Advances in Experimental Medicine and Biology

Volume 1459

Advances in Experimental Medicine and Biology provides a platform for scientific contributions in the main disciplines of the biomedicine and the life sciences. This series publishes thematic volumes on contemporary research in the areas of microbiology, immunology, neurosciences, biochemistry, biomedical engineering, genetics, physiology, and cancer research. Covering emerging topics and techniques in basic and clinical science, it brings together clinicians and researchers from various fields.

Advances in Experimental Medicine and Biology has been publishing exceptional works in the field for over 40 years, and is indexed in SCOPUS, Medline (PubMed), EMBASE, Reaxys, EMBiology, the Chemical Abstracts Service (CAS), and Pathway Studio.

2022 CiteScore: 6.2

Tilman Borggrefe
Benedetto Daniele Giaimo
Editors

Transcription factors in blood cell development

Springer

Editors
Tilman Borggrefe
Institute of Biochemistry
University of Giessen
Gießen, Hessen, Germany

Benedetto Daniele Giaimo
Institute of Biochemistry
University of Giessen
Gießen, Hessen, Germany

ISSN 0065-2598 ISSN 2214-8019 (electronic)
Advances in Experimental Medicine and Biology
ISBN 978-3-031-62730-9 ISBN 978-3-031-62731-6 (eBook)
https://doi.org/10.1007/978-3-031-62731-6

This Springer imprint is published by the registered company Springer Nature Switzerland AG
The registered company address is: Gewerbestrasse 11, 6330 Cham, Switzerland

Preface

Dear reader,

It is our pleasure to present this book on "Transcription Factors in Blood Cell Development" to you. Already in the 1990s, many of lineage specific transcription factors (TFs) had been discovered, long before the stem cell pluripotency TFs. Their exquisite specific expression, combined with techniques such as *Electrophoretic Mobility Shift Assay* (EMSA) or luciferase assays, has paved the way. Since then, the importance of such TFs has been highlighted using knockout or later conditional knockout mouse models and linking them to human diseases such as leukemia. Our book is a concerted effort to bring mechanistic insights, genome-wide data, and disease relevance together, focusing individual chapters on a particular TF.

Some of these TFs are master regulators of particular hematopoietic lineages. For example, early B-cell factor (EBF), BOB.1, or E2A are good examples for a lymphocyte specific TF, EKLF, or GATA1 more for the erythroid/megakaryocytic lineages.

All of them have been studied for decades in mouse but also human systems. Other TFs have been identified in the context of leukemia and/or as leukemic fusion-proteins. For several of these groups of different TFs, there will be examples in the current book.

The introduction of chromatin-immunoprecipitation coupled to high-throughput sequencing (ChIP-Seq) has tremendously advanced the research on hematopoietic TFs. Suddenly, it was possible to map genome-wide the exact binding sites. Subsequently, it was possible to evolve networks, such as the interdependence of TFs, by discovering feedback loops or feed-forward loops. Moreover, it was also evident that there are genomic sites that are bound by multiple TFs. Seemingly discoveries for basic research could be used for diagnostic purposes, but also in part to better define therapies, as described for some leukemic fusion proteins, for example, PML-RAR.

This book summarizes molecular aspects of TFs in the context of blood cell development and leukemia. We hope that it will inspire young scientists as well as scientists from other research areas, who may use the book as a starting point for further investigations or translational endeavors.

Gießen, Hessen, Germany

Tilman Borggrefe
Benedetto Daniele Giaimo

Contents

Contributors

Sajesan Aryal Department of Medicine, Division of Hematology/Oncology, University of Alabama at Birmingham Heersink School of Medicine, Birmingham, AL, USA
O'Neal Comprehensive Cancer Center, University of Alabama at Birmingham Heersink School of Medicine, Birmingham, AL, USA

Annika C. Betzler Department of Oto-Rhino-Larnygology, Ulm University Medical Center, Ulm, Germany

Core Facility Immune Monitoring, Ulm University, Ulm, Germany

James J. Bieker Department of Cell, Developmental, and Regenerative Biology, Icahn School of Medicine at Mount Sinai, New York, NY, USA

Mark Boothby Departments of Pathology, Microbiology, Immunology (Molecular Pathogenesis Division), Vanderbilt University Medical Center, Nashville, TN, USA
Medicine (Rheumatology and Immunology Division), Vanderbilt University Medical Center, Nashville, TN, USA
Vanderbilt Institute for Infection, Inflammation, Immunity (VI4), Nashville, TN, USA

Cornelia Brunner Department of Oto-Rhino-Larnygology, Ulm University Medical Center, Ulm, Germany
Core Facility Immune Monitoring, Ulm University, Ulm, Germany

Susan Chan Institut de Génétique et de Biologie Moléculaire et Cellulaire (IGBMC), Illkirch-Graffenstaden, France
Institut National de la Santé et de la Recherche Médicale (INSERM), U1258, Illkirch-Graffenstaden, France
Centre National de la Recherche Scientifique (CNRS), UMR7104, Illkirch-Graffenstaden, France
Université de Strasbourg, Illkirch-Graffenstaden, France

Sung Hoon Cho Departments of Pathology, Microbiology, Immunology (Molecular Pathogenesis Division), Vanderbilt University Medical Center, Nashville, TN, USA
Vanderbilt Institute for Infection, Inflammation, Immunity (VI4), Nashville, TN, USA

Stella T. Chou Department of Pediatrics, Division of Hematology, University of Pennsylvania Perelman School of Medicine, Children's Hospital of Philadelphia, Philadelphia, PA, USA
Department of Pathology and Laboratory Medicine, Division of Transfusion Medicine, University of Pennsylvania Perelman School of Medicine, Children's Hospital of Philadelphia, Philadelphia, PA, USA

Mary Louise Clarke Department of Biomedical Sciences, College of Medicine & Health, University of Birmingham, Edgbaston, Birmingham, UK
Department of Cancer & Genomic Sciences, College of Medicine & Health, University of Birmingham, Edgbaston, Birmingham, UK

Johannes Fechner Department of Eukaryotic Genetics, Institute of Biomedical Genetics, University of Stuttgart, Stuttgart, Germany

Jon Frampton Department of Cancer & Genomic Sciences, College of Medicine & Health, University of Birmingham, Edgbaston, Birmingham, UK

Bettina Maria Fuglerud Department of Biosciences, University of Oslo, Oslo, Norway

Odd Stokke Gabrielsen Department of Biosciences, University of Oslo, Oslo, Norway

Aishwarya Gurumurthy Department of Pathology, University of Michigan Medical School, Ann Arbor, MI, USA

Philippe Kastner Institut de Génétique et de Biologie Moléculaire et Cellulaire (IGBMC), Illkirch-Graffenstaden, France
Institut National de la Santé et de la Recherche Médicale (INSERM), U1258, Illkirch-Graffenstaden, France
Centre National de la Recherche Scientifique (CNRS), UMR7104, Illkirch-Graffenstaden, France
Université de Strasbourg, Illkirch-Graffenstaden, France
Faculté de Médecine, Université de Strasbourg, Strasbourg, France

Chara Konstantakopoulou Institute for Fundamental Biomedical Research, Biomedical Sciences Research Center Alexander Fleming, Vari, Greece
Department of Antibody Research Materials, Genmab B.V., Utrecht, The Netherlands

Jörn Lausen Department of Eukaryotic Genetics, Institute of Biomedical Genetics, University of Stuttgart, Stuttgart, Germany

Roza Berhanu Lemma Department of Biosciences, University of Oslo, Oslo, Norway
Centre for Molecular Medicine Norway (NCMM), Nordic EMBL Partnership, University of Oslo, Oslo, Norway

Rui Lu Department of Medicine, Division of Hematology/Oncology, University of Alabama at Birmingham Heersink School of Medicine, Birmingham, AL, USA
O'Neal Comprehensive Cancer Center, University of Alabama at Birmingham Heersink School of Medicine, Birmingham, AL, USA

Rebecca A. Luchtel Division of Hematology and Oncology, Department of Medicine, Northwestern University, Chicago, IL, USA

Kazuko Miyazaki Laboratory of Immunology, Institute for Life and Medical Sciences, Kyoto University, Kyoto, Japan

Masaki Miyazaki Laboratory of Immunology, Institute for Life and Medical Sciences, Kyoto University, Kyoto, Japan

Alexander C. Monovich Department of Pathology, University of Michigan Medical School, Ann Arbor, MI, USA

Stuart H. Orkin Dana-Farber/Boston Children's Cancer and Blood Disorders Center, Boston, MA, USA
Harvard Medical School and HHMI, Boston, MA, USA

Elvira Pelosi Department of Oncology, Istituto Superiore di Sanità, Rome, Italy

Marc Pfefferlé Department of Internal Medicine, Spital Limmattal, Schlieren, Switzerland

Sjaak Philipsen Department of Cell Biology, Erasmus MC, Rotterdam, The Netherlands

Russell J. H. Ryan Department of Pathology, University of Michigan Medical School, Ann Arbor, MI, USA

Mikael Sigvardsson Department of Biomedical and Clinical Sciences, Linköping University, Linköping, Sweden
Division of Molecular Hematology, Lund University, Lund, Sweden

Kaoru Takasaki Department of Pediatrics, Division of Hematology, University of Pennsylvania Perelman School of Medicine, Children's Hospital of Philadelphia, Philadelphia, PA, USA

Ugo Testa Department of Oncology, Istituto Superiore di Sanità, Rome, Italy

Florence Vallelian Department of Internal Medicine, University of Zurich and University Hospital of Zurich, Zurich, Switzerland

Mihalis Verykokakis Institute for Fundamental Biomedical Research, Biomedical Sciences Research Center Alexander Fleming, Vari, Greece

Pengcheng Zhang Department of Medicine, Division of Hematology/Oncology, University of Alabama at Birmingham Heersink School of Medicine, Birmingham, AL, USA
O'Neal Comprehensive Cancer Center, University of Alabama at Birmingham Heersink School of Medicine, Birmingham, AL, USA

Ge Zheng Dana-Farber/Boston Children's Cancer and Blood Disorders Center, Boston, MA, USA
Harvard Medical School and HHMI, Boston, MA, USA

Abbreviations

2-OHG	2-Hydroxyglutarate
4-OHT	4-Hydroxy-tamoxifen
4-OI	4-Octyl itaconate
5′-UTR	5′-Untranslated region
5hmC	5-Hydromethylcytosine
5mC	5-Methylcytosine
5PRA	5-Phosphoribosylamine
6-OHDA	6-Hydroxydopamine
15d-PGJ_2	15-Deoxy-delta-12,14-prostaglandin J2
αGalCer	Alpha-galactosylceramide
β-TrCP	Beta-transducin repeats-containing proteins
β-YAC	Beta-globin cluster yeast artificial chromosome transgene
A2AR	A2A adenosine receptor
aa	Amino acids
Ab	Antibody
ABC	Activated B-cell-like
ABC-DLBCL	Diffuse large B-cell lymphoma, activated B-cell subtype
ABCA1	ATP-binding cassette transporter 1
ABCG5	ATP-binding cassette g5
ABCG8	ATP-binding cassette g8
ABE	Adenine base editor
ABF-1	Activated B-cell factor-1
ABL1	Abelson tyrosine kinase 1 / Abelson leukemia virus proto-oncogene 1
ACAD10	Acyl-coenzyme A dehydrogenase family member 10
ACAD12	Acyl-coenzyme A dehydrogenase family member 12
ACC	Adenoid cystic carcinoma
ACH	Active chromatin hub
ACOD1	Aconitate decarboxylase 1
ACOT7	Acyl-CoA thioesterase 7
ACSS2	Acyl-CoA synthetase short-chain family member 2
AD	Activation domain
ADA	Adenosine deaminase

ADP	Adenosine diphosphatase
AF-1	Activation function-1
AF-2	Activation function-2
AFF	ALF transcription elongation factor
Ag	Antigen
AGM	Aorto-gonadal-mesonephric
AHR	Aryl hydrocarbon receptor
Aicda	Activation induced cytidine deaminase
AID	Activation-induced cytidine deaminase
AKR	Aldo-keto reductase
AKT	AKT serine/threonine kinase
AL	Acute leukemia
ALDA	Fructose-bisphosphate aldolase A
ALDH1	Aldehyde dehydrogenase 1
ALK	Anaplastic lymphoma kinase
ALI	Acute lung injury
ALL	Acute lymphoblastic leukemia
ALS	Amyotrophic lateral sclerosis
AML	Acute myeloid leukemia
AML1	Acute myeloid leukemia 1 (former name of RUNX1)
AMV	Avian myeloblastosis virus
ANKRD17	Ankyrin repeat domain 17
Aop	Anterior open (ETV6 / ETV7 homolog)
AP1	Activator protein 1
APC	Antigen-presenting cell
APL	Acute promyelocytic leukemia
APOBEC	Apolipoprotein B mRNA-editing enzyme, catalytic polypeptide
ApoE	Apolipoprotein E
AR	Adenosine receptor
ARE	Antioxidant-responsive elements
ARG1	Arginase-1
ARID1B	AT-rich interaction domain 1B
AS-C T3	Achaete-scute T3
ASC	Antibody-secreting cell
ASE	Anti-silencer element
ASH2L	ASH2 like, histone lysine methyltransferase complex subunit
ATAC-Seq	Assay for transposase-accessible chromatin sequencing
ATF	Activating transcription factor
ATF3	Activating transcription factor 3
ATF4	Activating transcription factor 4
ATF6	Activating transcription factor 6
ATG5	Autophagy-related protein 5
ATLL	Adult T cell leukemia/lymphoma
ATM	Ataxia Telangiectasia

ATO	Arsenic trioxide
ATP	Adenosine triphosphate
ATRA	All-trans retinoic acid
B-ALL	B-cell acute lymphoblastic leukemia
B29	B-cell-specific glycoprotein B29
BACH1	BTB domain and CNC homolog 1
Bach2	Broad complex-tramtrack-bric a brac and Cap'n'collar homolog 2
BAF	BRG1/BRM associated factor
BAFF	B-cell activating factor
BAFFR	B-cell activating factor receptor
BAL	Bronchoalveolar lavage
BATF	Basic leucine zipper transcription factor, ATF-like
BATF3	Basic leucine zipper ATF-like transcription factor 3
BCL	B-cell chronic lymphocytic leukemia /lymphoma
BCL2	B-cell lymphoma 2
BCL2L11/BIM	BCL2-like 11 / BCL2 interacting mediator of cell death
BCL2L14	BCL2-like 14
BCL6	B-cell lymphoma 6
BCL9	B-cell lymphoma 9
BCL11A	BCL11 transcription factor A
BCL11B	BCL11 transcription factor B
BCL2L1/BCLX$_L$	BCL2 Like 1 / B-cell lymphoma-extra large
BCMA	B-cell maturation factor
BCOR	BCL6 corepressor
BCP-ALL	B-cell precursor acute lymphoblastic leukemia
BCR	B-cell receptor
BCR	Breakpoint cluster region activator of RhoGEF and GTPase
BCRP	Breast cancer-resistant protein
BENC	Blood enhancer cluster
BET	Bromodomain and extra-terminal
BFU-E	Burst-forming unit–erythroid
BH3	Homology (domain)-3
bHLH	Basic helix-loop-helix
BHLHE40	Basic helix-loop-helix family member e40
BioID	Biotin identification
BKLF	Basic Krüppel-like factor
BL	Burkitt lymphoma
Blimp1	B-lymphocyte-induced maturation protein 1
BLP	B-primed lymphoid progenitor
BM	Bone marrow
BMDC	Bone marrow-derived dendritic cell
BMDM	Bone marrow-derived macrophage
BMF	BCL2 modifying factor
BMP4	Bone morphogenetic protein 4

BNIP3	BCL2/adenovirus E1B 19 kDa protein-interacting protein 3
BOB.1/OBF.1	B-cell Oct binding factor 1/Oct-binding factor 1
BORIS (CTCFL)	Brother of the regulator of imprinted sites
Bp	Base pair
BRCA1	Breast cancer gene 1
BRD	Bromodomain
BRD2	Bromodomain-containing protein 2
BRD3	Bromodomain-containing protein 3
BRD4	Bromodomain-containing protein 4
BRD7	Bromodomain-containing protein 7
BRDT	Bromodomain-containing protein T
BRG1	Brahma-related gene 1
BRM	Brahma
BST1	BM stroma cell antigen 1
BTB/POZ/ZF	Bric-à-brac, tramtrack, broad complex/Pox virus zinc finger
BTD box	Buttonhead box
BTK	Bruton's tyrosine kinase
Btla	B- and T-lymphocyte attenuator
bZIP	Basic region, leucine zipper
c-JUN	Transcription factor Jun
c-MPL/c-Mpl	Myeloproliferative leukemia protein
c-MYC	Cellular myelocytomatosis
C-TAD	Carboxy (C) terminus transactivation domain
C-terminal	Carboxy terminal
C-ZF	Carboxy (C) terminus zinc finger
C/EBPα	CCAAT/enhancer-binding protein alpha
C/EBPβ	CCAAT/enhancer-binding protein beta
Ca^{2+}	Calcium
CA-AML	Cytogenetically abnormal acute myeloid leukemia
cAMP	Cyclic adenyl monophosphate
CABIN1	Calcineurin binding protein 1
CALCOCO2/NDP52	Calcium-binding, coiled-coil domain-containing protein 2
CaMKII	Calcium/calmodulin-dependent protein kinase II
CaN	Calcineurin
CAR	Chimeric antigen receptor
CAR-T	Chimeric antigen receptor T
CAR-iMac	CAR-macrophages
CARD	Caspase recruitment domain
CASP1	Caspase 1
CASP8AP2	Caspase 8 associated protein 2
CAT	Catalase
CBE	Cytidine base editor
CBF	Core-binding factor
CBFβ	Core binding factor subunit beta

CBFA2	Core binding factor A2
CBFA2T3 (ETO2)	CBFA2/RUNX1 partner transcriptional co-repressor 3
CBP	CREB-binding protein
CBR	Carbonyl reductase
CBX5	Chromobox protein homolog 5
CC	Coiled-coil domain
CCL2	CC chemokine ligand 2
CCL3	CC chemokine ligand 3
CCL4	CC chemokine ligand 4
CCL5	CC chemokine ligand 5
CCL11	CC chemokine ligand 11
CCL17	CC chemokine ligand 17
CCR4	C-C motif chemokine receptor 4
CCR7	C-C chemokine receptor type 7
CD	Cluster of differentiation
CD3	Cluster of differentiation 3
Cd3d	CD3 delta subunit of T-cell receptor complex
Cd19	Cluster of differentiation 19
CD20	Cluster of differentiation 20
CD21	Cluster of differentiation 21
CD28	Cluster of differentiation 28
CD34	Cluster of differentiation 34
Cd79a	Cluster of differentiation 79a
Cd79b	Cluster of differentiation 79b
CDA	Congenital dyserythropoietic anemia
cDC	Classical dendritic cell / Conventional dendritic cell
cDCs1	Classical dendritic cell 1 / Conventional dendritic cell 1
cDCs2	Classical dendritic cell 2 / Conventional dendritic cell 2
CDC73	Cell division cycle 73
CDDO-Im	1-[2-cyano-3-,12-dioxooleana-1,9(11)-dien-28-oyl]imidazole
CDDO-Me	2-cyano-3,12-dioxo-oleana-1,9(11)-dien-28-oic acid methyl ester
CDK9	Cyclin-dependent kinase 9
CDK12	Cyclin-dependent kinase 12
CDKN1A	Cyclin-dependent kinase inhibitor 1A
CDKN1B	Cyclin-dependent kinase inhibitor 1B
CDL	Central disordered/linker region of ETV6
cDNA	Complementary DNA
CDX2	Caudal type homeobox 2
CDX4	Caudal type homeobox 4
CEBPA	CCAAT/Enhancer binding protein alpha
CEBPB	CCAAT/Enhancer binding protein beta
CEBPE	CCAAT/Enhancer binding protein epsilon

CEP	Congenital erythropoietic porphyria
cFos	FOS proto-oncogene, AP-1 transcription factor subunit
CFU-E	Colony forming unit-erythroid
cGAS	Cyclic GMP–AMP synthase
CHD3/Mi-2α	Chromodomain helicase DNA binding protein 3
CHD4/Mi-2β	Chromodomain helicase DNA binding protein 4
CHIP	Carboxyl terminus of hsc70-interacting protein
ChIP	Chromatin immunoprecipitation
ChIP-qPCR	Chromatin immunoprecipitation followed by quantitative polymerase chain reaction
ChIP-Seq	Chromatin immunoprecipitation followed by deep sequencing
cHL	Classical Hodgkin's lymphoma
CIA	Collagen-induced arthritis
CID	C-terminal inhibitory domain of ETV6
Cish	Cytokine inducible SH2-containing protein
CKII	Casein kinase II
CLP	Common lymphoid progenitor
CLRF2	Cytokine receptor-like factor 2
CLTC	Clathrin heavy chain
CM	Cardiac myxoma
CMA	Chaperone-mediated autophagy
CMAP	Connectivity map
CML	Chronic myelogenous leukemia
CMML	Chronic myelomonocytic leukemia
CMP	Common myeloid progenitor
CN-AML	Cytogenetically normal AML
CNC	Cap'n'Collar
CNOT3	CCR4-NOT transcription complex subunit 3
CNSHA	Congenital non-spherocytic hemolytic anemia
CO	Carbon monoxide
COP1	COP1 E3 ubiquitin ligase
COPD	Chronic obstructive pulmonary disease
CoV	Coronavirus
CR	Complete remission
CRABP1	Cellular retinoic acid binding protein I
CRBN	Cereblon
CRD	C-terminal regulatory domain
CRE	Cis-regulatory region / cis-regulatoty element
CREB	cAMP response element-binding protein
CREBBP	CREB binding protein
CRISPR	Clustered regularly interspaced short palindromic repeats
CRLF2	Cytokine receptor-like factor 2
CRM1	Chromosome region maintenance 1
cRNA	Circular RNA
CSF1	Colony stimulating factor 1

CSF1R	Colony stimulating factor 1 receptor
CSF2	Colony stimulating factor 2
CSR	Class-switch recombination
CTBP	Carboxy-terminal binding protein
CTBP1	Carboxy-terminal binding protein 1
CTCF	CCCTC-binding factor
CTCFL (BORIS)	CCCTC-binding factor like
CTD	C-terminal domain
CTIP1	COUP-TF-interacting protein 1
CTL	Cytotoxic T lymphocyte
CTLA4	Cytotoxic T lymphocyte antigen 4
CUL-4A	Culin 4A
CUL1	Cullin 1
CUL3	Cullin 3
CUL4	Cullin 4
CUT&RUN	Cleavage under targets and release using nuclease
CUT&TAG	Cleavage under targets and tagmentation
CVID	Common variable immunodeficiency
CXCL1	Chemokine (C-X-C motif) ligand 1
CXCL2	Chemokine (C-X-C motif) ligand 2
CXCL9	Chemokine (C-X-C motif) ligand 9
CXCL10	Chemokine (C-X-C motif) ligand 10
CXCL12	Chemokine (C-X-C motif) ligand 12
CXCL13	Chemokine C-X-C motif ligand 13
Cxcr4	C-X-C motif chemokine receptor 4
CXCR5	C-X-C motif chemokine receptor 5
CYP	Cytochrome P450
D	Diversity
DA	Dopaminergic
DAMP	Damage-associated molecular pattern
DAR	DNA binding through RARE
DAZAP1	DAZ associated protein 1
DBA	Diamond-Blackfan anemia
DBD	DNA-binding domain
dbSNP	Database for SNPs in the human genome
DC	Dendritic cell
DCreg	Regulatory dendritic cell
DD	Dimerization domain
DDB	Damaged DNA binding protein
DDX5	DEAD-box helicase 5
DEG	Differentially expressed gene
DEK	DEK proto-oncogene
DEM	Diethyl maleate
Dex-Mes	Dexamethasone 21-mesylate
DFS	Disease-free survival
DHODH	Dihydroorotate dehydrogenase
DHS	DNase I–hypersensitive site

dKO	Double knockout
DLBCL	Diffuse large B-cell lymphoma
DMF	Dimethyl fumarate
DMS	Demarcation membrane system
DMTN	Dematin
DN	Double negative ($CD4^{-}CD8^{-}$)
DN3	Double negative maturation stage 3 of thymocytes (negative for both CD4 and CD8)
DNA	Deoxyribonucleic acid
DNase2	Deoxyribonuclease 2, lysosomal
DNMT1	DNA methyltransferase 1
DNMT3A	DNA methyltransferase 3 alpha
DNTT	DNA nucleotidylexotransferase
DOT1L	DOT1-like histone lysine methyltransferase
DOTCom	DOT1L/KMT4 containing complex
DP	Double positive ($CD4^{+}CD8^{+}$)
DPF2	Double PHD fingers 2
DR	Direct repeat
DS	Down syndrome
DSB	Double-strand DNA break / Double-stranded break
dTAG	degradation tag
DU.582	Duke university 528
DUX4	Double homeobox four gene /Double homeobox 4
DYRK1A	Dual-specificity tyrosine-phosphorylation regulated kinase 1A
ER	ETV6::RUNX1
E1A	Adenovirus early region 1A
E2a	E2 and E47
E2A	Transcription factor E2 alpha / Transcription factor 3
E2F	E2F transcription factor
E4BP4	E4 binding protein 4
E47	E2A immunoglobulin enhancer-binding factors E12/E47 (transcription factor 3, TCF3)
EAE	Experimental autoimmune encephalomyelitis
EAF	ELL associated factor
EB	Elongation block
EBaso	Early basophilic erythroblast
EBF1	Early B-cell factor 1
EBI2	Epstein–Barr virus-induced gene 2
EBNA-1	Epstein–Barr nuclear antigen 1
E box	Ephrussi box
EBV	Epstein-Barr virus
EC	Epithelial cell
ECAR	Extracellular acidification rate
EGCG	Epigallocatechin-3-gallate
EGF	Epidermal growth factor

EGR1	Early growth response 1
EGR2	Early growth response 2
EGR3	Early growth response 3
EICE	ETS/IRF composite element
eIF2	Eukaryotic initiation factor 2
eIF2α	Eukaryotic initiation factor 2 alpha
eIF4	Eukaryotic initiation factor 4
EKLF	Erythroid Krüppel-like factor
ELAM	Endothelial leukocyte adhesion molecule
ELF1	E74-like ETS transcription factor 1
ELK1	ETS transcription factor ELK1
ELL	Elongation factor for RNA polymerase
ELOVL7	Elongation of very long chain fatty acids protein 7
EMSA	Electrophoretic mobility shift assay
ENCODE	Encyclopedia of DNA elements consortium
ENO1	Enolase 1
ENO4	Enolase 4
ENU	Ethylnitrosurea
EP300	E1A binding protein p300
ePBGD	Erythroid porphobilinogen deaminase
EPO	Erythropoietin
EPOR	Erythropoietin receptor
EpRE	Electrophile-responsive element
ERα	Estrogen receptor alpha
Erag	Enhancer of Rag
ERG	ETS-related gene
ERK1/MEK1	Extracellular-signal regulated kinase 1 and mitogen-activated protein kinase 1
ERK2	Extracellular-signal regulated kinase 2
ERK5	Extracellular-signal regulated kinase 5
Ery	Erythroid progenitor
ESC	Embryonic stem cell
ESR1/ERα	Estrogen receptor 1, estrogen receptor alpha
ET	Embryo transfer
ETO	Eight-twenty-one
ETO2 (CBFA2T3)	CBFA2/RUNX1 partner transcriptional co-repressor 3
ETP	Early T-progenitor
ETP-ALL	Early T precursor acute lymphoblastic leukemia
ETS	Erythroblastosis virus 26 transformation-specific
ETS1	ETS proto-oncogene 1
ETV6	ETS variant transcription factor 6
ETV6-RUNX1	ETS variant transcription factor 6-Runt related transcription factor 1 fusion gene
ETV7	ETS variant transcription factor 7
EVI9	Ecotropic viral integration site 9
EWS	Ewing sarcoma RNA binding protein 1 (encoded by EWSR1 gene)

EWSR1 Ewing sarcoma RNA binding protein 1
Ex Embryonic day x
EYA2 Eyes absent homolog 2
EZH2 Enhancer of zeste homolog 2
FAB French American-British
FACS Fluorescence-activated cell sorting
FADS1 Fatty acid desaturase 1
FATP-1 Fatty acid transport protein 1
FBL6 F-box and leucine-rich repeat protein 6 (former name of FBXL6)
FBXL6 F-box and leucine rich repeat protein 6
FDC Follicular dendritic cell
FIH Factor inhibiting HIF
FISH Fluorescent in situ hybridization
FKBP12 FK506-binding protein 12
FL Follicular lymphoma
FLASH/CASP8AP2 Caspase 8 associated protein 2
Fl-TAL1 Full-length TAL1
FLI1 Friend leukemia integration 1
FLT3 Fms-like tyrosine kinase 3
FLT3-ITD FMS-like tyrosine kinase 3-internal tandem duplication
FOE Friend of EKLF
FOG1 Friend of GATA1
Fos Fos proto-oncogene, AP-1 transcription factor subunit
Fos-B FBJ murine osteosarcoma viral oncogene homolog B
FOX Forkhead box transcription factor
FOXA Forkhead box A
FOXJ2 Forkhead box J2
FOXN3 Forkhead box N3
FOXO1 Forkhead box O1
FOXO3 Forkhead box O3
FOXP1 Forkhead box P1 transcription factor
FOXP3 Forkhead box P3
FPN Ferroportin
FTH Ferritin heavy chain 1
FTL Ferritin light chain
FUS Fused in sarcoma
G-CSF Granulocyte colony-stimulating factor
G1HE GATA1 hematopoietic enhancer
G1HRD GATA1 hematopoietic regulatory domain
G1ME Gata1^{-} megakaryocyte erythroid
G6PD Glucose-6-phosphate dehydrogenase
GABARAPL1 Gamma-aminobutyric acid receptor-associated protein-like 1

GABPα	GA binding protein transcription factor subunit alpha
GABPA	GA binding protein transcription factor subunit alpha
GAMES	Genetic analysis of multiple sclerosis in Europeans
GATA1	GATA-binding factor 1 / GATA-binding protein 1
GATA1s	GATA1 short (isoform)
GATA2	GATA-binding factor 2 / GATA-binding protein 2
GATA3	GATA-binding factor 3 / GATA-binding protein 3
GATAD2A	GATA zinc finger domain containing 2A
GATAD2B	GATA zinc finger domain containing 2B
GC	Germinal center
GC-Tfh	Follicular helper (fh) T-cells (Tfh), including those in germinal center
GCB	Germinal center B-cell-like
GCLC	Glutamate-cysteine ligase catalytic
GCLM	Glutamate-cysteine ligase modulator
GFI1	Growth factor independent 1 transcriptional repressor
GFI1B	Growth factor independent 1B transcriptional repressor
GFP	Green fluorescent protein
GFRA2	GDNF family receptor 2
GIST	Gastrointestinal stromal tumor
GKLF	Gut Krüppel-like factor
GLI	Glioma-associated oncogene homolog
GLUT	Glucose transporter
GLUT1	Glucose transporter 1
GM-CSF	Granulocyte-macrophage colony stimulating factor
GMP	Granulocyte/macrophage progenitor
GO	Gemtuzumab ozigomicin
GoF	Gain-of-function
GP1BA/Gp1ba	Glycoprotein 1b subunit alpha
GP1BB/Gp1bb	Glycoprotein 1b subunit beta
GPA	Glycophorin A
GPI1	Glucose phosphate isomerase 1
GPIIB/Gpiib	Glycoprotein IIb
Gpr183	G-protein coupled receptor 183
GPV/ Gpv	Glycoprotein V
GPVI/Gpvi	Glycoprotein VI
GPIX/Gpix	Glycoprotein IX
GPS2	G-protein pathway suppressor 2
GPX	Glutathione peroxidase
GPX2	Glutathione peroxidase 2
GRB2	Growth factor receptor bound protein 2
GRB10	Growth factor receptor bound protein 10

GRN	Gene regulatory network
gRNA	Guide RNA
GSEA	Gene set enrichment analysis
GSH	Glutathione
GSK3	*Glycogen synthase kinase-3*
GSK3β	*Glycogen synthase kinase-3 beta*
GSR1	Glutathione reductase 1
GSSG	Glutathione disulfide
GST	Glutathione S-transferase
GSX2/GSH2	GS homeobox 2
GLUT	Glucose transporter
GTF	General transcription factor
GWAS	Genome-wide association studies
H-L(3)MBT	Human lethal(3)malignant brain tumor-like
H_2O_2	Hydrogen peroxide
H3	Histone 3
H3K4	Histone 3 lysine 4
H3K4me1	Histone 3 lysine 4 monomethylation
H3K4me3	Histone 3 lysine 4 trimethylation
H3K9ac	Histone 3 lysine 9 acetylation
H3K9me1	Histone 3 lysine 9 monomethylation
H3K9me2	Histone 3 lysine 9 dimethylation
H3K27	Histone 3 lysine 27
H3K27ac	Histone 3 lysine 27 acetylation
H3K27me3	Histone 3 lysine 27 trimethylation
H3K36me3	Histone 3 lysine 36 trimethylation
H3K79	Histone 3 lysine 79
H3K79me2	Histone 3 lysine 79 dimethylation
H3R17	Histone 3 arginine 17
H4	Histone 4
HAT	Histone acetyltransferase
Hb	Hemoglobin
HBA	Hemoglobin, subtype alpha
HbA	Adult hemoglobin
HBB	Hemoglobin, subunit beta
HbF	Fetal hemoglobin
HBG	Hemoglobin, subtype gamma
HBG1	Hemoglobin subunit gamma 1
HBG2	Hemoglobin subunit gamma 2
HBO1	H4-specific histone acetylase
HbS	Sickle hemoglobin
HBS1L	HBS1 like translational GTPase
HBZ	HTLV-1 basic leucine zipper
HCC	Hepatocellular carcinoma
HCP1	Heme carrier protein 1
HCV	Hepatitis C virus
HD	Homeodomain
HDAC	Histone deacetylase

HDAC1	Histone deacetylase 1
HDAC2	Histone deacetylase 2
HDAC3	Histone deacetylase 3
HDAC4	Histone deacetylase 4
HDAC5	Histone deacetylase 5
HDAC7	Histone deacetylase 7
HDAC9	Histone deacetylase 9
HE	Hemogenic endothelial
HEB/TCF12	Transcription factor 12
HER2	Human epidermal growth factor receptor 2
hESC	Human embryonic stem cell
HF1	A 28-base-pair element in the rat cardiac myosin light-chain-2 gene
HFD	High-fat diet
Hgb F	Fetal hemoglobin
Hi-C	Analysis for genome-wide chromatin interactions
HIC2	Hypermethylated in cancer 2
HIF	Hypoxia-inducible factor
HIF1α	Hypoxia-inducible factor 1 alpha
HIF2α	Hypoxia-inducible factor 2 alpha
HIRA	Histone cell cycle regulator homolog A
HIV-1	Human immunodeficiency virus 1
HK1	Hexokinase 1
HK2	Hexokinase 2
HLH	Helix-loop-helix
HLXB9	Homeobox HB9 (former name of MNX1)
HMGB3	High-mobility group box 3
HMOX1	Heme-oxygenase 1
HMT	Histone methyltransferase
HNE	4-Hydroxynonenal
HNRNPF	Heterogeneous nuclear ribonucleoprotein F
HNRNPH1	Heterogeneous nuclear ribonucleoprotein H1
HNRNPH/F	HNRNPH1 and HNRNPF
HNRNPM	Heterogeneous nuclear ribonucleoprotein M
HNRNPUL1	Heterogeneous nuclear ribonucleoprotein U like 1
HOCT1	Organic cation uptake transporter 1
HOCT2	Organic cation uptake transporter 2
HOTAIR	HOX transcript antisense RNA
HOTTIP	HOXA transcript at distal tip
HOX	Homeobox
HOX-A	HOX cluster A
HOXA9	Homeobox A9
HOXA13	Homeobox A13
HOXB9	Homeobox B9
HPC	Hematopoietic progenitor cell
HPFH	Hereditary persistence of fetal hemoglobin
HRD1	ERAD-associated E3 ubiquitin-protein ligase 1

HRI	Heme-regulated elF2α kinase / Heme-regulated inhibitor
HRS	Hodgkin and Reed-Sternberg cells
HS	Hypersensitivity sites
HS I, II, III	DNase I hypersensitivity region I, II, III
HSC	Hematopoietic stem cell
HSCT	Hematopoietic stem cell transplantation
hSET1	Human SET domain containing 1A, histone lysine methyltransferase
HSPC	Hematopoietic stem/progenitor cell
HSV	Herpes simplex virus
HTH	Helix-turn-helix
HTLV-1	Human T-cell leukemia virus type 1
HUDEP	Human umbilical cord blood-derived erythroid progenitor
HUGO	Human genome organization
HX	HOX-specific hexapeptide
HXR9	HXR9 peptide
iAMP21	Intrachromosomal amplification of chromosome 21 (B-ALL subtype)
IBD	Inflammatory bowel disease
ICAM4	Intercellular adhesion molecule 4, Landsteiner-Wiener blood group
ICH	Intracerebral hemorrhage
ICOS	Inducible T-cell co-stimulator
ID1	Inhibitor of DNA binding 1
ID2	Inhibitor of DNA binding 2
IDA	Idarubicin
iDC	Immature dendritic cell
IDH1	Isocitrate dehydrogenase 1
IDH2	Isocitrate dehydrogenase 2
IDR	Intrinsically disordered protein regions
IE	Erythroid promoter/first exon
IEGs	Immediate early genes
IFIT2	Interferon-induced protein with tetratricopeptide repeats 2
IFN	Interferon
IFNβ	Interferon beta
IFNγ	Interferon gamma
Ifng	Interferon gamma
Ifngr1	Interferon gamma receptor 1
IFR	Interferon response factor
IFR2BP	Interferon response 2 binding protein
Ig	Immunoglobulin
IGF	Insulin-like growth factor
IGF2BP1	Insulin like growth factor 2 mRNA-binding protein 1
IgH	Immunoglobulin heavy (chain)

IgL	Immunoglobulin light (chain)
IκB	Inhibitors of NF-κB
IkBα	Inhibitors of NF-κB alpha
IK2F1/3	IKAROS family zinc finger 1/3
IKAROS	IKAROS family zinc finger
IKK	IκB kinase
IKZF1	Ikaros family zinc finger 1
IL	Interleukin
IL1β	Interleukin 1 beta
IL2	Interleukin 2
IL3	Interleukin 3
IL4	Interleukin 4
IL5	Interleukin 5
IL6	Interleukin 6
IL7	Interleukin 7
IL7R	Interleukin 7 receptor
IL7Ra	Interleukin 7 receptor subunit alpha
IL8	Interleukin 8
IL10	Interleukin 10
IL12	Interleukin 12
IL12B	Interleukin 12 beta
IL12R	Interleukin 12 receptor
IL13	Interleukin 13
IL17	Interleukin 17
IL18	Interleukin 18
IL21	Interleukin 21
IL23	Interleukin 23
IL23R	Interleukin 23 receptor
ILC	Innate lymphoid cell
In(Lu)	Inhibitor of Lutheran antigen
iNKT	Invariant natural killer-T
iNOS	Inducible NO synthase
IPF	Idiopathic pulmonary fibrosis
iPSC	Induced pluripotent stem cell
IPT/TIG	Immunoglobulin-like, plexins, transcription factors/transcription factor immunoglobulin
IRE1	Inositol-requiring enzyme 1α
IRF1	Interferon regulatory factor 1
IRF3	Interferon regulatory factor 3
IRF4	Interferon regulatory factor 4
IRF7	Interferon regulatory factor 7
IRF8	Interferon regulatory factor 8
IRX5	Iroquois-class homeodomain protein 5
ISRE	Interferon-stimulated response element
IT	Testis promoter/first exon
ITGA2B/Itga2b	Integrin subunit alpha IIb
iTreg	Induced regulatory T-cells
J	Joining

JAK	Janus kinase
JAK2	Janus kinase 2
JMJD1A	Jumonji domain containing 1A
JNK	c-Jun N-terminal kinase
JUNB	JunB proto-oncogene, AP-1 transcription factor subunit
KAT	Lysine acetyltransferase
KAT2B	Lysine acetyltransferase 2B
KAT3A	Lysine acetyltransferase 3A
KAT3B	Lysine acetyltransferase 3B
KAT5	Lysine acetyltransferase 5
KAT6A	Lysine acetyltransferase 6A
kB	Kilobase
kDa	Kilodalton
KDM1A	Lysine demethylase 1A
KDM1B	Lysine demethylase 1B
KDM5C	Lysine demethylase 5C
κE2	Immunoglobulin kappa chain enhancer E2
KEAP1	Kelch-like ECH-associated protein 1
KG	Ketoglutarate
KIT	KIT proto-oncogene, receptor tyrosine kinase
KIX	Kinase-inducible interacting
KLF1	Krüppel-like factor 1
KLF2	Krüppel-like factor 2
KLF3	Krüppel-like factor 3
KLF4	Krüppel-like factor 4
KLF5	Krüppel-like factor 5
KLF6	Krüppel-like factor 6
KLF9	Krüppel-like factor 9
KLF13	Krüppel-like factor 13
KLF14	Krüppel-like factor 14
KLF15	Krüppel-like factor 15
KLF16	Krüppel-like factor 16
KLF17	Krüppel-like factor 17
KLF18	Krüppel-like factor 18
KLFxPy	Krüppel-like factor (x) pseudogene (y)
KLRG1	Killer cell lectin-like receptor subfamily G member 1
KMT1C	Lysine methyltransferase 1C
KMT2A	Lysine methyltransferase 2A
KMT2B	Lysine methyltransferase 2B
KMT2C	Lysine methyltransferase 2C
KMT2D	Lysine methyltransferase 2D
KMT2F	Lysine methyltransferase 2F
KMT4	Lysine methyltransferase 4
KO	Knockout
KPNB1	Karyopherin subunit beta 1

L3MBTL1	L3MBTL histone methyl-lysine binding protein 1 [formerly H-L(3)MBT]
LAG3	Lymphocyte activation gene 3
LAZ3	Lymphoma associated zinc finger 3
LBaso	Late basophilic erythroblast
LBD	Ligand-binding domain
Lck	Lymphoid-specific protein-tyrosine kinase
LCMV	Lymphocytic choriomeningitis virus
LCR	Locus control region
LDB1	LIM domain-binding protein 1
LDL	Low-density lipoprotein
LDLR	Low-density lipoprotein receptor
LEDGF	Lens epithelium-derived growth factor (also referred to as PSIP1)
LEF1	Lymphoid enhancer-binding factor 1
LIM	LIN-11, Isl-1 and MEC-3 protein interaction domain composed of two zinc fingers
Lin^-	Lineage marker-negative
LIN28B	Lin-28 homolog B
LKLF	Lung Krüppel-like factor
LMO1	LIM domain only 1
LMO2	LIM domain only 2
LMO4	LIM domain only 4
LMPP	Lympho-myeloid primed progenitor / Lymphoid-primed multipotent progenitor
LN	Lymph node
lncRNA	Long non-coding RNA
LoF	Loss-of-function
LPS	Lipopolysaccharide
LRF	Leukemia/lymphoma related factor
LSC	Leukemic stem cell
LSD1	Lysine-specific histone demethylase 1A
LSK	Lineage markers, Sca-1, cKit
LT-HSC	Long-term hematopoietic stem cells
$LT\alpha_1\beta_2$	Lymphotoxin-$\alpha_1\beta_2$
LTi	Lymphoid tissue inducer
Lu	Lutheran blood group antigen
LTR	Long terminal repeat
LW	Landsteiner-Wiener
LYF-1	Lymphocyte factor 1
LYL1	Lymphoblastic leukemia 1
LYN	LYN proto-oncogene
M/LN-eo-TK	Myeloid and lymphoid neoplasms with eosinophilia and tyrosine kinase fusion genes
MADS	MCM-Agamous-Definiens-serum
MAF	MAF bZIP transcription factor
MAFF	MAF bZIP transcription factor F
MAFG	MAF bZIP transcription factor G

MAFK	MAF bZIP transcription factor K
MAFK	MAF bZIP transcription factor K
MAIT	Mucosa-associated invariant T
mAKAP	Myo-A kinase-anchored protein
MAML1	Mastermind like transcriptional coactivator 1
MAP4K4	Mitogen-activated protein kinase 4
MAPK	Mitogen-activated protein kinase
MARCO	Macrophage receptor with a collagenous structure
MARV	Marburg virus
MAZ	Myc-associated zinc finger protein
MB	MEF2D-BCL9
Mb1	Myoglobin 1
MBD2	Methyl-CpG binding domain protein 2
MBD3	Methyl-CpG binding domain protein 3
MBNL3	Muscleblind like splicing regulator 3
MBTD1	Mbt domain containing 1
MCD	MYD88-CD79B mutation cluster
MCH	Mean corpuscular hemoglobin
MCP-1	Monocyte chemotactic protein-1
MCS	Multi-species conserved sequence
MCV	Mean corpuscular volume
MDM	Monocyte-derived macrophage
MDM2	Mouse double minute 2, human homolog of; p53-binding protein, proto-oncogene
MDP	Monocyte/dendritic progenitor
MDR	Multidrug resistance-associated protein
MDS	Myelodysplastic syndrome
MDSC	Myeloid-derived suppressor cell
ME1	Malic enzyme 1
MED1	Mediator complex subunit 1
MED13L	Mediator complex subunit 13L
MED14	Mediator complex subunit 14
MEF2A	Myocyte-specific enhancer factor 2A
MEF2B	Myocyte-specific enhancer factor 2B
MEF2C	Myocyte-specific enhancer factor 2C
MEF2D	Myocyte-specific enhancer factor 2D
MEIS1	Myeloid ecotropic viral integration site homolog 1 / Myeloid ecotropic insertion site 1
MEK1	Mitogen-activated protein kinase 1
MEL	Murine erythroleukemia / mouse erythroleukemia cells
MEP	Megakaryocytic/erythroid progenitor
MH	MEF2D-HNRNPUL1
MHC	Major histocompatibility complex
MHCII	Major histocompatibility complex class II
MHRA	Medicines and Healthcare products Regulatory Agency

Mi-2α/CHD3	Chromodomain helicase DNA binding protein 3
Mi-2β/CHD4	Chromodomain helicase DNA binding protein 4
MIP-2	Macrophage inflammatory protein 2
miRNA	microRNA
MITF	Melanocyte inducing transcription factor
MIU	Motive interacting with ubiquitin
ML-DS	Myeloid leukemia of Down syndrome
MLL	Mixed-lineage leukemia
MLL1	Mixed lineage leukemia 1
MLL2	Mixed lineage leukemia 2
MLL3	Mixed lineage leukemia 3
MLL4	Mixed lineage leukemia 4
MLLr	MLL rearrangement
MM	Multiple myeloma
MMF	Monomethyl fumarate
MMP	Matrix metallopeptidase
MMP3	Matrix metallopeptidase 3
MMP9	Matrix metallopeptidase 9
MN1	Meningioma-1 proto-oncogene
MNX1	Motor neuron and pancreas homeobox 1
moDC	Monocyte-derived dendritic cell
MOG	Myelin oligodendrocyte glycoprotein
Mono	Monocyte
MORE	More of PORE
MOZ	Monocytic leukemia zinc finger protein
MPAL	Mixed phenotype acute leukemias
MPECs	Memory precursor cells
MPN	Myeloproliferative neoplasm
MPO	Myeloperoxidase
MPP	Multipotent progenitor
MPP4	Multipotent progenitor 4
MPTP	1-Methl-4-phenyl-1,2,3,6-tetrahydropyridine
MR1	Major histocompatibility complex, class I-related
MRE	MEF2D response element
MRF	Myogenesis regulator family
MRF4	Myogenic regulatory factor 4
MRL-lpr	Murphy Roths large lymphoproliferation mouse strain
mRNA	Messenger ribonucleic acid
MS	Multiple sclerosis
MSA	Multiple sequence alignment
MSCV	Mean sphered corpuscular volume
mSIN3A	SIN3 transcription regulator family member A
MSK1	Mitogen- and stress-activated kinase 1
MST	Mammalian STE20-like kinase
MST1	Mammalian STE20-like kinase 1
MST2	Mammalian STE20-like kinase 2
MSU	Monosodium urate

MTA1	Metastasis-associated protein 1
MTA2	Metastasis-associated protein 2
MTA3	Metastasis-associated protein 3
Mtb	Mycobacterium tuberculosis
mtDNA	Mitochondrial DNA
mTEC	Medullary thymic epithelial cell
MTHFD2	Methylenetetrahydrofolate dehydrogenase 2
mtHsp70	Mitochondrial heat shock protein 70
mTORC	Mechanistic target of rapamycin complex
mTORC1	Mechanistic target of rapamycin kinase complex 1
MYB	MYB proto-oncogene / Myeloblastosis viral oncogene homolog
MYC	MYC proto-oncogene / Myelocytomatosis viral oncogene homolog
MYD88	Myeloid differentiation primary response protein 88
MYF5	Myogenic factor 5
MYF6	Myogenic factor 6
MYOD	Myogenic differentiation 1
MYOG	Myogenin
MZ	Marginal zone
MZ B	Marginal zone B-cell
N-TAD	Amino terminus transactivation domain
N-terminal	Amino-terminal
N-ZF	Amino terminus zinc finger
NADPH	Reduced nicotinamide adenine dinucleotide phosphate
Nan	Neonatal anemia
NASH	Non-alcoholic steatohepatitis
NB	Nuclear bodies
NCOA2	Nuclear receptor coactivator 2
NCOA3	Nuclear receptor coactivator 3
NCOA6	Nuclear receptor coactivator 6
NCoR	Nuclear receptor corepressor
NCOR1	Nuclear receptor corepressor 1 (formerly N-CoR)
NCOR2	Nuclear receptor corepressor 2 (aka, SMRT)
ncRNA	Noncoding RNA
ND	N-terminal disordered region of ETV6
ND6	NADH dehydrogenase subunit 6
Neh	NRF2-ECH homology
NES	Nuclear export signal
NET	Neutrophil extracellular trap
NF-E2	Nuclear factor, erythroid 2
NF-kB	Nuclear factor kappa chain enhancer of activated B-cells
NF-Y	Nuclear transcription factor Y
NFAT	Nuclear factor of activated T-cells
NFE2/Nfe2	Nuclear factor, erythroid 2

NFIA	Nuclear factor I A
NFIX	Nuclear factor I X
NGS	Next generation sequencing
NHEJ	Non-homologous end joining
NHL	Non-Hodgkin lymphoma
NIPBL	Nipped-B-Like Protein
NK	Natural killer
NKT	Natural killer T
NLRP3	Nucleotide-binding oligomerization domain leucine rich repeat and pyrin domain-containing 3
NLS	Nuclear localization signal
NO	Nitric oxide
NOD	Non-obese diabetic
NOS2	Nitric oxide synthase 2
NPM1	Nucleophosmin
NPM1c	NPM1 gene mutation variant
NQO1	NAD(P)H quinone dehydrogenase 1
NRAS	Neuroblastoma RAS viral oncogene homolog
NRD	Negative regulatory domain
NRF1	Nuclear respiratory factor 1
NRF2	Nuclear factor erythroid 2-related factor 2 / NFE2-related factor 2
NRSF/REST	RE1 silencing transcription factor
nTreg	Natural regulatory T-cells
NTRK3	Neurotrophic receptor tyrosine kinase 3
NUP98	Nucleoporin 98
NUP98-HOXA9	Nucleoporin 98–Homeobox A9 fusion
NuRD	Nucleosome remodeling and deacetylase
Nurr77	Nuclear receptor 4A1
NuRSERY	Nuclear remodeling shuttle erythroid
NUTM1	Nuclear protein in testis midline carcinoma 1
NTRK3	Neurotrophic receptor tyrosine kinase 3
OBF1	Oct binding factor 1
OCA-B	Octamer coactivator from B-cells
OCR	Oxygen consumption rate
OCT1	Octamer transcription factor 1
OCT2	Octamer transcription factor 2
OCT4/POU5F1	POU class 5 homeobox 1, octamer motif binding factor 4
ODD	Oxygen-sensitive degron domain
OG	Oxoglutarate
OGD/R-injured neurons	Oxygen-glucose deprivation/reperfusion-injured neurons
OMP1	Olfactory marker protein 1
ORF	Open reading frame
Ortho	Orthochromatic erythroblast
OS	Overall survival
OVA	Ovalbumin

oxLDLs	Oxidized low-density lipoproteins
P-TEFb	Positive transcription elongation factor b
P/CAF	EP300/CBP-associated factor
P4.2	Human erythrocyte protein 4.2
P9	Postnatal day 9
P34	BOB.1/OBF.1 protein of 34 kDa
P35	BOB.1/OBF.1 protein of 35 kDa
P38	p38 Mitogen-activated protein kinases
P40	BOB.1/OBF.1 protein of 40 kDa
P53	Tumor protein P53
P66 (GATAD2A)	GATA zinc finger domain containing 2A
PAK2	P21 (RAC1) activated kinase 2
PAK4	P21 (RAC1) activated kinase 4
PAX5	Paired box 5
PAX5alt	PAX5-altered (B-ALL subtype)
PBC	Primary biliary cirrhosis
PBM	Protein-binding microarray
PBMC	Peripheral blood mononuclear cell
PBP/Pbp	Platelet basic protein
PBX1	Pre-B-cell leukemia homeobox 1
PBX2	Pre-B-cell leukemia homeobox 2
PBX3	Pre-B-cell leukemia homeobox 3
PC	Plasma cell
PC1	Principal component 1
PCET	EP300/KAT3B, CBP/KAT3A and ETO target
PD	Parkinson's disease
PD1	Programmed death 1
pDC	Plasmacytoid dendritic cell
Pdcd1	Programmed cell death 1
PDGFRB	Platelet-derived growth factor receptor beta
PDX	Patient-derived xenograft
PERK	Protein kinase R (PKR)-R-like endoplasmic reticulum kinase
PEST	Proline (P), glutamic acid (E), serine (S) and threonine (T)
PETN	Phosphatase and tensin homolog
PF4/Pf4	Platelet factor 4
PFKFB2	6-phosphofructo-2-kinase/fructose-2,6-biphosphatase 2
PFKFB4	6-phosphofructo-2-kinase/fructose-2,6-biphosphatase 4
PGA2	Prostaglandin A2
PGC-1α	Peroxisome proliferator-activated receptor gamma coactivator 1 alpha
PGD	Phosphogluconate dehydrogenase
PGE_2	Prostaglandin E2
PGF	6-phosphogluconate dehydrogenase
PGR	Progesterone receptor

PHD	Prolyl hydroxyl dioxygenase / Plant homeodomain
PHD-BD-PWWP	Plant homeodomain-bromodomain-Pro Trp Pro
PI	Phosphatidyl-inositol
PI3K-AKT	Phosphatidylinositol 3-kinase /protein kinase B
PI3K	Phosphoinositide-3-kinase / Phosphatidylinositol-3-kinase
PIAS1	Protein inhibitor of activated STAT 1
PIAS3	Protein inhibitor of activated STAT3
PIK3C3	Phosphatidylinositol 3-kinase catalytic subunit type 3
pIpC	Polyinosinic-polycytidylic acid
PITX1	Paired-like homeodomain 1
PKA	Protein kinase A
PKACA	cAMP-dependent protein kinase A
PKC	Protein kinase C
PKCA	Protein kinase C A
PKCθ	Protein kinase C θ
PKM2	Pyruvate kinase M2
PKR	Protein kinase R
PLZF	Promyelocytic leukemia zinc finger
PMA	Phorbol 12-myristate-13-acetate
PML	Promyelocytic leukemia protein
PML-RARA	Promyelocytic leukemia protein–Retinoic acid receptor alfa
PNT	Pointed (domain)
pO_2	Partial pressure of oxygen
POK	POZ domain Krüppel-like zinc finger
POKEMON	POZ and Krüppel erythroid myeloid ontogenic factor
Poly	Polychromatic erythroblast
POMP	Proteasome maturation protein
PORE	Palindromic OCT factor recognition element
Pou2af1	POU domain class 2-associating factor 1
POU_H	POU-homeodomain
POU_S	POU-specific domain
PP	Peyer's patch
PPAR	Peroxisome proliferator-activated receptor
PPARα	Peroxisome proliferator-activated receptor alpha
PPARβ/δ	Peroxisome proliferator-activated receptor beta/delta
PPARγ	Peroxisome proliferator-activated receptor gamma
PPI	Protein-protein interaction
PP2A-B55δ	Protein phosphatase 2A regulatory subunit B55 Delta
PP2B	Protein phosphatase 2B
PPAR	Peroxisome proliferator-activated receptor
PPAT	Phosphoribosyl pyrophosphate amidotransferase

PPI	Protein interaction
PPM1B	Protein phosphatase, Mg^{2+}/Mn^{2+} dependent 1B
PPP	Pentose phosphate pathway
PPP3	Protein phosphatase 3
PPRE	Peroxisome proliferator-activated receptor response element
PRC1	Polycomb repressive complex 1
PRC2	Polycomb repressive complex 2
Prdm1	Positive regulatory domain 1
PRDM1	PR/SET domain 1 / PR containing 1, with zinc finger domain / Positive regulatory domain 1
PRDM16	PR/SET domain 16
PRDX / PRX	Peroxiredoxin
PRDX1 / PRX1	Peroxiredoxin 1
PRDX2 / PRX2	Peroxiredoxin 2
pre-B-cells	B-cell precursors
pre-BCR	Pre-B-cell receptor
pre-TCR	Pre-Tα chain paired with a functional TCR β chain
Prep1–2	PKNOX family transcription factors 1-2
prepro-B-cell	Very early stage of B-cell development
PRMT	Protein arginine methyltransferase
PRMT1	Protein arginine methyltransferase 1
PRMT4	Protein arginine methyltransferase 4
PRMT5	Protein arginine methyltransferase 5
PRMT6	Protein arginine methyltransferase 6
pro-B-cells	B-cell progenitors
PRO-Seq	Precision run-on sequencing
ProE	Pro-erythroblast
PROTAC	Proteolysis targeting chimera
PRRs	Pattern recognition receptors
PSGL-1	P-selectin glycoprotein ligand 1
PSIP1	PC4 and SRSF1 interacting protein 1
PSM	Proteasome component-encoding
PSMA1	Proteasome 20S subunit alpha 1
PSMA4	Proteasome 20S subunit alpha 4
PSMB3	Proteasome subunit beta type-3
PSMB5	Proteasome subunit beta type-5
PSMB6	Proteasome subunit beta type-6
PSMC1	Proteasome 26S subunit, ATPase 1
PSMC3	Proteasome 26S subunit, ATPase 3
PSMD14	Proteasome 26S subunit, non-ATPase 14
PTD	Partial tandem duplication
PTEN	Phosphatase and tensin homolog
PTHrP	Parathyroid hormone-related protein
PTM	Post-translational modification
pTreg	Induced regulatory T-cells

PU.1/SPI1	Purine-rich box 1 / Hematopoietic transcription factor PU.1 / Spleen focus forming virus (SFFV) proviral integration protein (encoded by SPI1)
PUM1	Pumilio RNA-binding family member 1
PUS	Pseudo uridine synthetase
pVHL	von Hippel-Lindau protein
PYCARD/ASC	Apoptosis-associated speck-like protein containing caspase recruitment domain (CARD)
qPCR	Quantitative polymerase chain reaction
R-TEn	Rag-T-cell enhancer
R-TEn-p1-E-box$^{mut/mut}$	E-box mutations in the R-TEn region
R1	Repeat 1
R1B	Rag B-cell enhancer 1
R1pro	Rag1 promoter
R1pro-E-box$^{mut/mut}$	E-box mutation in R1pro
R2	Repeat 2
R2B	Rag B-cell enhancer 2
R3	Repeat 3
RA	Retinoic acid
RAD21	Double-strand-break repair protein Rad21 homolog
RAD23	RAD23 homolog, nucleotide excision repair protein
RAG	Recombination activating gene
RAG1	Recombination activating gene 1
RAG2	Recombination activating gene 2
RAR	Retinoic acid receptor
RARA	Retinoic acid receptor alfa
RARB	Retinoic acid recetpor beta
RARE	Retinoic acid response elements
RARG	Retinoic acid receptor gamma
RARS	Refractory anemia with ring sideroblasts
RAS	Rat sarcoma virus GTPase
RB/RB1	Retinoblastoma transcriptional corepressor 1
RBAP46 (RBBP7)	Retinoblastoma binding protein P46
RBAP48 (RBBP4)	Retinoblastoma binding protein P48
RBBP4	Retinoblastoma binding protein 4
RBBP7	Retinoblastoma binding protein 7
RBC	Red blood corpuscules
RBCC	RINGB-box-coiled coil
RBFOX2	RNA binding Fox-1 Homolog 2
RBM15	RNA binding motif protein 15
RBPJ	Recombination signal binding protein for immunoglobulin kappa J region
RBX1	RING-box protein 1
RCC	Renal cell carcinoma
RD2	Repressor domain 2
RDW	Red blood cell distribution width

REL	REL proto-oncogene, NF-kB subunit
REMs	regulatory elements
RhAG	Rhesus-associated glycoprotein
RHD	Rhesus blood group, D antigen
RIG-I	Retinoic acid-inducible gene I
RIOK2	RIO kinase 2
RLR	REL homology region
RNA	Ribonucleic acid
RNA-Seq	RNA followed by deep sequencing
RNAPII	RNA polymerase II
RNF4	RING finger protein 4
RNP	Ribonucleoprotein
ROR	Retinoid orphan receptor
RORγt	Retinoic acid receptor-related orphan receptor gamma t / Retinoic acid-related orphan receptor gamma t
Rorc	Retinoic acid receptor-related orphan receptor gamma
ROS	Reactive oxygen species
ROSE	Ranking of super-enhancer
RSL3	RAS-selective lethal 3
RSS	Recombination signal sequence
RSV	Respiratory syncytial virus
RT-PCR	Reverse transcriptase-polymerase chain reaction
RTM	Resident tissue macrophage
RUNX	Runt-related transcription factor
RUNX1	Runt-related transcription factor 1
RUNX2	Runt-related transcription factor 2
RUNX3	Runt-related transcription factor 3
RXR	Retinoid X receptor
RXRα	Retinoid X receptor alpha
S1PR	Sphingosine-1-phosphate receptor
SAGA	Spt-Ada-Gcn5-acetyltransferase complex, SAGA coactivator complex
SALL4	Spalt-like transcription factor 4
SAM	S-adenosyl methionine
SARS-CoV-2	Severe acute respiratory syndrome coronavirus-2
SATB1	Special AT-rich sequence-binding protein-1
Sca1	Stem cell antigen 1 (encoded by Ly6a)
SCD	Sickle cell disease
SCL/TAL1	Stem cell leukemia / T-cell acute leukemia 1
SCN	Suprachiasmatic nucleus
SCUBE1	Signal peptide, CUB domain, EGF-like 1
SE	Super-enhancer
SEC	Super elongation complex
Selplg	Selectin P ligand
SENP	Sentrin (SUMO) specific protease
SENP1	Sentrin-(SUMO)-specific protease 1

SENP3	SUMO specific peptidase 3
SETBP1	SET binding protein 1
SF3B1	Splicing factor 3b subunit 1
SFFV	Spleen focus forming virus
SFN	Sulforaphane
SHM	Somatic hypermutation
shmiR	shRNA embedded within an microRNA architecture
shRNA	Short hairpin RNA / Small hairpin RNA
SIAH	Seven in absentia homolog
SIM	SUMO-interacting motif
SIN	Septation initiation network
SIN3A	SIN3 transcription regulator family member A (formerly mSin3A)
siRNA	Small interfering RNA
Sis	Silencer in the intervening sequence
SKP1	S-phase kinase-associated protein 1
SKP2	S-phase kinase associated protein 2
SL-1	Serum response factor-associated factor SL1
SLAM	Signaling lymphocytic activation molecule
SLAM-seq	Thiol (SH)-linked alkylation for the metabolic sequencing of RNA
Slc2a1	Solute carrier family 2 member 1
Slc2a3	Solute carrier family 2 member 3
SLC7A11	Solute carrier family 7 member 11
SLE	Systemic lupus erythematosus
SLM	Short linear motif
SLO	Secondary lymphoid organ
SLUG	Snail family transcriptional repressor 2
SMAD1	SMAD family member 1
sMAF	Small musculoaponeurotic fibrosarcoma
sMAFF	MAF BZIP Transcription Factor F
sMAFG	MAF BZIP Transcription Factor G
sMAFK	MAF BZIP Transcription Factor K
SMARCA2	SWI/SNF related, matrix associated, actin dependent regulator of chromatin, subfamily A, member 2
SMARCA4	SWI/SNF related, matrix associated, actin dependent regulator of chromatin, subfamily A, member 4
SMARCC1	SWI/SNF related, matrix associated, actin dependent regulator of chromatin subfamily c member 1
SMC3	Structural maintenance of chromosomes 3
SMRT	Silencing mediator for retinoid and thyroid hormone receptors (aka NCOR2)
SNAI1	Snail family transcriptional repressor 1
SNP	Single nucleotide polymorphism
snRNA	Small nuclear RNA
Socs1	Suppressor of cytokine signaling 1

Socs3	Suppressor of cytokine signaling 3
SOD1	Superoxide dismutase 1
SOD3	Superoxide dismutase 3
SOX2	SRY-box transcription factor 2
SOX4	SRY-box transcription factor 4
SOX11	SRY-box transcription factor 11
SP	Single-positive (positive for either CD4 or CD8)
SP1	Specificity protein 1
SP2	Specificity protein 2
SP3	Specificity protein 3
SP4	Specificity protein 4
SP5	Specificity protein 5
SP6	Specificity protein 6
SP7	Specificity protein 7
SP8	Specificity protein 8
SP9	Specificity protein 9
SPI1/PU.1	SPI1 proto-oncogene / Spleen focus forming virus (SFFV) proviral integration protein / Hematopoietic transcription factor PU.1 / Purine-rich box 1
SPIB	Spi-B transcription factor
SPR-1	Sp1-related 1
SPR-2	Sp1-related 2
SQSTM1/ P62	Sequestosome 1
SRAF	SUMO-regulated activation function
SRF	Serum response factor
SRXN1	Sulfiredoxin
SS	Sjogren's syndrome
SS18	Synovial sarcoma translocation, chromosome 18
ST-HSC	Short-term hematopoietic stem cell
STAG2	Cohesin subunit SA-2
STAT	Signal transducer and activator of transcription
STAT1	Signal transducer and activator of transcription 1
STAT3	Signal transducer and activator of transcription 3
STAT5	Signal transducer and activator of transcription 5
STAT5A	Signal transducer and activator of transcription 5A
STAT6	Signal transducer and activator of transcription 6
sTIM-3	Soluble TIM-3
STING	Stimulator of interferon gene
STL	Several sesquiterpene lactones
SUMO	Small ubiquitin like modifier / Sumoylation
SUMO-1	Small ubiquitin like modifier 1
SUMO-2	Small ubiquitin like modifier 2
SUMO-3	Small ubiquitin like modifier 3
SUZ12	Suppressor of zeste 12 protein homolog
SWI/SNF	SWItch/sucrose non-fermentable
SYK	Spleen tyrosine kinase
SYNRG	Synergin gamma

T-ALL	T-cell acute lymphoblastic leukemia
T-BET / TBX21	T-box expressed in T-cells / T-box transcription factor 21
t-BHQ	Tert-butylhydroquinone
T1D	Type 1 diabetes
T21	Trisomy 21
TAD	Transcription activation domain / Transactivation domain
TAF3/TFIID	TATA-box binding protein associated factor 3/ Transcription initiation factor TFIID subunit 3
TAF4	TATA-box binding protein associated factor 4
TAF9	TATA-box binding protein associated factor 9
TAF12	TATA-box binding protein associated factor 12
TAF15	TATA-Box Binding Protein Associated Factor 15
TAL1	T-cell acute leukemia 1 / T-cell acute lymphocytic leukemia 1
TALDO1	Transaldolase 1
TALE	Three-amino-acid-loop-extension
TALEN	Transcription activator-like effector nucleases
TAM	Transient abnormal myelopoiesis
TAT	Cationic, cell-penetrating peptide sequence
TAX	Transactivator protein/p40
tBHQ	tert-Butylhydroquinone
TBX21	T-box transcription factor 21
TCA	Tricarboxylic acid cycle
TCF-1	T-cell factor 1
TCF3	Transcription factor 3 (encodes E2A)
TCF3-PBX1	TCF3-PBX1 Fusion Gene
TCF4	Transcription factor 4
TCF7	Transcription factor 7
TCF12	Transcription factor 12
TCR	T-cell receptor
TEL	Translocation ETS leukemia (former name of ETV6)
TEL2	Translocation ETS leukemia 2 (former name of ETV7)
TET	Tet methylcytosine dioxygenase / Ten-eleven translocation
TET2	Tet methylcytosine dioxygenase 2 / Ten-eleven translocation 2
TET3	Tet methylcytosine dioxygenase 3 / Ten-eleven translocation 3
TF	Transcription factor
Tfb1PH	Tfb1 pleckstrin homology domain
Tfh	T follicular helper
TFIID	Transcription factor II D
Tfr	T follicular regulatory
TFRC	Transferrin receptor

tg	Transgene
TGFβ	Transforming growth factor beta
Th	T-helper
Th9	T-helper 9
Th17	T-helper 17
THC5	Thrombocytopenia 5
THPOK	T-helper inducing POZ-Kruppel like factor
TIE2	TEK receptor tyrosine kinase
TIE2Cre	Cre recombinase under the control of TIE2 regulatory Elements
TIF1γ/TRIM33	Transcriptional intermediary factor 1 gamma, tripartite motif containing 33
TIG	Transcription factor immunoglobulin
TIM-3	T-cell immunoglobulin and mucin-domain containing-3
TIM4	T-cell immunoglobulin and mucin-domain containing 4
TIN	Tumor-infiltrating neutrophil
TIP60	TAT-interacting protein 60kDa
TK	Tyrosine kinase
TKT	Transketolase
TLR	Toll-like receptor
TLR2	Toll-like receptor 2
TLR3	Toll-like receptor 3
TRL4	Toll-like receptor 4
TLR7	Toll-like receptor 7
TLX1	T-cell leukemia homeobox 1
TMD	Transient myeloproliferative disorder
TME	Tumor microenvironment
TNBC	Triple-negative breast cancer
TNF	Tumor necrosis factor
TNF-α	Tumor necrosis factor alpha
TNPO3	Transportin 3
TP53	Tumor protein 53 / Tumor antigen P53
TPA	12-O-Tetradecanoylphorbol-13-acetate
TPEX	Precursor exhausted T-cells
tpms	Transcripts per million
TRADD	Tumor necrosis factor 1-associated via death domain
Traf6	TNF Receptor Associated Factor 6
Treg	Regulatory T-cell
TREM1	Triggering receptor I on myeloid cells
TREM2	Triggering receptor II on myeloid cells
TRIM33/ TIF1γ	Tripartite motif containing 33 / Transcriptional intermediary factor 1 gamma
TRK	Tropomyosin receptor kinase
TRPS1	Transcriptional repressor GATA binding 1
tTreg	Thymus-derived regulatory T-cells

TSA	Trichostatin A
TSHZ1	Teashirt zinc finger homeobox 1
TSLP	Thymic stromal lymphopoietin
TSS	Transcription start site
TXN	Thioredoxin
TXNRD11	Thioredoxin reductase 1
Ubx	Ultrabithorax
UBC9/UBE2I	Ubiquitin carrier protein 9 / ubiquitin conjugating enzyme E2 I
UBE2I/ UBC9	Ubiquitin conjugating enzyme E2 I / Ubiquitin carrier protein 9
UCB	Umbilical cord blood
UCP3	Uncoupling protein 3
UDP	Uridine diphosphate
UGT	Uridine diphosphate-glucuronosyltransferase
ULK1	Unc-51-like kinase 1
UPR	Unfolded protein response
UROS	Uroporphyrinogen III synthase
USF1	Upstream stimulation factor 1
USF2	Upstream stimulation factor 2
USP9X	Ubiquitin-specific peptidase 9 X-linked
UTR	Untranslated region
UV	Ultraviolet
UVB	Ultraviolet B
UXT	Ubiquitously expressed prefoldin like chaperone
V	Variable
VAV	Vav guanine nucleotide exchange factor
Vav-iCre	VAV gene-driven Cre recombinase
VCAM-1	Vascular cell adhesion molecule-1
VCP	VCP alosin-containing protein
VEGF	Vascular endothelial growth factor
VEGFA	Vascular endothelial growth factor A
V_H	Immunoglobulin heavy chain variable gene segment
VHL	von Hippel-Lindau
VPA	Valproic acid
Vpreb	V-Set pre-B cell surrogate light chain 1
VSVΔ51	Vesicular stomatitis virus Δ51
VWF/Vwf	Von Willebrand factor
WDR	WD repeat-containing protein
WDR3	WD repeat-containing protein 3
WDR5	WD repeat-containing protein 5
WFA	Withaferin A
WGD	Whole-genome duplication
WHO	World Health Organization
Wnt	Wingless-related integration site
WT	Wild-type
WT1	Wilms tumor transcription factor 1

XBP1	X-box binding protein 1
XPO1	Exportin 1
XLTT	X-linked thrombocytopenia with thalassemia
YAC	Yeast artificial chromosome
Yan	Former name of Aop
YEATS	YEATS domain containing
YS	Yolk sac
ZBP89	Zinc finger DNA-binding protein 89
ZBTB7A	Zinc finger and BTB (Broad-Complex, Tramtrack, and Bric a brac) domain containing 7A
Zbtb16	Zinc finger and BTB domain-containing protein
ZBTB17	Zinc finger and BTB domain containing 17
ZEB1	Zinc finger E-box binding homeobox 1
ZEB2	Zinc finger E-box binding homeobox 2
ZF	Zinc finger
ZFPM1	Zinc finger protein, friend of GATA family member 1
ZMYND8	Zinc finger MYND-type containing 8
ZNF	Zinc finger factor
ZnF23	Zinc finger protein 2 and 3
ZNF362	Zinc finger protein 362
ZNF384	Zinc finger protein 384
ZNF410	Zinc finger protein 410
ZNF423	Zinc finger protein 423
ZnF456	Zinc finger protein 4, 5 and 6
ZNF521	Zinc finger protein 521
Znfn1a1	Zinc finger protein, subfamily 1A, 1

Part I

Stem/Progenitor Cells

1 MYB: A Key Transcription Factor in the Hematopoietic System Subject to Many Levels of Control

Roza Berhanu Lemma, Bettina Maria Fuglerud, Jon Frampton, and Odd Stokke Gabrielsen

Abstract

MYB is a master regulator and pioneer factor highly expressed in hematopoietic progenitor cells (HPCs) where it contributes to the reprogramming processes operating during hematopoietic development. MYB plays a complex role being involved in several lineages of the hematopoietic system. At the molecular level, the *MYB* gene is subject to intricate regulation at many levels through several enhancer and promoter elements, through transcriptional elongation control, as well as post-transcriptional regulation. The protein is modulated by post-translational modifications (PTMs) such as SUMOylation restricting the expression of its downstream targets. Together with a range of interaction partners, cooperating transcription factors (TFs) and epigenetic regulators, MYB orchestrates a fine-tuned symphony of genes expressed during various stages of haematopoiesis. At the same time, the complex MYB system is vulnerable, being a target for unbalanced control and cancer development.

Keywords

MYB · c-Myb · Transcription factor · Pioneer factor · Enhancer · Epigenetic regulator · Target genes · SUMO

R. B. Lemma
Department of Biosciences, University of Oslo, Oslo, Norway

Centre for Molecular Medicine Norway (NCMM), Nordic EMBL Partnership, University of Oslo, Oslo, Norway

B. M. Fuglerud · O. S. Gabrielsen (✉)
Department of Biosciences, University of Oslo, Oslo, Norway
e-mail: o.s.gabrielsen@ibv.uio.no

J. Frampton
Department of Cancer & Genomic Sciences, College of Medicine & Health, University of Birmingham, Edgbaston, Birmingham, UK

Abbreviations

AML	acute myeloid leukaemia
AMV	avian myeloblastosis virus
ATAC-Seq	assay for transposase-accessible chromatin sequencing
ATF3 and ATF4	activating transcription factor 3 and 4
BACH1	BTB domain and CNC homolog 1
BCL2L11/BIM	BCL2-like 11 (apoptosis facilitator)
BENC	Blood ENhancer Cluster

T. Borggrefe, B. D. Giaimo (eds.), *Transcription factors in blood cell development*, Advances in Experimental Medicine and Biology 1459, https://doi.org/10.1007/978-3-031-62731-6_1

BHLHE40	basic helix-loop-helix family member e40
BMF	Bcl2 modifying factor
BORIS	brother of the regulator of imprinted sites
bZIP	basic region, leucine zipper
CDK9 and CDK12	cyclin-dependent kinase 9 and 12
C/EBPβ	CCAAT/enhancer-binding protein beta
CHD3/Mi-2α	chromodomain helicase DNA binding protein 3
ChIP	chromatin immunoprecipitation
ChIP-Seq	chromatin immunoprecipitation followed by deep sequencing
CML	chronic myelogenous leukaemia
CRD	C-terminal regulatory domain
CREBBP/CBP/KAT3A	CREB-binding protein, histone acetyltransferase CBP
CRISPR	clustered regularly interspaced short palindromic repeats
cRNA	circular RNA
CTCF	CCCTC-binding factor
CTCFL	CCCTC-binding factor like
DBD	DNA-binding domain
DHS	DNase I-hypersensitive site
DN3	double-negative maturation stage 3 of thymocytes (negative for both CD4 and CD8)
EB	elongation block
ENU	ethylnitrosurea
EP300/KAT3B	E1A binding protein p300, histone acetyltransferase p300
ESR1/ERα	estrogen receptor 1, estrogen receptor alpha
FLASH/CASP8AP2	caspase 8 associated protein 2
FOXA	forkhead box A
FOXN3	forkhead box N3
GATA1 and GATA2	GATA-binding protein 1 and 2
GATAD2A	GATA zinc finger domain containing 2A
GRB10	growth factor receptor bound protein 10
HAT	histone acetyltransferase
HbF	foetal haemoglobin
HBG1 and HBG2	haemoglobin subunit gamma 1 and 2
HBS1L	HBS1-like translational GTPase
HDAC	histone deacetylase
HOXB9	homeobox B9
HPC	hematopoietic progenitor cell
HRI	heme-regulated elF2α kinase
HSC	hematopoietic stem cell
HSPC	hematopoietic stem/progenitor cell
HTH	helix-turn-helix
IRF1	interferon regulatory factor 1
JUNB	JunB proto-oncogene, AP-1 transcription factor subunit
KDM1B	lysine demethylase 1B
KIT	KIT proto-oncogene, receptor tyrosine kinase
KIX	kinase-inducible interacting
KLF1 and KLF4	Krüppel-like factor 1 and 4
LDB1	LIM domain binding 1
LMO2	LIM domain only 2
lncRNA	long-noncoding RNA
MAF	MAF bZIP transcription factor

MAFK	MAF bZIP transcription factor K
MAML1	mastermind-like transcriptional coactivator 1
MAP4K4	mitogen-activated protein kinase kinase kinase kinase 4
MAZ	Myc-associated zinc finger protein
MBTD1	mbt domain containing 1
MED1, MED13L and MED14	mediator complex subunit 1, 13L and 14
MITF	melanocyte inducing transcription factor
MOZ/KAT6A	lysine acetyltransferase 6A
MYC	MYC proto-oncogene
ncRNAs	noncoding RNAs
NF-E2	nuclear factor, erythroid 2
NFIA	nuclear factor I A
NF-κB	nuclear factor kappa chain enhancer of activated B cells
NRD	negative regulatory domain
NRSF/REST	RE1 silencing transcription factor
OCT4/POU5F1	POU class 5 homeobox 1, octamer motif-binding factor 4
PAK2	p21 (RAC1) activated kinase 2
PGR	progesterone receptor
PIAS1	protein inhibitor of activated STAT 1
PITX1	paired like homeodomain 1
PPI	protein-protein interaction
PRMT	protein arginine methyltransferase
PTM	post-translational modification
qPCR	quantitative polymerase chain reaction
R1	repeat 1
R2	repeat 2
R3	repeat 3
RB/RB1	retinoblastoma transcriptional corepressor 1
REL	REL proto-oncogene, NF-kB subunit
RUNX1	Runt-related transcription factor 1
SAGA	Spt-Ada-Gcn5-acetyltransferase complex, SAGA coactivator complex
SENP	sentrin-(SUMO)-specific protease
SIM	SUMO-interacting motif
SKP2	S-phase kinase-associated protein 2
SLM	short linear motif
SMARCC1	SWI/SNF-related, matrix-associated, actin-dependent regulator of chromatin subfamily c member 1
SNAI1	snail family transcriptional repressor 1
SNP	single-nucleotide polymorphism
SOX2	SRY-box transcription factor 2
SP	single-positive maturation stage of thymocytes (positive for either CD4 or CD8)
SPI1/PU.1	hematopoietic transcription factor PU.1/spleen focus-forming virus (SFFV) proviral integration protein
SRAF	SUMO-regulated activation function

STAT5A	signal transducer and activator of transcription 5A
SUMO	small ubiquitin-like modifier
TAD	transactivation domain
TAF4 and TAF12	TATA-box binding protein associated factor 4 and 12
T-ALL	T-cell acute lymphoblastic leukaemia
TAL1	T-cell acute leukaemia 1
TET2	tet methylcytosine dioxygenase 2
TF	transcription factor
TIF1γ/TRIM33	transcriptional intermediary factor 1 gamma, tripartite motif containing 33
TPEX cells	precursor exhausted T-cells
Treg	regulatory T-cell
TSA	trichostatin A
TSS	transcription start site
UBC9/UBE2I	ubiquitin-conjugating enzyme E2 I
UTR	untranslated region
UXT	ubiquitously expressed prefoldin-like chaperone
WT	wild type

1.1 Introduction

The transcription factor (TF) MYB[1] is highly expressed in hematopoietic progenitor cells (HPCs) and plays a key role in regulating expression of genes involved in their differentiation and proliferation [1–9]. In haematopoiesis, MYB is required for the normal development of progenitor cells, and its down-regulation is necessary for their terminal differentiation. Studies in mice have shown that MYB is essential for normal haematopoiesis [10]. More specifically, it is required for the development of hematopoietic precursors rather than for their generation [11–13]. In situations where high MYB levels are maintained, normal hematopoietic differentiation is suppressed, and leukemic transformation may be promoted [1, 5]. This is the case in many human lymphoid and myeloid acute leukemias [14].

MYB appears to be involved at multiple stages of hematopoiesis, including crucial steps during T- and B-cell development [15–17]. MYB is required for differentiation to the pro-B-cell stage, survival during the pro-B-cell stage and the pro-B to pre-B-cell transition during B lymphopoiesis [18]. Several of the target genes acting as downstream mediators of the MYB-regulated function in these processes have been identified [19].

MYB is required during multiple stages of T-cell development [15, 16] and is essential for generating and maintaining stem cells in the $CD8^+$ T-cell memory compartment [20]. MYB is also involved in the gene regulatory network controlling the differentiation and function of thymus-derived effector regulatory T-cells (Treg), in which it regulates a large proportion of genes specific to these cells [21]. MYB is also involved in gene regulation in T cells via Notch signalling through the recruitment of RUNX1 and MYB to enhancer chromatin during thymocyte development [22]. Recently, MYB was found to orchestrate yet another subpopulation of T cells, precursor exhausted T-cells (TPEX) cells. When a pathogen- or tumor cell-derived antigen cannot be cleared, T cells form a stem-like compartment with the so-called TPEX cells. Tsui et al. [23] demonstrated that MYB controls the formation, maintenance, and therapeutic function of these cells. TPEX cells constitute a stem-like population that is central to the maintenance of long-term antiviral immunity and responsiveness to immunotherapy.

Due to its role in lineage determination and control of other transcription factors (TFs), MYB has been described as a master regulator [24–27]. In this chapter, we will focus on the

[1]For the sake of simplicity, we have used the capitalised human designation MYB throughout this review, as we assume that studies of Myb/c-Myb in non-human vertebrates are also relevant for human MYB. The same applies to other factors.

normal functions of MYB as a TF, on what controls its expression, and what its downstream targets are. Established knowledge is treated briefly to accommodate more of recent findings. The accompanying Chap. 15 will focus on the role of MYB in acute myeloid leukaemia (AML).

1.2 The MYB Protein and Its Functional Domains

Traditionally, the MYB protein (640 amino acids in humans) has been described as being composed of three functional regions: (I) a highly conserved DNA-binding domain (DBD) located in the N-terminal region of the protein, (II) a centrally located transactivation domain (TAD), and (III) a negative regulatory domain (NRD) in the C-terminal region [28] (Fig. 1.1a).

The importance of each of these three regions was illustrated by the outcome of two screening studies in mice, where genome-wide ethylnitrosourea (ENU) mutagenesis was employed to identify genes controlling specific hematopoietic processes [29, 30]. Both studies hit the *MYB* gene as the main affected locus. One study identified a point mutation in *MYB* that causes thrombocytosis, megakaryocytosis, anaemia, lymphopenia and lack of eosinophils [29]. This mutation, M303V, affects the TAD of MYB. The other study identified two point mutations (D152V and D384V) that were able to rescue mice from a myeloproliferative syndrome with supra-physiological expansion of megakaryocytes and platelet production [30]. D152V is located in the DBD and D384V in the NRD region of MYB.

According to the recent AlphaFold-model [31, 32] of the human MYB protein (https://alphafold.ebi.ac.uk/entry/P10242), its largest folded region is the DBD. In addition, the model suggests four additional smaller folded regions in the TAD and NRD regions, respectively (Fig. 1.1c). The majority of the protein is being modelled as disordered.

1.2.1 DNA-Binding Domain

The highly conserved DBD is composed of three imperfect repeats, R1, R2 and R3, each of 52–53 amino acids and with a characteristic tryptophan-rich signature motif (Fig. 1.1b). The R1, R2 and R3 repeats each contain three alpha-helices of which the last two form a helix-turn-helix (HTH)-related motif (Fig. 1.2a) [33–35]. The two last repeats constitute a minimal DBD sufficient for sequence-specific DNA binding, while R1 appears to have no specific interactions with DNA (Fig. 1.2a, lower panel) [35, 36]. Despite the close similarity between the repeats, they show dynamic differences. While R1 and R3 are tightly folded in solution [37, 38], several lines of evidence suggest that R2 is less ordered and does not generate a full HTH motif until it interacts with DNA [39, 40].

The DNA motif bound by the MYB DBD (MYB recognition element) was defined from in vitro studies as YAACNG, [41, 42], or a more refined version thereof [43, 44]. Using a dedicated yeast screening system, an extended optimal MYB recognition element consensus was defined closely mimicking known strong response elements and where a sequence of 11 nucleotides influence activity including a T-rich flanking sequence [45]. The motif most frequently used in chromatin, as deduced from ChIP-Seq peaks, is consistent with the general motif (Fig. 1.2b). Since R2+R3 are responsible for this recognition, the MRE mirrors this and appears to be bipartite. The first half-site [YAAC] is absolutely required for DNA binding, while the second half-site [NGHH] can adopt a range of sequences mainly affecting the half-life of the protein/DNA complex in vitro [46]. The orientation of the MYB-DNA complex is such that the first half-site interacts mainly with R3 and the less well-defined second half-site has mainly specific contacts with the R2 subdomain [35, 46].

In addition to its function as a recognition domain for a specific DNA sequence motif, the DBD is also able to associate directly with histones, which may play a role in chromatin inter-

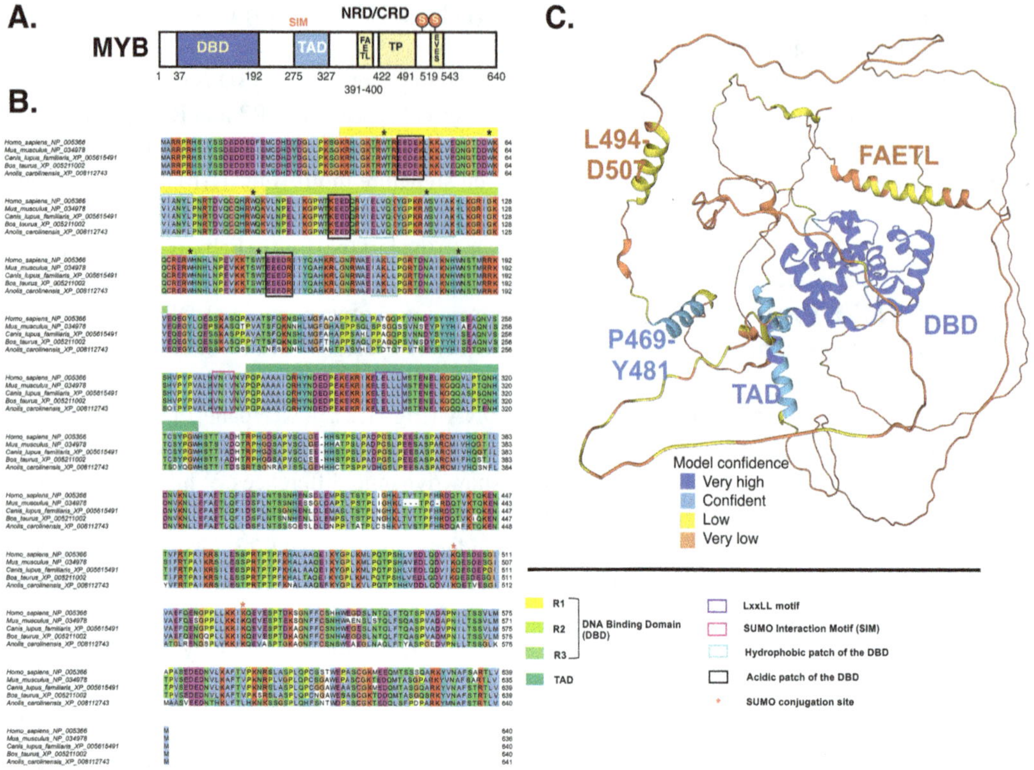

Fig. 1.1 The MYB protein with its domains. (**b**) A simplified sketch of MYB with its classical domain divisions. (**b**) Multiple sequence alignment (MSA) of selected MYB homologs in vertebrates. The DNA-binding domain (DBD) and its repeats R1, R2 and R3 are represented by bars with a yellow to green colour palette above the sequences. The signature Trp residues within each repeat are indicated with a black star (*). The transactivation domain (TAD) is represented by a green bar above the sequences. The LxxLL motif, SIM (SUMO Interaction Motif) and the hydrophobic and acidic patches of the DBD are marked with violet, purple, cyan and orange-yellow boxes, respectively. The two SUMO conjugation sites (K503 and K527) are highlighted with a red star (*). The MSA was generated using the MAFFT algorithm accessible through Jalview. The resulting alignment was visualized in Jalview v.2.10 [177]. ClustalX colouring scheme was used for displaying the MSA. **C.** The AlphaFold model [31, 32] of the human MYB protein (https://alphafold.ebi.ac.uk/entry/P10242)

action and decompaction. The repeats in the DBD are closely related to the chromatin-interacting SANT domain, found in other chromatin-regulating proteins and involved in recognition of histone tails [47, 48]. In line with this, the MYB DBD has been found to interact with histone tails [49, 50] and facilitate histone tail acetylation by recruitment of the HAT EP300/KAT3B [49]. Both the R2 and the R3 repeats seem to interact individually with H3, suggesting that MYB has at least two histone-binding modules within its DBD [51]. Moreover, the D152V mutant mentioned above interferes with this ability to bind histones. D152 is located in an acidic patch of the R3 repeat within the first α-helix of the repeat (Fig. 1.2a), and the D152V mutation weakens the binding to H3 by impairing the R3 histone-binding module while having no effect on DNA binding [51]. As a consequence, the ability of MYB to increase chromatin accessibility is impaired by the D152V mutation.

1.2.2 Transactivation Domain (TAD)

For human MYB, the exact borders of its TAD have not been defined, but based on early studies where a series of deletions were analysed, the

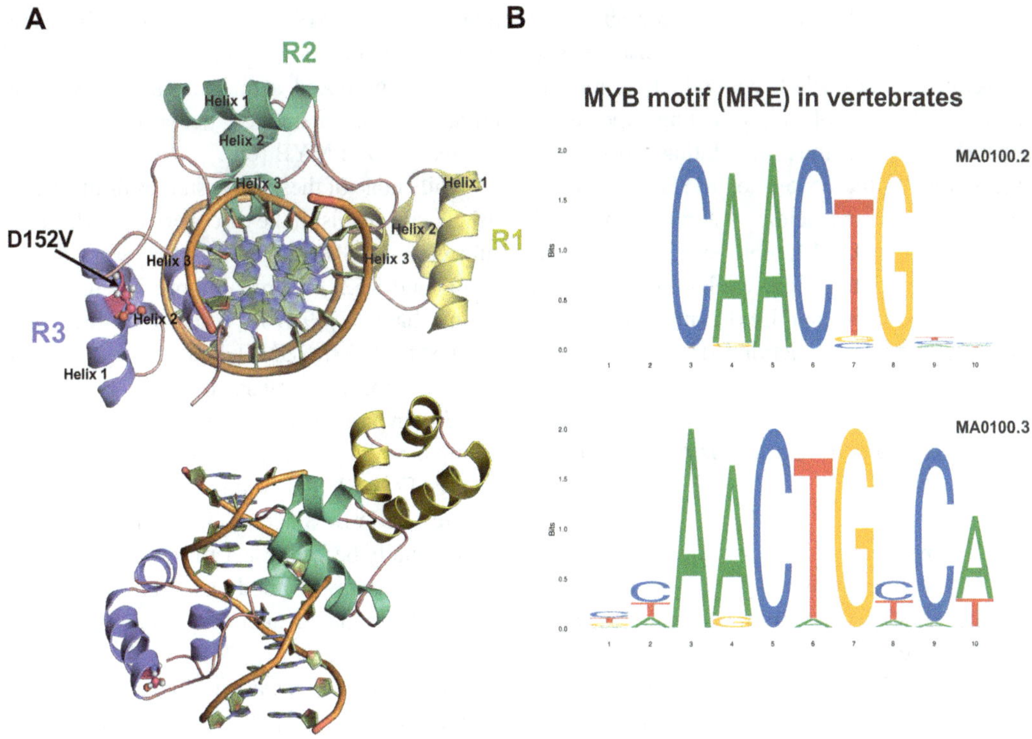

Fig. 1.2 Structural depiction of MYB DBD in complex with the DNA. (**b**) The structural depiction was obtained by fusing the NMR structure of the human MYB containing R2 and R3 in complex with DNA (**PDB ID: 1MSF**) with that of the crystal structure of mouse MYB as a ternary complex between the DNA and C/EBPβ **PDB ID: 1H88**. The human MYB R2 is displayed in green cyan colour, and the human MYB R3 is displayed in a slate blue colour, whereas the mouse R1 is displayed in olive yellow colour. Each repeat forms an HTH-related structure with three α-helices. The fusion between human MYB and mouse MYB structures was made using the *ce align* algorithm of pyMol. The irrelevant parts from the ternary structure PDB ID: 1H88 are hidden to only show part of the repeat that was missing from the PDB ID: 1MSF. PyMOL v.2.1.1 was used to render the structure. (**b**). MYB motif obtained from JASPAR database [178]. The two MYB motifs from the core collection of JASPAR, MA0199.2 and MA0100.3 are displayed. The MYB JASPAR motifs reflect the two MREs

TAD in MYB was localized to be in the amino acid residue range of 275–327 [52] (Fig. 1.1a and b). The TAD of MYB is acidic and contains a short linear motif (SLM) in the form of a LxxLL motif (aa 298–302) that is involved in the interaction with the KIX domain of the coactivators CREBBP/CBP/KAT3A and EP300/KAT3B [53–57].

According to the AlphaFold model [31, 32] of the MYB protein (https://alphafold.ebi.ac.uk/entry/P10242) (Fig. 1.1c), this central region contains an α-helix P289-G311, spanning the LxxLL motif. It has been suggested that KIX binding and folding are coupled leading to the formation of a helix upon interaction of TAD with the KIX domain of CREBBP/CBP/KAT3A and EP300/KAT3B [58, 59]. Consistent with this, recent dynamic modelling concluded that binding of a short stretch of amino acids in the TAD region of MYB (especially the LxxLL motif) establishes initial native contacts that prime the binding and folding of the rest of the peptide [60].

Is then the LxxLL motif the only feature important for TAD function or does the acidic patch also play a role? A recent mutational analysis addressed this question and observed strong effects exerted by both acidic replacements and mutations in the LxxLL motif [61]. This behaviour seems to fit the recently proposed SLM-exposure model of Slater et al. [62]. Their general TAD model comprises a specific hydrophobic SLM embedded in a disordered acidic region. The idea is that the intrinsic disorder and the acidic residues help keeping the hydrophobic motifs exposed to solvent, providing exposure to binding partners, thus facilitating coactivator interaction. In the context of MYB, this means that the disordered acidic context in the TAD region keeps the LxxLL motif exposed and available for KIX interaction, explaining why both acidic replacements and LxxLL mutations could totally abolish TAD function [61].

1.2.3 Negative Regulatory Domain (NRD)

The designation of most of the C-terminal half of the MYB protein as a 'negative regulatory domain' (NRD) comes from early studies showing that C-terminal deletions of chicken, mouse and human MYB greatly increased the activity of MYB in transactivation assays [28, 52, 63]. Moreover, the oncogenically activated versions of chicken MYB, avian myeloblastosis virus (AMV) v-MYB and E26 v-MYB, had large parts of this region deleted [64].

Detailed deletion studies of chicken MYB located the NRD function to two essentially non-overlapping regions of the protein corresponding to amino acids 425–464 and 499–558 [63, 65]. Although predicted to be largely disordered, the AlphaFold model of MYB [31, 32] predicts three smaller helices in the NRD, at T381-S401 (spanning the FAETL-motif), at P469-Y481 and at L494-D507, where the P469-Y481 has the best model confidence of the three (Fig. 1.1c).

No clear picture of the mechanism of negative regulation has been found, although functional effects have been attributed to specific subdomains termed FAETL and EVES [66, 67]. Fu and Lipsick defined a 10-amino-acid subregion (the 'FAETL' motif, 391–400 in human MYB) to be required for both transcriptional activation and oncogenic transformation by AMV v-MYB [67]. A helix-breaking insertion mutation in this region resulted in a large increase in transcriptional activation [68]. Dash et al. defined a conserved stretch of about 25 amino acids, termed the EVES motif, spanning 519–543 in human MYB, as an important region [66]. Based on yeast two-hybrid data, they proposed a mechanism involving an intramolecular interaction between this region and the DBD of MYB [66].

While none of these early studies firmly established a mechanism, the discovery of NRD being subjected to SUMO conjugation at two sites (K503 and K527) is probably one key feature of the mechanism [69, 70]. When these sites were mutated to abolish SUMOylation, a large increase in MYB-dependent transactivation was observed [69, 70], in particular on complex promoters with multiple binding sites for MYB because of lost synergy [71]. SUMO conjugation is not a unique feature of MYB. Since its discovery in 1996, it has gradually become clear that SUMOylation is a post-translational modification (PTM) quite frequently found in nuclear proteins, including TFs and coactivators [72–76].

Concerning the negative effect of the NRD, Molværsmyr et al. [71] identified a SUMO-dependent switch in the function of the NRD. The activity of MYB is not only reduced when NRD is SUMOylated, but NRD switches into being activating when SUMOylation is abolished, providing the SUMO-negative MYB with two activation functions, one in TAD that is constitutively active and one in the NRD being SUMO-regulated (termed SRAF, SUMO-regulated activation function). This was the basis for a proposal of renaming NRD simply as CRD, C-terminal regulatory domain [71].

One of the proposed functions of SUMOylation is to modulate protein-protein interaction (PPI), where the SUMO moiety can interact with other proteins through their SUMO-interacting motifs (SIM) [77]. Based on the SIM consensus, Sæther et al. [78] identified a functional SIM just in front of the TAD of MYB (V_{267}NIV). Abolishing the SUMO-binding properties by mutation resulted in a large increase in the transactivation potential of MYB. The combination of SUMO conjugation and SUMO binding in the same protein might indicate an intramolecular interaction, but the data argued against such a model. However, it was more surprising that only loss of SUMO binding, and not SUMO conjugation, enhanced the myeloid transformational potential of MYB. The SIM mutated MYB conferred a higher proliferative ability than the wild type (WT) and caused an effective differentiation block [78].

1.3 Upstream of MYB: What Controls Its Expression

1.3.1 Enhancers Controlling MYB Expression

As a lineage cell-specific TF, regulating hematopoietic stem/progenitor cells (HSPCs), *MYB* expression is expected to be controlled by a set of lineage-specific enhancers. The nature and use of these enhancers controlling the *MYB* gene are not fully understood, but some key elements have been identified. An erythroid enhancer was first dissected in an extensive study by Stadhouders et al. [79]. MYB is a critical regulator of erythroid development, and the intergenic region (125 kb) on chromosome 6q23, between the *HBS1L* and *MYB* genes, was found to harbour two erythroid-specific enhancers located at −84 and −71 kb upstream of *MYB* TSS (Fig. 1.3). These distal enhancer elements were occupied by several erythroid TFs, and long-distance physical interaction with the *MYB* gene was detected (*see* Fig. 1.3b). Interestingly, this region contains common single-nucleotide polymorphism (SNP) variants that are associated with elevated levels of foetal haemoglobin (HbF) and alterations of other clinically important human erythroid traits. High foetal globin gene expression is a crucial feature decreasing the severity of β-thalassaemia and sickle-cell anaemia [80–83]. This is why this region has been studied in some detail. While the SNPs in the region have no effect on *HBS1L* expression, they clearly influence the expression of *MYB*, suggesting a causal link between sustained HbF production and reduced levels of MYB. Notably, the *HBS1L-MYB* region is composed of 98 DNase I-hypersensitive sites (DHSs) identified in erythroid precursors and thus may have an even more complex enhancer landscape [84].

All these DHSs were analysed by Canver et al. [84] in the immortalized human erythroid cell line HUDEP-2 using saturating-mutagenesis screens with multiple CRISPR-associated nucleases allowing them to identify putative regulatory elements that control *MYB* expression. Here they confirmed *MYB* enhancer function in the −84 kb region and identified a novel *MYB* enhancer at −36 kb (*see* Fig. 1.3). They also observed decreased *MYB* expression after targeting the sequences within the −126, −83, −71 and −7 kb regions, suggesting that these regions may also contain *MYB* enhancer activity, although the authors remained cautious due to the increased off-target cleavage potential caused by repetitive sequences in their proximity.

Recently, Matsumura et al. [85] identified two additional functional enhancers controlling lineage-specific expression of *MYB*. Using an in silico approach, exploring a combination of clusters of retroviral integration sites representing open accessible chromatin around the MYB locus and available ChIP-Seq data for hematopoietic TFs, they identified a total of 11 candidate regulatory elements. Two of these were validated as functional in a zebrafish model and in mice. The *MYB* -74 enhancer was found to be activated robustly in immature T and B cells in mice. During B-cell development, the -74 enhancer appeared to be required to turn on the pre-pro-B-cell stage. In the T-cell lineage, the enhancer was active in early T lineage progenitor stages, as well as in the double negative 3 (DN3) stage and the CD4/CD8 single positive (SP) stage. In both lineages, the enhancer activity declined with B- or T-cell maturation. The *MYB* -68 enhancer showed a distinct activity profile being induced in granulocyte-/macrophage-lineage progenitors. How this *MYB* −68 enhancer regulates the differentiation of mast cells and basophils was studied in quite some detail.

Another recent study [86] of the upstream region of the *MYB* gene identified two functional enhancer regions −34k and −88k in the in the human myeloid leukaemia cell lines K562, U937 and HL-60. These enhancers were defined by long-range interaction between the upstream regions and the *MYB* promoter (using 4C assays), enhancer marks (H3K4me1 and H3K27ac), ChIP-qPCR and luciferase assays. The activity of

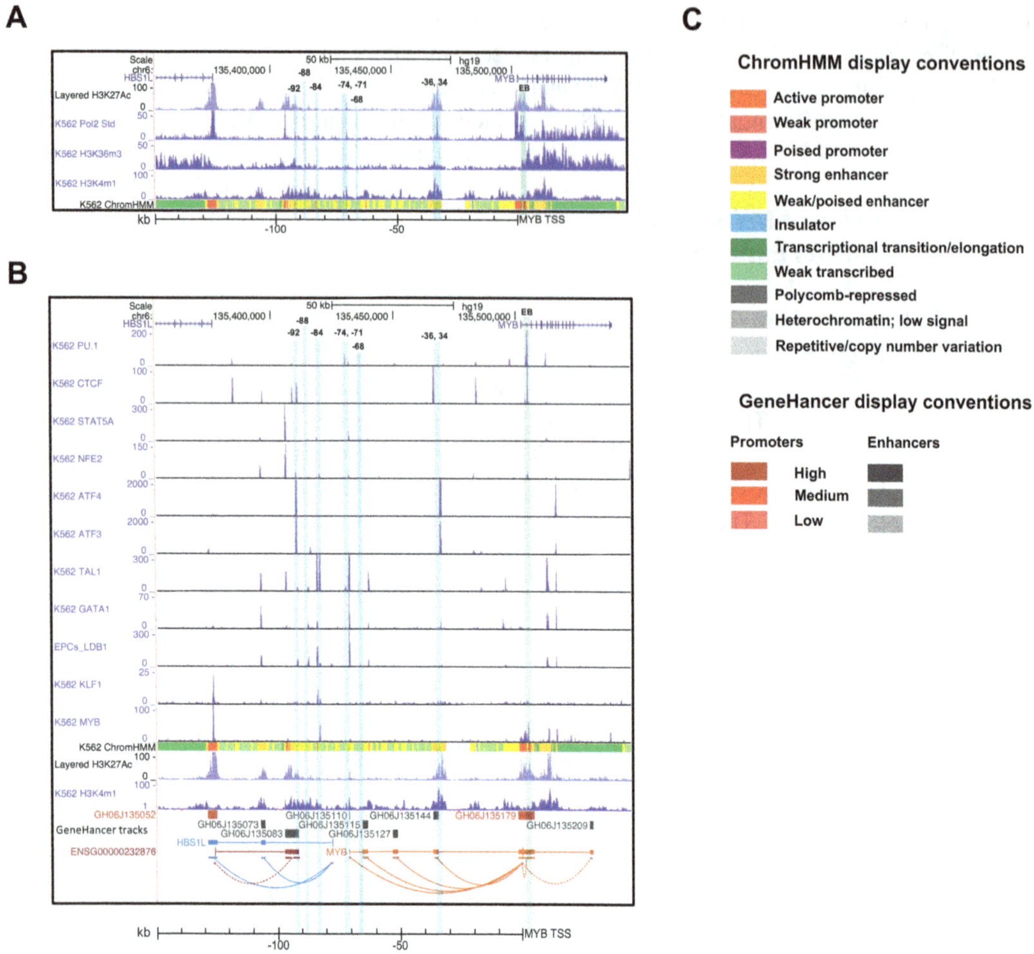

Fig. 1.3 Enhancers controlling MYB expression. (**a**) UCSC track showing ChIP-Seq profiles of various histone marks and RNA polII that align with important chromatin features such as enhancers and promoters. ChIP-seq occupancy for histone marks for H3K27ac, H3K36me3 and H3K4me1 along with the RNAPII ChIP-Seq occupancy is displayed. Chromatin segmentation data for K562 from chromHMM is also displayed in the UCSC genomic track display. Previously, well-characterized enhancers at the MYB-HBS1L locus identified from literature are highlighted with light green colour, and their genomic positions relative to MYB TSS are provided above each highlight. (**b**) UCSC track showing K562 ChIP-Seq profiles of various TFs that are known to regulate MYB expression along with MYB ChIP-Seq occupancy from K562 (GEO accession: GSE124541) from [150]. ChIP-Seq occupancy tracks for PU.1 (ENCODE-DCC accession, ENCFF562THD), CTCF (ENCODE-DCC accession, ENCFF468HJA), STAT5A (ENCODE-DCC accession, ENCFF746KSV), NF-E2 (ENCODE-DCC accession, ENCFF399JNI), ATF4 (ENCODE-DCC accession, ENCFF111CNS), ATF3 (ENCODE-DCC accession, ENCFF594WSC), TAL1 (ENCODE-DCC accession, ENCFF438KVV), GATA1 (ENCODE-DCC accession, ENCFF724KSO), LDB1 (GEO ID, GSM1273197) and KLF1 (ENCODE-DCC accession, ENCFF153SCC) along with public tracks for histone marks, chromHMM and GeneHancer regulatory elements (REMs) accessed through the UCSC genome browser were displayed. Display conventions for chromHMM and GeneHancer tracks are provided in panel **c**. Well-characterized enhancers at the MYB-HBS1L locus identified from literature are highlighted with light green colour, and their genomic positions relative to MYB TSS are provided above each highlight. In addition, we highlighted in light green the region within intron 1 of MYB gene, referred here as elongation block (EB). The GeneHancer tracks from the elite collection were displayed. (**c**) Display conventions for chromHMM and GeneHancer tracks

these enhancers declined when MYB was down-regulated during chemical-induced differentiation.

In a study of stress erythropoiesis where loss of β-globin triggers a rapid up-regulation of several globins, including very high levels of γ-globin, Boontanrart et al. [87] found that this regulation was mediated by enhancer elements in the *HBS1L-MYB* intergenic region, more precisely at -92 kb, and occurred through ATF4-mediated reprogramming (Fig. 1.3a and b). The conclusion was that this so-called β0-stress regulates γ-globin levels via MYB.

1.3.2 Specific TFs Regulating MYB Expression

Closely related to the identification of enhancers controlling *MYB* expression is the binding of key hematopoietic TFs to cis-regulatory elements in these enhancer regions. While the expression profiles of a range of hematopoietic TFs have been well characterized at the different stages of haematopoiesis [88–94], how these factors act and cooperate at the many regulatory elements controlling individual genes, such as the *MYB* gene, is far from fully elucidated.

As we have seen, the enhancers controlling *MYB* expression during erythropoiesis are relatively well dissected. This also applies to which erythroid TFs operate at these elements. For illustration, a selection of ChIP-Seq profiles for some of the relevant TFs (in K562 cells) is shown in Fig. 1.3b. Most of the factors shown are associated with positive regulation of *MYB* expression. TFs controlling erythropoiesis such as LDB1 (LIM domain-binding protein 1), GATA1 (GATA binding protein 1), TAL1 (T-cell acute leukaemia 1) and KLF1 (Krüppel-like factor 1) are found associated with the erythroid enhancers at -84 and -71 kb in primary human erythroid progenitors [79] (Fig. 1.3b). GATA1 is one of the key erythroid 'master regulators' and essential for erythropoiesis. TAL1 belongs to the family of basic helix–loop–helix (bHLH) transcription. It plays a critical role in early hematopoietic specification and is essential for development of all hematopoietic lineages (*see* Chap. 11). TAL1 and GATA1 often bind together at a compound motif [95]. Interestingly, most DNA-bound GATA1 and TAL1 proteins are contained within higher-order complexes (LDB1 complexes) that include the nuclear adapters LDB1 and LMO2 [96]. An important role of LDB1 complexes is to facilitate, via LDB1-mediated oligomerization, long-distance interactions, including the juxtaposition of erythroid promoters and enhancer elements. The LDB1 complex often colocalises and synergises with KLF1. KLF1 is an erythroid lineage restricted factor and one of the primary players in global erythroid gene regulation (*see* Chap. 10).

In the K562 cell line, dynamic binding of GATA1, TAL1 and CCAAT/enhancer-binding protein beta (C/EBPβ) was found at −34k and −88k during differentiation of the cells [86] (Fig. 1.3b).

Variants of the LDB1 complex are also found in non-erythroid hematopoietic lineages, in particular in hematopoietic stem cells (HSC) where GATA2 has replaced GATA1 [96]. This variant of the LDB1 complex appears to regulate a transcriptional program required for HSC maintenance [97]. This is part of the so-called 'GATA-switch' where GATA2, which is highly expressed in hematopoietic progenitors, is replaced by GATA1 during commitment of HSCs to the erythroid lineage [98].

MYB expression is also under the control of ATF4, which binds at an ATF4 consensus motif within the ATF4-controlled enhancer mentioned above [87] (*see* Fig. 1.3). ATF4 is a stress-induced basic leucine-zipper TF that dimerizes with other bZIP partners [99]. It is under translational control regulated by HRI (heme-regulated eIF2α kinase)-mediated phosphorylation of eIF2α [100]. The ATF4-*MYB* connection was demonstrated by MYB being down-regulated in ATF4 knockout cells, while overexpression of ATF4 rescued MYB levels [87]. This creates a regulatory pathway triggered by a loss of β-globin levels that leads to reduced ATF4 levels, which again decreases the levels of MYB, which finally causes an up-regulation of γ-globin and foetal haemoglobin.

Even if the regulatory system controlling *MYB* expression during erythropoiesis starts to be quite well defined, there are still elements bound by other TFs whose role is not fully explored in relation to *MYB*. One example shown in Fig. 1.3b is NF-E2, (nuclear factor, erythroid 2). The NF-E2 complex is essential for regulating erythroid and megakaryocytic maturation and binds together with MAFK or other small MAF proteins to the NF-E2 motif [101, 102].

How *MYB* expression is controlled in other lineages of the hematopoietic system is less well characterized, and its regulatory mechanisms there are incompletely defined.

The two non-erythroid enhancers identified by Matsumura et al. [85] are bound by multiple TFs in mouse hematopoietic progenitor HPC-7 cells, mouse erythroleukemia (MEL) cells and human T-cell acute lymphoblastic leukaemia (T-ALL) Jurkat cells. For the *MYB* -74 enhancer, activated in immature T and B cells, these TFs include TAL1, GATA3, RUNX1, TCF12 and TCF3. For the *MYB* -68 enhancer, induced in granulocyte-/macrophage-lineage progenitors, these TFs include TAL1, LMO2, GATA2, RUNX1, MEIS1, SPI1, FLI1 and GFI1B. The authors speculate that the key mast cell regulator MITF, which also binds to the *MYB* -68 enhancer, is another key element activating this enhancer. However, the specific role of these TFs on these enhancers was not further studied.

RUNX1 seems to play a role in maintaining *MYB* enhancer activity in T-ALL cells by binding to the two enhancer regions located at –92 kb and +15 kb of the mouse *MYB* gene [103]. This binding has also been shown to support the expression of the TAL1 oncogene, as part of an autoregulatory feedback loop involving GATA3 and MYB. *RUNX1* deletion in mouse T-ALL cells interferes with *MYB* enhancer activity.

In Fig. 1.3, in the lowest panel, we have also incorporated GeneHancer tracks [104], which identify putative human regulatory elements (enhancers and promoters) linked to their inferred target genes. The GeneHancer database was created by integrating >1 million regulatory elements from multiple genome-wide databases. The elements here are not limited to erythroid cells and thus illustrate that the regulation of MYB may have additional layers in other lineages of the hematopoietic system as well as in other tissues.

1.3.3 The MYB Promoter

The proximal promoter of the *MYB* gene harbours several GC boxes and binding sites for various transcriptional regulators, but despite some early studies [105–109], their joint action and relative importance have not been extensively studied. Here we will just point to some particular features of the proximal promoter, one being a highly conserved purine-rich region with three copies of four GGA repeats, [3x$(GGA)_4$], located 15-17 bps downstream of the TSS (on the bottom strand). Such sequences are known for being able to form unusual DNA structures, such as a i-motif structure [110]. A study from Palumbo et al. showed that this GGA repeat region of the murine *MYB* promoter was able to form special G-quadruplex structures [tetrad:heptad:heptad:tetrad (T:H:H:T)] [111]. Their functional analysis indicated that this unusual structure is both necessary for promoter activity and that it may act as a negative regulator of the *MYB* promoter. The mechanism proposed was that the MYC-associated zinc finger protein (MAZ) interacts with the T:H:H:T G-quadruplexes and represses the promoter activity. According to another study [112], the MAZ factor may also have a quite different role on the *MYB* promoter. This study found that a protein complex containing MAZ binds the *MYB* promoter in vivo and activates transcription via a binding site that overlaps an E2F element in the *MYB* promoter. The latter element would normally cause cell cycle repression by the RB proteins regulating the E2F family of TFs. The authors hypothesize that MAZ is essential to bypass *MYB* promoter repression by RB family members and to induce *MYB* expression. In their study, MAZ knockdown was found to inhibit *MYB* expression.

Overlapping with the GGA repeats is a binding site for the hematopoietic TF SPI1/PU.1, a purine-rich DNA core sequence 5′-GAGGAA-3′, also known as the PU box. Early studies showed that PU.1 negatively regulates the *MYB* promoter during granulocytic differentiation [105]. The logic proposed by the authors was that the ability of PU.1 to potentiate terminal myeloid differentiation involved down-regulation of *MYB* expression. However, ChIP-Seq data (*see* Fig. 1.3) suggest that SPI1/PU.1 has a more important binding site further down in intron 1 close to a regulatory site there (*see* below), at least in K562 cells. Its effect there has not, to our knowledge, been studied.

Another aspect of the proximal region is that *MYB* expression is regulated at the level of transcriptional elongation through an attenuation site or elongation block (EB) in the first intron 1700 bp downstream of TSS (marked EB in Fig. 1.3). The pause here is believed to be imposed by a stem-loop forming sequence followed by a poly (dT) tract [5, 113, 114]. This region was found to interact with the distal -36 and -81 kb elements [115]. Mapping of the phosphorylation status of RNAPII and elongation type of histone marks (H3K36me3) showed that the transition to productive transcription elongation occurs around the intronic site, which is located in the vicinity of a highly conserved CTCF binding site, suggesting some type of chromosome conformation being involved (Fig. 1.3). Release from this pause involves the positive elongation factor P-TEFb known to regulate pause release through phosphorylation of Ser2 in RNAPII [116]. P-TEFb was proposed to be recruited by ESR1/ERα in breast cancer cells [117], but it may also be recruited through looping involving the LDB1 complex, since the latter interacts with transcription elongation factors, like TIF1γ and CDK9 [115]. Members of the nuclear factor kappa chain enhancer of activated B-cells (NF-kB) family may also be involved [118, 119]. Furthermore, this mechanism may be exploited by MLL fusion oncoproteins, which are able to induce acute myeloid leukaemias by altering *MYB* expression. It was reported that the MLL-AF9 fusion protein regulates *MYB* transcription through this mechanism by recruiting the transcriptional regulator kinase CDK9 [120].

An intriguing aspect of the *MYB* proximal promoter is a switch in the use of TSS as seen particularly in some types of cancers [121, 122]. While the normal TSS (TSS1, RefSeq NM_005375) dominates in normal cells, an alternative TSS (TSS2) in intron 1 close to the start of exon 2 (GenBank X52126) has been found to dominate in acute lymphoblastic leukaemia (ALL) [121] and in adenoid cystic carcinoma (ACC) [122]. This leads to the expression of an N-terminally truncated MYB protein where the 20 first (and highly conserved) amino acids of the MYB protein are lost. Since this alternative TSS is downstream of the elongation block in intron 1, it would probably not be under the same elongation control. The consequence appears to be an altered set of genes regulated by the truncated MYB versus the normal MYB [122]. It is worth noticing that also the classical v-MYB encoded by the AMV has a large N-terminal deletion of 71 amino acids while keeping a functional DBD [64]. In ALL, the increased use of TSS2 was significantly associated with relapse and therapy-resistant disease [121] (*see* also the accompanying Chap. 15 on MYB in AML).

1.3.4 Other Regulatory Mechanisms

On top of the elaborate transcriptional control described above, several reports have established that *MYB* expression is also regulated at the post-transcriptional level by noncoding RNAs (ncRNAs), such as microRNAs, long-noncoding RNAs (lncRNAs) and circular RNAs (cRNAs) [123]. Xiao et al. first established that miR-150, expressed in mature lymphocytes, directly regulates MYB protein levels through two conserved miR-150 binding sites in the 3′-untraslated region (UTR) of the MYB mRNA [124]. Since this first example, several other studies have uncovered additional ncRNAs affecting MYB causing an intricate crosstalk between MYB and ncRNAs. This topic was recently reviewed [123] and will not be covered further here.

1.4 Downstream of MYB: Gene Programs that MYB Activates

1.4.1 MYB as a Cofactor in Epigenetic Writing

Specific histone PTMs are associated with enhancers and promoters and their state of activity [125, 126]. How these modifications are directed at these regions is not fully understood, but it is clear that the targeting of histone modifications involves direct interactions between the histone-modifying enzymes and the DNA-binding TFs. Several studies have shown links between MYB and histone tail modifications. One example is the recruitment of the HMT complex MLL1/KMT2A and PRMT4 to target genes by MYB, both leading to activation of MYB-dependent transcription [127, 128]. The chromatin remodeller CHD3/Mi-2α acts as a coactivator of MYB-dependent transcription [129]. However, the most extensively studied interaction of MYB with a histone-modifying enzyme is with the HATs EP300/KAT3B and CREBBP/CBP/KAT3A [53–57]. These HATs stimulate MYB-dependent transcriptional activation by acetylating the NRD of MYB and the neighbouring chromatin [130]. The main interaction surface has been mapped to the TAD of MYB and the KIX domain of the coactivators [53]. The C/H2 domain of CREBBP/CBP/KAT3A may further enhance the interaction [131]. The purpose of this interaction has been suggested to be a positioning of histones by MYB for EP300/KAT3B to acetylate [49]. Mo and co-workers showed that when the interaction with H3 is lost in MYB, the interaction with EP300/KAT3B remains intact, but H3 was not acetylated at lysines 18 and 23. The lost H3 interaction resulted in a block in activation of differentiation-associated genes which could be rescued by treating the cells with the HDAC inhibitor trichostatin A (TSA), suggesting that facilitating histone acetylation is a crucial part of MYB-dependent transcriptional activation. Later, the study of Mansour and co-workers found that somatically acquired MYB binding sites upstream of the *TAL1* oncogene in T-ALL cells result in MYB binding and recruitment of the HAT CREBBP/CBP/KAT3A, generating a super-enhancer defined by extensive acetylation of histone H3 (H3K27ac) as well as recruitment of core components of a major leukaemic transcriptional complex containing RUNX1, GATA3 and TAL1 itself [26]. Targeted deletion of the MYB binding site resulted in a collapse of the super-enhancer and completely abrogated endogenous *TAL1* expression, indicating that MYB acts in concert with CREBBP/CBP/KAT3A to regulate super-enhancer formation at genes critical for defining cell identity.

An additional aspect of the link between MYB, EP300/KAT3B and histone acetylation is the phenomenon of acetylation-induced chromatin dissociation of MYB, as observed by Fuglerud et al. [132], suggesting a model where MYB binds to regions of closed chromatin and then recruits histone acetyltransferases, thus facilitating histone acetylation, where MYB is acting as a cofactor for EP300/KAT3B at these sites. The resulting H3K27ac not only leads to chromatin opening but also causes the detachment of MYB from the acetylated chromatin, creating a dynamic occupancy of MYB in chromatin.

The recruitment of the MLL1/KTM2A complex by MYB was shown to occur through the adaptor protein MENIN at genes believed to be critical for transformation by MLL fusion proteins, such as HOXA9 and MEIS1 (more fully covered in the accompanying Chap. 15 on MYB in AML). Moreover, knockdown of MYB resulted in decreased *HOXA9* and *MEIS1* gene expression and a decrease in H3K4me3 at the *HOXA9* locus and globally in K562 cells [127]. The arginine methyltransferase PRMT4 was found to function as a coactivator of MYB target gene expression in hematopoietic cell lines and resulted in increased H3R17 methylation at the *mim*-1 promoter, but not the enhancer [128]. Recruitment of histone-modifying enzymes by MYB seems to primarily direct activating histone marks. In contrast, the interaction of MYB with the HAT TIP60/KAT5 has been reported to have a repressive role. The MYB/TIP60 complex was found to interact with HDAC1 and HDAC2, and the suppression of MYB-dependent activation was attributed to the action of the HDACs [133].

However, the effect on histone acetylation was not investigated in this study. Taken together, MYB appears to be a factor important for the recruitment of several chromatin modifiers to target regions and may function as a facilitator for epigenetic writers. As these modifiers are enzymes, MYB may be regarded as a 'cofactor' in epigenetic writing, allowing the enzymes to more efficiently modify chromatin.

1.4.2 Other Interaction Partners Cooperating with MYB

Several other proteins have been reported to interact with MYB and are modulating its role as a transcriptional regulator. These include some transcription factors, such as C/EBPβ [134, 135], and coactivators and histone-modifying enzymes as mentioned above. Several additional interaction partners of MYB have been identified. MYB cooperates with FLASH/CASP8AP2 in foci associated with active RNAPII, leading to enhancement of MYB-dependent gene activation [136]. FLASH/CASP8AP2 was found to interact with the DBD of MYB. Furthermore, two enzymes linking MYB to the SUMO system were identified as interaction partners, namely, the SUMO conjugation enzyme UBC9 and the SUMO ligase PIAS1 [70, 137, 138].

An unexpected specific interaction between MYB and the general transcription machinery was uncovered by Xu et al. [139, 140]. Using an RNA interference screen, the authors found that AML cells were hypersensitive to TAF12 knockdown. TAF12 is an essential component of the general transcription co-activating complexes TFIID and SAGA [141]. Because the phenotypic consequences of TAF12 depletion shared many features in common with genetic ablation of MYB, the authors investigated this MYB-TAF12 connection further. It turned out that genetic depletion of MYB phenocopied the effects of TAF12 depletion, and the authors identified a direct interaction between the TAF12/TAF4 histone-fold heterodimer and the TAD of MYB. Their findings suggested that MYB utilizes TAF12 and TAF4 to selectively associate with the TFIID complex. In support of this, their data showed that 90% of the genomic sites bound by MYB are co-occupied by TAF12 [139] (*see* also the accompanying Chap. 15 on MYB in AML).

Another specific link to the general transcription apparatus is the observation of a rapid collapse in the transcription of the *MYC* and *MYB* genes upon ablation of the MED14 subunit of the mediator in KBM7 cells, a near-haploid human cell line, which like K562 is a chronic myelogenous leukaemia (CML) cell line [142].

1.4.3 Pioneer TF

Compaction of DNA into chromatin limits the amount of free DNA that is available for many regulatory factors to bind. One model for targeting silent chromatin is by the action of pioneer TFs [143, 144]. This is a subclass of early developmental TFs that are proposed to associate with closed chromatin independently of other factors and are thus the first TFs to access a regulatory region. Upon binding, pioneer factors increase chromatin accessibility, thereby allowing access to other TFs and chromatin modifiers [145]. By changing the chromatin landscape and recruiting activators or repressors that by themselves are unable to engage with silent chromatin, the pioneer factors act as master regulators able to change downstream gene regulatory networks and cell identity [146]. Examples of pioneer factors include the forkhead box A (FOXA) factors, GATA factors, PU.1 as well as the reprogramming factors OCT4, SOX2 and KLF4 [145]. Fuglerud et al. concluded that MYB should be added to the group of pioneer factors being able to access closed chromatin at specific loci and thereby change gene programs during haematopoiesis. This was based on several characteristics typical of pioneer factors, and the observation that the pioneer activity could be specifically abrogated by the D152V mutant of MYB, a mutant with no effect on the DNA binding of MYB [51].

Several characteristic features of MYB are in line with how pioneer factors operate in general.

Nucleosome-binding affinities of the individual pioneer factors seem to correlate with their cell reprogramming potential [147], and their mechanism of action seems to involve interactions between pioneer factors and core histones [148]. In line with this, MYB interacts with core histones, and the interaction between the R3 repeat and core histones was abrogated by D152V [51]. A key feature of pioneer factors is their ability to increase chromatin accessibility [145]. ATAC-Seq data indicates the D152V mutation impairs the ability of MYB to cause chromatin opening at specific sites [51].

Pioneer factors are commonly found bound at enhancers, where they trigger transcriptional competency by priming the enhancer for activation [145, 149]. In K562 cells, MYB occupies a large number of sites in chromatin, the largest fraction of which is in enhancer regions (32.5% including strong and weak/poised enhancers), many of which are defined as super-enhancers [150]. Nearly all pioneer factors identified so far interact with EP300/KAT3B and CREBBP/CBP/KAT3A. Chromatin modifications and pioneer factors likely function synergistically, and pioneer factors seem to function as both readers and facilitators for the writers of the histone code. The HATs EP300/KAT3B and CREBBP/CBP/KAT3A are among the best studied interaction partners of MYB [53–57].

Pioneer factors play a key role in rewiring of gene networks during cellular differentiation. How does the pioneer function of MYB influence its downstream gene programs and which target genes it activates? In the case of MYB, the D152V mutant allowed this question to be specifically addressed. Transcriptome data from K562 cells indicates that the D152V mutation specifically affects MYB's ability to regulate genes involved in differentiation, causing failure in the ability of MYB to block induced differentiation of K562 cells into erythrocytes. The genes identified that do not require the pioneer function of MYB for regulation were almost exclusively involved in metabolic processes, including many housekeeping genes. This strongly suggests that the pioneer factor function of MYB is important for its role as a master regulator, associated with rewiring of gene programs, and thereby also connected to its functions in cancer.

1.4.4 Global Analysis of Target Genes

Several lessons were already learned from studies of the first identified MYB target gene, *mim*-1, which encodes a secreted protein expressed in the granules of promyelocytes [151]. First, oncogenic mutations in MYB may have specific effects regarding which target genes are activated by MYB. The *mim*-1 gene is activated by the normal MYB, but not by the AMV v-MYB. Second, the MYB-dependent transcription of the endogenous *mim*-1 gene involves chromatin opening at the *mim*-1 promoter [152]. Finally, combinatorial control, the lineage-restricted expression of *mim*-1 is due to a cooperation of MYB with C/EBPβ, which is highly expressed in the myelomonocytic lineage [153].

Several specific target genes have been identified during the years in studies where knockdown or induced expression of MYB were combined with various forms of genome-wide profiling. Examples of important hematopoietic target genes of MYB are *MYC*, *TAL1*, *GATA2*, *KIT* and *LMO2* [reviewed in [5, 7, 9, 154]]. Using knockdown and rescue, we and others identified a large group of putative target genes responding to MYB depletion in hematopoietic cell lines [25, 51, 155–157]. We have also used digital DNase footprinting combining the occurrence of the MYB binding signature with DNase accessibility in six different hematopoietic cell types, concluding that MYB is a TF with multifaceted target regulation depending on cell type [158].

However, to safely define a gene as a direct target gene requires a combination of ChIP and response to MYB knockdown [159]. The lack of ChIP-grade antibodies hampered for a long time the global analysis of MYB occupancy in chromatin. One of the first global ChIP-Seq analyses of MYB in hematopoietic cells therefore analysed occupancy of a truncated MYB fused to the ligand-binding domain of ESR1/ERα, using anti-ERα antibodies [155]. Recently, CRISPR-Cas9 was used to create an in-frame GFP insertion in the MYB locus of mouse MEL cells, allowing the use of the GFP moiety as an affinity tag for a CUT&RUN analysis uncovering MYB target genes [160]. An antibody against MYB phosphorylated in Ser11 has been successfully used in generating MYB ChIP-Seq profiles in some cell lines [26, 161]. The occupancy of full-length MYB in K562 cells independent of phosphorylation status was analysed using a 3×Ty1-tagged MYB [150]. The latter was accompanied by an extensive bioinformatic analysis.

In the latter study, MYB was found to occupy a great number of sites in K562 chromatin. These sites were classified according to their epigenetic profiles into five clusters, termed C1–C5: a promoter cluster (C1) with high level of H3K4me3 and strong ATAC signal, an enhancer cluster (C2) with the high level of the enhancer-marks H3K4me1and H3K27ac, an intragenic cluster (C3) enriched for H3K36me3 over a broad area, a cluster C4 corresponding to repressed chromatin with enrichment of H3K27me3 and finally a cluster of unmarked chromatin (C5) [150]. While the promoter cluster was dominated by housekeeping genes, the enhancer cluster (C2) was enriched for functions linked to the development of hematopoietic cells, reflecting that MYB occupies many cell type-specific enhancers in K562. Further analysis of colocalisation with other chromatin-bound factors revealed a 'promoter cluster' (like C1) and an 'enhancer cluster' (like C2), where the latter was characterized by colocalisation with TFs acting as hematopoietic master regulators (such as GATA1 and GATA2), as well as the HAT EP300/KAT3B enriched at enhancers [162].

The same study found that MYB is occupying nearly a third (31%) of the super-enhancers in K562, consistent with its role as a master regulator [150]. Super-enhancers are large clusters of enhancers working together to efficiently drive the expression of genes with a key role in defining cell identity [163, 164]. MYB may even operate as a nucleation site for the formation of super-enhancers, as illustrated by the report of Mansour et al. showing that somatic mutations that happen to create binding sites for MYB can lead to the formation of a super-enhancer upstream of the *TAL1* oncogene [26].

MYB is able to activate really long-range enhancers. The *MYC* gene was early on proposing to be a MYB target genes [165, 166]. It is responsive to MYB knockdown and rescue and requires the pioneer function of MYB [51]. However, while the promoter proximal region did not show convincing MYB occupancy, strong MYB peaks were first found in a super-enhancer region [150], named BENC (for Blood ENhancer Cluster), located 1.7 Mb downstream of the gene and known to control *MYC* expression [167]. Therefore, MYB does occupy key regulatory elements controlling *MYC* expression even if really long-range interactions are involved.

MYB may contribute to cell identity. In K562 cells, MYB is among the top eight most highly expressed TFs [150]. The others are GATA1, STAT5A, CTCFL (also known as BORIS), NF-E2, PITX1, HOXB9 and KLF1. As expected, several of the enhancers upstream of the genes encoding these TFs are occupied by a subset of these eight highly expressed TFs, including MYB. This self-reinforcing pattern supports the notion that MYB in K562 cells operates as a master regulator by cooperating with a subset of the other highly expressed TFs thus contributing to the cell type identity of K562.

The status as a master regulator should be reflected in it being able to activate target genes controlling the expression of other regulators in the hematopoietic system. The group of direct pioneer genes (i.e. requiring pioneer function of MYB) that are activated by MYB in K562 cells included several genes encoding TFs, chromatin modifiers as well as signalling proteins, consistent with a rewiring function [150]. The list included genes encoding components of transcriptional complexes, both specific ones (*LMO2* and *MAML1*) and more general ones (*TAF4*, *MED13L* and *CDK12*), several DNA-binding TFs (*MYC*, *STAT5A*, *PGR*, *GATAD2A*, *NFIA* and *SNAI1*) and also a few epigenetic regulators (*KDM1B*, *MBTD1* and *SMARCC1*). Finally, several genes encoding proteins involved in signalling pathways were present (*KIT*, *MAP4K4*, *PAK2*, *GRB10* and *SKP2*).

1.4.5 How SUMOylation Affects the MYB Target Genes

Since MYB is a TF subjected to SUMO-mediated regulation, through both SUMO conjugation and SUMO binding, a relevant question is how its SUMO status affects its downstream gene programs. To start to answer this, Lemma et al. performed RNA-Seq analysis in K562 cells after endogenous MYB knockdown and rescue using clones stably expressing either WT MYB or one of two MYB SUMO mutants, a double SUMO conjugation negative mutant (2KR-MYB) and a SIM mutant (ANAA-MYB) [168]. The results showed large effects of altering the SUMO status of MYB. A number of differentially regulated genes were observed, reflecting both quantitative and qualitative effects on the regulome of MYB. Strikingly, loss of SUMO conjugation led to up-regulation of many genes involved in the regulation of transcription and epigenetic modifications, suggesting that the master regulatory function of MYB seems to be restricted and kept in check by SUMOylation. To name a few examples, SUMO-dependent altered expression included target genes encoding key TFs such as *BHLHE40*, *ATF3*, *REL* and *FOXN3*, genes encoding transcriptional coactivators such as the mediator subunit *MED1*, epigenetic modifiers such as *TET2*, *MOZ/KAT6A* and *CREBBP/CBP/KAT3A*.

Somewhat unexpectedly, the two types of SUMO mutations of MYB, 2KR-MYB and ANAA-MYB showed quite similar changes in expression patterns despite having distinct pattern of SUMO association [168]. This probably relates to the concept of 'group SUMOylation' where clusters of proteins undergo SUMOylation in a concerted manner and where SUMO–SIM interactions contribute as a glue to stabilize specific protein clusters [reviewed in [169]]. Such a glue function has been suggested to play a role in stabilizing different chromatin states, thus preserving cell identities [170, 171].

Related to the role of SUMO in understanding MYB function, one specific target gene caught attention, the gene encoding the SUMO protease SENP1, a key enzyme that removes SUMO conjugation from SUMOylated proteins. Further studies of the MYB-*SENP1* link supported the existence of an autoactivation loop whereby MYB, through enhancing SENP1 levels (and reducing its repressive partner UXT), is itself being activated by a reduced level of repressive SUMOylation [168]. The implication is that MYB seems able not only to control a specific gene program directly but also to affect the SUMO landscape indirectly through *SENP1* regulation. Since MYB is overexpressed in many cancers including leukaemia, breast, and colorectal cancers [5, 7, 135, 172, 173], it may under these conditions be less SUMO-repressed because of this autoactivation phenomenon and the deregulation of *SENP1* expression. In this way, overexpressed MYB could be activated because of reduced SUMOylation levels, not unlike what happens to the classical v-MYB because of its deletions in NRD causing removal of its SUMO-conjugation sites. In support of this model, it was found that overexpressed MYB correlates positively with *SENP1* expression in leukaemia, breast and colorectal cancer patients [168]. Similarly, a high correlation was found in the same cohorts between MYB levels and targets enhanced by SUMO-negative MYB in K562. This is a new mechanism that may contribute to oncogenic activation of MYB.

1.4.6 Negatively Regulated Target Genes

The existence of target genes negatively regulated by MYB was first emphasized by Zhao et al. [155]. Their early global analysis uncovered that the transactivator MYB also functions to repress a large fraction of its direct targets, including several key regulators of myeloid differentiation, such as *SPI1/PU.1*, *RUNX1*, *JUNB* and *CEBPB*. A more recent example of negatively regulated target genes is during B-cell differentiation, where MYB regulates the intrinsic survival of $CD19^+$ pro-B-cells by repressing expression of the proapoptotic proteins BMF and BCL2L11/BIM [19]. Another interesting recent example of target genes that are negatively regulated by

MYB is the proteasome component-encoding (*PSM*) genes. Clarke et al. [174] observed altered proteosomal activity in MYB-insufficient mice creating an imbalance in proteosomal activity and perturbed proteostasis as a consequence of MYB insufficiency. This appears to be one key factor leading to age-related neoplasia as a direct consequence of the MYB insufficiency. In human K562 cells, several *PSM* genes are bound by MYB in their promoter regions and respond to MYB knockdown.

The negative correlation between MYB expression and HbF expression discussed earlier is yet an example of this negative regulation, a phenomenon that seems to be reflected also in the transcriptome data from K562 cells where both *HBG1* and *HBG2* are up-regulated upon MYB knockdown [51] (results not shown).

1.4.7 Cooperating TFs

A key feature of enhancers is their ability to function as integrated platforms for TF binding and combinatorial control where combinations of TFs cooperate with each other and with chromatin-modifying complexes to create the properly controlled activation of the linked promoters. A relevant question is therefore which TFs do MYB usually cooperate with. A well-studied classical cooperation is with C/EBPβ, where in particular the joint activation of the chicken *mim*-1 gene has been thoroughly studied [151–153, 175, 176]. A study of combinatorial TF modules within enhancers during erythropoiesis identified five distinct TF motif modules including expected associations such as GATA1 and TAL1, and some less expected associations such as NF-E2 and MYB [88].

In the global analysis of MYB occupancy in the K562 cell line, Lemma et al. [150] also analysed the co-occupancy between MYB and other TFs. The top five TFs most frequently found colocalized with MYB in K562 were ATF3, CTCF, IRF1, NRSF and BACH1 [150]. The two first were found in regions with genes enriched for functions linked to chromatin organization and expression, epigenetic or transcriptional regulations. The three others were linked to functions for transcriptional regulation, immune-related signalling and macromolecule/protein modifications, respectively. Neither these pairs nor the NF-E2-MYB pair has been subjected to in-depth studies. The preferred cooperating TFs will most probably differ between different lineages allowing the multifaceted regulation of MYB target genes depending on cell type.

Acknowledgements All downstream genome-wide analyses were performed on the Saga super computing resources (Project nn9374k) provided by UNINETT Sigma2—the National Infrastructure for High Performance Computing and Data Storage in Norway. We are grateful to all our previous co-authors for their indirect contribution to this review. Finally, we are thankful to Prof. Anthony Mathelier who provided support to R.B.L during the review process.

References

1. Greig KT, de Graaf CA, Murphy JM, Carpinelli MR, Pang SHM, Frampton J, Kile BT, Hilton DJ, Nutt SL (2010) Critical roles for c-Myb in lymphoid priming and early B-cell development. Blood 115:2796–2805. https://doi.org/10.1182/blood-2009-08-239210
2. Ramsay RG (2005) c-Myb a stem-progenitor cell regulator in multiple tissue compartments. Growth Factors 23:253–261. https://doi.org/10.1080/08977190500233730
3. Malaterre J, Mantamadiotis T, Dworkin S, Lightowler S, Yang Q, Ransome MI, Turnley AM, Nichols NR, Emambokus NR, Frampton J, Ramsay RG (2008) c-Myb is required for neural progenitor cell proliferation and maintenance of the neural stem cell niche in adult brain. Stem Cell 26:173–181. https://doi.org/10.1634/stemcells.2007-0293
4. Gonda TJ, Leo P, Ramsay RG (2008) Estrogen and MYB in breast cancer: potential for new therapies. Expert Opin Biol Ther 8:713–717. https://doi.org/10.1517/14712598.8.6.713
5. Ramsay RG, Gonda TJ (2008) MYB function in normal and cancer cells. Nat Rev Cancer 8:523–534. https://doi.org/10.1038/nrc2439
6. Oh IH, Reddy EP (1999) The myb gene family in cell growth, differentiation and apoptosis. Oncogene 18:3017–3033. https://doi.org/10.1038/sj.onc.1202839
7. Cicirò Y, Sala A (2021) MYB oncoproteins: emerging players and potential therapeutic targets in human cancer. Oncogene 10:19. https://doi.org/10.1038/s41389-021-00309-y

8. Liu X, Xu Y, Han L, Yi Y (2018) Reassessing the potential of Myb-targeted anti-cancer therapy. J Cancer 9:1259–1266. https://doi.org/10.7150/jca.23992
9. Wang X, Angelis N, Thein SL (2018) MYB—a regulatory factor in hematopoiesis. Gene 665:6–17. https://doi.org/10.1016/j.gene.2018.04.065
10. Mucenski ML, McLain K, Kier AB, Swerdlow SH, Schreiner CM, Miller TA, Pietryga DW Jr, Potter SS (1991) A functional c-myb gene is required for normal murine fetal hepatic hematopoiesis. Cell 65:677–689. https://doi.org/10.1016/0092-8674(91)90099-k
11. Sumner R, Crawford A, Mucenski M, Frampton J (2000) Initiation of adult myelopoiesis can occur in the absence of c-Myb whereas subsequent development is strictly dependent on the transcription factor. Oncogene 19:3335–3342. https://doi.org/10.1038/sj.onc.1203660
12. Emambokus N, Vegiopoulos A, Harman B, Jenkinson E, Anderson G, Frampton J (2003) Progression through key stages of haemopoiesis is dependent on distinct threshold levels of c-Myb. EMBO J 22:4478–4488. https://doi.org/10.1093/emboj/cdg434
13. Vegiopoulos A (2006) Coordination of erythropoiesis by the transcription factor c-Myb. Blood 107:4703–4710. https://doi.org/10.1182/blood-2005-07-2968
14. Westin EH, Gallo RC, Arya SK, Eva A, Souza LM, Baluda MA, Aaronson SA, Wong-Staal F (1982) Differential expression of the amv gene in human hematopoietic cells. Proc Nat Acad Sci 79:2194–2198. https://doi.org/10.1073/pnas.79.7.2194
15. Allen RD, Bender TP, Siu G (1999) c-Myb is essential for early T cell development. Genes Dev 13:1073–1078. https://doi.org/10.1101/gad.13.9.1073
16. Bender TP, Kremer CS, Kraus M, Buch T, Rajewsky K (2004) Critical functions for c-Myb at three checkpoints during thymocyte development. Nat Immunol 5:721–729. https://doi.org/10.1038/ni1085
17. Thomas MD, Kremer CS, Ravichandran KS, Rajewsky K, Bender TP (2005) c-Myb is critical for B cell development and maintenance of follicular B cells. Immunity 23:275–286. https://doi.org/10.1016/j.immuni.2005.08.005
18. Fahl SP, Crittenden RB, Allman D, Bender TP (2009) c-Myb is required for pro-B cell differentiation. J Immunol (Baltimore, Md: 1950) 183:5582–5592. https://doi.org/10.4049/jimmunol.0901187
19. Fahl SP, Daamen AR, Crittenden RB, Bender TP (2018) c-Myb Coordinates Survival and the Expression of Genes That Are Critical for the Pre-BCR Checkpoint. J Immunol (Baltimore, Md: 1950) 200:ji1302303-3463. https://doi.org/10.4049/jimmunol.1302303
20. Gautam S, Fioravanti J, Zhu W, Gall JBL, Brohawn P, Lacey NE, Hu J, Hocker JD, Hawk NV, Kapoor V, Telford WG, Gurusamy D, Yu Z, Bhandoola A, Xue H-H, Roychoudhuri R, Higgs BW, Restifo NP, Bender TP, Ji Y, Gattinoni L (2019) The transcription factor c-Myb regulates CD8+ T cell stemness and antitumor immunity. Nat Immunol 20:337–349. https://doi.org/10.1038/s41590-018-0311-z
21. Dias S, D'Amico A, Cretney E, Liao Y, Tellier J, Bruggeman C, Almeida FF, Leahy J, Belz GT, Smyth GK, Shi W, Nutt SL (2017) Effector eegulatory T cell differentiation and immune homeostasis depend on the transcription factor Myb. Immunity 46:78–91. https://doi.org/10.1016/j.immuni.2016.12.017
22. Rodríguez-Caparrós A, García V, Casal Á, López-Ros J, García-Mariscal A, Tani-ichi S, Ikuta K, Hernández-Munain C (2019) Notch signaling controls transcription via the recruitment of RUNX1 and MYB to enhancers during T cell development. J Immunol 202:2460–2472. https://doi.org/10.4049/jimmunol.1801650
23. Tsui C, Kretschmer L, Rapelius S, Gabriel SS, Chisanga D, Knöpper K, Utzschneider DT, Nüssing S, Liao Y, Mason T, Torres SV, Wilcox SA, Kanev K, Jarosch S, Leube J, Nutt SL, Zehn D, Parish IA, Kastenmüller W, Shi W, Buchholz VR, Kallies A (2022) MYB orchestrates T cell exhaustion and response to checkpoint inhibition. Nature 609:354–360. https://doi.org/10.1038/s41586-022-05105-1
24. Lefebvre C, Rajbhandari P, Alvarez MJ, Bandaru P, Lim WK, Sato M, Wang K, Sumazin P, Kustagi M, Bisikirska BC, Basso K, Beltrao P, Krogan N, Gautier J, Dalla-Favera R, Califano A (2010) A human B-cell interactome identifies MYB and FOXM1 as master regulators of proliferation in germinal centers. Mol Syst Biol 6:377–377. https://doi.org/10.1038/msb.2010.31
25. Lorenzo PI, Brendeford EM, Gilfillan S, Gavrilov AA, Leedsak M, Razin SV, Eskeland R, Sæther T, Gabrielsen OS (2011) Identification of c-Myb target genes in K562 cells reveals a role for c-Myb as a master regulator. Genes Cancer 2:805–817. https://doi.org/10.1177/1947601911428224
26. Mansour MR, Abraham BJ, Anders L, Berezovskaya A, Gutierrez A, Durbin AD, Etchin J, Lawton L, Sallan SE, Silverman LB, Loh ML, Hunger SP, Sanda T, Young RA, Look AT (2014) An oncogenic super-enhancer formed through somatic mutation of a noncoding intergenic element. Science 346:1373–1377. https://doi.org/10.1126/science.1259037
27. Capellera-Garcia S, Pulecio J, Dhulipala K, Siva K, Rayon-Estrada V, Singbrant S, Sommarin MNE, Walkley CR, Soneji S, Karlsson G, Raya Á, Sankaran VG, Flygare J (2016) Defining the minimal factors required for erythropoiesis through direct lineage conversion. Cell Rep 15:2550–2562. https://doi.org/10.1016/j.celrep.2016.05.027
28. Sakura H, Kanei-Ishii C, Nagase T, Nakagoshi H, Gonda TJ, Ishii S (1989) Delineation of three functional domains of the transcriptional activa-

tor encoded by the c-myb protooncogene. Proc Natl Acad Sci USA 86:5758–5762. https://doi.org/10.1073/pnas.86.15.5758
29. Sandberg ML, Sutton SE, Pletcher MT, Wiltshire T, Tarantino LM, Hogenesch JB, Cooke MP (2005) c-Myb and p300 regulate hematopoietic stem cell proliferation and differentiation. Dev Cell 8:153–166. https://doi.org/10.1016/j.devcel.2004.12.015
30. Carpinelli MR, Hilton DJ, Metcalf D, Antonchuk JL, Hyland CD, Mifsud SL, Rago LD, Hilton AA, Willson TA, Roberts AW, Ramsay RG, Nicola NA, Alexander WS (2004) Suppressor screen in Mpl-/- mice: c-Myb mutation causes supraphysiological production of platelets in the absence of thrombopoietin signaling. Proc Natl Acad Sci USA 101:6553–6558. https://doi.org/10.1073/pnas.0401496101
31. Jumper J, Evans R, Pritzel A, Green T, Figurnov M, Ronneberger O, Tunyasuvunakool K, Bates R, Žídek A, Potapenko A, Bridgland A, Meyer C, Kohl SAA, Ballard AJ, Cowie A, Romera-Paredes B, Nikolov S, Jain R, Adler J, Back T, Petersen S, Reiman D, Clancy E, Zielinski M, Steinegger M, Pacholska M, Berghammer T, Bodenstein S, Silver D, Vinyals O, Senior AW, Kavukcuoglu K, Kohli P, Hassabis D (2021) Highly accurate protein structure prediction with AlphaFold. Nature 596:583–589. https://doi.org/10.1038/s41586-021-03819-2
32. Varadi M, Anyango S, Deshpande M, Nair S, Natassia C, Yordanova G, Yuan D, Stroe O, Wood G, Laydon A, Žídek A, Green T, Tunyasuvunakool K, Petersen S, Jumper J, Clancy E, Green R, Vora A, Lutfi M, Figurnov M, Cowie A, Hobbs N, Kohli P, Kleywegt G, Birney E, Hassabis D, Velankar S (2021) AlphaFold protein structure database: massively expanding the structural coverage of protein-sequence space with high accuracy models. Nucleic Acids Res 50:D439–D444. https://doi.org/10.1093/nar/gkab1061
33. Frampton J, Gibson TJ, Ness SA, Döderlein G, Graf T (1991) Proposed structure for the DNA-binding domain of the Myb oncoprotein based on model building and mutational analysis. Protein Eng Des Sel 4:891–901. https://doi.org/10.1093/protein/4.8.891
34. Gabrielsen OS, Sentenac A, Fromageot P (1991) Specific DNA binding by c-Myb: evidence for a double helix-turn-helix-related motif. Science 253:1140–1143. https://doi.org/10.1126/science.1887237
35. Ogata K, Morikawa S, Nakamura H, Sekikawa A, Inoue T, Kanai H, Sarai A, Ishii S, Nishimura Y (1994) Solution structure of a specific DNA complex of the Myb DNA-binding domain with cooperative recognition helices. Cell 79:639–648. https://doi.org/10.1016/0092-8674(94)90549-5
36. Ness SA (1996) The Myb oncoprotein: regulating a regulator. Biochim Biophys Acta 1288:F123–F139. https://doi.org/10.1016/s0304-419x(96)00027-3
37. Ogata K, Hojo H, Aimoto S, Nakai T, Nakamura H, Sarai A, Ishii S, Nishimura Y (1992) Solution structure of a DNA-binding unit of Myb: a helix-turn-helix-related motif with conserved tryptophans forming a hydrophobic core. Proc Natl Acad Sci USA 89:6428–6432
38. Ogata K, Morikawa S, Nakamura H, Hojo H, Yoshimura S, Zhang R, Aimoto S, Ametani Y, Hirata Z, Sarai A (1995) Comparison of the free and DNA-complexed forms of the DNA-binding domain from c-Myb. Nat Struct Biol 2:309–320. https://doi.org/10.1038/nsb0495-309
39. Myrset AH, Bostad A, Jamin N, Lirsac PN, Toma F, Gabrielsen OS (1993) DNA and redox state induced conformational changes in the DNA-binding domain of the Myb oncoprotein. EMBO J 12:4625–4633
40. Zargarian L, Tilly VL, Jamin N, Chaffotte A, Gabrielsen OS, Toma F, Alpert B (1999) Myb-DNA recognition: role of tryptophan residues and structural changes of the minimal DNA binding domain of c-Myb. Biochemistry 38:1921–1929. https://doi.org/10.1021/bi981199j
41. Biedenkapp H, Borgmeyer U, Sippel AE, Klempnauer K-H (1988) Viral myb oncogene encodes a sequence-specific DNA-binding activity. Nat Cell Biol 335:835–837. https://doi.org/10.1038/335835a0
42. Nakagoshi H, Nagase T, Kanei-Ishii C, Ueno Y, Ishii S (1990) Binding of the c-myb proto-oncogene product to the simian virus 40 enhancer stimulates transcription. J Biol Chem 265:3479–3483. https://doi.org/10.1016/s0021-9258(19)39793-5
43. Howe KM, Watson RJ (1991) Nucleotide preferences in sequence-specific recognition of DNA by c-myb protein. Nucleic Acids Res 19:3913–3919. https://doi.org/10.1093/nar/19.14.3913
44. Weston K (1992) Extension of the DNA binding consensus of the chicken c-Myb and v-Myb proteins. Nucleic Acids Res 20:3043–3049. https://doi.org/10.1093/nar/20.12.3043
45. Berge T, Bergholtz SL, Andersson KB, Gabrielsen OS (2001) A novel yeast system for in vivo selection of recognition sequences: defining an optimal c-Myb-responsive element. Nucleic Acids Res 29:E99
46. Ording E, Kvåvik W, Bostad A, Gabrielsen OS (1994) Two functionally distinct half sites in the DNA-recognition sequence of the Myb oncoprotein. Eur J Biochem 222:113–120
47. Boyer LA, Latek RR, Peterson CL (2004) Opinion: The SANT domain: a unique histone-tail-binding module? Nat Rev Mol Cell Biol 5:158–163. https://doi.org/10.1038/nrm1314
48. Aasland R, Stewart AF, Gibson T (1996) The SANT domain: a putative DNA-binding domain in the SWI-SNF and ADA complexes, the transcriptional co-repressor N-CoR and TFIIIB. Trends Biochem Sci 21:87–88

49. Mo X, Kowenz-Leutz E, Laumonnier Y, Xu H, Leutz A (2005) Histone H3 tail positioning and acetylation by the c-Myb but not the v-Myb DNA-binding SANT domain. Genes Dev 19:2447–2457. https://doi.org/10.1101/gad.355405
50. Ko ER, Ko D, Chen C, Lipsick JS (2008) A conserved acidic patch in the Myb domain is required for activation of an endogenous target gene and for chromatin binding. Mol Cancer 7:77–77. https://doi.org/10.1186/1476-4598-7-77
51. Fuglerud BM, Lemma RB, Wanichawan P, Sundaram AYM, Eskeland R, Gabrielsen OS (2017) A c-Myb mutant causes deregulated differentiation due to impaired histone binding and abrogated pioneer factor function. Nucleic Acids Res 45:7681–7696. https://doi.org/10.1093/nar/gkx364
52. Kalkbrenner F, Guehmann S, Moelling K (1990) Transcriptional activation by human c-myb and v-myb genes. Oncogene 5:657–661
53. Zor T, de Guzman RN, Dyson HJ, Wright PE (2004) Solution structure of the KIX domain of CBP bound to the transactivation domain of c-Myb. J Mol Biol 337:521–534. https://doi.org/10.1016/j.jmb.2004.01.038
54. Parker D, Rivera M, Zor T, Henrion-Caude A, Radhakrishnan I, Kumar A, Shapiro LH, Wright PE, Montminy M, Brindle PK (1999) Role of secondary structure in discrimination between constitutive and inducible activators. Mol Cell Biol 19:5601–5607
55. Dai P, Akimaru H, Tanaka Y, Hou DX, Yasukawa T, Kanei-Ishii C, Takahashi T, Ishii S (1996) CBP as a transcriptional coactivator of c-Myb. Genes Dev 10:528–540. https://doi.org/10.1101/gad.10.5.528
56. Oelgeschläger M, Janknecht R, Krieg J, Schreek S, Luscher B (1996) Interaction of the co-activator CBP with Myb proteins: effects on Myb-specific transactivation and on the cooperativity with NF-M. EMBO J 15
57. Kasper LH, Boussouar F, Ney PA, Jackson CW, Rehg J, van Deursen JM, Brindle PK (2002) A transcription-factor-binding surface of coactivator p300 is required for haematopoiesis. Nat Cell Biol 419:738–743. https://doi.org/10.1038/nature01062
58. Suetaka S, Oka Y, Kunihara T, Hayashi Y, Arai M (2022) Rational design of a helical peptide inhibitor targeting c-Myb–KIX interaction. Sci Rep 12:816. https://doi.org/10.1038/s41598-021-04497-w
59. Thakur JK, Yadav A, Yadav G (2013) Molecular recognition by the KIX domain and its role in gene regulation. Nucleic Acids Res. https://doi.org/10.1093/nar/gkt1147
60. Herrera-Nieto P, Pérez A, Fabritiis GD (2023) Binding-and-folding recognition of an intrinsically disordered protein using online learning molecular dynamics. J Chem Theory Comput 19:3817–3824. https://doi.org/10.1021/acs.jctc.3c00008
61. Næs G, Storesund JO, Udayakumar P, Ledsaak M, Gabrielsen OS (2020) Dissecting the transactivation domain (tAD) of the transcription factor c-Myb to assess recent models of tAD function. FEBS Open Bio 10:2329–2342. https://doi.org/10.1002/2211-5463.12978
62. Staller MV, Holehouse AS, Swain-Lenz D, Das RK, Pappu RV, Cohen BA (2018) A high-throughput mutational scan of an intrinsically disordered acidic transcriptional activation domain. Cell Syst 6:444–455.e6. https://doi.org/10.1016/j.cels.2018.01.015
63. Dubendorff JW, Whittaker LJ, Eltman JT, Lipsick JS (1992) Carboxy-terminal elements of c-Myb negatively regulate transcriptional activation in cis and in trans. Genes Dev 6:2524–2535. https://doi.org/10.1101/gad.6.12b.2524
64. Lipsick JS, Wang DM (1999) Transformation by v-Myb. Oncogene 18:3047–3055. https://doi.org/10.1038/sj.onc.1202745
65. Wang DM, Dubendorff JW, Woo CH, Lipsick JS (1999) Functional analysis of carboxy-terminal deletion mutants of c-Myb. J Virol 73:5875–5886
66. Dash AB, Orrico FC, Ness SA (1996) The EVES motif mediates both intermolecular and intramolecular regulation of c-Myb. Genes Dev 10:1858–1869. https://doi.org/10.1101/gad.10.15.1858
67. Fu SL, Lipsick JS (1996) FAETL motif required for leukemic transformation by v-Myb. J Virol 70:5600–5610
68. Dubendorff JW, Lipsick JS (1999) Transcriptional regulation by the carboxyl terminus of c-Myb depends upon both the Myb DNA-binding domain and the DNA recognition site. Oncogene 18:3452–3460. https://doi.org/10.1038/sj.onc.1202679
69. Bies J, Markus J, Wolff L (2002) Covalent attachment of the SUMO-1 protein to the negative regulatory domain of the c-Myb transcription factor modifies its stability and transactivation capacity*. J Biol Chem 277:8999–9009. https://doi.org/10.1074/jbc.m110453200
70. Dahle Ø, Andersen TØ, Nordgård O, Matre V, Sal GD, Gabrielsen OS (2003) Transactivation properties of c-Myb are critically dependent on two SUMO-1 acceptor sites that are conjugated in a PIASy enhanced manner. Eur J Biochem 270:1338–1348. https://doi.org/10.1046/j.1432-1033.2003.03504.x
71. Molværsmyr A-K, Sæther T, Gilfillan S, Lorenzo PI, Kvaløy H, Matre V, Gabrielsen OS (2010) A SUMO-regulated activation function controls synergy of c-Myb through a repressor–activator switch leading to differential p300 recruitment. Nucleic Acids Res 38:4970–4984. https://doi.org/10.1093/nar/gkq245
72. Zhao X (2018) SUMO-mediated regulation of nuclear functions and signaling processes. Mol Cell 71:409–418. https://doi.org/10.1016/j.molcel.2018.07.027
73. Hendriks IA, D'Souza RCJ, Yang B, Vries MV, Mann M, Vertegaal ACO (2014) Uncovering global SUMOylation signaling networks in a site-specific manner. Nat Struct Mol Biol 21:927–936. https://doi.org/10.1038/nsmb.2890

74. Seeler J-S, Dejean A (2017) SUMO and the robustness of cancer. Nat Rev Cancer 17:184–197. https://doi.org/10.1038/nrc.2016.143
75. Kroonen JS, Vertegaal ACO (2020) Targeting SUMO signaling to wrestle cancer. Trend Cancer 7:496–510. https://doi.org/10.1016/j.trecan.2020.11.009
76. Hendriks IA, Lyon D, Young C, Jensen LJ, Vertegaal ACO, Nielsen ML (2017) Site-specific mapping of the human SUMO proteome reveals co-modification with phosphorylation. Nat Struct Mol Biol 24:325–336. https://doi.org/10.1038/nsmb.3366
77. Chang H-M, Yeh ETH (2020) SUMO: from bench to bedside. Physiol Rev 100:1599–1619. https://doi.org/10.1152/physrev.00025.2019
78. Sæther T, Pattabiraman DR, Alm-Kristiansen AH, Vogt-Kielland LT, Gonda TJ, Gabrielsen OS (2011) A functional SUMO-interacting motif in the transactivation domain of c-Myb regulates its myeloid transforming ability. Oncogene 30:212–222. https://doi.org/10.1038/onc.2010.397
79. Stadhouders R, Aktuna S, Thongjuea S, Aghajanirefah A, Pourfarzad F, van Ijcken W, Lenhard B, Rooks H, Best S, Menzel S, Grosveld F, Thein SL, Soler E (2014) HBS1L-MYB intergenic variants modulate fetal hemoglobin via long-range MYB enhancers. J Clin Invest 124:1699–1710. https://doi.org/10.1172/jci71520
80. Thein SL, Menzel S, Peng X, Best S, Jiang J, Close J, Silver N, Gerovasilli A, Ping C, Yamaguchi M, Wahlberg K, Ulug P, Spector TD, Garner C, Matsuda F, Farrall M, Lathrop M (2007) Intergenic variants of HBS1L-MYB are responsible for a major quantitative trait locus on chromosome 6q23 influencing fetal hemoglobin levels in adults. Proc Natl Acad Sci 104:11346–11351. https://doi.org/10.1073/pnas.0611393104
81. Lettre G, Sankaran VG, Bezerra MAC, Araújo AS, Uda M, Sanna S, Cao A, Schlessinger D, Costa FF, Hirschhorn JN, Orkin SH (2008) DNA polymorphisms at the BCL11A, HBS1L-MYB, and β-globin loci associate with fetal hemoglobin levels and pain crises in sickle cell disease. Proc Natl Acad Sci 105:11869–11874. https://doi.org/10.1073/pnas.0804799105
82. Ganesh SK, Zakai NA, van Rooij FJA, Soranzo N, Smith AV, Nalls MA, Chen M-H, Kottgen A, Glazer NL, Dehghan A, Kuhnel B, Aspelund T, Yang Q, Tanaka T, Jaffe A, Bis JCM, Verwoert GC, Teumer A, Fox CS, Guralnik JM, Ehret GB, Rice K, Felix JF, Rendon A, Eiriksdottir G, Levy D, Patel KV, Boerwinkle E, Rotter JI, Hofman A, Sambrook JG, Hernandez DG, Zheng G, Bandinelli S, Singleton AB, Coresh J, Lumley T, Uitterlinden AG, vanGils JM, Launer LJ, Cupples LA, Oostra BA, Zwaginga J-J, Ouwehand WH, Thein S-L, Meisinger C, Deloukas P, Nauck M, Spector TD, Gieger C, Gudnason V, van Duijn CM, Psaty BM, Ferrucci L, Chakravarti A, Greinacher A, O'Donnell CJ, Witteman JCM, Furth S, Cushman M, Harris TB, Lin J-P (2009) Multiple loci influence erythrocyte phenotypes in the CHARGE Consortium. Nat Genet 41:1191–1198. https://doi.org/10.1038/ng.466
83. Galarneau G, Palmer CD, Sankaran VG, Orkin SH, Hirschhorn JN, Lettre G (2010) Fine-mapping at three loci known to affect fetal hemoglobin levels explains additional genetic variation. Nat Genet 42:1049–1051. https://doi.org/10.1038/ng.707
84. Canver MC, Lessard S, Pinello L, Wu Y, Ilboudo Y, Stern EN, Needleman AJ, Galactéros F, Brugnara C, Kutlar A, McKenzie C, Reid M, Chen DD, Das PP, Cole MA, Zeng J, Kurita R, Nakamura Y, Yuan G-C, Lettre G, Bauer DE, Orkin SH (2017) Variant-aware saturating mutagenesis using multiple Cas9 nucleases identifies regulatory elements at trait-associated loci. Nat Genet 49:625–634. https://doi.org/10.1038/ng.3793
85. Matsumura T, Totani H, Gunji Y, Fukuda M, Yokomori R, Deng J, Rethnam M, Yang C, Tan TK, Karasawa T, Kario K, Takahashi M, Osato M, Sanda T, Suda T (2022) A Myb enhancer-guided analysis of basophil and mast cell differentiation. Nat Commun 13:7064. https://doi.org/10.1038/s41467-022-34906-1
86. Li M, Jiang P, Cheng K, Zhang Z, Lan S, Li X, Zhao L, Wang Y, Wang X, Chen J, Ji T, Han B, Zhang J (2021) Regulation of MYB by distal enhancer elements in human myeloid leukemia. Cell Death Dis 12:223. https://doi.org/10.1038/s41419-021-03515-z
87. Boontanrart MY, Schröder MS, Stehli GM, Banović M, Wyman SK, Lew RJ, Bordi M, Gowen BG, DeWitt MA, Corn JE (2020) ATF4 regulates MYB to increase γ-globin in response to loss of β-globin. Cell Rep 32:107993. https://doi.org/10.1016/j.celrep.2020.107993
88. Huang J, Liu X, Li D, Shao Z, Cao H, Zhang Y, Trompouki E, Bowman TV, Zon LI, Yuan G-C, Orkin SH, Xu J (2016) Dynamic control of enhancer repertoires drives lineage and stage-specific transcription during hematopoiesis. Dev Cell 36:9–23. https://doi.org/10.1016/j.devcel.2015.12.014
89. Benyoucef A, Haigh JJ, Brand M (2023) Unveiling the complexity of transcription factor networks in hematopoietic stem cells: implications for cell therapy and hematological malignancies. Front Oncol 13:1151343. https://doi.org/10.3389/fonc.2023.1151343
90. Wilson NK, Foster SD, Wang X, Knezevic K, Schütte J, Kaimakis P, Chilarska PM, Kinston S, Ouwehand WH, Dzierzak E, Pimanda JE, de Bruijn MFTR, Göttgens B (2010) Combinatorial transcriptional control in blood stem/progenitor cells: genome-wide analysis of ten major transcriptional regulators. Cell Stem Cell 7:532–544. https://doi.org/10.1016/j.stem.2010.07.016
91. Edginton-White B, Maytum A, Kellaway SG, Goode DK, Keane P, Pagnuco I, Assi SA, Ames L, Clarke M, Cockerill PN, Göttgens B, Cazier JB, Bonifer C (2023) A genome-wide relay of signalling-

responsive enhancers drives hematopoietic specification. Nat Commun 14:267. https://doi.org/10.1038/s41467-023-35910-9
92. Velten L, Haas SF, Raffel S, Blaszkiewicz S, Islam S, Hennig BP, Hirche C, Lutz C, Buss EC, Nowak D, Boch T, Hofmann W-K, Ho AD, Huber W, Trumpp A, Essers MAG, Steinmetz LM (2017) Human haematopoietic stem cell lineage commitment is a continuous process. Nat Cell Biol 19:271–281. https://doi.org/10.1038/ncb3493
93. Sugimura R, Jha DK, Han A, Soria-Valles C, da Rocha EL, Lu Y-F, Goettel JA, Serrao E, Rowe RG, Malleshaiah M, Wong I, Sousa P, Zhu TN, Ditadi A, Keller G, Engelman AN, Snapper SB, Doulatov S, Daley GQ (2017) Haematopoietic stem and progenitor cells from human pluripotent stem cells. Nature 545:432–438. https://doi.org/10.1038/nature22370
94. Yue Y, Sun Q, Chen L, Li G, Huang Q, Chen S (2023) Redefining hematopoietic progenitor cells and reaffirming lineage commitment by transcriptional atlas. biorxiv. https://doi.org/10.1101/2023.01.27.524347
95. Han GC, Vinayachandran V, Bataille AR, Park B, Chan-Salis KY, Keller CA, Long M, Mahony S, Hardison RC, Pugh BF (2016) Genome-wide organization of GATA1 and TAL1 determined at high resolution. Mol Cell Biol 36:157–172. https://doi.org/10.1128/mcb.00806-15
96. Love PE, Warzecha C, Li L (2014) Ldb1 complexes: the new master regulators of erythroid gene transcription. Trend Gen TIG 30:1–9. https://doi.org/10.1016/j.tig.2013.10.001
97. Li L, Jothi R, Cui K, Lee JY, Cohen T, Gorivodsky M, Tzchori I, Zhao Y, Hayes SM, Bresnick EH, Zhao K, Westphal H, Love PE (2011) Nuclear adaptor Ldb1 regulates a transcriptional program essential for the maintenance of hematopoietic stem cells. Nat Immunol 12:129–136. https://doi.org/10.1038/ni.1978
98. Bresnick EH, Lee H-Y, Fujiwara T, Johnson KD, Keles S (2010) GATA switches as developmental drivers. J Biol Chem 285:31087–31093. https://doi.org/10.1074/jbc.r110.159079
99. Wortel IMN, van der Meer LT, Kilberg MS, van Leeuwen FN (2017) Surviving stress: modulation of ATF4-mediated stress responses in normal and malignant cells. Trends Endocrinol Metab 28:794–806. https://doi.org/10.1016/j.tem.2017.07.003
100. Chen J-J, Zhang S (2022) Translational control by heme-regulated elF2α kinase during erythropoiesis. Curr Opin Hematol 29:103–111. https://doi.org/10.1097/moh.0000000000000704
101. Kim YW, Yun WJ, Kim A (2016) Erythroid activator NF-E2, TAL1 and KLF1 play roles in forming the LCR HSs in the human adult β-globin locus. Int J Biochem Cell Biol 75:45–52. https://doi.org/10.1016/j.biocel.2016.03.013
102. Hung H-L, Kim AY, Hong W, Rakowski C, Blobel GA (2001) Stimulation of NF-E2 DNA binding by CREB-binding protein (CBP)-mediated acetylation*. J Biol Chem 276:10715–10721. https://doi.org/10.1074/jbc.m007846200
103. Choi A, Illendula A, Pulikkan JA, Roderick JE, Tesell J, Yu J, Hermance N, Zhu LJ, Castilla LH, Bushweller JH, Kelliher MA (2017) RUNX1 is required for oncogenic Myb and Myc enhancer activity in T-cell acute lymphoblastic leukemia. Blood 130:1722–1733. https://doi.org/10.1182/blood-2017-03-775536
104. Fishilevich S, Nudel R, Rappaport N, Hadar R, Plaschkes I, Stein TI, Rosen N, Kohn A, Twik M, Safran M, Lancet D, Cohen D (2017) GeneHancer: genome-wide integration of enhancers and target genes in GeneCards. Database 2017:bax028. https://doi.org/10.1093/database/bax028
105. Bellon T, Perrotti D, Calabretta B (1997) Granulocytic differentiation of normal hematopoietic precursor cells induced by transcription factor PU.1 correlates with negative regulation of the c-myb promoter. Blood 90:1828–1839. https://doi.org/10.1182/blood.v90.5.1828
106. Nicolaides NC, Gualdi R, Casadevall C, Manzella L, Calabretta B (1991) Positive autoregulation of c-myb expression via Myb binding sites in the 5′ flanking region of the human c-myb gene. Mol Cell Biol 11:6166–6176. https://doi.org/10.1128/mcb.11.12.6166-6176.1991
107. Guerra J, Withers D, Boxer L (1995) Myb binding sites mediate negative regulation of c-myb expression in T- cell lines. Blood 86:1873–1880. https://doi.org/10.1182/blood.v86.5.1873.bloodjournal8651873
108. Nicolaides NC, Correa I, Casadevall C, Travali S, Soprano KJ, Calabretta B (1992) The Jun family members, c-Jun and JunD, transactivate the human c-myb promoter via an Ap1-like element. J Biol Chem 267:19665–19672. https://doi.org/10.1016/s0021-9258(18)41826-1
109. Campanero MR, Armstrong M, Flemington E (1999) Distinct cellular factors regulate the c-myb promoter through its E2F element. Mol Cell Biol 19:8442–8450. https://doi.org/10.1128/mcb.19.12.8442
110. Li H, Hai J, Zhou J, Yuan G (2016) The formation and characteristics of the i-motif structure within the promoter of the c-myb proto-oncogene. J Photochem Photobiol B 162:625–632. https://doi.org/10.1016/j.jphotobiol.2016.07.035
111. Palumbo SL, Memmott RM, Uribe DJ, Krotova-Khan Y, Hurley LH, Ebbinghaus SW (2008) A novel G-quadruplex-forming GGA repeat region in the c-myb promoter is a critical regulator of promoter activity. Nucleic Acids Res 36:1755–1769. https://doi.org/10.1093/nar/gkm1069
112. Álvaro-Blanco J, Urso K, Chiodo Y, Martín-Cortázar C, Kourani O, Arco PG, Rodríguez-Martínez M, Calonge E, Alcamí J, Redondo JM, Iglesias T, Campanero MR (2017) MAZ induces

MYB expression during the exit from quiescence via the E2F site in the MYB promoter. Nucleic Acids Res 45:9960–9975. https://doi.org/10.1093/nar/gkx641
113. Bender TP, Thompson CB, Kuehl WM (1987) Differential expression of c-myb mRNA in murine B lymphomas by a block to transcription elongation. Science 237:1473–1476. https://doi.org/10.1126/science.3498214
114. Hugo H, Cures A, Suraweera N, Drabsch Y, Purcell D, Mantamadiotis T, Phillips W, Dobrovic A, Zupi G, Gonda TJ, Iacopetta B, Ramsay RG (2006) Mutations in the MYB intron I regulatory sequence increase transcription in colon cancers. Genes Chromosom Cancer 45:1143–1154. https://doi.org/10.1002/gcc.20378
115. Stadhouders R, Thongjuea S, Andrieu-Soler C, Palstra R-J, Bryne JC, van den Heuvel A, Stevens M, de Boer E, Kockx C, van der Sloot A, van den Hout M, van Ijcken W, Eick D, Lenhard B, Grosveld F, Soler E (2011) Dynamic long-range chromatin interactions control Myb proto-oncogene transcription during erythroid development. EMBO J. https://doi.org/10.1038/emboj.2011.450
116. Chen FX, Smith ER, Shilatifard A (2018) Born to run: control of transcription elongation by RNA polymerase II. Nat Rev Mol Cell Biol 19:464–478. https://doi.org/10.1038/s41580-018-0010-5
117. Mitra P, Pereira LA, Drabsch Y, Ramsay RG, Gonda TJ (2012) Estrogen receptor- recruits P-TEFb to overcome transcriptional pausing in intron 1 of the MYB gene. Nucleic Acids Res. https://doi.org/10.1093/nar/gks286
118. Toth CR, Hostutler RF, Baldwin AS, Bender TP (1995) Members of the nuclear factor κB family transactivate the murine c-myb gene (∗). J Biol Chem 270:7661–7671. https://doi.org/10.1074/jbc.270.13.7661
119. Pereira LA, Hugo HJ, Malaterre J, Huiling X, Sonza S, Cures A, Purcell DFJ, Ramsland PA, Gerondakis S, Gonda TJ, Ramsay RG (2015) MYB elongation is regulated by the nucleic acid binding of NFκB p50 to the intronic stem-loop region. PLoS ONE 10:e0122919. https://doi.org/10.1371/journal.pone.0122919
120. Cao L, Mitra P, Gonda TJ (2019) The mechanism of MYB transcriptional regulation by MLL-AF9 oncoprotein. Sci Rep 9:20084. https://doi.org/10.1038/s41598-019-56426-7
121. Frerich CA, Sedam HN, Kang H, Mitani Y, El-Naggar AK, Ness SA (2019) N-terminal truncated Myb with new transcriptional activity produced through use of an alternative MYB promoter in salivary gland adenoid cystic carcinoma. Cancer 12:45. https://doi.org/10.3390/cancers12010045
122. Fehr A, Arvidsson G, Nordlund J, Lönnerholm G, Stenman G, Andersson MK (2023) Increased MYB alternative promoter usage is associated with relapse in acute lymphoblastic leukemia. Genes Chromosom Cancer. https://doi.org/10.1002/gcc.23151
123. Hu D, Shao W, Liu L, Wang Y, Yuan S, Liu Z, Liu J, Zhang J (2021) Intricate crosstalk between MYB and noncoding RNAs in cancer. Cancer Cell Int 21:653. https://doi.org/10.1186/s12935-021-02362-4
124. Xiao C, Calado DP, Galler G, Thai T-H, Patterson HC, Wang J, Rajewsky N, Bender TP, Rajewsky K (2007) MiR-150 controls B cell differentiation by targeting the transcription factor c-Myb. Cell 131:146–159. https://doi.org/10.1016/j.cell.2007.07.021
125. Zhao Z, Shilatifard A (2019) Epigenetic modifications of histones in cancer. Genome Biol 20:245–216. https://doi.org/10.1186/s13059-019-1870-5
126. Talbert PB, Henikoff S (2021) The Yin and Yang of histone marks in transcription. Annu Rev Genomics Hum Genet 22:1–24. https://doi.org/10.1146/annurev-genom-120220-085159
127. Jin S, Zhao H, Yi Y, Nakata Y, Kalota A, Gewirtz AM (2010) c-Myb binds MLL through menin in human leukemia cells and is an important driver of MLL-associated leukemogenesis. J Clin Invest 120:593–606. https://doi.org/10.1172/jci38030
128. Streubel G, Bouchard C, Berberich H, Zeller MS, Teichmann S, Adamkiewicz J, Müller R, Klempnauer K-H, Bauer U-M (2013) PRMT4 is a novel coactivator of c-Myb-dependent transcription in haematopoietic cell lines. PLoS Genet 9:e1003343. https://doi.org/10.1371/journal.pgen.1003343
129. Sæther T, Berge T, Ledsaak M, Matre V, Alm-Kristiansen AH, Dahle Ø, Aubry F, Gabrielsen OS (2007) The chromatin remodeling factor Mi-2α acts as a novel co-activator for human c-Myb*. J Biol Chem 282:13994–14005. https://doi.org/10.1074/jbc.m700755200
130. Tomita A, Towatari M, Tsuzuki S, Hayakawa F, Kosugi H, Tamai K, Miyazaki T, Kinoshita T, Saito H (2000) c-Myb acetylation at the carboxyl-terminal conserved domain by transcriptional co-activator p300. Oncogene 19:444–451. https://doi.org/10.1038/sj.onc.1203329
131. Sano Y, Ishii S (2001) Increased affinity of c-Myb for CREB-binding protein (CBP) after CBP-induced acetylation. J Biol Chem 276:3674–3682. https://doi.org/10.1074/jbc.m006896200
132. Fuglerud BM, Ledsaak M, Rogne M, Eskeland R, Gabrielsen OS (2018) The pioneer factor activity of c-Myb involves recruitment of p300 and induction of histone acetylation followed by acetylation-induced chromatin dissociation. Epigenetics Chromatin 11:35. https://doi.org/10.1186/s13072-018-0208-y
133. Zhao H, Jin S, Gewirtz AM (2012) The histone acetyltransferase TIP60 interacts with c-Myb and inactivates its transcriptional activity in human leukemia. J Biol Chem 287:925–934. https://doi.org/10.1074/jbc.m111.279950
134. Tahirov TH, Sato K, Ichikawa-Iwata E, Sasaki M, Inoue-Bungo T, Shiina M, Kimura K, Takata S, Fujikawa A, Morii H, Kumasaka T, Yamamoto M, Ishii S, Ogata K (2002) Mechanism of c-Myb-C/EBP beta cooperation from separated sites on a promoter. Cell 108:57–70

135. Klempnauer K-H (2023) C/EBPβ cooperates with MYB to maintain the oncogenic program of AML cells. Oncotarget 14:174–177. https://doi.org/10.18632/oncotarget.28377
136. Alm-Kristiansen AH, Saether T, Matre V, Gilfillan S, Dahle O, Gabrielsen OS (2008) FLASH acts as a co-activator of the transcription factor c-Myb and localizes to active RNA polymerase II foci. Oncogene 27:4644–4656. https://doi.org/10.1038/onc.2008.105
137. Alm-Kristiansen AH, Lorenzo PI, Molværsmyr A-K, Matre V, Ledsaak M, Sæther T, Gabrielsen OS (2011) PIAS1 interacts with FLASH and enhances its co-activation of c-Myb. Mol Cancer 10:21. https://doi.org/10.1186/1476-4598-10-21
138. Ledsaak M, Bengtsen M, Molværsmyr A-K, Fuglerud BM, Matre V, Eskeland R, Gabrielsen OS (2016) PIAS1 binds p300 and behaves as a coactivator or corepressor of the transcription factor c-Myb dependent on SUMO-status. Biochim Biophys Acta Gene Regul Mech 1859:705–718. https://doi.org/10.1016/j.bbagrm.2016.03.011
139. Xu Y, Milazzo JP, Somerville TDD, Tarumoto Y, Huang Y-H, Ostrander EL, Wilkinson JE, Challen GA, Vakoc CR (2018) A TFIID-SAGA perturbation that targets MYB and suppresses acute myeloid leukemia. Cancer Cell 33:13-28.e8. https://doi.org/10.1016/j.ccell.2017.12.002
140. Bell CC, Dawson MA (2018) TFIID and MYB share a therapeutic handshake in AML. Cancer Cell 33:1–3. https://doi.org/10.1016/j.ccell.2017.12.008
141. Bieniossek C, Papai G, Schaffitzel C, Garzoni F, Chaillet M, Scheer E, Papadopoulos P, Tora L, Schultz P, Berger I (2013) The architecture of human general transcription factor TFIID core complex. Nat Cell Biol 493:699–702. https://doi.org/10.1038/nature11791
142. Jaeger MG, Schwalb B, Mackowiak SD, Velychko T, Hanzl A, Imrichova H, Brand M, Agerer B, Chorn S, Nabet B, Ferguson FM, Müller AC, Bergthaler A, Gray NS, Bradner JE, Bock C, Hnisz D, Cramer P, Winter GE (2020) Selective Mediator dependence of cell-type-specifying transcription. Nat Genet 52:719–727. https://doi.org/10.1038/s41588-020-0635-0
143. Zaret KS (2020) Pioneer transcription factors initiating gene network changes. Ann Rev Gene 54. https://doi.org/10.1146/annurev-genet-030220-015007
144. Bulyk ML, Drouin J, Harrison MM, Taipale J, Zaret KS (2023) Pioneer factors—key regulators of chromatin and gene expression. Nat Rev Genet:1–7. https://doi.org/10.1038/s41576-023-00648-z
145. Zaret KS, Carroll JS (2011) Pioneer transcription factors: establishing competence for gene expression. Genes Dev 25:2227–2241. https://doi.org/10.1101/gad.176826.111
146. Iwafuchi-Doi M, Zaret KS (2016) Cell fate control by pioneer transcription factors. Development 143:1833–1837. https://doi.org/10.1242/dev.133900
147. Garcia MF, Moore CD, Schulz KN, Alberto O, Donague G, Harrison MM, Zhu H, Zaret KS (2019) Structural features of transcription factors associating with nucleosome binding. Mol Cell 75:921-932.e6. https://doi.org/10.1016/j.molcel.2019.06.009
148. Iwafuchi M, Cuesta I, Donahue G, Takenaka N, Osipovich AB, Magnuson MA, Roder H, Seeholzer SH, Santisteban P, Zaret KS (2020) Gene network transitions in embryos depend upon interactions between a pioneer transcription factor and core histones. Nat Genet 52:418–427. https://doi.org/10.1038/s41588-020-0591-8
149. Bulger M, Groudine M (2011) Functional and mechanistic diversity of distal transcription enhancers. Cell 144:327–339. https://doi.org/10.1016/j.cell.2011.01.024
150. Lemma RB, Ledsaak M, Fuglerud BM, Sandve GK, Eskeland R, Gabrielsen OS (2021) Chromatin occupancy and target genes of the haematopoietic master transcription factor MYB. Sci Rep 11:9008. https://doi.org/10.1038/s41598-021-88516-w
151. Ness SA, Marknell A, Graf T (1989) The v-myb oncogene product binds to and activates the promyelocyte-specific mim-1 gene. Cell 59:1115–1125. https://doi.org/10.1016/0092-8674(89)90767-8
152. Plachetka A, Chayka O, Wilczek C, Melnik S, Bonifer C, Klempnauer K-H (2008) C/EBPbeta induces chromatin opening at a cell-type-specific enhancer. Mol Cell Biol 28:2102–2112. https://doi.org/10.1128/mcb.01943-07
153. Mink S, Kerber U, Klempnauer KH (1996) Interaction of C/EBPbeta and v-Myb is required for synergistic activation of the mim-1 gene. Mol Cell Biol 16
154. George OL, Ness SA (2014) Situational awareness: regulation of the myb transcription factor in differentiation, the cell cycle and oncogenesis. Cancer 6:2049–2071. https://doi.org/10.3390/cancers6042049
155. Zhao L, Glazov EA, Pattabiraman DR, Al-Owaidi F, Zhang P, Brown MA, Leo PJ, Gonda TJ (2011) Integrated genome-wide chromatin occupancy and expression analyses identify key myeloid pro-differentiation transcription factors repressed by Myb. Nucleic Acids Res 39:4664–4679. https://doi.org/10.1093/nar/gkr024
156. Gambone JE, Dusaban SS, Loperena R, Nakata Y, Shetzline SE (2011) The c-Myb target gene neuromedin U functions as a novel cofactor during the early stages of erythropoiesis. Blood. https://doi.org/10.1182/blood-2009-09-242131
157. Zhao L, Ye P, Gonda TJ (2013) The MYB proto-oncogene suppresses monocytic differentiation of acute myeloid leukemia cells via transcriptional activation of its target gene GFI1. Oncogene. https://doi.org/10.1038/onc.2013.419
158. Bengtsen M, Klepper K, Gundersen S, Cuervo I, Drabløs F, Hovig E, Sandve GK, Gabrielsen OS, Eskeland R (2015) c-Myb binding sites in haematopoietic chromatin landscapes. Plos One

10:e0133280. https://doi.org/10.1371/journal.pone.0133280
159. Berge T, Matre V, Brendeford EM, Saether T, Lüscher B, Gabrielsen OS (2007) Revisiting a selection of target genes for the hematopoietic transcription factor c-Myb using chromatin immunoprecipitation and c-Myb knockdown. Blood Cell Mol Dis 39:278–286. https://doi.org/10.1016/j.bcmd.2007.05.007
160. Deleuze V, Garcia L, Rouaisnel B, Salma M, Kinoo A, Andrieu-Soler C, Soler E (2023) Efficient genome editing in erythroid cells unveils novel MYB target genes and regulatory functions. iScience 26:107641. https://doi.org/10.1016/j.isci.2023.107641
161. Abraham BJ, Hnisz D, Weintraub AS, Kwiatkowski N, Li CH, Li Z, Weichert-Leahey N, Rahman S, Liu Y, Etchin J, Li B, Shen S, Lee TI, Zhang J, Look AT, Mansour MR, Young RA (2017) Small genomic insertions form enhancers that misregulate oncogenes. Nat Commun 8:14385–14313. https://doi.org/10.1038/ncomms14385
162. Calo E, Wysocka J (2013) Modification of enhancer chromatin: what, how, and why? Mol Cell 49:825–837. https://doi.org/10.1016/j.molcel.2013.01.038
163. Whyte WA, Orlando DA, Hnisz D, Abraham BJ, Lin CY, Kagey MH, Rahl PB, Lee TI, Young RA (2013) Master transcription factors and mediator establish super-enhancers at key cell identity genes. Cell 153:307–319. https://doi.org/10.1016/j.cell.2013.03.035
164. Jia Y, Chng W-J, Zhou J (2019) Super-enhancers: critical roles and therapeutic targets in hematologic malignancies. J Hematol Oncol 12:77–17. https://doi.org/10.1186/s13045-019-0757-y
165. Cogswell JP, Cogswell PC, Cuddihy AM, Bender T, Engelke U, Marcu KB, Ting JP (1993) Mechanism of c-myc regulation by c-Myb in different cell lineages. Mol Cell Biol 13:2858–2869
166. Schmidt M, Nazarov V, Stevens L, Watson R, Wolff L (2000) Regulation of the resident chromosomal copy of c-myc by c-Myb is involved in myeloid leukemogenesis. Mol Cell Biol 20:1970–1981. https://doi.org/10.1128/mcb.20.6.1970-1981.2000
167. Bahr C, von Paleske L, Uslu VV, Remeseiro S, Takayama N, Ng SW, Murison A, Langenfeld K, Petretich M, Scognamiglio R, Zeisberger P, Benk AS, Amit I, Zandstra PW, Lupien M, Dick JE, Trumpp A, Spitz F (2018) A Myc enhancer cluster regulates normal and leukaemic haematopoietic stem cell hierarchies. Nat Cell Biol 553:515–520. https://doi.org/10.1038/nature25193
168. Lemma RB, Ledsaak M, Fuglerud BM, Rodríguez-Castañeda F, Eskeland R, Gabrielsen OS (2023) MYB regulates the SUMO protease SENP1 and its novel interaction partner UXT, modulating MYB target genes and the SUMO landscape. J Biol Chem 299:105062. https://doi.org/10.1016/j.jbc.2023.105062
169. Boulanger M, Chakraborty M, Tempé D, Piechaczyk M, Bossis G (2021) SUMO and transcriptional regulation: the lessons of large-scale proteomic, modifomic and genomic studies. Molecules 26:828. https://doi.org/10.3390/molecules26040828
170. Theurillat I, Hendriks IA, Cossec J-C, Andrieux A, Nielsen ML, Dejean A (2020) Extensive SUMO modification of repressive chromatin factors distinguishes pluripotent from somatic cells. Cell Rep 32:108146. https://doi.org/10.1016/j.celrep.2020.108146
171. Cossec J-C, Theurillat I, Chica C, Aguín SB, Gaume X, Andrieux A, Iturbide A, Jouvion G, Li H, Bossis G, Seeler J-S, Torres-Padilla M-E, Dejean A (2018) SUMO safeguards somatic and pluripotent cell identities by enforcing distinct chromatin states. Cell Stem Cell 23:742-757.e8. https://doi.org/10.1016/j.stem.2018.10.001
172. Kauraniemi P, Hedenfalk I, Persson K, Duggan DJ, Tanner M, Johannsson O, Olsson H, Trent JM, Isola J, Borg A (2000) MYB oncogene amplification in hereditary BRCA1 breast cancer. Cancer Res 60:5323–5328
173. Takao S, Forbes L, Uni M, Cheng S, Pineda JMB, Tarumoto Y, Cifani P, Minuesa G, Chen C, Kharas MG, Bradley RK, Vakoc CR, Koche RP, Kentsis A (2021) Convergent organization of aberrant MYB complex controls oncogenic gene expression in acute myeloid leukemia. elife 10:e65905. https://doi.org/10.7554/elife.65905
174. Clarke ML, Lemma RB, Walton DS, Volpe G, Noyvert B, Gabrielsen OS, Frampton J (2023) MYB insufficiency disrupts proteostasis in hematopoietic stem cells leading to age-related neoplasia. Blood 141:1858–1870. https://doi.org/10.1182/blood.2022019138
175. Burk O, Mink S, Ringwald M, Klempnauer KH (1993) Synergistic activation of the chicken mim-1 gene by v-myb and C/EBP transcription factors. EMBO J 12:2027–2038
176. Wilczek C, Chayka O, Plachetka A, Klempnauer K-H (2009) Myb-induced chromatin remodeling at a dual enhancer/promoter element involves non-coding rna transcription and is disrupted by oncogenic mutations of v-myb. J Biol Chem 284:35314–35324. https://doi.org/10.1074/jbc.m109.066175
177. Waterhouse AM, Procter JB, Martin DMA, Clamp M, Barton GJ (2009) Jalview Version 2—a multiple sequence alignment editor and analysis workbench. Bioinformatics 25:1189–1191. https://doi.org/10.1093/bioinformatics/btp033
178. Castro-Mondragon JA, Riudavets-Puig R, Rauluseviciute I, Berhanu Lemma R, Turchi L, Blanc-Mathieu R, Lucas J, Boddie P, Khan A, Manosalva Pérez N, Fornes O, Leung TY, Aguirre A, Hammal F, Schmelter D, Baranasic D, Ballester B, Sandelin A, Lenhard B, Vandepoele K, Wasserman WW, Parcy F, Mathelier A (2021) JASPAR 2022: the 9th release of the open-access database of transcription factor binding profiles. Nucleic Acids Res 50:gkab1113. https://doi.org/10.1093/nar/gkab1113

Part II

T-Cells

IKAROS Family Transcription Factors in Lymphocyte Differentiation and Function

2

Philippe Kastner and Susan Chan

Abstract

The IKAROS family of transcription factors comprises four zinc-finger proteins (IKAROS, HELIOS, AIOLOS, and EOS), which over the last decades have been established to be critical regulators of the development and function of lymphoid cells. These factors act as homo- or heterodimers and are involved both in gene activation and repression. Their function often involves cross-talk with other regulatory circuits, such as the JAK/STAT, NF-κB, and NOTCH pathways. They control lymphocyte differentiation at multiple stages and are notably critical for lymphoid commitment in multipotent hematopoietic progenitors and for T and B cell differentiation downstream of pre-TCR and pre-BCR signaling. They also control many aspects of effector functions in mature B and T cells. They are dysregulated or mutated in multiple pathologies affecting the lymphoid system, which range from leukemia to immunodeficiencies. In this chapter, we review the molecular and physiological function of these factors in lymphocytes and their implications in human pathologies.

Keywords

Ikaros family · Stem cells · B- and T-cell differentiation · Effector B- and T-cell functions · Leukemia · Inborn errors of immunity

Abbreviations

BCP-ALL	B-cell precursor acute lymphoblastic leukemia
BCR	B-cell receptor
BM	Bone marrow

P. Kastner (✉)
Institut de Génétique et de Biologie Moléculaire et Cellulaire (IGBMC), Illkirch-Graffenstaden, France

Institut National de la Santé et de la Recherche Médicale (INSERM), U1258, Illkirch-Graffenstaden, France

Centre National de la Recherche Scientifique (CNRS), UMR7104, Illkirch-Graffenstaden, France

Université de Strasbourg, Illkirch-Graffenstaden, France

Faculté de Médecine, Université de Strasbourg, Strasbourg, France
e-mail: scpk@igbmc.fr

S. Chan (✉)
Institut de Génétique et de Biologie Moléculaire et Cellulaire (IGBMC), Illkirch-Graffenstaden, France

Institut National de la Santé et de la Recherche Médicale (INSERM), U1258, Illkirch-Graffenstaden, France

Centre National de la Recherche Scientifique (CNRS), UMR7104, Illkirch-Graffenstaden, France

Université de Strasbourg, Illkirch-Graffenstaden, France
e-mail: scpk@igbmc.fr

T. Borggrefe, B. D. Giaimo (eds.), *Transcription factors in blood cell development*, Advances in Experimental Medicine and Biology 1459, https://doi.org/10.1007/978-3-031-62731-6_2

CLP	Common lymphoid precursor
CSR	Class-switch recombination
DBD	DNA-binding domain
DD	Dimerization domain
DN	Double-negative
Dntt	DNA nucleotidylexotransferase
DP	Double-positive
GoF	Gain-of-function
GWAS	Genome-wide association studies
HSPC	Hematopoietic stem/progenitor cell
LoF	Loss-of-function
MHC	Major histocompatibility complex
MM	Multiple myeloma
MPP	Multipotent progenitor
PTM	Posttranslational modification
RSS	Recombination signal sequence
T-ALL	T-cell acute lymphoblastic leukemia
TCR	T-cell receptor
TLR	Toll-like receptor
TF	Transcription factor
ZF	Zinc finger

2.1 Discovery of the IKAROS family

In the race to identify "master" transcription factors (TFs) of the lymphoid lineage in the late 1900s, two groups discovered a DNA-binding protein that was abundantly expressed in T cells and bound to the regulatory regions of genes important for T-cell function. This factor, named lymphocyte factor 1 (LYF-1) by the group of Stephen Smale [1, 2], was found to bind to the promoter of the *Dntt* gene, which encodes the DNA nucleotidylexotransferase enzyme that adds N-nucleotides to the V, D and J exons of the T and B cell receptors during gene recombination in immature, pre-T and pre-B cells. The same protein, named IKAROS by the group of Katia Georgopoulos [3], was reported to bind to an enhancer of the *Cd3d* gene, which encodes the delta subunit of the CD3 signaling complex in T cells. The IKAROS name prevailed, and it soon became apparent from mouse studies that IKAROS is a critical regulator of lymphoid cell development and hematopoiesis in general. Later studies revealed that IKAROS belonged to a small family of homologous factors that included HELIOS, AIOLOS, and EOS, all of which are key modulators of hematopoietic cell differentiation and function [4–7].

2.2 Overview and Molecular Functions

2.2.1 Gene Organization and Homologs

IKAROS, HELIOS, AIOLOS, and EOS are encoded by the *IKZF1-4* genes, respectively. They share a general domain organization (Fig. 2.1)—an acidic N-terminal region (20–25% glutamic or aspartic acid), followed by a DNA-binding domain (DBD) comprised of four C2H2 zinc fingers (ZF1-4) and a C-terminal dimerization domain (DD) containing two other C2H2 ZFs (ZF5 and −6) [5, 8]. The DBD and DD are highly conserved and account for most of the sequence homology between the family members. In contrast, the N-terminal region and the sequences between the DBD and DD are less conserved. How the C-terminal ZFs mediate protein-protein dimerization, rather than protein-DNA binding, is currently not well understood. While some amino acids critical for protein dimerization have been identified [9], the structural basis for ZF dimerization remains to be elucidated.

One striking feature of the *IKZF1-4* genes is that the coding exons are in-frame, which allows for alternative mRNA splicing and the generation of proteins with partial and/or dominant-negative properties [2, 8, 10]. Indeed, numerous isoforms of IKAROS family proteins have been described, in physiological and pathological conditions. In lymphocytes, for example, IKAROS is detected both as a full-length isoform (IKAROS-1) and a smaller protein that lacks part of the N-terminal region and ZF1 (IKAROS-2) [11–13]. In other hematopoietic cells, like macrophages, even smaller IKAROS isoforms have been found and

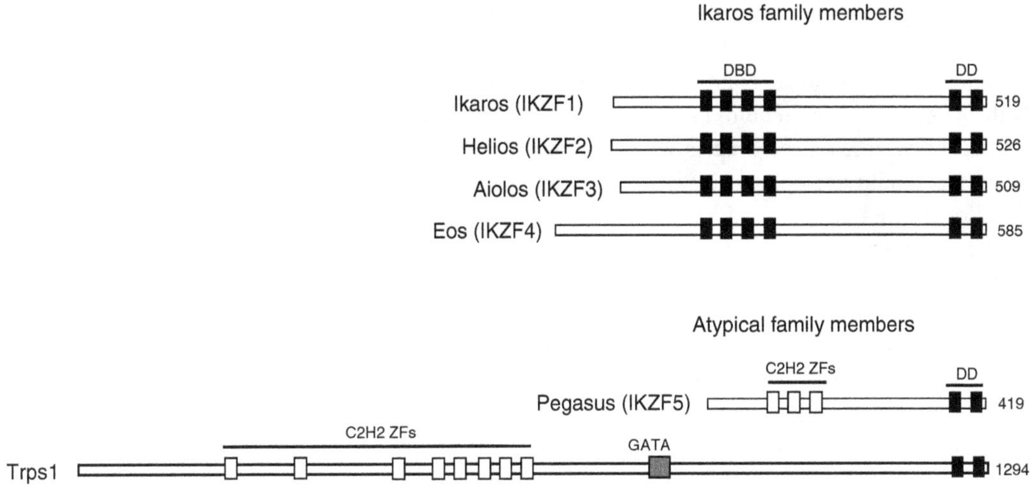

Fig. 2.1 Primary structure of IKAROS family members. Top: primary structure of the four core IKAROS family members. The black boxes correspond to C2H2 zinc fingers (ZFs), which are homologous among all four factors. Bottom: primary structure of the atypical family member PEGASUS and the transcriptional repressor GATA binding 1 (TRPS1). Black boxes indicate ZFs homologous to those of IKAROS, while the white and gray boxes correspond to ZFs unrelated to those of IKAROS

purported to be associated with maturation state [14]. How different isoforms contribute to IKAROS function remains unclear.

IKAROS family protein homologs containing the conserved DBD and DD are found in all vertebrates and tunicates [15], suggesting that the IKAROS family is present in chordates. In addition, IKAROS family proteins belong to a larger constellation of TFs, whose members exist in nearly all animals (with the possible exception of sponges, jellyfish, and corals), and whose structures are characterized by a DBD, containing varying numbers of C2H2 ZFs, and a C-terminal dimerization domain with two ZFs [16]. This constellation includes the *Drosophila melanogaster* Hunchback protein, an essential TF for anteroposterior axis development [17], and the *Caenorhabditis elegans* Ehn-3 and R08E3.4 [18]. Humans also have two members, PEGASUS and TRPS1, which are unrelated to the core IKAROS family proteins in their DBD [7, 9] (Fig. 2.1). PEGASUS contains a putative DBD with three ZFs, while TRPS1 has a DBD with 8 C2H2 ZFs and one GATA-type ZF. Only the four core IKAROS family proteins will be covered below.

2.2.2 DNA Binding

IKAROS family proteins have been reported to homo- and heterodimerize. Indeed, all possible combinations can form in vitro, and most likely in vivo, as immunoprecipitation of one member leads to the coprecipitation of other family members when co-expressed [4, 5]. It is currently unknown if or how the different dimers are functionally distinct.

What is clear is that all four family members possess highly conserved DNA-binding ZFs, suggesting that they recognize similar DNA motifs. Both in vitro DNA-binding experiments and bioinformatic motif enrichment analyses of data from chromatin immunoprecipitation followed by deep sequencing (ChIP-Seq) studies have consistently shown that IKAROS, HELIOS, and AIOLOS bind a core GGAA motif [19–23]. Beyond these four nucleotides, however, there is no clear consensus about the flanking sequences required for an IKAROS family TF target site. Since IKAROS family proteins bind DNA as dimers, high affinity binding sites are likely to contain two GGAA motifs in close proximity. This is true for some high affinity targets, like the synthetic BS4 oligonucleotide [8], or the physiological target sequence near the *Cish* gene promoter, where mutational analyses showed that both GGAA motifs are required for binding [24]. Interestingly, efficient binding is also observed when the motifs are pointed in different directions or spaced apart by differing nucleotide lengths. This suggests that binding is flexible and that the dimer recognizes sequences with different architectures [24]. There is also no clear understanding about the contribution of the flanking nucleotides to binding efficiency. In vitro target site selection for IKAROS or HELIOS homodimers have identified surrounding nucleotides responsible for high affinity binding [4, 8]. Curiously, these high affinity sequences, like the TGggaaT(A/G) one for IKAROS [8], are not often found at physiological targets, suggesting that the physiological target sequences have been selected for suboptimal binding.

In addition, posttranslational modifications (PTMs) of the IKAROS family proteins affect the binding capacity of these proteins. The groups of K. Georgopoulos and S. Dovat reported that IKAROS phosphorylation by casein kinase II (CKII) inhibits the binding of IKAROS to its target sequences [25, 26], and the CKII target residues in the N-terminal and inter-ZF regions are implicated. How phosphorylation events outside of the DBD interfere with DNA binding remains to be studied. Nonetheless, these observations suggested that CKII inhibitors may improve the survival of mice engrafted with B-cell leukemias bearing haploinsufficient *IKZF1* mutations by dephosphorylating and improving the function of the remaining wild-type (WT) IKAROS proteins [27].

The core GGAA motif also forms part of the recognition sequence of other TFs like those of the ETS, STAT (signal transducer and activator of transcription), and NFAT (nuclear factor of activated T cells) families, suggesting that IKAROS family proteins may compete with factors of other TF families for binding. Indeed, the Smale group reported that IKAROS competes with the ETS factor ELF1 to bind the *Dntt* gene, resulting in a switch from gene activation (ELF1) to repression (IKAROS) [28]. In addition, we have shown that the IKAROS target sequence in the *Cish* promoter coincides with a target sequence for STAT5, and that IKAROS binding leads to a widespread decrease of STAT5 binding in pre-B cells [24].

2.2.3 Repression and Activation

IKAROS family proteins act as transcriptional repressors or activators. IKAROS, AIOLOS, and HELIOS associate with the repressive Nucleosome Remodeling and Deacetylase (NuRD) complex and are thought to recruit or tether this complex to their target genes [20, 29, 30]. Additionally, IKAROS can interact with Polycomb Repressive Complex 2 (PRC2) in thymocytes to deposit or maintain the H3K27me3 repressive chromatin mark at other IKAROS targets [31]. A key feature of IKAROS family-mediated repression is the ability of these proteins to bind to γ-satellite sequences at centromeric heterochromatin [32]. It is currently thought that the concomitant binding to target genes poised for silencing and γ-satellite sequences triggers the repositioning of these genes to centromeric heterochromatin for permanent silencing [32–34].

IKAROS family factors also contribute to gene activation. In pre-B cells, Hu et al. showed that IKAROS-dependent activation of genes important for differentiation is mediated through the binding of IKAROS to super-enhancers also targeted by other B-cell TFs like EBF1, PAX5, and IRF4 [35]. IKAROS binding was required for promoting a permissive chromatin environment, as evidenced by reduced H3K27ac and H3K4me3 at these enhancers in IKAROS-deficient cells. Similarly, Ding et al. showed that IKAROS reexpression in an IKAROS-null T-cell leukemia cell line leads to a large-scale de novo formation of active enhancers, suggesting that IKAROS can act as a pioneering factor to promote chromatin opening [36].

An open question is how the target sites inform IKAROS family TFs to act as repressors or activators. One attractive hypothesis, which remains to be investigated, is that different target site architectures (i.e., orientation and/or spacing of the target motifs) may induce different conformations of the bound dimers which then recruit specific partners involved in repression or activation.

2.3 Physiological Functions

While this chapter focuses on the role of IKAROS TFs in lymphocytes, these proteins are also important differentiation factors in other hematopoietic and non-hematopoietic cells. In hematopoietic cells, IKAROS is required for early developmental decisions in dendritic cells [37, 38], the development of lymphoid tissue initiating cells [39], and the differentiation of erythrocytes and neutrophils [40, 41]. In non-hematopoietic cells, IKAROS, HELIOS, and EOS are important for the development of certain neuronal cells, notably in the retina and the striatum [42–46]. In addition, HELIOS is required for the maturation of outer hair cells in the cochlea, and HELIOS mutations may be involved in age-dependent hearing loss [47].

2.3.1 Hematopoietic Stem and Progenitor Cells

IKAROS and HELIOS are highly expressed in hematopoietic stem and progenitor cells (HSPCs), and both factors are critical in this compartment, as underscored by a > 100-fold reduction in stem cell activity in mice with constitutive dominant-negative DBD mutations in IKAROS, which inhibit the binding of IKAROS and its dimerization partners to target genes [48, 49], and a lesser reduction in animals with an *Ikzf1*-specific knockout (KO) mutation [48]. These studies suggested that IKAROS family members were essential for HSPC development and/or maintenance.

Other studies showed that IKAROS family proteins promote lymphoid commitment from HSPCs. Indeed, *Ikzf1* KO multipotent progenitor cells (MPPs) exhibit impaired expression of lymphocyte-specific genes and a differentiation block toward common lymphoid progenitors (CLPs) and mature B cells [50, 51]. In contrast, HSPCs deficient for Helios lose lymphoid priming potential by differentiating toward the megakaryocytic lineage [23]. Thus, IKAROS family proteins promote lymphoid priming both directly (IKAROS) and indirectly (HELIOS) in the HSPC compartment.

2.3.2 Lymphocyte Differentiation

T cells: Conventional αβ-lineage T cell development is marked by quality control checkpoints at the $CD4^-CD8^-$ double-negative (DN1–4) and $CD4^+$ $CD8^+$ double-positive (DP) thymocyte stages. They include the T cell receptor β-chain rearrangement in DN3a thymocytes, leading to the pre-T-cell receptor (pre-TCR) checkpoint (β-selection) that validates the presence of a functional TCR β-chain on the surface of DN3b thymocytes [52]; the interaction of the NOTCH1 receptor with DELTA-LIKE ligands expressed by cortical thymic stromal cells, which is critical for commitment of early thymic progenitors to the T-cell lineage, and to support differentiation, proliferation, and survival [53]; the progressive shutting off of NOTCH signaling after β-selection, to allow cell cycle exit for further differentiation and the rearrangement of the TCR α-chain; and the positive and negative selection events that occur at the DP to single-positive (SP) $CD4^+$ or $CD8^+$ thymocyte transition which depends on TCR signal strength.

IKAROS is implicated in setting the signaling thresholds of both the pre-TCR and the TCR. In its absence, *Rag1* KO thymocytes, normally blocked at the DN3a stage without a productive β-chain, differentiate into DN3b cells [54], suggesting that IKAROS is required at β-selection to eliminate cells with nonfunctional pre-TCRs. In addition, IKAROS-deficient thymocytes do not require major histocompatibility complex (MHC) molecules to develop past the DP stage and can develop into both SP $CD4^+$ and $CD8^+$ thymocytes even when expressing an incompatible MHC-restricted TCR, indicating that IKAROS is required for positive selection. Finally, mature IKAROS-deficient T cells proliferate robustly when activated via their TCR in the absence of co-stimulatory signals [55], indicating that IKAROS reduces the activation threshold of the TCR. These studies demonstrated that IKAROS is a negative regulator of pre-TCR and TCR activation, though the molecular mechanisms remain unclear.

What is more clear is that IKAROS is required for repressing the expression of genes activated by NOTCH1 signaling in DN3 thymocytes. Lack of IKAROS leads to increased responsiveness to NOTCH ligand-mediated stimulation and increased DN4 cell proliferation [56, 57]. In mice, IKAROS deficiency results in a rapid and aggressive NOTCH-mediated T-cell precursor-acute lymphoblastic lymphoma/leukemia (T-ALL) development that is 100% fatal [58, 59]. At the molecular level, IKAROS directly interacts with NOTCH1, and IKAROS inhibits the binding of the NOTCH1/RBPJ complex to common target genes [19], probably by steric hindrance. In addition, IKAROS recruits/tethers PRC2 to some NOTCH target genes like HES1 to epigenetically silence their expression in DN3 and DN4 cells. Indeed, IKAROS-mediated PRC2 silencing in thymocytes is key to its silencing of >500 genes, including hematopoietic progenitor cell-specific genes [31].

B cells: More than thymocytes, IKAROS is absolutely essential for the differentiation of the common B2 lineage of B cells in the bone marrow, where common lymphoid progenitors give rise to committed pro-B cells. $CD19^+$ pro-B cells rearrange the immunoglobulin heavy (IgH) chain locus and express the surrogate l5 note: the "l" in l5 should be the greek letter for lambda and Vpre-B light chains to form the pre-B cell receptor (pre-BCR) and become pre-B cells. Large pre-B cells undergo a transient proliferative phase that is dependent on pre-BCR signaling and IL7. Continuous pre-BCR signaling and migration from IL7-rich areas lead to cell cycle exit, germline Ig light (IgL) chain transcription and recombination, and the generation of small pre-B cells. Subsequent expression of the mature BCR marks the progression into the immature $CD19^+$ IgM^+ B-cell stage. Germline IKAROS null mice lack fetal and adult B lineage cells [60] due to the absence of CLPs [61, 62]. In addition, animals with a hypomorphic *Ikzf1*$^{L/L}$ mutation show a partial block in prB cell differentiation with the remaining cells displaying an easily activatable state, suggesting that IKAROS was required in B cell progenitors and mature B lymphocytes [63].

In pro-B cells, IKAROS is required for IgH rearrangement. IKAROS null HSPCs carrying an EBF1 transgene (tg) can bypass the differentiation block at the CLP stage and become $CD19^+$ pro-B-like cells that lack RAG1 and RAG2 expression; reintroduction of IKAROS in the pro-B cells rescues RAG1/2 expression and IgH rearrangement [62]. Similarly, *Pten* KO pro-B cells lacking IKAROS do not show IgH rearrangements, though proximal V_H rearrangement can be rescued by IKAROS reexpression and distal by IKAROS in conjunction with PAX5 [64]. While IKAROS-activated *Rag1/2* mRNA expression is important for IgH rearrangement, it is currently unclear if IKAROS directly controls other aspects of recombination. For example, IKAROS may directly activate the transcription of the V_H segments themselves, as well as the rearrangement frequency, as has been shown for IgL chain rearrangement in pre-B cells (*see* below).

Indeed, IKAROS is essential for multiple aspects of pre-B cell differentiation. Mutant mice in which *Ikzf1* is deleted at the late pro-B-cell stage (*Ikzf1*$^{f/f}$ x $Mb1\text{-}Cre^+$) show an absolute block in large pre-B cell development [65–67]. IKAROS is required to downregulate the mRNA expression of *Igll1* (encoding λ5) and initiate IgL chain germline expression and rearrangement. In addition, IKAROS is involved in the downregulation of IL7 receptor (IL7R) signaling through direct and indirect means. IKAROS directly represses the expression of genes encoding cell adhesion and integrin signaling proteins [66]; reduced adhesion promotes the movement of cells away from IL7-rich niches [68]. Further, IKAROS antagonizes the transcriptional activity of IL7/STAT5 signaling by competing with STAT5 for binding to common target elements [24]. IKAROS promotes IgL chain rearrangement by increasing *Rag1/2* mRNA levels and binds to the promoters and recombination signal sequence (RSS) regions of V_K genes [69, 70], where binding strength correlates with rearrangement frequency. IKAROS has also been shown to bind to a silencer element located between the Vκ and Jκ regions (named *Sis* for "silencer in the intervening sequence"), which is required to position the inactive allele to centromeric heterochromatin, prompting speculation about the role of *Sis* and IKAROS in allelic exclusion [71]. However, deletion of *Sis* did not impair allelic exclusion but instead skewed Igκ rearrangement toward increased use of proximal Vκ segments [72]. It remains however to be determined if IKAROS is implicated in this function. These various results suggested that IKAROS directly impacts the transcription and/or recombination of variable gene segments during VDJ recombination.

In addition to IKAROS, AIOLOS is also expressed during B cell differentiation. *Ikzf3* mRNA is detected in pro-B cells and increases with differentiation to reach maximal levels in immature $CD19^+$ IgM^+ B cells [21]. In contrast to IKAROS deficiency, B-cell differentiation occurs in the absence of AIOLOS, as all bone marrow (BM) B cell populations are detected in *Ikzf3* KO mice [73]. However, AIOLOS null pre-B cells do not efficiently downregulate *Igll1* mRNA expression [74]. Thus, both IKAROS and AIOLOS are required to downregulate pre-BCR signaling. In addition, germline $Ikzf3^{-/-}Ikzf1^{+/-}$ animals exhibit a differentiation block at the large pre-B cell stage [66], and mice expressing a G158R AIOLOS mutation, which may promote the formation of nonfunctional dimers upon heterodimerization with IKAROS, show a partial block in pre-B cell differentiation [75]. These studies highlight the importance of IKAROS family homo- and/or heterodimerization in early B cells, but it remains unknown how similar IKAROS and AIOLOS are in terms of molecular functions.

IKAROS also blocks the differentiation of B1 lineage cells. Animals harboring the hypomorphic $Ikzf1^{L/L}$ mutation have higher numbers of B1 progenitor ($CD93^+$ $CD19^+$ $B220^-$/lo IgM^-) and immature B1 cells, in the fetal liver and BM [13, 76], and interestingly, at the expense of B2 cells early in the postnatal period [76]. Thus, IKAROS may be involved in the cell fate choice between B1 and B2 cells.

2.3.3 Effector Cell Function

A number of studies have demonstrated that IKAROS family proteins have wide-ranging roles in regulating lymphocyte effector functions. Though an exhaustive review on this topic is beyond the scope of this chapter, we highlight some common themes regulated by IKAROS family proteins.

Humoral responses IKAROS and AIOLOS affect B cell antibody production both directly and indirectly. IKAROS is a direct regulator of Ig isotype selection during class-switch recombination (CSR) by repressing the transcriptional competition between constant region genes to control the timing and correct range of isotypes (IgG1, IgG2a/2b, IgG3, IgE, IgA) targeted by CSR [77]. On the other hand, AIOLOS is required for the generation of high-affinity plasma cells in the BM that contribute to serum antibody titers after a T cell-dependent immune response [78]. Both IKAROS and AIOLOS are required for the generation and/or maintenance of splenic marginal zone (MZ) B cells [13, 79], for reasons that remain unclear.

The precise functions of the IKAROS family proteins during a T-dependent B cell response are complex and include additional mechanisms in the T-helper (Th) cells themselves. For example, IKAROS indirectly promotes the humoral response by contributing to the polarization of naive $CD4^+$ Th cells into Th2 cells, which secrete IL4, IL5, and IL13, important for B cell help and CSR. In the absence of IKAROS, $CD4^+$ T cells skew toward the Th1 phenotype, due in part to an upregulation of the Th1-associated TF, T-BET, and the production of IFNγ, TNF-α, and IL2 [80, 81]. Similarly, EOS has also recently been shown to promote Th2 polarization [82]. AIOLOS also promotes the humoral response by contributing to the development of T follicular Th (Tfh) cells, required for germinal center formation [83]. In AIOLOS null mice, Tfh production is strongly reduced, and cytotoxic CD4+ T cell numbers are increased.

Lymphocyte activation IKAROS family proteins are required to boost the activation threshold of mature lymphocytes. T cells, expressing a dominant-negative IKAROS mutation that results in the loss of IKAROS family function, proliferate in response to low TCR activation without secondary signals and show higher resistance to signaling inhibitors [55]. Similarly, B cells carrying the hypomorphic *Ikzf1*$^{L/L}$ mutation are poised for activation [84]. When stimulated via their BCR, the mutant B cells grow bigger, proliferate earlier, and exhibit an activated transcriptional signature earlier than WT cells; this phenotype is linked to increased phosphorylation of kinases involved in BCR signaling, like ERK, AKT, and p38. Nonetheless, how IKAROS regulates the signaling cascades downstream of BCR and TCR engagement remains unknown.

IKAROS family proteins prevent T and B cells from developing an inflammatory phenotype. TCR activation of IKAROS-deficient naive T cells induces the expression of pro-inflammatory cytokines like granulocyte-macrophage colony-stimulating factor (GM-CSF), implicated in T cell-dependent inflammatory diseases like multiple sclerosis [12]. IKAROS null T cells exhibit large-scale changes in chromatin accessibility, with increased accessibility at places containing NF-κB (nuclear factor kappa chain enhancer of activated B cells) or STAT5 target motifs. In contrast, treatment of IKAROS-deficient T cells with STAT5 or NF-κB inhibitors prevents the ectopic expression of GM-CSF [12]. These results indicate that IKAROS antagonizes STAT5- and NF-κB activation in T cells. Interestingly, HELIOS also suppresses an inflammatory T cell phenotype in aging humans [85], and HELIOS levels decrease in naive CD4+ T cells with age. Pharmacological degradation of HELIOS leads to increased IL2 expression, STAT5 phosphorylation, and tissue infiltration in vivo. Thus STAT5 and NF-κB activation are closely associated with IKAROS or HELIOS deficiency in T cells.

Similarly, IKAROS deficiency in B cells results in an inflammatory phenotype that is associated with an enlarged spleen and loss of B cell anergy [86]. *Ikzf1* KO B cells lose their tolerance to self-antigens and secrete autoantibodies, which is linked to sustained NF-κB activation due to a reduction in the expression of negative feedback regulators of the Toll-like receptor (TLR) 4 signaling pathway. Macias-Garcia et al. showed that B1 B cells were the major producers of autoantibodies in IKAROS null animals, when stimulated with anti-IgM and anti-CD40 antibodies in the presence of lipopolysaccharides (LPSs) that signal through TLR4 [13]. Indeed, both *Ikzf1* KO B1 and B2 cells hyper-proliferate to TLR4 stimulation and show dysregulation of genes encoding both positive and negative regulators of BCR and TLR signaling.

Lastly, IKAROS family members may also limit the immune response by enhancing the function of regulatory T cells (Treg). Both HELIOS and EOS are highly expressed in Treg cells, and important for the optimal expression of Forkhead box P3 (FOXP3), the central TF for Treg cell development and function [22, 87]. Both have been reported to promote the anti-inflammatory activity of Treg cells, as cells rendered HELIOS deficient show a reduced capacity to rescue and suppress inflammation when injected into Scurfy mice lacking endogenous Treg cells [22, 88, 89]. Indeed, HELIOS-deficient Treg cells appear to be unstable under inflammatory conditions, a phenotype that is linked to increased STAT5 activity [22, 90].

Antagonism of STAT5 and NF-κB A common conclusion from the above studies is that IKAROS family proteins negatively regulate STAT5 and NF-κB activation (Table 2.1). This suggests that a central feature of IKAROS function is the antagonism of the STAT5 and NF-κB pathways. The precise mechanisms are not completely defined but may involve multiple layers that converge on the same end result, such as the transcriptional regulation of upstream regulators of STAT5 and NF-κB, as well as the competition for binding to common target sequences.

Table 2.1 Implication of dysregulated STAT5 or NF-kB activity in the function IKAROS family TFs

	Factor	Effect	Cellular system	Proposed mechanism	Reference
STAT5	IKAROS	Negative	Mouse pre-B cells	Inhibition of STAT5 binding to target sites; direct competition	[24]
	IKAROS	Negative	Human B-ALL	Targeting of common enhancers; transcriptional antagonism	[102]
	IKAROS	Negative	Mouse CD4+ T cells	Negative regulation of accessibility of genomic regions enriched for STAT5 target sites	[12]
	AIOLOS	Negative	Mouse Tfh cells	Decreased accessibility to STAT5 target sites	[83]
	AIOLOS	Negative	Mouse CD4+ T cells	Inhibition of IL2 expression	[127]
	HELIOS	Negative	Human CD4+ T cells	Inhibition of IL2 expression	[85]
	HELIOS	Negative	T cells from patients with Helios mutation	Inhibition of IL2 expression	[125]
	HELIOS	Negative	T cells from patients with HELIOS mutation	Inhibition of IL2 expression	[128]
	HELIOS	Positive	Mouse Treg cells	Decreased STAT5 activation in HELIOS KO Tregs upon IL2 stimulation	[22]
	EOS	Positive	Mouse CD4+ T cells	EOS-dependent tyrosine phosphorylation of STAT5; direct interaction with STAT5	[82]
NF-kB	IKAROS	Negative	Mouse CD4+ T cells	Negative regulation of accessibility of genomic regions enriched for STAT5 target sites	[12]
	IKAROS	Negative	Mouse follicular B cells	Repression of negative regulators of NF-kB activation	[86]
	AIOLOS	Positive	Human B-cell lymphoma	Stimulation of NF-kB activity by AIOLOS gain of function (GoF)	[114]
	IKAROS	Positive	Mouse macrophages	Promotes sustained NF-kB chromatin binding	[129]

2.4 Pathologies

Given their wide-ranging roles in lymphocyte development and function, it is not surprising that mutations or deletions of the genes encoding the IKAROS family members are associated with pathologies. Genome-wide association studies (GWAS) have linked the *IKZF1-4* genes to a number of diseases, mostly of autoimmune or inflammatory nature. Several family members are often associated with the same disease, suggesting cooperation (Table 2.2). However, specific family members are also linked to certain diseases. For example, only *IKZF4* (EOS) polymorphisms have been detected in autoimmune skin conditions like vitiligo and alopecia areata [91, 92]. IKAROS, HELIOS, and AIOLOS are also implicated in hematological cancers, where they mostly act as tumor suppressors but are sometimes oncogenic.

2.4.1 Cancers

IKZF1 loss-of-function (LoF) mutations are frequently detected in B-cell precursor-acute lymphoblastic leukemias (BCP-ALL). Somatic *IKZF1* deletions resulting in haploinsufficiency or the expression of a dominant-negative IKAROS (Ik6) are found in about 15% of pediatric and 50% of adult BCP-ALL cases (for reviews, see [93–96]). These deletions are linked to high-risk cases and used accordingly for patient stratification. *IKZF1* deletions are common in leukemias that exhibit an activated JAK/STAT pathway, like those carrying BCR-ABL translocations, activating JAK2 mutations, or amplifying CRLF2 mutations (which encodes the receptor for the TSLP cytokine) [97–101]. Since IKAROS antagonizes STAT5 function in both non-transformed and transformed lymphocytes,

Table 2.2 Immune system-related traits associated with *IKZF1-4* genes in GWAS studies

Trait	*IKZF1*	*IKZF2*	*IKZF3*	*IKZF4*
Acute lymphoblastic leukemia	x			
Allergic disease (asthma, hay fever, or eczema)	x		x	x
Asthma	x		x	x
Basophil count	x	x		
Crohn's disease	x		x	
Eosinophil count	x	x	x	x
Granulocyte count	x			
IgG glycosylation	x			
Inflammatory bowel disease	x		x	
Lymphocyte to monocyte ratio	x		x	
Monocyte count	x			
Multiple sclerosis	x	x	x	
Plasmacytoid dendritic cell count	x			
Rheumatoid arthritis	x			
Systemic lupus erythematosus	x	x	x	
Type I diabetes	x		x	
Eczema		x		
IL17 levels		x		
Systemic sclerosis		x		
Type II diabetes		x		
Atopic march/atopic asthma			x	
Primary biliary cholangitis			x	
Ulcerative colitis			x	
Alopecia areata				x
Vitiligo				x

Traits related to red blood cells, platelets, solid cancer, or cognitive functions are not represented. Data are from the GWAS collection database

these associations probably reflect an enhanced activation of the STAT pathway [24, 102]. *IKZF1* LoF mutations/deletions are also detected in BCP-ALL subtypes without activated STAT5, as well as other hematopoietic cell-derived leukemias such as acute myeloid leukemias or plasmacytoid dendritic cell neoplasms [103, 104]. Further, all BCP-ALL and myeloid leukemias with *IKZF1* LoF mutations display increased expression of stem cell-related target genes [105, 106]. As IKAROS represses stem cell-associated gene expression via PRC2 and/or other repressive complexes during early B and T cell development [20, 31, 34, 35], these leukemias may in addition be negatively affected by a reduction in IKAROS-mediated gene silencing. Lastly, IKAROS target genes encoding regulators of cell adhesion (e.g., focal adhesion kinase or p120-catenin) are upregulated in IKAROS-deficient BCP-ALL [66, 107, 108], allowing the leukemic cells to stay in the bone marrow and continue to receive IL7/STAT5 signals.

Interestingly, IKAROS and AIOLOS are required for cell growth and in multiple myeloma (MM), a cancer of plasmocytes, through a mechanism that remains unclear. The current standard treatment for MM comprises immunomodulatory drugs like lenalidomide [109]. Lenalidomide binds to the human (not mouse) cereblon E3 ubiquitin ligase to alter its structure, so that the lenalidomide-bound cereblon acquires the ability to target and degrade IKAROS and AIOLOS (but not HELIOS or EOS) through ZF2 of the DBD [110–112]. The discovery that IKAROS family proteins can be pharmacologically targeted has opened the way to developing novel immunomodulators for this family. For example, chemically induced changes to lenalidomide have allowed new forms (e.g., ALV2) to selectively degrade HELIOS [113], with broad implications for the treatment of other diseases. These compounds thus offer a new and powerful toolkit to study the function of IKAROS family proteins in human cells.

AIOLOS is also overexpressed in other mature B-cell malignancies, such as chronic lymphocytic leukemia, diffuse large B-cell lymphoma, mantle cell lymphoma, and Waldenström macroglobulinemia, and sometimes detected as a recurrent *IKZF3* L162R mutation [114]. The L162R mutation is gain-of-function (GoF) and increases the DNA-binding capacity of the mutant AIOLOS to some target genes to mimic AIOLOS overexpression. As a result, AIOLOS GoF promotes BCR signaling [114]. Curiously, although both IKAROS and AIOLOS are expressed in mature B cells, *IKZF1* mutations have not yet been described in B-cell lymphomas, possibly suggesting a functional divergence between IKAROS and AIOLOS in mature B cells.

Interestingly, IKAROS loss appears to promote distinct types of leukemias in human and mouse. In the mouse, loss of IKAROS is predominantly associated with acute T-cell leukemia, while *IKZF1* mutations are only rarely detected in human T-ALL [115]. This difference reflects different mechanisms of activation of the main oncogene for T-ALL, NOTCH1. In humans, NOTCH1 is mostly activated via point mutations that target its heterodimerization domain to make it independent from ligand [116]. In mice, its activation occurs mostly via RAG1/2-mediated deletion of its 5′ region (which includes the promoter and first exon), owing to the presence of cryptic recombination signals in the mouse, but not the human, sequence [117]. These deletions in turn leads to the awakening of an intragenic promoter that directs the generation of transcripts encoding N-terminally truncated, constitutively active proteins [118]. IKAROS strongly binds and represses this cryptic promoter in WT cells. Loss of IKAROS thus synergizes with the deletion of 5' *Notch1* sequences to lead to strong expression of truncated transcripts. Since 5' *NOTCH1* deletions do not occur in human T-ALL, there is also no need for loss of IKAROS as a mechanism for reinforcing NOTCH1 expression in these leukemias.

2.4.2 Germline Mutations and Link with Disease

A growing number of studies have identified germline mutations in *IKZF1*, *IKZF2*, and *IKZF3* in patients with immunodeficiencies (for more extensive review, *see* [119]). In most cases, these mutations are inherited and present as heterozygous alleles. The clinical picture is variable but falls within the "inborn errors of immunity" class and is characterized by dysregulated functions of B and T lymphocytes (Table 2.3). The defects in patients are generally in good agreement with the knowledge obtained from mutant mice. For example, loss of IKAROS function in humans is associated with impaired B cell development and autoimmunity. However, the genotype-phenotype relationship is often more complex, due to the outbred nature of human genetics. Thus, family members with the same *IKZF1* mutation may develop autoimmune syndromes, common variable immunodeficiency (CVID), BCP-ALL, or be asymptomatic [120–122].

Most of the mutations in the *IKZF* genes are LoF, though some GoF alterations have been described [123]. Mutations that result in large protein truncations probably act as null alleles, while missense mutations lead to the production of weakened isoforms that can still dimerize with, and inhibit, the products from unaffected *IKZF* alleles by potentially forming homo- and heterodimers. Understanding how these mutations work is therefore not straightforward. Many have been engineered in mice and mostly lead to defects that are consistent with those seen in patients, indicating that animal models represent a valuable tool to understand the dysfunctions associated with *IKZF* genetic defects.

The observations that most of the human *IKZF* mutations are missense and affect specific amino acids allow us to study the precise functions of various amino acids and domains of these factors. This is underscored by the distinct outcomes associated with mutations of two neighboring residues in AIOLOS (G159 and N160) that lead to distinct defects in patients and cognate mouse models ([124] for review). In addition, a growing number of identified mutations affect amino acids outside of the DBD and DD (notably for HELIOS, [125, 126]), which may offer clues into the role of poorly understood regions. Thus the studies of human mutations in animal models will greatly improve our understanding of the physiological and molecular functions of the IKAROS family TFs.

2.5 Open Questions

As of this writing, our general understanding of how IKAROS family proteins work remains fragmented, and key questions persist.

What is the specificity of each family member? Given that sequence differences between IKAROS, HELIOS, AIOLOS, and EOS are conserved across vertebrates, it is likely that each

Table 2.3 Germline mutation of *IKZF1-4* genes in human patients

Gene	Reference	Mutation type (monoallelic unless specific otherwise)	Main disease features
IKZF1	[120–122, 130]	- Various loss-of-function (LoF) missense mutations in DBD (affecting aa 139, 143, 147, 150, 162, 184, 210) - C-terminal truncations - Deletions (intragenic or entire gene)	Common variable immunodeficiency, recurrent infections, autoimmunity [immune thrombocytopenic purpura, IgA vasculitis, systemic lupus erythematosus (SLE)], acute lymphoblastic leukemia
IKZF1	[120, 122, 131]	Dimerization domain (DD) missense mutations (M494V; C467R, C467W, R502L)	Common variable immunodeficiency, recurrent infections, autoimmunity (immune thrombocytopenic purpura, IgA vasculitis, SLE), acute lymphoblastic leukemia
IKZF1	[122, 132]	Missense mutations in non-DNA-binding domain (DBD) and non-DD aa (R423C, A448R, G337R)	R423C: diamond-Blackfan anemia like syndrome G337R: inflammatory bowel disease A448R: hepatosplenomegaly, recurrent fever
IKZF1	[133]	Missense N159S, N159T (dominant-negative)	*Pneumocystis jirovecii* infections; low B cell, neutrophil, eosinophil, myeloid DC counts T cells with mostly naive phenotype, unable to evolve into effector memory T cells
IKZF1	[123]	Missense R183C (gain of function; GoF)	Allergy, autoimmunity and autoinflammation, plasma cell proliferation, T helper (Th) response skewing toward Th2, low regulatory T cell (Treg) numbers
IKZF2	[128]	C-terminal truncation (Y200X)	Combined immunodeficiency, chronic T-cell activation, increased inflammatory cytokine production, reduction of circulating mucosal associated invariant T cells
IKZF2	[125]	Missense I325V in inter-zinc finger (ZF) region (biallelic)	Recurrent respiratory infections, hypogammaglobulinemia, lower frequency of invariant natural killer-T (iNKT) cells, and mucosa-associated invariant T (MAIT) cells
IKZF2	[126]	C-terminal truncation (R291X) Missense in DBD (Y359C, V347M, R106W, N220S)	SLE, immune thrombocytopenia, Epstein-Barr virus (EBV)-associated hemophagocytic lymphohistiocytosis, hypogammaglobulinemia
IKZF2	[134]	Exon 5 duplication (dominant-negative) G153R (dominant-negative)	Immunodysregulation, craniofacial anomalies, hearing impairment, athelia (absence of nipples), developmental delay
IKZF3	[75]	Missense G159R	Profound B lymphopenia, hypogammaglobulinemia, lymphoma, skewing of T cells to memory and activated phenotypes, decreased T helper 17 (Th17) and increased Th1 helper T cells
IKZF3	[135]	Missense N160S	Normal to high B cell number, hypogammaglobulinemia

factor possesses specific molecular functions. What is the interplay between DNA binding and function? Depending on the target gene, IKAROS factors can either activate or repress transcription, suggesting that there are "instructions" embedded within the target sites to recruit the appropriate active or repressive complexes. What is the role of the PTMs? Indeed, 144 distinct PTMs have been identified for human IKAROS alone in large-scale proteomic analyses (PhosphositePlus repository; A. Aukenova, personal communication), suggesting that multiple signaling pathways modify and impact the function of IKAROS family proteins.

We anticipate that the future integration of large-scale molecular, genetic, and structural studies will allow us to gain a much clearer view of this essential family of TFs in lymphocyte development and function.

Acknowledgments We thank lab members for discussion. SC received funding from grant ANR-22-CE15-0017-01.

References

1. Lo K, Landau NR, Smale ST (1991) LyF-1, a transcriptional regulator that interacts with a novel class of promoters for lymphocyte-specific genes. Mol Cell Biol 11(10):5229–5243. https://doi.org/10.1128/mcb.11.10.5229-5243.1991
2. Hahm K, Ernst P, Lo K, Kim GS, Turck C, Smale ST (1994) The lymphoid transcription factor LyF-1 is encoded by specific, alternatively spliced mRNAs derived from the Ikaros gene. Mol Cell Biol 14(11):7111–7123. https://doi.org/10.1128/mcb.14.11.7111-7123.1994
3. Georgopoulos K, Moore DD, Derfler B (1992) Ikaros, an early lymphoid-specific transcription factor and a putative mediator for T cell commitment. Science 258(5083):808–812. https://doi.org/10.1126/science.1439790
4. Hahm K, Cobb BS, McCarty AS, Brown KE, Klug CA, Lee R, Akashi K, Weissman IL, Fisher AG, Smale ST (1998) Helios, a T cell-restricted Ikaros family member that quantitatively associates with Ikaros at centromeric heterochromatin. Genes Dev 12(6):782–796. https://doi.org/10.1101/gad.12.6.782
5. Morgan B, Sun L, Avitahl N, Andrikopoulos K, Ikeda T, Gonzales E, Wu P, Neben S, Georgopoulos K (1997) Aiolos, a lymphoid restricted transcription factor that interacts with Ikaros to regulate lymphocyte differentiation. EMBO J 16(8):2004–2013. https://doi.org/10.1093/emboj/16.8.2004
6. Kelley CM, Ikeda T, Koipally J, Avitahl N, Wu L, Georgopoulos K, Morgan BA (1998) Helios, a novel dimerization partner of Ikaros expressed in the earliest hematopoietic progenitors. Curr Biol 8(9):508–515. https://doi.org/10.1016/s0960-9822(98)70202-7
7. Perdomo J, Holmes M, Chong B, Crossley M (2000) Eos and pegasus, two members of the Ikaros family of proteins with distinct DNA binding activities. J Biol Chem 275(49):38347–38354. https://doi.org/10.1074/jbc.M005457200
8. Molnar A, Georgopoulos K (1994) The Ikaros gene encodes a family of functionally diverse zinc finger DNA-binding proteins. Mol Cell Biol 14(12):8292–8303. https://doi.org/10.1128/mcb.14.12.8292-8303.1994
9. McCarty AS, Kleiger G, Eisenberg D, Smale ST (2003) Selective dimerization of a C2H2 zinc finger subfamily. Mol Cell 11(2):459–470. https://doi.org/10.1016/s1097-2765(03)00043-1
10. Payne KJ, Nicolas JH, Zhu JY, Barsky LW, Crooks GM (2001) Cutting edge: predominant expression of a novel Ikaros isoform in normal human hemopoiesis. J Immunol 167(4):1867–1870. https://doi.org/10.4049/jimmunol.167.4.1867
11. Virely C, Moulin S, Cobaleda C, Lasgi C, Alberdi A, Soulier J, Sigaux F, Chan S, Kastner P, Ghysdael J (2010) Haploinsufficiency of the IKZF1 (IKAROS) tumor suppressor gene cooperates with BCR-ABL in a transgenic model of acute lymphoblastic leukemia. Leukemia 24(6):1200–1204. https://doi.org/10.1038/leu.2010.63
12. Bernardi C, Maurer G, Ye T, Marchal P, Jost B, Wissler M, Maurer U, Kastner P, Chan S, Charvet C (2021) CD4(+) T cells require Ikaros to inhibit their differentiation toward a pathogenic cell fate. Proc Natl Acad Sci USA 118(17). https://doi.org/10.1073/pnas.2023172118
13. Macias-Garcia A, Heizmann B, Sellars M, Marchal P, Dali H, Pasquali JL, Muller S, Kastner P, Chan S (2016) Ikaros is a negative regulator of B1 cell development and function. J Biol Chem 291(17):9073–9086. https://doi.org/10.1074/jbc.M115.704239
14. Klug CA, Morrison SJ, Masek M, Hahm K, Smale ST, Weissman IL (1998) Hematopoietic stem cells and lymphoid progenitors express different Ikaros isoforms, and Ikaros is localized to heterochromatin in immature lymphocytes. Proc Natl Acad Sci USA 95(2):657–662. https://doi.org/10.1073/pnas.95.2.657
15. Cupit PM, Hansen JD, McCarty AS, White G, Chioda M, Spada F, Smale ST, Cunningham C (2003) Ikaros family members from the agnathan Myxine glutinosa and the urochordate Oikopleura dioica: emergence of an essential transcription factor for adaptive

immunity. J Immunol 171(11):6006–6013. https://doi.org/10.4049/jimmunol.171.11.6006
16. Kerner P, Zelada Gonzalez F, Le Gouar M, Ledent V, Arendt D, Vervoort M (2006) The expression of a hunchback ortholog in the polychaete annelid Platynereis dumerilii suggests an ancestral role in mesoderm development and neurogenesis. Dev Genes Evol 216(12):821–828. https://doi.org/10.1007/s00427-006-0100-9
17. Pinnell J, Lindeman PS, Colavito S, Lowe C, Savage RM (2006) The divergent roles of the segmentation gene hunchback. Integr Comp Biol 46(4):519–532. https://doi.org/10.1093/icb/icj054
18. Large EE, Mathies LD (2010) hunchback and Ikaros-like zinc finger genes control reproductive system development in Caenorhabditis elegans. Dev Biol 339(1):51–64. https://doi.org/10.1016/j.ydbio.2009.12.013
19. Geimer Le Lay AS, Oravecz A, Mastio J, Jung C, Marchal P, Ebel C, Dembele D, Jost B, Le Gras S, Thibault C, Borggrefe T, Kastner P, Chan S (2014) The tumor suppressor Ikaros shapes the repertoire of notch target genes in T cells. Sci Signal 7 (317):ra28. https://doi.org/10.1126/scisignal.2004545
20. Zhang J, Jackson AF, Naito T, Dose M, Seavitt J, Liu F, Heller EJ, Kashiwagi M, Yoshida T, Gounari F, Petrie HT, Georgopoulos K (2011) Harnessing of the nucleosome-remodeling-deacetylase complex controls lymphocyte development and prevents leukemogenesis. Nat Immunol 13(1):86–94. https://doi.org/10.1038/ni.2150
21. Ferreiros-Vidal I, Carroll T, Taylor B, Terry A, Liang Z, Bruno L, Dharmalingam G, Khadayate S, Cobb BS, Smale ST, Spivakov M, Srivastava P, Petretto E, Fisher AG, Merkenschlager M (2013) Genome-wide identification of Ikaros targets elucidates its contribution to mouse B-cell lineage specification and pre-B-cell differentiation. Blood 121(10):1769–1782. https://doi.org/10.1182/blood-2012-08-450114
22. Kim HJ, Barnitz RA, Kreslavsky T, Brown FD, Moffett H, Lemieux ME, Kaygusuz Y, Meissner T, Holderried TA, Chan S, Kastner P, Haining WN, Cantor H (2015) Stable inhibitory activity of regulatory T cells requires the transcription factor Helios. Science 350(6258):334–339. https://doi.org/10.1126/science.aad0616
23. Cova G, Taroni C, Deau MC, Cai Q, Mittelheisser V, Philipps M, Jung M, Cerciat M, Le Gras S, Thibault-Carpentier C, Jost B, Carlsson L, Thornton AM, Shevach EM, Kirstetter P, Kastner P, Chan S (2021) Helios represses megakaryocyte priming in hematopoietic stem and progenitor cells. J Exp Med 218(10). https://doi.org/10.1084/jem.20202317
24. Heizmann B, Le Gras S, Simand C, Marchal P, Chan S, Kastner P (2020) Ikaros antagonizes DNA binding by STAT5 in pre-B cells. PLoS One 15(11):e0242211. https://doi.org/10.1371/journal.pone.0242211
25. Gomez-del Arco P, Maki K, Georgopoulos K (2004) Phosphorylation controls Ikaros's ability to negatively regulate the G(1)-S transition. Mol Cell Biol 24(7):2797–2807. https://doi.org/10.1128/MCB.24.7.2797-2807.2004
26. Gurel Z, Ronni T, Ho S, Kuchar J, Payne KJ, Turk CW, Dovat S (2008) Recruitment of ikaros to pericentromeric heterochromatin is regulated by phosphorylation. J Biol Chem 283(13):8291–8300. https://doi.org/10.1074/jbc.M707906200
27. Song C, Ge Z, Ding Y, Tan BH, Desai D, Gowda K, Amin S, Gowda R, Robertson GP, Yue F, Huang S, Spiegelman V, Payne JL, Reeves ME, Gurel Z, Iyer S, Dhanyamraju PK, Xiang M, Kawasawa YI, Cury NM, Yunes JA, McGrath M, Schramm J, Su R, Yang Y, Zhao Z, Lyu X, Muschen M, Payne KJ, Gowda C, Dovat S (2020) IKAROS and CK2 regulate expression of BCL-XL and chemosensitivity in high-risk B-cell acute lymphoblastic leukemia. Blood 136(13):1520–1534. https://doi.org/10.1182/blood.2019002655
28. Ernst P, Hahm K, Trinh L, Davis JN, Roussel MF, Turck CW, Smale ST (1996) A potential role for Elf-1 in terminal transferase gene regulation. Mol Cell Biol 16(11):6121–6131. https://doi.org/10.1128/MCB.16.11.6121
29. Kim J, Sif S, Jones B, Jackson A, Koipally J, Heller E, Winandy S, Viel A, Sawyer A, Ikeda T, Kingston R, Georgopoulos K (1999) Ikaros DNA-binding proteins direct formation of chromatin remodeling complexes in lymphocytes. Immunity 10(3):345–355. https://doi.org/10.1016/s1074-7613(00)80034-5
30. Sridharan R, Smale ST (2007) Predominant interaction of both Ikaros and Helios with the NuRD complex in immature thymocytes. J Biol Chem 282(41):30227–30238. https://doi.org/10.1074/jbc.M702541200
31. Oravecz A, Apostolov A, Polak K, Jost B, Le Gras S, Chan S, Kastner P (2015) Ikaros mediates gene silencing in T cells through Polycomb repressive complex 2. Nat Commun 6:8823. https://doi.org/10.1038/ncomms9823
32. Cobb BS, Morales-Alcelay S, Kleiger G, Brown KE, Fisher AG, Smale ST (2000) Targeting of Ikaros to pericentromeric heterochromatin by direct DNA binding. Genes Dev 14(17):2146–2160. https://doi.org/10.1101/gad.816400
33. Brown KE, Guest SS, Smale ST, Hahm K, Merkenschlager M, Fisher AG (1997) Association of transcriptionally silent genes with Ikaros complexes at centromeric heterochromatin. Cell 91(6):845–854. https://doi.org/10.1016/s0092-8674(00)80472-9
34. Liang Z, Brown KE, Carroll T, Taylor B, Vidal IF, Hendrich B, Rueda D, Fisher AG, Merkenschlager M (2017) A high-resolution map of transcriptional repression. elife 6. https://doi.org/10.7554/eLife.22767
35. Hu Y, Zhang Z, Kashiwagi M, Yoshida T, Joshi I, Jena N, Somasundaram R, Emmanuel AO, Sigvardsson M, Fitamant J, El-Bardeesy N, Gounari F, Van Etten RA, Georgopoulos K (2016) Superenhancer reprogramming drives a B-cell-

epithelial transition and high-risk leukemia. Genes Dev 30(17):1971–1990. https://doi.org/10.1101/gad.283762.116
36. Ding Y, Zhang B, Payne JL, Song C, Ge Z, Gowda C, Iyer S, Dhanyamraju PK, Dorsam G, Reeves ME, Desai D, Huang S, Payne KJ, Yue F, Dovat S (2019) Ikaros tumor suppressor function includes induction of active enhancers and super-enhancers along with pioneering activity. Leukemia 33(11):2720–2731. https://doi.org/10.1038/s41375-019-0474-0
37. Wu L, Nichogiannopoulou A, Shortman K, Georgopoulos K (1997) Cell-autonomous defects in dendritic cell populations of Ikaros mutant mice point to a developmental relationship with the lymphoid lineage. Immunity 7(4):483–492. https://doi.org/10.1016/s1074-7613(00)80370-2
38. Mastio J, Simand C, Cova G, Kastner P, Chan S, Kirstetter P (2018) Ikaros cooperates with Notch activation and antagonizes TGFbeta signaling to promote pDC development. PLoS Genet 14(7):e1007485. https://doi.org/10.1371/journal.pgen.1007485
39. Schjerven H, McLaughlin J, Arenzana TL, Frietze S, Cheng D, Wadsworth SE, Lawson GW, Bensinger SJ, Farnham PJ, Witte ON, Smale ST (2013) Selective regulation of lymphopoiesis and leukemogenesis by individual zinc fingers of Ikaros. Nat Immunol 14(10):1073–1083. https://doi.org/10.1038/ni.2707
40. Lopez RA, Schoetz S, DeAngelis K, O'Neill D, Bank A (2002) Multiple hematopoietic defects and delayed globin switching in Ikaros null mice. Proc Natl Acad Sci USA 99(2):602–607. https://doi.org/10.1073/pnas.022412699
41. Dumortier A, Kirstetter P, Kastner P, Chan S (2003) Ikaros regulates neutrophil differentiation. Blood 101(6):2219–2226. https://doi.org/10.1182/blood-2002-05-1336
42. Alsio JM, Tarchini B, Cayouette M, Livesey FJ (2013) Ikaros promotes early-born neuronal fates in the cerebral cortex. Proc Natl Acad Sci USA 110(8):E716–E725. https://doi.org/10.1073/pnas.1215707110
43. Boudreau-Pinsonneault C, David LA, Lourenco Fernandes JA, Javed A, Fries M, Mattar P, Cayouette M (2023) Direct neuronal reprogramming by temporal identity factors. Proc Natl Acad Sci USA 120(19):e2122168120. https://doi.org/10.1073/pnas.2122168120
44. Elliott J, Jolicoeur C, Ramamurthy V, Cayouette M (2008) Ikaros confers early temporal competence to mouse retinal progenitor cells. Neuron 60(1):26–39. https://doi.org/10.1016/j.neuron.2008.08.008
45. Martin-Ibanez R, Crespo E, Urban N, Sergent-Tanguy S, Herranz C, Jaumot M, Valiente M, Long JE, Pineda JR, Andreu C, Rubenstein JL, Marin O, Georgopoulos K, Mengod G, Farinas I, Bachs O, Alberch J, Canals JM (2010) Ikaros-1 couples cell cycle arrest of late striatal precursors with neurogenesis of enkephalinergic neurons. J Comp Neurol 518(3):329–351. https://doi.org/10.1002/cne.22215
46. Martin-Ibanez R, Pardo M, Giralt A, Miguez A, Guardia I, Marion-Poll L, Herranz C, Esgleas M, Garcia-Diaz Barriga G, Edel MJ, Vicario-Abejon C, Alberch J, Girault JA, Chan S, Kastner P, Canals JM (2017) Helios expression coordinates the development of a subset of striatopallidal medium spiny neurons. Development 144(8):1566–1577. https://doi.org/10.1242/dev.138248
47. Chessum L, Matern MS, Kelly MC, Johnson SL, Ogawa Y, Milon B, McMurray M, Driver EC, Parker A, Song Y, Codner G, Esapa CT, Prescott J, Trent G, Wells S, Dragich AK, Frolenkov GI, Kelley MW, Marcotti W, Brown SDM, Elkon R, Bowl MR, Hertzano R (2018) Helios is a key transcriptional regulator of outer hair cell maturation. Nature 563(7733):696–700. https://doi.org/10.1038/s41586-018-0728-4
48. Nichogiannopoulou A, Trevisan M, Neben S, Friedrich C, Georgopoulos K (1999) Defects in hemopoietic stem cell activity in Ikaros mutant mice. J Exp Med 190(9):1201–1214. https://doi.org/10.1084/jem.190.9.1201
49. Papathanasiou P, Attema JL, Karsunky H, Hosen N, Sontani Y, Hoyne GF, Tunningley R, Smale ST, Weissman IL (2009) Self-renewal of the long-term reconstituting subset of hematopoietic stem cells is regulated by Ikaros. Stem Cells 27(12):3082–3092. https://doi.org/10.1002/stem.232
50. Yoshida T, Ng SY, Zuniga-Pflucker JC, Georgopoulos K (2006) Early hematopoietic lineage restrictions directed by Ikaros. Nat Immunol 7(4):382–391. https://doi.org/10.1038/ni1314
51. Ng SY, Yoshida T, Zhang J, Georgopoulos K (2009) Genome-wide lineage-specific transcriptional networks underscore Ikaros-dependent lymphoid priming in hematopoietic stem cells. Immunity 30(4):493–507. https://doi.org/10.1016/j.immuni.2009.01.014
52. Rothenberg EV, Moore JE, Yui MA (2008) Launching the T-cell-lineage developmental programme. Nat Rev Immunol 8(1):9–21. https://doi.org/10.1038/nri2232
53. Thompson PK, Zuniga-Pflucker JC (2011) On becoming a T cell, a convergence of factors kick it up a Notch along the way. Semin Immunol 23(5):350–359. https://doi.org/10.1016/j.smim.2011.08.007
54. Winandy S, Wu L, Wang JH, Georgopoulos K (1999) Pre-T cell receptor (TCR) and TCR-controlled checkpoints in T cell differentiation are set by Ikaros. J Exp Med 190(8):1039–1048. https://doi.org/10.1084/jem.190.8.1039
55. Avitahl N, Winandy S, Friedrich C, Jones B, Ge Y, Georgopoulos K (1999) Ikaros sets thresholds for T cell activation and regulates chromosome propagation. Immunity 10(3):333–343. https://doi.org/10.1016/s1074-7613(00)80033-3

56. Kleinmann E, Geimer Le Lay AS, Sellars M, Kastner P, Chan S (2008) Ikaros represses the transcriptional response to Notch signaling in T-cell development. Mol Cell Biol 28(24):7465–7475. https://doi.org/10.1128/MCB.00715-08
57. Chari S, Winandy S (2008) Ikaros regulates Notch target gene expression in developing thymocytes. J Immunol 181(9):6265–6274. https://doi.org/10.4049/jimmunol.181.9.6265
58. Winandy S, Wu P, Georgopoulos K (1995) A dominant mutation in the Ikaros gene leads to rapid development of leukemia and lymphoma. Cell 83(2):289–299. https://doi.org/10.1016/0092-8674(95)90170-1
59. Dumortier A, Jeannet R, Kirstetter P, Kleinmann E, Sellars M, dos Santos NR, Thibault C, Barths J, Ghysdael J, Punt JA, Kastner P, Chan S (2006) Notch activation is an early and critical event during T-Cell leukemogenesis in Ikaros-deficient mice. Mol Cell Biol 26(1):209–220. https://doi.org/10.1128/MCB.26.1.209-220.2006
60. Wang JH, Nichogiannopoulou A, Wu L, Sun L, Sharpe AH, Bigby M, Georgopoulos K (1996) Selective defects in the development of the fetal and adult lymphoid system in mice with an Ikaros null mutation. Immunity 5(6):537–549. https://doi.org/10.1016/s1074-7613(00)80269-1
61. Allman D, Sambandam A, Kim S, Miller JP, Pagan A, Well D, Meraz A, Bhandoola A (2003) Thymopoiesis independent of common lymphoid progenitors. Nat Immunol 4(2):168–174. https://doi.org/10.1038/ni878
62. Reynaud D, Demarco IA, Reddy KL, Schjerven H, Bertolino E, Chen Z, Smale ST, Winandy S, Singh H (2008) Regulation of B cell fate commitment and immunoglobulin heavy-chain gene rearrangements by Ikaros. Nat Immunol 9(8):927–936. https://doi.org/10.1038/ni.1626
63. Kirstetter P, Thomas M, Dierich A, Kastner P, Chan S (2002) Ikaros is critical for B cell differentiation and function. Eur J Immunol 32(3):720–730. https://doi.org/10.1002/1521-4141(200203)32:3<720::AID-IMMU720>3.0.CO;2-P
64. Alkhatib A, Werner M, Hug E, Herzog S, Eschbach C, Faraidun H, Kohler F, Wossning T, Jumaa H (2012) FoxO1 induces Ikaros splicing to promote immunoglobulin gene recombination. J Exp Med 209(2):395–406. https://doi.org/10.1084/jem.20110216
65. Heizmann B, Kastner P, Chan S (2013) Ikaros is absolutely required for pre-B cell differentiation by attenuating IL-7 signals. J Exp Med 210(13):2823–2832. https://doi.org/10.1084/jem.20131735
66. Joshi I, Yoshida T, Jena N, Qi X, Zhang J, Van Etten RA, Georgopoulos K (2014) Loss of Ikaros DNA-binding function confers integrin-dependent survival on pre-B cells and progression to acute lymphoblastic leukemia. Nat Immunol 15(3):294–304. https://doi.org/10.1038/ni.2821
67. Schwickert TA, Tagoh H, Gultekin S, Dakic A, Axelsson E, Minnich M, Ebert A, Werner B, Roth M, Cimmino L, Dickins RA, Zuber J, Jaritz M, Busslinger M (2014) Stage-specific control of early B cell development by the transcription factor Ikaros. Nat Immunol 15(3):283–293. https://doi.org/10.1038/ni.2828
68. Fistonich C, Zehentmeier S, Bednarski JJ, Miao R, Schjerven H, Sleckman BP, Pereira JP (2018) Cell circuits between B cell progenitors and IL-7(+) mesenchymal progenitor cells control B cell development. J Exp Med 215(10):2586–2599. https://doi.org/10.1084/jem.20180778
69. Matheson LS, Bolland DJ, Chovanec P, Krueger F, Andrews S, Koohy H, Corcoran AE (2017) Local chromatin features including PU.1 and IKAROS binding and H3K4 methylation shape the repertoire of immunoglobulin Kappa genes chosen for V(D)J recombination. Front Immunol 8:1550. https://doi.org/10.3389/fimmu.2017.01550
70. Stadhouders R, de Bruijn MJ, Rother MB, Yuvaraj S, Ribeiro de Almeida C, Kolovos P, Van Zelm MC, van Ijcken W, Grosveld F, Soler E, Hendriks RW (2014) Pre-B cell receptor signaling induces immunoglobulin kappa locus accessibility by functional redistribution of enhancer-mediated chromatin interactions. PLoS Biol 12(2):e1001791. https://doi.org/10.1371/journal.pbio.1001791
71. Liu Z, Widlak P, Zou Y, Xiao F, Oh M, Li S, Chang MY, Shay JW, Garrard WT (2006) A recombination silencer that specifies heterochromatin positioning and ikaros association in the immunoglobulin kappa locus. Immunity 24(4):405–415. https://doi.org/10.1016/j.immuni.2006.02.001
72. Xiang Y, Zhou X, Hewitt SL, Skok JA, Garrard WT (2011) A multifunctional element in the mouse Igkappa locus that specifies repertoire and Ig loci subnuclear location. J Immunol 186(9):5356–5366. https://doi.org/10.4049/jimmunol.1003794
73. Wang JH, Avitahl N, Cariappa A, Friedrich C, Ikeda T, Renold A, Andrikopoulos K, Liang L, Pillai S, Morgan BA, Georgopoulos K (1998) Aiolos regulates B cell activation and maturation to effector state. Immunity 9(4):543–553. https://doi.org/10.1016/s1074-7613(00)80637-8
74. Thompson EC, Cobb BS, Sabbattini P, Meixlsperger S, Parelho V, Liberg D, Taylor B, Dillon N, Georgopoulos K, Jumaa H, Smale ST, Fisher AG, Merkenschlager M (2007) Ikaros DNA-binding proteins as integral components of B cell developmental-stage-specific regulatory circuits. Immunity 26(3):335–344. https://doi.org/10.1016/j.immuni.2007.02.010
75. Yamashita M, Kuehn HS, Okuyama K, Okada S, Inoue Y, Mitsuiki N, Imai K, Takagi M, Kanegane H, Takeuchi M, Shimojo N, Tsumura M, Padhi AK, Zhang KYJ, Boisson B, Casanova JL, Ohara O, Rosenzweig SD, Taniuchi I, Morio T (2021) A variant in human AIOLOS impairs adap-

tive immunity by interfering with IKAROS. Nat Immunol 22(7):893–903. https://doi.org/10.1038/s41590-021-00951-z
76. Simand C, Keime C, Caye A, Arfeuille C, Passet M, Kim R, Cave H, Clappier E, Kastner P, Chan S, Heizmann B (2022) Ikaros deficiency is associated with aggressive BCR-ABL1 B-cell precursor acute lymphoblastic leukemia independent of the lineage and developmental origin. Haematologica 107(1):316–320. https://doi.org/10.3324/haematol.2021.279125
77. Sellars M, Reina-San-Martin B, Kastner P, Chan S (2009) Ikaros controls isotype selection during immunoglobulin class switch recombination. J Exp Med 206(5):1073–1087. https://doi.org/10.1084/jem.20082311
78. Cortes M, Georgopoulos K (2004) Aiolos is required for the generation of high affinity bone marrow plasma cells responsible for long-term immunity. J Exp Med 199(2):209–219. https://doi.org/10.1084/jem.20031571
79. Cariappa A, Tang M, Parng C, Nebelitskiy E, Carroll M, Georgopoulos K, Pillai S (2001) The follicular versus marginal zone B lymphocyte cell fate decision is regulated by Aiolos, Btk, and CD21. Immunity 14(5):603–615. https://doi.org/10.1016/s1074-7613(01)00135-2
80. Quirion MR, Gregory GD, Umetsu SE, Winandy S, Brown MA (2009) Cutting edge: Ikaros is a regulator of Th2 cell differentiation. J Immunol 182(2):741–745. https://doi.org/10.4049/jimmunol.182.2.741
81. Thomas RM, Chen C, Chunder N, Ma L, Taylor J, Pearce EJ, Wells AD (2010) Ikaros silences T-bet expression and interferon-gamma production during T helper 2 differentiation. J Biol Chem 285(4):2545–2553. https://doi.org/10.1074/jbc.M109.038794
82. Tuazon JA, Read KA, Sreekumar BK, Roettger JE, Yaeger MJ, Varikuti S, Pokhrel S, Jones DM, Warren RT, Powell MD, Rasheed MN, Duncan EG, Childs LM, Gowdy KM, Oestreich KJ (2023) Eos promotes TH2 differentiation by interacting with and propagating the activity of STAT5. J Immunol 211(3):365–376. https://doi.org/10.4049/jimmunol.2200861
83. Read KA, Jones DM, Pokhrel S, Hales EDS, Varkey A, Tuazon JA, Eisele CD, Abdouni O, Saadey A, Leonard MR, Warren RT, Powell MD, Boss JM, Hemann EA, Yount JS, Xin G, Ghoneim HE, Lio CJ, Freud AG, Collins PL, Oestreich KJ (2023) Aiolos represses CD4(+) T cell cytotoxic programming via reciprocal regulation of T(FH) transcription factors and IL-2 sensitivity. Nat Commun 14(1):1652. https://doi.org/10.1038/s41467-023-37420-0
84. Heizmann B, Sellars M, Macias-Garcia A, Chan S, Kastner P (2016) Ikaros limits follicular B cell activation by regulating B cell receptor signaling pathways. Biochem Biophys Res Commun 470(3):714–720. https://doi.org/10.1016/j.bbrc.2016.01.060
85. Zhang H, Jadhav RR, Cao W, Goronzy IN, Zhao TV, Jin J, Ohtsuki S, Hu Z, Morales J, Greenleaf WJ, Weyand CM, Goronzy JJ (2023) Aging-associated HELIOS deficiency in naive CD4(+) T cells alters chromatin remodeling and promotes effector cell responses. Nat Immunol 24(1):96–109. https://doi.org/10.1038/s41590-022-01369-x
86. Schwickert TA, Tagoh H, Schindler K, Fischer M, Jaritz M, Busslinger M (2019) Ikaros prevents autoimmunity by controlling anergy and Toll-like receptor signaling in B cells. Nat Immunol 20(11):1517–1529. https://doi.org/10.1038/s41590-019-0490-2
87. Polak K, Marchal P, Taroni C, Ebel C, Kirstetter P, Kastner P, Chan S (2023) CD4(+) regulatory T cells lacking Helios and Eos. Biochem Biophys Res Commun 674:83–89. https://doi.org/10.1016/j.bbrc.2023.06.087
88. Sebastian M, Lopez-Ocasio M, Metidji A, Rieder SA, Shevach EM, Thornton AM (2016) Helios controls a limited subset of regulatory T cell functions. J Immunol 196(1):144–155. https://doi.org/10.4049/jimmunol.1501704
89. Gokhale AS, Gangaplara A, Lopez-Occasio M, Thornton AM, Shevach EM (2019) Selective deletion of Eos (Ikzf4) in T-regulatory cells leads to loss of suppressive function and development of systemic autoimmunity. J Autoimmun 105:102300. https://doi.org/10.1016/j.jaut.2019.06.011
90. Thornton AM, Shevach EM (2019) Helios: still behind the clouds. Immunology 158(3):161–170. https://doi.org/10.1111/imm.13115
91. Jin Y, Birlea SA, Fain PR, Ferrara TM, Ben S, Riccardi SL, Cole JB, Gowan K, Holland PJ, Bennett DC, Luiten RM, Wolkerstorfer A, van der Veen JP, Hartmann A, Eichner S, Schuler G, van Geel N, Lambert J, Kemp EH, Gawkrodger DJ, Weetman AP, Taieb A, Jouary T, Ezzedine K, Wallace MR, McCormack WT, Picardo M, Leone G, Overbeck A, Silverberg NB, Spritz RA (2012) Genome-wide association analyses identify 13 new susceptibility loci for generalized vitiligo. Nat Genet 44(6):676–680. https://doi.org/10.1038/ng.2272
92. Petukhova L, Duvic M, Hordinsky M, Norris D, Price V, Shimomura Y, Kim H, Singh P, Lee A, Chen WV, Meyer KC, Paus R, Jahoda CA, Amos CI, Gregersen PK, Christiano AM (2010) Genome-wide association study in alopecia areata implicates both innate and adaptive immunity. Nature 466(7302):113–117. https://doi.org/10.1038/nature09114
93. Kastner P, Dupuis A, Gaub MP, Herbrecht R, Lutz P, Chan S (2013) Function of Ikaros as a tumor suppressor in B cell acute lymphoblastic leukemia. Am J Blood Res 3(1):1–13
94. Olsson L, Johansson B (2015) Ikaros and leukaemia. Br J Haematol 169(4):479–491. https://doi.org/10.1111/bjh.13342
95. Marke R, van Leeuwen FN, Scheijen B (2018) The many faces of IKZF1 in B-cell precursor

acute lymphoblastic leukemia. Haematologica 103(4):565–574. https://doi.org/10.3324/haematol.2017.185603
96. Vairy S, Tran TH (2020) IKZF1 alterations in acute lymphoblastic leukemia: The good, the bad and the ugly. Blood Rev 44:100677. https://doi.org/10.1016/j.blre.2020.100677
97. Iacobucci I, Lonetti A, Messa F, Cilloni D, Arruga F, Ottaviani E, Paolini S, Papayannidis C, Piccaluga PP, Giannoulia P, Soverini S, Amabile M, Poerio A, Saglio G, Pane F, Berton G, Baruzzi A, Vitale A, Chiaretti S, Perini G, Foa R, Baccarani M, Martinelli G (2008) Expression of spliced oncogenic Ikaros isoforms in Philadelphia-positive acute lymphoblastic leukemia patients treated with tyrosine kinase inhibitors: implications for a new mechanism of resistance. Blood 112(9):3847–3855. https://doi.org/10.1182/blood-2007-09-112631
98. Iacobucci I, Storlazzi CT, Cilloni D, Lonetti A, Ottaviani E, Soverini S, Astolfi A, Chiaretti S, Vitale A, Messa F, Impera L, Baldazzi C, D'Addabbo P, Papayannidis C, Lonoce A, Colarossi S, Vignetti M, Piccaluga PP, Paolini S, Russo D, Pane F, Saglio G, Baccarani M, Foa R, Martinelli G (2009) Identification and molecular characterization of recurrent genomic deletions on 7p12 in the IKZF1 gene in a large cohort of BCR-ABL1-positive acute lymphoblastic leukemia patients: on behalf of Gruppo Italiano Malattie Ematologiche dell'Adulto Acute Leukemia Working Party (GIMEMA AL WP). Blood 114(10):2159–2167. https://doi.org/10.1182/blood-2008-08-173963
99. Harvey RC, Mullighan CG, Chen IM, Wharton W, Mikhail FM, Carroll AJ, Kang H, Liu W, Dobbin KK, Smith MA, Carroll WL, Devidas M, Bowman WP, Camitta BM, Reaman GH, Hunger SP, Downing JR, Willman CL (2010) Rearrangement of CRLF2 is associated with mutation of JAK kinases, alteration of IKZF1, Hispanic/Latino ethnicity, and a poor outcome in pediatric B-progenitor acute lymphoblastic leukemia. Blood 115(26):5312–5321. https://doi.org/10.1182/blood-2009-09-245944
100. Roberts KG, Morin RD, Zhang J, Hirst M, Zhao Y, Su X, Chen SC, Payne-Turner D, Churchman ML, Harvey RC, Chen X, Kasap C, Yan C, Becksfort J, Finney RP, Teachey DT, Maude SL, Tse K, Moore R, Jones S, Mungall K, Birol I, Edmonson MN, Hu Y, Buetow KE, Chen IM, Carroll WL, Wei L, Ma J, Kleppe M, Levine RL, Garcia-Manero G, Larsen E, Shah NP, Devidas M, Reaman G, Smith M, Paugh SW, Evans WE, Grupp SA, Jeha S, Pui CH, Gerhard DS, Downing JR, Willman CL, Loh M, Hunger SP, Marra MA, Mullighan CG (2012) Genetic alterations activating kinase and cytokine receptor signaling in high-risk acute lymphoblastic leukemia. Cancer Cell 22(2):153–166. https://doi.org/10.1016/j.ccr.2012.06.005
101. Mullighan CG, Miller CB, Radtke I, Phillips LA, Dalton J, Ma J, White D, Hughes TP, Le Beau MM, Pui CH, Relling MV, Shurtleff SA, Downing JR (2008) BCR-ABL1 lymphoblastic leukaemia is characterized by the deletion of Ikaros. Nature 453(7191):110–114. https://doi.org/10.1038/nature06866
102. Katerndahl CDS, Heltemes-Harris LM, Willette MJL, Henzler CM, Frietze S, Yang R, Schjerven H, Silverstein KAT, Ramsey LB, Hubbard G, Wells AD, Kuiper RP, Scheijen B, van Leeuwen FN, Muschen M, Kornblau SM, Farrar MA (2017) Antagonism of B cell enhancer networks by STAT5 drives leukemia and poor patient survival. Nat Immunol 18(6):694–704. https://doi.org/10.1038/ni.3716
103. Eckardt JN, Stasik S, Rollig C, Petzold A, Sauer T, Scholl S, Hochhaus A, Crysandt M, Brummendorf TH, Naumann R, Steffen B, Kunzmann V, Einsele H, Schaich M, Burchert A, Neubauer A, Schafer-Eckart K, Schliemann C, Krause SW, Herbst R, Hanel M, Hanoun M, Kaiser U, Kaufmann M, Racil Z, Mayer J, Oelschlagel U, Berdel WE, Ehninger G, Serve H, Muller-Tidow C, Platzbecker U, Baldus CD, Dahl A, Schetelig J, Bornhauser M, Middeke JM, Thiede C (2023) Mutated IKZF1 is an independent marker of adverse risk in acute myeloid leukemia. Leukemia. https://doi.org/10.1038/s41375-023-02061-1
104. Bastidas Torres AN, Cats D, Mei H, Fanoni D, Gliozzo J, Corti L, Paulli M, Vermeer MH, Willemze R, Berti E, Tensen CP (2020) Whole-genome analysis uncovers recurrent IKZF1 inactivation and aberrant cell adhesion in blastic plasmacytoid dendritic cell neoplasm. Genes Chromosomes Cancer 59(5):295–308. https://doi.org/10.1002/gcc.22831
105. Mullighan CG, Su X, Zhang J, Radtke I, Phillips LA, Miller CB, Ma J, Liu W, Cheng C, Schulman BA, Harvey RC, Chen IM, Clifford RJ, Carroll WL, Reaman G, Bowman WP, Devidas M, Gerhard DS, Yang W, Relling MV, Shurtleff SA, Campana D, Borowitz MJ, Pui CH, Smith M, Hunger SP, Willman CL, Downing JR, Children's Oncology G (2009) Deletion of IKZF1 and prognosis in acute lymphoblastic leukemia. N Engl J Med 360(5):470–480. https://doi.org/10.1056/NEJMoa0808253
106. Theocharides AP, Dobson SM, Laurenti E, Notta F, Voisin V, Cheng PY, Yuan JS, Guidos CJ, Minden MD, Mullighan CG, Torlakovic E, Dick JE (2015) Dominant-negative Ikaros cooperates with BCR-ABL1 to induce human acute myeloid leukemia in xenografts. Leukemia 29(1):177–187. https://doi.org/10.1038/leu.2014.150
107. Schjerven H, Ayongaba EF, Aghajanirefah A, McLaughlin J, Cheng D, Geng H, Boyd JR, Eggesbo LM, Lindeman I, Heath JL, Park E, Witte ON, Smale ST, Frietze S, Muschen M (2017) Genetic analysis of Ikaros target genes and tumor suppressor function in BCR-ABL1(+) pre-B ALL. J Exp Med 214(3):793–814. https://doi.org/10.1084/jem.20160049
108. Witkowski MT, Hu Y, Roberts KG, Boer JM, McKenzie MD, Liu GJ, Le Grice OD, Tremblay CS, Ghisi M, Willson TA, Horstmann MA, Aifantis I, Cimmino L, Frietze S, den Boer ML,

Mullighan CG, Smyth GK, Dickins RA (2017) Conserved IKAROS-regulated genes associated with B-progenitor acute lymphoblastic leukemia outcome. J Exp Med 214(3):773–791. https://doi.org/10.1084/jem.20160048
109. Cowan AJ, Green DJ, Kwok M, Lee S, Coffey DG, Holmberg LA, Tuazon S, Gopal AK, Libby EN (2022) Diagnosis and management of multiple myeloma: a review. JAMA 327(5):464–477. https://doi.org/10.1001/jama.2022.0003
110. Kronke J, Udeshi ND, Narla A, Grauman P, Hurst SN, McConkey M, Svinkina T, Heckl D, Comer E, Li X, Ciarlo C, Hartman E, Munshi N, Schenone M, Schreiber SL, Carr SA, Ebert BL (2014) Lenalidomide causes selective degradation of IKZF1 and IKZF3 in multiple myeloma cells. Science 343(6168):301–305. https://doi.org/10.1126/science.1244851
111. Lu G, Middleton RE, Sun H, Naniong M, Ott CJ, Mitsiades CS, Wong KK, Bradner JE, Kaelin WG Jr (2014) The myeloma drug lenalidomide promotes the cereblon-dependent destruction of Ikaros proteins. Science 343(6168):305–309. https://doi.org/10.1126/science.1244917
112. Fischer ES, Bohm K, Lydeard JR, Yang H, Stadler MB, Cavadini S, Nagel J, Serluca F, Acker V, Lingaraju GM, Tichkule RB, Schebesta M, Forrester WC, Schirle M, Hassiepen U, Ottl J, Hild M, Beckwith RE, Harper JW, Jenkins JL, Thoma NH (2014) Structure of the DDB1-CRBN E3 ubiquitin ligase in complex with thalidomide. Nature 512(7512):49–53. https://doi.org/10.1038/nature13527
113. Wang ES, Verano AL, Nowak RP, Yuan JC, Donovan KA, Eleuteri NA, Yue H, Ngo KH, Lizotte PH, Gokhale PC, Gray NS, Fischer ES (2021) Acute pharmacological degradation of Helios destabilizes regulatory T cells. Nat Chem Biol 17(6):711–717. https://doi.org/10.1038/s41589-021-00802-w
114. Lazarian G, Yin S, Ten Hacken E, Sewastianik T, Uduman M, Font-Tello A, Gohil SH, Li S, Kim E, Joyal H, Billington L, Witten E, Zheng M, Huang T, Severgnini M, Lefebvre V, Rassenti LZ, Gutierrez C, Georgopoulos K, Ott CJ, Wang L, Kipps TJ, Burger JA, Livak KJ, Neuberg DS, Baran-Marszak F, Cymbalista F, Carrasco RD, Wu CJ (2021) A hotspot mutation in transcription factor IKZF3 drives B cell neoplasia via transcriptional dysregulation. Cancer Cell 39 (3):380-393 e388. https://doi.org/10.1016/j.ccell.2021.02.003
115. Kastner P, Chan S (2011) Role of Ikaros in T-cell acute lymphoblastic leukemia. World J Biol Chem 2(6):108–114. https://doi.org/10.4331/wjbc.v2.i6.108
116. Weng AP, Ferrando AA, Lee W, Morris JP, Silverman LB, Sanchez-Irizarry C, Blacklow SC, Look AT, Aster JC (2004) Activating mutations of NOTCH1 in human T cell acute lymphoblastic leukemia. Science 306(5694):269–271. https://doi.org/10.1126/science.1102160
117. Ashworth TD, Pear WS, Chiang MY, Blacklow SC, Mastio J, Xu L, Kelliher M, Kastner P, Chan S, Aster JC (2010) Deletion-based mechanisms of Notch1 activation in T-ALL: key roles for RAG recombinase and a conserved internal translational start site in Notch1. Blood 116(25):5455–5464. https://doi.org/10.1182/blood-2010-05-286328
118. Jeannet R, Mastio J, Macias-Garcia A, Oravecz A, Ashworth T, Geimer Le Lay AS, Jost B, Le Gras S, Ghysdael J, Gridley T, Honjo T, Radtke F, Aster JC, Chan S, Kastner P (2010) Oncogenic activation of the Notch1 gene by deletion of its promoter in Ikaros-deficient T-ALL. Blood 116(25):5443–5454. https://doi.org/10.1182/blood-2010-05-286658
119. Kuehn HS, Boast B, Rosenzweig SD (2023) Inborn errors of human IKAROS: LOF and GOF variants associated with primary immunodeficiency. Clin Exp Immunol 212(2):129–136. https://doi.org/10.1093/cei/uxac109
120. Kuehn HS, Boisson B, Cunningham-Rundles C, Reichenbach J, Stray-Pedersen A, Gelfand EW, Maffucci P, Pierce KR, Abbott JK, Voelkerding KV, South ST, Augustine NH, Bush JS, Dolen WK, Wray BB, Itan Y, Cobat A, Sorte HS, Ganesan S, Prader S, Martins TB, Lawrence MG, Orange JS, Calvo KR, Niemela JE, Casanova JL, Fleisher TA, Hill HR, Kumanovics A, Conley ME, Rosenzweig SD (2016) Loss of B cells in patients with heterozygous nutations in IKAROS. N Engl J Med 374(11):1032–1043. https://doi.org/10.1056/NEJMoa1512234
121. Hoshino A, Okada S, Yoshida K, Nishida N, Okuno Y, Ueno H, Yamashita M, Okano T, Tsumura M, Nishimura S, Sakata S, Kobayashi M, Nakamura H, Kamizono J, Mitsui-Sekinaka K, Ichimura T, Ohga S, Nakazawa Y, Takagi M, Imai K, Shiraishi Y, Chiba K, Tanaka H, Miyano S, Ogawa S, Kojima S, Nonoyama S, Morio T, Kanegane H (2017) Abnormal hematopoiesis and autoimmunity in human subjects with germline IKZF1 mutations. J Allergy Clin Immunol 140(1):223–231. https://doi.org/10.1016/j.jaci.2016.09.029
122. Eskandarian Z, Fliegauf M, Bulashevska A, Proietti M, Hague R, Smulski CR, Schubert D, Warnatz K, Grimbacher B (2019) Assessing the functional relevance of variants in the IKAROS family zinc finger protein 1 (IKZF1) in a cohort of patients with primary immunodeficiency. Front Immunol 10:568. https://doi.org/10.3389/fimmu.2019.00568
123. Hoshino A, Boutboul D, Zhang Y, Kuehn HS, Hadjadj J, Ozdemir N, Celkan T, Walz C, Picard C, Lenoir C, Mahlaoui N, Klein C, Peng X, Azar A, Reigh E, Cheminant M, Fischer A, Rieux-Laucat F, Callebaut I, Hauck F, Milner J, Rosenzweig SD, Latour S (2022) Gain-of-function IKZF1 variants in humans cause immune dysregulation associated with abnormal T/B cell late differentiation. Sci Immunol 7(69):eabi7160. https://doi.org/10.1126/sciimmunol.abi7160

124. Yamashita M, Morio T (2022) AIOLOS variants causing immunodeficiency in human and mice. Front Immunol 13:866582. https://doi.org/10.3389/fimmu.2022.866582
125. Shahin T, Kuehn HS, Shoeb MR, Gawriyski L, Giuliani S, Repiscak P, Hoeger B, Yuce Petronczki O, Bal SK, Zoghi S, Dmytrus J, Seruggia D, Castanon I, Rezaei N, Varjosalo M, Halbritter F, Rosenzweig SD, Boztug K (2021) Germline biallelic mutation affecting the transcription factor Helios causes pleiotropic defects of immunity. Sci Immunol 6 (65):eabe3981. https://doi.org/10.1126/sciimmunol.abe3981
126. Shahin T, Mayr D, Shoeb MR, Kuehn HS, Hoeger B, Giuliani S, Gawriyski LM, Petronczki OY, Hadjadj J, Bal SK, Zoghi S, Haimel M, Jimenez Heredia R, Boutboul D, Triebwasser MP, Rialland-Battisti F, Costedoat Chalumeau N, Quartier P, Tangye SG, Fleisher TA, Rezaei N, Romberg N, Latour S, Varjosalo M, Halbritter F, Rieux-Laucat F, Castanon I, Rosenzweig SD, Boztug K (2022) Identification of germline monoallelic mutations in IKZF2 in patients with immune dysregulation. Blood Adv 6(7):2444–2451. https://doi.org/10.1182/bloodadvances.2021006367
127. Quintana FJ, Jin H, Burns EJ, Nadeau M, Yeste A, Kumar D, Rangachari M, Zhu C, Xiao S, Seavitt J, Georgopoulos K, Kuchroo VK (2012) Aiolos promotes TH17 differentiation by directly silencing Il2 expression. Nat Immunol 13(8):770–777. https://doi.org/10.1038/ni.2363
128. Hetemaki I, Kaustio M, Kinnunen M, Heikkila N, Keskitalo S, Nowlan K, Miettinen S, Sarkkinen J, Glumoff V, Andersson N, Kettunen K, Vanhanen R, Nurmi K, Eklund KK, Dunkel J, Mayranpaa MI, Schlums H, Arstila TP, Kisand K, Bryceson YT, Peterson P, Otava U, Syrjanen J, Saarela J, Varjosalo M, Kekalainen E (2021) Loss-of-function mutation in IKZF2 leads to immunodeficiency with dysregulated germinal center reactions and reduction of MAIT cells. Sci Immunol 6 (65):eabe3454. https://doi.org/10.1126/sciimmunol.abe3454
129. Oh KS, Gottschalk RA, Lounsbury NW, Sun J, Dorrington MG, Baek S, Sun G, Wang Z, Krauss KS, Milner JD, Dutta B, Hager GL, Sung MH, Fraser IDC (2018) Dual Roles for Ikaros in Regulation of Macrophage Chromatin State and Inflammatory Gene Expression. J Immunol 201(2):757–771. https://doi.org/10.4049/jimmunol.1800158
130. Dieudonne Y, Guffroy A, Vollmer O, Carapito R, Korganow AS (2019) IKZF1 loss-of-function variant causes autoimmunity and severe familial antiphospholipid syndrome. J Clin Immunol 39(4):353–357. https://doi.org/10.1007/s10875-019-00643-2
131. Kuijpers TW, Tromp SAM, van Leeuwen EMM, de Bree GJ (2022) Case report: a highly variable clinical and immunological presentation of IKAROS deficiency in a single family. Front Immunol 13:865838. https://doi.org/10.3389/fimmu.2022.865838
132. Brodie SA, Khincha PP, Giri N, Bouk AJ, Steinberg M, Dai J, Jessop L, Donovan FX, Chandrasekharappa SC, de Andrade KC, Maric I, Ellis SR, Mirabello L, Alter BP, Savage SA (2021) Pathogenic germline IKZF1 variant alters hematopoietic gene expression profiles. Cold Spring Harb Mol Case Stud 7(4). https://doi.org/10.1101/mcs.a006015
133. Boutboul D, Kuehn HS, Van de Wyngaert Z, Niemela JE, Callebaut I, Stoddard J, Lenoir C, Barlogis V, Farnarier C, Vely F, Yoshida N, Kojima S, Kanegane H, Hoshino A, Hauck F, Lhermitte L, Asnafi V, Roehrs P, Chen S, Verbsky JW, Calvo KR, Husami A, Zhang K, Roberts J, Amrol D, Sleaseman J, Hsu AP, Holland SM, Marsh R, Fischer A, Fleisher TA, Picard C, Latour S, Rosenzweig SD (2018) Dominant-negative IKZF1 mutations cause a T, B, and myeloid cell combined immunodeficiency. J Clin Invest 128(7):3071–3087. https://doi.org/10.1172/JCI98164
134. Mohajeri A, Vaseghi-Shanjani M, Rosenfeld JA, Yang GX, Lu H, Sharma M, Lin S, Salman A, Waqas M, Sababi Azamian M, Worley KC, Del Bel KL, Kozak FK, Rahmanian R, Biggs CM, Hildebrand KJ, Lalani SR, Nicholas SK, Scott DA, Mostafavi S, van Karnebeek C, Henkelman E, Halparin J, Yang CL, Armstrong L, Undiagnosed Diseases N, Care4Rare Canada C, Turvey SE, Lehman A (2023) Dominant negative variants in IKZF2 cause ICHAD syndrome, a new disorder characterised by immunodysregulation, craniofacial anomalies, hearing impairment, athelia and developmental delay. J Med Genet. https://doi.org/10.1136/jmg-2022-109127
135. Kuehn HS, Chang J, Yamashita M, Niemela JE, Zou C, Okuyama K, Harada J, Stoddard JL, Nunes-Santos CJ, Boast B, Baxter RM, Hsieh EWY, Garofalo M, Fleisher TA, Morio T, Taniuchi I, Dutmer CM, Rosenzweig SD (2021) T and B cell abnormalities, pneumocystis pneumonia, and chronic lymphocytic leukemia associated with an AIOLOS defect in patients. J Exp Med 218(12). https://doi.org/10.1084/jem.20211118

3 The Role of the Transcriptional Coactivator BOB.1/OBF.1 in Adaptive Immunity

Annika C. Betzler and Cornelia Brunner

Abstract

BOB.1/OBF.1 is a transcriptional coactivator involved in octamer-dependent transcription. Thereby, BOB.1/OBF.1 is involved in the transcriptional regulation of genes important for lymphocyte physiology. BOB.1/OBF.1-deficient mice reveal multiple B- and T-cell developmental defects. The most prominent defect of these mice is the complete absence of germinal centers (GCs) resulting in severely impaired T-cell-dependent immune responses. In humans, BOB.1/OBF.1 is associated with several autoimmune and inflammatory diseases but also linked to liquid and solid tumors. Although its role for B-cell development is relatively well understood, its exact role for the GC reaction and T-cell biology has long been unclear. Here, the contribution of BOB.1/OBF.1 for B-cell maturation is summarized, and recent findings regarding its function in GC B- as well as in various T-cell populations are discussed. Finally, a detailed perspective on how BOB.1/OBF.1 contributes to different pathologies is provided.

A. C. Betzler · C. Brunner (✉)
Department of Oto-Rhino-Larnygology, Ulm University Medical Center, Ulm, Germany

Core Facility Immune Monitoring, Ulm University, Ulm, Germany
e-mail: cornelia.brunner@uniklinik-ulm.de

Keywords

BOB.1/OBF.1 · B lymphocytes · T lymphocytes · Germinal center · Transcriptional regulation · Autoimmunity · Chronic inflammation · Lymphoma · Solid tumors

Abbreviations

ABF-1	Activated B-cell factor-1
Aicda	Activation-induced cytidine deaminase
AID	Activation-induced cytidine deaminase
ALK	Anaplastic lymphoma kinase
B29	B-cell-specific glycoprotein B29
BAFF	B-cell activating factor
BAFFR	B-cell activating factor receptor
Bcl2	B-cell lymphoma 2
Bcl6	B-cell lymphoma 6
BCMA	B-cell maturation factor
BCR	B-cell receptor
BL	Burkitt lymphoma
BM	Bone marrow
BOB.1/OBF.1	B-cell Oct binding factor 1/ Oct-binding factor 1
BTK	Bruton's tyrosine kinase
Btla	B- and T-lymphocyte attenuator

T. Borggrefe, B. D. Giaimo (eds.), *Transcription factors in blood cell development*, Advances in Experimental Medicine and Biology 1459, https://doi.org/10.1007/978-3-031-62731-6_3

Ca^{2+} Calcium
CD20 Cluster of differentiation 20
CD21 Cluster of differentiation 21
CD28 Cluster of differentiation 28
CD3 Cluster of differentiation 3
cHL Classical Hodgkin's lymphoma
CIA Collagen-induced arthritis
COPD Chronic obstructive pulmonary disease
CSR Class-switch recombination
CXCL13 Chemokine C-X-C motif ligand 13
CXCR5 C-X-C motif chemokine receptor 5
DC Dendritic cell
DEG Differentially expressed gene
DLBCL Diffuse large B-cell lymphoma
DOT1L DOT1-like histone lysine methyltransferase
EAE Experimental autoimmune encephalomyelitis
FDC Follicular dendritic cell
FL Follicular lymphoma
Foxo1 Forkhead box O1
FoxP3 Forkhead box P3
GAMES Genetic analysis of ultiple sclerosis in EuropeanS
GATA3 GATA Binding Protein 3
GC Germinal center
GIST Gastrointestinal stromal tumors
HER2 Human epidermal growth factor receptor 2
HRS Hodgkin and Reed-Sternberg cells
IBD Inflammatory bowel disease
Ifng Interferon gamma
IFNγ Interferon gamma
Ig Immunoglobulin
IgH Immunoglobulin heavy
IK2F1/3 IKAROS Family Zinc Finger 1/3
IL Interleukin
IPF Idiopathic pulmonary fibrosis
Irf4 Interferon regulatory factor 4
JMJD1A Jumonji domain containing 1A
LAZ3 Lymphoma-associated zinc finger 3
LN Lymph node
$LT\alpha_1\beta_2$ Lymphotoxin-$\alpha_1\beta_2$
MAPK Mitogen-activated protein kinases
Mef2b Myocyte-specific enhancer factor 2B
MM Multiple myeloma
MORE More of PORE
MRL-lpr Murphy Roths large lymphoproliferation mouse strain
MS Multiple sclerosis
Myc MYC Proto-oncogene
MZ B Marginal zone B cell
NFAT Nuclear factor of activated T cells
NF-κB Nuclear factor kappa chain enhancer of activated B cells
NHL Non-Hodgkin lymphoma
NOD Nonobese diabetic
OCA-B Octamer coactivator from B cells
OCT1 Octamer transcription factor 1
OCT2 Octamer transcription factor 2
p34 BOB.1/OBF.1 protein of 34 kDa
p35 BOB.1/OBF.1 protein of 35 kDa
p40 BOB.1/OBF.1 protein of 40 kDa
P9 Postnatal day 9
PAX5 Paired box 5
PBC Primary biliary cirrhosis
PI3-Akt Phosphatidylinositol 3-kinase / protein kinase B
PMA Phorbol 12-myristate-13-acetate
PORE Palindromic Oct-factor recognition element
Pou2af1 POU domain class 2-associating factor 1
POU_H POU homeodomain
POU_S POU-specific domain
PP Peyer's patch
PPI Protein-protein interaction
Prdm1 Positive regulatory domain 1
PU.1/SPI1 Hematopoietic transcription factor PU.1/Spleen focus forming virus (SFFV) proviral integration protein

RCC	Renal cell carcinoma
RNA-Seq	RNA followed by deep sequencing
RORγt	Retinoic acid receptor-related orphan receptor gamma t
RORγt	Retinoic acid-related orphan receptor γt
SHM	Somatic hypermutation
SIAH	Seven in absentia homolog
SLE	Systemic lupus erythematosus
SLO	Secondary lymphoid organ
SPI1/PU.1	Hematopoietic transcription factor PU.1/spleen focus-forming virus (SFFV) proviral integration protein
SPIB	Spi-B transcription factor
SS	Sjogren's syndrome
Syk	Spleen tyrosine kinase
T1D	Type 1 diabetes
Tbx21	T-box transcription factor 21
TCR	T-cell receptor
TF	Transcription factor
Tfh	T follicular helper
TGFβ	Transforming growth factor-β
Th	T-helper
TLR	Toll-like receptor
TNBC	Triple-negative breast cancer
Treg	Regulatory T cell
Wnt	Wingless-related integration site
WT	Wild type

3.1 The Transcriptional Coactivator BOB.1/OBF.1

BOB.1/OBF.1 (B-cell Oct binding factor 1/Oct-binding factor 1), also known as OCA-B (octamer coactivator from B cells), is encoded by the *Pou2af1* (POU domain class 2-associating factor 1) gene. BOB.1/OBF.1 is expressed in both B and T lymphocytes and acts as transcriptional coactivator of octamer-dependent transcription [1]. An involvement of a transcriptional coactivator during octamer-dependent transcription was first suggested in 1992 [2]. Three years later, the cDNA of the human transcriptional coactivator was isolated by three independent groups and alternatively named BOB.1 [3], OBF.1 [4], or OCA-B [5]. Shortly after, the murine homolog of the human BOB.1/OBF.1 was described [6, 7]. Since its discovery, research invested great efforts to resolve the function of BOB.1/OBF.1 in the murine and human immune system, both in physiological and pathological states. Relatively quickly it became clear that BOB.1/OBF.1 is important for B-cell development and function and in particular for the germinal center (GC) reaction [7–10]. Significantly later, a contribution of BOB.1/OBF.1 also in T-cell physiology became evident [11–13]. Further investigations also revealed an involvement of BOB.1/OBF.1 in various pathologies. Thus, BOB.1/OBF.1 plays a role in several B-cell malignancies. Its expression was described to be aberrantly regulated in GC-derived lymphomas but also in classical Hodgkin disease [14–16]. In addition, dysregulated BOB.1/OBF.1 expression drives the pathogenesis of autoimmune and chronic inflammatory diseases. Besides, an association of BOB.1/OBF.1 with different solid tumor entities was revealed. Recently, a 3-year-old patient was described with a homozygous mutation in the *POU2AF1* gene resulting in the absence of BOB.1/OBF.1 protein [17]. Deficiency of BOB.1/OBF.1 protein led to recurrent infections and a lack of GCs and immunoglobulins (Igs) as it is also described for BOB.1/OBF.1-deficient mice [17]. The regulation of BOB.1/OBF.1 as well as its specific role in B and T lymphocytes will be discussed in this chapter.

3.2 Role of BOB.1/OBF.1 in Octamer-Dependent Transcription

BOB.1/OBF.1 acts as transcriptional coactivator during octamer-dependent transcription. Octamer-dependent transcription is attained by transcription factors (TFs) that belong to the OCT family including OCT1 and OCT2. OCT1 is ubiquitously expressed, while OCT2 expression is found in cells of hematopoietic origin but also in neuronal cells [18, 19]. However, the OCT2 isoform found in neuronal cells differs from that expressed in lymphocytes [20]. These OCT-TFs bind to the octamer motif (5′-ATGCAAAT-3′), which can be found in regulatory elements of B- and T-cell-specific genes [21–23]. The octamer motif is preserved in all Ig heavy and light chain gene promoters as well as enhancer elements [21, 22] but was also found in regulatory regions of other B-cell-related genes including *B29*, *CD20*, *CD21*, *BTK*, and *PAX5* [24–27]. In addition, it is reported in promoters of ubiquitously expressed genes like small nuclear RNA (snRNA) genes or the histone *H2B* gene [28, 29]. BOB.1/OBF.1 interacts with OCT-TFs and enhances their binding selectivity to the octamer motif as well as their transcriptional activity [4, 30–32]. BOB.1/OBF.1 itself has only low DNA affinity but is recruited to the octamer site by its interaction with OCT-TFs [4, 30–32]. Thus, the ternary complex composed of BOB.1/OBF.1 together with OCT1/2 acts on promoters to induce expression of genes essential for lymphocyte development, signaling, and function [11, 13, 30, 32–35]. BOB.1/OBF.1 itself has two functional domains. BOB.1/OBF.1 uses its N-terminal domain to interact with OCT-TFs, while the C-terminal domain is intended to interact with the transactivation domains of OCT-TFs thereby driving transcription [3, 5, 6, 30]. OCT proteins harbor a POU domain, a bipartite DNA-binding segment, consisting of two structurally independent DNA-binding domains [DBD [36]]. These domains include the N-terminal POU-specific (POU_S) and the C-terminal POU homeodomain (POU_H), which are connected by a flexible linker [36]. The two subdomains bind the DNA major groove on opposite sides, as the POU_S domain contacts the 5′-ATGC and the POU_H the AAAT-3′ subsite [37]. BOB.1/OBF.1 can interact with OCT-TF monomers bound to the octamer motif. Thereby, BOB.1/OBF.1 contacts the POU_S, the POU_H, and the adenosine residue at position 5 of the octamer motif. In doing so, the transcriptional coactivator functions as a molecular clamp to keep these three molecules together [31, 32, 38]. Otherwise, OCT-TFs can bind to the octamer site as dimers. However, different dimer configurations can either favor or impede interaction with BOB.1/OBF.1. [39–42]. The so-called PORE sequence (Palindromic Oct-factor Recognition Element, 5′-ATTTGAAATGCAAAT-3′) favors BOB.1/OBF.1 binding, while the MORE (More of PORE, 5′-ATGCATATGCAT-3′) configuration hinders interaction with BOB.1/OBF.1 [41]. In the MORE configuration, bindings sites required for interaction with BOB.1/OBF.1 are blocked, as they are already used for dimerization [41].

3.3 Regulation of BOB.1/OBF.1 Expression

Two isoforms, namely, p34 and p35, have been described for the BOB.1/OBF.1 protein [4]. Both isoforms are encoded by the same mRNA but are translated from two different translation initiation codons [4]. The p35 isoform is generated from a p40 precursor isoform [43]. The p34 isoform is exclusively found in the nucleus, whereas the p35 isoform is additionally present in the cytoplasm but also membrane bound. BOB1/OBF.1- p35 is posttranslationally modified by myristoylation through the N-terminal glycine residue, which probably accounts for the predominant cytoplasmic and membrane localization of this isoform [43]. These differences in location and posttranslational modification (PTM) suggest that both isoforms carry out different functions. However, the precise function of both the cytoplasmic and also the membrane-bound p35 isoform is still unknown. Initially, both isoforms were considered to be evenly able to stimulate octamer-dependent transcription [4]. However, further research showed that it is predominantly the p34 isoform that activates transcription [43]. Due to its localization, it is hypothesized that the p35 isoform is rather involved in B-cell receptor (BCR) signaling, as an interaction of the p35 isoform with spleen tyrosine kinase (SYK) and GALECTIN-1 was demonstrated [44, 45]. The expression of BOB.1/OBF.1 is confined to lymphocytes. Both p34 and p35 isoforms are continuously expressed in B cells throughout all B-cell maturation stages [46]. However, BOB.1/OBF.1 expression levels vary at different B-cell developmental stages. The highest expression of BOB.1/OBF.1 is detected in GC B cells both in mice and humans [14, 47, 48]. Generally, its expression peaks in cycling low-density splenic B cells, whereas its expression is virtually absent in resting, high-density splenic B cells [14, 49]. In B cells, BOB.1/OBF.1 expression is regulated by canonical and noncanonical nuclear factor kappa chain enhancer of activated B cells (NF-κB) [50]. In this regard, it was shown that ligands stimulating Toll-like receptor (TLR)-2, -3, -4, -5, -6, -7, -8, and -9 were all able to induce BOB.1/OBF.1 as well as OCT2 expression in B cells [50]. Besides TLR ligands, also other inducers of the canonical NF-κB pathway including tumor necrosis factor α (TNF-α) but also inducers of the noncanonical NF-κB pathway like B-cell activating factor (BAFF) of the TNF family or lymphotoxinα1β2 (LTα1β2) induced BOB.1/OBF.1 expression in B cells [50]. In T cells, BOB.1/OBF.1 expression is regulated by NF-κB and nuclear factor of activated T cells (NFAT) [23]. Here, it was shown that expression of both BOB.1/OBF.1 and OCT2 depends on CALCINEURIN. CALCINEURIN dephosphorylates NFAT TFs, which facilitates their translocation into the nucleus thereby controlling the expression of genes required for T-cell physiology [51]. Moreover, CALCINEURIN participates in regulating T-cell receptor (TCR) induced NF-κB signaling and thereby again contributes to regulation of genes essential for T-cell function [51]. In both B and T cells, potential NF-κB and NFAT binding sites within the *Pou2af1* promoter have been identified contributing to its activity [23, 50]. Besides its regulation at the transcriptional level, expression levels of BOB.1/OBF.1 are also regulated via protein stability. BOB.1/OBF.1 interacts with known regulators of protein stability, namely, SIAH1 and SIAH2 (seven in absentia homolog) [49, 52]. Because of this interaction, BOB.1/OBF.1 is degraded in a proteasome-dependent manner [49, 52].

3.4 The Role of BOB.1/OBF.1 in B Cells

3.4.1 The Role of BOB.1/OBF.1 for Early B-Cell Maturation in the Bone Marrow

BOB.1/OBF.1 is expressed throughout complete B-cell ontogeny and is required for B-cell development at multiple stages. Accordingly, BOB.1/OBF.1-deficient mice reveal several defects in early and late B-cell development. BOB.1/OBF.1 expression is already essential during early B-cell maturation in the bone marrow (BM; Fig. 3.1). Analysis of BOB.1/OBF.1-deficient mice revealed a developmental block at the pro-/pre-B-cell stage [8, 53]. In detail, BOB.1/OBF.1 is critical for the transit of immature B cells from the BM to the periphery, since BOB.1/OBF.1-deficient mice feature reduced numbers of transitional and mature recirculating B cells in the BM [8, 53]. In addition, BOB.1/OBF.1-deficient B cells in the BM and in the periphery display increased apoptosis, which is presumably due to a reduction of the antiapoptotic protein B-cell lymphoma 2 (BCL2) [8, 54]. This assumption further grounded on findings showing that restoring BCL2 expression in BOB.1/OBF.1-deficient B cells rescues the number of transitional and recirculating B cells in the BM [54]. Moreover, surface expression of CD22 was found to be increased in cell populations of the BM in the absence of BOB.1/OBF.1 including pre-B, immature, and transitional B cells [55]. CD22 expression initiates at the pre-B-cell stages and increases throughout B-cell maturation in the BM [56]. However, its role for B-cell development in the BM is unknown. In mature B cells, CD22 is required for recirculating back from the periphery to the BM and is also known as a negative regulator of BCR signaling [57, 58]. To check whether the B developmental defects observed in the BM of BOB.1/OBF.1-deficient animals are a consequence of CD22 upregulation, BOB.1/OBF.1/CD22-double-deficient mice were analyzed [55]. Deficiency of both proteins normalized the transitional B-cell levels, but not the defect in recirculating B cells in the BM [55].

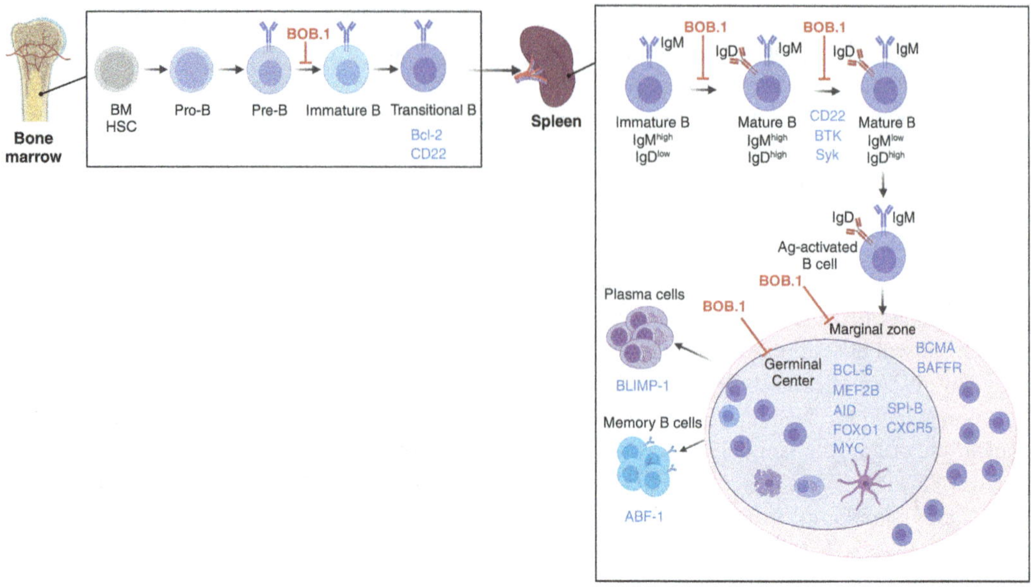

Fig. 3.1 BOB.1/OBF.1 is essential at several stages of B-cell development. BOB.1/OBF.1 deficiency leads to severe B-cell developmental defects in the bone marrow (BM). Here, BOB.1/OBF.1-deficient mice stuck in their development at the pro/pre-B-cell stage and the subsequent maturation is severely impaired. It is hypothesized that B-cell developmental defects in the BM could be due to BOB.1/OBF.1-mediated regulation of BCL2 and CD22. BOB.1/OBF.1 is also important for efficient B-cell maturation in the periphery. Here, BOB.1/OBF.1-deficient mice reveal increased levels of immature but decreased numbers of mature B cells. Defective BCR signaling in the absence of BOB.1/OBF.1 was reported which could be due to defective BTK or SYK signaling. BOB.1/OBF.1-deficient marginal zone (MZ) B cells are also defective in their development and function. Here, an influence of BOB.1/OBF.1 on BCMA and BAFFR expression is discussed. The hallmark of BOB.1/OBF.1-deficient animals is the complete absence of germinal centers (GCs). Recently, it was shown that BOB.1/OBF.1 controls the GC transcriptional program. Moreover, an involvement of BOB.1/OBF.1 in plasma and memory B-cell differentiation is postulated by regulating BLIMP-1 or ABF-1 expression. (This figure was created with BioRender.com)

Consequently, the defects of early B-cell maturation in the BM of BOB.1/OBF.1-deficient mice could partially be traced back to the upregulation of CD22 and the resulting hindering of BCR signaling.

3.4.2 The Role of BOB.1/OBF.1 for B-Cell Maturation in the Periphery

In addition, BOB.1/OBF.1 is essential at later stages during B-cell development in the periphery (Fig. 3.1). Specifically, BOB.1/OBF.1 is required for the maturation of follicular B cells, since BOB.1/OBF.1-deficient mice show increased levels of immature but decreased numbers of mature B cells [8, 9]. The number of marginal zone (MZ) B cells is also substantially diminished in the absence of the transcriptional coactivator [10]. BOB.1/OBF.1-deficient MZ B cells are beyond that also impaired in their function, as they are incapable of efficiently capturing the T-independent Ag TNP-Ficoll [10]. Interestingly, mRNA levels of the B-cell activating factor BAFF receptors, BAFFR and BCMA (B-cell maturation factor), are reduced upon BOB.1/OBF.1 deficiency [10]. Since BAFF/BAFFR or BCMA signaling is required for B-cell maturation and survival in the periphery, their reduced expression might contribute to the B-cell deficiency in BOB.1/OBF.1-deficient mice. In fact, some years later, it was shown that BOB.1/OBF.1 binds to an octamer motif upstream of the

BCMA transcription start site in multiple myeloma cells [59]. Further experiments confirmed that BOB.1/OBF.1 transactivates BCMA expression [59]. Consequently, mutation of the octamer-motif diminished BCMA transactivation [59]. For *Baff*, *Baffr*, or *Bcl2*, no direct interaction of BOB.1/OBF.1 with an octamer site within their promoters has yet been demonstrated.

Besides, BOB.1/OBF.1 is also involved in BCR signaling at different levels. BOB.1/OBF.1-deficient B cells reveal affected responses to BCR stimulation, as they fail to increase intracellular calcium (Ca^{2+}) levels [10]. This defect is probably due to the increased expression of aforementioned CD22 on BOB.1/OBF.1-deficient B cells, as BOB.1/OBF.1/CD22-double deficiency restores Ca^{2+} mobilization [55]. Findings demonstrating an involvement of BOB.1/OBF.1 in the regulation of Bruton's tyrosine kinase (BTK) and SYK, both important molecules for BCR signaling, further support the assumption that the transcriptional coactivator is implicated in BCR signaling [44, 60]. In this context, BOB.1/OBF.1 seems to play a dual role. On the one hand, it is involved in BCR signaling acting as a classical transcriptional coactivator on downstream target genes, including, e.g., BTK. The non-receptor tyrosine kinase BTK was shown to be a direct target of BOB.1/OBF.1 [60]. Both the murine and human *BTK* promoters feature a functional octamer motif to which BOB.1/OBF.1 and OCT1/2 can demonstrably bind to [60]. On the other hand, BOB.1/OBF.1 seems to be involved in BCR signaling in a nontranscriptional manner by directly interacting with key signaling components like SYK. In this regard, it is assumed that the cytoplasmic isoform of BOB.1/OBF.1 directly interacts with SYK thereby regulating its stability [44]. This is complemented by the suggestion that BOB.1/OBF.1 might regulate BCR signaling through an interaction with GALECTIN-1 [45]. GALECTIN-1 negatively regulates B-cell proliferation and tyrosine phosphorylation upon BCR stimulation. GALECTIN-1 levels were shown to be increased in BOB.1/OBF.1-deficient B cells, and a direct association of these two proteins was demonstrated [45].

3.4.3 The Contribution of BOB.1/OBF.1 to the Germinal Center Reaction

Besides all B-cell developmental defects in the BM and periphery that were discussed up to here, the most prominent feature of BOB.1/OBF.1-deficient mice is the complete lack of GCs. Consequently, these mice also feature dramatically reduced levels of isotype-switched Igs and post-GC B cells [7, 9]. However, for a long period of time, the precise contribution of BOB.1/OBF.1 to the GC reaction and more precisely BOB.1/OBF.1-regulated genes that could explain their complete absence were unknown. One of the main questions that was unresolved was at which stage of B-cell maturation—before or during the process of GC formation—BOB.1/OBF.1 expression is essential for proper GC formation.

Recent analyses of a conditional mouse system, allowing the deletion of BOB.1/OBF.1 at different timepoints throughout B-cell maturation, could show that BOB.1/OBF.1 expression is required during complete B-cell ontogeny for the formation of functional GCs [61]. By deleting BOB.1/OBF.1 specifically in GC B cells, it was proven that the failure to form GCs is a GC B-cell autonomous defect and cannot exclusively be explained by defective early B-cell maturation [61].

Over the years, a lot of efforts were made to identify BOB.1/OBF.1-regulated genes that could explain the absence of GCs. Along with reduced expression of the C-X-C motif chemokine receptor 5 (CXCR5) in BOB.1/OBF.1-deficient B cells, NF-κB, OCT2, and BOB.1/OBF.1 were shown to collectively regulate the promoter activity of *Cxcr5* in 1998 [62]. Recently, postnatal migration of lymphocytes into secondary lymphoid organs (SLOs) was examined in BOB.1/OBF.1-deficient animals [63]. On postnatal day 9 (P9), spleens of BOB.1/OBF.1-deficient animals revealed a considerable reduction of cellularity, which included decreased numbers of various B- and T-cell subpopulations but also diminished macrophages and dendritic cell (DCs) levels [63]. One main regulator of lymphocyte homing is the CXCL13-CXCR5 signaling axis

[64]. In mature SLOs, CXCR5-expressing B cells interact with CXCL13 secreting follicular DCs (FDCs), which result in the migration of B cells into the follicles [65, 66]. Therefore, impaired lymphocyte migration in BOB.1/OBF.1-deficient mice could either be due to defective CXCR5 expression of B cells or affected CXCL13 expression by FDCs. P9 BOB.1/OBF.1-deficient B cells exposed a severe reduction in the frequency of migration toward CXCL13 pointing to an impaired CXCR5 function in these cells [63]. Simultaneously, diminished FDC numbers and reduced CXCL13 expression could be observed in the absence of BOB.1/OBF.1 [63]. Consequently, reduced lymphocyte numbers of SLOs could be due to defective CXCL13-CXCR5 signaling in BOB.1/OBF.1-deficient mice. Impaired B-cell migration due to reduced CXCR5 and/or CXCL13 levels could also partially contribute to the dysfunction of GCs in the absence of BOB.1/OBF.1 but would not explain their complete absence. In 2006, SPIB, which is essential for GC formation and maintenance [67], was identified as BOB.1/OBF.1 target [68]. In 2021, one study identified several direct BOB.1/OBF.1 target genes regulating the GC response [34]. It was shown that BOB.1/OBF.1 together with OCT-TFs directly binds to and regulates genes encoding TFs relevant for the GC reaction including *Bcl6*, *Myc*, *Foxo1*, *Mef2b*, *Aicda*, and *Spi1* in primary B cells ([34] Fig. 3.1). BCL6 is the master regulator of the GC reaction and essential throughout all its phases. It is likely that BOB.1/OBF.1 induces BCL6 expression already during GC initiation. In the dark zone, where GC B-cell proliferation and somatic hypermutation (SHM) takes place, BOB.1/OBF.1 is involved in AID, BCL6, and FOXO1 induction [34]. In the light zone, where affinity maturation and class-switch recombination (CSR) occurs, BOB.1/OBF.1 might be involved in regulating FOXO1 and AID expression. MYC is another target gene of BOB.1/OBF.1, which is involved in regulating the reentry of light-zone GC B cells into the dark zone [69]. The study by Song et al. also revealed that BOB.1/OBF.1 is essential for the maintenance of the GC transcriptional program, since BOB.1/OBF.1 promotes BCL6 expression, but represses IRF4, whose expression would initiate the plasma cell program [34]. Another study suggested that various BOB.1/OBF.1 protein levels differentially influence cell-fate decisions in GCs during T-cell-dependent immune responses [70]. Physiological BOB.1/OBF.1 protein levels seem to promote plasma cell differentiation, while elevated BOB.1/OBF.1 expression seems to drive memory B-cell development by suppressing BCL6 and inducing ABF-1 expression [70]. In line, there is evidence that BOB.1/OBF.1 is involved in the induction of *Prdm1*, which encodes BLIMP-1 an essential regulator of plasma cell differentiation [35].

As a consequence of impaired GC function, BOB.1/OBF.1-deficient animals reveal dramatically reduced levels of secondary Igs. In that regard, it was demonstrated that the isotype switching process per se stays intact in the absence of BOB.1/OBF.1, but the transcription of switched Ig heavy (IgH) chain loci is reduced [71]. This was suggested to be due to defects in the function of 3′-IgH enhancer elements [72, 73].

3.5 The Role of BOB.1/OBF.1 in T Cells

BOB.1/OBF.1 has long been considered a B-cell-specific factor and was initially assumed to be irrelevant for T-cell physiology. However, in recent years, its role in T cells moved into the focus of research. In contrast to B cells, the role and function of BOB.1/OBF.1 in T cells are incompletely understood. In contrast to B cells, BOB.1/OBF.1 is not expressed in naïve T cells, but its expression can be induced in $CD4^+$ T cells upon treatment with PMA/ionomycin, αCD3/αCD28, or antigen receptor engagement [74–76]. Phosphorylation at Ser^{184} is required for inducible activation of BOB.1/OBF.1 in T cells [75]. Inducible activation of BOB.1/OBF.1 in T lymphocytes was first reported in 1997 [74, 75]. In 2003, BOB.1/OBF.1 expression was found in human T-cell lymphoma cell lines and primary samples [77]. By analyzing transgenic mice expressing BOB.1/OBF.1 predominantly in T cells, the TF SPIB was identified as direct BOB.1/OBF.1 target in thymocytes in 2006 [68]. Subsequent reevaluations of BOB.1/OBF.1-deficient mice revealed normal T-cell development in the thymus but reduced T-cell numbers, both of CD4⁺ and CD8⁺ T cells, in the periphery [11]. Further studies then showed that BOB.1/OBF.1 is relevant for CD4⁺ T-helper cell (Th) and memory T-cell development and function [11, 12, 78–80], which will be outlined in more details in the following.

3.5.1 The Role of BOB.1/OBF.1 in T-helper Cells

Since a reduction of peripheral T-cell numbers in BOB.1/OBF.1-deficient animals was reported, its role for different Th cell subsets was subsequently evaluated in more details. First defects in Th cells became evident after analyzing BOB.1/OBF.1-deficient mice infected with *Leishmania major* [11]. Thereby, BOB.1/OBF.1 was shown to control the Th1/Th2 cytokine balance as secretion of Th1 cytokine interferon γ (IFNγ) was nearly absent, while the secretion of Th2 cytokine IL13 was increased in the absence of BOB.1/OBF.1 [[11] Fig. 3.2]. As a consequence of this cytokine imbalance, BOB.1/OBF.1-deficient mice are not able to control *Leishmania major* infection [11]. Subsequent reconstitution experiments proved that this defect is CD4⁺ T-cell autonomous [11]. Further in vitro experiments demonstrated reduced production of Th1 cytokines IL2 and IFNγ, while Th2 cytokines IL4 and IL13 were produced in higher amounts in BOB.1/OBF.1-deficient polarized CD4⁺ T cells [11]. Consistently, BOB.1/OBF.1 was shown to activate *Ifng* and *Il2* promoters [11]. Multiple octamer motifs in the *Ifng* promoter could be identified, and binding of BOB.1/OBF.1 together with OCT-TFs could be demonstrated [11]. Moreover, BOB.1/OBF.1 and OCT2 activated the *Ifng* and *Il2* promoters. Later RNA followed by deep sequenc-

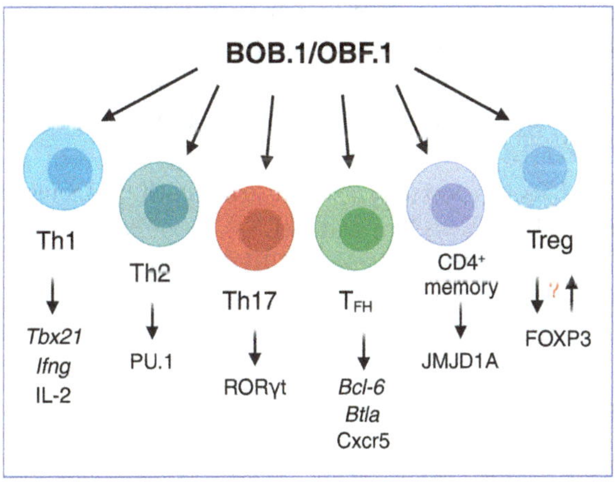

Fig. 3.2 BOB.1/OBF.1 is involved in CD4⁺ T-cell biology at multiple levels. BOB.1/OBF.1 regulates Th1/Th2 cytokine balance. BOB.1/OBF.1 was shown to transactivate the promoters of the Th1 cytokines *Ifng* and *Il2* promoters. Reduced *Tbx21* expression, the master TF of Th1 cytokines, was also reported in the absence of BOB.1/OBF.1. In Th2 cells, BOB.1/OBF.1 controls the TF PU.1, which interferes with the master Th2 regulator GATA3 thereby regulating Th2 cytokine release. BOB.1/OBF.1 also promotes Th17 cell differentiation by interacting with RORγt. BOB.1/OBF.1 is also essential for Tfh cell differentiation as BOB.1/OBF.1-deficient Tfh cells reveal impaired development. Here, BOB.1/OBF.1 was shown to regulate *Bcl6*, *Btla*, and Cxcr5 expression. BOB.1/OBF.1 is also required for CD4⁺ memory T-cell development and function. BOB.1/OBF.1 recruits the histone lysine demethylase JMJD1A to promoters of its target genes to keep them in a poised state for fast re-expression upon restimulation. BOB.1/OBF.1 was also associated with regulatory T-cell (Treg) function, but its role in this cell population is incompletely understood. So far, it is unclear whether and how BOB.1/OBF.1 expression affects the regulation of FOXP3, the master TF of Tregs, and vice versa. (This figure was created with BioRender.com)

ing (RNA-Seq) experiments identified more than 50 direct target genes of BOB.1/OBF.1 in association with OCT1 in restimulated CD4⁺ T cells [12]. These genes included among others multiple cytokines like *Il2* and *Ifng* as well as *Tbx21*, the master regulator of Th1 cells. Recently, a further study confirmed these results by observing reduced *Tbx21* and *Ifng* expression in BOB.1/OBF.1-deficient $CD4^+$ T cells by RNA-Seq [81]. Moreover, genes associated with IFNγ production and with response to IFNγ were differentially expressed in $CD4^+$ T cells in the absence of BOB.1/OBF.1 in this study [81]. On the other hand, no direct regulation of Th2 cytokine promoters by BOB.1/OBF.1 could be found. However, another mechanism by which BOB.1/OBF.1 seems to be involved in regulation of the expression of Th2 cytokines was described. This mechanism involves the TF PU.1, which is expressed in Th2 cells and alleviates Th2 cytokine expression by interfering with GATA3, the master regulator of Th2 cells [82]. Mechanistically, BOB.1/OBF.1-deficient T_H2 cells revealed reduced levels of the transcriptional factor PU.1 resulting in higher GATA3 activity and consequently higher Th2 cytokine release [11]. This mechanism was confirmed by showing that increased BOB.1/OBF.1 expression resulted in increased PU.1 levels and a suppression of both GATA3 expression and Th2 cytokine production [83]. Besides Th1 and Th2, a role for BOB.1/OBF.1 in Th17 differentiation and function was described [78, 79] (Fig. 3.2). Knockdown of BOB.1/OBF.1 led to the reduction of Th17 signature genes, while at the same time, the expression of genes relevant for other Th cell subsets was induced [78]. Polarization experiments suggested that BOB.1/OBF.1 impedes production of IL2, which is a known repressor of Th17 differentiation [78]. Another study confirmed that IL17 differentiation is affected in the absence of the transcriptional coactivator [79]. It could be demonstrated that BOB.1/OBF.1 interacts with the key Th17 TF RORγt, thereby regulating IL17 transcription [79]. In doing so, BOB.1/OBF.1 might contribute to T_H17-mediated autoimmune disorders, as BOB.1/OBF.1-deficient mice are resistant to experimental autoimmune encephalomyelitis (EAE) [79]. In line with the above described findings, BOB.1/OBF.1 expression was later on also confirmed in human Th1, Th2, and cells from tonsils and blood [80].

3.5.2 The Role of BOB.1/OBF.1 in T Follicular Helper (Tfh) Cells

Highest expression of BOB.1/OBF.1 in the T-cell compartment is found in T follicular helper (Tfh) cells in mice and humans [13, 80, 84]. However, the role of BOB.1/OBF.1 for Tfh cell development and function has long been controversially discussed. In 2012, it was first reported that BOB.1/OBF.1-deficient animals had reduced Tfh cells following infection with influenza virus [85]. However, subsequent reconstitution experiments performed in the same study pointed toward the conclusion that the observed Tfh cell defect might be B-cell mediated [85]. In 2014, BOB.1/OBF.1 deficiency was shown to prevent the differentiation of $CD4^+$ T cells into Tfh cells in an autoimmune mouse model [86]. Two years later, an in-depth study analyzing the contribution of BOB.1/OBF.1 for Tfh cell development was conducted [13]. Therein it was demonstrated that various Tfh cell subsets are significantly reduced in SLOs including spleen, lymph nodes (LNs), and Peyer's patches (PPs) of BOB.1/OBF.1-deficient mice after immunization [13]. Moreover, expression levels of CXCR5 on $CD4^+$ T cells were markedly reduced in the absence of BOB.1/OBF.1 similar to the reduction already described for B cells [13]. As it was already suggested that BOB.1/OBF.1 is involved in regulating *Cxcr5* promoter activities in B cells [62], these findings indicate that BOB.1/OBF.1 also promotes CXCR5 expression in $CD4^+$ T cells. By performing reconstitution experiments, the study further provided evidence that the observed impaired Tfh cell development in the absence of the transcriptional coactivator is a $CD4^+$ T-cell autonomous defect [13]. These results ruled out the possibility that reduced Tfh cell numbers are only a result of impaired B-T-cell interactions. Mechanistically, the study revealed that BOB.1/

OBF.1 together with OCT1/2 binds to and activates *Bcl6* and *Btla* promoters [13]. BCL6 is the master regulator of Tfh cell development. BCL6 is required already for initiation of Tfh cell differentiation, as its expression represses other TFs that would favor differentiation toward a Th cell phenotype [87]. Besides, its expression is also essential at later stages for Tfh cell maintenance [88]. BTLA and CXCR5 are both highly expressed in Tfh cells. Altogether, BOB.1/OBF.1 might be involved in Tfh cell differentiation by regulating among others the expression of BCL6, BTLAa, and CXCR5 (Fig. 3.2).

In the same year, another study contradicted these findings to some extent [80]. Here, CXCR5$^+$ PD-1$^+$ Tfh cells in BOB.1/OBF.1-deficient animals were analyzed, and an increase of this cell population in spleens of mutant mice could be observed [80]. As a result of these findings, authors concluded that the function of BOB.1/OBF.1 is rather the restriction of Tfh cell numbers than the promotion of their differentiation and development. In 2022, it was shown that Tfh differentiation is blocked at early maturation stages when BOB.1/OBF.1 is absent in hematopoietic cells [89]. However, a T-cell extrinsic role of BOB.1/OBF:1 for Tfh cell differentiation was proposed. In detail, it was assumed that BOB.1/OBF.1 expression in B cells enables Tfh cell differentiation [89].

Also in 2022, another study established a mouse model, which allowed the conditional deletion of BOB.1/OBF.1 precisely in T or more specifically in Tfh cells [81]. Significantly reduced levels of multiple Tfh cell subsets were found in spleens and LNs in both conditional knockouts (KOs) after immunization, supporting a T-cell intrinsic function of BOB.1/OBF.1 for Tfh cell development [81]. Particularly, the reduction of Tfh cells in mice in which BOB.1/OBF.1 was specifically deleted in this same cell population proves that impaired Tfh cell development is a Tfh cell intrinsic defect. These results show that decreased Tfh cell levels are not exclusively a consequence of reduced CD4$^+$ T cells prior to GC formation or defective CD4$^+$ T-cell differentiation into Tfh cells.

3.5.3 Role of BOB.1/OBF.1 in T cells for the GC Reaction

As described in previous subchapters, the most prominent defect of BOB.1/OBF.1-deficient mice is the complete absence of GCs. Long time, this defect was mainly attributed to the defective B-cell compartment, while the contribution of BOB.1/OBF.1 expression in T cells for the GC reaction remained unknown. To resolve the question whether and how BOB.1/OBF.1 expression in T cells contributes to the GC reaction, T-cell-specific *Pou2af1* conditional KO mice were analyzed. Here, mice in which the transcriptional coactivator was deleted either in all T cells or precisely in Tfh cells were immunized and analyzed regarding GC formation [81]. In doing so, it could be shown that deleting BOB.1/OBF.1 in T and Tfh cells led to impaired GC formation [81]. These findings clearly indicate that BOB.1/OBF.1 expression in T cells is also required for efficient GC formation. As a consequence, the contribution of BOB.1/OBF.1 to the GC reaction cannot be exclusively attributed to its expression in B cells. Moreover, this study provided data suggesting that BOB.1/OBF.1 expression in T cells is already required during GC initiation [81]. However, possible target genes of BOB.1/OBF.1 in CD4$^+$ T cells that might contribute to its regulatory function for GC formation remain to be determined.

3.5.4 Role of BOB.1/OBF.1 in CD4$^+$ Memory T Cells

In 2015, one study identified a requirement of BOB.1/OBF.1 for CD4$^+$ memory T-cell development and function ([12] Fig. 3.2). By analyzing the immune response to viral infection, the study revealed reduced memory T-cell numbers and impaired recall responses in the absence of BOB.1/OBF.1 [12]. Reconstitution experiments ruled out a B-cell-mediated effect on T-cell memory responses [12]. Further RNA-Seq experiments identified over 50 direct target genes of BOB.1/OBF.1 in association with OCT1 in restimulated CD4$^+$ T cells [12]. Generally, over

100 genes that are strongly induced after restimulation in the wild-type (WT) situation were markedly reduced in BOB.1/OBF.1-deficient cells [12]. These included among others genes encoding various cytokines, cytotoxic T lymphocyte effector molecules, and stem cell markers [12]. Expression of *Ifng* and *Il2*, which have already been previously reported to be diminished in BOB.1/OBF.1-deficient T cells [11], was also reduced upon restimulation compared to controls [12]. Further mechanistic investigations revealed that BOB.1/OBF.1 recruits the histone lysine demethylase JMJD1A to promoters of its target genes like *Il2* or *Ifng* [12]. JMJD1A promotes transcriptional activation as it demethylates H3K9me1 and H3K9me2, repressive histone marks impeding transcription [90]. The study could show that BOB.1/OBF.1 is essential for the association of JMJD1A with the *Il2* promoter, particularly in resting, but previously stimulated $CD4^+$ T cells [12]. BOB.1/OBF.1-deficient T cells fail to localize JMJD1A resulting in H3K9me2 accumulation and failure to express IL2. These findings in association with the observed impaired memory $CD4^+$ T-cell responses allow the conclusion that BOB.1/OBF.1 is required to keep key genes of $CD4^+$ memory T cells in a silent but poised state [12]. This is highly relevant to quickly reinduce gene expression at later stages for robust immunological memory responses. Therefore, BOB.1/OBF.1 is an important regulator of $CD4^+$ memory T-cell responses.

3.5.5 BOB.1/OBF.1 in Regulatory T Cells

As BOB.1/OBF.1 is essential for the GC reaction and important for $CD4^+$ T-cell biology, it is reasonable to assume that the transcriptional coactivator might also contribute to regulatory T- (Treg) and T follicular regulatory (Tfr) cell physiology. The task of Tregs is to maintain immune homeostasis and self-tolerance by suppressing the immune response. Tfr cells are a specialized subpopulation of Tregs that are key players during the GC reaction. Tfr cells can suppress the GC reaction to sustain the balance between immune activation and tolerance [91]. So far, very little is known about the function of BOB.1/OBF.1 in regulatory T-cell populations. Nevertheless, it is thought that BOB.1/OBF.1 has a role in these cells, as different studies reported altered *POU2AF1* expression in regulatory T-cell populations. One study showed that *POU2AF1* is upregulated in circulating human blood Tfr cells in comparison to Tfh and Treg cells [92]. Furthermore, *Pou2af1* expression was suppressed in stimulated murine $FOXP3^+$ cells compared to $FOXP3^-$ cells [93]. In addition, the *Pou2af1* promoter was occupied by FOXP3 indicating that FOXP3 might suppress BOB.1/OBF.1 expression in stimulated $CD4^+$ T cells [93]. On the other hand, $CD4^+$ T cells cultured together with Tregs revealed higher *Pou2af1* expression compared to T cells cultured without Tregs [94]. This finding could suggest that BOB.1/OBF.1 might be a potential transcriptional initiator of immune suppression. Others reported a downregulation of *POU2AF1* in Tregs isolated from patients with chronic infection compared to healthy individuals [95]. Coculture experiments of B cells together with Tfr cells revealed that one of the strongest downregulated genes by Tfr cells in B cells is *Pou2af1* [96]. Further investigations showed chromatin inaccessibility of *Pou2af1* during suppression by Tfr cells [96]. These findings suggest that *Pou2af1* expression in B cells might be suppressed by Tfr cells probably by epigenetic modification [96]. Deletion of BOB.1/OBF.1 in T cells by conditional mutagenesis resulted in enhanced splenic Treg cell levels [81]. In addition, mice showed an impaired antigen-specific immune response [81], which could indicate that BOB.1/OBF.1 expression in regulatory T-cell subsets might balance affinity maturation. The precise role of Tfr cells for affinity maturation is unclear. Initially, Tfr cells were considered to control antigen-specific antibody production, while more recent studies suggest Tfr cells to restrain affinity maturation [97].

In conclusion, the exact function of BOB.1/OBF.1 in regulatory T-cell populations particularly with regard to affinity maturation requires further investigations. Several studies could link T_{reg}/T_{fr} cells with BOB.1/OBF.1 expression, but so far it does not provide a clear picture how BOB.1/OBF.1 is regulated in these cells and whether and how it contributes to immune suppression.

3.6 BOB.1/OBF.1 in Disease

As BOB.1/OBF.1 is an important regulator of the adaptive immune response, it is not surprising that its aberrant expression contributes to several pathologies. A link between divergent BOB.1/OBF.1 expression and several autoimmune and chronic inflammatory disorders, B- and T-cell neoplasms as well as solid tumor entities, has been reported (Fig. 3.3). The contribution of the

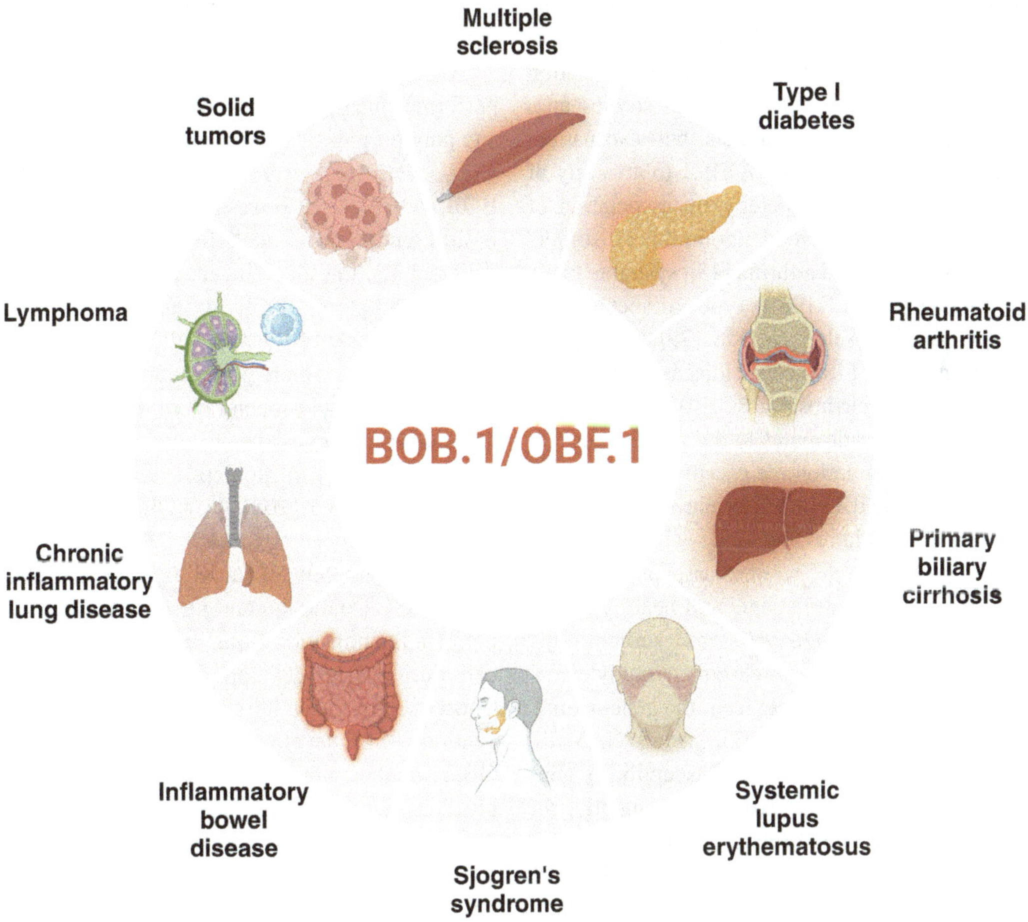

Fig. 3.3 BOB.1/OBF.1 expression is linked to diseases including autoimmunity, inflammation, liquid, and solid tumors. Multiple studies identified a contribution of BOB.1/OBF.1 to the pathology of various autoimmune and chronic inflammatory diseases. BOB.1/OBF.1 is a potential therapeutic target for the treatment of autoimmune disorders. BOB.1/OBF.1 is also aberrantly expressed in B- and T-cell neoplasms. It is highly expressed in GC-derived lymphomas and aberrantly silenced in classical Hodgkin lymphoma. Recently, an association of BOB.1/OBF.1 with various solid tumor entities was described. Therefore, BOB.1/OBF.1 has potential as diagnostic and prognostic biomarker in liquid and solid tumors. (This figure was created with BioRender.com)

transcriptional coactivator to these various diseases will be discussed in more details in the following.

3.7 BOB.1/OBF.1 in Autoimmunity and Chronic Inflammation

BOB.1/OBF.1 is a key transcriptional regulator of lymphocyte-specific genes. Transcriptional dysregulation can contribute to the development of autoimmunity and inflammation. Although the precise contribution of BOB.1/OBF:1 to the pathogenesis of autoimmunity and inflammation is incompletely understood, recent studies analyzing genetics, mouse models, but also patient material could link BOB.1/OBF.1 to a variety of disorders (Fig. 3.3). Thereby, an association of BOB.1/OBF.1 and several autoimmune disorders includes rheumatoid arthritis [48], systemic lupus erythematosus [SLE [98]], type 1 diabetes [T1D [99]], multiple sclerosis [MS [79]], Sjogren's syndrome (SS) [48], celiac disease [100], and primary biliary cirrhosis [PBC [101]]. Moreover, BOB.1/OBF.1 contributes to the pathogenesis of chronic inflammation, e.g., to inflammatory bowel disease [IBD [102]] or chronic inflammatory diseases of lungs and kidneys [103].

In 2003, a whole genome association screen of over 300 Belgian patients and healthy volunteers identified *POU2AF1* as susceptibility marker for MS [104]. A subsequent study evaluating all the data available from the genetic analysis of multiple sclerosis in EuropeanS (GAMES) project validated *POU2AF* as susceptibility gene for MS in Europeans [105]. Consequently, the contribution of BOB.1/OBF.1 to MS was further analyzed in a corresponding mouse model [79]. Here, the EAE model, a commonly used experimental model for MS, was applied to BOB.1/OBF.1-deficient animals [79]. EAE symptoms set in at later timepoints in the absence of the transcriptional coactivator compared to WT controls [79]. Additionally, symptoms are weaker as defined by the EAE score [79]. Lesser demyelination and decreased infiltration of inflammatory cells were also observed in BOB.1/OBF.1-deficient spinal cords [79]. Next the function of BOB.1/OBF.1 in Th17 cells was further addressed in this study, as Th17 cells and their secretion of IL17 are key mediators of MS pathology [106]. Indeed, authors could show that IL17 secretion is impaired in BOB.1/OBF.1-deficient $CD4^+$ T cells [79]. Mechanistically, BOB.1/OBF.1 was shown to interact with the ligand-binding domain of RORγt, the lineage defining TF of Th17 cells [79]. By interacting with RORγt, BOB.1/OBF.1 participates in the transcriptional regulation of *Il17* thereby playing a central role in Th17-mediated autoimmunity including MS.

Another big genome-wide association study performed in 2012 could associate OCT1-binding site polymorphisms with predisposition for SLE, T1D, PBC, and RA [107]. Contribution of BOB.1/OBF.1 to SLE was also further examined in different mouse models for the disease. Mice deficient for the zinc finger (ZF) TF AIOLOS develop symptoms similar to human SLE [108]. One study crossed AIOLOS- with BOB.1/OBF.1-deficient mice, which prevented SLE development. These double-mutant mice did not produce any autoantibodies, and no formation of immune-complex deposits in glomeruli of mice kidneys could be detected [108]. In addition, GCs that spontaneously develop in AIOLOS-single mutant mice were no longer found in double-mutant mice [108]. Double-mutant mice also revealed a block in the transition from pre-B to immature B cells during B-cell development in the BM [108]. Consequently, the authors concluded that the prevention of SLE development in AIOLOS/BOB.1/OBF.1-double mutant mice might be due to a block of autoreactive B-cell development [108]. A subsequent study by the same group this time crossed BOB.1/OBF.1-deficient mice with MRL-lpr mice, which is another well-described model for SLE [98]. Also, in the MRL-lpr model, additional BOB.1/OBF.1 deficiency prevented production of autoantibodies, development of hypergammaglobulinemia, immune-complex-driven glomerulonephritis, as well as early mortality [98]. In contrast to the previously studied AIOLOS/BOB.1/OBF.1-deficeint model, B-cell development was unaffected in the BOB-1/OBF.1-deficient MRL-lpr mice [98]. Instead, severely reduced levels of antibody secreting cells could be observed in these mice. Therefore, authors suggested that BOB.1/OBF.1 could contribute to SLE by promoting the generation of autoantibody-secreting cells. Even though the precise mechanism is not completely resolved, both studies provide evidence for an involvement of BOB.1/OBF.1 in SLE pathogenesis.

As previously mentioned, OCT1 binding site polymorphisms are also associated with a predisposition for T1D. Additionally, BOB.1/OBF.1 is critical for the CD4^{+} T-cell memory response, as described in more detail in the previous chapter [12], which can drive autoimmunity. In line, the Immunological Genome Project revealed enhanced BOB.1/OBF.1 expression levels in central memory but also pancreas-infiltrating, islet-reactive CD4^{+} T cells [99, 109]. Consequently, the contribution of BOB.1/OBF.1 was further evaluated in an appropriate mouse model. Therefore, BOB.1/OBF.1 was specifically deleted in T cells by conditional mutagenesis, and these mice were further backcrossed to nonobese diabetic (NOD) mice, a common animal model for T1D [99]. The study could show that NOD mice lacking BOB.1/OBF.1 expression in T cells were protected from spontaneous T1D indicating that BOB.1/OBF.1 expression in T cells is central for the pathogenesis of T1D [99]. TCR clonotype analysis revealed that BOB.1/OBF.1 ablation in T cells shifted the TCR landscape of islet effector/memory-like T cells to an anti-diabetogenic phenotype [99]. Autoreactive CD8^{+} T cells are still found in pancreatic LNs of T-cell conditional BOB.1/OBF.1-deficient mice. However, peptide stimulation experiments revealed that these CD8^{+} T cells fail to become activated and show reduced islet penetration [99]. In contrast, autoreactive CD4^{+} T-cell clones were present in conditional KO mice but were associated with anergic phenotypes [99]. As genetic ablation of BOB.1/OBF.1 in T cells impeded T1D development, the study next aimed to design a BOB.1/OBF.1 peptide inhibitor. The researchers managed to develop a membrane-permeable peptide inhibitor, which prevents the interaction of BOB.1/OBF.1 with its downstream target JMJD1A. As described in the previous subchapter, BOB.1/OBF.1 interacts with the histone lysine demethylase JMJD1A to keep previously activated genes in a poised state for rapid reexpression after antigen reencounter [12]. Logically, this interaction can be damaging in the context of autoimmunity. Administration of the newly designed inhibitor protected NOD mice from newly arisen T1D. Treatment normalized blood glucose levels and reduced CD8^{+} T-cell autoreactivity as well as the production of pro-inflammatory cytokines [99]. Collectively, this study demonstrates the potential of targeting BOB.1/OBF.1 for treatment of autoimmune disorders. Nevertheless, further studies are required to evaluate possible side effects of therapeutic BOB.1/OBF.1 inhibition particularly regarding its crucial function in B cells.

Another autoimmune disorder BOB.1/OBF.1 was described to contribute to its rheumatoid arthritis. Differential expression of BOB.1/OBF.1 in rheumatoid arthritis patients was described by several studies [48, 70, 110, 111]. Analysis of peripheral blood revealed lower *POU2AF1* mRNA expression in rheumatoid arthritis patients compared to healthy controls [111]. Moreover, a negative correlation between *POU2AF1* and disease stage was reported [111]. Another study could not detect differences in *POU2AF1* mRNA levels between rheumatoid arthritis and patients with spondyloarthritis, a chronic inflammatory joint disease [48]. In contrast, global transcriptome analysis revealed *POU2AF1* as one of the most upregulated genes in synovia of rheumatoid arthritis patients compared to controls [48]. In accordance, BOB.1/OBF.1 protein expression was also increased in rheumatoid arthritis synovial membrane sections [48]. These findings suggest a specific increase of BOB.1/OBF.1 expression at the key site of inflammation, the synovia, of rheumatoid arthritis patients. A further study also revealed significantly elevated *POU2AF1* mRNA levels in LNs of rheumatoid arthritis patients compared to the healthy control group further supporting an involvement of the transcriptional coactivator in rheumatoid arthritis pathogenesis [70]. To further evaluate the contribution of BOB.1/OBF.1 to the pathogenesis of rheumatoid arthritis, the collagen-induced arthritis (CIA) model, commonly used to study rheumatoid arthritis, was applied to BOB.1/OBF.1-deficient animals [48]. BOB.1/OBF.1-deficient mice did not produce antitype II collagen antibodies and were protected from arthritis [48]. Subsequent reconstitution experiments displayed that BOB.1/OBF.1 expression in B cells drives the autoimmune response and promotes arthritis development in the CIA mouse model [48]. Furthermore, the study provided evidence that enhanced BOB.1/OBF.1 expression is associated with the presence of ectopic GCs in rheumatoid arthritis synovitis but also in inflamed salivary glands of SS patients [48]. Overexpression experiments showed that high BOB.1/OBF.1 expression hinders B-cell differentiation into plasmablasts and instead drives differentiation into memory B cells during T-cell-dependent immune responses [70]. On the other hand, silencing BOB.1/OBF.1 expression by using short hairpin RNA impaired memory B-cell differentiation but promoted plasmablast formation [70]. Mechanistically, the study revealed that aberrantly high BOB.1/OBF.1 levels suppress BCL6 but induce ABF-1 expression, a TF favoring memory B-cell formation [70]. Therefore, it is possible that deregulated BOB.1/OBF.1 expression promotes accumulation of autoreactive memory B cells thereby driving autoimmunity. However, further studies are necessary to clarify the role of BOB.1/OBF.1 for autoimmunity.

Besides autoimmunity, BOB.1/OBF.1 also seems to contribute to chronic inflammatory diseases. With regard to IBD, mRNA levels of *POU2AF1* were reported to be upregulated in the ileum and adjacent mesenteric adipose tissue of patients with Crohn's disease [102]. A polymorphism in the *TNF* promoter resulting in higher TNF production was described to be associated with susceptibility for IBD [112]. In this context, it was shown that OCT1 specifically binds to this polymorphism next to a NF-κB-binding site [112]. Further experiments revealed OCT1 and NF-κB p65 interaction [112]. Therefore, it is possible that interaction of these two proteins on the *TNF* promoter might be relevant to IBD pathogenesis [112]. However, this assumption as well as a potential involvement of BOB.1/OBF.1 requires further investigations.

A meta-analysis of GWAS evaluated serum urea levels in Europeans and identified a polymorphism in the *POU2AF1* gene, which is thought to be associated with urea metabolism and kidney function [113]. Another study identified a further *POU2AF1* polymorphism associated with increased risk of low urine pH and kidney stones [114].

Several studies also describe a role of BOB.1/OBF.1 for various inflammatory lung diseases. Another meta-analysis of GWAS found a *POU2AF1* polymorphism associated with chronic obstructive pulmonary disease (COPD) [115]. Analysis of public data sets revealed *POU2AF1* downregulation in the blood of patients with end-stage chronic respiratory diseases including cystic fibrosis, idiopathic pulmonary fibrosis (IPF), and pulmonary hypertension compared to healthy controls [116–118]. In addition, *POU2AF1* was identified as predictive biomarker of bronchiolitis obliterans syndrome [116]. Downregulation of *POU2AF1* could predict the disease 6 months before its onset [116]. In silico analyses predicted *POU2AF1* as a regulator of the profibrotic transcriptional program in IPF [103]. Subsequently a possible link between BOB.1/OBF.1 and fibrosis was investigated by using the bleomycin mouse model, the most widely used pulmonary fibrosis model. Administration of bleomycin induces inflammation and fibrosis in murine lungs. BOB.1/OBF.1-deficient animals revealed less fibrotic severity in this model compared to WT controls [103]. Immunohistochemical staining of human lungs of IPF patients localized BOB.1/OBF.1 in lymphocyte aggregates found in fibrotic regions [103]. Additional RNA-Seq analysis revealed that BOB.1/OBF.1 is elevated in B cells that infiltrate lungs of IPF patients [103].

Conclusively, multiple studies identified a contribution of BOB.1/OBF.1 to various autoimmune and inflammatory diseases. Initial attempts have even been made to target BOB.1/OBF.1 as a therapeutic option. However, further research is needed to expand our understanding of how

BOB.1/OBF.1 drives autoimmunity and inflammation and how this knowledge could be used therapeutically.

3.8 BOB.1/OBF.1 in Lymphoma

Dysregulated BOB.1/OBF.1 expression is also associated with liquid tumors. As outlined in previous subchapters, BOB.1/OBF.1 is essential for the GC reaction. The GC reaction is orchestrated by a bunch of TFs that become activated or inactivated in distinct phases of the reaction. These transcriptional regulators can be hijacked to promote the development of GC-derived lymphomas. Most non-Hodgkin lymphomas (NHLs) derive from GC B cells including diffuse large B-cell lymphoma (DLBCL), follicular lymphoma (FL), and Burkitt lymphoma (BL) [119, 120]. As the transcriptional coactivator BOB.1/OBF.1 is critical for GC B cells, it is not surprising that its expression can also be detected in GC-derived lymphomas.

GC-derived B-cell lymphomas including BL, FL, and DLBCL revealed strong BOB.1/OBF.1 expression [14, 121, 122]. In non-GC-derived lymphomas, only low or no BOB.1/OBF.1 expression could be detected [14]. In general, this reflects the physiological situation where BOB.1/OBF.1 is highly expressed in cycling GC B cells, but low in resting B cells [14]. These findings do not prove a direct contribution of BOB.1/OBF.1 to B-cell lymphoma pathogenesis but still describe its potential as a possible diagnostic marker. Moreover, BOB.1/OBF.1 and OCT2 were identified as reliable markers in rare B-cell lymphomas that do not express conventional B-cell markers including anaplastic lymphoma kinase (ALK)-positive large B-cell lymphoma, plasmablastic lymphoma, and primary effusion lymphoma [123]. These findings strengthen the potential of BOB.1/OBF.1 as a diagnostic tool for B-cell lymphomas. Supportively, a requirement of BOB.1/OBF.1 for DLBCL cell proliferation and survival was reported [124, 125]. One study identified a DLBCL-specific super-enhancer located near the BOB.1/OBF.1 promoter controlling its expression [124]. Inhibition of this super-enhancer hindered DLBCL proliferation [124]. Later it was demonstrated that BOB.1/OBF.1 seems to mediate its effects via its interaction with OCT2 in DLBCL [125]. Therefore, targeting the BOB.1/OBF.1-OCT2 interaction could be a potent therapeutic strategy in DLBCL [125]. Most recently, it was demonstrated that the fast proliferation of GC-derived lymphoma cells is dependent on BOB.1/OBF.1 [34]. It could be shown that in the absence of the transcriptional coactivator, the expression of IRF4 is enhanced in lymphoma cells [34]. IRF4 promotes the GC exit, initiates plasma cell differentiation, and is a negative regulator of BL proliferation. Thus, BOB.1/OBF.1 represses IRF4 expression and thereby driving GC-derived lymphoma cell proliferation [34]. Interestingly, lymphoma cells with reduced BOB.1/OBF.1 expression adopt features associated with favorable prognosis suggesting that BOB.1/OBF.1 might be a valuable prognostic marker [34]. *POU2AF1* was also identified as a probable target for amplification in multiple myeloma (MM) [59]. MM is characterized by the malignant transformation of plasma cells. Here, knockdown of BOB.1/OBF.1 using small interfering RNA (siRNA) reduced MM cell proliferation [59]. On the other hand, ectopic BOB.1/OBF.1 expression increased MM cell growth [59]. Mechanistically, BOB.1/OBF.1 seems to drive myelomagenesis by transactivating BCMA [59]. It is hypothesized that BOB.1/OBF.1 is overexpressed by amplification in MM acting as a possible oncogene by transactivating BCMA thereby promoting the development of MM [59]. A further study identified BOB.1/OBF.1 as key regulator of MM [126]. Here, BOB.1/OBF.1 was shown to promote MM cell growth by interacting with TFs genes relevant for myelomagenesis, including *IRF4*, *PRDM1*, *IKZF1/3*, and *DOT1L* [126].

In contrast to the above-described NHLs, BOB.1/OBF.1 expression is silenced in classical Hodgkin lymphoma (cHL). The key feature of cHL is the presence of so-called Hodgkin and Reed-Sternberg cells (HRS), the malignant component of cHL [127]. These cells are also derived from GC B cells but lost expression of typical B-cell markers [128]. HRS cells also lack expression of the B-cell-related transcriptional regulators BOB.1/OBF.1 and OCT2 [15, 16, 129, 130]. There is evidence that the downregulation of B-cell-specific genes, including BOB.1/OBF.1 in HRS cells, is due to promoter hypermethylation [131]. HRS cells express rearranged Ig genes but are incapable of transcribing Ig [16]. Initially, it was thought that Ig expression is absent in cHD due to disrupting Ig gene mutations [132]. Later, another study found that HRS have an inactive Ig promoter which seems to be a result of BOB.1/OBF.1 and OCT2 downregulation [16]. Possibly, the silencing of BOB.1/OBF.1 combined with additional mutations might have a causative role in cHL.

Besides aberrant expression, BOB.1/OBF.1 chromosomal rearrangements or mutations involved in lymphomagenesis have been reported. Fusion of LAZ3/BCL6 and BOB.1/OBF.1 genes by chromosomal translocation has been revealed [133–135]. Additionally, a germline variation in the 3′-untranslated region (UTR) of the *POU2AF1* gene was shown to be associated with susceptibility to lymphoma [136]. In FL, *POU2F2*, encoding OCT2, and *POU2AF1* were also found to be significantly mutated [137].

Additionally, a chromosomal rearrangement of the *POU2AF1* gene was identified in one case of T-cell prolymphocytic leukemia [135]. Subsequent investigations revealed distinct expression of BOB.1/OBF.1 in most tumor cells in various T-cell lymphoma types [77]. However, it is unclear whether BOB.1/OBF.1 is directly involved in neoplastic transformation in T-cell lymphoma or whether its expression is just a secondary event but not involved in malignant transformation [77].

In total, all these studies underline the involvement of BOB.1/OBF.1 in lymphomagenesis, highlighting its diagnostic and therapeutic potential.

3.9 BOB.1/OBF.1 in Solid Tumors

Only recently, more and more studies have been published showing an association of BOB.1/OBF.1 with various solid tumor entities. Here, altered *POU2AF1* expression was linked to melanoma [138], breast cancer [139, 140], gastrointestinal stromal tumors (GISTs) [141], lung cancer [142], renal cell carcinoma (RCC) [143, 144], nasopharyngeal carcinoma [145, 146], and colorectal cancer [147].

POU2AF1 was associated with breast cancer in a study that could define three HER2-positive breast cancer subtypes based on their immune-related signatures [139]. The study identified a gene co-expression network that was associated with a $HER2^{+}$ subtype characterized by high immune cell infiltration resulting in better survival. These gene networks were significantly upregulated in this breast cancer subtype and involved eight TFs including *POU2AF1* [139]. A machine learning approach identified *POU2AF1* as novel potential prognostic biomarker for triple-negative breast cancer (TNBC) [140]. In this study, *POU2AF1* was differentially regulated in TNBC compared to other breast cancer subtypes. Moreover, patients with high *POU2AF1* expression showed longer metastasis-free survival compared to patients with low *POU2AF1* expression [140]. Further pathway enrichment analysis associated *POU2AF1* with signaling pathways including MAPK, PI3-Akt, Wnt, or TGFβ, which all contribute to metastasis [140].

POU2AF1 was also found to be differentially expressed in nasopharyngeal carcinoma compared to control samples [146]. Protein-protein interaction (PPI) networks were built with identified differentially expressed genes (DEGs), and here *POU2AF1* was among the 13 hub genes [146]. Another study aimed to detect abnormally methylated DEGs in nasopharyngeal carcinoma [145]. *POU2AF1* was found to be hypermethylated, and PPI network analysis identified the transcriptional coactivator as one of the top three hub genes [145]. Similar to breast cancer, high *POU2AF1* expression was associated with longer overall survival [145]. Thus, *POU2AF1* could also have potential as a prognostic biomarker for nasopharyngeal carcinoma.

Another study found *POU2AF1* to be significantly downregulated in high-risk GISTs compared to low-risk GISTs emphasizing its potential as prognostic biomarker [141]. Furthermore, *POU2AF1* was identified as cancer-related DEG in lung cancer [142]. In addition, its expression was significantly associated with better overall- and progression-free survival in lung adenocarcinoma [142]. In contrast, in CRC, several studies found a downregulation of *POU2AF1* in tumor compared to normal tissue [147–150]. Moreover, patients with lower *POU2AF1* expression had better survival compared to patients with high expression [147].

Two independent studies identified *POU2AF1* as top hub gene in RCC [143, 144]. Here, *POU2AF1* expression was associated with poor survival in RCC [143, 144]. In metastatic melanoma, POU2AF1 expression was found to be upregulated and described as independent risk factor [138].

A study analyzing whole cancer genomes detected mutation hotspots in the *POU2AF1* enhancer [151]. In three liver cancer patients, G > C mutations were found, while one head and neck squamous cell carcinoma patients revealed a G > A mutation [151]. Because the The G > A mutation creates binding motifs for TFs [151], the identified *POU2AF1* enhancer hotspot is likely to be a cancer driver [151].

Altogether, these studies link BOB.1/OBF.1 to different solid tumor entities and highlight its potential as diagnostic and prognostic biomarker. However, the role of BOB.1/OBF.1 in solid tumor is incompletely understood. Since the transcriptional coactivator is predominantly expressed in lymphocytes, it is reasonable to conclude that *POU2AF1* expression found in solid tumors arises from tumor-infiltrating immune cells. Yet, one study that tried to identify biomarkers for tumor purity detected a ten-gen set, which included *POU2AF1*, that was able to predict tumor purity in various tumor types [152]. Consequently, further research is required to clarify the role of BOB.1/OBF.1 in solid tumors.

Acknowledgments/Funding We would like to thank the Institute of Physiological Chemistry at the Ulm University (Head: Prof. Thomas Wirth) as well as the Department of Otorhinolaryngology at Ulm University Medical Center (Head: Prof. Dr. Thomas K. Hoffmann) for the for long-term and continuous research support. Our own research cited here was largely funded by the German Research Foundation (DFG; BR 2891/4-1 and BR 2891/8-1).

References

1. Brunner CWT (2006) BOB.1/OBF.1—a critical regulator of B cell function. Curr Immunol Rev 2(1):3–12
2. Luo Y, Fujii H, Gerster T, Roeder RG (1992) A novel B cell-derived coactivator potentiates the activation of immunoglobulin promoters by octamer-binding transcription factors. Cell 71(2):231–241
3. Gstaiger M, Knoepfel L, Georgiev O, Schaffner W, Hovens CM (1995) A B-cell coactivator of octamer-binding transcription factors. Nature 373(6512):360–362
4. Luo Y, Roeder RG (1995) Cloning, functional characterization, and mechanism of action of the B-cell-specific transcriptional coactivator OCA-B. Mol Cell Biol 15(8):4115–4124
5. Strubin M, Newell JW, Matthias P (1995) OBF-1, a novel B cell-specific coactivator that stimulates immunoglobulin promoter activity through association with octamer-binding proteins. Cell 80(3):497–506
6. Pfisterer P, Zwilling S, Hess J, Wirth T (1995) Functional characterization of the murine homolog of the B cell-specific coactivator BOB.1/OBF.1. J Biol Chem 270(50):29870–29880
7. Schubart DB, Rolink A, Kosco-Vilbois MH, Botteri F, Matthias P (1996) B-cell-specific coactivator OBF-1/OCA-B/Bob1 required for immune response and germinal centre formation. Nature 383:538
8. Hess J, Nielsen PJ, Fischer KD, Bujard H, Wirth T (2001) The B lymphocyte-specific coactivator BOB.1/OBF.1 is required at multiple stages of B-cell development. Mol Cell Biol 21(5):1531–1539
9. Nielsen PJ, Georgiev O, Lorenz B, Schaffner W (1996) B lymphocytes are impaired in mice lacking the transcriptional co-activator Bob1/OCA-B/OBF1. Eur J Immunol 26(12):3214–3218
10. Samardzic T, Marinkovic D, Nielsen PJ, Nitschke L, Wirth T (2002) BOB.1/OBF.1 deficiency affects marginal-zone B-cell compartment. Mol Cell Biol 22(23):8320–8331
11. Brunner C, Sindrilaru A, Girkontaite I, Fischer KD, Sunderkötter C, Wirth T (2007) BOB.1/OBF.1 controls the balance of TH1 and TH2 immune responses. EMBO J 26(13):3191–3202
12. Shakya A, Goren A, Shalek A, German CN, Snook J, Kuchroo VK et al (2015) Oct1 and OCA-B are selectively required for CD4 memory T cell function. J Exp Med 212(12):2115–2131
13. Stauss D, Brunner C, Berberich-Siebelt F, Hopken UE, Lipp M, Muller G (2016) The transcriptional coactivator Bob1 promotes the development of follicular T helper cells via Bcl6. EMBO J 35(8):881–898
14. Greiner A, Muller KB, Hess J, Pfeffer K, Muller-Hermelink HK, Wirth T (2000) Up-regulation of BOB.1/OBF.1 expression in normal germinal center B cells and germinal center-derived lymphomas. Am J Pathol 156(2):501–507

15. Re D, Müschen M, Ahmadi T, Wickenhauser C, Staratschek-Jox A, Holtick U et al (2001) Oct-2 and Bob-1 deficiency in Hodgkin and Reed Sternberg cells. Cancer Res 61(5):2080–2084
16. Stein H, Marafioti T, Foss H-D, Laumen H, Hummel M, Anagnostopoulos I et al (2001) Down-regulation of BOB.1/OBF.1 and Oct2 in classical Hodgkin disease but not in lymphocyte predominant Hodgkin disease correlates with immunoglobulin transcription. Blood 97(2):496–501
17. Kury P, Staniek J, Wegehaupt O, Janowska I, Eckenweiler M, Korinthenberg R et al (2021) Agammaglobulinemia with normal B-cell numbers in a patient lacking Bob1. J All Clin Immunol
18. Staudt LM, Singh H, Sen R, Wirth T, Sharp PA, Baltimore D (1986) A lymphoid-specific protein binding to the octamer motif of immunoglobulin genes. Nature 323(6089):640–643
19. He X, Treacy MN, Simmons DM, Ingraham HA, Swanson LW, Rosenfeld MG (1989) Expression of a large family of POU-domain regulatory genes in mammalian brain development. Nature 340(6228):35–41
20. Lillycrop KA, Latchman DS (1992) Alternative splicing of the Oct-2 transcription factor RNA is differentially regulated in neuronal cells and B cells and results in protein isoforms with opposite effects on the activity of octamer/TAATGARAT-containing promoters. J Biol Chem 267(35):24960–24965
21. Falkner FG, Zachau HG (1984) Correct transcription of an immunoglobulin kappa gene requires an upstream fragment containing conserved sequence elements. Nature 310(5972):71–74
22. Parslow TG, Blair DL, Murphy WJ, Granner DK (1984) Structure of the 5′ ends of immunoglobulin genes: a novel conserved sequence. Proc Natl Acad Sci U S A 81(9):2650–2654
23. Mueller K, Quandt J, Marienfeld RB, Weihrich P, Fiedler K, Claussnitzer M et al (2013) Octamer-dependent transcription in T cells is mediated by NFAT and NF-κB. Nucl Acid Res 41(4):2138–2154
24. Hermanson GG, Briskin M, Sigman D, Wall R (1989) Immunoglobulin enhancer and promoter motifs 5′ of the B29 B-cell-specific gene. Proc Natl Acad Sci U S A 86(19):7341–7345
25. Thévenin C, Lucas BP, Kozlow EJ, Kehrl JH (1993) Cell type- and stage-specific expression of the CD20/B1 antigen correlates with the activity of a diverged octamer DNA motif present in its promoter. J Biol Chem 268(8):5949–5956
26. Christensen SM, Martin BK, Tan SS, Weis JH (1992) Identification of sites for distinct DNA binding proteins including Oct-1 and Oct-2 in the Cr2 gene. J Immunol 148(11):3610–3617
27. Mahmoud MS, Kawano MM (1996) Cloning and analysis of the human Pax-5 gene promoter. Biochem Biophys Res Comm 228(1):159–164
28. Gunderson SI, Murphy JT, Knuth MW, Steinberg TH, Dahlberg JH, Burgess RR (1988) Binding of transcription factors to the promoter of the human U1 RNA gene studied by footprinting. J Biol Chem 263(33):17603–17610
29. Sive HL, Heintz N, Roeder RG (1986) Multiple sequence elements are required for maximal in vitro transcription of a human histone H2B gene. Mol Cell Biol 6(10):3329–3340
30. Gstaiger M, Georgiev O, van Leeuwen H, van der Vliet P, Schaffner W (1996) The B cell coactivator Bob1 shows DNA sequence-dependent complex formation with Oct-1/Oct-2 factors, leading to differential promoter activation. EMBO J 15(11):2781–2790
31. Cepek KL, Chasman DI, Sharp PA (1996) Sequence-specific DNA binding of the B-cell-specific coactivator OCA-B. Genes Dev 10(16):2079–2088
32. Chasman D, Cepek K, Sharp PA, Pabo CO (1999) Crystal structure of an OCA-B peptide bound to an Oct-1 POU domain/octamer DNA complex: specific recognition of a protein-DNA interface. Genes Dev 13(20):2650–2657
33. Brunner C, Laumen H, Nielsen PJ, Kraut N, Wirth T (2003) Expression of the aldehyde dehydrogenase 2-like gene is controlled by BOB.1/OBF.1 in B lymphocytes. J Biol Chem 278(46):45231–45239
34. Song S, Cao C, Choukrallah MA, Tang F, Christofori G, Kohler H et al (2021) OBF1 and Oct factors control the germinal center transcriptional program. Blood 137(21):2920–2934
35. Corcoran LM, Hasbold J, Dietrich W, Hawkins E, Kallies A, Nutt SL et al (2005) Differential requirement for OBF-1 during antibody-secreting cell differentiation. J Exp Med 201(9):1385–1396
36. Reményi A, Tomilin A, Schöler HR, Wilmanns M (2002) Differential activity by DNA-induced quarternary structures of POU transcription factors. Biochem Pharm 64(5-6):979–984
37. Klemm JD, Rould MA, Aurora R, Herr W, Pabo CO (1994) Crystal structure of the Oct-1 POU domain bound to an octamer site: DNA recognition with tethered DNA-binding modules. Cell 77(1):21–32
38. Sauter P, Matthias P (1998) Coactivator OBF-1 makes selective contacts with both the POU-specific domain and the POU homeodomain and acts as a molecular clamp on DNA. Mol Cell Biol 18(12):7397–7409
39. Botquin V, Hess H, Fuhrmann G, Anastassiadis C, Gross MK, Vriend G et al (1998) New POU dimer configuration mediates antagonistic control of an osteopontin preimplantation enhancer by Oct-4 and Sox-2. Genes Dev 12(13):2073–2090
40. Reményi A, Tomilin A, Pohl E, Lins K, Philippsen A, Reinbold R et al (2001) Differential dimer activities of the transcription factor Oct-1 by DNA-induced interface swapping. Mol Cell 8(3):569–580
41. Tomilin A, Reményi A, Lins K, Bak H, Leidel S, Vriend G et al (2000) Synergism with the Coactivator OBF-1 (OCA-B, BOB-1) Is Mediated by a Specific POU Dimer Configuration. Cell 103(6):853–864
42. Yeremenko N, Danger R, Baeten D, Tomilin A, Brouard S (2021) Transcriptional regulator BOB.1: molecular mechanisms and emerging role in chronic

inflammation and autoimmunity. Autoimmun Rev 20(6):102833
43. Yu X, Wang L, Luo Y, Roeder RG (2001) Identification and characterization of a novel OCA-B isoform: implications for a role in B cell signaling pathways. Immunity 14(2):157–167
44. Siegel R, Kim U, Patke A, Yu X, Ren X, Tarakhovsky A et al (2006) Nontranscriptional regulation of SYK by the coactivator OCA-B is required at multiple stages of B cell development. Cell 125(4):761–774
45. Yu X, Siegel R, Roeder RG (2006) Interaction of the B cell-specific transcriptional coactivator OCA-B and galectin-1 and a possible role in regulating BCR-mediated B cell proliferation. J Biol Chem 281(22):15505–15516
46. Schubart DB, Sauter P, Massa S, Friedl EM, Schwarzenbach H, Matthias P (1996) Gene structure and characterization of the murine homologue of the B cell-specific transcriptional coactivator OBF-1. Nucleic Acids Res 24(10):1913–1920
47. Qin XF, Reichlin A, Luo Y, Roeder RG, Nussenzweig MC (1998) OCA-B integrates B cell antigen receptor-, CD40L- and IL 4-mediated signals for the germinal center pathway of B cell development. EMBO J 17(17):5066–5075
48. Levels MJ, Van Tok MN, Cantaert T, Cañete JD, Kroese FG, Germar K et al (2017) The transcriptional coactivator Bob1 is associated with pathologic B cell responses in autoimmune tissue inflammation. Arthrit Rheum 69(4):750–762
49. Tiedt R, Bartholdy BA, Matthias G, Newell JW, Matthias P (2001) The RING finger protein Siah-1 regulates the level of the transcriptional coactivator OBF-1. EMBO J 20(15):4143–4152
50. Kilzheimer M, Quandt J, Langhans J, Weihrich P, Wirth T, Brunner C (2015) NF-κB-dependent signals control BOB.1/OBF.1 and Oct2 transcriptional activity in B cells. Eur J Immunol 45(12):3441–3453
51. Serfling E, Berberich-Siebelt F, Avots A, Chuvpilo S, Klein-Hessling S, Jha MK et al (2004) NFAT and NF-kappaB factors-the distant relatives. Int J Biochem Cell Biol 36(7):1166–1170
52. Boehm J, He Y, Greiner A, Staudt L, Wirth T (2001) Regulation of BOB.1/OBF.1 stability by SIAH. EMBO J 20(15):4153–4162
53. Schubart DB, Rolink A, Schubart K, Matthias P (2000) Cutting edge: lack of peripheral B cells and severe agammaglobulinemia in mice simultaneously lacking Bruton's tyrosine kinase and the B cell-specific transcriptional coactivator OBF-1. J Immun 164(1):18–22
54. Brunner C, Marinkovic D, Klein J, Samardzic T, Nitschke L, Wirth T (2003) B cell-specific transgenic expression of Bcl2 rescues early B lymphopoiesis but not B cell responses in BOB.1/OBF.1-deficient mice. J Exp Med 197(9):1205–1211
55. Samardzic T, Gerlach J, Muller K, Marinkovic D, Hess J, Nitschke L et al (2002) CD22 regulates early B cell development in BOB.1/OBF.1-deficient mice. Eur J Immunol 32(9):2481–2489
56. Stoddart A, Ray RJ, Paige CJ (1997) Analysis of murine CD22 during B cell development: CD22 is expressed on B cell progenitors prior to IgM. Int Immunol 9(10):1571–1579
57. Nitschke L, Carsetti R, Ocker B, Köhler G, Lamers MC (1997) CD22 is a negative regulator of B-cell receptor signalling. Curr Biol CB 7(2):133–143
58. Nitschke L, Floyd H, Ferguson DJ, Crocker PR (1999) Identification of CD22 ligands on bone marrow sinusoidal endothelium implicated in CD22-dependent homing of recirculating B cells. J Exp Med 189(9):1513–1518
59. Zhao C, Inoue J, Imoto I, Otsuki T, Iida S, Ueda R et al (2008) POU2AF1, an amplification target at 11q23, promotes growth of multiple myeloma cells by directly regulating expression of a B-cell maturation factor, TNFRSF17. Oncogene 27(1):63–75
60. Brunner C, Wirth T (2006) Btk expression is controlled by Oct and BOB.1/OBF.1. Nucleic Acids Res 34(6):1807–1815
61. Betzler AC, Fiedler K, Hoffmann TK, Fehling HJ, Wirth T, Brunner C (2021) BOB.1/OBF.1 is required during B-cell ontogeny for B-cell differentiation and germinal center function. Eur J Immunol
62. Wolf I, Pevzner V, Kaiser E, Bernhardt G, Claudio E, Siebenlist U et al (1998) Downstream activation of a TATA-less promoter by Oct-2, Bob1, and NF-kappaB directs expression of the homing receptor BLR1 to mature B cells. J Biol Chem 273(44):28831–28836
63. Betzler AC, Fiedler K, Kokai E, Wirth T, Hoffmann TK, Brunner C (2021) Impaired Peyer's patch development in BOB.1/OBF.1-deficient mice. Eur J Immunol 51(7):1860–1863
64. Ansel KM, Ngo VN, Hyman PL, Luther SA, Förster R, Sedgwick JD et al (2000) A chemokine-driven positive feedback loop organizes lymphoid follicles. Nature 406(6793):309–314
65. Forster R, Pabst O, Bernhardt G (2008) Homeostatic chemokines in development, plasticity, and functional organization of the intestinal immune system. Semin Immunol 20(3):171–180
66. Okada T, Ngo VN, Ekland EH, Forster R, Lipp M, Littman DR et al (2002) Chemokine requirements for B cell entry to lymph nodes and Peyer's patches. J Exp Med 196(1):65–75
67. Su GH, Chen HM, Muthusamy N, Garrett-Sinha LA, Baunoch D, Tenen DG et al (1997) Defective B cell receptor-mediated responses in mice lacking the Ets protein, Spi-B. Embo J 16(23):7118–7129
68. Bartholdy B, Du Roure C, Bordon A, Emslie D, Corcoran LM, Matthias P (2006) The Ets factor Spi-B is a direct critical target of the coactivator OBF-1. Proc Nat Acad Sci 103(31):11665–11670
69. Dominguez-Sola D, Victora GD, Ying CY, Phan RT, Saito M, Nussenzweig MC et al (2012) The proto-oncogene MYC is required for selection in the germinal center and cyclic reentry. Nat Immunol 13(11):1083–1091

70. Levels MJ, Fehres CM, van Baarsen LGM, van Uden NOP, Germar K, O'Toole TG et al (2019) BOB.1 controls memory B-cell fate in the germinal center reaction. J Autoimmun 101:131–144
71. Kim U, Qin XF, Gong S, Stevens S, Luo Y, Nussenzweig M et al (1996) The B-cell-specific transcription coactivator OCA-B/OBF-1/Bob-1 is essential for normal production of immunoglobulin isotypes. Nature 383(6600):542–547
72. Stevens S, Ong J, Kim U, Eckhardt LA, Roeder RG (2000) Role of OCA-B in 3′-IgH enhancer function. J Immunol 164(10):5306–5312
73. Tang H, Sharp PA (1999) Transcriptional Regulation of the Murine 3′ IgH Enhancer by OCT-2. Immunity 11(5):517–526
74. Sauter P, Matthias P (1997) The B cell-specific coactivator OBF-1 (OCA-B, Bob-1) is inducible in T cells and its expression is dispensable for IL-2 gene induction. Immunobiology 198(1-3):207–216
75. Zwilling S, Dieckmann A, Pfisterer P, Angel P, Wirth T (1997) Inducible expression and phosphorylation of coactivator BOB.1/OBF.1 in T cells. Science 277(5323):221–225
76. Moriuchi M, Moriuchi H (2001) Octamer transcription factors up-regulate the expression of CCR5, a coreceptor for HIV-1 entry *. J Biol Chem 276(12):8639–8642
77. Marafioti T, Ascani S, Pulford K, Sabattini E, Piccioli M, Jones M et al (2003) Expression of B-lymphocyte-associated transcription factors in human T-cell neoplasms. Am J Pathol 162(3):861–871
78. Yosef N, Shalek AK, Gaublomme JT, Jin H, Lee Y, Awasthi A et al (2013) Dynamic regulatory network controlling TH17 cell differentiation. Nature 496(7446):461–468
79. Ikegami I, Takaki H, Kamiya S, Kamekura R, Ichimiya S (2019) Bob1 enhances RORγt-mediated IL-17A expression in Th17 cells through interaction with RORγt. Biochem Biophys Res Commun 514(4):1167–1171
80. Yamashita K, Kawata K, Matsumiya H, Kamekura R, Jitsukawa S, Nagaya T et al (2016) Bob1 limits cellular frequency of T-follicular helper cells. Eur J Immunol 46(6):1361–1370
81. Betzler AC, Ezić J, Abou Kors T, Hoffmann TK, Wirth T, Brunner C (2022) T cell specific BOB.1/OBF.1 expression promotes germinal center response and T helper cell differentiation. Front Immunol 13:889564
82. Chang H-C, Zhang S, Thieu VT, Slee RB, Bruns HA, Laribee RN et al (2005) PU.1 expression delineates heterogeneity in primary Th2 cells. Immunity 22(6):693–703
83. Mattes J, Collison A, Plank M, Phipps S, Foster PS (2009) Antagonism of microRNA-126 suppresses the effector function of TH2 cells and the development of allergic airways disease. Proc Natl Acad Sci U S A 106(44):18704–18709
84. Locci M, Havenar-Daughton C, Landais E, Wu J, Kroenke Mark A, Arlehamn Cecilia L et al (2013) Human circulating PD-1+CXCR3-CXCR5+ memory Tfh cells are highly functional and correlate with broadly neutralizing HIV antibody responses. Immunity 39(4):758–769
85. Karnowski A, Chevrier S, Belz GT, Mount A, Emslie D, D'Costa K et al (2012) B and T cells collaborate in antiviral responses via IL-6, IL-21, and transcriptional activator and coactivator, Oct2 and OBF-1. J Experimen Med 209(11):2049–2064
86. Chevrier S, Kratina T, Emslie D, Karnowski A, Corcoran LM (2014) Germinal center-independent, IgM-mediated autoimmunity in sanroque mice lacking Obf1. Immunol Cell Biol 92(1):12–19
87. Choi J, Crotty S (2021) Bcl6-mediated transcriptional regulation of follicular helper T cells (TFH). Trends Immunol 42(4):336–349
88. Ise W, Inoue T, McLachlan JB, Kometani K, Kubo M, Okada T et al (2014) Memory B cells contribute to rapid Bcl6 expression by memory follicular helper T cells. Proc Nat Acad Sci 111(32):11792–11797
89. Lombard-Vadnais F, Lacombe J, Chabot-Roy G, Ferron M, Lesage S (2022) OCA-B does not act as a transcriptional coactivator in T cells. Immun Cell Biol 100(5):338–351
90. Goda S, Isagawa T, Chikaoka Y, Kawamura T, Aburatani H (2013) Control of histone H3 lysine 9 (H3K9) methylation state via cooperative two-step demethylation by Jumonji domain containing 1A (JMJD1A) homodimer. J Biol Chem 288(52):36948–36956
91. Chung Y, Tanaka S, Chu F, Nurieva RI, Martinez GJ, Rawal S et al (2011) Follicular regulatory T cells expressing Foxp3 and Bcl-6 suppress germinal center reactions. Nat Med 17(8):983–988
92. Wing JB, Kitagawa Y, Locci M, Hume H, Tay C, Morita T et al (2017) A distinct subpopulation of CD25(-) T-follicular regulatory cells localizes in the germinal centers. Proc Natl Acad Sci U S A 114(31):E6400–E64e9
93. Marson A, Kretschmer K, Frampton GM, Jacobsen ES, Polansky JK, MacIsaac KD et al (2007) Foxp3 occupancy and regulation of key target genes during T-cell stimulation. Nature 445(7130):931–935
94. Sukiennicki TL, Fowell DJ (2006) Distinct molecular program imposed on CD4+ T cell targets by CD4+CD25+ regulatory T cells. J Immunol (Baltimore, Md: 1950) 177(10):6952–6961
95. Metenou S, Coulibaly YI, Sturdevant D, Dolo H, Diallo AA, Soumaoro L et al (2014) Highly heterogeneous, activated, and short-lived regulatory T cells during chronic filarial infection. Eur J Immunol 44(7):2036–2047
96. Sage PT, Ron-Harel N, Juneja VR, Sen DR, Maleri S, Sungnak W et al (2016) Suppression by T(FR) cells leads to durable and selective inhibition of B cell effector function. Nat Immunol 17(12):1436–1446
97. Stebegg M, Kumar SD, Silva-Cayetano A, Fonseca VR, Linterman MA, Graca L (2018) Regulation of the germinal center response. Front Immunol 9:2469

98. Zuo J, Ge H, Zhu G, Matthias P, Sun J (2007) OBF-1 is essential for the generation of antibody-secreting cells and the development of autoimmunity in MRL-lpr mice. J Autoimmun 29(2-3):87–96
99. Kim H, Perovanovic J, Shakya A, Shen Z, German CN, Ibarra A et al (2021) Targeting transcriptional coregulator OCA-B/Pou2af1 blocks activated autoreactive T cells in the pancreas and type 1 diabetes. J Exp Med 218(3)
100. Trynka G, Hunt KA, Bockett NA, Romanos J, Mistry V, Szperl A et al (2011) Dense genotyping identifies and localizes multiple common and rare variant association signals in celiac disease. Nat Gen 43(12):1193–1201
101. Nakamura M, Nishida N, Kawashima M, Aiba Y, Tanaka A, Yasunami M et al (2012) Genome-wide association study identifies TNFSF15 and POU2AF1 as susceptibility loci for primary biliary cirrhosis in the Japanese population. Am J Human Gen 91(4):721–728
102. da Silva FAR, Pascoal LB, Dotti I, Setsuko Ayrizono ML, Aguilar D, Rodrigues BL et al (2020) Whole transcriptional analysis identifies markers of B, T and plasma cell signaling pathways in the mesenteric adipose tissue associated with Crohn's disease. J Translat Med 18(1):44
103. McDonough JE, Ahangari F, Li Q, Jain S, Verleden SE, Herazo-Maya J et al (2019) Transcriptional regulatory model of fibrosis progression in the human lung. JCI Insight 4(22)
104. Goris A, Sawcer S, Vandenbroeck K, Carton H, Billiau A, Setakis E et al (2003) New candidate loci for multiple sclerosis susceptibility revealed by a whole genome association screen in a Belgian population. J Neuroimmunol 143(1-2):65–69
105. Ban M, Booth D, Heard R, Stewart G, Goris A, Vandenbroeck K et al (2006) Linkage disequilibrium screening for multiple sclerosis implicates JAG1 and POU2AF1 as susceptibility genes in Europeans. J Neuroimmunol 179(1-2):108–116
106. Maddur MS, Miossec P, Kaveri SV, Bayry J (2012) Th17 cells: biology, pathogenesis of autoimmune and inflammatory diseases, and therapeutic strategies. Am J Pathol 181(1):8–18
107. Maurano MT, Humbert R, Rynes E, Thurman RE, Haugen E, Wang H et al (2012) Systematic localization of common disease-associated variation in regulatory DNA. Science 337(6099):1190–1195
108. Sun J, Matthias G, Mihatsch MJ, Georgopoulos K, Matthias P (2003) Lack of the transcriptional coactivator OBF-1 prevents the development of systemic lupus erythematosus-like phenotypes in aiolos mutant mice. J Immunol 170(4):1699–1706
109. Heng TSP, Painter MW, Elpek K, Lukacs-Kornek V, Mauermann N, Turley SJ et al (2008) The immunological genome project: networks of gene expression in immune cells. Nat Immunol 9(10):1091–1094
110. Macías-Segura N, Castañeda-Delgado JE, Bastian Y, Santiago-Algarra D, Castillo-Ortiz JD, Alemán-Navarro AL et al (2018) Transcriptional signature associated with early rheumatoid arthritis and healthy individuals at high risk to develop the disease. PLoS One 13(3):e0194205
111. Romo-García MF, Nava-Ramírez HS, Zapata-Zúñiga M, Macías-Segura N, Santiago-Algarra D, Castillo-Ortiz JD et al (2019) Evaluation of SUMO1 and POU2AF1 in whole blood from rheumatoid arthritis patients and at risk relatives. Int J Immunogen 46(2):59–66
112. van Heel DA, Udalova IA, De Silva AP, McGovern DP, Kinouchi Y, Hull J et al (2002) Inflammatory bowel disease is associated with a TNF polymorphism that affects an interaction between the OCT1 and NF(-kappa)B transcription factors. Human Mol Gen 11(11):1281–1289
113. Thio CHL, Reznichenko A, van der Most PJ, Kamali Z, Vaez A, Smit JH et al (2019) Genome-Wide Association Scan of Serum Urea in European Populations Identifies Two Novel Loci. Am J Nephrol 49(3):193–202
114. Benonisdottir S, Kristjansson RP, Oddsson A, Steinthorsdottir V, Mikaelsdottir E, Kehr B et al (2018) Sequence variants associating with urinary biomarkers. Human Mol Gene 28(7):1199–1211
115. Cho MH, McDonald ML, Zhou X, Mattheisen M, Castaldi PJ, Hersh CP et al (2014) Risk loci for chronic obstructive pulmonary disease: a genome-wide association study and meta-analysis. Lancet Respirat Med 2(3):214–225
116. Danger R, Royer PJ, Reboulleau D, Durand E, Loy J, Tissot A et al (2017) Blood gene expression predicts bronchiolitis obliterans syndrome. Front Immunol 8:1841
117. Chesné J, Danger R, Botturi K, Reynaud-Gaubert M, Mussot S, Stern M et al (2014) Systematic analysis of blood cell transcriptome in end-stage chronic respiratory diseases. PLoS One 9(10):e109291
118. Herazo-Maya JD, Noth I, Duncan SR, Kim S, Ma SF, Tseng GC et al (2013) Peripheral blood mononuclear cell gene expression profiles predict poor outcome in idiopathic pulmonary fibrosis. Sci Transl Med 5(205):205ra136
119. Basso K, Dalla-Favera R (2015) Germinal centres and B cell lymphomagenesis. Nat Rev Immunol 15(3):172–184
120. Mlynarczyk C, Fontán L, Melnick A (2019) Germinal center-derived lymphomas: The darkest side of humoral immunity. Immunol Rev 288(1):214–239
121. Loddenkemper C, Anagnostopoulos I, Hummel M, Jöhrens-Leder K, Foss H-D, Jundt F et al (2004) Differential Eμ enhancer activity and expression of BOB.1/OBF.1, Oct2, PU.1, and immunoglobulin in reactive B-cell populations, B-cell non-Hodgkin lymphomas, and Hodgkin lymphomas. J Pathol 202(1):60–69
122. Sáez A-I, Artiga M-J, Sánchez-Beato M, Sánchez-Verde L, García J-F, Camacho F-I et al (2002) Analysis of octamer-binding transcription factors Oct2 and Oct1 and their coactivator BOB.1/OBF.1 in Lymphomas. Mod Pathol 15(3):211–220

123. Yin L, Xu J, Li M, Reddy V, Zhou Q, Liu H et al (2016) Oct2 and Bob1 are sensitive and specific markers in lineage determination of B cell lymphomas with no expression of conventional B cell markers. Histopathol 69(5):775–783
124. Chapuy B, McKeown MR, Lin CY, Monti S, Roemer MG, Qi J et al (2013) Discovery and characterization of super-enhancer-associated dependencies in diffuse large B cell lymphoma. Cancer Cell 24(6):777–790
125. Hodson DJ, Shaffer AL, Xiao W, Wright GW, Schmitz R, Phelan JD et al (2016) Regulation of normal B-cell differentiation and malignant B-cell survival by OCT2. Proc Nat Acad Sci 113(14):E2039–E2E46
126. De Matos SR, Shirasaki R, Tang H, Yamano S, Barwick BG, Gandolfi S et al (2020) POU2AF1 as a master regulator of oncogenic transcription factor networks in myeloma. Blood 136(Supplement 1):18–19
127. Küppers R, Klein U, Schwering I, Distler V, Bräuninger A, Cattoretti G et al (2003) Identification of Hodgkin and Reed-Sternberg cell-specific genes by gene expression profiling. J Clin Invest 111(4):529–537
128. Weniger MA, Küppers R (2021) Molecular biology of Hodgkin lymphoma. Leukemia 35(4):968–981
129. Theil J, Laumen H, Marafioti T, Hummel M, Lenz G, Wirth T et al (2001) Defective octamer-dependent transcription is responsible for silenced immunoglobulin transcription in Reed-Sternberg cells. Blood 97(10):3191–3196
130. Hertel CB, Zhou XG, Hamilton-Dutoit SJ, Junker S (2002) Loss of B cell identity correlates with loss of B cell-specific transcription factors in Hodgkin/Reed-Sternberg cells of classical Hodgkin lymphoma. Oncogene 21(32):4908–4920
131. Ushmorov A, Leithäuser F, Sakk O, Weinhaüsel A, Popov SW, Möller P et al (2006) Epigenetic processes play a major role in B-cell-specific gene silencing in classical Hodgkin lymphoma. Blood 107(6):2493–2500
132. Kanzler H, Küppers R, Hansmann ML, Rajewsky K (1996) Hodgkin and Reed-Sternberg cells in Hodgkin's disease represent the outgrowth of a dominant tumor clone derived from (crippled) germinal center B cells. J Exp Med 184(4):1495–1505
133. Galiègue Zouitina S, Quief S, Hildebrand MP, Denis C, Lecocq G, Collyn-d'Hooghe M et al (1996) The B cell transcriptional coactivator BOB1/OBF1 gene fuses to the LAZ3/BCL6 gene by t(3;11)(q27;q23.1) chromosomal translocation in a B cell leukemia line (Karpas 231). Leukemia 10(4):579–587
134. Galiègue-Zouitina S, Quief S, Hildebrand MP, Denis C, Lecocq G, Collyn-d'Hooghe M et al (1995) Fusion of the LAZ3/BCL6 and BOB1/OBF1 genes by t(3; 11) (q27; q23) chromosomal translocation. Comptes rendus de l'Academie des sciences Serie III, Sciences de la vie 318(11):1125–1131
135. Yuille MA, Galiegue-Zouitina S, Hiorns LR, Jadayel D, De Schouwer PJ, Catovsky D et al (1996) Heterogeneity of breakpoints at the transcriptional co-activator gene, BOB-1, in lymphoproliferative disease. Leukemia 10(9):1492–1496
136. Zhai K, Chang J, Hu J, Wu C, Lin D (2017) Germline variation in the 3′-untranslated region of the POU2AF1 gene is associated with susceptibility to lymphoma. Mol Carcinogen 56(8):1945–1952
137. Krysiak K, Gomez F, White BS, Matlock M, Miller CA, Trani L et al (2017) Recurrent somatic mutations affecting B-cell receptor signaling pathway genes in follicular lymphoma. Blood 129(4):473–483
138. Ji C, Li Y, Yang K, Gao Y, Sha Y, Xiao D et al (2020) Identification of four genes associated with cutaneous metastatic melanoma. Open Med (Warsaw, Poland) 15(1):531–539
139. Huang L, Liu X, Li L, Wang L, Wu N, Liu Z (2022) Novel immune subtypes identification of HER2-positive breast cancer based on immunogenomic landscape. Med Oncol (Northwood, London, England) 39(5):92
140. Thalor A, Kumar Joon H, Singh G, Roy S, Gupta D (2022) Machine learning assisted analysis of breast cancer gene expression profiles reveals novel potential prognostic biomarkers for triple-negative breast cancer. Comp Struct Biotechnol J 20:1618–1631
141. Jin S, Zhu W, Li J (2018) Identification of key genes related to high-risk gastrointestinal stromal tumors using bioinformatics analysis. J Cancer Res Therap 14(Suppl):S243–S2s7
142. Ullah MA, Alam S, Moin AT, Ahamed T, Shohael AM (2023) Risk factors and actionable molecular signatures in COVID-19-associated lung adenocarcinoma and lung squamous cell carcinoma patients. Comp Biol Med 158:106855
143. Wan B, Liu B, Huang Y, Lv C (2020) Identification of genes of prognostic value in the ccRCC microenvironment from TCGA database. Mol Gene Genom Med 8(4):e1159
144. Zeng Q, Zhang W, Li X, Lai J, Li Z (2020) Bioinformatic identification of renal cell carcinoma microenvironment-associated biomarkers with therapeutic and prognostic value. Life Sci 243:117273
145. Han B, Yang X, Zhang P, Zhang Y, Tu Y, He Z et al (2020) DNA methylation biomarkers for nasopharyngeal carcinoma. PLoS One 15(4):e0230524
146. Zhang JZ, Wu ZH, Cheng Q (2019) Screening and identification of key biomarkers in nasopharyngeal carcinoma: evidence from bioinformatic analysis. Medicine 98(48):e17997
147. Pan Z, He Y, Zhu W, Xu T, Hu X, Huang P (2021) A dynamic transcription factor signature along the colorectal adenoma-carcinoma sequence in patients with co-occurrent adenoma and carcinoma. Front Oncol 11:597447

148. Pan J, Weng Z, Xue C, Lin B, Lin M (2021) The bioinformatics-based analysis identifies 7 immune-related genes as prognostic biomarkers for colon cancer. Front Oncol 11
149. Guo M, Li X, Li J, Li B (2023) Identification of the prognostic biomarkers and their correlations with immune infiltration in colorectal cancer through bioinformatics analysis and in vitro experiments. Heliyon 9(6):e17101
150. Liu Y, Li C, Dong L, Chen X, Fan R (2020) Identification and verification of three key genes associated with survival and prognosis of COAD patients via integrated bioinformatics analysis. Biosci Rep 40(9)
151. Juul RI, Nielsen MM, Juul M, Feuerbach L, Pedersen JS (2021) The landscape and driver potential of site-specific hotspots across cancer genomes. NPJ Genom Med 6(1):33
152. Li Y, Umbach DM, Bingham A, Li QJ, Zhuang Y, Li L (2019) Putative biomarkers for predicting tumor sample purity based on gene expression data. BMC Genom 20(1):1021

Key Functions of the Transcription Factor BCL6 During T-Cell Differentiation

4

Chara Konstantakopoulou
and Mihalis Verykokakis

Abstract

T lymphocytes consist of several subtypes with distinct functions that help to coordinate an immune response. They are generated within the thymus through a sequential developmental pathway that produces subsets with diverse antigen specificities and functions. Naïve T cells populate peripheral lymphoid organs and are activated upon foreign antigen encounter. While most T cells die soon after activation, a memory population survives and is able to quickly respond to secondary challenges, thus providing long-term immunity to the host. Although cell identity is largely stable and is instructed by cell-specific transcriptional programs, cells may change their transcriptional profiles to be able to adapt to new functionalities. Central to these dynamic processes are transcription factors, which control cell fate decisions, through direct regulation of gene expression. In this book chapter, we review the functions of the transcription factor B-cell lymphoma 6 (BCL6), which directs the fate of several lymphocyte subsets, including helper, cytotoxic, and innate-like T cells, but can also be involved in lymphomagenesis in humans.

Keywords

BCL6 · T lymphocytes · Memory T cells · Differentiation · Transcription

Abbreviations

AP1	Activator protein 1
BATF	Basic leucine zipper transcription factor, ATF-like
BCL6	B-cell lymphoma 6
BCOR	BCL6 corepressor
BLIMP1	B lymphocyte-induced maturation protein-1
OBF1	Oct-binding factor 1
BTB/POZ/ZF	Bric-à-brac, tramtrack, broad complex/Pox virus zinc finger
Bach2	Broad complex-tramtrack-bric a brac and Cap'n'collar homolog 2
CD	Cluster differentiation

C. Konstantakopoulou
Institute for Fundamental Biomedical Research, Biomedical Sciences Research Center Alexander Fleming, Vari, Greece

Department of Antibody Research Materials, Genmab B.V., Utrecht, The Netherlands

M. Verykokakis (✉)
Institute for Fundamental Biomedical Research, Biomedical Sciences Research Center Alexander Fleming, Vari, Greece
e-mail: verykokakis@fleming.gr

T. Borggrefe, B. D. Giaimo (eds.), *Transcription factors in blood cell development*, Advances in Experimental Medicine and Biology 1459, https://doi.org/10.1007/978-3-031-62731-6_4

CLP Common lymphoid progenitor
CTBP1 Carboxy-terminal binding protein 1
CUL1 Cullin 1
CUL3 Cullin 3
CXCR5 C-X-C motif chemokine receptor 5
CCR7 C-C chemokine receptor type 7
DBD DNA-binding domain
DLBCL Diffuse large B-cell lymphoma
DN Double negative
DP Double positive
E4BP4 E4-binding protein 4
EBI2 Epstein–Barr virus-induced gene 2
EGR2 Early growth response 2
EP300/KAT3B E1A-binding protein p300/lysine acetyltransferase 3B
EZH2 Enhancer of zeste homolog 2
FOXP3 Forkhead box P3
GC Germinal center
Gata3 GATA Binding Protein 3
Gpr183 G-protein-coupled receptor 183
HDAC Histone deacetylase
HSC Hematopoietic stem cell
Hk2 Hexokinase 2
ICOS Inducible T-cell co-stimulator
ID2 Inhibitor of DNA-binding 2
IFNγ Interferon gamma
IL Interleukin
Ifngr1 Interferon gamma receptor 1
Il12r Interleukin 12 receptor
Il23r Interleukin 23 receptor
KLRG1 Killer cell lectin-like receptor subfamily G member 1
KO Knockout
Klf2 Kruppel-like factor 2
LCMV Lymphocytic choriomeningitis virus
LEF1 Lymphoid enhancer-binding factor 1
Lck Lymphoid-specific protein-tyrosine kinase
MAIT Mucosal-associated invariant T
MPEC Memory precursor cell
MR1 Major histocompatibility complex, Class I-Related
MTA3 Metastasis-associated protein 3
NCoR1 Nuclear receptor corepressor 1
NKT Natural killer T
HOCT1/ HOCT2 Organic cation uptake transporter 1/ 2
OVA Ovalbumin
PD1 Programmed death 1
PEST Proline (P), glutamic acid (E), serine (S), and threonine (T)
PLZF Promyelocytic leukemia zinc finger
PSGL-1 P-selectin glycoprotein ligand 1
Pdcd1 Programmed cell death 1
Pkm2 Pyruvate kinase M2
Prdm1 PR containing 1, with zinc finger domain
RD2 Repressor domain 2
RORγt Retinoic acid receptor-related orphan receptor γt
RUNX2/3 Runt-related transcription factor 2/3
Rorc Retinoic acid receptor-related orphan receptor gamma
SATB1 Special AT-rich sequence-binding protein-1
SLAM Signaling lymphocytic activation molecule
SMRT Silencing mediator for retinoid and thyroid hormone receptors
STAT Signal transducer and activator of transcription
Selplg Selectin P ligand
Slc2a1 Solute carrier family 2 member 1
Slc2a3 Solute carrier family 2 member 3
Socs3 Suppressor of cytokine signaling 3

Socs1	Suppressor of cytokine signaling 1
Stat5	Signal transducer and activator of transcription 5
T-BET	T-box expressed in T cells
TCF-1	T-cell factor 1
TF	Transcription factor
THPOK	T-helper inducing POZ-Kruppel-like factor
TIM-3	T-cell immunoglobulin and mucin-domain containing-3
TNFα	Tumor necrosis factor-α
TSS	Transcription start site
Tbx21	T-box transcription factor 21
Tcf7	Transcription factor 7
Tfh	T follicular helper
Th	T helper
Treg cells	Regulatory T cell
Zbtb16	Zinc finger and BTB domain-containing protein

4.1 Introduction

The immune system is comprised of a multitude of cell types with distinct characteristics, which should be tightly coordinated in time and space in order for the organism to effectively respond to environmental threats. All immune cells derive from a single hematopoietic stem cell (HSC) that resides in the bone marrow and is able to differentiate into erythroid, myeloid, and lymphoid lineages. As HSCs differentiate toward the lymphoid lineage, they gradually lose their myeloid and erythroid potential [1]. Common lymphoid progenitors (CLPs) migrate to the thymus, where, under the strong influence of Notch signaling, they upregulate a T-cell gene expression program, while they turn off alternative lineage programs, such as those of B cells and innate lymphocytes [2]. Committed T-cell progenitors give rise to multiple T-cell lineages in the thymus, including helper CD4$^+$ and cytotoxic CD8$^+$ αβ T cells, as well as γδ T cells. In addition to these conventional T-cell lineages, rare unconventional T-cell subsets with innate characteristics, such as natural killer T (NKT) and mucosal-associated invariant T (MAIT) cells, are produced in the thymus, following a distinct developmental trajectory [3]. Naïve T cells that populate peripheral immune organs have the ability to differentiate into short-lived effector sublineages after activation by foreign antigens and cytokine stimulation. Eventually, a population of memory T cells will persist in the periphery, in some cases throughout life, which is significantly different from naïve T cells coming out of the thymus. In the case of NKT and MAIT cells, distinct differentiation programs unfold during development, without prior foreign antigen stimulation, thus leading to stable peripheral subsets [4].

Cell fate is determined by unique gene expression programs that direct the developmental pathway of precursor cells toward a specific lineage, as well as the homeostasis and function of the ensuing mature cells. At the same time, they repress those programs that may lead to the development of alternative lineages. In the context of T lymphopoiesis, progenitor cells pass through a series of intermediate developmental stages before eventually give rise to mature T cell. Therefore, the deployment of a specific gene expression program is a process that should be strictly coordinated in time and space, to ensure production of the appropriate range of T-cell subsets. While micro-environmental signals are largely responsible for the induction of lineage-specific gene expression patterns, the stability and feed-forward regulation of cell-intrinsic transcriptional networks are essential for the establishment and maintenance of a stable cell identity. Although there are many layers of gene expression control, transcriptional regulation is the initial step to processing the information encoded in the genome into functional units. Of utmost importance during this process are proteins called transcription factors (TFs), which have the unique ability to directly bind to gene regulatory elements in a DNA sequence-specific manner, while interacting with the general transcriptional machinery, thus controlling the transcriptional output of a gene. Small alterations in TF activity may lead to diversions from established developmental pathways, the severity of which is powerfully illustrated by several human disorders that have been associated with mutations in genes encoding for TFs [5, 6].

Earlier studies have suggested that a given gene expression program associated with a specific cell fate is determined by the activity of a critical TF, also known as a master regulator TF. According to this definition, master regulators are strictly expressed in a particular cell type and are both required and sufficient for programming the respective cell fate. A well-defined example of a master regulator is the TF promyelocytic leukemia zinc finger (PLZF), whose expression is specifically induced upon commitment to the NKT cell lineage, while it is not expressed in conventional T cells; in addition, genetic deletion of PLZF completely abolishes NKT cell development, while ectopic PLZF expression imposes innate-like properties in otherwise conventionally developing $CD4^+$ T cells [7, 8]. However, rather than acting alone, TFs operate in a combinatorial fashion to regulate cell fate decisions. For example, the TF B-cell lymphoma (BCL) 6 was discovered as a highly expressed gene in T follicular helper (Tfh) $CD4^+$ cells and, upon activation, directs naïve $CD4^+$ T cells toward the Tfh lineage, at the expense of alternative $CD4^+$ T-helper (Th) lineages [9–11] (Fig. 4.2). Surprisingly, low levels of BCL6 were also required for the development of the antagonistic Th1 lineage, through regulation of T-box expressed in T cells (T-BET) activity [12], whereas transient BCL6 expression was essential for NKT and Th1 cell development [13, 14]. This pleiotropy of functions is attributed to the wide range of molecular mechanisms and interacting partners utilized by BCL6 to control transcriptional activity. In this chapter, we will examine how BCL6 contributes to the development of T cells while exploring the molecular mechanisms involved in the control of BCL6 expression and functions.

4.2 BCL6: Discovery, Structure, and Regulation

BCL6 is a transcriptional repressor that was initially known for its high expression and oncogenic role in diffuse large B-cell lymphoma (DLBCL) [15]. BCL6 repressed the activity of tumor suppressor and cell cycle arrest genes, such as p53, thus allowing for the progression of the cell cycle even upon DNA damage [16–18]. Germline *Bcl6* knockout (KO) mice developed multiple anomalies, such as growth retardation, prominent myocarditis, and multisystemic inflammation which eventually led to early death, indicating that BCL6 is essential for normal growth of the organism [19]. Importantly, BCL6 can be targeted therapeutically against DLBCL, with small-molecule inhibitors or degraders [20].

Normally, BCL6 is highly expressed in germinal center (GC) B cells [19] and is required for their differentiation and survival. Importantly, BCL6 deletion not only abrogated the formation of GC B cells [21] but also led to the loss of Tfh cells [22]. Additionally, expression of BCL6 is of prominent importance in B lymphopoiesis in the bone marrow [23]; during transition from pro- to pre-B cells, pre-B-cell receptor (BCR) signaling induces high expression of *Bcl6* that in turn suppresses expression of DNA damage response genes, thus protecting developing B cells from apoptosis. However, this DNA damage tolerance is tightly regulated, through DNA damage-induced phosphorylation of BCL6, which leads to degradation by the ubiquitin-proteasome pathway [24, 25].

BCL6 is a 79-kD protein and belongs to the bric-à-brac, tramtrack, and broad complex (BTB)/Pox virus zinc finger (POZ/ZF) family of TFs. BCL6 binds directly to specific DNA sequences through its C-terminal domain, which includes six ZF DNA-binding domains (DBDs; Fig. 4.1). Although BCL6 is expressed in multiple lymphocyte subsets, including GC B cells and Tfh cells, it binds to distinct sets of distal and proximal gene regulatory elements in a cell type-specific manner [26]. In fact, genome-wide studies demonstrated that BCL6 binds to specific consensus sequences in Tfh cells, which are strikingly different from those in B cells [27], thus accounting for the distinct regulated gene set. Many BCL6 target genes are involved in fundamental biological processes in Tfh cells, including cell adhesion, calcium signaling, and purine metabolism, while BCL6 directly represses TFs that regulate differentiation of

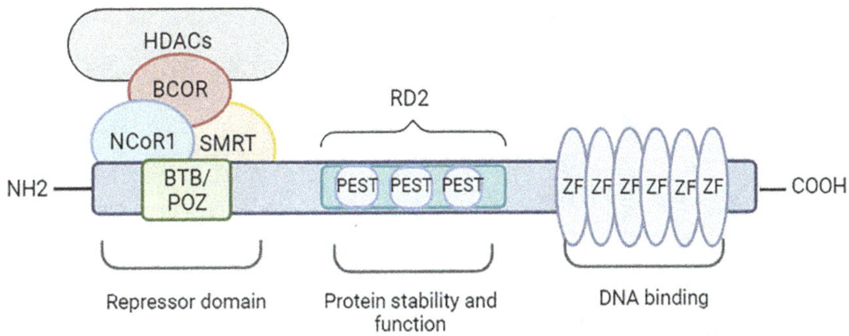

Fig. 4.1 BCL6 protein structure. BCL6 is a member of the BTB/POZ/ZF family of transcription factors (TFs). In the C-terminal domain, it contains six zinc finger (ZF) binding motifs, through which it binds to target DNA sequences. The middle region includes three PEST (proline (P), glutamic acid (E), serine (S), and threonine (T)) domains that are responsible for protein stability and function. Finally, the BTB domain, located at the N-terminal, facilitates protein-protein interactions, enabling interactions with other corepressors, such as the NCoR1 (nuclear receptor corepressor 1), NCoR2/silencing mediator for retinoid and thyroid hormone receptors (SMRT) and B-cell corepressor (BCOR). The middle region contains the RD2 (repressor domain 2) domain, which is an autonomous repressor domain that mediates recruitment of HDACs (histone deacetylases), thus suppressing transcription of target genes

alternative lineages. Surprisingly, BCL6 DNA-binding motifs were identified in only 10% of the BCL6-bound regions genome-wide, while motifs specific for other TFs were enriched, indicating that BCL6 may indirectly bind DNA [26]. Consistent with this hypothesis, BCL6 co-localizes in a subset of promoters and enhancers with other TFs, such as T-BET [12, 28] and activator protein 1 (AP1) [26], and represses their target genes.

The BTB domain, which is located at the N-terminal part of the protein, facilitates protein-protein interactions and is often involved in protein dimerization, indicating that BCL6 probably acts as a dimer (Fig. 4.1). Finally, the middle region includes three proline (P), glutamic acid (E), serine (S), and threonine (T) (PEST) domains that are responsible for protein stability and function [24]. BCL6 represses transcription through recruitment of histone deacetylases (HDACs) by two independent transcriptional repression domains located at the BTB and middle domains (repressor domain 2, RD2), which interact with distinct corepressor complexes. The BTB domain enables interactions with other corepressors, such as nuclear receptor corepressor (NCoR) 1 [29], NCoR2/silencing mediator for retinoid and thyroid hormone (SMRT) receptors [29, 30], and BCL6 corepressor (BCOR) [31], which are essential for BCL6-dependent repression, while the corepressor metastasis-associated protein 3 (MTA3) interacts with BCL6 via its middle region [32] and carboxy-terminal binding protein 1 (CTBP1) with both domains [33]. Therefore, it is no surprise that mutations interrupting these two domains were associated with completely distinct phenotypes in mice [34, 35].

BCL6 expression should be tightly regulated, as evident by the fact that small fluctuations in BCL6 levels may control activities related with different cell lineages and differentiation states. During T-cell development, *Bcl6* mRNA and protein are induced by pre-TCR (pre-Tα chain paired with a functional TCRβ chain) signaling during the $CD4^{-}CD8^{-}$ double-negative (DN) 3-to-4 transition and are at the highest level in $CD4^{+}CD8^{+}$ double-positive (DP) cells [36] (Fig. 4.2a). After positive selection, BCL6 level decreases, although it is sustained in a minor subset of early committed NKT precursors [13]. *Bcl6* mRNA is barely detectable in naive T cells; however, it is sharply induced early upon $CD8^{+}$ T-cell activation [37] and remains high in memory $CD8^{+}$ T cells but quickly declines in effector $CD8^{+}$ T cells (Fig. 4.2b). While *Bcl6* levels are maintained during Tfh differentiation, they are quickly down-regulated in Th1 differentiation through the

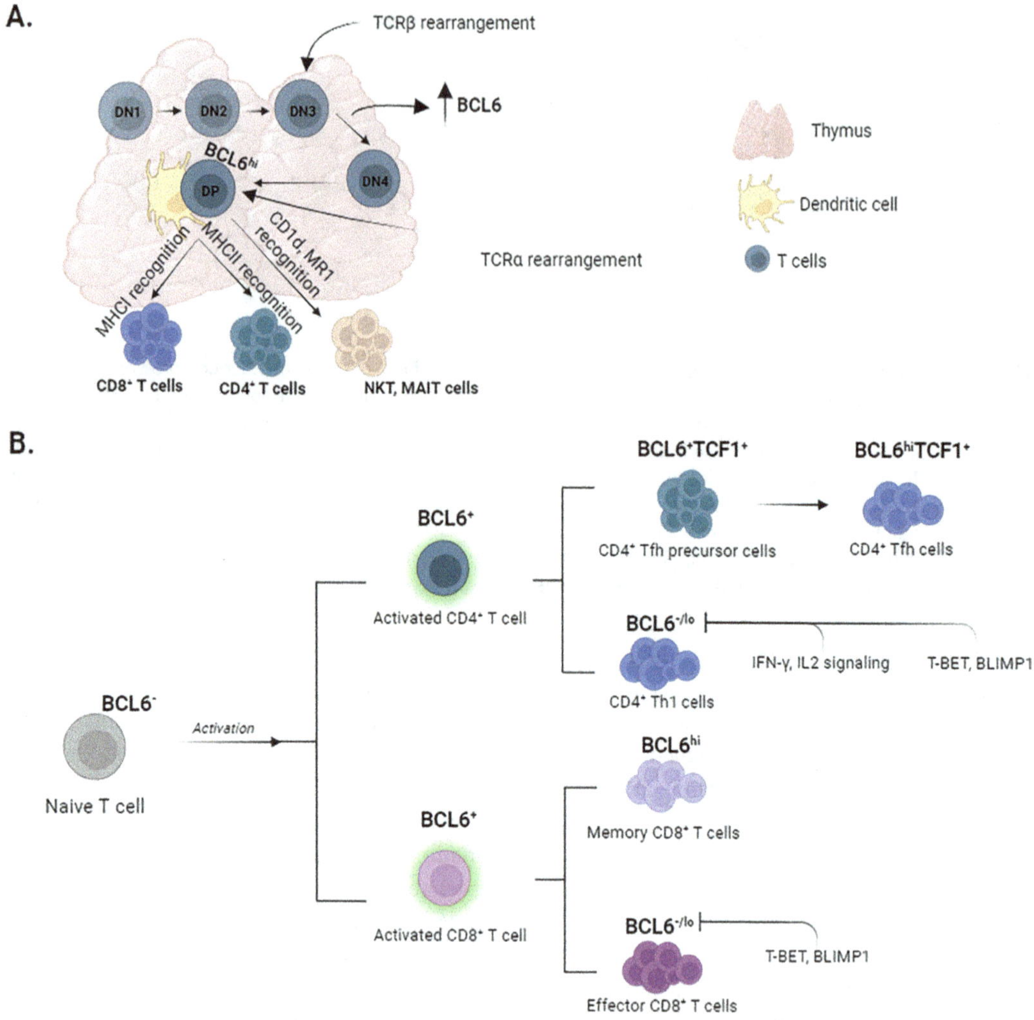

Fig. 4.2 BCL6 regulates multiple steps during T-cell development. (**a**) During thymic T-cell development, BCL6 expression is initiated in DN3-DN4 cells. BCL6 expression peaks at the DP stage, before turning off in thymic CD4$^+$ and CD8$^+$ T cells. In addition, although BCL6 is not detectable in the vast majority of natural killer T (NKT) cells, it is highly expressed in a minor immature subset. Accordingly, BCL6 is required for the development of innate-like T cells, such as NKT and MAIT cells, but not for conventional T cells. (**b**) BCL6 is expressed shortly after activation of CD4$^+$ and CD8$^+$ T cells, after infection. In CD4$^+$ T cells, sustained expression of BCL6 together with TCF-1 leads to the generation of Tfh cells, while downregulation of BCL6 enables differentiation of non-Tfh lineages. In CD8$^+$ T cells, BCL6 expression is high in memory precursor cells, while it is downregulated in short-lived effector cells

function of interferon gamma (IFNγ) and high interleukin (IL) 2 signaling and the activities of the transcription factors T-BET and B lymphocyte-induced maturation protein-1 (BLIMP1) (Fig. 4.2b). Similarly, *Bcl6* mRNA transcription is directly suppressed by T-BET and BLIMP1 in terminally differentiated CD8$^+$ effector T cells, indicating that the BCL6-T-BET-BLIMP1 axis is a broad molecular determinant of T-cell-mediated immunity [38].

BCL6 expression is also regulated through modifications of the protein at the post-transcriptional level and regulation of protein stability. E1A binding protein p300/lysine acetyltransferase 3B (EP300/KAT3B)-mediated acetylation of the BCL6 middle domain causes dissociation of BCL6 from its target genes in the nucleus and relief from transcriptional repression [39]. BCL6 is also inactivated in B cells through ubiquitin-mediated protein degradation by the E3 ligase cullin (CUL)1, after phosphorylation by mitogen-activated protein kinase (MAPK) and ataxia telangiectasia (ATM) in response to B-cell receptor (BCR) signaling [40, 41]. In developing T and NKT cells, BCL6 interacts with the ubiquitin ligase CUL3 [42, 43], and CUL3 deficiency resulted in increased expression of BCL6 and Tfh-associated genes in DP cells. Of note, this increase was observed at the transcriptional level and was mediated by basic leucine zipper transcription factor, ATF-like (BATF), which bound and regulated *Bcl6* promoter activity [42]. Interestingly, the *Bcl6* promoter is highly decorated with H3K27ac in DP cells and interacts with distal super enhancers at 250 kb and 500 kb upstream of the gene locus. These interactions are mediated by the genome organizer special AT-rich sequence-binding protein-1 (SATB1), which also interacts with CUL3 [44], indicating that CUL3 most likely regulates BCL6 expression through mechanisms other than protein ubiquitination and degradation.

4.3 BCL6 and CD4⁺ T Cells

In response to acute infection, naïve CD4⁺ T cells are activated and provide help to other lymphocytes in multiple ways: for example, they differentiate into BCL6^{hi} Tfh cells, which provide help to B cells, thus facilitating antibody production and affinity maturation, but they also differentiate into non-Tfh lineages, contributing to the coordination of innate and adaptive immune responses. Long-term CD4⁺ T-cell responses rely on the presence of memory precursor cells, which differentiate into effector CD4⁺ T cells upon secondary or persistent infection. Shortly after activation, CD4⁺ T cells segregate into BCL6⁺ and BCL6⁻ subsets, in response to IL2 and T-cell factor 1 (TCF-1) [45–47]; these early BCL6⁺TCF-1⁺ precursors are skewed toward Tfh differentiation at the expense of non-Tfh lineages, mainly through downregulation of the antagonistic TF BLIMP1 [48, 49]. However, in case of chronic antigen persistence, BCL6-expressing TCF-1^{hi} memory precursor cells have the ability to self-renew and give rise to secondary CD4⁺ T-cell effectors [48]. Similarly, Tfh generation depends on BCL6 and TCF-1 [50], thus indicating that common transcriptional networks may be shared during the differentiation of Tfh cells, CD4⁺ memory cells, and CD8⁺ memory cells (see below).

BCL6 is highly expressed in CD4⁺ Tfh cells and is necessary and sufficient for their proper development; generation of Tfh cells is completely abolished in the absence of BCL6, while enforced expression of BCL6 in CD4⁺ T cells induces Tfh differentiation [9–11]. Interestingly, temporal deletion of BCL6 in already established Tfh cells, using the CD4-Cre^{ERT2} mouse strain, resulted in Th1 differentiation, indicating that BCL6 is also important for the maintenance of the Tfh identity [51]. This is particularly important, because continuous antigen presentation is required to sustain BCL6 expression, Tfh cell expansion, and eventually the GC reaction [22, 51]. While T-cell receptor (TCR) signaling induces and maintains BCL6 expression, cytokine and inducible T-cell co-stimulator (ICOS) receptor signaling, through the signal transducer and activator of transcription (STAT) family of TFs, also contribute to BCL6 induction during early Tfh differentiation [52]. In turn, BCL6 controls a gene regulatory network that is important for the induction/maintenance of the Tfh fate but also suppression of alternative fates (see below).

The activity of several TFs that are essential for Tfh differentiation converges on the regulation of the *Bcl6* gene locus. TCF-1 directly binds to several regions proximal to the transcription start site (TSS) of the *Bcl6* gene and promotes BCL6 expression during early Tfh differentiation [53, 54]. Deletion of either TCF-1, or the related TF lymphoid enhancer-binding factor 1 (LEF1), diminishes *Bcl6* expression, while enforced expression of LEF1 in CD4⁺ T cells enhances *Bcl6* expression, even in cells with Th1 characteristics [54]. Moreover, enforced expression of BCL6 in TCF-1-deleted CD4⁺ T cells rescues Tfh differentiation, indicating that TCF-1 acts upstream of BCL6 [55]. TCF-1 binding sites on the *Bcl6* promoter are associated with high H3K27ac, which is positively correlated with gene expression. In Tfh cells, H3K27ac peaks, including those on the *Bcl6* regulatory region, were paradoxically correlated with genomic binding of Enhancer of Zeste homolog 2 (EZH2), a H3K27me3 methyltransferase with repressive activity; In fact, EZH2 co-localizes with TCF-1 on *Bcl6* gene promoter and is required for *Bcl6* expression [56]. Importantly, TCF-1 not only promotes *Bcl6* expression but also suppresses transcription of the PR containing 1, with zinc finger domain (*Prdm1*) (encoding for the antagonistic BLIMP1 protein), thus further contributing to high BCL6 levels in Tfh cells [55, 57].

Additional TFs have been shown to control Tfh differentiation through regulation of *Bcl6* gene expression. BATF is another positive regulator of *Bcl6* in Tfh cells, which binds directly to *Bcl6* regulatory loci [58], thus promoting Tfh differentiation; however, it is unclear whether it acts in coordination with the TCF1/LEF1-related network. T-helper inducing POZ-Kruppel-like factor (THPOK) binds to intron 1 of the *Bcl6* gene, while genetic deletion of THPOK in CD4$^+$ T cells results in compromised BCL6 expression [59]. Early growth response 2 (EGR2) directly regulates *Bcl6* expression, while *Egr2/Egr3*-deficient CD4$^+$ T cells fail to differentiate into Tfh. Re-expression of BCL6 in *Egr2/Egr3*-deleted cells rescued Tfh differentiation, indicating that EGR2/EGR3 are likely upstream of BCL6 [60]. The TF Oct-binding factor 1 (OBF1) is highly expressed in Tfh cells and regulates, together with organic cation uptake transporter 1 and 2 (HOCT1/HOCT2), Tfh differentiation through direct binding and promotion of BCL6 expression [61]. Although these TFs clearly impact how BCL6 is regulated in Tfh cells, how they coordinate their functions during Tfh differentiation is largely unexplored.

As a bona fide transcriptional repressor, BCL6 controls Tfh differentiation through repression of molecules that direct the development of alternative Th lineages. A primary direct target gene of BCL6 is *Prdm1*, which encodes for the antagonistic Th1-related TF BLIMP1. BLIMP1 expression is downregulated in Tfh cells, possibly suggesting that repression of BLIMP1 by BCL6 may be sufficient to unfold the Tfh cell program. However, CD4$^+$ T cells with simultaneous deletion of BCL6 and BLIMP1 still fail to generate Tfh cells [62], thus demonstrating that BCL6 regulates additional genes, beyond *Prdm1*, to control the Tfh fate. Consistent with this, in both human and mice Tfh cells, BCL6 binds to regulatory regions of several genes that encode for TFs critical for the development of non-Tfh lineages, such as T-box transcription factor 21 (*Tbx21*), GATA-binding protein 3 (*Gata3*), RAR-related orphan receptor gamma (*Rorc*), and signal transducer and activator of transcription (*Stat*) 5 [9, 10, 26, 27, 63, 64]. In addition to TFs, BCL6 directly regulates key cytokines and cytokine receptors, such as *Il17a*, IFNγ receptor 1 (*Ifngr1*), IL12 receptor (*Il12r*), and *Il23r* [9, 26], which are involved in the differentiation of non-Tfh cells. Furthermore, BCL6 binds and represses expression of C-C chemokine receptor type 7 (*Ccr7*) and selectin P ligand (*Selplg*) (which encodes for P-selectin glycoprotein ligand 1; PSGL-1), and G-protein-coupled receptor 183 (*Gpr183*) (encoding for Epstein–Barr virus-induced gene 2; EBI2), which regulate migration of T cells in extrafollicular zones of secondary lymphoid organs [26]. Importantly, BCL6 represses transcription of Kruppel-like factor 2 (*Klf2*) [62], which is essential for the expression of genes involved in T-cell migration and restricts Tfh differentiation [65]. Therefore, BCL6 serves as a hub for a transcriptional network that not only enforces the Tfh developmental gene program but also the proper localization of CD4$^+$ T cells in the follicular zone where Tfh cells are required to control GC formation and antibody production (Fig. 4.3).

Although high levels of BCL6 are essential for Tfh development, low expression of BCL6 contributes to development of antagonistic non-Tfh lineages. For example, when in low amounts during Th1 differentiation, BCL6 interacts with T-BET through its DBD [28]. As a result, T-BET sequesters BCL6 away from its cognate binding sites on the genome and instead, BCL6 is redirected to T-BET-bound genomic loci. This asso-

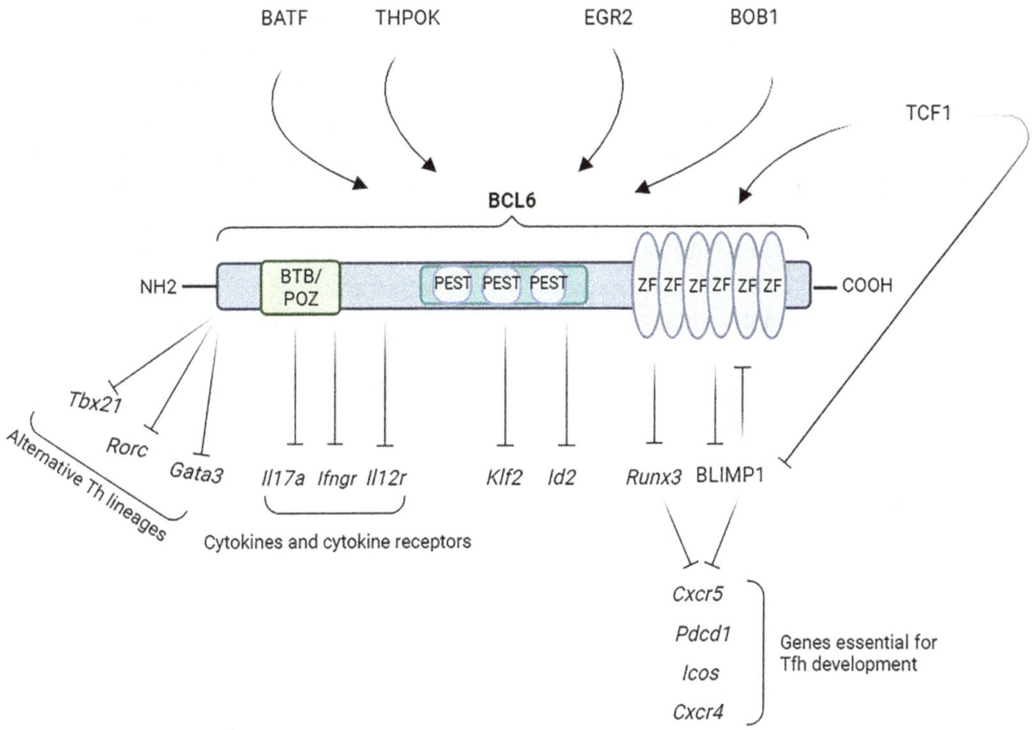

Fig. 4.3 BCL6 is the hub of a transcriptional network that controls differentiation of Tfh cells. BCL6 represses molecules that are associated with alternative lineages, including transcription factors (TFs; *Tbx21*, *Gata3*, and *Rorc*), cytokines, and cytokine receptors. In addition, BCL6 represses TFs that repress genes essential for Tfh development

ciation enables T-BET, which is a transcriptional activator, to repress target genes related to Th1 functions, such as suppressor of cytokine signaling (*Socs*) 1 and *Socs3*, which are inhibitors of IFNγ signaling, or transcription factor 7 (*Tcf7*), which is associated with alternative Th lineages [12]. Therefore, T-BET hijacks the ability of BCL6 to serve as a transcriptional repressor to suppress development of non-Th1 fate. BCL6 is also lowly expressed in regulatory T-cell (Treg) cells within the tumor microenvironment, and loss of BCL6 in Treg cells impaired their suppressive capacity. In a clinical setting, increased BCL6 expression in Treg cells serves as a poor prognostic marker in patients with colorectal cancer or melanoma, while Treg-specific deletion of BCL6 in combination with immune checkpoint inhibitors led to delayed tumor growth [66].

BCL6 represses a set of genes that would otherwise suppress molecules important for Tfh cell homing and function [50, 67]. C-X-C motif chemokine receptor 5 (CXCR5) is a key molecule in Tfh biology, because it directs T cells in the follicular zone and is regulated by multiple TFs. *Cxcr5* is repressed in non-Tfh cells by BLIMP1 and Runt-related transcription factor 2/3 (RUNX2/3) proteins, which bind directly to enhancer and promoter regions within the *Cxcr5* genetic locus [11, 62]. In addition, *Cxcr5* expression is repressed by inhibitor of DNA-binding 2 (ID2) which impairs E proteins family members from binding to their respective DNA-binding sites [68]. In addition to *Cxrc5*, BLIMP1, RUNX2/3, and ID2 repress other genes that are essential for Tfh development, such as programmed cell death 1 (*Pdcd1*) (encoding for PD1), *Icos*, and *Cxcr4* [62].

BCL6 directly represses *Prdm1*, *Runx3*, *Klf2*, and *Id2*, thus releasing their inhibitory effect on key Tfh-associated molecules. Importantly, BCL6 represses transcription of the microRNA miR-17-92, which downregulates *Cxcr5* expression [10]. BCL6 is thus a transcriptional repressor at the apex of a network that promotes Tfh development, not only by repressing the program of alternative fates but also by repressing expression of protein-coding genes and microRNAs that suppress Tfh cell differentiation (Fig. 4.3).

4.4 BCL6 and Innate-Like T Cells

Innate-like T cells, including NKT and MAIT cells, are a subtype of lymphocytes with memory characteristics. As their name implies, they share characteristics with both the innate and the adaptive immune cells; they constitutively express innate-related receptors, such as NK1.1, and a rearranged oligoclonal TCR on their surface, while they can secrete a variety of cytokines just hours after activation. NKT cells recognize lipid antigens in the context of CD1, while MAIT cells recognize vitamin metabolites through major histocompatibility complex and class I-related (MR1) presentation [69]. Based on the expression of signature TFs and cytokines, NKT cells are distinguished in different subsets, which develop in the thymus prior foreign antigen stimulation: NKT1 cells express T-BET and are predominantly IFNγ producers; NKT2 cells express high levels of GATA3 and IL4; and NKT17 cells are RORγt and IL17 expressors [70]. In addition to these subsets that are a product of thymic development, at least two other subsets are found in the periphery. NKT10 cells express the immunosuppressive cytokine IL10 and E4-binding protein 4 (E4BP4) (but not forkhead box P3; FOXP3) and reside mainly in the adipose tissue [71] or emerge after repeated activation with the NKT-specific ligand alpha-GalactosylCeramide (αGalCer) in the periphery [72].

Earlier studies have shown that a small subset of splenic NKT cells provided help to B cells within the follicular zone, after stimulation with aGalCer, in an IL21-dependent manner [73, 74]. Furthermore, they contributed to formation of early GCs and antibody affinity maturation; given the functional similarities with Tfh cells, this NKT cell subset was termed follicular helper NK (NKTfh). Consistent with their conventional Tfh counterparts, NKTfh cells upregulated BCL6, and their emergence was abrogated in the absence of BCL6 [73]. However, a recent study revealed additional functions for BCL6 during thymic NKT cell development [13]. NKT cells branch off conventional T-cell development at the DP stage, where TCR-CD1d interactions, assisted by signaling lymphocytic activation molecule (SLAM) signaling, commit DP cells to the NKT cell lineage. Upon selection, NKT cells follow a step-wise developmental pathway that culminates in the generation of mature cells. The earliest committed NKT progenitors that emerge immediately post-selection are known as stage 0 cells and begin to upregulate the NKT-specifying TF PLZF (encoded by zinc finger and BTB domain-containing protein 16; *Zbtb16*). Stage 0 cells begin to expand and give rise into intermediate developmental states, termed stage 1 and stage 2, at which point cells begin to express low levels of cytokines, especially IL4. Stage 3 cells represent mature NKT cells and may quickly secrete a broad range of cytokines [69] (Fig. 4.4a). During the stage 0-to-1 transition, cells deploy an innate-like program, characterized by upregulation of *Zbtb16*, *Tbx21*, and NK-related cell receptors, which is accompanied by an open chromatin state in the respective gene loci [13]. Although the vast majority of NKT cells do not express BCL6 at the steady state, stage 0 cells transiently express high levels of BCL6 (Fig. 4.4a). Conditional deletion of BCL6 at the DP stage of thymic development led to severe reduction in the frequency and number of thymic and peripheral NKT cells (Fig. 4.4b). Interestingly, the development of MAIT cells was also compromised, showing that BCL6 may be a global regulator of innate-like T-cell development. The few remaining NKT cells accumulated at stage 0, whereas the number of stage 1 cells was reduced, indicating that the developmental path-

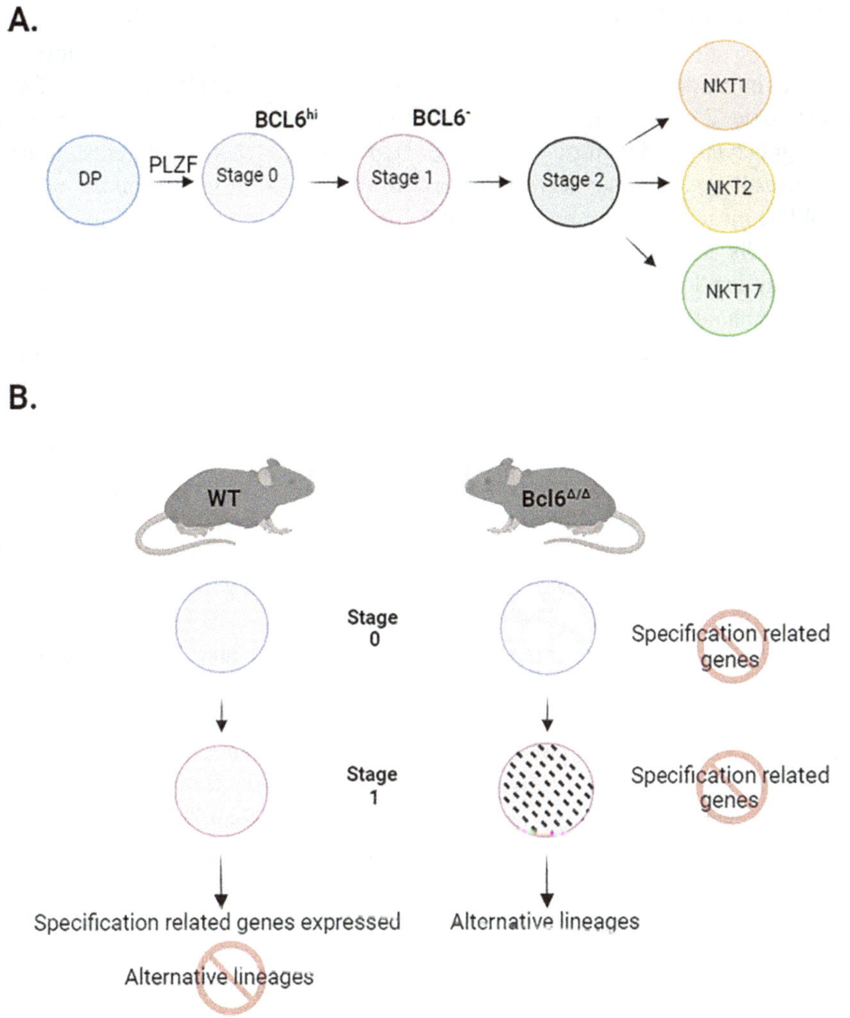

Fig. 4.4 BCL6 controls development of NKT cells. (**a**) NKT cells follow a step-wise developmental pathway during their thymic development. Stage 0 cells are early committed NKT cells that arise immediately after positive selection of DP cells and express high levels of BCL6. BCL6 expression is sharply lost in subsequent developmental stages and in mature NKT cells. (**b**) Loss of BCL6 in DP cells leads in accumulation of stage 0 NKT cells and loss of stage 1 and mature NKT cells. In addition, the remaining stage 1 cells do not upregulate genes associated with NKT cell differentiation, while they retain expression of genes associated with alternative lineages

way of NKT cells was blocked in the absence of BCL6 (Fig. 4.4b). Mechanistically, BCL6 was required for the upregulation of *Zbtb16* and its target genes, while it also repressed the premature induction of genes related to NKT cell maturation, such as T-BET. The remaining NKT cells acquired a naïve-like identity characterized by BTB domain and Cap'n'collar homolog 2 (*Bach2*) expression and reduced production of cytokines upon in vitro activation. BCL6 regulated the chromatin accessibility of the *Zbtb16* locus and other key genes for NKT cell differentiation. Finally, the few stage 1 NKT cells that developed in the absence of BCL6 largely maintained the stage 0-associated transcriptional program and chromatin accessibility profile. Therefore, these results indicate that BCL6 regulates the inception of the innate T-cell program, prior to PLZF induction (Fig. 4.4b).

4.5 BCL6 and CD8⁺ T-Cells

Upon microbial infection, CD8⁺ T cells differentiate into cytotoxic T cells, which kill infected cells through granzymes and perforin and secrete cytokines such as IFNγ and tumor necrosis factor α (TNFα). While most differentiating CD8⁺ cells act as effector cells and have a short lifespan, a few of them may live longer and persist in lymphoid or non-lymphoid organs and in the circulation, forming a memory T-cell pool. In addition to classical activated T-cell markers (CD44 and CD122), CD8⁺ memory cells express high levels of the IL7 receptor subunit alpha (IL7Ra) and low levels of killer cell lectin-like receptor subfamily G member 1 (KLRG1), while short-term effector CD8⁺ cells are IL7Ra^{low}KLRG1^{hi}. Early studies have shown that BCL6 is induced quickly after activation of CD8⁺ T cells [37] and the expansion of CD8⁺ T cells with a memory phenotype in neonatal mice was dependent on the expression of BCL6, which promoted their homeostatic proliferation under physiological conditions [75] (Fig. 4.2b). Importantly, transgenic BCL6 expression in T cells driven by the lymphoid-specific protein-tyrosine kinase (*Lck*) promoter resulted in increased numbers of CD44⁺CD8⁺ T cells for longer time than wild-type (WT) mice, showing that BCL6 expression was sufficient to sustain the memory CD8⁺ T-cell pool. BCL6 was also required and sufficient to promote the formation of antigen-specific memory CD8⁺ T cells after immunization with ovalbumin (OVA), thus indicating that BCL6 is important not only for the homeostatic maintenance of memory cells but also for long-term immunity. Consistent with this, expansion of CD8⁺ T cells after secondary boosting with OVA was more prominent in the presence of the BCL6 transgene, while the *Bcl6* mRNA was highly expressed in central memory CD8⁺ T cells but was barely detectable in effector memory and naïve CD8⁺ T cells [76].

Long-term immunity relies on the ability of memory precursor cells (MPECs) to resist contraction after infection and give rise to memory CD8⁺ T cells. While BCL6 accumulates in central memory CD8⁺ T cells, it is also expressed highly in IL7R⁺ MPECs [77]. More importantly, BCL6-deficient CD8⁺ T cells failed to differentiate into the IL7R^{hi}KLRG1^{lo} subset after acute lymphocytic choriomeningitis virus (LCMV) infection [77, 78]. In contrast, when *Bcl6* was deleted after acute LCMV infection to bypass BCL6 deficiency in MPECs, the number and cytotoxic activity of memory CD8⁺ T cells did not change, indicating that BCL6 is not required for the maintenance and function of the memory CD8⁺ T-cell pool [78]. Expression of BCL6 during acute viral infection was induced by STAT3 expression through signaling by IL21 and IL10 [77], while it was repressed by the STAT5-IL2 receptor signaling pathway [79]. High IL2 in the microenvironment promoted effector CD8⁺ T-cell differentiation by increasing expression of glycolytic genes, such as solute carrier family 2 member 1 (*Slc2a1*), and *Slc2a3*, pyruvate kinase M2 (*Pkm2*), and hexokinase 2 (*Hk2*), which are directly repressed by BCL6, at least during CD4⁺ Th1 differentiation [80]. Therefore, it is possible that *Bcl6* regulates the fate decision of activated T cells, by controlling a common metabolic reprogramming pathway, in both CD4⁺ and CD8⁺ T cells. Consistent with this hypothesis, expression of BCL6 was inversely correlated with that of BLIMP1 in effector and memory CD8⁺ T cells, while IL2 promoted BLIMP1 expression. As in B and CD4⁺ T cells, BCL6 was repressed by BLIMP1 in effector CD8⁺ T cells, and the absence of BLIMP1 led to increased BCL6 expression and memory formation [38].

Unlike acute viral infection, during chronic viral infection, CD8⁺ T cells acquire an exhausted phenotype, characterized by upregulation of inhibitory cell surface receptors and altered transcriptional program, which eventually leads to their compromised effector function. During chronic viral infection and cancer, CD8⁺ T cells differentiate into CXCR5⁺TCF-1^{hi} and CXCR5⁻TCF-1^{lo} subsets, with disparate functional properties; CXCR5⁺TCF-1^{hi} cells are characterized by a stem cell-like transcriptional signature and are more potent effectors than CXCR5⁻TCF-1^{lo} cells, which express higher levels of exhaustion markers, such as T-cell immunoglobulin and mucin-domain containing-3 (TIM-3), and mount weaker recall immune responses [81, 82]. TCF-1 directly binds to the *Bcl6* promoter and induces expression of BCL6 in TIM-3^{lo} cells while represses the expression of BLIMP1. Importantly, overexpression of BCL6 in P14 CD8^{+} T cells promoted the generation of CXCR5^{+}TIM-3^{lo} cells after infection, indicating that BCL6 alone is sufficient to induce a progenitor fate in antigen-specific CD8^{+} T cells after activation [82]. In acute LCMV infection, BCL6 binds to the *Tcf7* genomic locus and upregulates the expression of TCF-1 in KLRG1^{lo} CD8^{+} MPECs [78]. Overexpression of TCF1 in *Bcl6*-deficient CD8^{+} T cells restored the ability of MPECs to generate functional memory CD8^{+} T cells, indicating that TCF-1 and BCL6 form a positive feedback loop that stabilizes that memory CD8^{+} T-cell fate. Although it is still not clear how BCL6 promotes TCF-1 expression and which additional genes it may regulate, it is now evident that the TCF-1-BCL6 axis is a central node in the gene regulatory network that controls long-term immunity after infection.

In summary, recent data indicate that BCL6 has broad functions during T-cell differentiation. In addition to its established role as a signature transcription factor in Tfh cells, BCL6 is required for the generation of memory CD4^{+} and CD8^{+} T cells, while it is also required for the development of unconventional T cells with memory characteristics, such as NKT and MAIT cells. Interestingly, BCL6 is highly expressed in memory precursor cells and immature NKT cells, but its expression is significantly lower in more differentiated subsets, indicating that it supports a progenitor program with stem cell-like features; indeed, CD8 memory precursor cells share a transcriptional signature similar to that of HSCs, including expression of BCL6 and a subset of its target genes. A transcriptional network involving a positive feedback loop between BCL6 and TCF-1 and antagonistic interactions between BCL6, T-BET, and BLIMP1 is operational during the differentiation of both CD4^{+} and CD8^{+} T cells; however, it is not clear yet whether BCL6 participates in a similar network during NKT cell differentiation. Nonetheless, a core set of TFs, including BCL6, is broadly functioning to establish several memory-like T-cell fates. Given that BCL6 also regulates the fate of germinal center B cells, the coordinated function of this TF during an immune response across multiple lymphocyte subsets is critical for long-term immunity and host protection.

Acknowledgments MV was supported by a grant from the Hellenic Foundation for Research and Innovation (Grant No 486, ID_NKT). The authors declare no conflicts of interest.

References

1. de Pooter RF, Kee BL (2010) E proteins and the regulation of early lymphocyte development. Immunol Rev 238(1):93–109
2. Shin B, Rothenberg EV (2023) Multi-modular structure of the gene regulatory network for specification and commitment of murine T cells. Front Immunol 14:1108368
3. Ashby KM, Hogquist KA (2023) A guide to thymic selection of T cells. Nat Rev Immunol 23(10):697
4. Verykokakis M, Kee BL (2018) Transcriptional and epigenetic regulation of innate-like T lymphocyte development. Curr Opin Immunol 51:39–45
5. Lazarevic V, Glimcher LH (2011) T-bet in disease. Nat Immunol 12(7):597–606
6. Basso K, Dalla-Favera R (2010) BCL6: master regulator of the germinal center reaction and key oncogene in B cell lymphomagenesis. Adv Immunol 105:193–210
7. Savage AK, Constantinides M, G, Han J. et al (2008) The transcription factor PLZF directs the effector program of the NKT cell lineage. Immunity 29(3):391–403
8. Kovalovsky D, Uche O, U, Eladad S. et al (2008) The BTB-zinc finger transcriptional regulator PLZF controls the development of invariant natural killer T cell effector functions. Nat Immunol 9(9):1055–1064
9. Nurieva RI, Chung Y, Martinez GJ et al (2009) Bcl6 mediates the development of T follicular helper cells. Science 325(5943):1001–1005
10. Yu D, Rao S, Tsai LM et al (2009) The transcriptional repressor Bcl-6 directs T follicular helper cell lineage commitment. Immunity 31(3):457–468
11. Johnston RJ, Poholek AC, DiToro D et al (2009) Bcl6 and Blimp-1 are reciprocal and antagonistic regulators of T follicular helper cell differentiation. Science 325(5943):1006–1010
12. Oestreich KJ, Huang AC, Weinmann AS (2011) The lineage-defining factors T-bet and Bcl-6 collaborate

to regulate Th1 gene expression patterns. J Exp Med 208(5):1001–1013
13. Gioulbasani M, Galaras A, Grammenoudi S et al (2020) The transcription factor BCL-6 controls early development of innate-like T cells. Nat Immunol 21(9):1058–1069
14. Nakayamada S, Kanno Y, Takahashi H et al (2011) Early Th1 cell differentiation is marked by a Tfh cell-like transition. Immunity 35(6):919–931
15. Ye BH, Lo CF, Knowles DM et al (1993) Alterations of a zinc finger-encoding gene, BCL-6, in diffuse large-cell lymphoma. Science 262(5134):747–750
16. Cattoretti G, Pasqualucci L, Ballon G et al (2005) Deregulated BCL6 expression recapitulates the pathogenesis of human diffuse large B cell lymphomas in mice. Cancer Cell 7(5):445–455
17. Ranuncolo SM, Polo JM, Dierov J et al (2007) Bcl-6 mediates the germinal center B cell phenotype and lymphomagenesis through transcriptional repression of the DNA-damage sensor ATR. Nat Immunol 8(7):705–714
18. Ranuncolo SM, Wang L, Polo JM et al (2008) BCL6-mediated attenuation of DNA damage sensing triggers growth arrest and senescence through a p53-dependent pathway in a cell context-dependent manner. J Biol Chem 283(33):22565–22572
19. Dent AL, Shaffer AL, Yu X et al (1997) Control of inflammation, cytokine expression, and germinal center formation by BCL-6. Science 276 (5312):589–592
20. Gu H, He J, Li Y et al (2022) B-cell lymphoma 6 inhibitors: current advances and prospects of drug development for diffuse large B-cell lymphomas. J Med Chem 65(23):15559–15583
21. Ye BH, Cattoretti G, Shen Q et al (1997) The BCL-6 proto-oncogene controls germinal-centre formation and Th2-type inflammation. Nat Genet 16(2):161–170
22. Choi YS, Yang JA, Yusuf I et al (2013) Bcl6 expressing follicular helper CD4 T cells are fate committed early and have the capacity to form memory. J Immunol 190(8):4014–4026
23. Duy C, Yu JJ, Nahar R et al (2010) BCL6 is critical for the development of a diverse primary B cell repertoire. J Exp Med 207(6):1209–1221
24. Basso K, Dalla-Favera R (2012) Roles of BCL6 in normal and transformed germinal center B cells. Immunol Rev 247(1):172–183
25. Phan RT, Saito M, Kitagawa Y et al (2007) Genotoxic stress regulates expression of the proto-oncogene Bcl6 in germinal center B cells. Nat Immunol 8(10):1132–1139
26. Hatzi K, Nance PJ, Kroenke MA et al (2015) BCL6 orchestrates Tfh cell differentiation via multiple distinct mechanisms. J Exp Med 212(4):539–553
27. Liu X, Lu H, Chen T et al (2016) Genome-wide analysis identifies Bcl6-controlled regulatory networks during T follicular helper cell differentiation. Cell Rep 14(7):1735–1747
28. Oestreich KJ, Mohn SE, Weinmann AS (2012) Molecular mechanisms that control the expression and activity of Bcl-6 in TH1 cells to regulate flexibility with a TFH-like gene profile. Nat Immunol 13(4):405–411
29. Huynh KD, Bardwell VJ (1998) The BCL-6 POZ domain and other POZ domains interact with the co-repressors N-CoR and SMRT. Oncogene 17(19):2473–2484
30. Dhordain P, Albagli O, Lin RJ et al (1997) Corepressor SMRT binds the BTB/POZ repressing domain of the LAZ3/BCL6 oncoprotein. Proc Natl Acad Sci U S A 94(20):10762–10767
31. Huynh KD, Fischle W, Verdin E et al (2000) BCoR, a novel corepressor involved in BCL-6 repression. Genes Dev 14(14):1810–1823
32. Fujita N, Jaye DL, Geigerman C et al (2004) MTA3 and the Mi-2/NuRD complex regulate cell fate during B lymphocyte differentiation. Cell 119(1):75–86
33. Mendez LM, Polo JM, Yu JJ et al (2008) CtBP is an essential corepressor for BCL6 autoregulation. Mol Cell Biol 28(7):2175–2186
34. Huang C, Hatzi K, Melnick A (2013) Lineage-specific functions of Bcl-6 in immunity and inflammation are mediated by distinct biochemical mechanisms. Nat Immunol 14(4):380–388
35. Huang C, Gonzalez DG, Cote CM et al (2014) The BCL6 RD2 domain governs commitment of activated B cells to form germinal centers. Cell Rep 8(5):1497–1508
36. Solanki A, Yanez DC, Lau C et al (2020) The transcriptional repressor Bcl6 promotes pre-TCR-induced thymocyte differentiation and attenuates Notch1 activation. Development 147(19)
37. Yoshida K, Sakamoto A, Yamashita K et al (2006) Bcl6 controls granzyme B expression in effector CD8+ T cells. Eur J Immunol 36(12):3146–3156
38. Rutishauser RL, Martins GA, Kalachikov S et al (2009) Transcriptional repressor Blimp-1 promotes CD8(+) T cell terminal differentiation and represses the acquisition of central memory T cell properties. Immunity 31(2):296–308
39. Bereshchenko OR, Gu W, Dalla-Favera R (2002) Acetylation inactivates the transcriptional repressor BCL6. Nat Genet 32(4):606–613
40. Duan S, Cermak L, Pagan JK et al (2012) FBXO11 targets BCL6 for degradation and is inactivated in diffuse large B-cell lymphomas. Nature 481(7379):90–93
41. Niu H, Ye BH, Dalla-Favera R (1998) Antigen receptor signaling induces MAP kinase-mediated phosphorylation and degradation of the BCL-6 transcription factor. Genes Dev 12(13):1953–1961
42. Mathew R, Mao A, Chiang AH et al (2014) A negative feedback loop mediated by the Bcl6-cullin 3 complex limits Tfh cell differentiation. J Exp Med 211(6):1137–1151
43. Mathew R, Seiler MP, Scanlon ST et al (2012) BTB-ZF factors recruit the E3 ligase cullin 3

to regulate lymphoid effector programs. Nature 491(7425):618–621
44. Zelenka T, Klonizakis A, Tsoukatou D et al (2022) The 3D enhancer network of the developing T cell genome is shaped by SATB1. Nat Commun 13(1):6954
45. Baumjohann D, Okada T, Ansel KM (2011) Cutting ddge: distinct waves of BCL6 expression during T follicular helper cell development. J Immunol 187(5):2089–2092
46. DiToro D, Winstead CJ, Pham D et al (2018) Differential IL-2 expression defines developmental fates of follicular versus nonfollicular helper T cells. Science 361(6407)
47. Ichii H, Sakamoto A, Arima M et al (2007) Bcl6 is essential for the generation of long-term memory CD4+ T cells. Int Immunol 19(4):427–433
48. Xia Y, Sandor K, Pai JA et al (2022) BCL6-dependent TCF-1(+) progenitor cells maintain effector and helper CD4(+) T cell responses to persistent antigen. Immunity 55(7):1200–1215.e6
49. Ciucci T, Vacchio MS, Chen T et al (2022) Dependence on Bcl6 and Blimp1 drive distinct differentiation of murine memory and follicular helper CD4+ T cells. J Exp Med 219(1)
50. Choi J, Crotty S (2021) Bcl6-mediated transcriptional regulation of follicular helper T cells (T(FH)). Trends Immunol 42(4):336–349
51. Alterauge D, Bagnoli JW, Dahlstrom F et al (2020) Continued Bcl6 expression prevents the trans-differentiation of established Tfh cells into Th1 cells during acute viral infection. Cell Rep 33(1):108232
52. Choi YS, Kageyama R, Eto D et al (2011) ICOS receptor instructs T follicular helper cell versus effector cell differentiation via induction of the transcriptional repressor Bcl6. Immunity 34(6):932–946
53. Xu L, Cao Y, Xie Z et al (2015) The transcription factor TCF-1 initiates the differentiation of T(FH) cells during acute viral infection. Nat Immunol 16(9):991–999
54. Choi YS, Gullicksrud JA, Xing S et al (2015) LEF-1 and TCF-1 orchestrate T(FH) differentiation by regulating differentiation circuits upstream of the transcriptional repressor Bcl6. Nat Immunol 16(9):980–990
55. Wu T, Shin HM, Moseman AE et al (2015) TCF1 is required for the T follicular helper cell response to viral infection. Cell Rep 12(12):2099–2110
56. Li F, Zeng Z, Xing S et al (2018) Ezh2 programs T(FH) differentiation by integrating phosphorylation-dependent activation of Bcl6 and polycomb-dependent repression of p19Arf. Nat Commun 9(1):5452
57. Shao P, Li F, Wang J et al (2019) Cutting edge: tcf1 instructs T follicular helper cell differentiation by repressing blimp1 in response to acute viral infection. J Immunol 203(4):801–806
58. Ise W, Kohyama M, Schraml BU et al (2011) The transcription factor BATF controls the global regulators of class-switch recombination in both B cells and T cells. Nat Immunol 12(6):536–543
59. Vacchio MS, Ciucci T, Gao Y et al (2019) A thpok-directed transcriptional circuitry promotes Bcl6 and Maf expression to orchestrate T follicular helper differentiation. Immunity 51(3):465–478.e6
60. Ogbe A, Miao T, Symonds ALJ et al (2015) Early growth response genes 2 and 3 regulate the expression of Bcl6 and differentiation of T follicular helper cells. J Biol Chem 290(33):20455–20465
61. Stauss D, Brunner C, Berberich-Siebelt F et al (2016) The transcriptional coactivator Bob1 promotes the development of follicular T helper cells via Bcl6. Embo j 35(8):881–898
62. Choi J, Diao H, Faliti CE et al (2020) Bcl-6 is the nexus transcription factor of T follicular helper cells via repressor-of-repressor circuits. Nat Immunol 21(7):777–789
63. Kusam S, Toney LM, Sato H et al (2003) Inhibition of Th2 differentiation and GATA-3 expression by BCL-6. J Immunol 170(5):2435–2441
64. Mondal A, Sawant D, Dent AL (2010) Transcriptional repressor BCL6 controls Th17 responses by controlling gene expression in both T cells and macrophages. J Immunol 184(8):4123–4132
65. Lee JY, Skon CN, Lee YJ et al (2015) The transcription factor KLF2 restrains $CD4^+$ T follicular helper cell differentiation. Immunity 42(2):252–264
66. Li Y, Wang Z, Lin H et al (2020) Bcl6 preserves the suppressive function of regulatory T cells during tumorigenesis. Front Immunol 11:806
67. Sheikh AA, Groom JR (2021) Transcription tipping points for T follicular helper cell and T-helper 1 cell fate commitment. Cell Mol Immunol 18(3):528–538
68. Shaw LA, Belanger S, Omilusik KD et al (2016) Id2 reinforces TH1 differentiation and inhibits E2A to repress TFH differentiation. Nat Immunol 17(7):834–843
69. Bendelac A, Savage PB, Teyton L (2007) The biology of NKT cells. Annu Rev Immunol 25:297–336
70. Lee YJ, Holzapfel KL, Zhu J et al (2013) Steady-state production of IL-4 modulates immunity in mouse strains and is determined by lineage diversity of iNKT cells. Nat Immunol 14(11):1146–1154
71. Lynch L, Michelet X, Zhang S et al (2015) Regulatory iNKT cells lack expression of the transcription factor PLZF and control the homeostasis of T(reg) cells and macrophages in adipose tissue. Nat Immunol 16(1):85–95
72. Sag D, Krause P, Hedrick CC et al (2014) IL-10-producing NKT10 cells are a distinct regulatory invariant NKT cell subset. J Clin Invest 124(9):3725–3740
73. Chang PP, Barral P, Fitch J et al (2011) Identification of Bcl-6-dependent follicular helper NKT cells that provide cognate help for B cell responses. Nat Immunol 13(1):35–43
74. King IL, Fortier A, Tighe M et al (2011) Invariant natural killer T cells direct B cell responses to cognate lipid antigen in an IL-21-dependent manner. Nat Immunol 13(1):44–50

75. Ichii H, Sakamoto A, Hatano M et al (2002) Role for Bcl-6 in the generation and maintenance of memory CD8+ T cells. Nat Immunol 3(6):558–563
76. Ichii H, Sakamoto A, Kuroda Y et al (2004) Bcl6 acts as an amplifier for the generation and proliferative capacity of central memory CD8+ T cells. J Immunol 173(2):883–891
77. Cui W, Liu Y, Weinstein JS et al (2011) An interleukin-21-interleukin-10-STAT3 pathway is critical for functional maturation of memory CD8+ T cells. Immunity 35(5):792–805
78. Liu Z, Guo Y, Tang S et al (2019) Cutting edge: transcription factor BCL6 is required for the generation, but not maintenance, of memory CD8(+) T cells in acute viral infection. J Immunol 203(2):323–327
79. Pipkin ME, Sacks JA, Cruz-Guilloty F et al (2010) Interleukin-2 and inflammation induce distinct transcriptional programs that promote the differentiation of effector cytolytic T cells. Immunity 32(1):79–90
80. Oestreich KJ, Read KA, Gilbertson SE et al (2014) Bcl-6 directly represses the gene program of the glycolysis pathway. Nat Immunol 15(10):957–964
81. Im SJ, Hashimoto M, Gerner MY et al (2016) Defining CD8+ T cells that provide the proliferative burst after PD-1 therapy. Nature 537(7620):417–421
82. Wu T, Ji Y, Moseman EA et al (2016) The TCF1-Bcl6 axis counteracts type I interferon to repress exhaustion and maintain T cell stemness. Sci Immunol 1(6)

Part III

B-Cells

The Function of E2A in B-Cell Development

5

Masaki Miyazaki and Kazuko Miyazaki

Abstract

Helix-loop-helix (HLH) transcription factors (TFs) play a key role in various cellular differentiation and function through the regulation of enhancer activity. E2A, a member of the mammalian E-protein family (class I HLH protein), is well known to play an important role in hematopoiesis, especially in adaptive lymphocyte development. E2A instructs B- and T-cell lineage development through the regulation of enhancer activity for B- or T-cell signature gene expression, including *Rag1* and *Rag2* (*Rag1/2*) genes. In this chapter, we mainly focus on the function of E2A in B-cell development and on the roles of E2A in establishing the enhancer landscape through the recruitment of EP300/KAT3B, chromatin remodeling complex, mediator, cohesion, and TET proteins. Finally, we demonstrate how E2A orchestrates the assembly of the *Rag1/2* gene super-enhancer (SE) formation by changing the chromatin conformation across the *Rag* gene locus.

M. Miyazaki (✉) · K. Miyazaki
Laboratory of Immunology, Institute for Life and Medical Sciences, Kyoto University, Kyoto, Japan
e-mail: mmiyazaki@infront.kyoto-u.ac.jp

Keywords

E2A · B-cell development · Basic-helix-loop-helix transcription factor · Enhancer regulation · Adaptive lymphocyte · Rag1 and Rag2 expression

Abbreviations

HLH	Helix-loop-helix
TF	transcription factor
Rag1	recombination-activating gene 1; *Rag1/2*: *Rag1* and *Rag2*
PRRs	pattern recognition receptors
V	variable
D	diversity
J	joining
Ig	immunoglobulin
TCR	T-cell receptor
RSS	recombination signal sequence
Igh	Ig heavy chain
pro-B cells	B-cell progenitors
pre-BCR	pre-B-cell receptor
pre-B cells	B-cell precursors
Igl	Ig light chain

T. Borggrefe, B. D. Giaimo (eds.), *Transcription factors in blood cell development*, Advances in Experimental Medicine and Biology 1459, https://doi.org/10.1007/978-3-031-62731-6_5

DN	$CD4^-CD8^-$ double negative
DP	$CD4^+CD8^+$ double positive
PHD	plant homeodomain
HSPCs	hematopoietic stem/progenitor cells
BM	bone marrow
MPPs	multipotent progenitors
LMPPs	lymphoid-primed multipotent progenitors
CLPs	common lymphoid progenitors
GMPs	granulocyte/macrophage progenitors
ILCs	innate lymphoid cells
DCs	dendritic cells
Lin^-	lineage marker-negative
IL7R	IL7 receptor
BLPs	B-primed lymphoid progenitors
GFRA2	GDNF family receptor 2
BST1	BM stroma cell antigen 1
GRN	gene regulatory network
bHLH	basic helix-loop-helix
CRE	cis-regulatory element
pDCs	plasmacytoid DCs
cDC	classical DC
LTi	lymphoid tissue inducer
κE2	Ig kappa chain enhancer E2
AS-C T3	achaete-scute T3
EMSA	electrophoretic mobility shift assay
ESCs	embryonic stem cells
pre-pro-B cell	very early stage of B-cell development
WGD	whole-genome duplication
AD	activation domains
HAT	histone acetyltransferase
TET	ten-eleven translocation
PCET	EP300/KAT3B-CBP/KAT3A and ETO target
Mb1	myoglobin 1
Pax5	Paired box 5
BET	bromodomain and extra-terminal
IDRs	intrinsically disordered protein regions
Erag	enhancer of *Rag*
ASE	anti-silencer element
R-TEn	*Rag*-T-cell enhancer
R1B	*Rag* B-cell enhancer 1 *R2B*: *Rag* B-cell enhancer 2
R1pro	*Rag1* promoter
R-TEn-p1-E-box$^{mut/mut}$	E-box mutations in the *R-TEn* region
SE	super-enhancer
R1pro-E-box$^{mut/mut}$	E-box mutation in *R1pro*
BRD	bromodomain
HNRNPH1	heterogeneous nuclear ribonucleoprotein H1
HNRNPF	heterogeneous nuclear ribonucleoprotein F
TAF3/TFIID	TATA-box binding protein-associated factor 3/transcription initiation factor TFIID subunit 3
Vpreb	V-set pre-B-cell surrogate light chain 1
Cxcr4	C-X-C motif chemokine receptor 4
analysis for genome-wide chromatin interactions	Hi-C
fused in sarcoma	FUS
SATB1	special AT-rich sequence-binding protein 1
PMA	phorbol 12-myristate-13-acetate
BAF	BRG1/BRM-associated factor
Traf6	TNF receptor-associated factor 6

5.1 Introduction

Immune responses are mediated by adaptive and innate immune cells, and they cooperatively evoke immune responses aimed at protecting against pathogen invasion. Innate immune cells (macrophages and neutrophils) are activated by pattern recognition receptors (PRRs) that recognize common microbial components as danger signals. In contrast, adaptive immune cells, such as B cells and T cells, are stimulated through antigen-specific diverse receptors, which are generated by somatic recombination involving variable (V), diversity (D), and joining (J) gene segments of immunoglobulin (*Ig*) and T-cell receptor (*TCR*) genes. V(D)J recombination of *Ig* and *TCR* genes is a characteristic feature of adaptive lymphocytes that is mediated by the recombination-activating gene 1 (RAG1) and RAG2 protein complex [1–3]. The RAG1 and RAG2 complex recognize and cleave the recombination signal sequence (RSS) flanking the V, D, and J gene segments of the *Ig* and *TCR* gene loci. Following B-cell lineage commitment, D-J recombination of the Ig heavy (*Igh*) chain gene is initiated, and then V-DJ recombination occurs in B-cell progenitors (pro-B-cells) [4]. When the Ig heavy chain is paired with the surrogate light chain (l5 and VpreB), they form a pre-B-cell receptor (pre-BCR). Pre-BCR signaling induces cell expansion accompanied by IL7 signaling and differentiation into B-cell precursors (pre-B cells) [5]. If the first Ig heavy chain gene recombination fails, V-DJ recombination of the second allele of the *Igh* gene is attempted. In pre-B cells, V-J rearrangement of the Ig light (*Igl*) chain gene occurs to generate a pair of IgH and IgL for B-cell receptors (IgM expressing immature B-cells) [6]. Similarly, TCRβ and/or TCRγ/δ V(D)J recombination is initiated in immature $CD4^-CD8^-$ (double negative; DN) thymocytes. Following the success of productive *TCRβ* gene recombination in DN cells, they differentiate into $CD4^+CD8^+$ (double positive; DP) thymocytes and start V-J recombination of *TCRα* genes. DP cells that successfully express the TCRα/β complex undergo positive and negative selection to eliminate harmful T cells [7]. During these *Igh* and *Igl* gene recombination events, RAG1 binds and cleaves DNA [introduction of double-strand DNA breaks (DSBs) at RSSs]. RAG2 contains a plant homeodomain (PHD) finger in its C-terminus and recognizes hypermethylated histone H3K4 modification through this PHD, which enhances RAG1 binding and catalytic activity [3]. In both developing T and B cells, the RAG1 and RAG2 protein complex binds in a highly focal manner to a small region of D-J gene segments of the *Igh* and *TCRβ* genes and J gene segments of *Igκ* and *TCRα*, referred to as recombination centers. The recombination center coordinates V(D)J recombination by providing specific sites for capturing V-D-J gene segments [8]. Because RAG-mediated DNA cleavage can potentially lead to oncogenic translation and since a subset of leukemic cells express *Rag* genes [9–11], the expression of the *Rag1* and *Rag2* genes is limited and strictly controlled in a highly cell type-specific manner [6]. Only developing B and T cells [pro-B and pre-B cells in the BM, pro-T (DN3) and pre-T (DP) cells in the thymus] express RAG1/2. Hence, *Rag1/2* gene expression typically reflects the cell lineage commitment toward adaptive immune cells, and the regulatory mechanisms of *Rag1/2* gene expression should be controlled by TFs essential for B- and T-cell development [12].

During hematopoiesis in adults, hematopoietic stem/progenitor cells (HSPCs) in the BM have the ability to both reconstitute all hematopoietic cells and self-renew to maintain the hematopoietic system [13]. HSPCs can give rise to multipotent progenitors (MPPs) and then differentiate into lymphoid-primed multipotent progenitors (LMPPs) [14], which then differentiate into common lymphoid progenitors (CLPs) or granulocyte-macrophage progenitors (GMPs) [15, 16]. CLPs give rise to all lymphocytes, including B cells, T cells, natural killer (NK) cells, innate lymphoid cells (ILCs), and dendritic cells (DCs), while GMPs are restricted to myeloid cell pathways, such as granulocytes and macrophages. CLPs are defined as lineage marker-negative (Lin^-) $Flt3^{high}$ IL7 receptor (IL7R) ^{high} $Sca1^{low\ or\ -}$ KIT $^{low\ or\ -}$ cells by flow cytometric analysis. The restriction of lymphoid versus myeloid cell lineage choice during hematopoiesis in vivo was clarified by $IL7R^{Cre}$-mediated fate mapping and further high-resolution endogenous DNA Polylox barcoding [17–20]. As mentioned, CLPs include heterogeneous populations of each lymphoid lineage cell and are subdivided into Ly6D-positive (B-primed lymphoid progenitors: BLPs) and Ly6D-negative CLPs, and BLPs mainly generate B lineage cells after transplantation, although they do not express the B lineage marker CD19 [21, 22]. More recently, among Ly6D-expressing BLPs, GFRA2 and BST1 expression define a developmental trajectory to pro-B cells in CLPs, and BST1+ BLPs restrict other lymphoid lineages, including T-cell potential [23]. During early B-cell development, a gene regulatory network (GRN) of TFs, such as IKAROS, E2A, EBF1, PAX5, and FOXO1, which are essential for B-cell development, orchestrates the B-cell lineage commitment and the expression of lineage-specific genes, including *Rag1/2*, pre-BCR/BCR-associated molecules (l5 and VpreB, Iga, and Igb), and prevents hematopoietic malignancies [24].

E-proteins are basic helix-loop-helix (bHLH) TFs involved in the regulation of multiple hematopoietic development and include E2A (*Tcf3* gene; E12 and E47 protein), HEB (*Tcf12* gene), and E2-2 (*Tcf4* gene) [25]. E-proteins form either homodimers or heterodimers with other E-proteins or HLH proteins via the HLH dimerization domain and bind to the E-box motif (CANNTG) within the cis-regulatory element (CRE, enhancer region) of the target genes through the basic DNA-binding region. In contrast, ID-proteins (ID1-ID4) contain an HLH dimerization domain but lack the basic region for DNA binding. Therefore, when ID proteins form heterodimers with E-proteins, they antagonize the DNA binding of E-proteins and thereby function as negative regulators of E-proteins (Fig. 5.1a) [26]. It is well established that the transcriptional balance between E- and ID-proteins (the E-ID axis) controls developmental trajectories of lymphoid lineages and their activation, especially T cells, B cells, and DCs (Fig. 5.1b) [7, 24, 27, 28]. The E2A gene encodes the E47 and E12 proteins (splicing isoform of the E2A gene), and the E47 protein is essential for B-cell development [29]. E2A also acts in pro-T cells along with HEB to activate T-cell signature gene expression and simultaneously suppress ILC development in the thymus [7, 30–33]. E2A and HEB also regulate the positive selection of DP cells in the thymus [34], and HEB is also needed for invariant natural killer T (iNKT)-cell development [35, 36]. In addition, E2-2 controls type I interferon-producing plasmacytoid DCs (pDCs) and is essential for the maintenance of pDC identity, while ID2 regulates antigen-presenting classical DC (cDC) development by neutralizing E2-2 activity [37–40]. ID2 is also well known as a critical regulator of the development of all ILC subsets, including ILC1-3, NK cell, and lymphoid tissue inducer (LTi) cells [41–43]. In this chapter, we will focus on the roles of E2A in B-cell development, molecular mechanisms that underpin enhancer activity, and the regulation of *Rag1/Rag2* gene expression.

5.2 E2A Functions in B-Cell Lineage Commitment and Development

5.2.1 Discovery of E2A (E47 and E12) and E2A Knockout Mouse

E12 and E47 were first identified as a specific binding factors to Ig kappa chain enhancer E2 (κE2) containing an E-box, and they were found to form heterodimers with other HLH proteins, such as MYOD or achaete-scute T3 (AS-C T3) by electrophoretic mobility shift assay (EMSA) [44, 45]. E12 was clone number #12, and E47 was clone #47, and further investigation clarified that E12 and E47 were encoded in the E2A gene and generated by differential splicing of exons encoding the DNA-binding domain [46]. Indeed, E12 and E47 were the first E-proteins identified. Interestingly, it was found that the splicing factors HNRNPH1 and HNRNPF (HNRNPH/F) regulate E2A splicing in embryonic stem cells (ESCs), and a high level of HNRNPH/F leads to higher E12 expression, and decreasing HNRNPH/P during differentiation causes switching to E47 expression (alternative splicing switch of E2A) [47]. Following the discovery of the *E2A* gene, an *E2A*-deficient mice were generated that failed to generate B cells and lacked *Ig* DJ rearrangements, accompanied by severe impairments in *Rag1*, *Mb1*, *CD19*, and *Pax5* expression [48, 49]. Interestingly, B-cell development in *E47-*

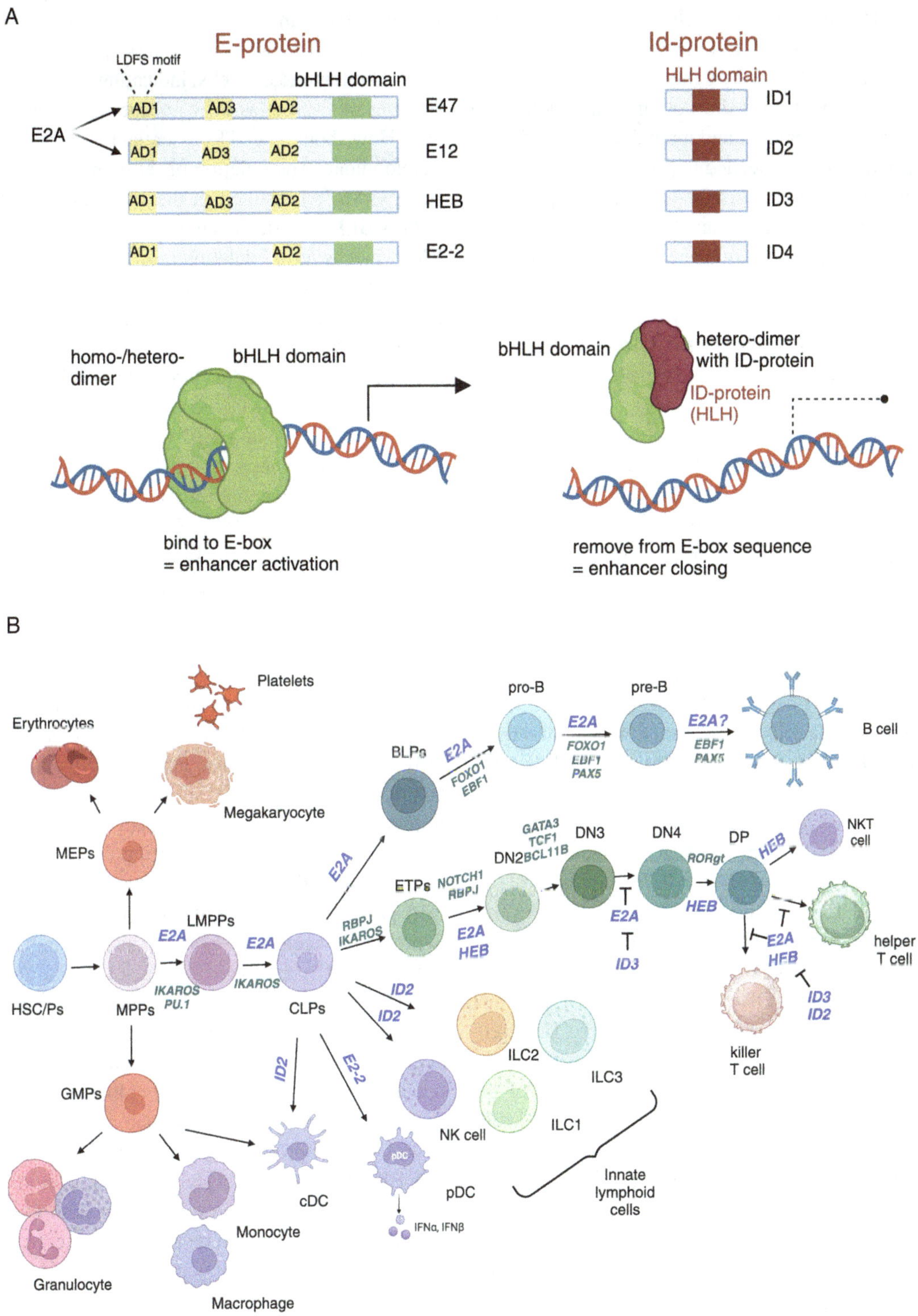

Fig. 5.1 (**a**) Mammalian E-protein family and ID-proteins (**b**) The role of E-proteins and ID-proteins in hematopoiesis

ablated mice was severely blocked at the very early stage of B-cell development (pre-pro-B cell) as well as *E2A*-deficient mice [29], while *E12*-deficient mice did not show any defects in B-cell development and maintenance [29, 50]. In contrast, E12 is needed to promote receptor editing of *Igl* gene rearrangement upon exposure to self-antigen, and E12 and E47 act to sequentially regulate *Ig* light chain gene rearrangement [50].

5.2.2 E2A Expression and Function During B-Cell Development

What is the expression profile of E2A during adaptive lymphocyte (B- and T-cell) and innate lymphoid cell (ILC) development? By analyzing the E2A-GFP fusion protein reporter mouse, it appeared that E2A protein levels were already high in HSCs and increased in LMPPs, CLPs, and BLPs and became highest in pro-B cells, the committed stage of the B-cell lineage. After pro-B cells, the E2A level gradually declines in immature B-cell populations (pre-B and IgM+ immature B cells) and is very low in mature IgD+ B cells in the BM [33, 51]. Interestingly, during ILC development, the E2A level remains mid to low in the ILC precursor population ($\alpha4\beta7^{hi}CD25^{-}$), showing an inverse correlation between E2A and ID2 or IL7R expression [33]. Similar to B-cell development, the E2A level is high in early T progenitors (ETPs), highest in committed DN2 thymocytes, and remains high until the resting DP compartment [30]. Consistent with these expression patterns of the E2A protein, E2A plays roles in early hematopoiesis. E2A/E47 is needed for the maintenance of the HSC pool and the transition from HSC to LMPPs and CLPs [52–55]. CLPs are a heterogeneous population, and Ly6D-expressing BLPs are primarily B lineage restricted [21]. Although E2A-deficient mice showed fewer CLPs (approximately half), BLPs are severely impaired in E2A null mice [21]. Consistent with these findings, the deletion of E2A and HEB after competitive BM transplantation resulted in a complete loss of B-cell development in the BM accompanied by severe defects in BLPs, while impairment in LMPPs and CLPs was mild to moderate in E2A/HEB double-deficient cells, indicating cell intrinsic roles of E-proteins in B lineage commitment [33]. How about other TFs essential for the B-cell development? The deletion of the *Foxo1* gene in HSPCs (*Vav*^{Cre}) resulted in the accumulation of Ly6D+ BLPs and defects in CD19^{+} B cells in the BM, whereas *Foxo1* deletion in BLPs/pro-B cells (*Mb1*^{Cre}) resulted in the accumulation of pro-B cells accompanied by lower expression of *E2A* and *Ebf1* [56, 57]. Similar to *Foxo1* decifiency, *Ebf1* deletion led to a developmental block at the transition from BLPs to pro-B cells [56]. Furthermore, E2A and EBF1 regulate FOXO1 expression, and these three TF regulatory networks regulate the enhancer landscape for B-cell lineage commitment [58]. Curiously, not only BLPs and pro-B cells but also DN2 and DN3 thymocytes were severely affected in the absence of E2A and HEB genes without affecting ETPs in the thymus, indicating that ETPs are able to migrate into the thymus in the absence of E2A and HEB and that E-proteins are essential for not only B-cell but also T-cell lineage commitment [33]. This clearly demonstrates that E2A plays a central role in adaptive lymphocyte lineage commitment. Interestingly, HEB could functionally replace the E2A gene in supporting B-cell development when HEB was driven by the endogenous E2A promoter, indicating that divergence among E-proteins in cell type-specific function is partially defined by their gene expression [59]. From these observations, it is suggested that gene duplication from ancestral genes may have occurred in part to permit genes to be placed under novel regulatory elements [46]. Indeed, a reconstructed maximum likelihood phylogenetic tree of three E-protein homologs for E2A, E2-2, and HEB shows three clades for these in jawed vertebrates, while lamprey E-protein homologs are distinct from those three in jawed vertebrates [60]. From this observation, it was suggested that these paralogs were generated through the two widely recognized rounds of whole-genome duplication (WGD) in vertebrates [61]. Hence, it is plausible that ancestral jawed vertebrates expressed four E-protein paralogs and that one paralog was lost early in evolution prior to the divergence of jawed vertebrates [60].

5.2.3 The Molecular Function of E2A in B-Cell Development

In addition to the bHLH domain, E-proteins contain three highly conserved activation domains (AD1-3) [62–64]. AD1 is located at the N-terminus of E2A; AD3 and AD2 are in the middle of the E2A protein; and the bHLH DNA binding and dimerization domain is located at the C-terminus (Fig. 5.1a) [64]. AD1 of E2A was first identified to activate target gene transcription through the recruitment of histone acetyltransferases (HATs) EP300/KAT3B and CBP/KAT3A, and AD1 also contributes to binding to the transcriptional repressor ETO protein to suppress target gene expression [62, 65, 66]. In addition, AD2 also interacts with P300/KAT3B and CBP/KAT3A or ETO2 cooperatively with AD1 to promote or repress the transcriptional activity of target genes [67–69]. It was reported that E-proteins enhance the recruitment of TAF3/TFIID to the promoter through AD3 binding, thereby promoting target gene transcription [70].

A genome-wide analysis of TF occupancies and histone modification has advanced our understanding of transcriptional regulation in B-cell lineage commitment [58]. During the transition from pre-pro-B cells (*Ebf1*-deficient cells) to pro-B cells (*Rag1*-deficient cells), E2A occupancies are dynamically increased during developmental progression, and E2A occupancies in pro-B cells are associated with monomethylation of histone H3K4 (H3K4me1) across putative enhancer regions. In addition, E2A binding is associated with RUNX1 and ETS motifs in pre-pro-B cells, while E2A binding across the putative enhancers specific to pro-B cells (cluster II and III) is associated with EBF1 motifs. Interestingly, E2A and EBF1 bound to the putative enhancer regions of B-cell-specific genes, such as *Cd19*, *Cd79a (Mb1)*, *Cd79b*, *Pax5*, *Vpreb1/2/3*, and *Cxcr4*, and they also bound to the *Foxo1* locus to induce its expression. From these genome-wide analyses of TFs, it was demonstrated a global network of TFs, BCR signaling, and survival factors that orchestrates B-cell specificity [58]. Subsequently, they further analyzed the whole-genome chromatin structure upon B-cell lineage commitment by using Hi-C methods [71]. By analyzing Hi-C data, it was shown that genomes are organized into two compartments: transcriptionally repressive and permissive chromatin compartments defined by the principal component 1 (PC1) value [72]. Notably, the *Ebf1* gene locus repositioned from the lamina to the nuclear interior during the transition from pre-pro-B cells to pro-B cells [71]. In this study, E2A bound at the chromatin loop attachment regions, together with EP300/KAT3B and PU.1, associated with intra- and interdomain interactions that are developmentally regulated. After B lineage commitment, EBF1, E2A, and other TFs switched compartments to establish new chromatin interactions linked to a B-cell signature gene. Another prominent chromatin switched region involved the *Igκ* locus. Specifically, in pro-B cells, the *Igκ* locus showed an intricate and elaborate pattern of interactions of chromatin loops. Interestingly, these long-range chromatin loops were closely associated with E2A occupancy, suggesting that E2A is involved in tethering Vκ segments across vast genomic distance [24, 71]. These results strongly support the previous model that E2A directly recruits EP300/KAT3B and CBP/KAT3A HATs to enhancer regions to activate target gene expression. Furthermore, by analyzing whole-genome DNA methylation status, B lineage commitment was accompanied by large-scale changes in not only chromatin structures but also DNA cytosine modification at CpG islands associated with enhancer, promoter, and anchor regions [73]. Hypomethylated DNA regions were found exclusively in the permissive chromatin compartment and were associated with active enhancer or anchor regions folding the genome [73]. How were the CpG islands in the regulatory regions hypomethylated during B-cell development? Is E2A involved in the process of demethylation? Ten-eleven translocation

(TET) enzymes facilitate DNA demethylation, and deletions of TET2 and TET3 in mice at the early B-cell stage using *Mb1*[Cre] blocked B-cell development at the pro-B stage, accompanied by decreased *Irf4* expression and impaired *Igκ* recombination [74]. TET2/3 double knockout (dKO) pro-B cells showed increased CpG methylation at the *Igκ* enhancers (iEκ, 3'Eκ, and dEκ) where E2A and PU.1 bind, as well as decreased chromatin accessibility at enhancer regions. Surprisingly, E2A and PU.1 directly interact with TET2, as shown by an immunoprecipitation assay, indicating that lineage-specific TFs such as E2A directly recruit TET proteins to remove DNA methylation [74]. Consistent with this, CpG islands at the *Rag* gene enhancer were hypermethylated when E2A binding was specifically blocked by generating E-box mutations in this enhancer region, indicating that E2A recruits TET protein to the lymphocyte-specific enhancer to eliminate CpG DNA methylation [75]. These data suggest that E2A instructs the adaptive lymphoid lineage, especially from multipotent progenitors to the B-cell lineage in the BM. Indeed, it was demonstrated that E2A-deficient BM progenitor cells showed self-renewal activity without losing multipotent differentiation potential in vitro [76], and in addition, blocking E2A activity by expressing the E-protein antagonist ID2 in hematopoietic progenitor cells (HPCs) was able to expand these progenitor cells with multipotency and readily differentiate upon withdrawal of ID2 expression into the B-cell lineage [77]. Interestingly, a subset of lymphoid and myeloid enhancers was already poised or active enhancer status, indicating multilineage priming of enhancers preceding commitment to the lymphoid ore myeloid lineage cells [77]. Consistent with this important finding, it was shown that the status of enhancers for lymphoid, myeloid, and erythroid lineages were already poised or active in HSCs and MPPs, and hematopoiesis progresses through gain and loss of these lineage-specific enhancers, associated with lineage-specific TF regulation [78]. Therefore, E2A primarily instructs and specifies the adaptive lymphoid lineage by activating these poised enhancers for the B-cell lineage and by suppressing myeloid or erythroid enhancers. Indeed, B-cell differentiation in E47-deficient fetal liver progenitor cells can be rescued by forced EBF1 expression but not PAX5 in vitro, suggesting that E2A acts upstream of EBF1 [79]. How does E2A activate enhancer regions and evolve the enhancer landscape for the adaptive lymphoid lineage? The AD1 helix domain contains a novel four-amino-acid-sequence LDFS motif that is present in class I HLH TF including E2A, HEB, and E2-2, and the LDFS motif stimulates transcription by recruiting the SAGA histone acetyltransferase complex [80]. Another group called these conserved motifs PCET (EP300/KAT3B, CBP/KAT3A, and ETO target) motifs, which are important for interacting with EP300 /KAT3B and ETO [64, 66]. As mentioned above, the AD1 domain also recruits acetyltransferases EP300/KAT3B and CBP/KAT3A [64–66]. Histone H3K27 acetylation (H3K27ac) is recognized by bromodomain and extra-terminal (BET) histone reader protein, and the bromodomain protein Brahma-related gene 1 (BRG1) complex binds to the E2A-occupied enhancer region to evict nucleosomes to provide chromatin accessibility to other B-cell-specific TFs [81]. Recently, a prion-like domain in the C-terminal domain (CTD) of EBF1 has been shown to have liquid-liquid phase separation formation ability, which enables the recruitment of the RNA-binding protein FUS and BRG1 [82]. In addition to EBF1, intrinsically disordered protein regions (IDRs) are highly frequently found to be enriched in E2A, which also contribute to the formation of phase separation. Collectively, E2A and EBF1 cooperatively facilitate chromatin opening through the recruitment of BRG1 and act as pioneer factors to initiate phase-separated super-enhancer (SE) formation to promote B-cell signature gene expression. To instruct SE formation for B-cell lineage commitment, large-scale chromatin conformation changes are needed. Since blocking E2A binding to the *Rag* enhancer led to a failure of SE formation accompanied by a loss of cohesion complex recruitment across the *Rag* gene locus [75], E2A seems to directly orchestrate chromatin

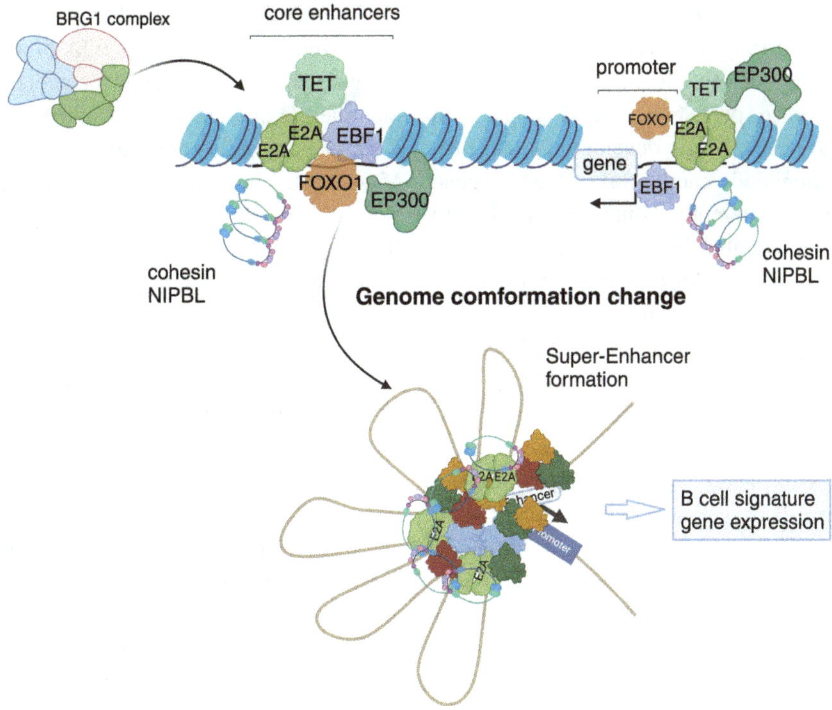

Fig. 5.2 E2A organizes the chromatin conformation and super-enhancer (SE) formation to induce the B-cell signature gene expression

conformation changes (Fig. 5.2). According to the integrative analysis of genome-wide Hi-C and ChIP-Seq data in pro-B cells, E2A occupancy is closely associated with EP300/KAT3B, PU.1, EBF1, CTCF, and cohesin complex component (RAD21) at chromatin loop attachment regions (anchor region) [71]. In particular, switching nuclear compartments between multipotent progenitors and pro-B cells, which include the *Ebf1* gene locus, establishes new intra- and interdomain interactions for the B-cell signature gene expression program [46, 71]. Notably, while pre-pro-B cells mainly exhibited adjacent chromatin looping across the *Igκ* gene locus, pro-B cells showed an intricate and elaborating chromatin interactions, and many interactions between *Igκ* intronic enhancers and genomic elements were closely associated with E2A occupancy, consistent with the initial discovery of E2A occupancy at the *Igκ* enhancer binding [71].

5.3 The Regulation of Rag1/2 Expression and E2A Function

5.3.1 B- and T-Cell-Specific Rag Gene Enhancers

As mentioned in the introduction, RAG-mediated V(D)J recombination of *Ig* and *TCR* genes is a characteristic feature of adaptive lymphocytes, and the regulation of *Rag1/2* genes is a hallmark of adaptive lymphoid lineage commitment [7]. *Rag* gene expression is restricted to developing B and T cells, and there are two waves of Rag expression [6, 83]. The first wave of Rag1/2 expression initiates D-J and V-DJ recombination of the *Igh* gene in pro-B cells. *Rag1/2* gene expression is transiently downregulated after successful *Igh* gene rearrangement during the developmental transition from pro-B to pre-B cells. In pre-B cells, *Rag1/2* genes are reexpressed, leading to immunoglobulin light chain (IgL) recombination. Following successful recombination of IgL and assembly of Ig, *Rag1/2* expression is repressed. Since persistent *Rag1/2* expression results in aberrant thymic development and profound immunodeficiency [84], *Rag1/2* gene expression is stringently controlled in B- and T-cell development [6]. Whereas both developing B and T cells express *Rag1/2* genes, they use distinct cell type-specific enhancers for their expression. It was reported that the B-cell-specific enhancer (enhancer of *Rag*; *Erag*) is 23 kb upstream of the *Rag2* gene promoter and that the deletion of *Erag* resulted in a reduction in *Rag1/2* gene expression and a moderate developmental block at pro-B stages without affecting T-cell development [85]. It was shown that E2A bound to the *Erag* enhancer in pro-B cells and was involved in *Erag* enhancer activity in nonlymphoid cells [85]. Following the identification of *Erag*, it has been reported that FOXO1 positively regulates the *Erag* enhancer and that GFI1B and c-MYB negatively regulate the *Erag* region in cell lines [57, 86–88]. Regarding the T-cell-specific enhancer, an anti-silencer element (*ASE*), which is located 73 kb upstream of the *Rag2* promoter and 8 kb in length, was identified [89]. The deletion of *ASE* in mice showed a reduction in *Rag1/2* expression in DN3 and DP thymocytes but not in developing B cells in the BM [89]. ASE-deficient mice showed a defect in the positive selection of DP cells but not in the β-selection of DN3 cells [89]. Since *Erag* deletion resulted in moderate impairment of B-cell development, additional enhancer or other regulatory mechanisms need to be determined [6]. In addition, it remains unclear how these enhancers regulate *Rag1* and *Rag2* gene loci in coordination and how their expression is restrained in innate immune cells. Exploring these molecular mechanisms surely shed light on the question of how lymphoid progenitors commit to the adaptive lymphocyte lineage.

By analyzing the E2A occupancies and ATAC-Seq data of DN3 and pro-B cells in the *Rag1/2* gene loci, two B-cell-specific *Rag* gene enhancers (*Rag* B-cell enhancer 1/2: *R1B* and *R2B*) and one T-cell-specific *Rag* gene enhancer [*Rag*-T-cell enhancer (*R-TEn*)] were identified. Furthermore, one common E2A-binding region adjacent to the *Rag1* gene promoter was also identified (*Rag1* promoter: *R1pro*) [75]. Indeed, *R-TEn* is included in the 8 kb *ASE*; *R1B* is located 15 kb upstream of the *Rag1* promoter; and *R2B* partially overlaps with the *Erag* enhancer (Fig. 5.3a). *R1B/R2B* and *R-TEn* directly loop uniquely to the *Rag1/2* gene promoter regions and form distinct genome structures between B and T cells. B- or T-cell-specific TFs [PAX5, ETS1, IKAROS, and IRF4 (B-cell TFs) and BCL11B, TCF1, RUNX1, IKAROS, and GATA3 (T-cell TFs)] bind to these enhancer regions in a cell type-specific manner. The *R-TEn*, *R2B*, and *R1pro* regions are SEs accompanied by elaborate chromatin interactions. While the deletion of either *R1B* or *R2B* in mice showed mild to moderate impairments in B-cell development, *R1B/R2B* double deletion in mice led to a severe developmental block at the pro-B-cell stage, indicating redundancy among these two B-cell enhancers [12, 75]. As well as *ASE*-deficient mice, deletion of *R-TEn* in mice resulted in defects in the positive selection of DP cells and the β-selection of DN3 cells. The developmental blocks in *R1B/R2B* or *R-TEn* deletion mice were caused by a failure of *Rag1/2* expression and impaired *TCR* and *IgH* gene recombination [75].

5.3.2 E2A/E-Proteins Regulate Rag Gene Enhancer and Rag1 Promoter Activities

What TF regulates the *Rag* enhancer? While *Rag1/2* gene expression and chromatin accessibility of *R-TEn* were not affected in *Tcf1*- or *Bcl11b*-deficient thymocytes, the deletion of E2A and HEB in DP cells led to a significant down-regulation of *Rag1/2* expression, accompanied by a loss of chromatin accessibility in the *R-TEn* and *Rag1* promoter regions, indicating that E2A regulates and maintains *Rag* gene enhancer activity. In addition, it was reported that the chromatin organizer SATB1 bound to the *ASE* and *Rag1* promoter to facilitate inclusion of the *Rag1* and *Rag2* gene loci in the active chromatin hub [90].

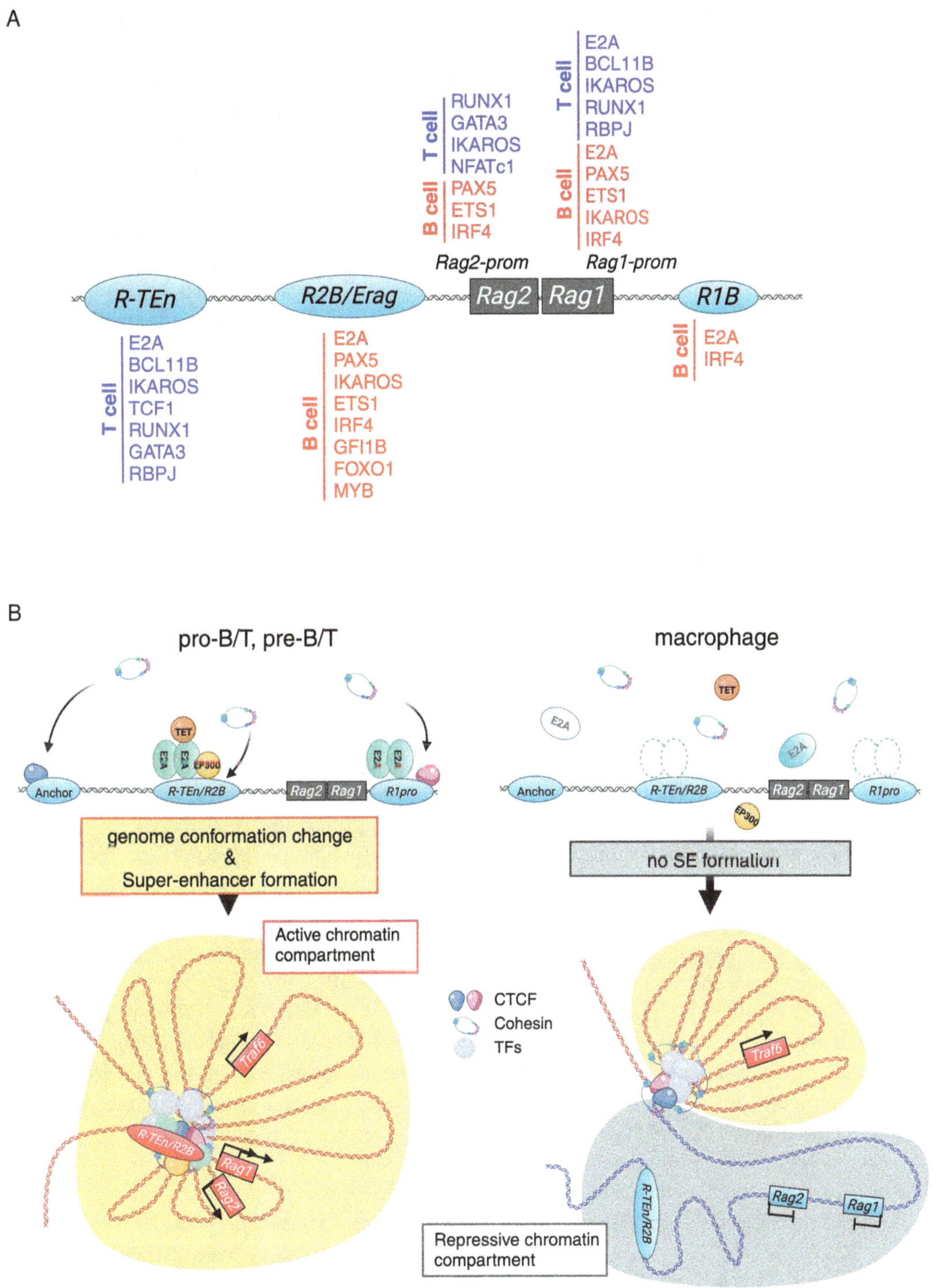

Fig. 5.3 The *Rag* gene locus and its cis-regulatory elements. (**a**) A schematic diagram of the *Rag1* and *Rag2* gene locus and B-cell-specific (*R1B* and *R2B*) and T-cell-specific (*R-TEn*) enhancer regions. (**b**) Regulation of B- and T-cell-specific *Rag* gene enhancers and *Rag* gene locus mediated by E2A/E-protein

They also reported GATA3 and E2A as critical regulators of *ASE* [91]. Finally, blocking the E2A/E-protein binding to this enhancer by generating mice with E-box mutations in the *R-TEn* region (*R-TEn-p1-E-box*$^{mut/mut}$) resulted in impaired *Rag1/2* expression and developmental block at the positive selection of DP thymocytes, as seen in the entire enhancer deletion mice (*R-TEn*$^{del/del}$), indicating that the E2A/E-protein regulates this cell type-specific *Rag1/2* enhancer [75]. Furthermore, blocking E2A binding to this enhancer led to the complete loss of chromatin accessibility and chromatin interactions across the entire *Rag1/2* gene locus accompanied by a loss of cohesin complex recruitment. This suggests that the E2A orchestrates cohesin-mediated looping and SE formation, collaborating with SATB1 (chromatin hub) to activate *Rag* gene expression [12, 90]. In addition to the *Rag* enhancer, E2A commonly binds to the *R1pro* region, which is located in close proximity from the *Rag1* promoter. In early studies, it was demonstrated that RUNX1 and E2A regulate *Rag1*-promoter activity by using a DP cell line [91]. E-box mutation in *R1pro* (*R1pro-E-box*$^{mut/mut}$) resulted in a severe developmental block at both β-selection of thymocytes and pre-BCR selection of pro-B cells, caused by defects in V(D)J recombination of *TCRβ* and *IgH* genes, similar to *Rag1*- or *Rag2*-deficient mice. Surprisingly, *Rag2* expression in DN3 and pro-B cells from *R1pro-E-box*$^{mut/mut}$ mice is normal, and the chromatin accessibilities of the *R1B/R2B* and *R-TEn* regions are not affected by a loss of *Rag1* promoter activity, indicating that cell type-specific enhancers and *Rag2* expression are independently regulated by *Rag1* promoter activity [75]. This outcome clearly demonstrates that E-proteins regulate *Rag1* promoter activity in the adaptive lymphoid lineage. However, *Rag2* promoter activity seems to be more complicatedly regulated.

The *Rag2* promoter has two CREs (D3 and Ep) to which C/EBP and NFkB bind to activate the *Rag2* promoter [92, 93]. In addition, GATA3, PAX5, c-MYB, and LEF1 bind to the *Rag2* promoter region [94, 95]. However, the regulation of *Rag2* promoter activity and responsible TFs in vivo remains to be determined. Finally, how is *Rag1/2* expression suppressed at positive selection? It was reported that signal-dependent suppression of *Rag1/2* gene expression and the disassembly of *Rag1* enhancer-promoter interactions depend on IKAROS. Blocking IKAROS binding to the *Rag1* promoter by mutation (*Rag1p* mIkaros) or *Ikaros* deletion (*Ikzf1* KO) in VL3-3M2 DP cell lines caused upregulation of *Rag1* gene expression, and the suppression of *Rag1/2* expression mediated by strong signals (PMA and ionomycin) was substantially impaired in *Rag1p* mIkaros and *Ikzf1* KO cells, accompanied by a loss of disassembly of enhancer-promoter interactions [91]. However, the suppression of *Rag1/2* gene expression upon mature BCR signaling remains undetermined.

Taken together, upon lineage commitment to the adaptive lymphoid lineage from multipotent progenitors, E2A binds to cell type-specific *Rag* gene enhancers (*R1B/R2B* and *R-TEn*) and the *R1pro* region independently and then recruits CBP/KAT3A and EP300/KAT3B, which induces a high level of histone H3K27ac across enhancer regions. The chromatin remodeling complex BAF, including BRG1 and BRM, recognizes acetylated histones through the bromodomain (BRD) and evicts nucleosomes across these enhancer regions [96]. E2A directly recruits TET proteins to remove CpG DNA methylation across the enhancer regions [74]. In fact, hypomethylation of CpG islands across the *R1B*, *R2B*, and *R-TEn* enhancers was observed in pro-B/pre-B and DN3/DP cells, respectively, and blocking E2A binding to *R-TEn* resulted in a failure to remove CpG island methylation across the *R-TEn* region, indicating that E2A is needed for the recruitment of TET proteins to the cell type-specific enhancer regions [75]. Since blocking E2A binding to *R-TEn* led to a loss of chromatin interactions across the entire *Rag* gene locus accompanied by a failure of cohesin complex recruitment, E2A instructs SE formation by recruiting the cohesin complex [12, 75]. In contrast, in innate immune cells such as macrophages, there is no topologically associated domain (TAD) formation across the *Rag* gene locus, and *R1D* seems to function as an insulator to sequester the *Rag1/2* gene away from the active neighboring *Traf6* gene to prevent aberrant *Rag1/2* gene expression (Fig. 5.3b) [7, 12].

In summary, E2A instructs adaptive immunity through orchestrating B- or T-cell-specific enhancer landscapes cooperatively with other E-proteins, B- or T-cell-specific TFs, transcription mediators, chromatin remodeling factors, and TET proteins, by which E-protein/E2A simultaneously suppresses the innate immune cell lineage.

Acknowledgment We deeply thank C. Murre for reviewing the manuscript and for kind suggestions. Funding: This work was supported by the KAKENHI (23H02734 to M.M.) and Japan Agency for Medical Research and Development (AMED) (22gm6110030h0004 to M.M.) and Japan Science and Technology Agency (JST) (JPMJPR2388 to M.M.).

References

1. Schatz DG, Oettinger MA, Baltimore D (1989) The V(D)J recombination activating gene, RAG-1. Cell 59(6):1035–1048
2. Oettinger MA, Schatz DG, Gorka C, Baltimore D (1990) RAG-1 and RAG-2, adjacent genes that synergistically activate V(D)J recombination. Science 248(4962):1517–1523
3. Schatz DG, Ji Y (2011) Recombination centers and the orchestration of V(D)J recombination. Nat Rev Immunol 11(4):251–263
4. Vettermann C, Schlissel MS (2010) Allelic exclusion of immunoglobulin genes: models and mechanisms. Immunol Rev 237(1):22–42. https://doi.org/10.1111/j.1600-065X.2010.00935.x
5. Mandel EM, Grosschedl R (2010) Transcription control of early B cell differentiation. Curr Opin Immunol 22(2):161–167. https://doi.org/10.1016/j.coi.2010.01.010
6. Kuo TC, Schlissel MS (2009) Mechanisms controlling expression of the RAG locus during lymphocyte development. Curr Opin Immunol 21(2):173–178. https://doi.org/10.1016/j.coi.2009.03.008
7. Hidaka R, Miyazaki K, Miyazaki M (2022) The E-Id Axis Instructs Adaptive Versus Innate Lineage Cell Fate Choice and Instructs Regulatory T Cell Differentiation. Front Immunol 13:890056. https://doi.org/10.3389/fimmu.2022.890056
8. Ji Y, Resch W, Corbett E, Yamane A, Casellas R, Schatz DG (2010) The in vivo pattern of binding of RAG1 and RAG2 to antigen receptor loci. Cell 141(3):419–431. https://doi.org/10.1016/j.cell.2010.03.010
9. Umiel T, Pattengale P, Weinberg K (1993) Recombination activating gene-1 (RAG-1) expression in all differentiation stages of B-lineage precursor acute lymphoblastic leukemia. Leukemia 7(3):435–440
10. Han Q, Ma J, Gu Y, Song H, Kapadia M, Kawasawa YI et al (2019) RAG1 high expression associated with IKZF1 dysfunction in adult B-cell acute lymphoblastic leukemia. J Cancer 10(16):3842–3850. https://doi.org/10.7150/jca.33989
11. Aref S, Khaled N, Menshawy NE, Sabry M, Agder MA (2020) Clinical value of RAG1 expression and IKZF1 deletions in Philadelphia negative pediatric B cell precursor acute lymphoblastic leukemia. Pediatr Hematol Oncol 37(5):380–389. https://doi.org/10.1080/08880018.2020.1767739
12. Miyazaki K, Miyazaki M (2021) The interplay between chromatin architecture and lineage-specific transcription factors and the regulation of Rag gene expression. Front Immunol 12:659761. https://doi.org/10.3389/fimmu.2021.659761
13. Spangrude GJ, Heimfeld S, Weissman IL (1988) Purification and characterization of mouse hematopoietic stem cells. Science 241(4861):58–62. https://doi.org/10.1126/science.2898810
14. Adolfsson J, Månsson R, Buza-Vidas N, Hultquist A, Liuba K, Jensen CT et al (2005) Identification of Flt3+ lympho-myeloid stem cells lacking erythro-megakaryocytic potential a revised road map for adult blood lineage commitment. Cell 121(2):295–306. https://doi.org/10.1016/j.cell.2005.02.013
15. Kondo M, Weissman IL, Akashi K (1997) Identification of clonogenic common lymphoid progenitors in mouse bone marrow. Cell 91(5):661–672. https://doi.org/10.1016/s0092-8674(00)80453-5
16. Akashi K, Traver D, Miyamoto T, Weissman IL (2000) A clonogenic common myeloid progenitor that gives rise to all myeloid lineages. Nature 404(6774):193–197. https://doi.org/10.1038/35004599
17. Schlenner SM, Madan V, Busch K, Tietz A, Laufle C, Costa C et al (2010) Fate mapping reveals separate origins of T cells and myeloid lineages in the thymus. Immunity 32(3):426–436. https://doi.org/10.1016/j.immuni.2010.03.005
18. Busch K, Klapproth K, Barile M, Flossdorf M, Holland-Letz T, Schlenner SM et al (2015) Fundamental properties of unperturbed haematopoiesis from stem cells in vivo. Nature 518(7540):542–546. https://doi.org/10.1038/nature14242
19. Pei W, Feyerabend TB, Rössler J, Wang X, Postrach D, Busch K et al (2017) Polylox barcoding reveals haematopoietic stem cell fates realized in vivo. Nature 548(7668):456–460. https://doi.org/10.1038/nature23653
20. Shang F, Rodewald HR (2022) Toward the dissection of hematopoietic stem cell fates and their determinants. Curr Opin Genet Dev 75:101945. https://doi.org/10.1016/j.gde.2022.101945
21. Inlay MA, Bhattacharya D, Sahoo D, Serwold T, Seita J, Karsunky H et al (2009) Ly6d marks the earliest stage of B-cell specification and identifies the branch-

point between B-cell and T-cell development. Genes Dev 23(20):2376–2381. https://doi.org/10.1101/gad.1836009
22. Mansson R, Zandi S, Welinder E, Tsapogas P, Sakaguchi N, Bryder D et al (2010) Single-cell analysis of the common lymphoid progenitor compartment reveals functional and molecular heterogeneity. Blood 115(13):2601–2609. https://doi.org/10.1182/blood-2009-08-236398
23. Jensen CT, Åhsberg J, Sommarin MNE, Strid T, Somasundaram R, Okuyama K et al (2018) Dissection of progenitor compartments resolves developmental trajectories in B-lymphopoiesis. J Exp Med 215(7):1947–1963. https://doi.org/10.1084/jem.20171384
24. Aubrey M, Warburg ZJ, Murre C (2022) Helix-loop-helix proteins in adaptive immune development. Front Immunol 13:881656. https://doi.org/10.3389/fimmu.2022.881656
25. Kee BL (2009) E and ID proteins branch out. Nat Rev Immunol 9(3):175–184. https://doi.org/10.1038/nri2507
26. Murre C (2005) Helix-loop-helix proteins and lymphocyte development. Nat Immunol 6(11):1079–1086. https://doi.org/10.1038/ni1260
27. Lazorchak A, Jones ME, Zhuang Y (2005) New insights into E-protein function in lymphocyte development. Trends Immunol 26(6):334–338. https://doi.org/10.1016/j.it.2005.03.011
28. Miyazaki K, Miyazaki M, Murre C (2014) The establishment of B versus T cell identity. Trends Immunol 35(5):205–210
29. Bain G, Robanus Maandag EC, te Riele HP, Feeney AJ, Sheehy A, Schlissel M et al (1997) Both E12 and E47 allow commitment to the B cell lineage. Immunity 6(2):145–154. https://doi.org/10.1016/s1074-7613(00)80421-5
30. Miyazaki M, Rivera RR, Miyazaki K, Lin YC, Agata Y, Murre C (2011) The opposing roles of the transcription factor E2A and its antagonist Id3 that orchestrate and enforce the naive fate of T cells. Nat Immunol 12(10):992–1001. https://doi.org/10.1038/ni.2086
31. Braunstein M, Anderson MK (2011) HEB-deficient T-cell precursors lose T-cell potential and adopt an alternative pathway of differentiation. Mol Cell Biol 31(5):971–982. https://doi.org/10.1128/mcb.01034-10
32. Braunstein M, Anderson MK (2012) HEB in the spotlight: transcriptional regulation of T-cell specification, commitment, and developmental plasticity. Clin Dev Immunol 678705(10):22
33. Miyazaki M, Miyazaki K, Chen K, Jin Y, Turner J, Moore AJ et al (2017) The E-Id protein axis specifies adaptive lymphoid cell identity and suppresses thymic innate lymphoid cell development. Immunity 46(5):818–834.e4
34. Jones ME, Zhuang Y (2007) Acquisition of a functional T cell receptor during T lymphocyte development is enforced by HEB and E2A transcription factors. Immunity 27(6):860–870. https://doi.org/10.1016/j.immuni.2007.10.014
35. D'Cruz LM, Knell J, Fujimoto JK, Goldrath AW (2010) An essential role for the transcription factor HEB in thymocyte survival, Tcra rearrangement and the development of natural killer T cells. Nat Immunol 11(3):240–249. https://doi.org/10.1038/ni.1845
36. Roy S, Zhuang Y (2015) Orchestration of invariant natural killer T cell development by E and Id proteins. Crit Rev Immunol 35(1):33–48. https://doi.org/10.1615/critrevimmunol.2015012207
37. Cisse B, Caton ML, Lehner M, Maeda T, Scheu S, Locksley R et al (2008) Transcription factor E2-2 is an essential and specific regulator of plasmacytoid dendritic cell development. Cell 135(1):37–48. https://doi.org/10.1016/j.cell.2008.09.016
38. Ghosh HS, Cisse B, Bunin A, Lewis KL, Reizis B (2010) Continuous expression of the transcription factor e2-2 maintains the cell fate of mature plasmacytoid dendritic cells. Immunity 33(6):905–916. https://doi.org/10.1016/j.immuni.2010.11.023
39. Grajkowska LT, Ceribelli M, Lau CM, Warren ME, Tiniakou I, Nakandakari Higa S et al (2017) Isoform-specific expression and feedback regulation of E protein TCF4 control dendritic cell lineage specification. Immunity 46(1):65–77. https://doi.org/10.1016/j.immuni.2016.11.006
40. Verykokakis M, Zook EC, Kee BL (2014) ID'ing innate and innate-like lymphoid cells. Immunol Rev 261(1):177–197. https://doi.org/10.1111/imr.12203
41. Yokota Y, Mansouri A, Mori S, Sugawara S, Adachi S, Nishikawa S et al (1999) Development of peripheral lymphoid organs and natural killer cells depends on the helix-loop-helix inhibitor Id2. Nature 397(6721):702–706. https://doi.org/10.1038/17812
42. Serafini N, Vosshenrich CA, Di Santo JP (2015) Transcriptional regulation of innate lymphoid cell fate. Nat Rev Immunol 15(7):415–428. https://doi.org/10.1038/nri3855
43. Cherrier DE, Serafini N, Di Santo JP (2018) Innate lymphoid cell development: a T cell perspective. Immunity 48(6):1091–1103. https://doi.org/10.1016/j.immuni.2018.05.010
44. Murre C, McCaw PS, Baltimore D (1989) A new DNA binding and dimerization motif in immunoglobulin enhancer binding, daughterless, MyoD, and myc proteins. Cell 56(5):777–783. https://doi.org/10.1016/0092-8674(89)90682-x
45. Murre C, McCaw PS, Vaessin H, Caudy M, Jan LY, Jan YN et al (1989) Interactions between heterologous helix-loop-helix proteins generate complexes that bind specifically to a common DNA sequence. Cell 58(3):537–544. https://doi.org/10.1016/0092-8674(89)90434-0
46. Murre C (2019) Helix-loop-helix proteins and the advent of cellular diversity: 30 years of discovery. Genes Dev 33(1–2):6–25. https://doi.org/10.1101/gad.320663.118
47. Yamazaki T, Liu L, Lazarev D, Al-Zain A, Fomin V, Yeung PL et al (2018) TCF3 alternative splicing

controlled by hnRNP H/F regulates E-cadherin expression and hESC pluripotency. Genes Dev 32(17–18):1161–1174. https://doi.org/10.1101/gad.316984.118
48. Bain G, Maandag EC, Izon DJ, Amsen D, Kruisbeek AM, Weintraub BC et al (1994) E2A proteins are required for proper B cell development and initiation of immunoglobulin gene rearrangements. Cell 79(5):885–892
49. Zhuang Y, Soriano P, Weintraub H (1994) The helix-loop-helix gene E2A is required for B cell formation. Cell 79(5):875–884
50. Beck K, Peak MM, Ota T, Nemazee D, Murre C (2009) Distinct roles for E12 and E47 in B cell specification and the sequential rearrangement of immunoglobulin light chain loci. J Exp Med 206(10):2271–2284. https://doi.org/10.1084/jem.20090756
51. Zhuang Y, Jackson A, Pan L, Shen K, Dai M (2004) Regulation of E2A gene expression in B-lymphocyte development. Mol Immunol 40(16):1165–1177
52. Yang Q, Kardava L, St Leger A, Martincic K, Varnum-Finney B, Bernstein ID et al (2008) E47 controls the developmental integrity and cell cycle quiescence of multipotential hematopoietic progenitors. J Immunol 181(9):5885–5894. https://doi.org/10.4049/jimmunol.181.9.5885
53. Dias S, Månsson R, Gurbuxani S, Sigvardsson M, Kee BL (2008) E2A proteins promote development of lymphoid-primed multipotent progenitors. Immunity 29(2):217–227. https://doi.org/10.1016/j.immuni.2008.05.015
54. Semerad CL, Mercer EM, Inlay MA, Weissman IL, Murre C (2009) E2A proteins maintain the hematopoietic stem cell pool and promote the maturation of myelolymphoid and myeloerythroid progenitors. Proc Natl Acad Sci U S A 106(6):1930–1935. https://doi.org/10.1073/pnas.0808866106
55. Borghesi L, Aites J, Nelson S, Lefterov P, James P, Gerstein R (2005) E47 is required for V(D)J recombinase activity in common lymphoid progenitors. J Exp Med 202(12):1669–1677. https://doi.org/10.1084/jem.20051190
56. Mansson R, Welinder E, Åhsberg J, Lin YC, Benner C, Glass CK et al (2012) Positive intergenic feedback circuitry, involving EBF1 and FOXO1, orchestrates B-cell fate. Proc Natl Acad Sci U S A 109(51):21028–21033. https://doi.org/10.1073/pnas.1211427109
57. Dengler HS, Baracho GV, Omori SA, Bruckner S, Arden KC, Castrillon DH et al (2008) Distinct functions for the transcription factor Foxo1 at various stages of B cell differentiation. Nat Immunol 9(12):1388–1398. https://doi.org/10.1038/ni.1667
58. Lin YC, Jhunjhunwala S, Benner C, Heinz S, Welinder E, Mansson R et al (2010) A global network of transcription factors, involving E2A, EBF1 and Foxo1, that orchestrates B cell fate. Nat Immunol 11(7):635–643. https://doi.org/10.1038/ni.1891
59. Zhuang Y, Barndt RJ, Pan L, Kelley R, Dai M (1998) Functional replacement of the mouse E2A gene with a human HEB cDNA. Mol Cell Biol 18(6):3340–3349. https://doi.org/10.1128/mcb.18.6.3340
60. Yoshikawa G, Miyazaki K, Ogata H, Miyazaki M (2021) The evolution of Rag gene enhancers and transcription factor E and Id proteins in the adaptive immune system. Int J Mol Sci 22(11). https://doi.org/10.3390/ijms22115888
61. Simakov O, Marlétaz F, Yue JX, O'Connell B, Jenkins J, Brandt A et al (2020) Deeply conserved synteny resolves early events in vertebrate evolution. Nat Ecol Evol 4(6):820–830. https://doi.org/10.1038/s41559-020-1156-z
62. Aronheim A, Shiran R, Rosen A, Walker MD (1993) The E2A gene product contains two separable and functionally distinct transcription activation domains. Proc Natl Acad Sci U S A 90(17):8063–8067. https://doi.org/10.1073/pnas.90.17.8063
63. Massari ME, Jennings PA, Murre C (1996) The AD1 transactivation domain of E2A contains a highly conserved helix which is required for its activity in both Saccharomyces cerevisiae and mammalian cells. Mol Cell Biol 16(1):121–129. https://doi.org/10.1128/mcb.16.1.121
64. Liang JJ, Peng H, Wang JJ, Liu XH, Ma L, Ni YR et al (2021) Relationship between the structure and function of the transcriptional regulator E2A. J Biol Res (Thessalon) 28(1):15. https://doi.org/10.1186/s40709-021-00146-5
65. Bradney C, Hjelmeland M, Komatsu Y, Yoshida M, Yao TP, Zhuang Y (2003) Regulation of E2A activities by histone acetyltransferases in B lymphocyte development. J Biol Chem 278(4):2370–2376. https://doi.org/10.1074/jbc.M211464200
66. Zhang J, Kalkum M, Yamamura S, Chait BT, Roeder RG (2004) E protein silencing by the leukemogenic AML1-ETO fusion protein. Science 305(5688):1286–1289. https://doi.org/10.1126/science.1097937
67. Quong MW, Massari ME, Zwart R, Murre C (1993) A new transcriptional-activation motif restricted to a class of helix-loop-helix proteins is functionally conserved in both yeast and mammalian cells. Mol Cell Biol 13(2):792–800. https://doi.org/10.1128/mcb.13.2.792-800.1993
68. Denis CM, Langelaan DN, Kirlin AC, Chitayat S, Munro K, Spencer HL et al (2014) Functional redundancy between the transcriptional activation domains of E2A is mediated by binding to the KIX domain of CBP/p300. Nucleic Acids Res 42(11):7370–7382. https://doi.org/10.1093/nar/gku206
69. Bradney C, Hjelmeland M, Komatsu Y, Yoshida M, Yao TP, Zhuang Y (2003) Regulation of E2A activities by histone acetyltransferases in B lymphocyte development. J Biol Chem 278(4):2370–2376
70. Chen WY, Zhang J, Geng H, Du Z, Nakadai T, Roeder RG (2013) A TAF4 coactivator function for E proteins that involves enhanced TFIID binding. Genes Dev 27(14):1596–1609. https://doi.org/10.1101/gad.216192.113

71. Lin YC, Benner C, Mansson R, Heinz S, Miyazaki K, Miyazaki M et al (2012) Global changes in the nuclear positioning of genes and intra- and interdomain genomic interactions that orchestrate B cell fate. Nat Immunol 13(12):1196–1204. https://doi.org/10.1038/ni.2432
72. Lieberman-Aiden E, van Berkum NL, Williams L, Imakaev M, Ragoczy T, Telling A et al (2009) Comprehensive mapping of long-range interactions reveals folding principles of the human genome. Science 326(5950):289–293. https://doi.org/10.1126/science.1181369
73. Benner C, Isoda T, Murre C (2015) New roles for DNA cytosine modification, eRNA, anchors, and superanchors in developing B cell progenitors. Proc Natl Acad Sci U S A 112(41):12776–12781. https://doi.org/10.1073/pnas.1512995112
74. Lio CW, Zhang J, González-Avalos E, Hogan PG, Chang X, Rao A (2016) Tet2 and Tet3 cooperate with B-lineage transcription factors to regulate DNA modification and chromatin accessibility. Elife:5. https://doi.org/10.7554/eLife.18290
75. Miyazaki K, Watanabe H, Yoshikawa G, Chen K, Hidaka R, Aitani Y et al (2020) The transcription factor E2A activates multiple enhancers that drive Rag expression in developing T and B cells. Sci Immunol 5(51). https://doi.org/10.1126/sciimmunol.abb1455
76. Ikawa T, Kawamoto H, Wright LY, Murre C (2004) Long-term cultured E2A-deficient hematopoietic progenitor cells are pluripotent. Immunity 20(3):349–360. https://doi.org/10.1016/s1074-7613(04)00049-4
77. Mercer EM, Lin YC, Benner C, Jhunjhunwala S, Dutkowski J, Flores M et al (2011) Multilineage priming of enhancer repertoires precedes commitment to the B and myeloid cell lineages in hematopoietic progenitors. Immunity 35(3):413–425. https://doi.org/10.1016/j.immuni.2011.06.013
78. Lara-Astiaso D, Weiner A, Lorenzo-Vivas E, Zaretsky I, Jaitin DA, David E et al (2014) Immunogenetics. Chromatin state dynamics during blood formation. Science 345(6199):943–949. https://doi.org/10.1126/science.1256271
79. Seet CS, Brumbaugh RL, Kee BL (2004) Early B cell factor promotes B lymphopoiesis with reduced interleukin 7 responsiveness in the absence of E2A. J Exp Med 199(12):1689–1700. https://doi.org/10.1084/jem.20032202
80. Massari ME, Grant PA, Pray-Grant MG, Berger SL, Workman JL, Murre C (1999) A conserved motif present in a class of helix-loop-helix proteins activates transcription by direct recruitment of the SAGA complex. Mol Cell 4(1):63–73. https://doi.org/10.1016/s1097-2765(00)80188-4
81. Bossen C, Murre CS, Chang AN, Mansson R, Rodewald HR, Murre C (2015) The chromatin remodeler Brg1 activates enhancer repertoires to establish B cell identity and modulate cell growth. Nat Immunol 16(7):775–784. https://doi.org/10.1038/ni.3170
82. Wang Y, Zolotarev N, Yang CY, Rambold A, Mittler G, Grosschedl R (2020) A prion-like domain in transcription factor EBF1 promotes phase separation and enables B cell programming of progenitor chromatin. Immunity. https://doi.org/10.1016/j.immuni.2020.10.009
83. Wilson A, Held W, MacDonald HR (1994) Two waves of recombinase gene expression in developing thymocytes. J Exp Med 179(4):1355–1360. https://doi.org/10.1084/jem.179.4.1355
84. Wayne J, Suh H, Sokol KA, Petrie HT, Witmer-Pack M, Edelhoff S et al (1994) TCR selection and allelic exclusion in RAG transgenic mice that exhibit abnormal T cell localization in lymph nodes and lymphatics. J Immunol 153(12):5491–5502
85. Hsu LY, Lauring J, Liang HE, Greenbaum S, Cado D, Zhuang Y et al (2003) A conserved transcriptional enhancer regulates RAG gene expression in developing B cells. Immunity 19(1):105–117. https://doi.org/10.1016/s1074-7613(03)00181-x
86. Amin RH, Schlissel MS (2008) Foxo1 directly regulates the transcription of recombination-activating genes during B cell development. Nat Immunol 9(6):613–622. https://doi.org/10.1038/ni.1612
87. Schulz D, Vassen L, Chow KT, McWhirter SM, Amin RH, Möröy T et al (2012) Gfi1b negatively regulates Rag expression directly and via the repression of FoxO1. J Exp Med 209(1):187–199. https://doi.org/10.1084/jem.20110645
88. Timblin GA, Xie L, Tjian R, Schlissel MS (2017) Dual mechanism of Rag gene repression by c-Myb during pre-B cell proliferation. Mol Cell Biol 37(12). https://doi.org/10.1128/mcb.00437-16
89. Yannoutsos N, Barreto V, Misulovin Z, Gazumyan A, Yu W, Rajewsky N et al (2004) A cis element in the recombination activating gene locus regulates gene expression by counteracting a distant silencer. Nat Immunol 5(4):443–450. https://doi.org/10.1038/ni1053
90. Hao B, Naik AK, Watanabe A, Tanaka H, Chen L, Richards HW et al (2015) An anti-silencer- and SATB1-dependent chromatin hub regulates Rag1 and Rag2 gene expression during thymocyte development. J Exp Med 212(5):809–824. https://doi.org/10.1084/jem.20142207
91. Naik AK, Byrd AT, Lucander ACK, Krangel MS (2019) Hierarchical assembly and disassembly of a transcriptionally active RAG locus in CD4(+)CD8(+) thymocytes. J Exp Med 216(1):231–243. https://doi.org/10.1084/jem.20181402
92. Yu W, Misulovin Z, Suh H, Hardy RR, Jankovic M, Yannoutsos N et al (1999) Coordinate regulation of RAG1 and RAG2 by cell type-specific DNA elements 5' of RAG2. Science 285(5430):1080–1084. https://doi.org/10.1126/science.285.5430.1080
93. Monroe RJ, Chen F, Ferrini R, Davidson L, Alt FW (1999) RAG2 is regulated differentially in B and T cells by elements 5′ of the promoter. Proc Natl

Acad Sci U S A 96(22):12713–12718. https://doi.org/10.1073/pnas.96.22.12713
94. Kishi H, Wei XC, Jin ZX, Fujishiro Y, Nagata T, Matsuda T et al (2000) Lineage-specific regulation of the murine RAG-2 promoter: GATA-3 in T cells and Pax-5 in B cells. Blood 95(12):3845–3852
95. Wu CX, Zhao WP, Kishi H, Dokan J, Jin ZX, Wei XC et al (2004) Activation of mouse RAG-2 promoter by Myc-associated zinc finger protein. Biochem Biophys Res Commun 317(4):1096–1102. https://doi.org/10.1016/j.bbrc.2004.03.159
96. Nguyen UT, Bittova L, Müller MM, Fierz B, David Y, Houck-Loomis B et al (2014) Accelerated chromatin biochemistry using DNA-barcoded nucleosome libraries. Nat Methods 11(8):834–840. https://doi.org/10.1038/nmeth.3022

6 Hypoxia and the Hypoxia-Inducible Factors in Lymphocyte Differentiation and Function

Mark Boothby and Sung Hoon Cho

Abstract

Molecular oxygen doubles as a biomolecular building block and an element required for energy generation and metabolism in aerobic organisms. A variety of systems in mammalian cells sense the concentration of oxygen to which they are exposed and are tuned to the range present in our blood and tissues. The ability to respond to insufficient O_2 in tissues is central to regulation of erythroid lineage cells, but challenges also are posed for immune cells by a need to adjust to very different oxygen concentrations. Hypoxia-inducible factors (HIFs) provide a major means of making such adjustments. For adaptive immunity, lymphoid lineages are initially defined in bone marrow niches; T lineage cells arise in the thymus, and B cells complete maturation in the spleen. Lymphocytes move from these first stops into microenvironments (bloodstream, lymphatics, and tissues) with distinct oxygenation in each. Herein, evidence pertaining to functions of the HIF transcription factors (TFs) in lymphocyte differentiation and function is reviewed. For the CD4+ and CD8+ subsets of T cells, the case is very strong that hypoxia and HIFs regulate important differentiation events and functions after the naïve lymphocytes emerge from the thymus. In the B lineage, the data indicate that HIF1 contributes to a balanced regulation of B-cell fates after antigen (Ag) activation during immunity. A model synthesized from the aggregate literature is that HIF in lymphocytes generally serves to modulate function in a manner dependent on the molecular context framed by other TFs and signals.

Keywords

Hypoxia · Hypoxia-inducible factor · Immunity · Lymphocyte · Helper T cell · CTL · Immune memory · Tumor · B cell · Germinal center

M. Boothby (✉)
Departments of Pathology, Microbiology, Immunology (Molecular Pathogenesis Division), Vanderbilt University Medical Center, Nashville, TN, USA

Medicine (Rheumatology and Immunology Division), Vanderbilt University Medical Center, Nashville, TN, USA

Vanderbilt Institute for Infection, Inflammation, Immunity (VI4), Nashville, TN, USA
e-mail: mark.boothby@vumc.org

S. H. Cho
Departments of Pathology, Microbiology, Immunology (Molecular Pathogenesis Division), Vanderbilt University Medical Center, Nashville, TN, USA

Vanderbilt Institute for Infection, Inflammation, Immunity (VI4), Nashville, TN, USA

T. Borggrefe, B. D. Giaimo (eds.), *Transcription factors in blood cell development*, Advances in Experimental Medicine and Biology 1459, https://doi.org/10.1007/978-3-031-62731-6_6

Abbreviations

Ab	Antibody
Ag	antigen
AID	gene name *Aicda* (activation-induced cytidine deaminase)
ASC	antibody-secreting cell
AR	adenosine receptor
A2AR	A2A adenosine receptor
BCL2, BCL6	B-cell lymphoma (protein number 2, 6)
BCR	B-cell receptor
BNIP	BCL2/adenovirus E1B 19 kDa protein-interacting protein
CAR	chimeric antigen receptor
CD	cluster of differentiation
CLP	common lymphoid progenitor
CoV	coronavirus
CRISPR	clustered regularly interspaced short palindromic repeats
CTL	cytotoxic T lymphocyte
CTLA-4	cytotoxic T-lymphocyte antigen 4
EAE	experimental autoimmune encephalomyelitis
EC	epithelial cell
ECAR	extracellular acidification rate
EPO	erythropoietin
FOX	forkhead box transcription factor
GC	germinal center
G-CSF	granulocyte colony-stimulating factor
GLUT	glucose transporter
HIF	hypoxia-inducible factor
HSC	hematopoietic stem cell
IFN	interferon
Ig	immunoglobulin
IL	interleukin
iNKT	invariant natural killer-T
KG	ketoglutarate
KLRG1	killer cell lectin-like receptor subfamily G member 1
LAG-3	lymphocyte activation gene 3
LCMV	lymphocytic choriomeningitis virus
LoF	loss of function
LSK	lineage markers, Sca-1, cKit
MAIT	mucosa-associated invariant T
MHC	major histocompatibility complex
Mtb	*Mycobacterium tuberculosis*
mTEC	medullary thymic epithelial cell
mTORC	mechanistic target of rapamycin complex
MZ	marginal zone
NRF2	nuclear factor erythroid 2-related factor 2
OG	oxoglutarate
PC	plasma cell
PD1	programmed death 1
PHD	prolyl hydroxyl dioxygenase
PI	phosphatidyl-inositol
PI3K	PI 3-kinase
RBC	red blood corpuscle
ROR	retinoid orphan receptor
S1PR	sphingosine-1-phosphate receptor
SARS	severe adult respiratory syndrome
SHM	somatic hypermutation
STAT	signal transducer and activator of transcription
TCR	T-cell receptor
TF	transcription factor
Tfh	follicular Th
Treg	regulatory T cell
Th	T-helper
TGF	transforming growth factor
VEGF	vascular endothelial growth factor
(p)VHL	von Hippel-Lindau (protein)

6.1 The Molecular and Cellular Physiology of Responding to Oxygen Concentration

Complex metazoan organisms require oxygen distribution to tissues and their cellular components. In most cases, this need is met with a specialized circulatory system. In mammals—which are the main reference set for this review—oxygen is delivered via a specialized pool of red blood corpuscules (RBC), packets of hemoglobin that develop from erythroid progenitors in the bone marrow. Erythroid precursor production in turn represents a branch-point in differentiation potential for the hematopoietic stem cells (HSCs). After its uptake onto hemoglobin in RBC, molecular oxygen (O_2) is delivered from the circulation across the endothelium and progressively consumed in traversing layers of cells in a tissue. Hypoxia needs to be considered at both the level of whole-body physiology and that of each type of cell in its microenvironment(s). To assure enough oxygenation, regulation of overall RBC and hemoglobin biomass is crucial. For these purposes, a small set of specialized cells in the kidney senses insufficient O_2 delivery rates. A shortfall leads to increased gene transcription, production, and release of the hormone erythropoietin (EPO) (reviewed in [1, 2]). EPO stimulates expansion of the erythroid progenitor mass in the marrow, in part by directing increased allocation of stem cells to this branch of hematopoiesis [3]. Analyses of the EPO promoter identified a regulatory element that, in turn, led to identification of hypoxia-inducible transcription factors (HIFs) [4].

The molecular aspects of how HIFs are induced by hypoxia are both beautiful as biochemical processes and essential for consideration of their roles in lymphocyte differentiation and function. These transcription factors (TFs) consist of two heterodimeric proteins, HIF1 and HIF2, each of which consists of one α- and one β-subunit (reviewed in [5, 6]). The HIF β-subunit, which also is shared with other TFs such as the aryl hydrocarbon receptor, requires an α-partner, HIF1α or HIF2α (encoded in the mouse by the *Hif1a* and *Epas1* genes, respectively). The primary sequence of each α-subunit contains a pair of two closely spaced proline residues that can undergo hydroxylation catalyzed by any of three similar enzymes termed prolyl hydroxyl dioxygenases (PHDs 1–3) [5–7]. These enzymes use O_2 and the metabolite alpha-ketoglutarate (αKG, alternatively termed 2-OG, i.e., 2-oxoglutarate) as substrates along with vitamin C (ascorbic acid) and Fe^{+++} as cofactors in a reaction that adds a hydroxyl group to two alpha subunit prolines and converts the αKG to succinate. Two hydroxyprolines are recognized by the von Hippel-Lindau protein (pVHL), an E3 component that directs rapid K48-linked polyubiquitylation and proteasomal degradation of the HIFα subunits. This mechanism of posttranslational stabilization of α-subunits is a major means of controlling the amounts of heterodimeric HIF and HIF2 that translocate to the nucleus, bind to DNA recognition sequences in chromatin, and induce expression of EPO, and many other "HIF target genes" are that [5, 6]. A second hydroxylation reaction restrains the transcriptional activity of HIFs by targeted posttranslational modifications (PTMs) of the trans-activation domain [5, 6]. This reaction—catalyzed by a 2OG-dependent dioxygenase termed FIH, factor inhibiting HIF, or HIF asparaginyl hydroxylase—also uses molecular oxygen to restrain activation of HIF-regulated genes.

The precise molecular details of how these enzymes mediate the induction of a response to hypoxia include a beautiful merger of evolution and biochemistry. An elegant mode of sensing relies on the Km of the PHD enzymes, because these are "tuned" in a way that reduces their capacity to hydroxylate HIFα subunit prolines in the physiological range of intracellular oxygen concentrations [5, 6]. Thus, as the partial pressure of oxygen (pO_2) falls low enough, HIF are stabilized. Of note, regulation is more complex, and even basic regulation may not be limited to this mechanism. As one nexus of HIF stabilization with intermediary metabolism, succinate and then fumarate—downstream from αKG in the Krebs Cycle—can impose feedback inhibition on PHD enzymatic activity, thereby reducing prolyl hydroxylation and pVHL-mediated α-subunit degradation [5–7]. Another general means of increasing HIF stabilization and transcriptional drive is through the capacity of reactive oxygen species (ROS) or specifically hydrogen peroxide (H_2O_2) to attenuate or block activity of the HIF hydroxylases ([8, 9], reviewed in [10]). As will be seen in sections below, T- and B-cell antigen (Ag) receptor engagement (i.e., stimulation during the initiation of adaptive immune responses) lead to HIF stabilization and nuclear induction even in conditions of normoxia or the hyperoxia of a tissue culture incubator [11, 12]. While it is not worked out which mechanism(s) lead to sus-

ments. Given the nature of the scientific enterprise, however, it is unlikely the basis for the starkly different results on this fundamental issue will be resolved.

Because of potential clinical scenarios, a variant question bears consideration: namely, whether HIFs might matter in HSCs under conditions outside the normal physiological range.

For instance, inhibitors of PHD enzymatic activity are approved as agents for mobilizing RBC production in the setting of anemia (reviewed in [37]). In this regard, it has been reported that PHD inhibitors can enhance mobilization of HSCs in a HIF1-dependent manner, and that this dependence also applied to the capacity of G-CSF (granulocyte colony-stimulating factor) to assist HSC mobilization [38, 39]. Other work indicates that HSCs' maintenance is altered under conditions of HIF hyper-stabilization after either treatment with PHD inhibitors or genetic inactivation of PHD2, both in normal homeostasis and under conditions of stress [40]. Unpublished evidence indicates that the entire process of marrow development is profoundly disturbed if pVHL is inactivated in hematopoietic cells (for instance, as part of our experiments reported in [41]). More recent work is notable in providing evidence that inactivation of the gene encoding the transcriptional repressor ID2 shifts a crucial balance in hematopoietic stem/progenitor cells (HSPCs) so as to favor activation and progress at the expense of long-term maintenance of the self-renewal [42]. As noted above, sufficient enforcement of quiescence is supported by hypoxia, arguably in a HIF-dependent manner. Indeed, the lack of ID2 was associated with lower HIF stabilization at steady-state along with "increased cycling, ROS production, mitochondrial activation, ATP production, and DNA damage." Importantly for the present issue, enforcement of HIF1 stabilization mitigated the defects noted above, and evidence was presented to the effect that ID2 (which can act in mature lymphocytes) can interfere with the capacity of pVHL to direct proteasomal degradation of HIF1α [42]. All together, the evidence of this recent work pointed to a model in which ID2 and HIF1 operated in a positive loop to participate in titrating activation and developmental progression versus renewal of HSCs.

6.2.2 Ontogenic Progression of B Lineage Cells to Yield the Pre-immune Repertoire

CLPs derived from the HSC progress through intermediate stages as lymphoblasts and then prolymphocytes, which then appear to partition between remaining in the bone marrow versus entering the arteriovenous circulation. In the marrow, precursors may start to commit to the B lymphoid lineage (Fig. 6.1) and then progress through stages in which a membrane-anchored B-cell Ag receptor is assembled from the antibody (Ab)-encoding heavy and then light chain genes. The membrane-anchored immunoglobulin (Ig) heavy (H) chain forms during a pro-B-cell stage and, paired with a surrogate light (L) chain, constitutively signals so that cells can advance to a rapidly proliferating early pre-B status. If rearrangement at the Ig L chain locus generates a partner for the H chain, the complete Ag receptor complex, including invariant signaling chains, is displayed on the surface of an immature or "transitional" B cell. These not fully matured B cells leave the bone marrow and complete maturation in the spleen.

As noted above, there are hypoxic zones in or throughout the bone marrow. Indeed, an important paper showed that developing B cells in the marrow do experience hypoxia [43]. To what extent does the functional insufficiency of oxygen lead to essential functions of HIF TFs long this series of stages in B-cell differentiation? Early papers provided evidence that original means of gene-targeted (unconditional) LoF for HIF-1α reduced the production of B cells in a manner that would appear to be lymphoid lineage-intrinsic [44, 45]. This impediment was imputed to a stage of rapid proliferation at which glycolytic metabolism—which HIF1 tends to promote—was particularly important [45]. However, a wrinkle was added by work in which the inactivation of the *Hif1a* gene was achieved in a B lineage-specific manner by introgression of the widely used *mb1*-Cre (*Cd79a*-Cre) transgene [12]. This paper sought to emphasize a different aspect of B-cell biology—a subset of B lineage cells with suppressive or inhibitory properties linked to IL10 production—and reported that the phenotypic progression of ontogeny in the marrow was normal. One potential reason for the findings might be if BCR signaling were haplo-insufficient for mb1 (*Cd79a*) and MB1 levels lower on the transgenic B cells, since MB1 is an essential signaling component of the BCR complex. In unpublished evidence with LoF in work for [46], we observed rapid perturbations of B lineage ontogeny in the marrow that were consistent with a functional redundancy of HIF1 and HIF2. Specifically, *Rosa26*-CreERT2-driven deletions distorted late pre-B to immature B as well as the balance between T1 and T2 populations within a week but only if both *Hif1a* and *Epas1* were inactivated (unpublished observations). The lack of a clear way to reconcile these streams of evidence means that, as with long-term self-renewal of HSCs, the role of HIFs during normal physiological ontogeny is less definitively established. On the whole, however, the evidence weighs heavily on the pan of the balance that favors a contribution of HIF1 to the regulated progression of normal B-cell development.

As with HSCs, a variant of the question needs also to be considered. Specifically, what are effects of excessive HIF stabilization, either due to sustained hypoxemia (low oxygen tension in the blood, for instance, due to lung pathology), impaired PHD hydroxylation of the α-subunits, or stabilization due to genetic inactivation of the *Vhl* gene? In this case, as with HSCs, the data in a beautiful study show that persistent stabilization or hyper-induction of HIFs changes marrow progression of the B lineage via a HIF-dependent mechanism [43]. Mechanistically, this effect appears to involve a capacity of prolonged or excessive HIF1 stabilization to promote increased apoptosis of the developing repertoire through a mechanism dependent on BIM (*Bcl2l11*), a pro-apoptotic BCL2-homology (Domain)-3-only (BH3) member of the *BCL2*-gene superfamily [43]. Of note, the data also pointed to some changes in the pre-immune repertoire of Ag receptors among the B cells that reached maturity. This finding suggests that HIF involvement in some way influences the spectrum of Ag specificities prior to an immune challenge—be it through changing the set point for negative selection or subtly influencing the variable (Ig V) gene usage during receptor generation. Of note, too, follow-up work provided evidence that sustained hypoxemia in patients suffering infection by the severe acute respiratory syndrome coronavirus-2 (SARS-CoV-2) yielded congruent findings the mechanistic basis of which was illuminated using mimetic conditions in noninfected mouse models [47]. In particular, not only did the numbers of circulating B cells in hypoxemic patients decrease acutely (i.e., in a matter of weeks) but also hypoxia or persistent HIF stabilization (via B lineage-specific inactivation of the *Vhl* gene) markedly lowered B cells in secondary lymphoid organs (spleen, lymph nodes) [47]. Of note, these experimental interventions in mice reduced numbers at both the very early (pro-pre-B stage) and a later apoptosis-prone T2 transitional B-cell stage. These results reinforce findings in an earlier paper, in which use of *Cd19*-Cre (which acts as B-cells reach the semi-mature transitional stages) drove inactivation of *Vhl* [48]. The mechanistic focus in this paper was on the HIF-driven shift to a reductive metabolism of glutamine in which its conversion to citrate and subsequent contribution to palmitoylation of procaspase-8 caused undue activation of caspase-mediated apoptosis. Computational analyses of the gene expression signatures in comparing mice and human subjects suggest a similarity of the two models [47]. For arithmetic reasons, it is likely that both decreased differentiation and premature death were operative in the hypoxemic adult patients, in that the decrease in circulating B cells appears too large to be imputed solely to less effective differentiation. Notwithstanding this aspect of the important findings, the best working inference at present, in combining the bodies of evidence, is that ontogenic progression of B-cell development is HIF-dependent in normal and pathological physiology.

6.2.3 Development of the Naïve, Pre-immune T-Cell Repertoire

By contrast to the B lineage, in which semi-matured cells develop in the marrow, some of the circulating CLP population may differentiate into natural killer (NK) cells, whereas another fraction enters the thymus (Fig. 6.2). This organ is specialized for production of a pre-immune repertoire of mature T lymphocytes that bear Ag receptors able to interact with the particular individual's polymorphic variants of proteins encoded in the major histocompatibility complex (MHC). A first step is for a Notch-regulated progressive commitment to T lineage identity [49]. During the evolution from seeding to commit-

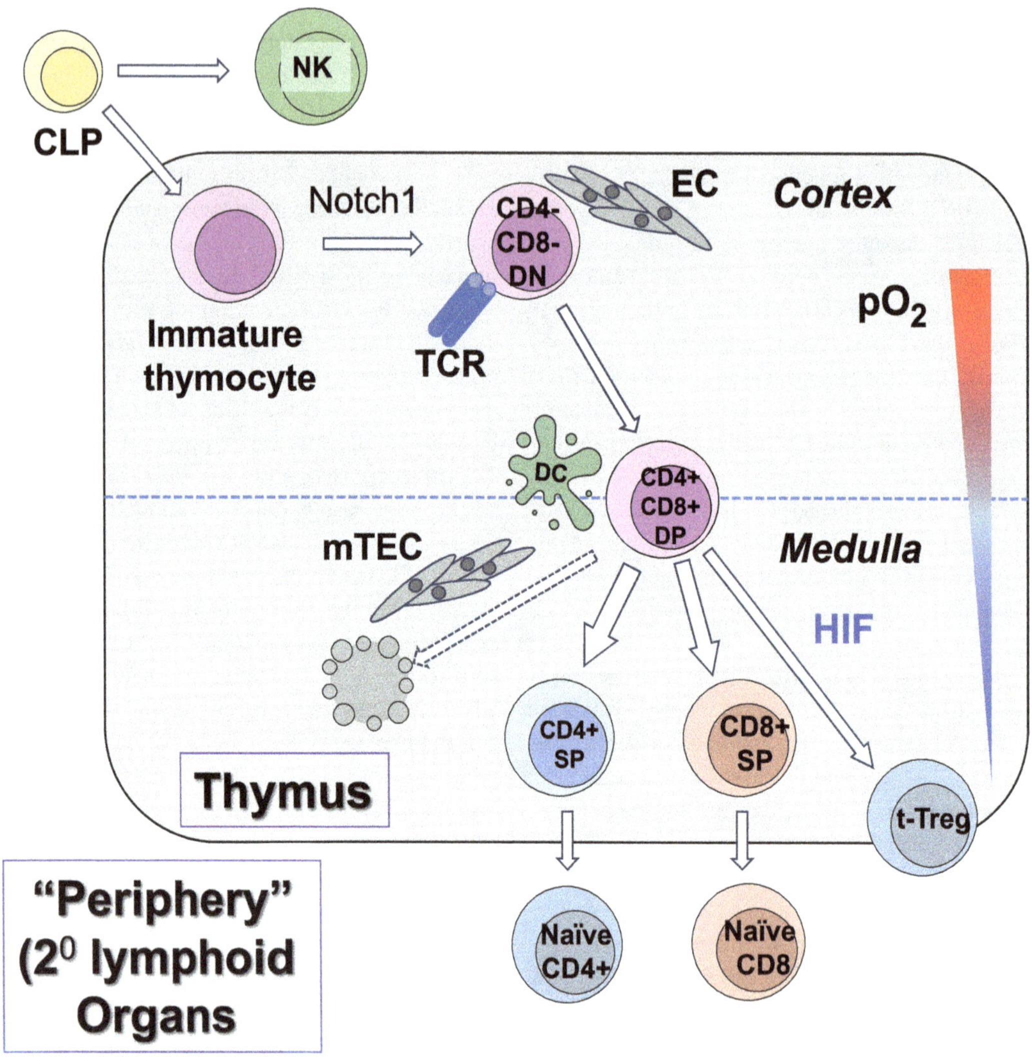

Fig. 6.2 Establishment and ontogenic progression of major T-cell lineages in the thymus. As discussed in the text, CLPs can generate thymus-independent NK cells or, under the influence of Notch signaling in the thymus, can develop and cement T lineage identity, leading to $CD4^+$ $CD8^+$ DP thymocytes that express a TCR. These DP cells then undergo positive and negative forms of selection through interactions with thymic epithelial cells (ECs) and dendritic cells (DCs), resulting in production of naïve $CD4^+$, $CD8^+$, and thymus-derived ("natural") Treg (tTreg) expressing FOXP3. While the significance is not clear, there is more hypoxia in the thymic medulla, in which specialized medullary thymic epithelial cells (mTECs) execute additional pruning of the repertoire of TCR (negative selection)

ment, these early phases of immature T cells move from a more central point of entry to a paracortical region, after which their daughter cells then move through the thymic cortex toward a cortico-medullary junction. During this phase, the T lineage identity is established by success in sequential recombination events that allow formation of a pre-TCR (a pre-Tα chain paired with a functional TCR β chain), followed by surface-expressed heterodimeric (αβ or γδ) T-cell Ag receptor components of a stoichiometric signaling complex that includes invariant CD3 chains. The surface display of a functional TCR complex is essential for immature thymocytes to differentiate into mature T cells, both by surviving apoptotic checkpoints and through reprogramming of gene expression. The nature of the randomly generated TCR specificity and its molecular "fit" with proteins encoded on thymic cells (cortical and medullary epithelial cells (ECs); dendritic cells (DCs) positioned at the cortico-medullary junction) determine a choice between a $CD4^+$ ($CD8^-$) fate or the reciprocal $CD8^+$ ($CD4^-$) state. One additional branch-point in the development of mature T-cell identity is one in which the forkhead box TF FOXP3 becomes highly expressed in a minority subset of thymic $CD4^+$ ($CD8^-$) cells. FOXP3 directs a profound reprogramming of gene expression and cellular metabolism that are part of the regulatory T cell (Treg) identity [50, 51]. These thymus-derived Treg cells are of central importance to direct and indirect inhibition of "conventional" CD4 and CD8 T cells and thereby are vital for what has termed "dominant tolerance" [50, 51]. The relationships among hypoxia, HIFs, and these T-cell subsets will be covered in a later section.

Early work using distinct approaches—including both direct measurements with an O_2-sensing microelectrode [44] and pimonidazole staining [52]—identified substantial hypoxia in the thymus. The intravital labeling of functionally hypoxic cells with pimonidazole adducts was most concentrated and apparent in the medulla—a site at which a phase of negative selection is imposed by specialized thymic medullary ECs—and was sparse in the cortex. Of note, the latter of these papers provided evidence that hypoxia could reduce the rates of apoptotic death of cultured thymocytes, the vast majority of which are TCR^{med} $CD4^+$ $CD8^+$ (double-positive, DP) cells that undergo the processes of positive and negative thymic selection in the cortex and at the cortico-medullary junction, respectively. Inasmuch as these processes are central to shaping the pre-immune T-cell repertoire, these findings raised important questions about HIF.

Notwithstanding the potential biological impact of the answering the questions about HIF (−1) function in thymic T-cell production under physiological conditions, the answers are somewhat mixed and unsatisfying. This question was secondary in the paper with a crucial genetic approach, but it was notable that inactivation of *Hif1a* coding capacity with a Cre-driving transgene active starting at an early stage of thymocyte ontogeny reduced thymocyte numbers to ½ of control numbers [53]. The findings in this paper indicated that HIF1 promotes the population of the thymus in a manner affecting the DP (TCR-expressing $CD4^+$ $CD8^+$ cells) and more matured single-positive ($CD4^+$ $CD8^-$ or $CD4^-$ $CD8^+$) thymocytes to similar extents. A conundrum was posed by the earlier observation that failed to find increased expression of hypoxia-induced genes, even when measuring *Bnip3*, a HIF-dependent mRNA whose characteristics suited it to being increased even if only a fraction of the thymocytes were hypoxic [52].

The main focus of the work with cell lineage-restricted LoF early in the T lineage was on sustained (hyper-)induction of HIFs, which was achieved by the use of the same proximal *Lck* promoter-driven Cre expression to inactivate the *Vhl* gene. This experimental approach documented that, in principle, HIF-dependent gene expression could readily be induced in thymocytes [53]. This finding indicated that the target genes could be turned on as long as HIF stabilization was persistent and at a high enough level, addressing the paradox that a substantial fraction of thymocytes is hypoxic yet such gene expression could not be detected. Although the biochemical basis for this set of observations was not resolved, it highlights an important issue that recurs in much of the work with differentiation and function among different types and subsets of mature lymphocyte. There are at least two possibilities—neither of them mutually exclusive of the other. One is that in a particular individual thymocyte, the degree and duration of hypoxia are limited in time and intensity to a level whereat HIF induction is generally insufficient to trigger changes in gene transcription. The other derives from the often-overlooked mass of insight into regulation of gene expression, in which generally no TF is an island unto itself. Thus, the constellation and concentrations of TFs that define the identity of different stages of thymocyte may supersede HIFs and prevent their capacity to increase expression of their usual target genes. As an example of this principle, there is evidence that TFs such as T-BET (i.e., T-box transcription factor 21, expressed in T cells, encoded by *Tbx21*) and BCL6 interfere with a HIF-driven metabolic reprogramming [54].

In any event, the biological consequence of lacking pVHL from early in T lineage development was a massive reduction in thymocyte cellularity in a manner that impeded differentiation or survival past the $CD4^-$ $CD8^-$ stage (when *Vhl* gene inactivation would be executed) and thereby reduced the TCR-expressing $CD4^+$ $CD8^+$ DP cells and their singly $CD4^+$ and $CD8^+$ descendants [53]. The effect was imputed to increased apoptosis, but the mechanism for the observed effects was not defined. One possibility is that the metabolic program imposed by sustained and high-level HIF1—would tend to reduce oxidative processes and favor the glycolytic pathway (reviewed in [55]). On the surface, this possibility might seem at odds with the requirement of $CD4^-$ $CD8^-$ thymocytes for sufficient glycolytic flux stimulated by Notch in a manner mediated by complex 1 of the mechanistic target of rapamycin complex 1 (mTORC1) [56]. However, many biological processes require being in a "Goldilocks" zone, so that activity that is too sustained or at too high a level might be deleterious. What was established is that the loss of thymocytes caused by lack of pVHL was substantially mitigated by concurrent inactivation of the *Hif1a* gene (i.e., *Vhl* Δ/Δ *Hif1a* Δ/Δ), so that cellularity was at the reduced level of isolated HIF1 deficiency [53]. These findings indicate that HIF1 stabilization that resulted from the lack of pVHL was a major cause of the drastically reduced population of thymocytes beyond the double-negative (DN) stage. Nonetheless, a challenge to this concept is posed by a later publication in which a separate team of researchers reported far more modest effects after *Cd4*-Cre-driven deletion of the conditional *Vhl* allele (*Vhl* f/f) [57]. Specifically, normal frequencies and numbers of each major thymocyte subset along with normal numbers of $CD4^+$ T cells were reported, albeit with a decrease in the mature $CD8^+$ T-cell compartment. A likely explanation for this stark difference in results is that *Cd4*-Cre is known to delete later in thymocyte progression than the *Lck*-Cre used in the earlier work; by inference, the loss of pVHL expression was most crucial at an early stage of $CD4^-$ $CD8^+$ (DN) thymocytes. An implication is that prolonged use of compounds that inhibit PHD enzymatic activity could have similar effects during the phase when thymopoiesis produces new T cells.

6.3 Influences of Hypoxia and HIFs During and After Activation of Naïve T Cells: Balancing Functional Subsets

The output of a pre-immune repertoire of T cells sets the stage for the myriad functions of these lymphocytes in managing the relationship between microbes and complex metazoans seeking to maximize reproductive fitness. Included in this output are types of T cells that are "less conventional" than the canonical model of thymic positive and negative selection sketched in the preceding section—including "invariant NK T-cells" (in mice, termed iNKT) and mucosa-associated invariant T (MAIT) cells. All of these branches of T lineage identity include subsets that circulate through or establish relatively persistent residency in tissues, both at sites of inflammation and in anatomical regions most closely interfacing with the microbial universe. In addition, however, most include phases of passing through the arterial circulation before and/or after residence in a secondary lymphoid organ (spleen or lymph node) and in some cases moving from tissue interstitial into the lymphatic vessels and thoracic duct. The concentration of oxygen varies among these sites from a range at which HIF1α and HIF2α are rapidly degraded to levels at which hypoxic HIF induction is expected [27–29, 44]. In addition to this likelihood of hypoxic stabilization of HIFα subunits to form transcriptionally active αβ dimers, TCR-mediated T-cell activation drives normoxic HIF induction.

Some of the earliest work on the relationship among hypoxia, HIF, and T cells revealed that the transduction of signals from the TCR after its stimulation is altered, as is the balance between two HIF1α isoforms [58–61]. These points, while not pursued deeply in later work, are important in that the exact mix of quantitative signaling along pathways that are induced by the Ag receptor and costimulatory receptors is a major determinant of the differentiation trajectories of T cells ([62, 63], reviewed in [64, 65]).

6.4 CD8 T Cells

After initial activation and proliferation, CD8 T cells differentiate into cytolytic effectors and ultimately—in parallel or as linear descendants of effectors—into at least three classes of memory lymphocyte (effector memory, the most actively recirculating; central, tending to be in secondary lymphoid organs; and tissue-resident) (reviewed in [66–70]). The main focus of work relating to function during this phase of the lineage is on the capacity to kill target cells bearing antigenic peptide in an MHC class I protein. In evolutionary terms, these targets might be host cells infected with an intracellular microbe such as a virus or intracellular bacterium. However, the capacity of cytolytic attack to aid in control or elimination of cancer cells is a major driver of research in this area (Fig. 6.3). Additional fates for CD8 T cells after TCR stimulation by peptide: MHC-I complexes appear to include some form of ineffectual effector function termed exhaustion [71, 72]. As for each of the cell types in which HIF functions are explored, a distinction needs to be made between effects of pVHL elimination—with constitutive stabilization, usually at higher levels than what is achieved with hypoxia—and HIF LoF on its own.

Benchmark analyses of cytotoxic T lymphocyte (CTL) development and function focused on a cancer model and provided evidence of a HIF1-dependent mechanism that restrained the effector actions of CD8 T cells via increased expression of two ecto-enzymes that convert extracellular ATP to adenosine and act on adenosine receptors on the T cells [73–76] (Fig. 6.3). Of note, the HIF1—A2a adenosine receptor (A2AR) connection—appears sufficiently robust as to support moving an adenosine receptor inhibitor into the clinic [76]. This body of work has been independently confirmed and embellished or extended by several groups. One line of papers showed that induction of HIF1, but not HIF2, mediated the reduced differentiation, interferon gamma (IFNγ) expression, and antitumor activity of CD8 T cells caused by intratumoral hypoxia [77]. Intriguingly, in addition to mediation of increased glycolytic gene expression, HIF1 also supported induction of the vascular endothelial growth factor VEGF-A, whose expression by CD8 T cells made a modest contribution toward the HIF-dependent antitumor function [78]. Of note, the same research teams provided evidence of a HIF-dependent accumulation of the metabolite 2-hydroxyglutarate (2-OHG)—especially the S-enantiomer—in activated CD8 T cells [78]. The high concentrations of 2-OHG in CD8 T cells provide a link to the altered gene regulation [79], because this oncometabolite—originally prominent as a product of mutations altering intermediary metabolism—inhibits several dioxygenase enzymes that edit chromatin methylation [80–83]. Further follow-up analyses have more recently provided evidence that part of the function of HIF—which is preponderantly A2AR-mediated [76, 84]—is on mitochondrial function and has experimentally confirmed the requirement for one of the ATP-processing ecto-enzymes, CD39 [85, 86] as shown in the earlier papers [73–76]. Cell-intrinsic impacts on hypoxia and HIF1 during T-cell activation and function

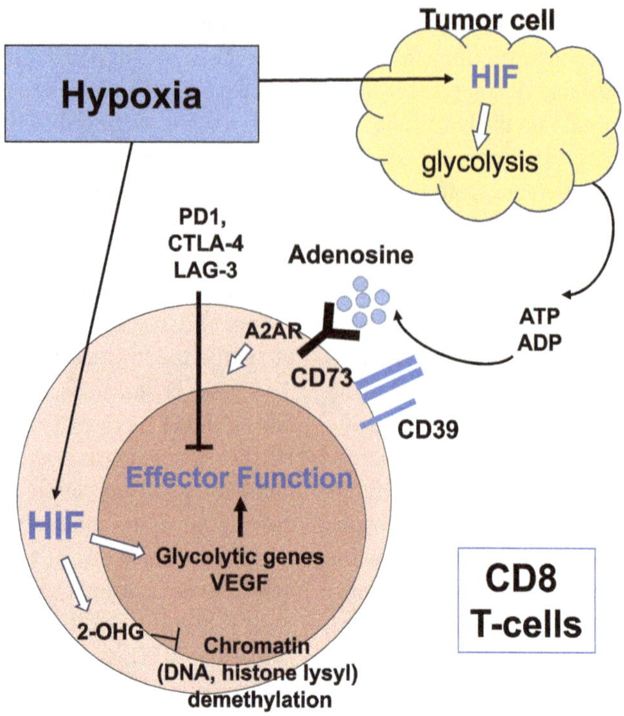

Fig. 6.3 HIF stabilization, for instance, in an oxygen-poor tumor microenvironment, acts in CD8 T cells to alter their physiology and function. As discussed in the main text, increased expression of the purine-processing ecto-enzymes in CD39 and CD73 leads to increased processing of extracellular ATP and ADP into AMP and then adenosine, which then acts on A2 adenosine receptors (A2AR) on the activated CD8 T cell to reduce their functional vigor. HIF also leads to 2-hydroxyglutarate (2-OHG) generation and shifts the balance between effector and memory functions

also have been analyzed in models of infection or inflammation. Here also hypoxia inhibited T cells by HIF-dependent mechanisms [87–89]. Although the potential for distinct differences to arise in chronic infection or inflammation when compared to acute models (reviewed in [90]), a coherent and replicated body of evidence indicates that HIF1 mediates the acquisition of key effector functions of CD8 T cells during and after their activation by Ag.

6.4.1 HIF in Engineered Settings: Bypassing pVHL-Mediated Degradation; Adoptive Cell Therapy

A special case pertaining to functions of HIF in T cells arises in considering experiments with non-physiological regulation of stabilization and induction. As opposed to persistent HIF stabilization by elimination of pVHL, α-subunit prolyl hydroxylation has been disrupted by conditional Cre-mediated inactivation of the three genes encoding PHD enzymes with an approach that impacted all CD4 subsets (covered below) as well as CD8 T cells [91]. This work—which focused on use of lung cancer model—provided evidence that these enzymes restrain CD8 T cells as persistent HIF stabilization increased antitumor activity and CTL function. One notable model-dependent difference from the bulk of other results was that evidence of substantial nonredundant HIF2 functions although HIF1 had a more potent function.

As in analyses of cell development and differentiation, gene-targeted inactivation of *Vhl* gene coding provides a simplified approach to testing effects of HIFs within particular cell lineages. Inactivation of the *Vhl* gene function has been performed using *dLck*-iCre (a codon-optimized version of Cre), whose expression of the recombinase at the end of thymic positive selection [92] tends to minimize developmental effects of the LoF (e.g., [93, 94]). Using this transgene, CD8 T cells were found to have increased effector function and to cause more immunopathology in mouse models of acute and chronic models of infection with lymphocytic choriomeningitis virus (LCMV) [95]. This HIF1 (but not −2)-dependent process was associated with increased proton secretion and decreased respiration (O_2 consumption) as well as a shift in ostensible subsets of CD8 effectors to a profile more enriched in the marker killer cell lectin-like receptor subfamily G member 1 (KLRG1) [95]. In follow-up work, this group more formally tested the capacity of the pVHL CTL to form protective and function memory T cells [96]. The findings provided evidence that memory CTL well able to generate recall effectors could form efficiently despite a persistent increase in glucose-stimulated extracellular acidification rate (ECAR)—presumed (but not shown) to be heightened glycolysis with less glucose oxidation [96]. This finding countered a previous inference that the transition to memory involved decreasing glycolytic flux [97]. In continuing this line of experiments using the same genetic intervention, elimination of pVHL enhanced the carcinoma-directed cytotoxic activity and restrained tumor growth in transplantable models [96, 98]. This increased functionality was lost when the genes encoding HIF1α and HIF2α were inactivated along with the elimination of pVHL [98]. In addition to confirming earlier analyses covered in the previous paragraph, this study developed evidence that the control of HIF activity by VHL regulates the balance of memory subsets derived from CD8 T cells and regulates the effectiveness of a memory CTL response. Thus, CD8 T cells lacking pVHL not only had a gene signature and other markers used as surrogates for the tissue-resident type of memory T cell (i.e., Trm) but also were shown to provide anamnestic protection in tumor rechallenge experiments [98]. This analysis suggests that sustained enforcement of HIF stabilization in CD8 T cells generated for adoptive immunotherapy might enhance their therapeutic efficacy. Parallel work used both *Vhl* gene inactivation and delivery of unmutated or modified HIF1α and HIF2α into activated CD8 T cells [99]. In addition to reinforcing the conclusion that the absence of pVHL can enhance the activity of transferred CTL, this study [99] provided evidence for a capacity of HIF2 to contribute to regulation, so that not all aspects of the lymphocyte biology are simply HIF1-dependent. In practice, this concept remains only a possibility. Such cellular therapies involve mixed CD4 and CD8 T cells and, in the case of different chimeric Ag receptor (CAR) T-cell products or models, many combinations and permutations of signaling components. These complexities will pose major obstacles to movement past basic experimental models with transplanted tumors. Nonetheless, follow-up work has indicated that a reduced oxygen tension during initial post-activation culture of CD8$^+$ CAR T cells—which would stabilize HIFs—increased their antitumor activity [100]. Moreover, the molecular biology of HIF regulation may be exploited in a different manner by using the oxygen-sensitive degron domain (ODD) as a molecular switch to constrain the CAR activity in normoxic environments, so that local tumor hypoxia might limit T-cell attack [101–103]. Since various normal tissue microenvironments are hypoxic (reviewed in [30, 31]), this intriguing approach also will require more extensive refinement. Current CAR T-cell products involve a mixture of CD4$^+$ and CD8$^+$ T cells, increasing the need to understand functions of HIF in CD4$^+$ subset generation and function.

6.5 CD4 T-Cell Subsets

CD4 T cells emerge in the periphery after processes of thymic positive and negative selection and include those termed "conventional" (selected via an αβ TCR, among other characteristics), the thymus-derived or natural Treg (tTreg or nTreg) lineage (Fig. 6.2). Other thymus-derived T cells (e.g., γδ T cells; iNKT; MAIT) are not dealt with here due to the paucity of studies relating their programming to HIFs. Progeny of each cell from the pool of naïve conventional CD4 T cells can acquire diverse functional features after activation by Ag and proliferation. These characteristics fit patterns that are termed helper (Th) subsets—Th1, −2, −9, −17, 22, follicular Th (Tfh)—or an induced Treg program (pTreg, also termed iTreg). Although originated based on patterns of cytokine gene expression, key functional features include distinct expression of receptors for chemoattractant cytokines (chemokines). One foundational concept that probably links to HIFs in the overall context of other TFs is that quantitative balancing of metabolic programs is distinct in each subset [104, 105] (reviewed in [106, 107]). Although viewed as differentiated states, a substantial capacity for flexibility (more commonly termed plasticity) ([108] reviewed in [109]) and altered gene expression or function in vivo bears emphasizing due to literature documenting HIF-mediated changes induced by hypoxia.

6.5.1 Balancing Pathogenic and Homeostatic Th17 Cells Against Suppressive Treg Cells

One cluster of fate potentials for the activated CD4 T cells centers on development of a capacity to produce IL17, leading to designation of these subsets as Th17 cells (Fig. 6.4a). Transforming growth factor-beta (TGFβ) contributes to driving the differentiation of two different classes of Th17 but also is central to generation of the suppressive pTreg cells (reviewed in [110–112]) (Fig. 6.4b). The gateway to exploration of HIF was opened by papers reporting that hypoxia stimulated in vitro Th17 differentiation, in part via a mechanism drawing on the HIF-dependent induction of receptors for extracellular adenosine generated by ATP release [113–117]. Later papers explored the mechanisms by which HIF1 weighs in the balance between Th17 and pTreg fates, with rigorous distinction of CD4 T-cell subsets in mind [118–121]. Thus, HIF1 was found to be a key mediator of overall Th17 differentiation, with this work focusing on a direct transcriptional effect on induction of a TF central to Th17 identity, i.e., the retinoid orphan receptor (ROR) RORγt [118, 119, 121]. Of note, the proinflammatory cytokine IL6 strongly promoted HIF1α stabilization by a mechanism dependent on the TF signal transducer and activator of transcription (STAT)3. An independent line of investigation found a similar dependence of Th17 but not pTreg on HIF1 but developed the mechanistic aspect of this balance through identification of mTORC1 activation in the developing Th17 and an increase in glycolysis [119]. Thus, the inhibitor of glycolysis and flux past junctions onto branches before pyruvate generation decreased Th17 differentiation in vitro while increasing pTreg frequencies and mitigated incidence and clinical severity of Th17-dependent disease model [119]. An important recognition subsequent to these studies is one that distinguishes homeostatic Th17—which maintain barrier homeostasis in microbiome-exposed sites such as the intestines—from a pathogenic sub-subset [122]. Thus, one unresolved issue in all the work is if the HIF1 requirement applies to both the pathogenic and homeostatic Th17 sub-subsets. Although the in vitro experiments were performed with conditions ill suited to generating pathogenic Th17 cells, the use of the mouse model experimental autoimmune encephalomyelitis (EAE) would suggest that HIF1 is important for the pathogenic Th17 cells. An important consideration related to the interplay between the environment and the generality (or lack thereof) of findings is that FOXP3 in itself appears to reprogram metabolic function [123], though FOXP3-independent reprogramming also is documented [124]. While not directly testing the role of HIF1 (or HIF2), elegant work has used CRISPR-Cas9-driven inactivation of genes encoding enzymes along the glycolytic pathway that support this inference [125]. Results of this latter paper indicated that glycolysis is essential for the pathogenic (colitis; EAE) Th17 sub-subset but not the homeostatic counterparts. This distinction depended on whether or not the interruption of conventional glycolysis could be bypassed by diverting products of the oxidative pentose

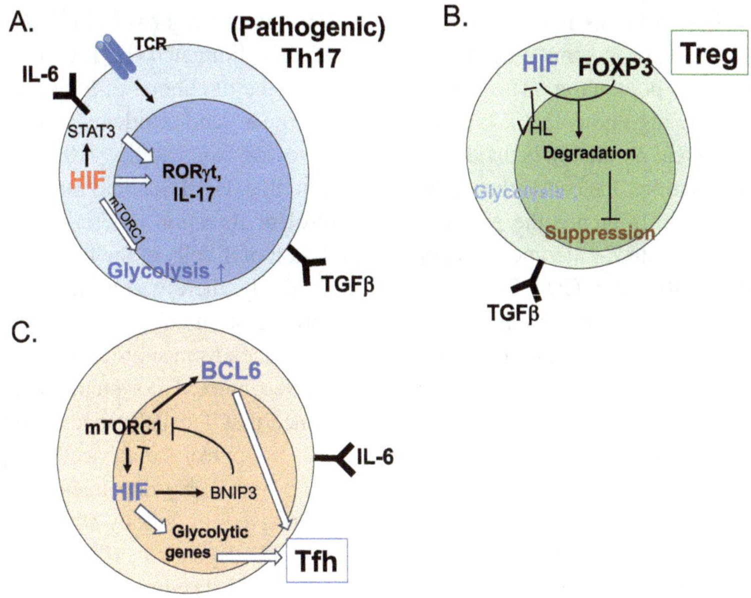

Fig. 6.4 HIF effects and influence on selected subsets of CD4$^+$ T cells. Shown are the three types of CD4$^+$ T cells in which HIF has been most studied, (**a**) Th17 effectors, (**b**) Treg, and (**c**) follicular helper T cells that stimulate B cells prior to and after entry into the germinal center reaction. (**a**) As discussed in the text, both STAT3 and increased glycolytic activity have been identified as mechanisms by which HIF can regulate the balance between Th17 and (**b**) Treg cell fates after TCR-mediated activation of naïve CD4$^+$ T cells. Detailed discussion and citations are in the body of the text. (**c**) Using distinct and different models and time points, experiments have shown that HIF either promotes or restrains the Tfh population or its function after an immune challenge elicits GC. In any case, HIF promotes glycolysis—which is important for Tregs—probably in part via a mechanism involving BCL2/adenovirus E1B 19 kDa protein-interacting protein (BNIP)3. A speculative thesis consistent with the disparate data is that the impact of HIF on proliferation, differentiation, and function of Tfh is modulated by a balance among at least this transcription factor (TF), the repressor BCL6, and mTORC1

phosphate pathway into a reductive continuation that allowed for pyruvate generation. Importantly, hypoxia was identified as a guide to this differential requirement.

6.5.2 Follicular Helpers of B Cells in Ab Responses

Follicular helper (fh) T cells (Tfh), including those in germinal centers (GC) (GC-Tfh), represent another specialized subset of CD4 T cells, one that is essential for elicitation of optimal Ab responses while minimizing the risk of developing sustained production of pathological Ab directed against "self" proteins (reviewed in [126]). Several papers have provided evidence of hypoxia in the follicle and especially in most germinal centers, motivating investigation of HIF functions in development or function of Tfh cells [41, 46, 127, 128] (Fig. 6.4c). An important function of HIF1 in promoting Tfh development and function, and its contribution to increasing an Ab response after immunization, was shown in two concurrent studies [46, 129]. Nonetheless, another approach and body of work yielded evidence of HIF1 tending to restrain the capacity to develop or function as Tfh or GC-Tfh cells [130]. Underscoring the likely complexities of the regulatory balance and systems, in each body of data the HIF effect, could be linked to regulation of mTORC1. Another work is potentially in line with a repressive function of HIF in that elimination of A2AR from CD4 T cells transferred into T-cell-deficient mice led to a 30% increase in Tfh after immunization [131]. Moreover, elimination of pVHL during thymic ontogeny was found to cause a dramatic decrease in Tfh cell number and function [57]. These findings indicate that with superinduction of HIF, or in some other settings, it restrains the Tfh population. The apparent conundrum probably highlights that a key aspect of HIF in physiological regulation is that HIFs do not operate in isolation. Instead, other TFs or modifiers are likely to influence the probability spectrum of "differentiated" outcomes in a population of progeny cells (here, CD4 T cells). The functional characteristics of the Tfh and GC-Tfh subsets revolve around substantially increased or very high expression of the transcriptional repressor B-cell lymphoma protein 6 BCL6 [126]. Although not directly analyzed in the setting of help to B cells, this protein—whose direct actions on chromatin mostly repress transcription—reportedly interferes with HIF in the metabolic regulation of Th1 cells [54]. Other works indicated that the ability to increase rates of glucose uptake or glycolytic flux was vital for Tfh cells [132, 133], and the context dependency of the impact of phosphoinositol-3-kinase (PI3K) activity on Tfh has been noted [134]. Accordingly, one straightforward model to reconcile the results is that the precise balance between transcriptional drive of HIF and the repressive brake of BCL6 determines the response of glycolysis (increase vs not) and thereby the numbers and function of Tfh and GC-Tfh cells. Given that Tfh appears able to form memory T-cell populations with some plasticity [135], whether or how HIF influence the persistence or functions upon reactivation of memory Tfh are intriguing but unanswered questions.

6.5.3 Other T Helper Subsets

Notwithstanding the potential importance of the issues in medical or evolutionary terms, considerably less is known about HIF functions in the differentiation of other T helper subsets or their maintenance as contributors to immune memory. Informative results stem from work with a mouse model of *Mycobacterium tuberculosis* (Mtb) infection in which T lineage-restricted inactivation from the outset of a pre-immune peripheral (CD4, CD8) T-cell repertoire was achieved using two different Cre driver alleles (*Cd4*- and *dLck*-iCre) that eliminated HIF1, pVHL, or both [136]. Despite the likelihood of hypoxia in tuberculous granulomas and evidence that IFNγ-induced nitric oxide can stabilize HIF1α, these means of eliminating HIF1 had no substantive effect on control of infection (bacterial burden) or accumulation of T cells in the lungs of infected mice. Intriguingly, there was a meaningful increase in the proliferation rate of the HIF-deficient T cells. Parallel work from another group, albeit restricted to an elegant ex vivo system, provided important evidence that T cells' capacity to act on macrophages to control the intracellular pathogens can involve both IFNγ-dependent and granulocyte-macrophage colony-stimulating factor (GM-CSF)-dependent, IFNγ-independent mechanisms [137]. The latter was HIF1-dependent, as is the macrophage contribution to antimycobacterial control [138], implying that there may be meaningful in vivo functions of HIF in the type 1 helper response. Of note, the exact mix and balance among the different T-cell subsets in which HIF1 has functional impacts could yield a null result with respect to bacterial burden and lung T-cell numbers in a setting where the impact on Th1 cells was offset by altered CD8 or Treg. Moreover, problems of Th1 cell generation or the response on the IFNγ cytokine/receptor axis are heavily represented in analyses of monogenic mutations in humans that manifest as difficulty controlling attenuated mycobacteria (reviewed in [139]). Accordingly, the more granular details not yet investigated may uncover as yet elusive functions of HIF-1 in Th1 cells, especially in people with subtly hypomorphic variants of the receptors for interleukin (IL)-12 or IFNγ.

The work in the mouse model of tuberculosis mostly used T-cell type-specific elimination of pVHL for constitutive HIFα-subunit stabilization, potentially modeling the effects of persistent hypoxia or normoxic HIF stabilization via treatment with PHD inhibitors. Such inactivation of the *Vhl* gene caused substantial increases in bacterial burden due to dramatic reductions in the recruitment of T cells to the infected lungs [136]. Notably, there was a dramatic further decrease in the Mtb-specific IFNγ$^+$ (Th1) cells among the

CD4 T cells recovered from lungs, and the CD8 population also had less IFNγ$^+$ cells. This latter finding represents an important counterpoint to what may seem the opposite result in papers discussed in the section on pVHL elimination in CD8 T cells [95]. In any event, the effects of *Vhl* gene inactivation were reversed by concurrent inactivation of *Hif1a*, akin to the other work. The profound differences among these results with CTL presumably underscore how the function of HIF1 in T-cell differentiation and function can depend on the exact clinical or biological context (e.g., intracellular bacterium versus viral pathogen or cancer cells).

6.6 Regulation of Mature B Cells and Their Descendants by Hypoxia and HIF

The mature naïve B lymphocytes of the B2 lineage that populate secondary lymphoid organs (SLO)—and other anatomic sites such as the peritoneum and various lymphoid aggregates—are thought of most in terms of their development into Ab-secreting cells. Of note, B cells do have additional functions such as Ag processing and presentation (the internalization of polypeptides, their degradation into a quasi-random set of peptides, and capture of these peptides on class II MHC-encoded proteins) and secretion of immune regulatory or pro-inflammatory cytokines. Although not considered further in this review linking to HIF TFs and hypoxia, it should be noted that some portion of circulating Ab derive from other sources, including marginal zone (MZ) B cells and the distinct developmental branch termed the B1 B lymphocyte (reviewed in [140]). The canonical process leading to development of Ab secretion involves conversion of follicular or MZ B cells into plasmablasts (still in cell cycle) and then plasma cells (PC)—a distinct cell type in which major aspects of the transcriptional landscape have been revised, RNA splicing has been reprogrammed to a secreted form of heavy chain mRNA (reviewed in [141]). Most commonly, these processes arise in follicular B cells, which reside in distinct zones in SLO, after their particular BCR encounters a molecule it recognizes (Fig. 6.5). Concurrent with B-cell activation, helper (CD4$^+$) T cells may also be activated by specific peptide: MHC-II complexes on Ag-presenting cells (APCs) such as DCs. In this setting, both B and T cells may move to be together at the T:B border rather than in the primary follicle. High-affinity and/or multivalent interactions between BCR and Ag can drive extra-follicular differentiation into Ab-secreting PC.

Importantly, however, activated B cells whose Ag receptor has a lower affinity or avidity can be recruited into a secondary follicle (reviewed in [142]), a new micro-anatomic locale along with an activated CD4$^+$ helper T cell. In this germinal center (GC) niche, B cells increase their levels of the TF BCL6 as well as AID, an activation-induced cytidine deaminase that uniquely mediates biochemical processes to cause somatic hypermutation (SHM) at the Ab gene-encoding locus (Fig. 6.5). This quasi-random introduction of nucleotide mutations in the rapidly proliferating GC B-cell population is central to a capacity to diversify the sequences of BCR in progeny cells. In cycles that involve movement between a dark zone, with rapid proliferation, and a light zone, GC B cells compete with one another for requisite new rounds of stimulation via interaction with GC T follicular helper T cells. The nature of this competition favors BCR with higher affinity for the cognate Ag, so that the overall process tends to select for the GC B-cell progeny with higher-affinity BCR. Among the progeny, some will revert to a quiescent state as memory B cells, whereas others will differentiate into Ab-secreting PC. There are many functional and biological implications of this complex iterative process that depends on levels of AID. These include *(i)* a broader array of Ag variants can be recognized upon a later encounter (e.g., infection with a SARS-CoV-2 variant whose external Spike protein has a somewhat different sequence) and *(ii)* higher affinities of Ab can be elicited over time after infection or vaccination, and/or by repeated exposures to Ag. In addition, it appears that the GC-derived population of plasmablasts likely is more efficient at generating long-lived

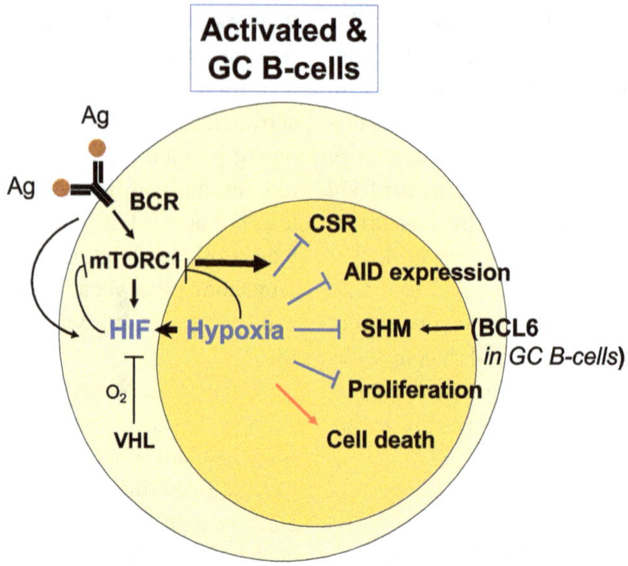

Fig. 6.5 B cells, activated by antigen (Ag) binding to the BCR, increase mTORC1 activity—which promotes proliferation and the activation-induced cytidine deaminase AID, which mediates Ig class-switch recombination (CSR) and, especially in GC B-cells, somatic hypermutation (SHM). Aspects of HIF physiology operate in a reciprocal balance in that mTORC1 can increase HIF levels but HIF stabilization restrains mTORC1, probably in part by blunting increases in the capacity for amino acid uptake [41]. In GC B cells, BCL6 expression is particularly high, in some way fostering SHM, and potentially altering the impact of HIFs as discussed in text

populations of PC at tissue sites (bone marrow or near mucosal surfaces), thereby sustaining Ab concentrations in the circulation years after the immune exposure that elicited B-cell differentiation into PC.

The nature of the GC reaction and of AID-dependent SHM provide a framework for consideration of how hypoxia and HIF TFs influence mature B cells and their PC descendants during and after immunization. Earlier papers had used oxygen-sensing microprobes to provide indications that the spleen has areas with oxygen concentrations (O_2 tension) in the range that would stabilize HIF [28, 44]. Motivated by this finding, initial studies explored the microanatomic distribution of hypoxia using intravital labeling of viable cells with pimonidazole or a similar compound [41, 127, 128]. These papers, using several types of immunization, all found that most (but not all) GC were prominently labeled. Of note, even though the GC-phenotype B cells exhibited greatest adduct formation (i.e., hypoxia), the B-cell follicle had a higher signal than other parts of the spleen or white pulp in the one evaluable case [41]. By this generally accepted criterion for detection of a functionally hypoxic state, then both Tfh cells (as noted in the previous section) and GC B experience an imbalance between O_2 delivery rates and consumption. Moreover, at least three papers have presented evidence that HIF1 levels were higher in GC B cells than non-GC counterparts [41, 128, 143], and in vitro cultures at an oxygen tension (1%) mimicking that at which the intravital tagging of cells with iminidazole adducts would occur [41]. These findings suggest that hypoxia stabilizes HIF1α in most GC of immunized mice. Alternatively, it is possible that in addition to the observed hypoxia stimulation of Ag receptors (BCR and TCR in B and T cells, respectively) drives HIF induction in the secondary follicles (i.e., the GC)—as noted in the section above that touched on the biochemistry of HIF induction. Indeed, an inference to this effect was reported based on analyses of immunofluorescent images after concurrent staining for pimonidazole and HIF1α [143]. Finally, recent work potentially connected the relationship among sufficiency of oxygen supply, GC versus extrafollicular B-cell generation of Ab-secreting cells, and Ab response. Specifically, the enteric pathogen *Salmonella typhimurium* was found to elicit myeloid cell recruitment that led to an environmental hypoxia and collapse of GC [144], which would then favor lower-affinity Ab with less SHM [142]. Although the function of HIF was not analyzed in this work, it is reminiscent of foundational studies of a function for HIF1 in protecting epithelial integrity during inflammation [145].

As with the other cell types discussed above, cell type-specific inactivation of the gene encoding pVHL has been a convenient experimental approach to investigate how HIF induction can influence B-cell physiology. Work on these lines shows that such HIF stabilization induced prior to immunization, blunted activity of mTORC1, and reduced B-cell proliferation, expression of AID, and inclusion of at least one type of Ag-binding B cells in the GC [41]. The decreased proliferation of pVHL-depleted GC B cells was consistent with the earliest work on HIF in B cells [146] and in itself would be expected to decrease class-switching, GC size, and SHM [142, 147]. Indeed, defects were manifested by altered Ab class switching that reduced the most pro-inflammatory isotype of IgG (IgG2c) and by impedance to the increase in affinity with time or Agn rechallenge (i.e., there was decreased affinity maturation) [41]. Combined LoF approaches led to data indicating that the effects of persistent HIF hyperinduction due to inactivation of *Vhl* gene expression were mediated by HIFs, most likely just HIF1 [41, 47]. Potential limitations of this evidence include what probably were supraphysiological levels of stabilized HIF1 in pVHL-deficient B cells, the unremitting nature of the HIF induction, and the use of gene inactivation prior to B-cell activation (discussed two paragraphs later). Notwithstanding these caveats, a vitally important study used mice and SARS-CoV-2 patients to investigate how sustained hypoxemia affected B cells and Ab. This work provided evidence of decreased somatic hypermutation and reduced Ab affinities in humans, consistent with the impact of B-cell type-specific *Vhl* gene inactivation [47].

These findings and the existence of at least two mechanisms by which HIFα subunits can be stabilized in B cells under physiological conditions are pertinent to consideration of the mechanisms by which HIF functions in the mature lymphocyte. To the extent the body of papers with evidence that HIF1 deficiency alters B-cell differentiation is correct, consideration of the question needs to be restricted to methods of gene inactivation that defer loss of function until after the formation of a normal mature preimmune repertoire and population. Published analyses using the conditional ("floxed") *Hif1a* alleles showed that germinal centers and Ab responses derived from HIF1-depleted B cells were reduced [143]. Further, the evidence suggested that there was a direct effect on the GC B cells [143]. This evidence indicates that HIF1 is required in B cells for generation of a normal GC reaction.

Importantly, B cells divide extensively after initial Ag-induced activation and T-cell help, and the clonally expanded population does not start to move into nascent GC until after day 3 postimmunization [148]. Some of the earliest work already suggested that HIF1 has a complex set of effects on cell cycle arrest and regulators [146]. While those in vitro data were generated using mice with a constitutive inactivation of both *Hif1a* alleles in all cells, increased death and decreased rates of population growth also have been observed in experiments that used conditional LoF [41, 143]. As such, one interpretation of the data derived from immunization experiments is that HIF1 expression in B cells was important prior to their differentiation into GC B cells driven by heightened expression of BCL6. On this view, the impact of HIF on GC B-cells under physiological conditions might be indirect and relate to their history prior to GC entry rather than a direct function. Indeed, an alternative possibility has been mooted based on experiments using transfers of B cells with a restricted BCR repertoire into recipient mice whose B cells do not enter GC. In this work, it was reported that a gene set enrichment analysis (GSEA) failed to find increased expression of a hypoxia signature or of increased glucose-dependent proton excretion [149].

The authors lean to the view that HIF1 does act within GC B cells. Broader analyses of RNA expression patterns in GC B cells in mice with normal BCR repertoires, using independent data sets of other publications, provided evidence of hypoxia-related signatures [150]. Work of other labs has validated a relatively new transgenic means of driving expression of the tamoxifen-activated recombinase CreER^{T2} in germinal centers, i.e., the regulatory elements of the G protein-coupled sphingosine phosphate receptor S1PR2 [151–153]. *S1pr2* gene expression is low until the development of GC B-cell identity and localization, so 4-hydroxy-tamoxifen (4-OHT) injection at a time after coalescence of GC is considered exclusively to yield recombination in this subset of B cells. GC B-cell numbers and the IgM Ab-secreting cells (ASC) and Ab-derived from them were all reduced (~50%) when *Hif1a* deletion in B cells was driven by *S1pr2*-CreER^{T2} (Cho SH and Boothby M, unpublished observations). These data indicate that HIF1 functions directly in GC B cells for their homeostasis and capacity to generate Ab-secreting plasmablasts/plasma cells. While the mechanisms accounting for the capacity of HIF1 in B cells to aid the generation of Ag-specific Ab have not been elucidated, published work [154] and preprint data [155] indicate that an enhanced rate of glucose uptake and metabolism likely is one important part of the mechanism. Thus, the glucose transporter GLUT1—encoded by a HIF1 target gene (*Slc2a1*)—is increased on GC B cells, which take up glucose several (~2.5)-fold more rapidly than resting controls [155]. Initial work suggested that B lineage-specific inactivation of the *Slc2a1* gene decreased an Ab response after immunization, but the use of *Cd19*-Cre in this work also lowered the number of B cells to an extent quantitatively the same as the effect on Ab, and the effect on specific cell types derived from the pre-immune mature B cells was not explored [154]. In work extending this analysis of a HIF target gene in B cells, reduced glucose flux in the activated B cells decreased the initial early population of GC B cells, directly affected GC B cells, and in turn decreased the efficiency with which they could generate PC [155]. Parallel and independent work using a *Cd23*-Cre transgene to drive inactivation of *Slc2a1* and deplete GLUT1 from predominantly the naïve follicular B-cell population also identified defects in the initial GC formation and in the generation of PC (Pracht K., personal communication, and [156]). In addition to this metabolic mechanism, in vitro data suggest that HIF1 may also promote B-cell migration [157], processes that are crucial for entry into the secondary follicle and for cycling within the GC [142]. Taken together, the data indicate that HIF—induced either by BCR stimulation of hypoxia—acts in activated and GC B cells to promote the population size and differentiation into PC.

6.7 Outlook

Because oxygen delivery rates vary yet are essential in mammals and myriad other organisms, hypoxia response mechanisms—and the HIF TFs foremost among them—are essential for adjusting cellular and organismal physiology and function. The papers noted here—a necessarily incomplete a survey of the topic—lead to a conclusion that HIF, and especially HIF1, are vital as regulators of the lymphocytic lineages and subsets that are the core of adaptive immunity. However, this overview of the literature provides multiple instances in which the effect of inactivating *Hif1a*, the gene encoding the α-subunit of HIF1, differs among studies. Apart from "fluctuation around a mean" in which there actually is no function of HIF1, the far more likely explanation is that HIF's impact on functional outputs is highly dependent on precise conditions and the quantitative network connecting HIFs, other TFs, and a framework imposed by a particular nutrient and metabolic microenvironment. This synthetic model fits with the fact that selective pressures in evolution tend to promote flexibility in and fine-tuning of outputs as central features of the immune system and probably of hematopoietic production. Apart from the broad sense of the model, the narrow point—that the quantitative impact across expressed genes and HIF target genes will vary according to cell type and effects of other TFs—seems self-evident based on what is known. A corollary is that relying on "HIF (or hypoxia) gene signatures" falls short on rigor. Accordingly, there is an urgent needed tool to develop and propagate a "HIF sensor mouse" based on a design validated in cell lines [158] and in vivo [159]. In such an approach, the level of expression of a fluorophore is controlled by HIF-responsive promoter elements and an oxygen degron domain, so that its emission intensity is proportional to the transcriptional impact of HIF1 and HIF2 in each cell. In addition to facilitating intravital imaging of the degrees of HIF stabilization, the stability is sufficient to allow application of single-cell techniques to lymphocytes (and many other cells) ex vivo. More broadly, analogous biosensors of TFs and metabolic states will advance understanding at molecular level of variance in the impact of HIFs on lymphocyte subsets, as will the development of systems for experimentally controlled fluctuations in HIF in vivo rather just stable LoF. At a point where PHD inhibitors have entered the clinic—and the world of competitive athletics—important issues of how manipulating HIF activity affects human health remain to be explored.

References

1. Haase VH (2013) Regulation of erythropoiesis by hypoxia-inducible factors. Blood Rev 27:41–53
2. Semenza GL (2023) Regulation of erythropoiesis by the hypoxia-inducible Ffctor pathway: Effects of genetic and pharmacological perturbations. Annu Rev Med 74:307–319
3. Tsiftsoglou AS (2021) Erythropoietin (EPO) as a key regulator of erythropoiesis, bone remodeling and endothelial transdifferentiation of multipotent mesenchymal stem cells (MSCs): Implications in regenerative medicine. Cells 10:2140. https://doi.org/10.3390/cells10082140
4. Semenza GL, Nejfelt MK, Chi SM, Antonarakis SE (1991) Hypoxia-inducible nuclear factors bind to an enhancer element located 3′ to the human erythropoietin gene. Proc Natl Acad Sci USA 88:5680–5684
5. Semenza GL (2001) HIF-1 and mechanisms of hypoxia sensing. Curr Opin Cell Biol 12:167–171
6. Kaelin WG Jr, Ratcliffe PJ (2008) Oxygen sensing by metazoans: the central role of the HIF hydroxylase pathway. Mol Cell 30:393–402
7. Takeda K, Aguila HL, Parikh NS, Li X, Lamothe K, Duan LJ, Takeda H, Lee FS, Fong GH (2008) Regulation of adult erythropoiesis by prolyl hydroxylase domain proteins. Blood 111:3229–3235
8. Bell EL, Klimova TA, Eisenbart J, Moraes CT, Murphy MP, Budinger GR, Chandel NS (2007) The Qo site of the mitochondrial complex III is required for the transduction of hypoxic signaling via reactive oxygen species production. J Cell Biol 177:1029–1036
9. Chua YL, Dufour E, Dassa EP, Rustin P, Jacobs HT, Taylor CT, Hagen T (2010) Stabilization of hypoxia-inducible factor-1alpha protein in hypoxia occurs independently of mitochondrial reactive oxygen species production. J Biol Chem 285:31277–31284
10. Hagen T (2012) Oxygen versus reactive oxygen in the regulation of HIF-1α: The balance tips. Biochem Res Int 2012:436981. https://doi.org/10.1155/2012/436981
11. Nakamura H, Makino Y, Okamoto K, Poellinger L, Ohnuma K, Morimoto C, Tanaka H (2005) TCR engagement increases hypoxia-inducible factor-1 alpha protein synthesis via rapamycin-sensitive pathway under hypoxic conditions in human peripheral T-cells. J Immunol 174:7592–7599
12. Meng X, Grötsch B, Luo Y, Knaup KX, Wiesener MS, Chen XX, Jantsch J, Fillatreau S, Schett G, Bozec A (2018) Hypoxia-inducible factor-1α is a critical transcription factor for IL10-producing B-cells in autoimmune disease. Nat Commun 9:251. https://doi.org/10.1038/s41467-017-02683-x
13. Wheeler ML, Defranco AL (2012) Prolonged production of reactive oxygen species in response to B-cell receptor stimulation promotes B-cell activation and proliferation. J Immunol 189:4405–4416
14. Toth RK, Warfel NA (2017) Strange bedfellows: nuclear factor, erythroid 2-like 2 (Nrf2) and hypoxia-inducible factor 1 (HIF-1) in tumor hypoxia. Antioxidants (Basel) 6:27. https://doi.org/10.3390/antiox6020027
15. Ohl K, Tenbrock K (2021) Oxidative stress in SLE T-Cells, Is NRF2 really the target to treat? Front Immunol 12(633845). https://doi.org/10.3389/fimmu.2021.633845
16. Kojima H, Kadono K, Hirao H, Dery KJ, Kupiec-Weglinski JW (2023) CD4+ T-Cell NRF2 signaling improves liver transplantation outcomes by modulating T-Cell activation and differentiation. Antioxid Redox Signal 38:670–683
17. Zhao Z, Wang Y, Gao Y, Ju Y, Zhao Y, Wu Z, Gao S, Zhang B, Pang X, Zhang Y, Wang W (2023) The PRAK-NRF2 axis promotes the differentiation of Th17 cells by mediating the redox homeostasis and glycolysis. Proc Natl Acad Sci USA 120:e2212613120
18. Akashi K, Traver D, Kondo M, Weissman IL (1999) Lymphoid development from hematopoietic stem cells. Int J Hematol 69:217–226

19. Kondo M (2010) Lymphoid and myeloid lineage commitment in multipotent hematopoietic progenitors. Immunol Rev 238:37–46
20. Lévesque JP, Winkler IG (2011) Hierarchy of immature hematopoietic cells related to blood flow and niche. Curr Opin Hematol 18:220–225
21. Höfer T, Busch K, Klapproth K, Rodewald HR (2016) Fate mapping and quantitation of hematopoiesis in vivo. Annu Rev Immunol 34:449–478
22. Nazaraliyev A, Richard E, Sawai CM (2020) In-vivo differentiation of adult hematopoietic stem cells from a single-cell point of view. Curr Opin Hematol 27:241–247
23. Suda T, Takubo K, Semenza GL (2011) Metabolic regulation of hematopoietic stem cells in the hypoxic niche. Cell Stem Cell 9:298–310
24. Spencer JA, Ferraro F, Roussakis E, Klein A, Wu J, Runnels JM, Zaher W, Mortensen LJ, Alt C, Turcotte R, Yusuf R, Côté D, Vinogradov SA, Scadden DT, Lin CP (2014) Direct measurement of local oxygen concentration in the bone marrow of live animals. Nature 508:269–273
25. Itkin T, Gur-Cohen S, Spencer JA, Schajnovitz A, Ramasamy SK, Kusumbe AP, Ledergor G, Jung Y, Milo I, Poulos MG, Kalinkovich A, Ludin A, Kollet O, Shakhar G, Butler JM, Rafii S, Adams RH, Scadden DT, Lin CP, Lapidot T (2016) Distinct bone marrow blood vessels differentially regulate haematopoiesis. Nature 532:323–328
26. Christodoulou C, Spencer JA, Yeh SA, Turcotte R, Kokkaliaris KD, Panero R, Ramos A, Guo G, Seyedhassantehrani N, Esipova TV, Vinogradov SA, Rudzinskas S, Zhang Y, Perkins AS, Orkin SH, Calogero RA, Schroeder T, Lin CP, Camargo FD (2020) Live-animal imaging of native haematopoietic stem and progenitor cells. Nature 578:278–283
27. Star-Lack JM, Adalsteinsson E, Adam MF, Terris DJ, Pinto HA, Brown JM, Spielman DM (2000) In vivo 1H MR spectroscopy of human head and neck lymph node metastasis and comparison with oxygen tension measurements. Am J Neuroradiol 21:183–193
28. Braun RD, Lanzen JL, Snyder SA, Dewhirst MW (2001) Comparison of tumor and normal tissue oxygen tension measurements using OxyLite or microelectrodes in rodents. Am J Physiol Heart Circ Physiol 280:H2533–H2544
29. Lanzen J, Braun RD, Klitzman B, Brizel D, Secomb TW, Dewhirst MW (2006) Direct demonstration of instabilities in oxygen concentrations within the extravascular compartment of an experimental tumor. Cancer Res 66:2219–2223
30. Taylor CT, Colgan SP (2017) Regulation of immunity and inflammation by hypoxia in immunological niches. Nat Rev Immunol 17:774–785
31. Burrows N, Maxwell PH (2017) Hypoxia and B-cells. Exp Cell Res 356:197–203
32. Kandarakov O, Belyavsky A, Semenova E (2022) Bone marrow niches of hematopoietic stem and progenitor cells. Int J Mol Sci 23:4462. https://doi.org/10.3390/ijms23084462
33. Takubo K, Goda N, Yamada W, Iriuchishima H, Ikeda E, Kubota Y, Shima H, Johnson RS, Hirao A, Suematsu M, Suda T (2010) Regulation of the HIF-1alpha level is essential for hematopoietic stem cells. Cell Stem Cell 7:391–402
34. Vukovic M, Sepulveda C, Subramani C, Guitart AV, Mohr J, Allen L, Panagopoulou TI, Paris J, Lawson H, Villacreces A, Armesilla-Diaz A, Gezer D, Holyoake TL, Ratcliffe PJ, Kranc KR (2016) Adult hematopoietic stem cells lacking Hif-1α self-renew normally. Blood 127:2841–2846
35. Yang W, Cong Y (2021) Gut microbiota-derived metabolites in the regulation of host immune responses and immune-related inflammatory diseases. Cell Mol Immunol 18:866–877
36. Guitart AV, Subramani C, Armesilla-Diaz A, Smith G, Sepulveda C, Gezer D, Vukovic M, Dunn K, Pollard P, Holyoake TL, Enver T, Ratcliffe PJ, Kranc KR (2013) Hif-2α is not essential for cell-autonomous hematopoietic stem cell maintenance. Blood 122:1741–1745
37. Sanghani NS, Haase VH (2019) Hypoxia-inducible factor activators in renal anemia: current clinical experience. Adv Chronic Kidney Dis 26:253–266
38. Forristal CE, Winkler IG, Nowlan B, Barbier V, Walkinshaw G, Levesque JP (2013) Pharmacologic stabilization of HIF-1α increases hematopoietic stem cell quiescence in vivo and accelerates blood recovery after severe irradiation. Blood 121:759–769
39. Forristal CE, Nowlan B, Jacobsen RN, Barbier V, Walkinshaw G, Walkley CR, Winkler IG, Levesque JP (2015) HIF-1α is required for hematopoietic stem cell mobilization and 4-prolyl hydroxylase inhibitors enhance mobilization by stabilizing HIF-1α. Leukemia 29:1366–1378
40. Singh RP, Franke K, Kalucka J, Mamlouk S, Muschter A, Gembarska A, Grinenko T, Willam C, Naumann R, Anastassiadis K, Stewart AF, Bornstein S, Chavakis T, Breier G, Waskow C, Wielockx B (2013) HIF prolyl hydroxylase 2 (PHD2) is a critical regulator of hematopoietic stem cell maintenance during steady-state and stress. Blood 121:5158–5166
41. Cho SH, Raybuck AL, Stengel K, Wei M, Beck TC, Volanakis E, Thomas JW, Hiebert S, Haase VH, Boothby MR (2016) Germinal centre hypoxia and regulation of antibody qualities by a hypoxia response system. Nature 537:234–238
42. Jakubison BL, Sarkar T, Gudmundsson KO, Singh S, Sun L, Morris HM, Klarmann KD, Keller JR (2022) ID2 and HIF-1α collaborate to protect quiescent hematopoietic stem cells from activation, differentiation, and exhaustion. J Clin Invest 132:e152599. https://doi.org/10.1172/JCI152599
43. Burrows N, Bashford-Rogers RJM, Bhute VJ, Peñalver A, Ferdinand JR, Stewart BJ, Smith JEG, Deobagkar-Lele M, Giudice G, Connor TM, Inaba A, Bergamaschi L, Smith S, Tran MGB, Petsalaki E, Lyons PA, Espeli M, Huntly BJP, Smith KGC,

Cornall RJ, Clatworthy MR, Maxwell PH (2020) Dynamic regulation of hypoxia-inducible factor-1α activity is essential for normal B-cell development. Nat Immunol 21:1408–1420

44. Caldwell CC, Kojima H, Lukashev D, Armstrong J, Farber M, Apasov SG, Sitkovsky MV (2001) Differential effects of physiologically relevant hypoxic conditions on T lymphocyte development and effector functions. J Immunol 167:6140–6149
45. Kojima H, Kobayashi A, Sakurai D, Kanno Y, Hase H, Takahashi R, Totsuka Y, Semenza GL, Sitkovsky MV, Kobata T (2010) Differentiation stage-specific requirement in hypoxia-inducible factor-1alpha-regulated glycolytic pathway during murine B-cell development in bone marrow. J Immunol 184:154–163
46. Cho SH, Raybuck AL, Blagih J, Kemboi E, Haase VH, Jones RG, Boothby MR (2019) Hypoxia-inducible factors in CD4+ T-cells promote metabolism, switch cytokine secretion, and T-cell help in humoral immunity. Proc Natl Acad Sci USA 116:8975–8984
47. Kotagiri P, Mescia F, Hanson AL, Turner L, Bergamaschi L, Peñalver A, Richoz N, Moore SD, Ortmann BM, Dunmore BJ, Morgan MD, Tuong ZK, Cambridge Institute of Therapeutic Immunology and Infectious Disease-National Institute of Health Research (CITIID-NIHR) COVID BioResource Collaboration, Göttgens B, Toshner M, Hess C, Maxwell PH, Clatworthy MR, Nathan JA, Bradley JR, Lyons PA, Burrows N, Smith KGC (2022) The impact of hypoxia on B-cells in COVID-19. EBio Med 77:103878. https://doi.org/10.1016/j.ebiom.2022.103878
48. Xu S, Huo J, Huang Y, Aw M, Chen S, Mak S, Yip LY, Ho YS, Ng SW, Tan AH, Lee A, Ou X, Lam KP (2019) von Hippel-Lindau protein maintains metabolic balance to regulate the survival of naive B lymphocytes. iScience 17:379–392
49. Thompson PK, Zúñiga-Pflücker JC (2011) On becoming a T-cell, a convergence of factors kick it up a Notch along the way. Semin Immunol 23:350–359
50. Fontenot JD, Rudensky AY (2005) A well adapted regulatory contrivance: regulatory T-cell development and the forkhead family transcription factor Foxp3. Nat Immunol 6:331–337
51. Savage PA, Klawon DEJ, Miller CH (2020) Regulatory T-cell development. Annu Rev Immunol 38:421–453
52. Hale LP, Braun RD, Gwinn WM, Greer PK, Dewhirst MW (2001) Hypoxia in the thymus: role of oxygen tension in thymocyte survival. Am J Physiol Heart Circ Physiol 282:H1467–H1477
53. Biju MP, Neumann AK, Bensinger SJ, Johnson RS, Turka LA, Haase VH (2004) Vhlh gene deletion induces Hif-1-mediated cell death in thymocytes. Mol Cell Biol 24:9038–9047
54. Oestreich KJ, Read KA, Gilbertson SE, Hough KP, McDonald PW, Krishnamoorthy V, Weinmann AS (2014) Bcl-6 directly represses the gene program of the glycolysis pathway. Nat Immunol 15:957–964
55. Kierans SJ, Taylor CT (2021) Regulation of glycolysis by the hypoxia-inducible factor (HIF): implications for cellular physiology. J Physiol 599:23–37
56. Ciofani M, Zúñiga-Pflücker JC (2005) Notch promotes survival of pre-T-cells at the beta-selection checkpoint by regulating cellular metabolism. Nat Immunol 6:881–888
57. Zhu Y, Zhao Y, Zou L, Zhang D, Aki D, Liu YC (2019) The E3 ligase VHL promotes follicular helper T-cell differentiation via glycolytic-epigenetic control. J Exp Med 216:1664–1681
58. Lukashev D, Caldwell C, Ohta A, Chen P, Sitkovsky M (2001) Differential regulation of two alternatively spliced isoforms of hypoxia-inducible factor-1 alpha in activated T lymphocytes. J Biol Chem 276:48754–48763
59. Makino Y, Nakamura H, Ikeda E, Ohnuma K, Yamauchi K, Yabe Y, Poellinger L, Okada Y, Morimoto C, Tanaka H (2003) Hypoxia-inducible factor regulates survival of antigen receptor-driven T-cells. J Immunol 171:6534–6540
60. Neumann AK, Yang J, Biju MP, Joseph SK, Johnson RS, Haase VH, Freedman BD, Turka LA (2005) Hypoxia inducible factor 1 alpha regulates T-cell receptor signal transduction. Proc Natl Acad Sci USA 102:17071–11706
61. Lukashev D, Klebanov B, Kojima H, Grinberg A, Ohta A, Berenfeld L, Wenger RH, Ohta A, Sitkovsky M (2006) Cutting edge: hypoxia-inducible factor 1alpha and its activation-inducible short isoform I.1 negatively regulate functions of CD4+ and CD8+ T lymphocytes. J Immunol 177:4962–4965
62. Chang CF, D'Souza WN, Ch'en IL, Pages G, Pouyssegur J, Hedrick SM (2012) Polar opposites: Erk direction of CD4 T-cell subsets. J Immunol 189:721–731
63. Tubo NJ, Pagán AJ, Taylor JJ, Nelson RW, Linehan JL, Ertelt JM, Huseby ES, Way SS, Jenkins MK (2013) Single naive CD4+ T cells from a diverse repertoire produce different effector cell types during infection. Cell 153:785–796
64. Constant SL, Bottomly K (1997) Induction of Th1 and Th2 CD4+ T-cell responses: the alternative approaches. Annu Rev Immunol 15:297–322
65. Tubo NJ, Jenkins MK (2014) TCR signal quantity and quality in CD4+ T-cell differentiation. Trends Immunol 35:591–596
66. Masopust D, Soerens AG (2019) Tissue-resident T-cells and other resident leukocytes. Annu Rev Immunol 37:521–546
67. Behr FM, Parga-Vidal L, Kragten NAM, van Dam TJP, Wesselink TH, Sheridan BS, Arens R, van Lier RAW, Stark R, van Gisbergen KPJM (2020) Tissue-resident memory CD8+ T-cells shape local and systemic secondary T-cell responses. Nat Immunol 21:1070–1081
68. Turner SJ, Bennett TJ, La Gruta NL (2021) CD8+ T-cell memory: The why, the when, and the how.

Cold Spring Harb Perspect Biol 13(5):a038661. https://doi.org/10.1101/cshperspect.a038661

69. Pace L (2021) Temporal and epigenetic control of plasticity and fate decision during CD8+ T-cell memory differentiation. Cold Spring Harb Perspect Biol 13:a037754. https://doi.org/10.1101/cshperspect.a037754
70. Muroyama Y, Wherry EJ (2021) Memory T-cell heterogeneity and terminology. Cold Spring Harb Perspect Biol 13:a037929. https://doi.org/10.1101/cshperspect.a037929. PMID: 33782027
71. Philip M, Schietinger A (2022) CD8+ T-cell differentiation and dysfunction in cancer. Nat Rev Immunol 22:209–223
72. van der Heide V, Humblin E, Vaidya A, Kamphorst AO (2022) Advancing beyond the twists and turns of T-cell exhaustion in cancer. Sci Transl Med 14:eabo4997. https://doi.org/10.1126/scitranslmed.abo4997
73. Ohta A, Ohta A, Madasu M, Kini R, Subramanian M, Goel N, Sitkovsky M (2009) A2A adenosine receptor may allow expansion of T-cells lacking effector functions in extracellular adenosine-rich microenvironments. J Immunol 183:5487–5493
74. Hatfield SM, Kjaergaard J, Lukashev D, Belikoff B, Schreiber TH, Sethumadhavan S, Abbott R, Philbrook P, Thayer M, Shujia D, Rodig S, Kutok JL, Ren J, Ohta A, Podack ER, Karger B, Jackson EK, Sitkovsky M (2014) Systemic oxygenation weakens the hypoxia and hypoxia inducible factor 1α-dependent and extracellular adenosine-mediated tumor protection. J Mol Med (Berl) 92:1283–1292
75. Hatfield SM, Kjaergaard J, Lukashev D, Schreiber TH, Belikoff B, Abbott R, Sethumadhavan S, Philbrook P, Ko K, Cannici R, Thayer M, Rodig S, Kutok JL, Jackson EK, Karger B, Podack ER, Ohta A, Sitkovsky MV (2015) Immunological mechanisms of the antitumor effects of supplemental oxygenation. Sci Transl Med 7:277ra30. https://doi.org/10.1126/scitranslmed.aaa1260
76. Kjaergaard J, Hatfield S, Jones G, Ohta A, Sitkovsky M (2018) A_{2A} adenosine receptor gene deletion or synthetic A_{2A} antagonist liberate tumor-reactive CD8+ T-cells from tumor-induced immunosuppression. J Immunol 201:782–791
77. Palazon A, Tyrakis PA, Macias D, Veliça P, Rundqvist H, Fitzpatrick S, Vojnovic N, Phan AT, Loman N, Hedenfalk I, Hatschek T, Lövrot J, Foukakis T, Goldrath AW, Bergh J, Johnson RS (2017) An HIF-1α/VEGF-A axis in cytotoxic T-cells regulates tumor progression. Cancer Cell 32:669–683.e5
78. Tyrakis PA, Palazon A, Macias D, Lee KL, Phan AT, Veliça P, You J, Chia GS, Sim J, Doedens A, Abelanet A, Evans CE, Griffiths JR, Poellinger L, Goldrath AW, Johnson RS (2016) S-2-hydroxyglutarate regulates CD8+ T-lymphocyte fate. Nature 540:236–241
79. Ford BR, Vignali PDA, Rittenhouse NL, Scharping NE, Peralta R, Lontos K, Frisch AT, Delgoffe GM, Poholek AC (2022) Tumor microenvironmental signals reshape chromatin landscapes to limit the functional potential of exhausted T-cells. Sci Immunol 7:eabj9123. https://doi.org/10.1126/sciimmunol.abj9123
80. Xu W, Yang H, Liu Y, Yang Y, Wang P, Kim SH, Ito S, Yang C, Wang P, Xiao MT, Liu LX, Jiang WQ, Liu J, Zhang JY, Wang B, Frye S, Zhang Y, Xu YH, Lei QY, Guan KL, Zhao SM, Xiong Y (2011) Oncometabolite 2-hydroxyglutarate is a competitive inhibitor of α-ketoglutarate-dependent dioxygenases. Cancer Cell 19:17–30
81. Hancock RL, Dunne K, Walport LJ, Flashman E, Kawamura A (2015) Epigenetic regulation by histone demethylases in hypoxia. Epigenomics 7:791–811
82. Waitkus MS, Diplas BH, Yan H (2018) Biological role and therapeutic potential of IDH mutations in cancer. Cancer Cell 34:186–195
83. Lio CJ, Rao A (2019) TET enzymes and 5hmC in adaptive and innate immune systems. Front Immunol 10:210. https://doi.org/10.3389/fimmu.2019.00210
84. Ohta A, Madasu M, Subramanian M, Kini R, Jones G, Choukèr A, Ohta A, Sitkovsky M (2014) Hypoxia-induced and A2A adenosine receptor-independent T-cell suppression is short-lived and easily reversible. Int Immunol 26:83–91
85. Scharping NE, Rivadeneira DB, Menk AV, Vignali PDA, Ford BR, Rittenhouse NL, Peralta R, Wang Y, Wang Y, DePeaux K, Poholek AC, Delgoffe GM (2021) Mitochondrial stress induced by continuous stimulation under hypoxia rapidly drives T-cell exhaustion. Nat Immunol 22:205–215
86. Vignali PDA, DePeaux K, Watson MJ, Ye C, Ford BR, Lontos K, McGaa NK, Scharping NE, Menk AV, Robson SC, Poholek AC, Rivadeneira DB, Delgoffe GM (2023) Hypoxia drives CD39-dependent suppressor function in exhausted T-cells to limit antitumor immunity. Nat Immunol 24:267–279
87. Ohta A, Diwanji R, Kini R, Subramanian M, Ohta A, Sitkovsky M (2011) In vivo T-cell activation in lymphoid tissues is inhibited in the oxygen-poor microenvironment. Front Immunol 2:27. https://doi.org/10.3389/fimmu.2011.00027
88. Thiel M, Caldwell CC, Kreth S, Kuboki S, Chen P, Smith P, Ohta A, Lentsch AB, Lukashev D, Sitkovsky MV (2007) Targeted deletion of HIF-1alpha gene in T-cells prevents their inhibition in hypoxic inflamed tissues and improves septic mice survival. PLoS One 2:e853. https://doi.org/10.1371/journal.pone.0000853. PMID: 17786224
89. Georgiev P, Belikoff BG, Hatfield S, Ohta A, Sitkovsky MV, Lukashev D (2013) Genetic deletion of the HIF-1α isoform I.1 in T-cells enhances antibacterial immunity and improves survival in a murine peritonitis model. Eur J Immunol 43:655–666
90. Chung HK, McDonald B, Kaech SM (2021) The architectural design of CD8+ T-cell responses in acute and chronic infection: Parallel structures with divergent fates. J Exp Med 218:e20201730. https://doi.org/10.1084/jem.20201730

91. Doedens AL, Phan AT, Stradner MH, Fujimoto JK, Nguyen JV, Yang E, Johnson RS, Goldrath Clever D, Roychoudhuri R, Constantinides MG, Askenase MH, Sukumar M, Klebanoff CA, Eil RL, Hickman HD, Yu Z, Pan JH, Palmer DC, Phan AT, Goulding J, Gattinoni L, Goldrath AW, Belkaid Y, Restifo NP (2016) Oxygen sensing by T-cells establishes an immunologically tolerant metastatic niche. Cell 166:1117–1131.e14
92. Zhang DJ, Wang Q, Wei J, Baimukanova G, Buchholz F, Stewart AF, Mao X, Killeen N (2005) Selective expression of the Cre recombinase in late-stage thymocytes using the distal promoter of the Lck gene. J Immunol 174:6725–6731
93. Lee K, Gudapati P, Dragovic S, Spencer C, Joyce S, Killeen N, Magnuson MA, Boothby M (2010) Mammalian target of rapamycin protein complex 2 regulates differentiation of Th1 and Th2 cell subsets via distinct signaling pathways. Immunity 32:743–753
94. Lee K, Nam KT, Cho SH, Gudapati P, Hwang Y, Park DS, Potter R, Chen J, Volanakis E, Boothby M (2012) Vital roles of mTOR complex 2 in Notch-driven thymocyte differentiation and leukemia. J Exp Med 209:713–728
95. Doedens AL, Phan AT, Stradner MH, Fujimoto JK, Nguyen JV, Yang E, Johnson RS, Goldrath AW (2013) Hypoxia-inducible factors enhance the effector responses of CD8(+) T-cells to persistent antigen. Nat Immunol 14:1173–1182
96. Phan AT, Doedens AL, Palazon A, Tyrakis PA, Cheung KP, Johnson RS, Goldrath AW (2016) Constitutive glycolytic metabolism supports CD8^{+} T-cell effector memory differentiation during viral infection. Immunity 45:1024–1037
97. Pearce EL, Walsh MC, Cejas PJ, Harms GM, Shen H, Wang LS, Jones RG, Choi Y (2009) Enhancing CD8 T-cell memory by modulating fatty acid metabolism. Nature 460:103–107
98. Liikanen I, Lauhan C, Quon S, Omilusik K, Phan AT, Bartrolí LB, Ferry A, Goulding J, Chen J, Scott-Browne JP, Yustein JT, Scharping NE, Witherden DA, Goldrath AW (2021) Hypoxia-inducible factor activity promotes antitumor effector function and tissue residency by CD8+ T-cells. J Clin Invest 131:e143729. https://doi.org/10.1172/JCI143729
99. Veliça P, Cunha PP, Vojnovic N, Foskolou IP, Bargiela D, Gojkovic M, Rundqvist H, Johnson RS (2021) Modified Hypoxia-Inducible Factor expression in CD8^{+} T-cells increases antitumor efficacy. Cancer Immunol Res 9:401–414
100. Cunha PP, Minogue E, Krause LCM, Hess RM, Bargiela D, Wadsworth BJ, Barbieri L, Brombach C, Foskolou IP, Bogeski I, Velica P, Johnson RS (2023) Oxygen levels at the time of activation determine T-cell persistence and immunotherapeutic efficacy. elife 12:e84280. https://doi.org/10.7554/eLife.84280
101. Juillerat A, Marechal A, Filhol JM, Valogne Y, Valton J, Duclert A, Duchateau P, Poirot L (2017) An oxygen sensitive self-decision making engineered CAR T-cell. Sci Rep 7:39833. https://doi.org/10.1038/srep39833
102. Kosti P, Opzoomer JW, Larios-Martinez KI, Henley-Smith R, Scudamore CL, Okesola M, Taher MYM, Davies DM, Muliaditan T, Larcombe-Young D, Woodman N, Gillett CE, Thavaraj S, Maher J, Arnold JN (2021) Hypoxia-sensing CAR T-cells provide safety and efficacy in treating solid tumors. Cell Rep Med 2:100227
103. He H, Liao Q, Zhao C, Zhu C, Feng M, Liu Z, Jiang L, Zhang L, Ding X, Yuan M, Zhang X, Xu J (2021) Conditioned CAR-T-cells by hypoxia-inducible transcription amplification (HiTA) system significantly enhances systemic safety and retains antitumor efficacy. J Immunother Cancer 9:e002755. https://doi.org/10.1136/jitc-2021-002755
104. Michalek RD, Gerriets VA, Jacobs SR, Macintyre AN, MacIver NJ, Mason EF, Sullivan SA, Nichols AG, Rathmell JC (2011) Cutting edge: distinct glycolytic and lipid oxidative metabolic programs are essential for effector and regulatory CD4+ T-cell subsets. J Immunol 186:3299–3303
105. Cao Y, Rathmell JC, Macintyre AN (2014) Metabolic reprogramming towards aerobic glycolysis correlates with greater proliferative ability and resistance to metabolic inhibition in CD8 versus CD4 T-cells. PLoS One 9:e104104
106. MacIver NJ, Michalek RD, Rathmell JC (2013) Metabolic regulation of T lymphocytes. Annu Rev Immunol 31:259–283
107. Chapman NM, Boothby MR, Chi H (2020) Metabolic coordination of T-cell quiescence and activation. Nat Rev Immunol 20:55–70
108. Adeeku E, Gudapati P, Mendez-Fernandez Y, Van Kaer L, Boothby M (2008) Flexibility accompanies commitment of memory CD4 lymphocytes derived from IL-4 locus-activated precursors. Proc Natl Acad Sci USA 105:9307–9312
109. Zhu J, Paul WE (2010) Heterogeneity and plasticity of T helper cells. Cell Res 20:4–12
110. Kimura A, Kishimoto T (2010) IL-6: regulator of Treg/Th17 balance. Eur J Immunol 40:1830–1835
111. Josefowicz SZ, Lu LF, Rudensky AY (2012) Regulatory T-cells: mechanisms of differentiation and function. Annu Rev Immunol 30:531–564
112. Moreau JM, Velegraki M, Bolyard C, Rosenblum MD, Li Z (2022) Transforming growth factor-β1 in regulatory T-cell biology. Sci Immunol. 7:eabi4613. https://doi.org/10.1126/sciimmunol.abi4613
113. Ben-Shoshan J, Maysel-Auslender S, Mor A, Keren G, George J (2008) Hypoxia controls CD4+CD25+ regulatory T-cell homeostasis via hypoxia-inducible factor-1alpha. Eur J Immunol 38:2412–2418
114. Yin Y, Wen S, Li G, Wang D (2009) Hypoxia enhances stimulating effect of amyloid beta peptide (25-35) for interleukin 17 and T helper lymphocyte subtype 17 upregulation in cultured peripheral blood mononuclear cells. Microbiol Immunol 53:281–286
115. Yin Y, Li G (2011) Hypoxia induces T Helper 17 cell upregulation in cultured peripheral blood mono-

nuclear cells from chronic stage patients of severe cerebral infarction. Microbiol Immunol 55:130–134
116. Wilson JM, Kurtz CC, Black SG, Ross WG, Alam MS, Linden J, Ernst PB (2011) The A2B adenosine receptor promotes Th17 differentiation via stimulation of dendritic cell IL-6. J Immunol 186:6746–6752
117. Ikejiri A, Nagai S, Goda N, Kurebayashi Y, Osada-Oka M, Takubo K, Suda T, Koyasu S (2012) Dynamic regulation of Th17 differentiation by oxygen concentrations. Int Immunol 24:137–146
118. Dang EV, Barbi J, Yang HY, Jinasena D, Yu H, Zheng Y, Bordman Z, Fu J, Kim Y, Yen HR, Luo W, Zeller K, Shimoda L, Topalian SL, Semenza GL, Dang CV, Pardoll DM, Pan F (2011) Control of T(H)17/T(reg) balance by hypoxia-inducible factor 1. Cell 146:772–784
119. Shi LZ, Wang R, Huang G, Vogel P, Neale G, Green DR, Chi H (2011) HIF1alpha-dependent glycolytic pathway orchestrates a metabolic checkpoint for the differentiation of TH17 and Treg cells. J Exp Med 208:1367–1376
120. Clambey ET, McNamee EN, Westrich JA, Glover LE, Campbell EL, Jedlicka P, de Zoeten EF, Cambier JC, Stenmark KR, Colgan SP, Eltzschig HK (2012) Hypoxia-inducible factor-1 alpha-dependent induction of FoxP3 drives regulatory T-cell abundance and function during inflammatory hypoxia of the mucosa. Proc Natl Acad Sci USA 109:E2784–E2793. https://doi.org/10.1073/pnas.1202366109
121. Wang H, Flach H, Onizawa M, Wei L, McManus MT, Weiss A (2014) Negative regulation of Hif1a expression and TH17 differentiation by the hypoxia-regulated microRNA miR-210. Nat Immunol 15:393–401
122. Bettelli E, Carrier Y, Gao W, Korn T, Strom TB, Oukka M, Weiner HL, Kuchroo VK (2006) Reciprocal developmental pathways for the generation of pathogenic effector TH17 and regulatory T-cells. Nature 441:235–238
123. Angelin A, Gil-de-Gómez L, Dahiya S, Jiao J, Guo L, Levine MH, Wang Z, Quinn WJ 3rd, Kopinski PK, Wang L, Akimova T, Liu Y, Bhatti TR, Han R, Laskin BL, Baur JA, Blair IA, Wallace DC, Hancock WW, Beier UH (2017) Foxp3 reprograms T-cell metabolism to function in low-glucose, high-lactate environments. Cell Metab 25:1282–1293.e7
124. Charbonnier LM, Cui Y, Stephen-Victor E, Harb H, Lopez D, Bleesing JJ, Garcia-Lloret MI, Chen K, Ozen A, Carmeliet P, Li MO, Pellegrini M, Chatila TA (2019) Functional reprogramming of regulatory T-cells in the absence of Foxp3. Nat Immunol 20:1208–1219
125. Wu L, Hollinshead KER, Hao Y, Au C, Kroehling L, Ng C, Lin WY, Li D, Silva HM, Shin J, Lafaille JJ, Possemato R, Pacold ME, Papagiannakopoulos T, Kimmelman AC, Satija R, Littman DR (2020) Niche-selective inhibition of pathogenic Th17 cells by targeting metabolic redundancy. Cell 182:641–654.e20
126. Vinuesa CG, Linterman MA, Yu D, MacLennan IC (2016) Follicular helper T-cells. Annu Rev Immunol 34:335–368
127. Jellusova J, Cato MH, Apgar JR, Ramezani-Rad P, Leung CR, Chen C, Richardson AD, Conner EM, Benschop RJ, Woodgett JR, Rickert RC (2017) Gsk3 is a metabolic checkpoint regulator in B-cells. Nat Immunol 18:303–312
128. Abbott RK, Thayer M, Labuda J, Silva M, Philbrook P, Cain DW, Kojima H, Hatfield S, Sethumadhavan S, Ohta A, Reinherz EL, Kelsoe G, Sitkovsky M (2016) Germinal center hypoxia potentiates immunoglobulin class switch recombination. J Immunol 197:4014–4020
129. Dong L, He Y, Zhou S, Cao Y, Li Y, Bi Y, Liu G (2019) HIF1α-dependent metabolic signals control the differentiation of follicular helper T-cells. Cells 8:1450
130. Huang B, Phelan JD, Preite S, Gomez-Rodriguez J, Johansen KH, Shibata H, Shaffer AL 3rd, Xu Q, Jeffrey B, Kirby M, Anderson S, Yang Y, Gossa S, McGavern DB, Staudt LM, Schwartzberg PL (2022) In vivo CRISPR screens reveal a HIF-1α-mTOR-network regulates T follicular helper versus Th1 cells. Nat Commun 13:805. https://doi.org/10.1038/s41467-022-28378-6
131. Abbott RK, Silva M, Labuda J, Thayer M, Cain DW, Philbrook P, Sethumadhavan S, Hatfield S, Ohta A, Sitkovsky M (2017) The GS protein-coupled A2a adenosine receptor controls T-cell help in the Germinal Center. J Biol Chem 292:1211–1217
132. Zeng H, Cohen S, Guy C, Shrestha S, Neale G, Brown SA, Cloer C, Kishton RJ, Gao X, Youngblood B, Do M, Li MO, Locasale JW, Rathmell JC, Chi H (2016) mTORC1 and mTORC2 kinase signaling and glucose metabolism drive follicular helper T-cell differentiation. Immunity 45:540–554
133. Zou X, Choi SC, Zeumer-Spataro L, Scindia Y, Moser EK, Morel L (2022) Metabolic regulation of follicular helper T-cell differentiation in a mouse model of lupus. Immunol Lett 247:13–21
134. Preite S, Huang B, Cannons JL, McGavern DB, Schwartzberg PL (2019) PI3K Orchestrates T follicular helper cell differentiation in a context-dependent manner: implications for autoimmunity. Front Immunol 9:3079. https://doi.org/10.3389/fimmu.2018.03079
135. Künzli M, Schreiner D, Pereboom TC, Swarnalekha N, Litzler LC, Lötscher J, Ertuna YI, Roux J, Geier F, Jakob RP, Maier T, Hess C, Taylor JJ, King CG (2020) Long-lived T follicular helper cells retain plasticity and help sustain humoral immunity. Sci Immunol. 5:eaay5552. https://doi.org/10.1126/sciimmunol.aay5552
136. Liu R, Muliadi V, Mou W, Li H, Yuan J, Holmberg J, Chambers BJ, Ullah N, Wurth J, Alzrigat M, Schlisio S, Carow B, Larsson LG, Rottenberg ME (2022) HIF-1 stabilization in T-cells hampers the control of Mycobacterium tuberculosis infection. Nat Commun 13:5093. https://doi.org/10.1038/s41467-022-32639-9
137. Van Dis E, Fox DM, Morrison HM, Fines DM, Babirye JP, McCann LH, Rawal S, Cox JS, Stanley SA (2022) IFN-γ-independent control of M. tuber-

culosis requires CD4 T-cell-derived GM-CSF and activation of HIF-1α. PLoS Pathog 18:e1010721. https://doi.org/10.1371/journal.ppat.1010721
138. Braverman J, Sogi KM, Benjamin D, Nomura DK, Stanley SA (2016) HIF-1α is an essential mediator of IFN-γ-dependent immunity to Mycobacterium tuberculosis. J Immunol 197:1287–1297
139. Bustamante J, Boisson-Dupuis S, Abel L, Casanova JL (2014) Mendelian susceptibility to mycobacterial disease: genetic, immunological, and clinical features of inborn errors of IFN-γ immunity. Semin Immunol 26:454–470
140. Boothby MR, Brookens SK, Raybuck AL, Cho SH (2022) Supplying the trip to antibody production-nutrients, signaling, and the programming of cellular metabolism in the mature B lineage. Cell Mol Immunol 19:352–369
141. Boothby MR, Hodges E, Thomas JW (2019) Molecular regulation of peripheral B-cells and their progeny in immunity. Genes Dev 33:26–48
142. Victora GD, Nussenzweig MC (2022) Germinal centers. Annu Rev Immunol 40:413–442
143. Li L, Feng C, Qin J, Li D, Liu M, Han S, Zheng B (2021) Regulation of humoral immune response by HIF-1α-dependent metabolic reprogramming of the germinal center reaction. Cell Immunol 367:104409. https://doi.org/10.1016/j.cellimm.2021.104409
144. Biram A, Liu J, Hezroni H, Davidzohn N, Schmiedel D, Khatib-Massalha E, Haddad M, Grenov A, Lebon S, Salame TM, Dezorella N, Hoffman D, Abou Karam P, Biton M, Lapidot T, Bemark M, Avraham R, Jung S, Shulman Z (2022) Bacterial infection disrupts established germinal center reactions through monocyte recruitment and impaired metabolic adaptation. Immunity 55:442–458.e8
145. Campbell EL, Bruyninckx WJ, Kelly CJ, Glover LE, McNamee EN, Bowers BE, Bayless AJ, Scully M, Saeedi BJ, Golden-Mason L, Ehrentraut SF, Curtis VF, Burgess A, Garvey JF, Sorensen A, Nemenoff R, Jedlicka P, Taylor CT, Kominsky DJ, Colgan SP (2014) Transmigrating neutrophils shape the mucosal microenvironment through localized oxygen depletion to influence resolution of inflammation. Immunity 40:66–77
146. Goda N, Ryan HE, Khadivi B, McNulty W, Rickert RC, Johnson RS (2003) Hypoxia-inducible factor 1alpha is essential for cell cycle arrest during hypoxia. Mol Cell Biol 23:359–369
147. Gitlin AD, Shulman Z, Nussenzweig MC (2014) Clonal selection in the germinal centre by regulated proliferation and hypermutation. Nature 509:637–640
148. Roco JA, Mesin L, Binder SC, Nefzger C, Gonzalez-Figueroa P, Canete PF, Ellyard J, Shen Q, Robert PA, Cappello J, Vohra H, Zhang Y, Nowosad CR, Schiepers A, Corcoran LM, Toellner KM, Polo JM, Meyer-Hermann M, Victora GD, Vinuesa CG (2019) Class-switch recombination occurs infrequently in germinal centers. Immunity 51:337–350.e7
149. Weisel FJ, Mullett SJ, Elsner RA, Menk AV, Trivedi N, Luo W, Wikenheiser D, Hawse WF, Chikina M, Smita S, Conter LJ, Joachim SM, Wendell SG, Jurczak MJ, Winkler TH, Delgoffe GM, Shlomchik MJ (2020) Germinal center B-cells selectively oxidize fatty acids for energy while conducting minimal glycolysis. Nat Immunol 21:331–342
150. Boothby MR, Raybuck A, Cho SH, Stengel KR, Haase VH, Hiebert S, Li J (2021) Over-generalizing about GC (hypoxia): Pitfalls of limiting breadth of experimental systems and analyses in framing informatics conclusions. Front Immunol 12:664249. https://doi.org/10.3389/fimmu.2021.664249
151. Green JA, Cyster JG (2012) S1PR2 links germinal center confinement and growth regulation. Immunol Rev 247:36–51
152. Ise W, Fujii K, Shiroguchi K, Ito A, Kometani K, Takeda K, Kawakami E, Yamashita K, Suzuki K, Okada T, Kurosaki T (2018) T follicular helper cell-germinal center B-cell interaction strength regulates entry into plasma cell or recycling germinal center cell fate. Immunity 48:702–715.e4
153. Viant C, Weymar GHJ, Escolano A, Chen S, Hartweger H, Cipolla M, Gazumyan A, Nussenzweig MC (2020) Antibody affinity shapes the choice between memory and germinal center B-cell fates. Cell 183:1298–1311
154. Caro-Maldonado A, Wang R, Nichols AG, Kuraoka M, Milasta S, Sun LD, Gavin AL, Abel ED, Kelsoe G, Green DR, Rathmell JC (2014) Metabolic reprogramming is required for the antibody production that is suppressed in anergic but exaggerated in chronically BAFF-exposed B-cells. J Immunol 192:3626–3636
155. Brookens SK, Cho SH, Paik Y, Meyer K, Raybuck AL, Park C, Greenwood DL, Rathmell JC, Boothby MR (2023) Plasma cell differentiation, antibody quality, and initial germinal center B-cell population depend on glucose influx rate. BioRxiv 2023:557599
156. Bierling T, Gumann A, Ottmann S, Schulz S, Van Schie K, Weckwerth L, Wichert M, Kuwert F, Hauke M, Freudenreich T, Toes R, Mielenz D, Jack H, Pracht K (2023) Abstract P141. Eur J Immunol 53(Suppl. 2):138. https://doi.org/10.1002/eji.202370300
157. Hilchey SP, Palshikar MG, Emo JA, Li D, Garigen J, Wang J, Mendelson ES, Cipolla V, Thakar J, Zand MS (2020) Cyclosporine a directly affects human and mouse B-cell migration in vitro by disrupting a HIF-1α-dependent, O_2 sensing, molecular switch. BMC Immunol 21:13. https://doi.org/10.1186/s12865-020-0342-8
158. Burr SP, Costa AS, Grice GL, Timms RT, Lobb IT, Freisinger P, Dodd RB, Dougan G, Lehner PJ, Frezza C, Nathan JA (2016) Mitochondrial protein lipoylation and the 2-oxoglutarate dehydrogenase complex controls HIF1α stability in aerobic conditions. Cell Metab 24:740–752
159. Wang S, Raybuck A, Shiuan E, Cho SH, Wang Q, Brantley-Sieders DM, Edwards D, Allaman MM, Nathan J, Wilson KT, DeNardo D, Zhang S, Cook R, Boothby M, Chen J (2020) Selective inhibition of mTORC1 in tumor vessels increases antitumor immunity. JCI Insight 5:e139237

Early B-Cell Factor 1: An Archetype for a Lineage-Restricted Transcription Factor Linking Development to Disease

7

Mikael Sigvardsson

Abstract

The development of highly specialized blood cells from hematopoietic stem cells (HSCs) in the bone marrow (BM) is dependent upon a stringently orchestrated network of stage- and lineage-restricted transcription factors (TFs). Thus, the same stem cell can give rise to various types of differentiated blood cells. One of the key regulators of B-lymphocyte development is early B-cell factor 1 (EBF1). This TF belongs to a small, but evolutionary conserved, family of proteins that harbor a Zn-coordinating motif and an IPT/TIG (immunoglobulin-like, plexins, transcription factors/transcription factor immunoglobulin) domain, creating a unique DNA-binding domain (DBD). EBF proteins play critical roles in diverse developmental processes, including body segmentation in the *Drosophila melanogaster* embryo, and retina formation in mice. While several EBF family members are expressed in neuronal cells, adipocytes, and BM stroma cells, only B-lymphoid cells express EBF1. In the absence of EBF1, hematopoietic progenitor cells (HPCs) fail to activate the B-lineage program. This has been attributed to the ability of EBF1 to act as a pioneering factor with the ability to remodel chromatin, thereby creating a B-lymphoid-specific epigenetic landscape. Conditional inactivation of the *Ebf1* gene in B-lineage cells has revealed additional functions of this protein in relation to the control of proliferation and apoptosis. This may explain why EBF1 is frequently targeted by mutations in human leukemia cases. This chapter provides an overview of the biochemical and functional properties of the EBF family proteins, with a focus on the roles of EBF1 in normal and malignant B-lymphocyte development.

Keywords

Early b-cell factor · Transcription factors · Gene regulatory networks · B-lymphocyte development

Abbreviations

BM	Bone marrow
EBF1	Early B-cell factor 1
TF	Transcription factors
IPT/TIG	Immunoglobulin-like, plexins, transcription factors/transcription factor immunoglobulin

M. Sigvardsson (✉)
Department of Biomedical and Clinical Sciences, Linköping University, Linköping, Sweden

Division of Molecular Hematology, Lund University, Lund, Sweden
e-mail: mikael.sigvardsson@med.lu.se

T. Borggrefe, B. D. Giaimo (eds.), *Transcription factors in blood cell development*, Advances in Experimental Medicine and Biology 1459, https://doi.org/10.1007/978-3-031-62731-6_7

HPCs	Hematopoietic progenitor cells
OMP1	Olfactory marker protein 1
HLH	Helix-loop-helix
DBD	DNA-binding domain
TAD	Transcription activation domain
BRG1	Brahma-related gene 1
PUS	Pseudouridine synthetase
IKZF1	Ikaros family zinc finger 1
KLF4	Krüppel-like factor 4
EGR1	Early growth response 1
RBPJ	Recombination signal binding protein for immunoglobulin kappa J region
Rag1	Recombination-activating gene 1
Dntt	DNA nucleotidylexotransferase
FOXO1	Forkhead box O1
GRN	Gene regulatory networks
IL7	Interleukin 7
CLP	Common lymphoid progenitor
JAK-STAT	Janus kinase-signal transducer and activator of transcription
ZBTB17	Zinc finger and BTB domain containing 17
PAX5	Paired box 5
pre-BCR	pre-B-cell receptor
B-ALL	B-cell acute lymphoblastic leukemia
IgH	Immunoglobulin heavy
LoF	Loss-of-function
CLRF2	Cytokine receptor-like factor 2
PDGFRB	Platelet-derived growth factor receptor beta
ZNF	Zinc finger factor

7.1 The EBF Protein Family Represents Ancient Regulators of Development

The cloning of early B-cell factor 1 (EBF1) was reported in two papers that identified a protein that bound to the promoter of the *olfactory marker protein 1* (*OMP1*) gene [1] as well as the B-lineage-restricted CD79a (Mb-1) promoter [2]. This protein was accordingly named OLF1 or EBF1 and, for a time, was denoted as O/E-1 before it was eventually assigned the official name of EBF1. Subsequent reports identified three additional members of this family in mice, EBF2 (mMot1/O/E-3), EBF3 (O/E-2) [3–5], and O/E-4 [6], all of which display a high degrees of similarity in the DNA-binding and dimerization domains. Even though EBFs 1, 2, and 3 would appear to be largely functionally redundant [7], their distinct expression patterns indicate biologically unique functions for these factors. For instance, while B-lymphoid cells express only EBF1, bone marrow (BM) stroma cells express both EBF1 and EBF2 [8]. The identification of Collier, an EBF homolog from *Drosophila melanogaster* [9], implied that EBF proteins are part of an evolutionarily conserved family, with members identified in numerous species (Fig. 7.1) [1, 2, 10–14]. This indicates that the EBF family proteins have their evolutionary origin before the divergence of insects and mammals, which occurred about 500 million years ago. The difficulties experienced in identifying EBF-related proteins in yeast and plants suggest that this protein family arose in the earliest common animal ancestors.

Even though the expression of EBF family proteins is restricted, these factors exert important functions in a variety of tissues and organ systems, as well as in the embryonal development processes of several species. These include segmentation [15] and appropriate muscle development [16] in *Drosophila melanogaster* embryos, as well as neurogenesis in *Xenopus laevis* [11, 17] and *Caenorhabditis elegans* embryos [14]. EBF proteins are also expressed during mouse embryonic development [3, 5], wherein they are important for appropriate neuronal [18, 19], retinal [20], and muscle development [21]. These proteins are also involved in adipogenesis, in that they drive fibroblasts towards adipogenic cell fates [7, 22]. EBF proteins are also involved in the regulation and formation of functional stroma cells [23, 24] and osteoblasts [25] in the BM. While the EBF family of proteins exerts diverse functions during development, this chapter focuses on how EBF1 acts as a key regulator of normal and malignant B-lymphocyte development.

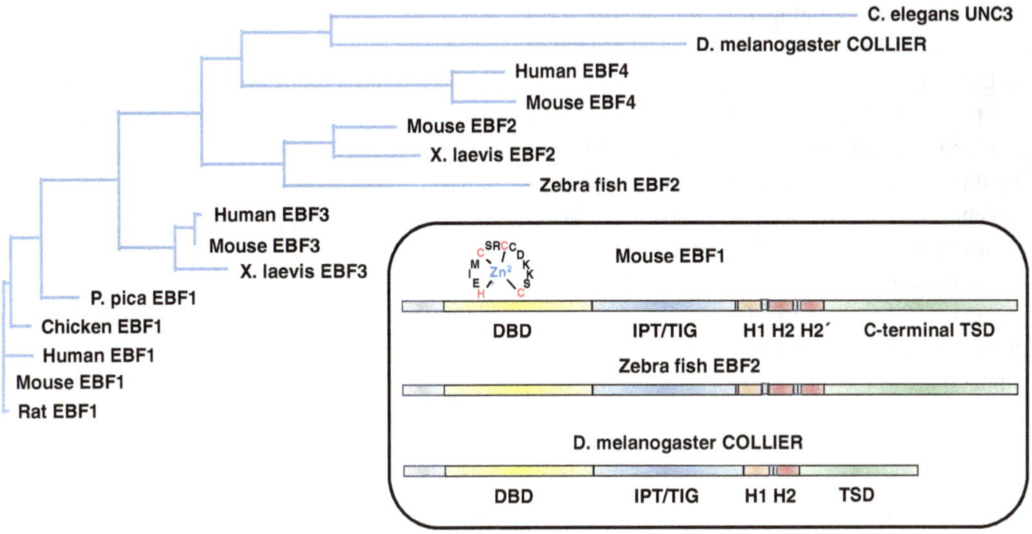

Fig. 7.1 Phylogenetic tree of some EBF family members, displaying the genetic relationships between the EBF proteins from different species. The tree was constructed using the ClustalW software and gene sequences derived from the EBI database. The box contains schematic drawings of EBF proteins from different species with specific domains indicated

7.2 The EBF Protein Family Has Unique Biochemical Features

The EBF family of proteins displays several unique features but also shares homologies with other transcription factor (TF) families (Fig. 7.1). While the EBF proteins harbor a novel type of helix-loop-helix (HLH) motif that resembles the second helix of classical basic-helix-loop-helix (bHLH) proteins such as TCF3 (transcription factor 3, E12/47) and myogenic differentiation 1 (MyoD1) [26, 27], they carry a completely different type of DNA-binding domain (DBD). Even the most evolutionary distal EBF family proteins display >80% homology in the DBD, pointing to the importance of this structure. The DBD is composed of an amino-terminal cysteine- and histidine-rich domain that harbors a Zn^{2+} coordination motif, which contains three cysteines and one histidine [28]. The core sequence of this motif, H-X_3-C-X_2-C-X_5-C [28], creates a small Zn^{2+}-stabilized loop [29, 30] that is conserved across the EBF protein family. The EBF protein contains a second conserved region with structural similarity to the IPT/TIG (immunoglobulin-like, plexins, transcription factors/transcription factor immunoglobulin) domains [29, 30] that are present in the DNA-binding regions of the TFs NFAT (nuclear factor of activated T cells) and NF-κB (nuclear factor kappa chain enhancer of activated B cells) [31]. The IPT domain forms an immunoglobulin-like fold of β-sheets that is connected by loops and is involved in DNA binding [30] as well as protein/protein interactions. This includes interactions with ZNF423 (zinc finger protein 423) [32] and ZNF521 (zinc finger protein 521) [33, 34], as well as Transportin 3 (TNPO3) [35] and CCR4-NOT transcription complex subunit 3 (CNOT3) [36]. The IPT domain is also found in plexin proteins that are involved in axonal guidance [37, 38]. This process is reported to involve EBF family proteins [14], and the IPT domain may integrate semaphorin signaling [37] into the gene regulatory networks (GRNs) in migrating neurons.

The EBF proteins bind variants of the palindromic DNA sequence 5′-ATTCCCN-NGGGAAT-3′ [39] as homodimers or heterodimers [2, 5]. Dimerization is dependent upon the atypical HLH motif, which is composed of two regions related to helix 2 in classical bHLH proteins [26]. The classical HLH motif is composed of two dissimilar amphipathic helices connected by a loop [26]. However, EBF-1 was first reported to lack any sequence corresponding to helix 1, but to instead harbor a duplication of helix 2, resulting in a domain that was capable of mediating dimerization when coupled to a heterologous protein [28]. The duplication of helix 2 is found exclusively in vertebrate EBF proteins, and not in *Caenorhabditis elegans Unc-3* [14] or *Drosophila melanogaster* Collier [9]. Dimerization of these proteins has been suggested to instead depend on a putative helix 1 that is present in all the EBF family proteins [9]. However, deletion of this region in the mouse protein does not appear to have any dramatic effect on the DNA-binding capability of EBF1 [2]. The dimerization appears to be essential for high-affinity binding but not for target site recognition, as an EBF-1 variant that is missing a functional HLH domain can bind to the consensus motif with low affinity [2, 28, 30].

EBF-1 has been reported to harbor two distinct transcription activation domains (TADs): TAD I, a context-dependent domain in the DBD, and TAD II, a C-terminal domain that is located downstream of the dimerization motif [28]. The TADII domain, which contains multiple serine/threonine/proline residues, has the ability to activate transcription when fused to a GAL-4 DBD [28]. This region is poorly conserved among the EBF family members, with only about 50% homology across EBFs 1, 2, and 3 [2, 5, 6]. However, the fact that the C-terminals of Collier and Unc-3 are serine-rich [9, 14] indicates that even though the primary amino acid sequences diverge, the basic features of this region are conserved. This region has been reported to harbor a prion-like domain and to interact with the chromatin remodeler Brahma-related gene 1 (BRG1), as well as with the RNA-binding protein PUS (pseudouridine synthetase), underlining the capability of EBF1 to act as a pioneering TF in early B-cell development [40].

Therefore, even though the EBF protein family displays homologies with other TFs, it displays many unique features. These include the structures of the DBD and the dimerization domain. Interestingly, even though the EBF proteins carry a unique DBD, their DNA target sites overlap to some extent with those of other factors involved in lymphocyte development, such as Ikaros family zinc finger 1 (IKZF1), Krüppel-like factor 4 (KLF4), early growth response 1 (EGR1), and recombination signal binding protein for immunoglobulin kappa J region (RBPJ). This has interesting implications for lymphocyte development, as these factors may compete for EBF1 binding to target sites during B-cell development [35, 41, 42].

7.3 EBF1 in B-Cell Development

Our current understanding of the role and function of EBF1 in B-cell development has evolved in parallel with the acquisition of detailed understanding of the developmental trajectories of lymphoid cells in mice. It has been reported that while GATA-binding factor 1 (GATA1) expression in early progenitor cells is associated with a myeloid/erythroid cell fate, the expression of SPI1 (also known as PU.1) is a marker of cells that are entering developmental pathways to generate lymphoid/myeloid lineages [43]. Loss of myeloid potential and entry into the lymphoid pathway appears to depend on the dosage of SPI1, with high levels promoting development into myeloid lineages and low levels resulting in a lymphoid cell fate [44]. This phenomenon is explained by SPI1 acting in a regulatory network in which this TF activates the gene that encodes the growth factor independent 1 transcriptional repressor (GFI1) [45]. GFI1 has the ability to suppress transcription from the gene that encodes the Zn finger protein IKZF1 [45], a TF of crucial importance for entry into lymphoid developmental pathways [46, 47]. IKZF1 acts in conjunction with E-proteins, e.g., TCF3, TCF4, and TCF12 [26], to activate lymphoid-associated genes such as *recombination-activating gene 1* (*Rag1*) and *DNA nucleotidylexotransferase* (*Dntt*) [46, 48–50]. E-proteins also activate transcription of the gene encoding the TF Forkhead box O1 (FOXO1) [51], which plays essential roles in early B-cell development [52, 53]. FOXO1 activates transcription of the *Rag* genes [52] and also targets the *Ebf1* gene to create an autoregulatory loop and initiate transcription of a B-lymphoid-restricted program [54].

A linkage has also been defined between *Ebf1* expression and active interleukin (IL)-7 signaling already in the classical common lymphoid progenitor (CLP) compartment [55, 56]. Functional IL7 signaling is essential for early B-cell development [56–59] and has been suggested to directly influence the activation of the *Ebf1* gene via induction of the Janus kinase-signal transducer and activator of transcription (JAK-STAT) pathway [60]. Moreover, ectopic expression of EBF1 partially rescues B-cell development in IL7 [55] or zinc finger and BTB domain containing 17 (ZBTB17)-deficient mice [61], further strengthening the linkage between EBF1 and extracellular signaling in early B-cell development.

At the time of the seminal paper describing the phenotype of the *Ebf1*- knockout (KO) mice [62], early B-cell development was coarsely defined based on the combined expression of B220 and CD43 [63]. As the BM of EBF1-deficient animals was found to contain a significant number of B220$^+$CD43$^+$ cells, albeit without detectable expression of B-lineage-restricted genes [62], it was concluded that EBF1 was essential for the pro-B-cell stage. However, as the limitations associated with using B220 and CD43 as markers of pro-B cells became increasingly apparent [64–70] and as *Ebf1* expression was detected already within the classical CLP compartment [67–70], it became obvious that EBF1 had a prominent role in the establishment of the earliest B-lineage progenitors [69, 70]. *Ebf1*$^{-/-}$ mice have an expanded LY6D$^-$GFRA2$^-$ CLP compartment, while an absolute majority of the LY6D$^+$ and all the GFRA2$^+$ cells are absent [69]. The few LY6D$^+$ cells that are detected lack transcription of B-lineage-restricted target genes, highlighting the essential role of EBF1 in the establishment of the B-lineage transcriptional program (Fig. 7.2) [69]. Even though many of the regulatory elements that can be annotated to B-lineage-restricted genes are accessible in multipotent progenitor cells [71], activation of the lineage-specific transcriptional program involves changes in chromosome nuclear positioning [72]. To some extent, this appears to depend on the C-terminal domain of EBF1 [73]. It has been suggested that this region interacts with chromatin remodelers such as BRG1 [40, 71, 74], and it has been reported to be of relevance for the formation of the phase-separated structures at EBF1 target genes [40], a process that is believed to be of importance for functional transcription.

While the endogenous levels of EBF1 expression in early progenitors are sufficient to activate a substantial fraction of the B-lineage transcriptional program [75–77], these levels are not sufficient to enable EBF1 to mediate lineage restriction in the absence of the paired domain TF paired box 5 (PAX5) [75, 78–81]. Even though ectopic EBF1 expression results in lineage restriction in progenitor cells [82–84], PAX5 appears to be critical for preventing lineage plasticity in normal B-cell development [75, 78, 79]. This is because EBF1 and PAX5 create a positive feedback loop in which the two proteins activate each other's coding genes (Fig. 7.2) [84, 85]. Loss of EBF1 in committed cells [86] or trans-heterozygote loss of *Ebf1* and *Pax5* has also been reported to result in dedifferentiation and loss of B-lineage identity, further highlighting the importance of the EBF1-PAX5 feedback loop in B-lymphocyte development.

In addition to the feedback loops, EBF1 appears to be a key component of coordinated feed-forward loops (Fig. 7.2) [87]. Such loops are created when two developmentally regulated TFs share regulatory elements and target genes, such that they cooperatively activate transcriptional programs during cell differentiation. EBF1 shares a high number of target genes with several other factors in the network, including TCF3, PAX5, and FOXO1 [71, 88–92], driving the differentiation pathway forward. The target genes for EBF1 in B-cell progenitors include several genes for components of the pre-B-cell receptor (pre-BCR), such as *Igll1*, *VpreB1*, *Cd79a*, and *Cd79b* [2, 12, 89, 91, 93, 94]. The proteins encoded by these genes form a complex with a newly rearranged immunoglobulin heavy chain (IgH), so as to create a functional receptor on the B-cell progenitors (reviewed in [95]). The pre-

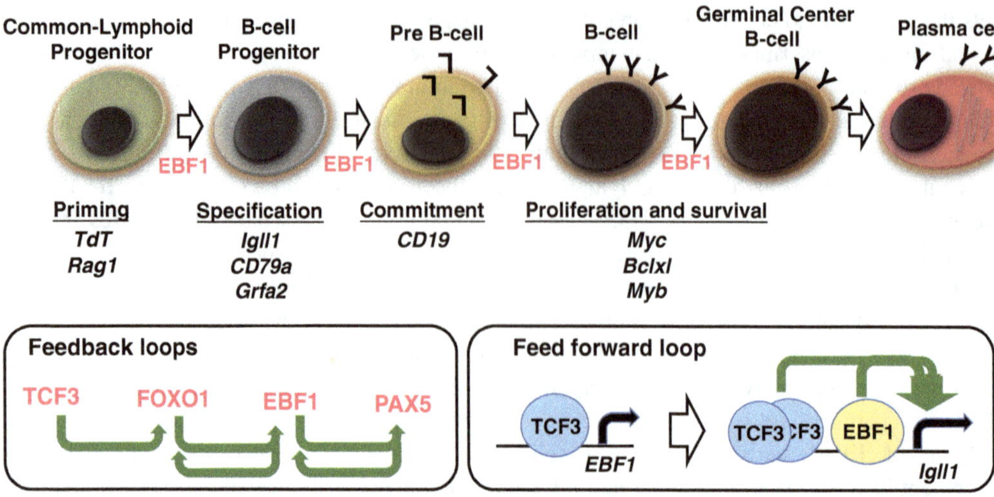

Fig. 7.2 Schematic drawing over B-cell development and the function of EBF1 in the different stages of B-lymphocyte development. The left box shows examples of autoregulatory feed-back loops where lineage-restricted TFs regulate each other's transcription. The right box exemplifies a feed-forward loop where TCF3 activates transcription of *Ebf1* allowing both factors to collaboratively activate a downstream target gene

BCR acts in synergy with signaling through the IL7 receptor to promote a proliferative burst of cells that have undergone functional Ig heavy-chain recombination [96]. Through a change in the signaling state of the pre-BCR [97, 98], activation of the AKT signaling pathway is disrupted, resulting in increased levels of nuclear FOXO1 [99]. This induces cell cycle blockage and reinitiation of *Rag* expression [99], both of which events are crucial for Ig light-chain recombination [52, 53, 100]. At this stage of development, the ability of EBF1 to drive transcription from target genes, such as *Igll1*, is disrupted by competitive binding of IKZF proteins to the target promoter [42], thereby restricting the production of surrogate light-chain genes to the progenitor compartment. This is in line with the observation that EBF1 displays a partly different target gene spectra in early B-cell development, as compared to in the mature B cell [101, 102]. Inactivation of the *Ebf1* gene in mature cells results in defective terminal differentiation and germinal center formation [101, 102]. These defects have been observed in all peripheral B-cell populations, including follicular, germinal center, and B1 B cells [101, 102]. While the defects include impaired expression of mRNA encoding several signal transducers, surface receptors, and TFs, the major phenotype appears to be related to defective proliferation [101, 102]. This is likely a consequence of EBF1 directly controlling the transcription of genes involved in cell survival, proliferation, and metabolism [101–104].

7.4 Ebf1: A Dose-Dependent Tumor Suppressor in Leukemia

While the essential role of EBF1 in normal B-cell development is well-established, large-scale genome analyses have identified *EBF1* mutations in about 7% of human B-cell acute lymphoblastic leukemia (B-ALL) cases (Table 7.1) [105, 106]. This malignancy is characterized by uncontrolled proliferation of progenitor cells in the BM, disrupting normal blood cell development. The link between development and leukemia was established already in 1844 when Alfred Donné reported the observation of immature blast cells in the blood samples from some of his patients [107]. The majority (65%–70%) of childhood B-ALL cases and about 50% of the cases in adults lack detectable expression of IgH chain protein, classifying them as B-cell precursor ALL [108, 109]. Pre-B ALL, defined based on expres-

Table 7.1 Examples of genetic alterations involving *EBF1* in human B-lineage malignancies

Malignant phenotype	Genetic alteration	Co-mutation	Comments	References
B-lineage ALL	Deletion		In 4%–6% of the pediatric B-ALL cases	[105] [106]
			In 4%–7% of the adult B-ALL cases	[147] [148]
			Enriched (25%) in relapsing cases of childhood B-ALL	[149]
		t(12:21) (p13;q22) ETV6-RUNX	Combined with ETV6-RUNX1. In 11% of ETV6-RUNX1 B-ALL cases	[105]
	Del (5) (q33;q33)		EBF1-PDGFRB fusion	[117] [150]
Mixed phenotype ALL	Deletion		Rare but detected	[151] [152]
B-cell lymphoma	t(5;14) (q33;q32)		Translocation to IG-heavy chain locus. Overexpression.	[121]

sion of the intracellular IgH chain, constitutes about 25% of the cases, while only 2%–5% of childhood B-ALLs are Ig surface-positive [108, 109]. This basic division of the different classes of B-ALL can be significantly refined; taking into consideration the morphologic, genetic, and clinical data, it is currently possible to define 28 variants of B-ALL [110]. Overall, this emphasizes the close linkage between development and disease in B-ALL.

With such a close linkage between development and disease, it is not surprising that mutations have been detected in the genes for several developmentally important TFs, in addition to *EBF1*. These include *PAX5*, *IKZF1* (IKAROS), and *TCF3* [105, 106]. Even though germline variants of *IKZF1* [111, 112] and *PAX5* [113, 114] have been reported to predispose for leukemia development, TF genes are mainly targeted by somatic mutations. As *EBF1* is mutated in about 7%, *PAX5* in 30%, and *IKZF1* in 23% of human B-ALL cases [105, 106], targeting developmentally important TFs would appear to be a central part of the oncogenic process in cases of B-ALL. The genetic alterations are frequently loss-of-function (LoF) mutations that appear in combination with translocations, resulting in the expression of fusion proteins such as *BCR-ABL* and *ETV-RUNX1* [105, 106, 115] or the overexpression of signal transducers or cytokine receptors, such as cytokine receptor-like factor 2 (CLRF2) [114]. In some cases, TF mutations appear to result in a "double hit," combining loss of TF function with hyperactivation of kinase activity. This is exemplified by translocations that result in the formation of a PAX5/JAK2 fusion protein [116] or an internal deletion in chromosome 5, creating a fusion protein comprising EBF1 and the protein kinase platelet-derived growth factor receptor beta (PDGFRB) [117]. This fusion protein is detected in 8% of cases diagnosed as Philadelphia-like ALL [118]. The fusion protein retains both the dimerization domains and DNA-binding regions of EBF1 but appears to be unable to activate target genes [119]. Instead, the dimerization domains of the EBF1 fusion component appear to result in dimerization and hyperactivation of PDGFRB [119].

Even though EBF1 is involved in the formation of fusion proteins [105, 106, 117], and in B-cell lymphoma, translocation to the IgH chain locus [120, 121], the most common genetic alteration is heterozygote LoF mutation [105, 106]. This does not completely align with the classical two-hit hypothesis that stipulates that for a tumor to arise, both alleles of a tumor suppressor need to be inactivated [122]. This could be explained by dominant-negative effects of the mutated proteins, in similarity to what has been reported for P53 [123, 124]. As EBF1 binds to DNA as a dimer [28], it is possible that a mutation in the

DBD generates a dominant-negative protein. However, in most B-ALL cases that have a mutated *Ebf1* gene, the genetic abnormality results in monoallelic loss of a significant portion of the gene, resulting in that no truncated protein is generated [105, 106]. Furthermore, increased expression of the EBF1 repressors zinc finger factor (ZNF) 423 [125] and ZNF521 [126] has been reported in B-ALL cases. Therefore, it appears that EBF1 acts as a dose-dependent tumor suppressor in B-cell development.

Developmental biology contains examples of how gene dosage impacts biological processes. A classic example of this is sex determination in the *Drosophila melanogaster* embryo, as this is not determined by the presence of a Y chromosome, but rather by the number of X chromosomes [127]. Interestingly, hyperdiploid and near-haploid B-ALLs show similar gene expression patterns, which may be explained by the fact that while hyperdiploid tumor cells often gain chromosomes 10, 14, 18, and 21, near-haploid cells often lose one copy of all chromosomes but these four [115]. The findings that heterozygous inactivation of the *Spi1* [44], *Ikzf1* [47], or *Bcl11a* [128] gene impairs the formation of B-lymphoid cells support the general notion that TF dosages are highly significant for B-lymphocyte development. EBF1 itself appears to act in a dose-dependent manner, as the loss of one allele of *Ebf1* has profound effects on early B-cell development [129–131]. The effect of monoallelic loss of *Ebf1* is exacerbated if it occurs concomitantly with mutations in the genes for other TFs, such as TCF3 [129, 130], RUNX1 [132], or PAX5 [133]. This may be a consequence of the regulatory networks creating self-enforcing positive-feedback loops that control the levels of the participating TFs. This is exemplified by the finding that EBF1 activates *Pax5* transcription, thereby increasing the level of PAX5 protein, which acts as a positive regulator of the *Ebf1* gene, to create a reinforcing loop [84, 85]. While this arrangement creates a powerful system for progression along a developmental trajectory, the system may be sensitive to perturbation, as reduced functionality of one component could impair the functions of the other factors within the loop. This is potentially reflected by the finding that B-cell leukemia develops in several mouse models where these networks are disrupted by trans-heterozygous inactivation of TF-encoding genes [100, 133–136]. Disruptions of regulatory TF loops can also be found in human B-lineage leukemias, as *PAX5* or *EBF1* deletions are frequently found together with additional genetic aberrations involving TFs, such as t(12;21)(p13;q22) (ETV6-RUNX1) or t(1;19)(q23;p13) (*TCF3-PBX1*) [105, 106] or mutations in *IKZF1* [105, 106, 137].

The essential roles of these TFs in normal cell differentiation and the finding that restored PAX5 expression in a mouse model of B-ALL is sufficient to circumvent the differentiation block [138] indicate that TF mutations contribute to the transformation process mainly by impairing differentiation. However, the role of lineage-restricted TFs is likely to be much more complex [139], given that humans who carry B-ALL-predisposing mutations in the *PAX5* gene do not show any impairment of B-cell development before the onset of disease [113]. This question has been addressed in a mouse model that expressed a constitutively active version of STAT5 [140]. Expressing this transgene on an *Ebf1*$^{+/-}$ or a *Pax5*$^{+/-}$ background resulted in a substantial proportion of the animals developing B-ALL [140]. However, when crossed to RAG1-deficient mice, these animals rarely developed leukemia [140]. Given that RAG1 deficiency, resulting in a complete block of early B-cell development, could not compensate for a reduced TF dosage and these animals rarely developed disease [140], it appears that EBF1 and PAX5 contribute to transformation through additional mechanisms.

During normal lymphocyte development in the mouse, IKZF1 has been proposed to play an important role in the control of genes that are involved in the regulation of metabolism and regulation of proliferation [141]. In light of the finding that B-cell progenitors are normally metabolically constrained, loss of IKZF1 can lead to the deregulation of metabolic pathways, conferring on the cells the ability to respond to proliferation signals in a pathologic manner [142,

143]. Similar functions have been proposed for PAX5 [142, 143] and EBF1 [103], revealing that the activities of lineage-restricted TFs are not limited to the regulation of lineage-restricted genes. This is further supported by the finding that conditional inactivation of *Ebf1* in CD19⁺ pro-B cells results in impaired survival and proliferation [101]. This has been attributed to the ability of EBF1 to regulate the expression of *c-Myb*, *BclXL* [101], and *Myc* [103, 104].

The regulation of the *Myc* gene in normal B-cell progenitors represents an interesting example of how a regulatory loop can result in lineage-restricted TFs gaining control over broadly expressed genes [104]. Even though EBF1 is critical for activation of the *Pax5* gene [62, 70], short-term loss of nuclear EBF1 does not result in reduced expression of PAX5 [104]. This becomes highly relevant because while EBF1 activates *Myc* transcription, the same gene is repressed by PAX5 [103, 104], thereby creating a balanced regulatory network for the control of proliferation and apoptosis of pro-B cells [104]. Thus, when there is loss of the activator, EBF1, in combination with maintained expression of the repressor, PAX5, the cells display cell cycle arrest, which can be rescued by ectopic expression of *Myc* [104]. It should be noted that malignant transformation results in loss of EBF1 dependency [101], likely as a result of deregulated transcriptional control of *Myc* [103]. Both EBF1 and PAX5 have also been reported to control the activities of the RAG proteins [144, 145]. This may be of high importance, as misdirected RAG-mediated recombination events have been proposed as common mechanisms for genetic abnormalities in certain types of human leukemias [115, 146]. Therefore, lineage-restricted TFs may contribute to malignant transformation via multiple, different molecular mechanisms that are not directly related to cellular differentiation per se.

7.5 Concluding Remarks

Even though EBF1 is one of the most intensively characterized regulators of stage- and lineage-specific transcription in the hematopoietic system, the functional complexity of this protein in normal and malignant B-cell development means that EBF1 still harbors many secrets. Even though many target genes for EBF1 have been identified [70, 93], none of these genes is likely to be sufficient to substitute for EBF1 in B-cell development or to explain the early developmental block observed in *Ebf1*-deficient mice. One possibility is that EBF1 has the ability, via the C-terminal TAD, to reconfigure chromatin in the early progenitors [73], such that the function would be to create a permissive epigenetic landscape rather than to activate any specific target gene. However, even in the absence of this region, B-lymphoid cells are generated, albeit in lower numbers [73]. It is puzzling that a lineage-restricted TF can take over the regulation of such central processes as proliferation and apoptosis. Even though the regulation of the *Myc* gene by EBF1 provides insights into how a lineage-restricted regulatory network can control an ubiquitously expressed gene [104], it remains unclear as to whether this is a general phenomenon or a unique feature of the *Myc* gene. Ongoing investigations of the functions and roles of EBF1 in B-cell development will provide further insights into the roles of GRNs in cell differentiation.

Acknowledgments The author wishes to thank Vincent Collins for constructive comments and suggestions. This work was supported by grants from the Swedish Cancer Society (20-1153), the Swedish Childhood Cancer Foundation (2022-0019), and the Swedish Research Council (2021-02379).

References

1. Wang MM, Reed RR (1993) Molecular cloning of the olfactory neuronal transcription factor Olf-1 by genetic selection in yeast. Nature 364:121–126
2. Hagman J, Belanger C, Travis A et al (1993) Cloning and functional characterization of early B-cell factor, a regulator of lymphocyte-specific gene expression. Genes Dev 7:760–773
3. Garel S, Marin F, Mattei MG et al (1997) Family of Ebf/Olf-1-related genes potentially involved in neuronal differentiation and regional specification in the central nervous system. Dev Dyn 210(3):191–205
4. Malgaretti N, Pozzoli O, Bosetti A et al (1997) Mmot1, a new helix-loop-helix transcription factor gene displaying a sharp expression bound-

ary in the embryonic mouse brain. J Biol Chem 272(28):17632–17639
5. Wang SS, Tsai RYL, Reed RR (1997) The characterization of the Olf-1/EBF-like HLH transcription factor family: implications in olfactory gene regulation and neuronal development. J Neurosci 17(11):4149–4158
6. Wang SS, Betz AG, Reed RR (2002) Cloning of a novel Olf-1/EBF-like gene, O/E-4, by degenerate oligo- based direct selection. Mol Cell Neurosci 20(3):404–414
7. Jimenez MA, Akerblad P, Sigvardsson M et al (2007) Critical role for Ebf1 and Ebf2 in the adipogenic transcriptional cascade. Mol Cell Biol 27(2):743–757
8. Lagergren A, Mansson R, Zetterblad J et al (2007) The Cxcl12, periostin, and Ccl9 genes are direct targets for early B-cell factor in OP-9 stroma cells. J Biol Chem 282(19):14454–14462
9. Crozatier M, Valle D, Dubois L et al (1996) Collier, a novel regulator of Drosophila head development, is expressed in a single mitotic domain. Curr Biol 6(6):707–718
10. Bally-Cuif L, Dubois L, Vincent A (1998) Molecular cloning of Zcoe2, the zebrafish homolog of Xenopus Xcoe2 and mouse EBF-2, and its expression during primary neurogenesis. Mech Dev 77(1):85–90
11. Dubois L, Bally-Cuif L, Crozatier M et al (1998) XCoe2, a transcription factor of the Col/Olf-1/EBF family involved in the specification of primary neurons in Xenopus. Curr Biol 8(4):199–209
12. Gisler R, Jacobsen S-E, Sigvardsson M (2000) Cloning of human early B cell factor and identification of target genes suggest a conserved role for EBF in B cell development. Blood 96:1457–1464
13. Nieminen P, Liippo J, Lassila O (2000) Pax-5 and EBF are expressed in committed B-cell progenitors prior to the colonization of the embryonic bursa of fabricius. Scand J Immunol 52(52):465–469
14. Prasad BC, Ye B, Zackhary R et al (1998) unc-3, a gene required for axonal guidance in Caenorhabditis elegans, encodes a member of the O/E family of transcription factors. Development 125(8):1561–1568
15. Crozatier M, Valle D, Dubois L et al (1999) Head versus trunk patterning in the Drosophila embryo; collier requirement for formation of the intercalary segment. Development 126(19):4385–4394
16. Crozatier M, Vincent A (1999) Requirement for the Drosophila COE transcription factor Collier in formation of an embryonic muscle: transcriptional response to notch signalling. Development 126(7):1495–1504
17. Garcia-Dominguez M, Poquet C, Garel S et al (2003) Ebf gene function is required for coupling neuronal differentiation and cell cycle exit. Development 130(24):6013–6025
18. Garel S, Marin F, Grosschedl R et al (1999) Ebf1 controls early cell differentiation in the embryonic striatum. Development 126(23):5285–5294
19. Corradi A, Croci L, Broccoli V et al (2003) Hypogonadotropic hypogonadism and peripheral neuropathy in Ebf2-null mice. Development 130(2):401–410
20. Jin K, Jiang H, Mo Z et al (2010) Early B-cell factors are required for specifying multiple retinal cell types and subtypes from postmitotic precursors. J Neurosci 30(36):11902–11916
21. Jin S, Kim J, Willert T et al (2014) Ebf factors and MyoD cooperate to regulate muscle relaxation via Atp2a1. Nat Commun 5:3793
22. Akerblad P, Lind U, Liberg D et al (2002) Early B-cell factor (o/e-1) is a promoter of adipogenesis and involved in control of genes important for terminal adipocyte differentiation. Mol Cell Biol 22(22):8015–8025
23. Derecka M, Herman JS, Cauchy P et al (2020) EBF1-deficient bone marrow stroma elicits persistent changes in HSC potential. Nat Immunol 21(3):261–273
24. Qian H, Badaloni A, Chiara F et al (2013) Molecular characterization of prospectively isolated multipotent mesenchymal progenitors provides new insight into the cellular identity of mesenchymal stem cells in mouse bone marrow. Mol Cell Biol 33(4):661–677
25. Kieslinger M, Folberth S, Dobreva G et al (2005) EBF2 regulates osteoblast-dependent differentiation of osteoclasts. Dev Cell 9(6):757–767
26. Murre C (2019) Helix-loop-helix proteins and the advent of cellular diversity: 30 years of discovery. Genes Dev 33(1–2):6–25
27. Murre C, McCaw PS, Baltimore D (1989) A new DNA binding and dimerisation motif in immunoglobulin enhancer binding, daughterless, myoD and myc proteins. Cell 56:777–783
28. Hagman J, Gutch MJ, Lin H et al (1995) EBF contains a novel zinc coordination motif and multiple dimerization and transcriptional activation domains. EMBO J 14(12):2907–2916
29. Siponen MI, Wisniewska M, Lehtio L et al (2010) Structural determination of functional domains in early B-cell Factor (EBF) family of transcription factors reveals similarities to rel DNA-binding proteins and a novel dimerization motif. J Biol Chem 285(34):25875–25879
30. Treiber N, Treiber T, Zocher G et al (2010) Structure of an Ebf1: DNA complex reveals unusual DNA recognition and structural homology with Rel proteins. Genes Dev 24(20):2270–2275
31. Aravind L, Koonin E (1999) Gleaning non-trivial structural, functional and evolutionary information about proteins by iterative database searches. J Mol Biol 287:1023–1040
32. Tsai RY, Reed RR (1997) Cloning and functional characterization of Roaz, a zinc finger protein that interacts with O/E-1 to regulate gene expression: implications for olfactory neuronal development. J Neurosci 17(11):4159–4169

33. Warming S, Liu P, Suzuki T et al (2003) Evi3, a common retroviral integration site in murine B-cell lymphoma, encodes an EBFAZ-related Kruppel-like zinc finger protein. Blood 101(5):1934–1940
34. Bond HM, Mesuraca M, Carbone E et al (2004) Early hematopoietic zinc finger protein (EHZF), the human homologue to mouse Evi3, is highly expressed in primitive human hematopoietic cells. Blood 103(6):2062–2070
35. Bayer M, Boller S, Ramamoothy S et al (2022) Tnpo3 enables EBF1 function in conditions of antagonistic Notch signaling. Genes Dev 36(15–16):901–915
36. Yang CYRS, Boller S, Rosenbaum M et al (2016) Interaction of CCR4-NOT with EBF1 regulates gene-specific transcription and mRNA stability in B lymphopoiesis. Genes Dev 30(15):2310–2324
37. Tamagnone L, Comoglio PM (2000) Signalling by semaphorin receptors: cell guidance and beyond. Trends Cell Biol 10(9):377–383
38. Artigiani S, Comoglio PM, Tamagnone L (1999) Plexins, semaphorins, and scatter factor receptors: a common root for cell guidance signals? IUBMB Life 48(5):477–482
39. Travis A, Hageman J, Hwang L et al (1993) Purification of early-B-cell factor and characterization of its DNA-binding specificity. Mol Cell Biol 13:3392–3400
40. Wang Y, Zolotarev N, Yang CY et al (2020) A prion-like domain in transcription factor EBF1 promotes phase separation and enables B cell programming of progenitor chromatin. Immunity 53(6):1151–1167
41. Smith EM, Akerblad P, Kadesch T et al (2005) Inhibition of EBF function by active Notch signaling reveals a novel regulatory pathway in early B-cell development. Blood 106(6):1995–2001
42. Thompson EC, Cobb BS, Sabbattini P et al (2007) Ikaros DNA-binding proteins as integral components of B cell developmental stage specific regulatory circuits. Immunity 26(3):335–344
43. Arinobo Y, Mizuno S, Chong Y et al (2007) Reciprocal activation of Gata-1 and PU.1 marks initial specification of hematopoietic stem cells into myeloerythroid and myelolymphoid lineages. Cell Stem Cell 1:416–427
44. DeKoter RP, Singh H (2000) Regulation of B lymphocyte and macrophage development by graded expression of PU.1. Science 288(5470):1439–1441
45. Spooner CJ, Cheng JX, Pujadas E et al (2009) A recurrent network involving the transcription factors PU.1 and Gfi1 orchestrates innate and adaptive immune cell fates. Immunity 31(4):576–586
46. Ng SY, Yoshida T, Zhang J et al (2009) Genome-wide lineage-specific transcriptional networks underscore Ikaros-dependent lymphoid priming in hematopoietic stem cells. Immunity 30(4):493–507
47. Ferreiros-Vidal I, Carroll T, Taylor B et al (2013) Genome-wide identification of Ikaros targets elucidates its contribution to mouse B-cell lineage specification and pre-B-cell differentiation. Blood 121(10):1769–1782
48. Semerad CL, Mercer EM, Inlay MA et al (2009) E2A proteins maintain the hematopoietic stem cell pool and promote the maturation of myelolymphoid and myeloerythroid progenitors. Proc Natl Acad Sci USA 106(6):1930–1935
49. Yang Q, Kardava L, St Leger A et al (2008) E47 controls the developmental integrity and cell cycle quiescence of multipotential hematopoietic progenitors. J Immunol 181(9):5885–5894
50. Dias S, Mansson R, Gurbuxani S et al (2008) E2A proteins promote development of lymphoid-primed multipotent progenitors. Immunity 29(2):217–227
51. Welinder E, Mansson R, Mercer EM et al (2011) The transcription factors E2A and HEB act in concert to induce the expression of FOXO1 in the common lymphoid progenitor. Proc Natl Acad Sci USA 108(42):17402–17407
52. Amin RH, Schlissel MS (2008) Foxo1 directly regulates the transcription of recombination-activating genes during B cell development. Nat Immunol 9(6):613–622
53. Dengler HS, Baracho GV, Omori SA et al (2008) Distinct functions for the transcription factor Foxo1 at various stages of B cell differentiation. Nat Immunol 9(12):1388–1398
54. Mansson R, Welinder E, Ahsberg J et al (2012) Positive intergenic feedback circuitry, involving EBF1 and FOXO1, orchestrates B-cell fate. Proc Natl Acad Sci USA 109(51):21028–21033
55. Dias S, Silva H Jr, Cumano A et al (2005) Interleukin-7 is necessary to maintain the B cell potential in common lymphoid progenitors. J Exp Med 201(6):971–979
56. Tsapogas P, Zandi S, Ahsberg J et al (2011) IL-7 mediates Ebf-1-dependent lineage restriction in early lymphoid progenitors. Blood 118(5):1283–1290
57. von Freeden-Jeffry U, Vieira P, Lucian LA et al (1995) Lymphopenia in interleukin (IL)-7 gene-deleted mice identifies IL-7 as a nonredundant cytokine. J Exp Med 181(4):1519–1526
58. Peschon JJ, Morrissey PJ, Grabstein KH et al (1994) Early lymphocyte expansion is severely impaired in interleukin 7 receptor-deficient mice. J Exp Med 180(5):1955–1960
59. Goetz CA, Harmon IR, O'Neil JJ et al (2004) STAT5 activation underlies IL7 receptor-dependent B cell development. J Immunol 172(8):4770–4778
60. Roessler S, Gyory I, Imhof S et al (2007) Distinct promoters mediate the regulation of Ebf1 gene expression by interleukin-7 and Pax5. Mol Cell Biol 27(2):579–594
61. Kosan C, Saba I, Godmann M et al (2010) Transcription factor miz-1 is required to regulate interleukin-7 receptor signaling at early commitment stages of B cell differentiation. Immunity 33(6):917–928
62. Lin H, Grosschedl R (1995) Failure of B-cell differentiation in mice lacking the transcription factor EBF. Nature 376(6537):263–267

63. Hardy RR, Carmack CE, Shinton SA et al (1991) Resolution and characterization of pro-B and pre-pro-B cell stages in normal mouse bone marrow. J Exp Med 173(5):1213–1225
64. Rolink A, ten Boekel E, Melchers F et al (1996) A subpopulation of B220+ cells in murine bone marrow does not express CD19 and contains natural killer cell progenitors. J Exp Med 183(1):187–194
65. Alberti-Servera L, von Muenchow L, Tsapogas P et al (2017) Single-cell RNA sequencing reveals developmental heterogeneity among early lymphoid progenitors. EMBO J 36(24):3619–3633
66. Balciunaite G, Ceredig R, Massa S et al (2005) A B220+ CD117+ CD19- hematopoietic progenitor with potent lymphoid and myeloid developmental potential. Eur J Immunol 35(7):2019–2030
67. Mansson R, Zandi S, Anderson K et al (2008) B-lineage commitment prior to surface expression of B220 and CD19 on hematopoietic progenitor cells. Blood 112(4):1048–1055
68. Mansson R, Zandi S, Welinder E et al (2010) Single-cell analysis of the common lymphoid progenitor compartment reveals functional and molecular heterogeneity. Blood 115(13):2601–2609
69. Jensen CT, Ahsberg J, Sommarin MNE et al (2018) Dissection of progenitor compartments resolves developmental trajectories in B-lymphopoiesis. J Exp Med 215(7):1947–1963
70. Zandi S, Mansson R, Tsapogas P et al (2008) EBF1 is essential for B-lineage priming and establishment of a transcription factor network in common lymphoid progenitors. J Immunol 181(5):3364–3372
71. Strid T, Okuyama K, Tingvall-Gustafsson J et al (2021) B lymphocyte specification is preceded by extensive epigenetic priming in multipotent progenitors. J Immunol 206(11):2700–2713
72. Lin YC, Benner C, Mansson R et al (2012) Global changes in the nuclear positioning of genes and intra- and interdomain genomic interactions that orchestrate B cell fate. Nat Immunol 13(12):1196–1204
73. Boller S, Ramamoorthy S, Akbas D et al (2016) Pioneering activity of the C-terminal domain of EBF1 shapes the chromatin landscape for B cell programming. Immunity 44(3):527–541
74. Gao H, Lukin K, Ramirez J et al (2009) Opposing effects of SWI/SNF and Mi-2/NuRD chromatin remodeling complexes on epigenetic reprogramming by EBF and Pax5. Proc Natl Acad Sci USA 106(27):11258–11263
75. Zandi S, Ahsberg J, Tsapogas P et al (2012) Single-cell analysis of early B-lymphocyte development suggests independent regulation of lineage specification and commitment in vivo. Proc Natl Acad Sci USA 109(39):15871–15876
76. Nutt SL, Urbanek P, Rolink A et al (1997) Essential functions of Pax5 (BSAP) in pro-B cell development: difference between fetal and adult B lymphopoiesis and reduced V-to-DJ recombination at the IgH locus. Genes Dev 11(4):476–491
77. Urbánek P, Wang Z-Q, Fetka I et al (1994) Complete block of early B cell differentiation and altered patterning of the posterior midbrain in mice lacking Pax5/BSAP. Cell 79:901–912
78. Nutt SL, Heavey B, Rolink AG et al (1999) Commitment to the B-lymphoid lineage depends on the transcription factor Pax5. Nature 401(6753):556–562
79. Rolink AG, Nutt SL, Melchers F et al (1999) Long-term in vivo reconstitution of T-cell development by Pax5- deficient B-cell progenitors. Nature 401(6753):603–606
80. Souabni A, Cobaleda C, Schebesta M et al (2002) Pax5 promotes B lymphopoiesis and blocks T cell development by repressing Notch1. Immunity 17(6):781–793
81. Mikkola I, Heavey B, Horcher M et al (2002) Reversion of B cell commitment upon loss of Pax5 expression. Science 297(5578):110–113
82. Lenaerts A, Kucinski I, Deboutte W et al (2022) EBF1 primes B-lymphoid enhancers and limits the myeloid bias in murine multipotent progenitors. J Exp Med 219(11):e2021243
83. Zhang Z, Cotta CV, Stephan RP et al (2003) Enforced expression of EBF in hematopoietic stem cells restricts lymphopoiesis to the B cell lineage. EMBO J 22(18):4759–4769
84. Pongubala JM, Northrup DL, Lancki DW et al (2008) Transcription factor EBF restricts alternative lineage options and promotes B cell fate commitment independently of Pax5. Nat Immunol 9(2):203–215
85. Decker T, Pasca di Magliano M et al (2009) Stepwise activation of enhancer and promoter regions of the B cell commitment gene Pax5 in early lymphopoiesis. Immunity 30(4):508–520
86. Nechanitzky R, Akbas D, Scherer S et al (2013) Transcription factor EBF1 is essential for the maintenance of B cell identity and prevention of alternative fates in committed cells. Nat Immunol 14(8):867–875
87. Mangan S, Alon U (2003) Structure and function of the feed-forward loop network motif. Proc Natl Acad Sci USA 100(21):11980–11985
88. Sigvardsson M (2000) Overlapping expression of early B-cell factor and basic helix-loop- helix proteins as a mechanism to dictate B-lineage-specific activity of the lambda5 promoter. Mol Cell Biol 20(10):3640–3654
89. Sigvardsson M, O'Riordan M, Grosschedl R (1997) EBF and E47 collaborate to induce expression of the endogenous immunoglobulin surrogate light chain genes. Immunity 7(1):25–36
90. Fitzsimmons D, Hodsdon W, Wheat W et al (1996) Pax-5 (BSAP) recruits Ets proto-oncogene family proteins to form functional ternary complexes on a B-cell-specific promoter. Genes Dev 10(17):2198–2211
91. Lin YC, Jhunjhunwala S, Benner C et al (2010) A global network of transcription factors, involving

E2A, EBF1 and Foxo1, that orchestrates B cell fate. Nat Immunol 11(7):635–643
92. Maier H, Colbert J, Fitzsimmons D et al (2003) Activation of the early B-cell-specific mb-1 (Ig-alpha) gene by Pax-5 is dependent on an unmethylated Ets binding site. Mol Cell Biol 23(6):1946–1960
93. Treiber T, Mandel EM, Pott S et al (2010) Early B cell factor 1 regulates B cell gene networks by activation, repression, and transcription- independent poising of chromatin. Immunity 32(5):714–725
94. Akerblad P, Rosberg M, Leanderson T et al (1999) The B29 (immunoglobulin beta-chain) gene is a genetic target for early B-cell factor. Mol Cell Biol 19(1):392–401
95. Melchers F (2015) Checkpoints that control B cell development. J Clin Invest 125(6):2203–2210
96. Erlandsson L, Licence S, Gaspal F et al (2005) Both the pre-BCR and the IL-7Ralpha are essential for expansion at the pre-BII cell stage in vivo. Eur J Immunol 35(6):1969–1976
97. Mandal M, Powers SE, Ochiai K et al (2009) Ras orchestrates exit from the cell cycle and light-chain recombination during early B cell development. Nat Immunol 10(10):1110–1117
98. Herzog S, Hug E, Meixlsperger S et al (2008) SLP-65 regulates immunoglobulin light chain gene recombination through the PI(3)K-PKB-Foxo pathway. Nat Immunol 9(6):623–631
99. Ochiai K, Maienschein-Cline M, Mandal M et al (2012) A self-reinforcing regulatory network triggered by limiting IL-7 activates pre-BCR signaling and differentiation. Nat Immunol 13(3):300–307
100. Lu R, Medina K, Lancki D et al (2003) IRF-4,8 orchestrate the pre-B-to-B transition in lymphocyte development. Genes Dev 17:1703–1708
101. Gyory I, Boller S, Nechanitzky R et al (2012) Transcription factor Ebf1 regulates differentiation stage-specific signaling, proliferation, and survival of B cells. Genes Dev 26(7):668–682
102. Vilagos B, Hoffmann M, Souabni A et al (2012) Essential role of EBF1 in the generation and function of distinct mature B cell types. J Exp Med 209(4):775–792
103. Ramamoorthy S, Kometani K, Herman JS et al (2020) EBF1 and Pax5 safeguard leukemic transformation by limiting IL-7 signaling, Myc expression, and folate metabolism. Genes Dev 34(21–22):1503–1519
104. Somasundaram R, Jensen CT, Tingvall-Gustafsson J et al (2021) EBF1 and PAX5 control pro-B cell expansion via opposing regulation of the Myc gene. Blood 137(22):3037–3049
105. Mullighan CG, Goorha S, Radtke I et al (2007) Genome-wide analysis of genetic alterations in acute lymphoblastic leukaemia. Nature 446(7137):758–764
106. Kuiper RP, Schoenmakers EF, van Reijmersdal SV et al (2007) High-resolution genomic profiling of childhood ALL reveals novel recurrent genetic lesions affecting pathways involved in lymphocyte differentiation and cell cycle progression. Leukemia 21(6):1258–1266
107. Donné A (1844) Cours de microscopie complémentaire des études médicales. Anatomie microscopique et physiologie des fluides de l'Economie. Ballière, Paris, pp 135–136
108. Jennings CD, Foon KA (1997) Recent advances in flow cytometry: application to the diagnosis of hematologic malignancy. Blood 90(8):2863–2892
109. Craig FE, Foon KA (2008) Flow cytometric immunophenotyping for hematologic neoplasms. Blood 111(8):3941–3967
110. Arber DA, Orazi A, Hasserjian RP et al (2022) International consensus classification of myeloid neoplasms and acute leukemias: integrating morphologic, clinical, and genomic data. Blood 140(11):1200–1228
111. Churchman ML, Qian M, Te Kronnie G et al (2018) Germline genetic IKZF1 variation and predisposition to childhood acute lymphoblastic leukemia. Cancer Cell 33(5):937–948
112. Trevino LR, Yang W, French D et al (2009) Germline genomic variants associated with childhood acute lymphoblastic leukemia. Nat Genet 41(9):1001–1005
113. Shah S, Schrader KA, Waanders E et al (2013) A recurrent germline PAX5 mutation confers susceptibility to pre-B cell acute lymphoblastic leukemia. Nat Genet 45(10):1226–1231
114. Gu Z, Churchman ML, Roberts KG et al (2019) PAX5-driven subtypes of B-progenitor acute lymphoblastic leukemia. Nat Genet 51(2):296–307
115. Brady SW, Roberts KG, Gu Z et al (2022) The genomic landscape of pediatric acute lymphoblastic leukemia. Nat Genet 54(9):1376–1389
116. Jurado S, Fedl AS, Jaritz M et al (2022) The PAX5-JAK2 translocation acts as dual-hit mutation that promotes aggressive B-cell leukemia via nuclear STAT5 activation. EMBO J 41(7):e108397
117. Roberts KG, Morin RD, Zhang J et al (2012) Genetic alterations activating kinase and cytokine receptor signaling in high-risk acute lymphoblastic leukemia. Cancer Cell 22(2):153–166
118. Schwab C, Ryan SL, Chilton L et al (2016) EBF1-PDGFRB fusion in pediatric B-cell precursor acute lymphoblastic leukemia (BCP-ALL): genetic profile and clinical implications. Blood 127(18):2214–2218
119. Welsh SJ, Churchman ML, Togni M et al (2018) Deregulation of kinase signaling and lymphoid development in EBF1-PDGFRB ALL leukemogenesis. Leukemia 32(1):38–48
120. Iida S, Rao PH, Ueda R et al (1999) Chromosomal rearrangement of the PAX-5 locus in lymphoplasmacytic lymphoma with t(9;14)(p13;q32). Leuk Lymphoma 34(1–2):25–33
121. Bouamar H, Abbas S, Lin AP et al (2013) A capture-sequencing strategy identifies IRF8, EBF1, and APRIL as novel IGH fusion partners in B-cell lymphoma. Blood 122(5):726–733

122. Knudson AG Jr (1971) Mutation and cancer: statistical study of retinoblastoma. Proc Natl Acad Sci USA 68(4):820–823
123. Hinds P, Finlay C, Levine AJ (1989) Mutation is required to activate the p53 gene for cooperation with the ras oncogene and transformation. J Virol 63(2):739–746
124. Finlay CA, Hinds PW, Tan TH et al (1988) Activating mutations for transformation by p53 produce a gene product that forms an hsc70-p53 complex with an altered half-life. Mol Cell Biol 8(2):531–539. https://doi.org/10.1128/mcb.8.2.531-539.1988
125. Harder L, Eschenburg G, Zech A et al (2013) Aberrant ZNF423 impedes B cell differentiation and is linked to adverse outcome of ETV6-RUNX1 negative B precursor acute lymphoblastic leukemia. J Exp Med 210(11):2289–2304
126. Yamasaki N, Miyazaki K, Nagamachi A et al (2010) Identification of Zfp521/ZNF521 as a cooperative gene for E2A-HLF to develop acute B-lineage leukemia. Oncogene 29(13):1963–1975
127. Bridges CB (1916) Non-disjunction as proof of the chromosome theory of heredity (concluded). Genetics 1(2):107–163
128. Yu Y, Wang J, Khaled W et al (2012) Bcl11a is essential for lymphoid development and negatively regulates p53. J Exp Med 209(13):2467–2483
129. O'Riordan M, Grosschedl R (1999) Coordinate regulation of B cell differentiation by the transcription factors EBF and E2A. Immunity 11(1):21–31
130. Ahsberg J, Ungerback J, Strid T et al (2013) Early B-cell factor 1 regulates the expansion of B-cell progenitors in a dose-dependent manner. J Biol Chem 288(46):33449–33461
131. Lukin K, Fields S, Guerrettaz L et al (2011) A dose-dependent role for EBF1 in repressing non-B-cell-specific genes. Eur J Immunol 41(6):1787–1793
132. Lukin K, Fields S, Lopez D et al (2010) Compound haploinsufficiencies of Ebf1 and Runx1 genes impede B cell lineage progression. Proc Natl Acad Sci USA 107(17):7869–7874
133. Prasad MA, Ungerback J, Ahsberg J et al (2015) Ebf1 heterozygosity results in increased DNA damage in pro-B cells and their synergistic transformation by Pax5 haploinsufficiency. Blood 125(26):4052–4059
134. Heltemes-Harris LM, Hubbard GK, LaRue RS et al (2021) Identification of mutations that cooperate with defects in B cell transcription factors to initiate leukemia. Oncogene 40(43):6166–6179
135. Jo SH, Schatz JH, Acquaviva J et al (2010) Cooperation between deficiencies of IRF-4 and IRF-8 promotes both myeloid and lymphoid tumorigenesis. Blood 116(15):2759–2767
136. Sokalski KM, Li SK, Welch I et al (2011) Deletion of genes encoding PU.1 and Spi-B in B cells impairs differentiation and induces pre-B cell acute lymphoblastic leukemia. Blood 118 (10):2801-2808
137. Okuyama K, Strid T, Kuruvilla J et al (2019) PAX5 is part of a functional transcription factor network targeted in lymphoid leukemia. PLoS Genet 15(8):e1008280
138. Liu GJ, Cimmino L, Jude JG et al (2014) Pax5 loss imposes a reversible differentiation block in B-progenitor acute lymphoblastic leukemia. Genes Dev 28(12):1337–1350
139. Sigvardsson M (2023) Transcription factor networks link B-lymphocyte development and malignant transformation in leukemia. Genes Dev 37(15–16):703–723
140. Heltemes-Harris LM, Willette MJ, Ramsey LB et al (2011) Ebf1 or Pax5 haploinsufficiency synergizes with STAT5 activation to initiate acute lymphoblastic leukemia. J Exp Med 208(6):1135–1149
141. Zhang J, Jackson AF, Naito T et al (2011) Harnessing of the nucleosome-remodeling-deacetylase complex controls lymphocyte development and prevents leukemogenesis. Nat Immunol 13(1):86–94
142. Chan LN, Chen Z, Braas D et al (2017) Metabolic gatekeeper function of B-lymphoid transcription factors. Nature 542(7642):479–483
143. Xiao G, Chan LN, Klemm L et al (2018) B-cell-specific diversion of glucose carbon utilization reveals a unique vulnerability in B cell malignancies. Cell 173(2):470–484
144. Timblin GA, Schlissel MS (2013) Ebf1 and c-Myb repress rag transcription downstream of Stat5 during early B cell development. J Immunol 191(9):4676–4687
145. Zhang Z, Espinoza CR, Yu Z et al (2006) Transcription factor Pax5 (BSAP) transactivates the RAG-mediated V(H)-to-DJ(H) rearrangement of immunoglobulin genes. Nat Immunol 7(6):616–624
146. Papaemmanuil E, Rapado I, Li Y et al (2014) RAG-mediated recombination is the predominant driver of oncogenic rearrangement in ETV6-RUNX1 acute lymphoblastic leukemia. Nat Genet 46(2):116–125
147. Paulsson K, Cazier JB, Macdougall F et al (2008) Microdeletions are a general feature of adult and adolescent acute lymphoblastic leukemia: Unexpected similarities with pediatric disease. Proc Natl Acad Sci USA 105(18):6708–6713
148. Safavi S, Hansson M, Karlsson K et al (2015) Novel gene targets detected by genomic profiling in a consecutive series of 126 adults with acute lymphoblastic leukemia. Haematologica 100(1):55–61
149. Yang JJ, Bhojwani D, Yang W et al (2008) Genome-wide copy number profiling reveals molecular evolution from diagnosis to relapse in childhood acute lymphoblastic leukemia. Blood 112(10):4178–4183
150. Lengline E, Beldjord K, Dombret H et al (2013) Successful tyrosine kinase inhibitor therapy in a refractory B-cell precursor acute lymphoblastic leukemia with EBF1-PDGFRB fusion. Haematologica 98(11):e146–e148
151. Yan L, Ping N, Zhu M et al (2012) Clinical, immunophenotypic, cytogenetic, and molecular genetic features in 117 adult patients with mixed-phenotype acute leukemia defined by WHO-2008 classification. Haematologica 97(11):1708–1712
152. Alexander TB, Gu Z, Iacobucci I et al (2018) The genetic basis and cell of origin of mixed phenotype acute leukaemia. Nature 562(7727):373–379

Part IV

Myeloid Cells

8 Transcription Factor NRF2 in Shaping Myeloid Cell Differentiation and Function

Marc Pfefferlé and Florence Vallelian

Abstract

NFE2-related factor 2 (NRF2) is a master transcription factor (TF) that coordinates key cellular homeostatic processes including antioxidative responses, autophagy, proteostasis, and metabolism. The emerging evidence underscores its significant role in modulating inflammatory and immune processes. This chapter delves into the role of NRF2 in myeloid cell differentiation and function and its implication in myeloid cell-driven diseases. In macrophages, NRF2 modulates cytokine production, phagocytosis, pathogen clearance, and metabolic adaptations. In dendritic cells (DCs), it affects maturation, cytokine production, and antigen presentation capabilities, while in neutrophils, NRF2 is involved in activation, migration, cytokine production, and NETosis. The discussion extends to how NRF2's regulatory actions pertain to a wide array of diseases, such as sepsis, various infectious diseases, cancer, wound healing, atherosclerosis, hemolytic conditions, pulmonary disorders, hemorrhagic events, and autoimmune diseases. The activation of NRF2 typically reduces inflammation, thereby modifying disease outcomes. This highlights the therapeutic potential of NRF2 modulation in treating myeloid cell-driven pathologies.

Keywords

NFE2-related factor 2 (NRF2) · Innate immunity · Inflammation · Myeloid cells · Myeloid differentiation · Macrophages · Neutrophils · Dendritic cells

Abbreviations

15d-PGJ_2	15-Deoxy-delta-12,14 prostaglandin J2
4-OI	4-Octyl itaconate
5PRA	5-Phosphoribosylamine
ABCA1	ATP-binding cassette transporter 1
ACAD10	Acyl-coenzyme A dehydrogenase family member 10
ACAD12	Acyl-coenzyme A dehydrogenase family member 12
ACOT7	Acyl-CoA thioesterase 7

M. Pfefferlé
Department of Internal Medicine, Spital Limmattal, Schlieren, Switzerland

F. Vallelian (✉)
Department of Internal Medicine, University of Zurich and University Hospital of Zurich, Zurich, Switzerland
e-mail: florence.vallelian@usz.ch

T. Borggrefe, B. D. Giaimo (eds.), *Transcription factors in blood cell development*, Advances in Experimental Medicine and Biology 1459, https://doi.org/10.1007/978-3-031-62731-6_8

ACSS2	Acyl-CoA synthetase short-chain family member 2
AHR	Aryl hydrocarbon receptor
AKRs	Aldo-keto reductases
ALDA	Fructose-bisphosphate aldolase A
ALI	Acute lung injury
AP-1	Activator protein 1
APC	Antigen-presenting cell
ApoE	Apolipoprotein E
ARE	Antioxidant-responsive elements
ARG1	Arginase-1
ATF4	Activating transcription factor 4
ATF6	Activating transcription factor 6
ATG5	Autophagy-related protein 5
BAF	BRG-/BRM-associated factor
BAL	Bronchoalveolar lavage
BCRP	Breast cancer-resistant protein
BM	Bone marrow
BMDC	Bone marrow-derived dendritic cell
BMDM	Bone marrow-derived macrophage
BRCA1	Breast cancer gene 1
bZIP	Basic region, leucine zipper
C/EBPβ	CCAAT/enhancer-binding protein beta
CALCOCO2/NDP52	Calcium-binding, coiled-coil domain-containing protein 2
CAR-iMac	CAR-macrophages
CASP1	Caspase 1
CBP/KAT3A	CREB-binding protein/lysine acetyltransferase 3A
CBR	Carbonyl reductase
CCL11	CC chemokine ligand 11
CCL17	CC chemokine ligand 17
CCL2	CC chemokine ligand 2
CCL3	CC chemokine ligand 3
CCL4	CC chemokine ligand 4
CCL5	CC chemokine ligand 5
CD	Cluster of differentiation
cDC	Conventional dendritic cell
cDC	Conventional dendritic cell
CDDO-Im	1-[2-cyano-3-,12-dioxooleana-1,9(11)-dien-28-oyl]imidazole
CDDO-Me	2-Cyano-3,12-dioxo-oleana-1,9(11)-dien-28-oic acid methyl ester
cGAS	Cyclic GMP-AMP synthase
c-JUN	Transcription factor Jun
CLP	Common lymphoid progenitor
CMP	Common myeloid progenitor
c-MYC	Cellular myelocytomatosis oncogene
CO	Carbon monoxide
COPD	Chronic obstructive pulmonary disease
CSF1	Colony-stimulating factor 1
CXCL1	Chemokine (C-X-C motif) ligand 1
CXCL10	Chemokine (C-X-C motif) ligand 10
CXCL12	Chemokine (C-X-C motif) ligand 12
CXCL2	Chemokine (C-X-C motif) ligand 2
CXCL9	Chemokine (C-X-C motif) ligand 9
CYP	Cytochrome P450
DAMP	Damage-associated molecular pattern
DC	Dendritic cell
DCreg	Regulatory dendritic cell
DDB	Damaged DNA-binding protein
DEM	Diethyl maleate
Dex-Mes	Dexamethasone 21-mesylate
DMF	Dimethyl fumarate
DPF2	Double PHD fingers 2
EAE	Experimental autoimmune encephalomyelitis
eIF2α	Eukaryotic initiation factor 2α
ELOVL7	Elongation of very-long-chain fatty acids protein 7
ENO1	Enolase 1
ENO4	Enolase 4
EpRE	Electrophile-responsive element
ERK	Extracellular signal-regulated protein kinase

FADS1	Fatty acid desaturase 1
FATP-1	Fatty acid transport protein 1
FPN	Ferroportin
FTH1	Ferritin heavy chain 1
FTL	Ferritin light chain
GABARAPL1	Amma-aminobutyric acid receptor-associated protein-like 1
GCLC	Glutamate-cysteine ligase catalytic
GM-CSF	Granulocyte-macrophage colony stimulating factor
GMP	Granulocyte/macrophage progenitor
GSH	Glutathione
GSK3	Glycogen synthase kinase 3
GSSG	Glutathione disulfide
GST	Glutathione S-transferase
HCC	Hepatocellular carcinoma
HCP1	Heme carrier protein 1
HCV	Hepatitis C virus
HDAC3	Histone deacetylase 3
HFD	High-fat diet
HMOX1	Heme-oxygenase 1
HNE	4-Hydroxynonenal
HRD1	ERAD-associated E3 ubiquitin-protein ligase 1
HSC	Hematopoietic stem cell
ICH	Intracerebral hemorrhage
iDC	Immature dendritic cell
IFIT2	Interferon-induced protein with tetratricopeptide repeats 2
IFN	Interferon
IFNβ	Interferon beta
IFNγ	Interferon gamma
IKK	IκB kinase
IL10	Interleukin 10
IL12	Interleukin= 12
IL12B	Interleukin 12 subunit beta
IL18	Interleukin 18
IL1β	Interleukin 1 beta
IL23	Interleukin 23
IL4	Interleukin 4
IL6	Interleukin 6
IL8	Interleukin 8
iNOS	Inducible NO synthase
iPSC	Induced pluripotent stem cell
IRE1	Inositol-requiring enzyme 1α
IRF3	Interferon regulatory factor 3
IRF7	Interferon regulatory factor 7
ISRE	Interferon-stimulated response element
IκB	Inhibitors of NF-κB
JNK	c-Jun N-terminal kinase
KO	Knockout
LDL	Low-density lipoprotein
LDLR	Low-density lipoprotein receptor
MARCO	Macrophage receptor with collagenous structure
MARV	Marburg virus
MDM	Monocyte-derived macrophage
MDP	Monocyte/dendritic progenitor
MDR	Multidrug resistance-associated protein
MDSC	Myeloid-derived suppressor cell
MEF2D	Myocyte-specific enhancer factor 2D
MHC	Major histocompatibility complex (MHC) class II (MHC II)
MHCII	Major histocompatibility complex (MHC) class II
MMF	Monomethyl fumarate
MMP	Matrix metallopeptidase
moDC	Monocyte-derived dendritic cell
MOG	myelin oligodendrocyte glycoprotein
MPO	Myeloperoxidase
MSK1	Mitogen- and stress-activated kinase 1
MST	Mammalian STE20-like kinase (MST)
MSU	Monosodium urate
mtDNA	Mitochondrial DNA
MYD88	Myeloid differentiation primary response protein 88 (MyD88)
NASH	Nonalcoholic steatohepatitis
Neh	NRF2-ECH homology
NES	Nuclear export signal
NET	Neutrophil extracellular trap
NF-κB	Nuclear factor kappa-light-chain-enhancer of activated B cells
NLRP3	Nucleotide-binding oligomerization domain leucine-rich repeat and pyrin domain-containing 3
NO	Nitric oxide
NOS2	Nitric oxide synthase 2
NQO1	NAD(P)H quinone dehydrogenase 1
NRF1	Nuclear respiratory factor 1
NRF2	NFE2-related factor 2
OCR	Oxygen consumption rate
OxLDLs	Oxidized low-density lipoproteins
p38	p38 mitogen-activated protein kinases
p53	Tumor protein P53

PBMC	Peripheral blood mononuclear cell
PERK	PKR-like ER kinase
PFKFB2	6-Phosphofructo-2-kinase/fructose-2,6-biphosphatase 2
PFKFB4	6-Phosphofructo-2-kinase/fructose-2,6-biphosphatase 4
PGA2	Prostaglandin A2
PGD	Phosphogluconate dehydrogenase
PGE2	Prostaglandin E2 (PGE2)
PGF	6-Phosphogluconate dehydrogenase
PKC	Protein kinase C
PKR	Protein kinase R
PMA	Phorbol 12-myristate-13-acetate
POMP	Proteasome maturation protein
PPAR	Peroxisome proliferator-activated receptor
PPAT	Phosphoribosyl pyrophosphate amidotransferase
PPI	Protein interaction
PPP	Pentose phosphate pathway
PPRE	Peroxisome proliferator-activated receptor response element
PRDX1	Peroxiredoxin-1
PRR	Pathogen recognition receptor
PSMA1	Proteasome 20S subunit alpha 1
PSMA4	Proteasome 20S subunit alpha 4
PSMB3	Proteasome subunit beta type-3
PSMB5	Proteasome subunit beta type-5
PSMB6	Proteasome subunit beta type-6
PSMC1	Proteasome 26S subunit, ATPase 1
PSMC3	Proteasome 26S subunit, ATPase 3
PSMD14	Proteasome 26S subunit, non-ATPase 14 Ribose-5-phosphate (R5P)
RBX	RING-box protein
RIG-I	Retinoic acid-inducible gene I
ROS	Reactive oxygen species
RSL3	RAS-selective lethal 3
RSV	Respiratory syncytial virus
RTM	Resident tissue macrophage
RXR	Retinoid X receptor
SFN	Sulforaphane
siRNA	Small interfering RNA
sMAF	Small musculoaponeurotic fibrosarcoma
sMAFF	MAF BZIP transcription factor F
sMAFG	MAF BZIP transcription factor G
sMAFK	MAF BZIP transcription factor K
SP1	Specificity protein 1
SQSTM1/P62	Sequestosome 1
STAT1	Signal transducer and activator of transcription 1
sTIM-3	Soluble TIM-3
STING	Stimulator of interferon gene
SWI/SNF	Switch/sucrose non-fermentable
t-BHQ	Tert-butylhydroquinone
TCA	Tricarboxylic acid cycle
Th	T-helper cell
TIM4	T-cell immunoglobulin and mucin domain containing 4
TIN	Tumor-infiltrating neutrophil
TKT	Transketolase
TLR	Toll-like receptor
TME	Tumor microenvironment
TNF	Tumor necrosis factor
Treg	Regulatory T cell
TREM1	Triggering receptor I on myeloid cells
TREM2	Triggering receptor II on myeloid cells
TSS	Transcription start sites
UCP3	Uncoupling protein 3
ULK1	Unc-51-like kinase 1
UPR	Unfolded protein response
UV	Ultraviolet
UVB	Ultraviolet B
VSVΔ51	Vesicular stomatitis virus Δ51
WDR	WD repeat-containing protein
XBP1	X-box-binding protein 1
β-TrCP	Beta-transducin repeats-containing proteins (β-TrCP)

8.1 Protein Structure

NFE2-related factor 2 (NRF2) is a member of the Cap'n'Collar (CNC) family of basic region, leucine zipper (bZIP) transcription factors (TFs) first described in 1994 [1]. The human NRF2 protein spans 605 amino acids and comprises 7 preserved domains, referred to as the NRF2-ECH homology (Neh) domains. These domains are conserved across species and play crucial roles in protein functionality [2]. Neh1 domain contains the CNC-bZIP domain, which allows the protein to interact with MAF proteins [3]. Neh2 domain regulates the activity of NRF2 by interacting with its negative regulator, Kelch-like ECH-associated protein 1 (KEAP1). KEAP1 is a component of the cullin 3 (CUL3)-based E3 ubiquitin ligase complex, which targets NRF2 for degradation by recruiting CUL3 and the small RING finger protein, the RING of Cullin 1 (CUL1), also known as RBX1, to form an active E3 ligase complex [4]. The NRF2-KEAP1 interaction is mediated by the DLG and ETGE motifs within the N-terminal Ned2 domain [5]. Two KEAP1 molecules bind to a single NRF2 molecule to form a trimer [6]. This two-site association is required for the ubiquitination of lysine residues in the Neh2 domain of NRF2 and rapid proteasome-mediated degradation of the protein [7–9]. Neh3-5 domains of NRF2 are transcriptional activation domains (TADs) that recruit coactivators to enhance its transcriptional activity [10, 11]. Neh5 domain also contains a nuclear export signal (NES), allowing NRF2 to shuttle between the nucleus and cytoplasm [12]. Neh6 is a serine-rich domain that interacts with the Beta-transducin repeats-containing proteins (β-TrCP) ubiquitin ligase complex. Phosphorylation of specific serine residues in Neh6 leads to the ubiquitination and subsequent degradation of NRF2 [13]. This domain contributes to the fine-tuning of NRF2 protein levels. Finally, Neh7 is a TAD that interacts with the retinoid X receptor alpha (RXRα) and mediates the synergistic activation of target genes when NRF2 forms heterodimers with RXRα. This interaction represses the transcriptional activity of NRF2 [14, 15].

8.2 NRF2 Regulation

Given the wide range of processes controlled by NRF2, its activity is tightly regulated at multiple levels.

8.2.1 Transcriptional Regulation

In addition to its self-regulation, NRF2 can be transcribed through the binding of several other TFs including aryl hydrocarbon receptor (AHR) [16], peroxisome proliferator-activated receptor γ (PPARγ) [17], nuclear factor kappa-light-chain-enhancer of activated B-cells (NF-κB) [18], specificity protein 1 (SP1) [19], tumor protein P53 (p53) [19], myocyte-specific enhancer factor 2D (MEF2D) [20], transcription factor Jun (c-JUN), cellular myelocytomatosis oncogene (c-MYC) [21], and breast cancer gene 1 (BRCA1) [22] to its promoter. In addition, epigenetic changes such as methylation of the NRF2 promoter in CpG islands, histone 3 (H3) methylation, and histone 4 (H4) acetylation can affect NRF2 transcriptional regulation [23].

8.2.2 Posttranscriptional Regulation

8.2.2.1 Proteasomal Degradation

Under homeostatic conditions, NRF2 is constitutionally expressed but rapidly degraded by the ubiquitin-proteasome system in the cytoplasm, resulting in low cellular levels of NRF2. NRF2 stability is regulated by binding to KEAP1. However, when KEAP1-NRF2 binding is disrupted, NRF2 is activated through derepression, where its components interfere with KEAP1-NRF2 binding. Therefore, it is more accurate to refer to the NRF2-KEAP1 sensor system. The murine KEAP1 domain comprises 25 cysteine residues that are capable of reacting with different electrophiles with varying degrees of sensitivity [24–26]. There are three major cysteine sensors on KEAP1, namely, Cys151, Cys273, and Cys288, and four classes of chemical inducers of NRF2 [26]. Class I and II inducers use Cys151 and Cys288, respectively; class III requires all three residues; and class IV inducers function independently of all three cysteine residues. The most common NRF2 inducers identified to date are listed in Table 8.1. These inducers can be electrophilic agents, natural or synthetic compounds, metal ions, or reactive oxygen species (ROS).

Interestingly, most of the compounds inhibit newly synthesized NRF2 proteins from binding to single KEAP1 proteins, rather than disrupting the existing NRF2-KEAP1 trimers [27].

Table 8.1 Inducers of NRF2

Class	Inducers
Class I	Diethyl maleate (DEM), dimethyl fumarate (DMF), nitric oxide (NO), tert-butylhydroquinone (t-BHQ), sulforaphane (SFN), 1-[2-cyano-3-,12-dioxooleana-1,9(11)-dien-28-oyl]imidazole (CDDO-Im)
Class II	OA-NO_2, 4-hydroxynonenal (HNE), AS^{3+}
Class III	15-deoxy-delta-12,14-prostaglandin J2 (15d-PGJ_2)
Class IV	Prostaglandin A2 (PGA2), $Cd2^+$, $Zn2^+$, dexamethasone 21-mesylate (Dex-Mes), H_2O_2

8.2.2.2 miRNA

Many small interfering RNAs (siRNAs) have also been shown to regulate NRF2 protein level by directly targeting the NRF2 mRNA and the mRNA coding for proteins that modulate NRF2 level and function [28].

8.2.3 Nuclear Translocation and Association with sMAF

KEAP1 is predominantly localized to the perinuclear cytoplasm [29], where it forms loose interactions with the cytoskeleton [30, 31]. Once the interaction between NRF2 and KEAP1 is disrupted, NRF2 translocates to the nucleus and heterodimerizes with small musculoaponeurotic fibrosarcoma (sMAF) through its bZip domains. MAFs can be classified into large and small, including small MAF BZIP transcription factor F (sMAFF), small MAF BZIP transcription factor G (sMAFG), and MAF BZIP transcription factor K (sMAFK), based on their molecular size. Only large MAFs can homodimerize with one another, whereas small MAFs solely facilitate gene transactivation when bound to specific CNC proteins such as NRF2 [32]. NRF2-sMAFs interaction is indispensable for NRF2 binding to DNA. This interaction also plays a role in preserving the stability of nuclear localization of the complex by concealing a redox-insensitive nuclear export signal (NES) motif within the Neh1 domain of NRF2 (AA 537–546) [33, 34].

8.2.4 DNA Binding

NRF2-sMAF heterodimers bind to antioxidant-responsive elements (ARE) [35] also known as electrophile-responsive elements (EpREs) [36]. These elements have a consensus sequence of TGACNNNGC [37] and are present in the promoters of specific genes. By binding to these motifs, NRF2-sMAF heterodimers facilitate the recruitment of RNA polymerase (RNAP) and other transcriptional coactivators to initiate mRNA synthesis and subsequent translation of downstream genes, ultimately contributing to the cellular antioxidant response.

8.2.5 Deactivation

KEAP1-CUL3-RBX1 captures and degrades newly synthesized cytoplasmic NRF2 proteins before the nuclear translocation of the TF [8]. Although this ubiquitinase system primarily regulates the global cellular activity of NRF2, KEAP1 complexes can also translocate to the nucleus. Here, they dissociate NRF2 from ARE and guide it back to the cytosol for degradation using a NES. This intricate process highlights the multifaceted nature of the KEAP1-CUL3-RBX1 system in the maintenance of cellular homeostasis [38].

In addition to KEAP1, three other E3 ubiquitin ligase complexes mediate the ubiquitylation and proteasomal degradation of NRF2 after activation: βTrCP-S-phase kinase-associated protein 1 (SKP1)-CUL1-RBX1 [39], WD repeat-containing protein 3 (WDR3)-CUL4-damaged DNA-binding protein (DDB) 1 [40], and ERAD-associated E3 ubiquitin-protein ligase 1 (HRD1) [41].

While the KEAP1 ubiquitinase system is responsible for the constitutive degradation of NRF2 in the cytoplasm under basal conditions and to some extent after activation, the βTrCP-SKP1-CUL1-RBX1, WDR3-CUL4-DDB, and HDR1 complexes can degrade NRF2 in the cytoplasm or the nucleus during oxidative conditions independently of KEAP1, thereby ensuring a tight control over the duration and intensity of

NRF2-mediated transcriptional responses by ensuring its termination [39, 40, 42].

8.2.6 Phosphorylation

Various kinases can phosphorylate NRF2, resulting in diverse effects on its stability [43]. Phosphorylation by extracellular signal-regulated protein kinases 1, 2, and 5 (ERK1, EKR2, and ERK5); c-Jun N-terminal kinase (JNK); PI3K-AKT; protein kinase C (PKC); casein kinase II (CKII); and protein kinase R (PKR)-R-like endoplasmic reticulum kinase (PERK) increases NRF2 stability [44]. Conversely, phosphorylation of NRF2 by p38 mitogen-activated protein kinases (p38) and glycogen synthase kinase 3 (GSK3) leads to its destabilization [39].

8.3 NRF2 Target Genes and Functions Common to All Cell Types/Core Pancellular Functions

8.3.1 NRF2 Regulates Antioxidant Stress Response and Drug Detoxification

NRF2-sMAF plays a vital role in maintaining cellular redox equilibrium. By activating the expression of ARE-containing cytoprotective genes, NRF2 ensures the coordinated transcription of various antioxidative enzymes and genes involved in phase I, II, and III xenobiotic detoxification processes [45]. NRF2 induces the expression of virtually all key antioxidative pathways. This includes the genes responsible for the glutathione (GSH) and thioredoxin (TXN) antioxidant systems, along with genes involved in the regeneration of NADPH and the elimination of ROS [46].

8.3.1.1 Glutathione (GSH) and Thioredoxin (TXN) Antioxidant System

NRF2 regulates the production of GSH by controlling the expression of genes encoding glutamate-cysteine ligase catalytic (GCLC) and modulator (GCLM) subunits [47] and by increasing the import of cystine through the modulation of the *SLC7A11* gene encoding the XCT subunit of the xc$^-$ system. The reduction of cystine generates cysteine, and its reaction with glutamate is the rate-limiting step in GSH synthesis [48]. NRF2 also supports GSH maintenance by inducing the expression of glutathione peroxidase 2 (GPX2) and several glutathione S-transferases (GSTs) [49]. These enzymes work together to oxidize GSH to oxidized glutathione (GSSG) and effectively detoxify ROS. NRF2 not only controls the regeneration of new GSH by inducing the transcription of glutathione reductase 1 (GSR1), which converts GSSG back to GSH in an NADPH-dependent manner [50], but also enhances the expression of key enzymes involved in the thioredoxin-based antioxidant system. This system, consisting of thioredoxin, thioredoxin reductase 1 (TXNRD1), and sulfiredoxin (SRXN1), is essential for the reduction of oxidized protein thiols [51–53]. Additionally, the redox cycling enzymes peroxiredoxin (PRX), glutathione peroxidase (GPX), superoxide dismutase 1 (SOD1), and catalase (CAT) are key NRF2 targets [54].

8.3.1.2 NADPH Production

NRF2 supports the production of NAPDH by inducing the main NADPH-regulating enzymes: glucose-6-phosphate dehydrogenase (G6PD), 6-phosphogluconate dehydrogenase (PGF), isocitrate dehydrogenase 1 (IDH1), and malic enzyme 1 (ME1) [55–58].

8.3.1.3 Xenobiotic Detoxification Processes

Phase I enzymes catalyze oxidative, reductive, and hydrolytic reactions of xenobiotics. These include genes encoding NAD(P)H quinone dehydrogenase 1 (NQO1), carbonyl reductases (CBRs), aldo-keto reductases (AKRs), aldehyde dehydrogenase 1 (ALDH1), and certain cytochrome P450 oxidoreductases (CYPs) [45, 59]. Phase II enzymes are responsible for conjugation reactions and include enzymes such as GST, UDP-glucuronosyltransferase (UGT), and UDP-glucuronic acid synthesis enzymes [59]. Phase III enzymes are drug efflux transporters that export conjugated metabolites after phase I and II reactions. These include multidrug resistance-associated proteins (MDR), breast cancer-resistant proteins (BCRP), and ATP-binding cassette g5 (ABCG5) and g8 (ABCG8) [59]. Apart from upregulating genes of all aforementioned encoding phase I, II and III enzymes, NRF2 plays a significant role in heme and iron metabolism which will be discussed in detail below.

8.3.2 NRF2 Regulates Metabolic Reprogramming

8.3.2.1 Upregulation of Aerobic Glycolysis

NRF2 plays an important role in glucose metabolism and has a profound impact on cellular antioxidative processes. NRF2 enhances the expression of GLUT1, a glucose transporter that increases cellular glucose uptake [60]. Additionally, it upregulates several glycolytic enzymes such as hexokinase 1 and 2 (HK1/2), glucose phosphate isomerase 1 (GPI1), 6-phosphofructo-2-kinase/fructose-2,6-biphosphatase 2 (PFKFB2), 6-phosphofructo-2-kinase/fructose-2,6-biphosphatase 4 (PFKFB4), fructose-bisphosphate aldolase A (ALDA), enolase 1 (ENO1), enolase 4 (ENO4), and pyruvate kinase M2 (PKM2) [60]. These enzymes work together to facilitate glucose import during glycolytic reactions. This increased glucose import not only contributes to energy production but also supports the maintenance of glycolytic intermediates necessary for anabolic reactions [56].

8.3.2.2 Upregulation of the Pentose Phosphate Pathway

NRF2 induces the expression of G6PD and phosphogluconate dehydrogenase (PGD), enzymes that regulate the diversion of the first glycolytic intermediate, glucose-6-phosphate, toward the oxidative phase of the pentose phosphate pathway (PPP). These enzymes produce NADPH, a reducing agent that is essential for the functioning of antioxidative enzymes and for the synthesis of lipids and nucleotides [56]. NRF2 also promotes the upregulation of two enzymes involved in the non-oxidative phase of the PPP, namely, transketolase (TKT) and transaldolase 1 (TALDO1) [56].

8.3.2.3 Upregulation of the *De Novo* Purine Biosynthesis Pathway

Ribose-5-phosphate (R5P), a precursor in the biosynthesis of nucleotides and a product of PPP, is catalyzed to 5-phosphoribosylamine (5PRA) by phosphoribosyl pyrophosphate amidotransferase (PPAT). This conversion is considered a rate-limiting step in purine synthesis. Methylenetetrahydrofolate dehydrogenase 2 (MTHFD2) provides one-carbon units for purine biosynthesis from serine amino acids derived from glycolytic intermediates [61]. Transcription of both PPAT and MTHFD2 is upregulated by NRF2.

8.3.2.4 NRF2 Regulates Lipid Metabolism

NRF2 regulates the degradation and synthesis of triglycerides/phospholipids, lipid transport, and fatty acid oxidation, all of which vary depending on the organ and context. Key genes involved in lipid metabolism are targeted and activated by NRF2, such as genes encoding elongation of very-long-chain fatty acids protein 7 (ELOVL7), acyl-CoA synthetase short-chain family member 2 (ACSS2), acyl-CoA thioesterase 7 (ACOT7), fatty acid desaturase 1 (FADS1), acyl-coenzyme A dehydrogenase family member 10 (ACAD10), and acyl-coenzyme A dehydrogenase family member 12 (ACAD12) [60]. This has been demonstrated in the esophagus of mice with a deficiency in KEAP1, resulting in elevated intracellular levels of phospholipids and long-chain free fatty acids [60]. Constitutive NRF2 activation in hepatocytes [62] has led to similar observations. Studies using a genetic knockout (KO) of NRF2 have shown organ-dependent effects. *Nrf2*-deficient C57BL/6 mice exhibited a tendency toward increased hepatic triglyceride levels compared to their wild-type counterparts when fed a high-fat diet (HFD) along with dysregulated fatty acid β-oxidation-related gene expression [63]. *Nrf2*-deficient mice displayed reduced adipose tissue mass, smaller adipocyte formation, and resistance to weight gain and obesity caused by HFD [64]. These findings suggested that NRF2 plays a crucial role in regulating lipid metabolism. The authors also highlighted the varied and context-dependent effects of NRF2 on triglyceride dynamics, adipose tissue properties, and complex networks of lipid-related processes in various cellular environments [64].

8.3.3 NRF2 Regulates Autophagy

The relationship between NRF2 and autophagy, which controls the degradation of defective proteins and organelles, is well established. NRF2 enhances autophagy by coordinating the transcription of key genes such as the ones encoding sequestosome 1 (SQSTM1/P62), calcium-binding, coiled-coil domain-containing protein 2 (CALCOCO2/NDP52), unc-51-like kinase 1 (ULK1), autophagy-related protein 5 (ATG5), and gamma-aminobutyric acid receptor-associated protein-like 1 (GABARAPL1) [65]. Deficient or insufficient autophagy can result in increased levels of oxidized proteins and organelles. This, in turn, activates NRF2. Additionally, inadequate autophagy leads to the accumulation of the cargo protein P62, which has been shown to sequester KEAP1. This accumulation also triggers NRF2 activation [66].

8.3.4 NRF2 Participates in the Unfolded Protein Response and Proteostasis

Ensuring the maintenance of a healthy cellular proteome involves proper synthesis and folding of new proteins, as well as a surveillance system that recognizes and eliminates dysfunctional proteins [67]. This process, referred to as proteostasis, relies on the unfolded protein response (UPR) which consists of three signaling arms that sense misfolded protein-associated stress: (1) inositol-requiring enzyme 1α (IRE1)-X-box binding protein 1 (XBP1), (2) PKR-like ER kinase (PERK)-eukaryotic initiation factor 2α (eIF2α) -activating transcription factor 4 (ATF4), and (3) activating transcription factor 6 (ATF6). All converge on the transcription of a battery of genes to restore endoplasmic reticulum (ER) homeostasis and alleviate cellular stress caused by the accumulation of misfolded proteins [67]. NRF2 participates in this complex process through various mechanisms.

- Integration with the UPR: Accumulation and aggregation of misfolded proteins in the ER results in excessive production of ROS from the mitochondria, which leads to the activation of NRF2 [68]. Furthermore, NRF2 can be phosphorylated and activated by PERK, a stress-responsive kinase that is a part of the UPR [69].
- Regulation of UPR genes: Activation of NRF2 stimulates the expression of genes involved in the UPR [62] such as those encoding XBP1 and ATF4 [70]. NRF2 interacts and associates with ATF4 to alleviate proteotoxic stress [71]. This indicates that NRF2 not only responds to UPR signals but also actively contributes to the regulation of UPR-related genes, thereby enhancing the cell's ability to cope with ER stress.
- Proteasome regulation: NRF2 is involved in controlling the basal and inducible expression of genes encoding multiple subunits of the 20S proteasome in both mice and humans. These include genes for proteasome 20S subunit alpha 1 (PSMA1), proteasome 20S subunit alpha 4 (PSMA4), proteasome subunit beta type-3 (PSMB3), proteasome subunit beta type-5 (PSMB5), proteasome subunit beta type-6 (PSMB6), proteasome 26S subunit ATPase 1 (PSMC1), proteasome 26S subunit ATPase 3 (PSMC3), and proteasome 26S subunit non-ATPase 14 (PSMD14) [72, 73]. Additionally, NRF2 binds to the promoter and induces the expression of the *proteasome maturation protein* (*POMP*) gene, which is involved in proteasome assembly [74].

8.3.5 NRF2 Regulates Mitochondrial Physiology and Biogenesis

NRF2 plays a role in mitochondrial biogenesis by controlling the transcription of *nuclear respiratory factor 1* (*NRF1*) [75] and *peroxisome proliferator-activated receptor gamma coactivator 1 alpha* (*PGC-1α*) [76]. Not only does NRF2 promote the generation of new mitochondria but also enhances the activity of existing mitochondria. This is achieved through the upregulation of various metabolic pathways (mentioned before), including glycolysis, the PPP, amino acid metabolism, and glutaminolysis. Consequently, there is an increased influx of substrates into the tricarboxylic acid cycle (TCA) cycle, leading to enhanced mitochondrial function [77].

Mouse embryonic fibroblasts with constitutive genetic activation of NRF2 exhibit higher basal mitochondrial membrane potential (ΔΨm) and ATP levels, while KO cells demonstrate the opposite phenotype [78]. To counterbalance these changes and prevent excessive ROS formation as a by-product of oxidative phosphorylation during mitochondrial respiration, NRF2 induces the expression of *uncoupling protein 3* (*UCP3*) [79]. Furthermore, NRF2 activation helps preserve mitochondrial DNA (mtDNA) integrity, which is crucial for preventing cell death and inflammation during increased mitochondrial activity [80].

8.4 NRF2 in Myeloid Cells and Dendritic Cells

In addition to their involvement in essential cellular processes such as survival, response to oxidative stress, metabolism, mitochondrial activity, autophagy, and proteostasis, NRF2 has been found to play a role in leukocyte differentiation and function.

Myeloid cells commonly refer to a heterogeneous group of immune cells derived from a common progenitor in the bone marrow (BM), the common myeloid progenitor (CMP), which arises early in hematopoiesis and segregates from a common lymphoid progenitor (CLP). CMP differentiation further branches into granulocyte/monocyte progenitors (GMPs) and monocyte/dendritic progenitors (MDPs) [81]. In turn, GMPs foster two differentiation pathways, one for granulocytes and one for monocytes, whereas MDPs give rise mainly to conventional dendritic cells (cDCs) and only marginally to monocytes [81]. Under appropriate stimulation conditions, such as during inflammation, monocytes can further differentiate into monocyte-derived macrophages (MDMs) [82] and monocyte-derived dendritic cells (moDCs) [83]. Therefore, myeloid cells include monocytes, cDCs, and granulocytes, such as neutrophils, as all these cells are thought to arise from CMP differentiation [84]. Although most resident tissue macrophage (RTMs) populations are derived from embryonic hematopoiesis [85], macrophages have historically remained included in the myeloid cell compartment. Thus, the remainder of this chapter uses the abovementioned definition of myeloid cells.

8.4.1 NRF2 and Myeloid Cell Differentiation

Hematopoietic stem cells (HSCs), known for their sensitivity to ROS, reside in antioxidative niches within the BM, where oxygen levels are low [86]. A low oxygen level state is key for maintaining stem cell quiescence in HSC through hypoxia-inducible factor 1 alpha (HIF1α)-sustained activation [87]. When HSCs are exposed to increasing ROS concentrations, it triggers their fate commitment. This observation, first made in *Drosophila melanogaster* [88], suggests that intracellular ROS accumulation, as a consequence of stress and/or inflammation, acts as a signaling mechanism that promotes stem cell differentiation [86, 89]. NRF2, as a pivotal modulator of the redox homeostasis, is essential in maintaining a balance between HSC quiescence and self-renewal [90, 91]. *Nrf2*-deficient mice had fewer HSCs in the quiescent G_0 state, including long-term HSC (LT-HSCs). Despite their hyperproliferative potential, these *Nrf2* KO HSCs did not have a competitive advantage when transplanted into primary recipients. This was likely due to inefficient homing and retention in the BM, possibly caused by reduced expression of the homing receptor CXCR4, independent of ROS concentration [90]. CXCR4 transcription is directly regulated by the binding of NRF2 to its promoter [91]. Although essential for maintaining HSC function, *Nrf2*-deficient HSPCs displayed similar myeloid and lymphoid commitment [91]. Recent studies have linked NRF2 to the double PHD fingers 2 (DPF2), a component of the BRG-/BRM-associated factor (BAF) (also known as switch/sucrose nonfermentable (SWI/SNF)) chromatin-remodeling complex, essential for the maintenance of HSCs [92]. Hematopoietic-specific *Dpf2* KO mice developed uncontrolled HSC proliferation with leukopenia alongside severe anemia and systemic inflammation due to excessive myeloid-cell lineage commitment, impaired macrophage polarization, and unrestrained T-helper (Th) cell activation [93]. Further investigations revealed that the absence of DPF2 results in improper binding of NRF2 to its target genes, disrupting antioxidative and anti-inflammatory transcriptional regulation in HSCs [94]. Interestingly, pharmacological reactivation of NRF2 by oral administration of 1-[2-cyano-3-,12-dioxooleana-1,9(11)-dien-28-oyl]imidazole (CDDO-Im) reversed the inflammatory phenotype and lethality in *Dpf2* KO mice [93].

In conditional *Keap1* KO mice, NRF2 activation was shown to skew the differentiation of HSCs toward the granulocyte-monocyte lineage. The differentiation trajectory was restored in mice where both *Keap1* and *Nrf2* were knocked out [95], underscoring the priming role of NRF2 in granulocyte-monocyte development. Similar observations were made in a study by Vallelian et al., who studied the effects of NRF2 activation on myeloid cell differentiation in granulocyte-macrophage colony stimulating factor (GM-CSF)-stimulated murine BM cell cultures. Exposure to heme, an important NRF2-activating metabolite, or to pharmacological NRF2 activators (ML-334 and RA-839) caused myeloid progenitors to deviate from their differentiation trajectories [96]. DC differentiation was abandoned in favor of enhanced differentiation toward antioxidant, iron-recycling macrophages, an effect reversed by the genetic KO of NRF2. The cells resulting from these cultures lost their ability to activate antigen-specific CD4$^+$ T cells [96]. Heme-induced NRF2 activation in myeloid progenitors may support physiological adaptations to hemolytic stress. Consequently, transgenic mice with constitutive intravascular hemolysis displayed lower levels of cDCs in the organs of extramedullary hematopoiesis, accompanied by a reduced capacity for antigen presentation and CD4$^+$ T-cell priming [96]. This may partially explain the intrinsic immune deficiency of individuals with genetic hemolytic diseases.

Taken together, these data indicate that the intricate role of NRF2 in HSCs extends beyond maintaining redox balance and affects HSC fate determination by influencing quiescence, self-renewal, and lineage specification (Fig. 8.1).

8.4.2 NFR2 Regulates Inflammatory Pathways

NRF2 interacts with many core inflammatory pathways, whether at the level of pathogen sensing, signaling cascades, and TFs. Although most of these interactions have been investigated in macrophages, they are also applicable to all innate immune cells [97].

8.4.2.1 Toll-Like Receptor Signaling

Innate immune cells are equipped with pathogen recognition receptors (PRRs), including Toll-like receptors (TLRs). These receptors recognize conserved microbial peptides and glycans or foreign genetic material, as well as damage-associated molecular patterns (DAMPs) that trigger inflammatory signaling cascades. Recent studies have elucidated the interplay between TLR activation and NRF2 [98]. The TLR agonists lipopolysaccharide (LPS), peptidoglycan, poly I:C, and resiquimod for TRL4, TLR2, TLR3, and TLR7, respectively, directly induced NRF2 by increasing the levels of the autophagy adapter protein P62, a known degrader of KEAP1 [99, 100]. Conversely, NRF2 fine-tuned TLR4 expression, while NRF2 silencing upregulated TLR4 protein levels [101].

8.4.2.2 NF-κB Signaling

NF-κB is a master TF regulating key aspects of both innate and adaptive immunity, as well as cell proliferation, apoptosis, cell migration, and

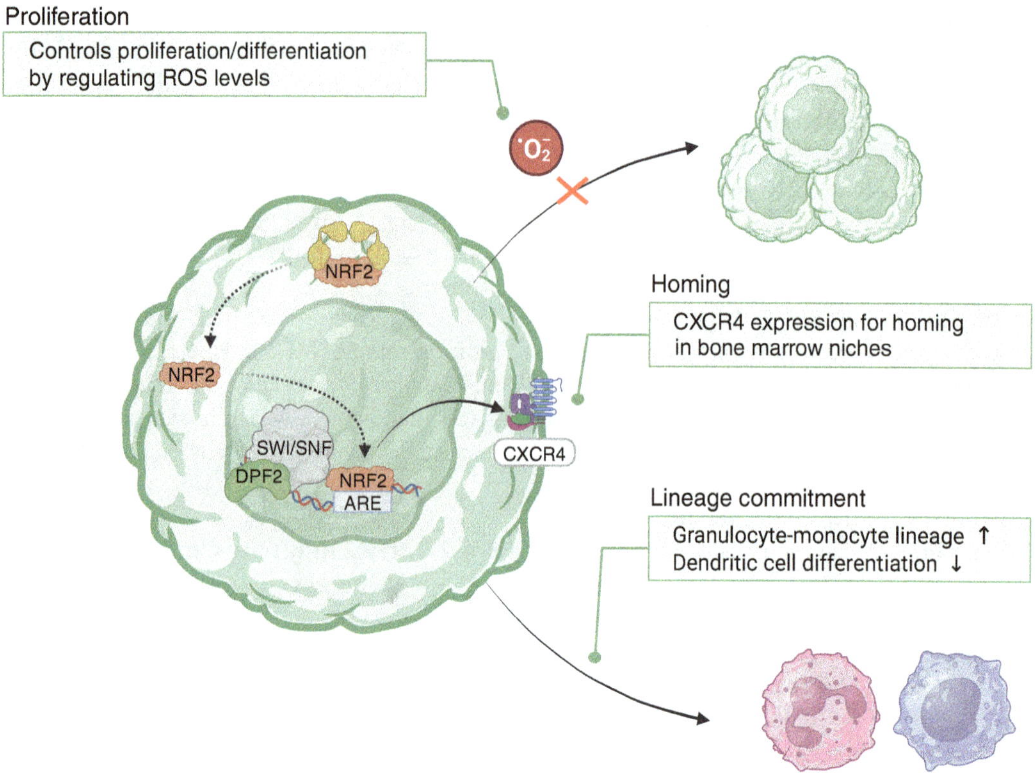

Fig. 8.1 NRF2's role on myeloid cell differentiation

angiogenesis [102, 103]. One of the most significant and well-documented functions of NF-κB is regulating the expression of proinflammatory cytokines and chemokines, particularly in macrophages. Its activation occurs when specific inhibitors known as inhibitors of NF-κB (IκB) are degraded after phosphorylation by the IκB kinase (IKK) complex. NRF2 and NF-κB possess an antagonistic relationship. NRF2 is negatively regulated by the P65 subunit of NF-κB which can form a stabilizing association with KEAP1 [104] but also through deprivation of CREB-binding protein/lysine acetyltransferase 3A (CBP/KAT3A) and recruitment of histone deacetylase 3 (HDAC3) to ARE regions [105]. Conversely, NRF2 can directly inhibit inflammatory NF-κB signaling. This is achieved by blocking the activity of IKK kinase and thereby preventing the degradation of IkBα, which is required for NF-κB activation [106]. Beyond these direct interactions, NF-κB is also susceptible to ROS-mediated activation (through alternative IκBα phosphorylation) or inhibition [107]. Thus, the reduction in ROS by NRF2 can indirectly affect NF-κB activation. It has also been shown that KEAP1 inhibits NF-κB signaling by promoting the proteasomal degradation of IκB [108].

8.4.2.3 Inflammasome

Inflammasomes are cytoplasmic multimeric protein complexes composed of cysteine proteases, a sensor protein, and an adaptor protein PYCARD/ASC (apoptosis-associated speck-like protein-containing caspase recruitment domain (CARD)). Many sensor proteins have been discovered in recent years and have been shown to detect different threats, ranging from a variety of endogenous or exogenous stimuli [109–112]. Upon detection of a threat by a sensor protein, the proteins assemble to form the inflammasome complex. This assembly ultimately leads to the activation of caspase effector proteins, which can cleave and activate the proprotein cytokines interleukin 1 beta (IL1β) and interleukin 18

(IL18) [109–112], or initiate the process of pyroptosis.

Data on the interplay between NRF2 and inflammasomes are complex and involve many levels of interaction [113]. One of the first interactions involves the nucleotide-binding oligomerization domain leucine-rich repeat and pyrin domain-containing 3 (NLRP3) sensor, which detects ROS. NRF2 not only naturally limits the intracellular accumulation of ROS and, by extension, the activation of the NLRP3 inflammasome, as shown in human THP1-derived macrophages [114], but also suppresses its expression. Exposure of keratinocytes or THP1 macrophages to several NRF2 activators (SFN, tBHQ, DMF, or 15d-PGJ_2) before ultraviolet B (UVB) irradiation inhibits inflammasome activation as demonstrated by reduced levels of processed caspase 1 (CASP1) and mature IL1β and IL18. This effect, however, is suspected to be independent of NRF2 expression, as the time of exposure to NRF2 activators before UVB irradiation was only 30 min, which is too early to elicit any transcriptional changes. Moreover, protein synthesis blockade by cycloheximide did not prevent inflammasome inhibition by NRF2 activators [115]. Conversely, NRF2 appears to be required for fully activating the sensor protein. Upon exposure to NLRP3 activators (including alum, silica, alhydrogel, monosodium urate, ATP, and nigericin), NRF2-deficient macrophages displayed reduced levels of PYCARD/ASC and decreased maturation and secretion of CASP1 and IL1β [116]. Other studies reported comparable results. When NRF2 was suppressed using siRNA and then stimulated with UV for NLRP3 inflammasome activation, there was a reduction in IL1β production [115]. Mechanistically, a direct physical interaction between the TF and the inflammasome pathway has been proposed [115], supported by the observation that NRF2 could co-localize with PYCARD/ASC proteins [116]. Interestingly, NLRP3 inflammasome activation seems to downregulate NRF2 expression. UVB irradiation of human primary keratinocytes, while inducing strong CASP1 activity, led to a notable decrease in NRF2 protein levels and a reduction in NRF2 target gene expression [115]. Similar results were observed when keratinocytes were treated with NLRP3 activators, such as nigericin [115]. Based on these observations, the authors suggested that the NLRP3 inflammasome might lead to NRF2 protein degradation. This hypothesis was strengthened by findings that NRF2 degradation by UVB or nigericin was prevented when *PYCARD* or *NAP1L3* was knocked out, or when cells were treated with ROS blockers [115].

8.4.2.4 Antiviral Functions

Type I interferons (IFNs), which comprise a group of 13 cytokines, are essential for orchestrating the host defense against viruses. They initiate an antiviral transcriptional program in infected and uninfected cells, thereby hampering viral replication [117]. NRF2 inhibits the expression of Type I IFNs in various cell types including macrophages. A comparison of the transcriptomes of BM-derived macrophages (BMDMs) and splenic tissues isolated from *Nrf2* KO and wild-type mice revealed upregulated basal expression of pathways associated with type I IFN in the absence of NRF2 [118]. In studies conducted on fibroblasts, NRF2 deficiency led to elevated activation of interferon regulatory factor 3 (IRF3), the main TF governing the transcription of type I IFN genes [106]. In line with these findings, the exposure of HEK293T cells infected with stomatitis virus to the NRF2 activator sulforaphane led to a dose-dependent reduction in IFN-stimulated response element (ISRE) binding. This, in turn, decreased the expression of key antiviral cytokines, such as CC chemokine ligand 5 (CCL5), chemokine (C-X-C motif) ligand 10 (CXCL10), stimulator of interferon genes (STING), retinoic acid-inducible gene I (RIG-I), and interferon regulatory factor 7 (IRF7). This observation was attributed to diminished IRF3 phosphorylation, nuclear translocation, and binding to its cognate binding site. The underlying mechanism is believed to be associated with enhanced cellular autophagy programs [119]; a recent proteomic study demonstrated suppressed interferon pathways in LPS-challenged BMDMs exposed to the KEAP1 inhibitor 4-octyl itaconate (4-OI), resulting in decreased expression of genes encoding interferon beta (IFNβ) and interferon-induced protein with tetratricopeptide repeats 2 (IFIT2) [120].

The cyclic GMP-AMP synthase (cGAS)-STING is a key component of the cytosolic DNA-sensing system. This system detects both foreign and aberrant genetic material and initiates a defensive immune response [121, 122]. In humans, NRF2 negatively regulates the STING pathway. This was evidenced by elevated levels of STING mRNA, and protein in the NRF2 silenced human epithelial cell line A549. Consequently, these cells exhibited decreased infectivity to the herpes simplex virus (HSV), indicating that the NRF2-STING increases susceptibility to DNA viruses [123].

8.4.3 NRF2 in Macrophages

Macrophages are a heterogeneous group of innate immune cells that are widely distributed in organisms. Their multifaceted functions are dictated by cues from their direct microenvironment and range from pathogen sensing and elimination to orchestrating inflammation, antigen presentation, tissue remodeling, repair, and the control of lipid and glucose metabolism [85]. To fulfill these roles, macrophages can adopt sub-identities, termed polarization, depending on their origin, tissue of residence, or local signals [85]. In addition to its essential function in maintaining macrophage homeostasis by regulating antioxidative responses, metabolic balance, and cytoprotection, NRF2 also plays a role in unique macrophage activities, such as cytokine production, pathogen defense, phagocytosis, and detoxification (Fig. 8.2) [97, 124].

8.4.3.1 Cytokine and Chemokine Production

A pivotal study by Kobayashi et al. unveiled a direct interaction between NRF2 and the transcription of several proinflammatory cytokines [125]. In this study, BMDMs were differentiated from mice with constitutional *Nrf2* deficiency or conditional KO for *Keap1* in myeloid cells, specifically *Keap1*$^{flox/flox}$ crossed with LysM-Cre mice, hereafter referred to as KpCKO mice. While NRF2 activation did not affect the differentiation of BMDMs, researchers observed a profound suppression of M1-associated genes in KpCKO cells. These genes include genes encoding for interleukin 6 (IL6), IL1β, nitric oxide synthase 2 (NOS2), and interleukin 12 subunit beta (IL12B), which were suppressed when stimulated with LPS and IFNγ. Other M1-associated genes such as tumor necrosis factor (TNF) and IRF1, as well as M2-associated genes such as arginase-1 (ARG1) and CCL17, were not affected by NRF2 activation. Further analysis using NRF2 ChIP-Seq revealed that NRF2 binding to regions upstream of the abovementioned genes suppressed proinflammatory genes, along with 198 other M1-associated genes. Interestingly, most of these binding peaks were predominantly observed in macrophages, but not in nonmyeloid cells such as Hepa1 cells [37]—in contrast to other antioxidative NRF2 targets that had binding peaks in both cell types. ARE motifs can be found upstream of certain genes, suggesting that NRF2 binds either directly or through its association with other proteins. Experiments using the NRF2 pharmacological activator diethyl maleate (DEM) revealed that the inhibition of M1-associated genes coincided with the early induction of NRF2-target genes, indicating that these effects are direct consequences of NRF2 activation and genome binding. This transcriptional repression was independent of the ROS levels. ChIP-qPCR experiments using an RNA polymerase II (RNAPII)-specific antibody demonstrated a significant reduction in RNAPII binding to the transcription starting sites (TSS) of IL6 and IL1β in LPS-stimulated BMDMs treated with DEM. Interestingly, mRNA stability and LPS-induced histone modifications of newly transcribed-M1-associated genes were not affected by DEM-mediated NRF2 activation, nor was the binding of the P65 (NF-κB) or CCAAT/enhancer-binding protein beta (C/EBPβ) and c-JUN ((activator protein 1 (AP1)) to the TSS of these genes. These findings demonstrate that NRF2 interferes with the transcription initiation and RNAPII recruitment to *Il6* and *Il1b* TSSs independently of the recruitment of proinflammatory TFs.

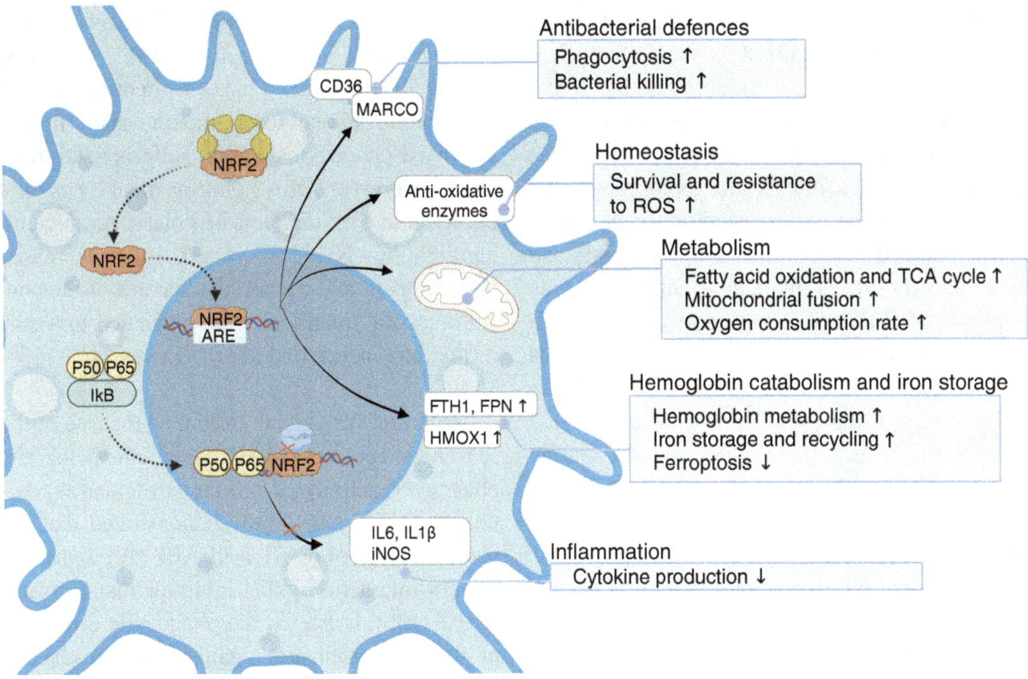

Fig. 8.2 NRF2's functions in macrophages

Luciferase reporter experiments were performed in 293T cells using a luciferase vector ligated 1 kb upstream of the *Il6* gene. Transfection with the active form of NRF2 (NRF2^{T80R}) mildly increased the luciferase activity, which was abolished when the ARE motif in the *Il6* promoter was mutated. Transfection with P65 increased the luciferase reporter activity. However, cotransfection of P65 and NRF2^{T80R} in 293T cells resulted in reduced activity, even with a mutated ARE motif. This suggests that NRF2 hinders *Il6* gene expression by disrupting P65-mediated transcriptional activation, independent of ARE binding. The data, along with previous ChIP-Seq and ChIP-qPCR results, imply that NRF2 regulates proinflammatory gene transcription, by associating with recruited NF-κB upstream of these genes, thereby affecting NF-κB signaling and RNAPII activity.

Using induced human pluripotent stem cell (iPSC)-derived engineered CAR-macrophages (CAR-iMac), researchers could demonstrate that NRF2 activation via CRISPR-mediated KEAP1 deletion represses expression of genes encoding proinflammatory cytokines IL6, IL1β, CXCL9, and CXCL10 upon IFNγ and LPS co-challenge [126]. Conversely, inhibition of NRF2 by reduction of intracellular levels of itaconate, an intrinsic metabolite inhibitor of KEAP1 [127], resulted in increased expression of the same cytokines [126].

8.4.3.2 Phagocytosis and Clearance of Bacteria

Macrophages are the principal phagocytic cells of the body. NRF2 activation can modulate the phagocytic capacity; however, the precise mechanism remains unclear. NRF2 indirectly supports phagocytic activity by maintaining redox balance. Phagocytosis is dependent on dynamic cytoskeletal changes. Since oxidative stress can affect the stability and function of the cytoskeleton, NRF2's role in protecting against oxidative damage supports the rearrangement necessary for phagocytosis [128].

The direct augmentation of phagocytosis by NRF2 is mediated by the induction of specific scavenger receptors. In this context, both CD36 and macrophage receptor with collagenous structure (MARCO), recognized as scavenger receptors, are susceptible to modulation by NRF2 signaling, which subsequently influences their phagocytic activity. Expression of the gene encoding CD36, a membrane glycoprotein that functions as a scavenger receptor for apoptotic cells, oxidized phospholipids, lipoproteins, and pathogens [129], is markedly enhanced upon exposure of RAW264.7 murine macrophage cell line or BMDMs to the NRF2 activator DEM. This induction arises from the binding of NRF2 to ARE motifs in the promoter region of the first exon of the gene [130]. Furthermore, in murine peritoneal macrophages, the NRF2-CD36 signaling axis controls non-opsonic phagocytosis of *Plasmodium falciparum*-parasitized erythrocytes [131]. Supporting these observations, macrophages treated with curcumin exhibited elevated non-opsonic phagocytosis of *Plasmodium falciparum*, an effect attributed to the NRF2-dependent upregulation of CD36 [132]. In vitro studies on RAW264.7 macrophages revealed that inhibition of NRF2 through binding of soluble TIM-3 (sTIM-3) reduces the uptake and killing of *Listeria Monocytogenes* by downregulating the scavenger receptor CD36 [133]. Mechanistically, sTIM-3 exposure inhibited NRF2 from entering the nucleus and promoted its ubiquitination and subsequent degradation, which, in turn, reduced *Cd36* expression [133]. Corresponding in vivo experiments have echoed these findings, showing that mice overexpressing TIM-3 challenged with intravenous *Listeria monocytogenes* have diminished peritoneal macrophage *Cd36* expression, elevated bacterial counts in the peritoneal fluid and spleen homogenate, and the largest liver foci of infection [133].

MARCO, a class A scavenger receptor, mediates the binding and uptake of gram-positive and gram-negative bacteria, oxidized low-density lipoprotein (LDL), and environmental particles by macrophages [134, 135]. The expression of this receptor is also linked to augmentation of phagocytosis. Ex vivo studies on alveolar macrophages from patients with chronic obstructive pulmonary disease (COPD) treated with sulforaphane revealed an increase in phagocytosis and clearance of *Haemophilus influenzae* and *Pseudomonas aeruginosa* [136]. Similarly, murine alveolar macrophages from mice exposed to cigarette smoke for 6 months also showed improved phagocytosis upon sulforaphane treatment. However, this improvement was not observed in *Nrf2*-deficient alveolar macrophages, highlighting NRF2's crucial role. Increased phagocytosis and bacterial clearance following sulforaphane treatment were attributed to a corresponding increase in *MARCO* expression [134, 135].

Recently, Wang et al. examined the role of the scavenger receptor MARCO in enhancing the phagocytic activity of murine peritoneal macrophages [137]. This study demonstrated that the administration of IR-61, a KEAP1-NRF2 protein-protein interaction (PPI) inhibitor that activates NRF2 [138], bolstered the phagocytic efficiency of these macrophages against latex beads and *Escherichia coli*. Moreover, it augmented the bactericidal activity against *Staphylococcus aureus* and *Escherichia coli* [137]. These augmented phagocytosis and bactericidal functions were NRF2-dependent. When NRF2 was silenced using siRNAs, the enhancing effects of IR-61 on macrophages were nullified.

Interestingly, BV-2 cells, which are used as in vitro models for microglia, also show increased phagocytosis of zymosan bioparticles and damaged erythrocytes upon upregulation of NRF2 through monascin or Xuezhikang exposure [139].

8.4.3.3 Metabolism, Antioxidative, and Detoxification Functions

Activation of NRF2 markedly enhances the antioxidative capacity of macrophages, thereby directly influencing their function and survival. In murine BMDMs, protein kinases mammalian STE20-like (MST) 1 and 2 sense ROS, specifically phagosomal and mitochondrial ROS. This prompts the activation of NRF2 via the phosphorylation of KEAP1, resulting in the expression of cytoprotective antioxidative genes. Double KO (dKO) of *Mst1/Mst2* led to premature macrophage aging and increased susceptibility to ROS-induced macrophage death. This phenotype can be rescued by the pharmacological activation of NRF2, underscoring the role of this TF in cytoprotection [140]. Similarly, exposure of RAW264.7 macrophages to low doses of curcumin has been found to increase the expression of a battery of antioxidative enzymes. This enhances their resistance to H_2O_2- and ROS-induced cell death [141]. In line with these data, proteomic analysis of BMDMs with genetically activated NRF2 revealed an elevation in proteins linked to oxidative stress regulation, whereas metabolomic studies further confirmed increased concentrations of antioxidant metabolites [120]. Similar observations were made in a transcriptomic analysis of BMDMs exposed to pharmacological activators of NRF2, in which pathways related to ROS detoxification were greatly enriched [142].

The metabolic state of macrophages is closely related to their polarization state. By influencing metabolic pathways, NRF2 indirectly affects macrophage polarization, thereby affecting the inflammatory response. As mentioned above, NRF2 promotes the PPP, ensuring a steady supply of NADPH for antioxidant defense; modulates mitochondrial function; enhances oxidative phosphorylation, which is characteristic of anti-inflammatory macrophages; and regulates genes involved in lipid and amino acid metabolism. Ryan et al. conducted a comprehensive metabolomic analysis of BMDMs with varying NRF2 status: wild-type, *Nrf2*-deficient (NRF2 inactive), or *Keap1*-deficient (NRF2 active) [120]. The metabolome profiles differed markedly between the groups. In *Keap1*-deficient BMDMs, there was a surge in fatty acid oxidation, as evidenced by elevated levels of specific carnitines. Simultaneously, the TCA cycle (fumarate, malate, and 2-ketoglutarate), central to cellular energy production, is suppressed, while bioenergetic compounds (NAD^+, creatine, and phosphocreatine) increase, either unstimulated or after LPS challenge [120]. Respirometry analysis of the oxygen consumption rate (OCR) demonstrated increased basal respiration rates in conjunction with increased ATP production [120]. These metabolic adaptations were accompanied by higher levels of mitochondrial fusion proteins in NRF2-activated macrophages. Under conditions of NRF2 suppression, the main alteration was an increase in TCA cycle activity with a decrease in the OCR compared to the wild-type [120].

In LPS-stimulated *Keap1* KO BMDMs, the metabolic response shifted toward glycolysis and amino acid and lipid metabolism. These changes were mirrored by increased extracellular acidification rates [120]. The mitochondria also displayed striking morphological changes, with a switch from fragmented to a large elongation of the organelle. In contrast, *Nrf2*-deficient BMDMs showed the opposite effects under LPS challenge, with suppressed glycolysis and negligible mitochondrial alterations [120].

8.4.3.4 Modulation of PPARγ Activity

Peroxisome proliferator-activated receptors (PPARs), comprising the three isoforms PPARα, PPARβ/δ, and PPAR γ, are a family of nuclear receptors widely expressed in all mammalian tissues. These receptors are involved in many cellular processes including differentiation, metabolism, and oncogenesis [143]. Particularly, PPARγ and NRF2 have been shown to interact with each other [144]. PPARγ is expressed in adipose tissue, skeletal muscles, intestine tissues, and immune cells [145] and is implicated in modulation of genes linked to oxidative defenses, lipid uptake, and inflammation.

The NRF2 and PPARγ pathways are connected by positive feedback loops. The expression of the *PPARG* (encoding for PPARγ) gene is positively regulated by NRF2 [146] as several ARE motifs are present in its promoter [17, 147]. *Nrf2*-deficient mice exhibit compromised expression of PPARγ [147]. Conversely, the expression of NRF2 seems to also be positively regulated by PPARγ as several PPAR response element (PPRE) motifs are present in the promoter of the *Nrf2* gene [148] and KO of *Pparg* in mice led to a reduced NRF2 expression [45]. Moreover, it is plausible that the two TFs collaborate on similar target genes to achieve maximum expression, as some genes such as GST possess both ARE and PPRE sequences in their promoter regions [148, 149] as does the gene of the scavenger receptor CD36.

Upon binding to the retinoid X receptor (RXR) [150], PPARγ positively regulates genes involved in lipid uptake such as CD36 and fatty acid transport protein 1 (FATP-1) [145, 151]. This is in particular relevant in the pathophysiology of atherosclerosis as foam cell formation is dependent on PPARγ induction and PPARγ-deficient macrophages are incapable of uptaking oxidized low-density lipoproteins (OxLDLs). Additionally, cholesterol uptake and efflux in macrophages is also controlled by PPARγ-mediated expression of cholesterol efflux cassettes of the ATP-binding cassette transporter 1 (ABCA1) family [152]. PPARγ also modulates the inflammatory functions of macrophages. *Pparg*-deficient murine BMDMs challenged with LPS display increased production of proinflammatory cytokines TNF, IL1β, IL6, and IL12 and a reduced production of anti-inflammatory cytokine interleukin 10 (IL10) compared to wild-type cells [153].

8.4.3.5 Coordination of Heme and Iron Metabolism

Macrophages are the central hub responsible for phagocytosis of senescent or damaged red blood cells, supporting heme detoxification and iron recycling [154]. Red blood cell toxins, hemoglobin, and heme accumulate in macrophages as a result of erythrophagocytosis; receptor-mediated endocytosis when heme is bound to carrier proteins such as haptoglobin, hemopexin, or albumin; or transfer by specialized importers such as heme carrier protein 1 (HCP1) [155, 156]. The iron atom in heme molecules is highly reactive and can interact with H_2O_2 to form hydroxyl radicals (HO•), thereby mediating significant intracellular oxidative stress. This oxidative stress is sufficient to disrupt the KEAP1-NRF2 association, leading to the stabilization and nuclear translocation of NRF2 [157]. Once activated, NRF2 plays an essential role in the coordination of hemoglobin catabolism and iron recycling [158, 159].

Heme oxygenase 1 (HMOX1) is the rate-limiting enzyme that catalyzes heme catabolism and functions by cleaving the alpha-mesocarbon of Fe-protoporphyrin-IX using three O_2 molecules [160, 161]. While many redundant pathways exist that lead to the upregulation of *HMOX1/Hmox1*, NRF2 significantly enhances its expression by binding to ARE motifs on the promoter of the gene [142, 162]. Additionally, NRF2 coordinates the expression of all the antioxidative enzymes in this process, mitigating oxidative stress-related damage. This catabolic activity results in the formation of biliverdin, carbon monoxide (CO), and ferrous iron. Ferrous iron is either stored in macrophages bound to ferritin, which consists of heavy-chain (FTH) and light-chain (FTL) subunits, or exported by ferroportin (FPN). NRF2 has been shown to bind to ARE motifs in the promoter regions of genes encoding ferritin H in immortalized human cell lines and to induce its transcription [163]. This finding is further supported by bulk RNA-sequencing data from murine BMDMs, where NRF2 activators amplify ferritin H expression [142]. In murine BMDMs and peritoneal macrophages, NRF2 activation by electrophilic compounds increases the expression of ferroportin, an effect not observed in *Nrf2* KO [164].

In addition, the upregulation of NRF2 induces resilience against ferroptosis, a cellular death process distinct from apoptosis, necrosis, necrotic apoptosis, or autophagy, and is associated with intracellular iron accumulation and, consequently, high oxidative stress derived from ROS production by Fe^{2+}-mediated Fenton-like oxidative reactions [165, 166]. In vitro studies using human peripheral blood mononuclear cells (PBMCs) showed that exposure to the ferroptosis-inducing agent RSL3 (RAS-selective lethal 3) leads, on average, to fivefold more cellular death in cells silenced for NRF2 compared to the control [167]. Similarly, in THP1 cells that differentiated into a macrophage-like state with phorbol 12-myristate-13-acetate (PMA), LPS-induced ferroptosis was rescued by exposure to the NRF2 activator 4-octyl itaconate (4-OI), and silencing of *NRF2* abolished this effect [168].

In summary, NRF2 upregulation plays a pivotal role in macrophages during hemoglobin and heme catabolism, orchestrating the expression of

HMOX1, antioxidative enzymes, and key iron-handling genes. NRF2's influence extends to ferritin and ferroportin, which determine iron storage and export, and enhances macrophage resistance to ferroptosis.

8.4.4 NRF2 in Dendritic Cells

DCs are a heterogeneous group of immune cells that originate from a hematopoietic lineage distinct from monocytes and neutrophils [81]. The cDCs are professional antigen-presenting cells (APCs). Their primary functions are to detect pathogens, take up antigens, and subsequently present these antigens to unprimed lymphocytes, thereby initiating adaptive immune responses. While research on NRF2's role in DCs is not as extensive as that in macrophages, there is a growing recognition of its impact on their maturation, cytokine production, and antigen presentation capabilities (Fig. 8.3).

8.4.4.1 Dendritic Cell Maturation, Costimulatory Receptor Expression, and T-Cell Activation

DCs mature in response to pathogens and foreign substances. This maturation process involves the upregulation of surface molecules and the release of signaling molecules, enhancing their ability to present antigens and activate the adaptive immune system, which is crucial for mounting an effective response against the identified threat.

Purified human monocytes were cultured in a medium supplemented with GM-CSF and IL4 to stimulate moDC differentiation [169]. The addition of the NRF2 activator DMF to the medium blocked DC maturation, as demonstrated by the reduced expression of genes encoding for CD1a, CD40, CD80, CD86, and HLA-DR, culminating in impaired T-cell activation capability [169]. In similar experiments, human monocyte-derived immature DCs (iDCs) incubated with the NRF2 activator monomethyl fumarate (MMF) showed

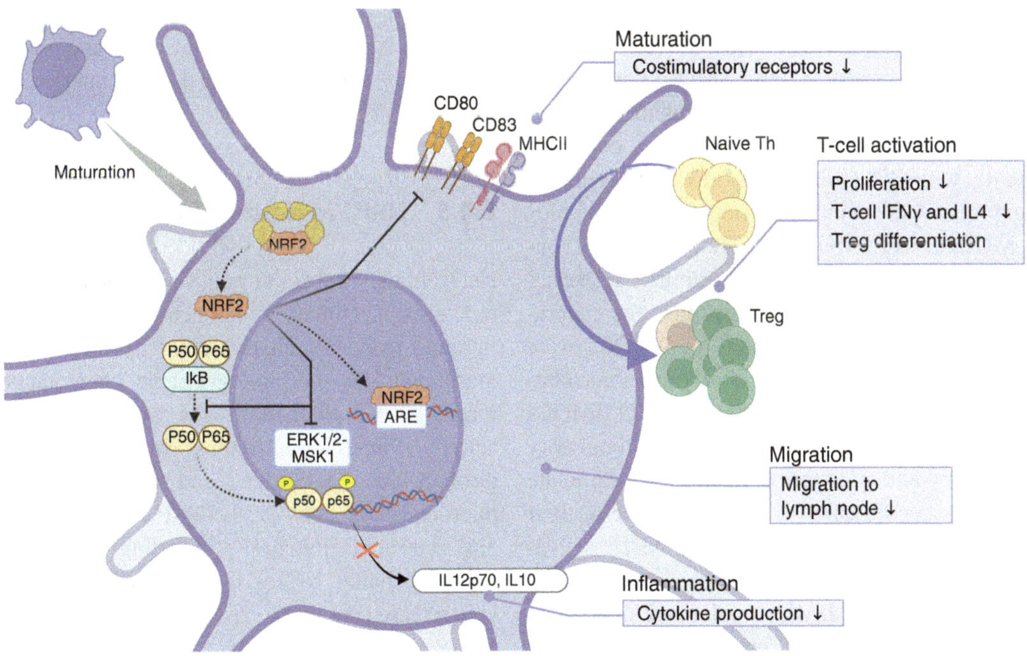

Fig. 8.3 NRF2's functions in dendritic cells

reduced expression of genes encoding the costimulatory receptors CD83, CD40, and CD80 compared to control DCs upon LPS challenge [170]. Researchers have demonstrated that MMF prevents DC maturation and polarization irrespective of the polarization factor involved. Whether polarized with LPS/IFNγ/pertussis toxin for cDCs1, LPS/prostaglandin E2 (PGE_2) for cDCs2, or LPS/IL10 for DCreg, all MMF-DC cultures exhibited reduced IL12p70 and IL10 production after maturation [170]. This inhibitory effect of MMF was also evident when DCs were matured in the presence of TLR ligands. Additionally, while $CD4^+$ naïve T-cells and MMF-DC co-cultures showed similar T-cell proliferation in comparison to control untreated DCs [170], the production of IFNγ and IL4 by these primed T cells was markedly reduced in co-cultures with MMD-DCs compared to co-cultures with control DCs [170]. This observation was even more pronounced under conditions of primed lymphocyte restimulation using LPS or anti-CD3 and anti-CD28 monoclonal antibodies. The same dysfunctional phenotype was observed after exposure to high doses of DMF [171].

Further studies in murine in vitro models validated and expanded on these findings. BM-derived dendritic cells (BMDCs) exposed to MMF from the start of the differentiation process (adjunction to the medium on day 0, concomitantly with GM-CSF and IL4) exhibited reduced expression of major histocompatibility complex (MHC) class II (MHCII), CD80, and CD86, coupled with impaired migratory ability. However, when NRF2 signaling was disrupted, the maturation and migration capacities of BMDCs were restored. A study by Yeang et al. further corroborated these results, noting that *Nrf2* KO BMDCs display increased expression of costimulatory receptors and a superior capacity to stimulate antigen-specific $CD8^+$ T cells compared to their wild-type counterparts [172, 173]. NRF2 activation also skews the type of T-cell response that DCs can generate. BMDCs polarized in the presence of MMF supported regulatory T-cell (Treg) proliferation compared to control BMDC cocultures [173]. In a recent study, Vallelian et al. demonstrated that heme-NRF2 signaling in BMDCs suppresses $CD4^+$ T-cell activation and proliferation by redirecting BM differentiation routes. Additionally, interruption of NRF2 signaling restored the priming function of BM cells [96].

8.4.4.2 Cytokine Production

NRF2 activators like DMF and MMF affect cytokine production in DCs. When human iDCs were treated with MMF, a notable decrease in the production of cytokines IL12p70 and IL10 was observed after LPS or LPS/IFNγ challenge [170]. This reduction in cytokine release has been linked to diminished NF-κB activation. Furthermore, in human DCs, high-dose DMF exposure resulted in decreased IL12 and IL6 production, potentially due to suppressed extracellular signal-regulated kinases 1/2 (ERK1/2)-mitogen- and stress-activated kinase 1 (MSK1) signaling that affects P65 phosphorylation and its subsequent translocation to the nucleus. Moreover, there was a reduced phosphorylation of H3 by MSK1, which interfered with local DNA relaxation and the accessibility of P65 to gene promoters such as *IL6*, *IL12B*, and *IL8* (encoding for interleukin 8 (IL8)) [171]. In addition, even in the absence of any inflammatory trigger, MMF-treated BMDCs produced significantly lower levels of the proinflammatory cytokines IL6, IL12, and interleukin 23 (IL23) [173].

8.4.5 NRF2 in Neutrophils

Neutrophils, a subset of granulocytic leukocytes, represent a critical component of the innate immune response, characterized by their polymorphonuclear morphology, rapid chemotactic migration to inflamed tissues, and phagocytic activity, thereby contributing to host defense through the engulfment and degradation of microbial pathogens [174]. NRF2 impacts various neutrophil functions, including activation, oxidative burst, NETosis, migration, and cytokine release (Fig. 8.4).

8.4.5.1 Activation and Survival

A recent investigation showed that NRF2 activation in murine BM-derived $Ly6G^+$ neutrophils by

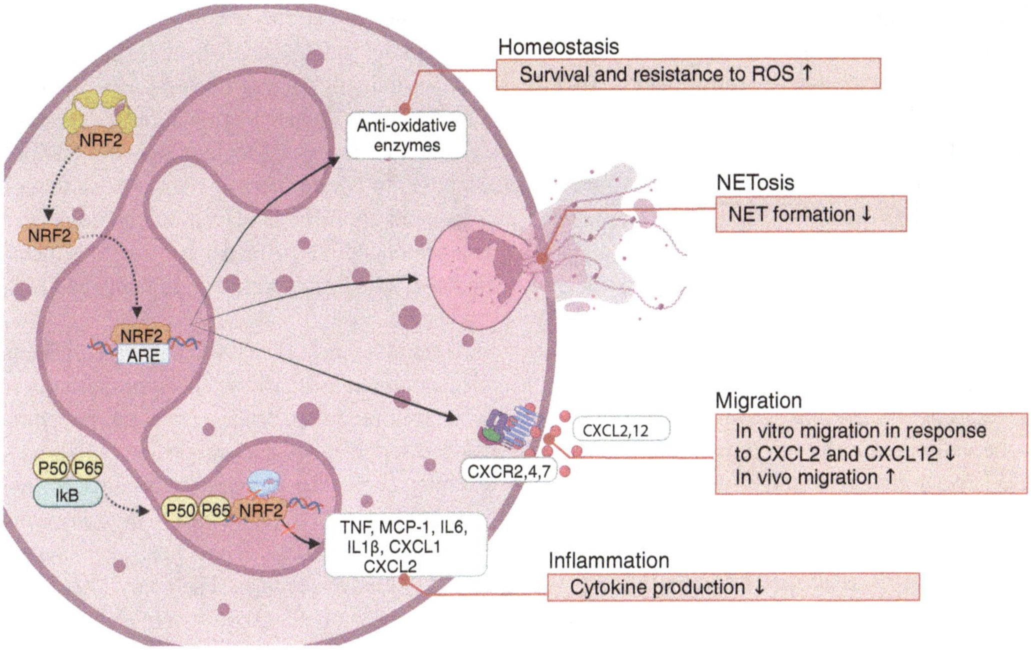

Fig. 8.4 NRF2's functions in neutrophils

exposure to synthetic 2-cyano-3,12-dioxo-oleana-1,9(11)-dien-28-oic acid methyl ester (CDDO-Me) promotes cell survival and reduces cell necrosis [175]. Similarly, the high-persistence and long-term survival of tumor-infiltrating neutrophils (TINs) were attributed to NRF2 upregulation through aconitate decarboxylase 1 (ACOD1), augmenting the resistance against ROS-induced cell death and ferroptosis [176].

8.4.5.2 Oxidative Burst and Neutrophil Extracellular Trap (NET) Formation

Neutrophils undergo a unique cellular death process termed NETosis, during which they release a mesh of cytoskeletal proteins, histones, granzymes, and nuclear components such as neutrophil elastase and myeloperoxidase (MPO), which create neutrophil extracellular traps (NETs) [177]. NETs are considered a cornerstone of the innate immune extracellular defense mechanisms, given their ability to trap and kill bacteria and degrade toxic factors [178]. NRF2 activation by curcumin has been shown to hinder NETosis in neutrophils by decreasing ROS accumulation and oxidative burst, a process essential for proper NET formation [179].

8.4.5.3 Cytokine Production

Research on neutrophils has also revealed the importance of NRF2 in the suppression of proinflammatory cytokines. In vivo LPS stimulation of peritoneal neutrophils from *Nrf2*- deficient mice resulted in a significantly increased production of TNF, IL6, monocyte chemotactic protein-1 (MCP-1), and macrophage inflammatory protein 2 (MIP-2), along with elevated ROS levels [180]. Similarly, *Nrf2*-deficient murine neutrophils stimulated with the TLR2 agonist zymosan showed increased expression of genes encoding proinflammatory cytokines TNF, IL1β, and IL6 and CCL3, CXCL1, and CXCL2 [181].

8.4.5.4 Migration

An in vitro transwell migration assay revealed that *Nrf2*-deficient murine neutrophils had decreased migration in response to the chemokines CXCL2 and CXCL12 compared to wild-type neutrophils [181]. Interestingly, opposite findings were observed in vivo. When challenged with dinitrochlorobenzene-induced skin inflammation, *Nrf2*-deficient mice display increased chemokine recruitment (CCL2, CCL4, CCL11) and a higher influx of neutrophils at the inflamed site than wild-type mice [182]. Similar observations have been made in other disease models, such as LPS-induced endotoxemic shock [180] or pneumonia [183]. These contradictory findings between in vitro and in vivo results may stem from elevated oxidative damage in a system-wide *Nrf2* KO. This can amplify the production of inflammatory mediators, potentially counterbalancing the inherently reduced migration propensity of NRF2-lacking neutrophils.

8.5 Implications of Modulating NRF2 in Myeloid Cells for Disease Pathogenesis

NRF2's numerous functions in myeloid cells ultimately dictate their behavior and influence the development of many diseases.

8.5.1 Sepsis and Extracellular Bacterial Infections

Sepsis is a life-threatening organ dysfunction caused by a dysregulated host response to infection [184]. It is initially characterized by an excessive innate immune response, leading to a cytokine storm and a hypermetabolic catabolic state, which increases metabolic substrate availability [185]. Metabolic alterations occur as the disease progresses. A key aspect of this metabolic shift is mitochondrial dysfunction, now recognized as a key driver of sepsis-induced cellular damage. This dysfunction is marked by an impaired TCA cycle and intensification of glycolysis [186]. Mitochondrial damage is largely attributed to the accumulation of ROS and reduced transcription of genes encoding proteins essential for cellular respiration [187]. The cellular ATP generation machinery is impaired due to inadequate cellular oxygen utilization despite sufficient supply.

The role of NRF2 in sepsis is dual, depending on the phase and context of the disease. NRF2 prevents oxidative stress-related tissue damage, helps to control hyperinflammation by attenuating cytokine release, and reprograms metabolic responses to foster energy production. Although timely and appropriate activation can be protective, prolonged NRF2 activation may have detrimental effects by suppressing immunity and ROS production.

Thimmulappa et al. showed that *Nrf2*-deficient mice exhibited a significantly higher mortality in two sepsis models: cecal ligation and puncture-induced septic shock and intravenous LPS endotoxemic shock. The increased mortality rate has been attributed to a pronounced surge in proinflammatory cytokines in the lungs and uncontrolled oxidative damage. The study also found elevated levels of NF-κB protein in the peritoneal macrophages of these mice [188]. Further investigation revealed that this effect was partially mediated by an overactive TLR4-myeloid differentiation primary response protein 88 (MyD88) pathway. Remarkably, ablation of the NADPH subunit GP91 (phox) rescued increased mortality, presumably by reducing LPS-induced ROS generation in macrophages [189]. In contrast, pharmacological activation of NRF2 in septic mice with bacterial lung infection using the heptamethine dye RI-61 strongly activated NRF2 in macrophages, enhancing bacterial clearance at infection sites and improving survival rates [137].

Macrophages, which are a significant source of cytokines, play a central role in sepsis protection through NRF2 [190]. Nevertheless, although active NRF2 macrophages mitigate excessive inflammation, this suppression can compromise defenses against pathogens. Recent studies demonstrated that NRF2 activation in liver macrophages can promote bacterial growth and lower survival rates. Mice subjected to intratracheal *Klebsiella pneumoniae* delivery prior to the transfusion of damaged erythrocytes showed increased mortality and extrapulmonary bacterial proliferation [191]. Upon engulfment of stressed erythrocytes, heme-NRF2 signaling in liver macrophages markedly suppressed signal transducer and activator of transcription 1 (STAT1) expression and IFN-related pathways, essential for control of *Klebsiella pneumoniae* [142, 191]. Furthermore, NRF2-induced antioxidative pathways, while protecting against oxidative damage, can inadvertently create conditions that favor bacterial growth due to the paucity of ROS [192].

8.5.2 Intracellular Infection

A study that investigated the early response of macrophages to intracellular infection by bacteria of the *Mycobacterium tuberculosis* complex, emphasized the significant contribution of NRF2 to the pathophysiology of tuberculosis [193]. During the early stages of infection, alveolar macrophages showed a surprising increase in NRF2 activation, leading to a dampened inflammatory response. This adaptation contributed to reduced intrinsic macrophage defense mechanisms, providing a favorable niche for bacterial growth. In addition, NRF2 activation blocked the death of infected macrophages, allowing the bacteria to replicate intracellularly. However, when *Nrf2* was specifically knocked out in macrophages and DCs, the benefits to the pathogen were reversed. The *Nrf2*-deficient cells had a stronger proinflammatory gene signature, increased expression of MHCII, and higher rates of cellular death upon infection, resulting in a significantly reduced mycobacterial burden [193].

The impact of NRF2 on viral infections is complex and varies among different viruses. NRF2 deficiency can worsen certain infections due to increased ROS-related tissue damage, as observed in respiratory syncytial virus (RSV) [194, 195] and influenza virus infections [196, 197]. In contrast, NRF2 may repress antiviral defenses. In a murine model of vaginal herpes infection, *Nrf2*-deficient mice showed reduced viral replication accompanied by increased levels of type I IFNs in vaginal washes, correlating with better clinical outcomes [118]. NRF2 deficiency also led to increased survival in mice infected with the intraperitoneal *Marburg virus* (MARV) [198]. Interestingly, MARV might hijack the NRF2-KEAP1 signaling system to promote NRF2 activation and ensure its survival [199]. Similarly, activation of NRF2 by sulforaphane increased vesicular stomatitis virus Δ51 (VSVΔ51) replication and sensitized cancer cells to viral oncolysis [119]. Inhibition of NRF2 using siRNA in hepatocellular carcinoma (HCC) cell lines infected with hepatitis C virus (HCV) led to reduced viral titers [200]. Mounting an appropriate antiviral response is closely associated with intracellular ROS levels. In human MoDCs exposed to dengue virus, decreasing ROS through inhibition of the NOX-complex led to reduced IFNγ production and diminished control over virus replication. Similar results were observed when NRF2 was inhibited by siRNA [201].

The role of NRF2 in infectious diseases and sepsis is dichotomous and varies depending on the replication site of the pathogen. NRF2 active macrophages can be protective, aid in the clearance of extracellular pathogens, or be detrimental by suppressing IFN pathways, thus potentially promoting the replication of intracellular pathogens.

8.5.3 Cancer

Mutations in the NRF2-KEAP1 sensing system of tumor cells have been linked to both protumorigenic and antitumorigenic effects [202] by promoting cytoprotection or inhibiting ROS-mediated protumoral mutations [203–205]. However, the impact of NRF2 on myeloid cells within the tumor microenvironment (TME) remains poorly explored. The TME is composed of various myeloid cells, including tumor-associated macrophages, myeloid-derived suppressor cells (MDSCs), myeloid-derived DCs, and TINs. These cells significantly influence tumor cell behavior and are potential therapeutic targets. Tumor cells release metabolites and signaling molecules that influence their surroundings. For example, the release of lactate by tumor cells shapes the phenotype of neighboring macrophages toward a protumoral phenotype [206, 207]. THP1 cells, PBMC-derived macrophages co-cultured with hepatocellular carcinoma cells (HepG2 and Huh 7) or pancreatic cancer cells (SUIT2 and Panc-1), showed increased NRF2 activation driven by tumor-derived lactate. This

NRF2 activation led to a protumoral, anti-inflammatory M2-like phenotype, with these macrophages secreting VEGF in quantities sufficient to induce an epithelial-mesenchymal transition in cancer cells [208].

Researchers explored the role of itaconate, a metabolite produced within the TME [209]. This metabolite, secreted by tumor-associated macrophages, MDSCs and TINs, acts as an inducer of NRF2 [127]. Itaconate was found to inhibit $CD8^+$ T-cell function and enhance TIN survival, thereby promoting tumor progression. In melanoma and murine mammary tumor models, mice deficient in ACOD1, the enzyme-mediating itaconate synthesis, showed reduced tumor growth and increased sensitivity to immune checkpoint blockade [210]. Wang et al. observed that *Acod1*-deficient CAR-iMac effectively killed tumor cells in vitro and suppressed tumor growth in ovarian and pancreatic cancer mouse models. This was attributed to low itaconate-NRF2 signaling [126]. Furthermore, another mechanism of NRF2 activation in the myeloid cells within the TME involves the phagocytosis of extravasated erythrocytes, a phenomenon commonly observed in tumors due to the heightened permeability and fragility of intratumoral vessels, leading to microhemorrhages [211]. A recent investigation demonstrated how intratumoral bleeding leads to the development of tumor-promoting macrophages [212]. Through single-cell RNA sequencing and experiments in 3D cell cultures, this study found that heme-NRF2 signaling in myeloid cells leads to the formation of procancerous tumor-associated macrophages. These cells contributed to cancer invasion and metastasis by facilitating epithelial-mesenchymal transition. Also, NRF2-activated macrophages were resistant against tumoricidal reprogramming by IFNγ and anti-CD40 antibodies. Conditional *Keap1* KO mice-bearing colon adenocarcinoma tumors (MC38) were unresponsive to anti-CD40 immunotherapy. Notably, when NRF2 signaling in macrophages was disrupted, these cells enhanced their capacity to kill tumor cells [212]. These findings align with previous observations connecting HMOX1-expressing macrophages to cancer escape. Macrophages characterized by high HMOX-1 levels and low expression of genes encoding for MHCII have been identified in various tumor types and were associated with protumoral activity, as their blockade reduced metastasis and improved the response to immunotherapy [213–215].

8.5.4 Wound Healing

Wound healing, a complex physiological process involving cellular, molecular, and biochemical events, is significantly modulated by NRF2 [216].

Wound macrophages orchestrate tissue repair by eliminating debris and facilitating hematoma resolution, which are essential early steps in wound healing. Through red blood cells and heme metabolism, NRF2 signaling imparts a profound anti-inflammatory phenotype to infiltrating macrophages, fostering both tissue repair and resolution of inflammation [142, 211]. Maassen et al. observed that itaconate-mediated activation of NRF2 steers human monocyte-derived macrophages toward $TGF\beta^+$ reparative macrophages. This phenotypic switch was accompanied by a suppression of LPS-induced collagen uptake [217].

In healthy individuals, wound healing typically progresses without medical intervention. However, in the context of diabetes, continual hyperglycemia disrupts this mechanism through glycosylation of proteins, thereby undermining the functionality of endothelial, stromal, and immune cells. Diabetic wounds exhibit dysregulated blood perfusion, increased inflammation, and impaired cell proliferation and migration [218]. In particular, macrophages adopt improper phenotypic changes that result in reduced matrix metallopeptidase (MMP) levels and an inability to trigger proper tissue remodeling, angiogenesis, and reepithelialization [219]. In rat peritoneal macrophages, exposure to hyperglycemic conditions inhibited NRF2 activation, increasing ROS and proinflammatory cytokine levels. Both NRF2 overexpression and pharmacological activation reversed the deleterious effects of hyperglycemia. Based on these observations, systemic treatment with the NRF2 activator DMF led to accelerated wound healing in diabetic rats compared to the control through the modulation of inflammation and oxidative stress [220]. Similar observations have been made using different NRF2 agonists, either systemically or locally [221–225]. In most of these studies, the mechanism involved heightened macrophage wound repair orchestration capabilities, with overall reduced inflammation and oxidative stress.

Myocardial infarction represents another context in which NRF2 modulation has a significant impact on wound healing. Cardiac tissue macrophages are associated with tissue repair and limited adverse tissue remodeling [226, 227]. Single-cell RNA sequencing data of myocardial infarct tissues from both mice and humans identified a subset of resident tissue macrophages, defined by high NRF2 activation and low expression of interferon-stimulated genes [228]. Interestingly, pharmacological inhibition of IFN signaling reduced inflammation and adverse ventricular remodeling and improved survival in mice [229, 230]. These findings suggest that NRF2 activity in cardiac macrophages protects against pathological remodeling and inflammation, ultimately leading to beneficial clinical outcomes.

8.5.5 Atherosclerosis

Atherosclerosis, defined as a chronic inflammatory disease of the arterial intima with accumulation of lipids and consequent oxidative stress, is profoundly affected by NRF2 activity. Macrophages in atherosclerotic lesions take up oxidized lipids, including oxLDLs, leading to the transformation of cholesterol-laden proinflammatory foam cells that direct the remodeling of the vascular wall and plaque development.

Increased atherosclerosis has been observed in *Nrf2* KO BM-transplanted atherogenic low-density lipoprotein receptor (*Ldlr*)-deficient mice fed a high-fat diet for 6 weeks [231]. Peritoneal macrophages from these *Nrf2*-deficient mice exposed to modified LDLs showed higher expression of proinflammatory cytokines and increased lipid uptake [231]. Extending the high-fat diet duration to 7 months in *Ldlr* conditional *Nrf2* double KO mice, another group observed a significant increase in atherosclerotic lesions and more severe liver inflammation and fibrosis compared to wild-type mice for NRF2 [232]. These findings align with more recent evidence, suggesting that the transcriptional inhibition of NRF2 via ERK5 phosphorylation contributes to the development and progression of atherosclerosis by conferring proliferative capacities to macrophages [233]. These findings collectively indicate that the absence of NRF2 in myeloid cells worsens the early and advanced stages of the disease. However, contrasting results have been reported in studies using apolipoprotein E (*ApoE*)-deficient (*Apoe* KO) mice on a high-fat diet, indicating a need for further research [234–236]. Considering the distinct anti-atherogenic effects of ApoE and the abnormal lipoprotein profile seen in ApoE-deficient mice, experts recommend validating these findings in more physiologically relevant atherosclerosis models.

8.5.6 Hemolysis

Hemolysis is characterized by the premature destruction of red blood cells and occurs either intravascularly or through erythrophagocytosis by macrophages in the liver and spleen (extravascular hemolysis). The uncontrolled breakdown of erythrocytes releases hemoglobin and heme, leading to nitric oxide depletion in the vasculature and oxidative stress [155, 156, 237–239]. Compartmentalized hemolysis occurs locally upon hemorrhage and hematoma formation, such as in subarachnoid hemorrhages or advanced atherosclerotic lesions [155, 211, 240]. While the orchestration of antioxidative defenses by NRF2 is essential in all cell types during hemolysis, NRF2 also plays an important role in the myeloid compartment, particularly in macrophages, which are the central hub for safe heme catabolism and iron recycling [241]. Under conditions of heme stress, colony-stimulating factor 1 (CSF1)-NRF2 signaling was shown to expand the scavenger macrophage pool by promoting monocyte differentiation into FPN⁺ T-cell immunoglobulin and mucin domain containing 4 (TIM4) macrophages. This specialized subset of macrophage transiently accumulated in the liver, carrying out heme detoxification and iron recycling to support erythropoiesis [242]. In sickle cell disease, the NRF2-CSF1 axis redirected the differentiation routes of monocytes, favoring the generation of blood patrolling monocytes [243]. These monocytes constitute a protective myeloid population, preventing vaso-occlusive disease [244].

Studying how hemoglobin exposure modifies macrophage polarization, Boyle et al. identified a subset of antioxidative atheroprotective $CD163^{high}$ $HLA\text{-}DR^{low}$ macrophage in the vicinity of intra-atherosclerotic plaque hemorrhages [245]. They reproduced this phenotype in vitro by exposing PBMC-derived macrophages to hemoglobin, observing an increased IL10 to TNF ratio upon LPS treatment and decreased ROS levels in response to opsonized zymosan, a prototypical macrophage oxidative burst stimulant [245]. This shift toward an anti-inflammatory and antioxidative phenotype was linked to NRF2 as the NRF2 activator tert-butylhydroquinone (t-BHQ) drove similar phenotypic changes [157]. Subsequent studies have found a comparable transformation in the liver macrophages of mice with spherocytosis and sickle cell disease [142]. In these mice, ongoing erythrophagocytosis by liver macrophages and intracellular heme accumulation led to their transformation into an idiosyncratic phenotype, termed erythrophagocyte, with high expression of genes related to heme metabolism and iron recycling (*Hmox1*, *Fth1*, and *Ftl1*), antioxidative enzymes *Nqo1* and peroxiredoxin-1 (*Prdx1*), enhanced expression of the scavenger receptor *Marco*, and low expression of genes related to MHCII. Upregulated pathways in erythrophagocytes were associated with antioxidative responses, heme metabolism, oxidative phosphorylation, and glycolysis, whereas downregulated pathways were related to immunoinflammatory responses, such as the IFNγ response. This phenotypic switch confered protection against macrophage-driven inflammatory liver diseases, such as CD40-driven necro-hepatitis or nonalcoholic steatohepatitis (NASH) [142, 246]. Therapeutic reprogramming of liver macrophages by heme-albumin complexes or anti-erythrocyte antibody administration to tumor-bearing mice abolishes liver toxicity induced by an immunotherapeutic anti-CD40 antibody [247, 248]. Collectively, NRF2 enforces the reprogramming of macrophages to enable the mounting of an appropriate antioxidative, iron-recycling anti-inflammatory response in order to mitigate the oxidative effects of cell-free hemoglobin. Although the role of NRF2 signaling in reducing immunity could be viewed as detrimental to patients with systemic hemolytic diseases, it has likely evolved to suppress harmful inflammation and autoimmunity during wound healing [247, 248].

8.5.7 Lung Diseases

Long-term exposure to cigarette smoke in COPD leads to oxidative stress, cellular damage, and chronic inflammation, resulting in small airway disease and pulmonary emphysema. Given the crucial role of oxidative stress in the pathogenesis of COPD, NRF2 has emerged as a promising therapeutic candidate [249]. *Nrf2*-deficient mice exposed for 6 months to cigarette smoke show earlier and more severe emphysema than their wild-type littermates [250], a finding also observed in murine elastase-induced emphysema models [251]. Wild-type mice fed a diet containing the potent NRF2 activator CDDO-Im and exposed to cigarette smoke showed significantly reduced lung oxidative stress, alveolar cell apoptosis, and pulmonary hypertension [250]. However, clinical translation to humans has posed challenges and yielded unsatisfactory results. Daily oral administration of the NRF2 activator sulforaphane did not result in significant changes in the severity of the disease or inflammation of NRF2-target gene expression in alveolar macrophages, possibly due to incomplete NRF2 activation as a result of the diet [252]. A recent study highlighted the clinical translation potential of NRF2 activators in lung disease. In patients with COPD, both alveolar and circulating peripheral blood monocyte-derived macrophages display impaired redox balance and profound depletion of energy reserves, with impaired mitochondrial respiration in favor of glycolysis, correlating with ineffective efferocytosis. Remarkably, ex vivo activation of NRF2 with the pharmacological activator KI-696 not only restored the metabolic defects of macrophages but also rescued COPD-associated transcriptome changes [253].

In OVA-challenge murine models of allergen-driven asthmatic inflammation, *Nrf2*-deficient mice pronounced mucus cell hyperplasia and infiltration of eosinophils compared to wild-type littermates, correlating with an increased expression of genes for IL4 and IL13 in bronchoalveolar lavage (BAL) [254].

Evidence suggests that NRF2 plays an essential role in the pathophysiology of acute lung injury (ALI). *Nrf2*-deficient mice succumbed to butylated hydroxytoluene-induced ALI at doses tolerated by wild-type mice [17]. A similar observation was made in a model of ALI induced by intraperitoneal LPS administration [188]. Although no clinical studies have been conducted to translate these findings to humans, results showed that NRF2 gene polymorphisms increase the risk of ALI after major trauma [255].

NRF2 activation by intraperitoneal administration of epigallocatechin-3-gallate (EGCG) in rats protected from an intratracheal bleomycin-induced pulmonary fibrosis with markedly reduced ROS generation, histologically observable pulmonary fibrosis, and collagen levels in the lungs [256].

8.5.8 Intracerebral Hemorrhage

In mice, the pharmacological activation of NRF2 through oral administration of Xuezhikang or monascin in murine models of collagenase-induced intracerebral hemorrhage (ICH) has shown clear beneficial effects. The treatment led to measurable improvements in neurological deficits and decreased intracerebral hemoglobin levels. The therapeutic effects were largely attributed to the functional transformation of microglia and increased phagocytic responses. NRF2 agonists increased the expression of triggering receptor II on myeloid cells (TREM2), CD206, and brain-derived neurotrophic factor, while inhibiting the expression of triggering receptor I on myeloid cells (TREM1), CD80, and TNF. Further, increased phagocytic capabilities were observed in BV-2 microglia-like cells exposed to the same NRF2 pharmacological activators, as detailed above [139], supporting the notion that NRF2 enhances the phagocytic function of myeloid cells. In a related study, induction of erythrophagocytosis through the administration of recombinant human cerebral dopamine neurotrophic factor was shown to promote hematoma resolution in a mouse model of ICH [257]. Consistent with the study of Pfefferle et al. [142], heme-NRF2 signaling reprogrammed macrophages/microglia toward an anti-inflammatory and antioxidative state, promoting neurological recovery [257].

8.5.9 Autoimmune Diseases

In autoimmune diseases, a critical aspect is the presentation of self-antigens by DCs to naïve T cells, leading to an improper adaptive immune response against self-components [258]. Treatment with the NRF2 activators DMF and MMF improved clinical scores in models of murine experimental autoimmune encephalomyelitis (EAE). This improvement was associated with suppressed expression of inflammatory cytokines in the lymph nodes and increased GATA3 expression [259]. The efficacy of these agents was supported by the observation that transferring T cells co-cultured with DMF-treated BMDCs to EAE mice suppress disease development [259]. Further research in different EAE models reinforced the therapeutic potential of NRF2 pathway modulation. *Keap1* KO mice with EAE exhibited improved clinical scores and reduced IL6 expression [125]. In myelin oligodendrocyte glycoprotein (MOG)-induced EAE models, *Nrf2* deficiency was linked to higher levels of mature DCs in the spinal cord, while MMF treatment on days 2 and 7 after disease induction led to significant disease amelioration and reduced spinal cord levels of inflammatory CD4+ T-cell subsets [173]. In the clinical context, particularly for patients with relapsing remitting multiple sclerosis, NRF2 activator DMF has been established as first-line therapy. Its effectiveness has been attributed to several mechanisms, but a key aspect relies on the inhibition of proinflammatory pathways in myeloid cells [260, 261]. This mechanism aligns with the observed effects of NRF2 in mouse models of EAE.

The protection conferred by NRF2 activation in autoimmunity has also been observed in other autoimmune models. In regulatory T-cell-deficient scurfy mice, which progressively develop lethal multiorgan inflammation, systemic genetic NRF2 activation through the reduction of *Keap1* expression [262, 263] improved tissue inflammation and reduced lethality [264]. Interestingly, this protective effect requires systemic NRF2 activation, as targeted activation in myeloid cells or DCs was insufficient. Treatment of mice with the pharmacological NRF2 activator, A-1396076, improved the clinical score in different autoimmune models, including antibody-mediated arthritis, lupus nephritis, and sterile ovalbumin lung inflammation models [265]. Treatment with A-1396076 improved clinical outcomes and reduced the number of mature DCs in BAL of mice sensitized with intranasal OVA/LPS [265]. The effectiveness of A-139607 in autoimmune arthritis was contingent on the timing of administration. It was beneficial when administered simultaneously with or before antigen sensitization but showed no clinical effect when given after antibody challenge, emphasizing a specific disruption of the antigen presentation process by NRF2 [265]. Similar effects were observed in hemolytic mice. In these mice, elevated exposure to the prooxidants hemoglobin and heme results in widespread NRF2 activation. Remarkably, splenic DCs exhibited diminished capacity to prime CD4+ T cells [96].

8.6 Conclusion

NRF2 has emerged as a pivotal player in the regulation of myeloid cell biology, transcending its traditional role in orchestrating antioxidative responses, metabolic regulation, protein homeostasis, and autophagy. Accumulating evidence suggests that the activation of NRF2 not only promotes the differentiation of myeloid cells but also significantly influences their functions. NRF2 achieves this by interfering with critical inflammatory pathways, which includes the suppression of TLR expression, attenuation of inflammasome activity, and the inhibition of NF-κB signaling cascades. In macrophages, NRF2 enhances phagocytosis and bacterial clearance capabilities. The TF also prompts a metabolic reprogramming that leads to increased oxygen consumption, augmented activity of the TCA cycle, and higher ATP production. The upregulation of antioxidant genes by NRF2 grants myeloid cells increased resistance to ROS-induced cytotoxicity. Additionally, NRF2 upregulates the heme catabolic machinery, further supporting cellular protection mechanisms. In DCs, the activation of NRF2 has been observed to hinder cellular maturation and compromise the efficiency of antigen presentation, which in turn diminishes the DCs' capacity to activate T cells. For neutrophils, NRF2 promotes cell survival and reduces necrosis, including NETosis. NRF2 activation also interferes with neutrophil migration and impairs protease secretion. The array of phenotypic and functional modifications in myeloid cells induced by NRF2 has tangible implications for disease progression and outcomes. As such, there is a compelling case for further investigation into the therapeutic potential of NRF2-targeting strategies in the context of myeloid cell-mediated diseases.

References

1. Moi P, Chan K, Asunis I et al (1994) Isolation of NF-E2-related factor 2 (Nrf2), a NF-E2-like basic leucine zipper transcriptional activator that binds to the tandem NF-E2/AP1 repeat of the beta-globin locus control region. Proc Natl Acad Sci U S A 91:9926–9930
2. Canning P, Sorrell FJ, Bullock AN (2015) Structural basis of Keap1 interactions with Nrf2. Free Radic Biol Med 88:101–107

3. Motohashi H, Katsuoka F, Engel JD, Yamamoto M (2004) Small Maf proteins serve as transcriptional cofactors for keratinocyte differentiation in the Keap1-Nrf2 regulatory pathway. Proc Natl Acad Sci U S A 101:6379–6384
4. Furukawa M, Xiong Y (2005) BTB protein Keap1 targets antioxidant transcription factor Nrf2 for ubiquitination by the Cullin 3-Roc1 ligase. Mol Cell Biol 25:162–171
5. Tong KI, Katoh Y, Kusunoki H et al (2006) Keap1 recruits Neh2 through binding to ETGE and DLG motifs: characterization of the two-site molecular recognition model. Mol Cell Biol 26:2887–2900
6. Iso T, Suzuki T, Baird L, Yamamoto M (2016) Absolute Amounts and Status of the Nrf2-Keap1-Cul3 Complex within Cells. Mol Cell Biol 36:3100–3112
7. McMahon M, Itoh K, Yamamoto M, Hayes JD (2003) Keap1-dependent Proteasomal Degradation of Transcription Factor Nrf2 Contributes to the Negative Regulation of Antioxidant Response Element-driven Gene Expression*. J Biol Chem 278:21592–21600
8. Zhang DD, Lo S-C, Cross JV et al (2004) Keap1 is a redox-regulated substrate adaptor protein for a Cul3-dependent ubiquitin ligase complex. Mol Cell Biol 24:10941–10953
9. Baird L, Llères D, Swift S, Dinkova-Kostova AT (2013) Regulatory flexibility in the Nrf2-mediated stress response is conferred by conformational cycling of the Keap1-Nrf2 protein complex. Proc Natl Acad Sci U S A 110:15259–15264
10. Nioi P, Nguyen T, Sherratt PJ, Pickett CB (2005) The carboxy-terminal Neh3 domain of Nrf2 is required for transcriptional activation. Mol Cell Biol 25:10895–10906
11. Katoh Y, Itoh K, Yoshida E et al (2001) Two domains of Nrf2 cooperatively bind CBP, a CREB binding protein, and synergistically activate transcription. Genes Cells 6:857–868
12. Li W, Yu S-W, Kong A-NT (2006) Nrf2 Possesses a Redox-sensitive Nuclear Exporting Signal in the Neh5 Transactivation Domain*. J Biol Chem 281:27251–27263
13. McMahon M, Thomas N, Itoh K et al (2004) Redox-regulated turnover of Nrf2 is determined by at least two separate protein domains, the redox-sensitive Neh2 degron and the redox-insensitive Neh6 degron. J Biol Chem 279:31556–31567
14. Wang XJ, Hayes JD, Henderson CJ, Wolf CR (2007) Identification of retinoic acid as an inhibitor of transcription factor Nrf2 through activation of retinoic acid receptor alpha. Proc Natl Acad Sci U S A 104:19589–19594
15. Wang H, Liu K, Geng M et al (2013) RXRα inhibits the NRF2-ARE signaling pathway through a direct interaction with the Neh7 domain of NRF2. Cancer Res 73:3097–3108
16. Miao W, Hu L, Scrivens PJ, Batist G (2005) Transcriptional regulation of NF-E2 p45-related factor (NRF2) expression by the aryl hydrocarbon receptor-xenobiotic response element signaling pathway: direct cross-talk between phase I and II drug-metabolizing enzymes. J Biol Chem 280:20340–20348
17. Cho H-Y, Gladwell W, Wang X et al (2010) Nrf2-regulated PPAR{gamma} expression is critical to protection against acute lung injury in mice. Am J Respir Crit Care Med 182:170–182
18. Rushworth SA, Zaitseva L, Murray MY et al (2012) The high Nrf2 expression in human acute myeloid leukemia is driven by NF-κB and underlies its chemo-resistance. Blood 120:5188–5198
19. Tung M-C, Lin P-L, Wang Y-C et al (2015) Mutant p53 confers chemoresistance in non-small cell lung cancer by upregulating Nrf2. Oncotarget 6:41692–41705
20. Nagar S, Noveral SM, Trudler D et al (2017) MEF2D haploinsufficiency downregulates the NRF2 pathway and renders photoreceptors susceptible to light-induced oxidative stress. Proc Natl Acad Sci U S A 114:E4048–E4056
21. DeNicola GM, Karreth FA, Humpton TJ et al (2011) Oncogene-induced Nrf2 transcription promotes ROS detoxification and tumorigenesis. Nature 475:106–109
22. Kang HJ, Hong YB, Kim HJ et al (2011) Detoxification: a novel function of BRCA1 in tumor suppression? Toxicol Sci 122:26–37
23. Yu S, Khor TO, Cheung K-L et al (2010) Nrf2 expression is regulated by epigenetic mechanisms in prostate cancer of TRAMP mice. PLoS One 5:e8579
24. Zhang DD, Hannink M (2003) Distinct cysteine residues in Keap1 are required for Keap1-dependent ubiquitination of Nrf2 and for stabilization of Nrf2 by chemopreventive agents and oxidative stress. Mol Cell Biol 23:8137–8151
25. Takaya K, Suzuki T, Motohashi H et al (2012) Validation of the multiple sensor mechanism of the Keap1-Nrf2 system. Free Radic Biol Med 53:817–827
26. Saito R, Suzuki T, Hiramoto K et al (2016) Characterizations of Three Major Cysteine Sensors of Keap1 in Stress Response. Mol Cell Biol 36:271–284
27. Abiko Y, Miura T, Phuc BH et al (2011) Participation of covalent modification of Keap1 in the activation of Nrf2 by tert-butylbenzoquinone, an electrophilic metabolite of butylated hydroxyanisole. Toxicol Appl Pharmacol 255:32–39
28. Kurinna S, Werner S (2015) NRF2 and microRNAs: new but awaited relations. Biochem Soc Trans 43:595–601
29. Watai Y, Kobayashi A, Nagase H et al (2007) Subcellular localization and cytoplasmic complex status of endogenous Keap1. Genes Cells 12:1163–1178
30. Kang M-I, Kobayashi A, Wakabayashi N et al (2004) Scaffolding of Keap1 to the actin cytoskeleton con-

trols the function of Nrf2 as key regulator of cytoprotective phase 2 genes. Proc Natl Acad Sci U S A 101:2046–2051
31. Nguyen T, Nioi P, Pickett CB (2009) The Nrf2-antioxidant response element signaling pathway and its activation by oxidative stress. J Biol Chem 284:13291–13295
32. Kannan MB, Solovieva V, Blank V (2012) The small MAF transcription factors MAFF, MAFG and MAFK: current knowledge and perspectives. Biochim Biophys Acta 1823:1841–1846
33. Li W, Jain MR, Chen C et al (2005) Nrf2 Possesses a redox-insensitive nuclear export signal overlapping with the leucine zipper motif. J Biol Chem 280:28430–28438
34. Li W, Yu S, Liu T et al (2008) Heterodimerization with small Maf proteins enhances nuclear retention of Nrf2 via masking the NESzip motif. Biochim Biophys Acta 1783:1847–1856
35. Rushmore TH, Pickett CB (1990) Transcriptional regulation of the rat glutathione S-transferase Ya subunit gene. Characterization of a xenobiotic-responsive element controlling inducible expression by phenolic antioxidants. J Biol Chem 265:14648–14653
36. Friling RS, Bensimon A, Tichauer Y, Daniel V (1990) Xenobiotic-inducible expression of murine glutathione S-transferase Ya subunit gene is controlled by an electrophile-responsive element. Proc Natl Acad Sci U S A 87:6258–6262
37. Hirotsu Y, Katsuoka F, Funayama R et al (2012) Nrf2–MafG heterodimers contribute globally to antioxidant and metabolic networks. Nucleic Acids Res 40:10228–10239
38. Sun Z, Zhang S, Chan JY, Zhang DD (2007) Keap1 controls postinduction repression of the Nrf2-mediated antioxidant response by escorting nuclear export of Nrf2. Mol Cell Biol 27:6334–6349
39. Chowdhry S, Zhang Y, McMahon M et al (2013) Nrf2 is controlled by two distinct β-TrCP recognition motifs in its Neh6 domain, one of which can be modulated by GSK-3 activity. Oncogene 32:3765–3781
40. Lo JY, Spatola BN, Curran SP (2017) WDR23 regulates NRF2 independently of KEAP1. PLoS Genet 13:e1006762
41. Wu T, Zhao F, Gao B et al (2014) Hrd1 suppresses Nrf2-mediated cellular protection during liver cirrhosis. Genes Dev 28:708–722
42. Hayes JD, Chowdhry S, Dinkova-Kostova AT, Sutherland C (2015) Dual regulation of transcription factor Nrf2 by Keap1 and by the combined actions of β-TrCP and GSK-3. Biochem Soc Trans 43:611–620
43. He F, Ru X, Wen T (2020) NRF2, a transcription factor for stress response and beyond. Int J Mol Sci 21. https://doi.org/10.3390/ijms21134777
44. Liu T, Lv Y-F, Zhao J-L et al (2021) Regulation of Nrf2 by phosphorylation: Consequences for biological function and therapeutic implications. Free Radic Biol Med 168:129–141
45. Hayes JD, Dinkova-Kostova AT (2014) The Nrf2 regulatory network provides an interface between redox and intermediary metabolism. Trends Biochem Sci 39:199–218
46. Tonelli C, Chio IIC, Tuveson DA (2018) Transcriptional Regulation by Nrf2. Antioxid Redox Signal 29:1727–1745
47. Moinova HR, Mulcahy RT (1999) Up-regulation of the human gamma-glutamylcysteine synthetase regulatory subunit gene involves binding of Nrf-2 to an electrophile responsive element. Biochem Biophys Res Commun 261:661–668
48. Sasaki H, Sato H, Kuriyama-Matsumura K et al (2002) Electrophile response element-mediated induction of the cystine/glutamate exchange transporter gene expression. J Biol Chem 277:44765–44771
49. Chanas SA, Jiang Q, McMahon M et al (2002) Loss of the Nrf2 transcription factor causes a marked reduction in constitutive and inducible expression of the glutathione S-transferase Gsta1, Gsta2, Gstm1, Gstm2, Gstm3 and Gstm4 genes in the livers of male and female mice. Biochem J 365:405–416
50. Harvey CJ, Thimmulappa RK, Singh A et al (2009) Nrf2-regulated glutathione recycling independent of biosynthesis is critical for cell survival during oxidative stress. Free Radic Biol Med 46:443–453
51. Hawkes H-JK, Karlenius TC, Tonissen KF (2014) Regulation of the human thioredoxin gene promoter and its key substrates: a study of functional and putative regulatory elements. Biochim Biophys Acta 1840:303–314
52. Sakurai A, Nishimoto M, Himeno S et al (2005) Transcriptional regulation of thioredoxin reductase 1 expression by cadmium in vascular endothelial cells: role of NF-E2-related factor-2. J Cell Physiol 203:529–537
53. Abbas K, Breton J, Planson A-G et al (2011) Nitric oxide activates an Nrf2/sulfiredoxin antioxidant pathway in macrophages. Free Radic Biol Med 51:107–114
54. He F, Antonucci L, Karin M (2020) NRF2 as a regulator of cell metabolism and inflammation in cancer. Carcinogenesis 41:405–416
55. Lee J-M, Calkins MJ, Chan K et al (2003) Identification of the NF-E2-related factor-2-dependent genes conferring protection against oxidative stress in primary cortical astrocytes using oligonucleotide microarray analysis. J Biol Chem 278:12029–12038
56. Mitsuishi Y, Taguchi K, Kawatani Y et al (2012) Nrf2 redirects glucose and glutamine into anabolic pathways in metabolic reprogramming. Cancer Cell 22:66–79
57. Thimmulappa RK, Mai KH, Srisuma S et al (2002) Identification of Nrf2-regulated genes induced by the chemopreventive agent sulforaphane by oligonucleotide microarray. Cancer Res 62:5196–5203

58. Wu KC, Cui JY, Klaassen CD (2011) Beneficial role of Nrf2 in regulating NADPH generation and consumption. Toxicol Sci 123:590–600
59. Wu KC, Cui JY, Klaassen CD (2012) Effect of graded Nrf2 activation on phase-I and -II drug metabolizing enzymes and transporters in mouse liver. PLoS One 7:e39006
60. Fu J, Xiong Z, Huang C et al (2019) Hyperactivity of the transcription factor Nrf2 causes metabolic reprogramming in mouse esophagus. J Biol Chem 294:327–340
61. Tibbetts AS, Appling DR (2010) Compartmentalization of Mammalian folate-mediated one-carbon metabolism. Annu Rev Nutr 30:57–81
62. He F, Antonucci L, Yamachika S et al (2020) NRF2 activates growth factor genes and downstream AKT signaling to induce mouse and human hepatomegaly. J Hepatol 72:1182–1195
63. Tanaka Y, Ikeda T, Yamamoto K et al (2012) Dysregulated expression of fatty acid oxidation enzymes and iron-regulatory genes in livers of Nrf2-null mice. J Gastroenterol Hepatol 27:1711–1717
64. Pi J, Leung L, Xue P et al (2010) Deficiency in the Nuclear Factor E2-related Factor-2 Transcription Factor Results in Impaired Adipogenesis and Protects against Diet-induced Obesity *. J Biol Chem 285:9292–9300
65. Pajares M, Jiménez-Moreno N, García-Yagüe ÁJ et al (2016) Transcription factor NFE2L2/NRF2 is a regulator of macroautophagy genes. Autophagy 12:1902–1916
66. Komatsu M, Kurokawa H, Waguri S et al (2010) The selective autophagy substrate p62 activates the stress responsive transcription factor Nrf2 through inactivation of Keap1. Nat Cell Biol 12:213–223
67. Hetz C, Zhang K, Kaufman RJ (2020) Mechanisms, regulation and functions of the unfolded protein response. Nat Rev Mol Cell Biol 21:421–438
68. Pajares M, Cuadrado A, Rojo AI (2017) Modulation of proteostasis by transcription factor NRF2 and impact in neurodegenerative diseases. Redox Biol 11:543–553
69. Cullinan SB, Zhang D, Hannink M et al (2003) Nrf2 is a direct PERK substrate and effector of PERK-dependent cell survival. Mol Cell Biol 23:7198–7209
70. Glover-Cutter KM, Lin S, Blackwell TK (2013) Integration of the unfolded protein and oxidative stress responses through SKN-1/Nrf. PLoS Genet 9:e1003701
71. He CH, Gong P, Hu B et al (2001) Identification of activating transcription factor 4 (ATF4) as an Nrf2-interacting protein. Implication for heme oxygenase-1 gene regulation. J Biol Chem 276:20858–20865
72. Kwak M-K, Wakabayashi N, Greenlaw JL et al (2003) Antioxidants enhance mammalian proteasome expression through the Keap1-Nrf2 signaling pathway. Mol Cell Biol 23:8786–8794
73. Kapeta S, Chondrogianni N, Gonos ES (2010) Nuclear erythroid factor 2-mediated proteasome activation delays senescence in human fibroblasts. J Biol Chem 285:8171–8184
74. Jang J, Wang Y, Kim H-S et al (2014) Nrf2, a regulator of the proteasome, controls self-renewal and pluripotency in human embryonic stem cells. Stem Cells 32:2616–2625
75. Piantadosi CA, Carraway MS, Babiker A, Suliman HB (2008) Heme oxygenase-1 regulates cardiac mitochondrial biogenesis via Nrf2-mediated transcriptional control of nuclear respiratory factor-1. Circ Res 103:1232–1240
76. Gureev AP, Shaforostova EA, Popov VN (2019) Regulation of Mitochondrial Biogenesis as a Way for Active Longevity: Interaction Between the Nrf2 and PGC-1α Signaling Pathways. Front Genet 10:435
77. Singh A, Happel C, Manna SK et al (2013) Transcription factor NRF2 regulates miR-1 and miR-206 to drive tumorigenesis. J Clin Invest 123:2921–2934
78. Holmström KM, Baird L, Zhang Y et al (2013) Nrf2 impacts cellular bioenergetics by controlling substrate availability for mitochondrial respiration. Biol Open 2:761–770
79. Anedda A, López-Bernardo E, Acosta-Iborra B et al (2013) The transcription factor Nrf2 promotes survival by enhancing the expression of uncoupling protein 3 under conditions of oxidative stress. Free Radic Biol Med 61:395–407
80. Hu Q, Ren J, Li G et al (2018) The mitochondrially targeted antioxidant MitoQ protects the intestinal barrier by ameliorating mitochondrial DNA damage via the Nrf2/ARE signaling pathway. Cell Death Dis 9:403
81. Liu Z, Gu Y, Chakarov S et al (2019) Fate Mapping via Ms4a3-Expression History Traces Monocyte-Derived Cells. Cell 178:1509–1525.e19
82. Ginhoux F, Jung S (2014) Monocytes and macrophages: developmental pathways and tissue homeostasis. Nat Rev Immunol 14:392–404
83. Schlitzer A, McGovern N, Ginhoux F (2015) Dendritic cells and monocyte-derived cells: Two complementary and integrated functional systems. Semin Cell Dev Biol 41:9–22
84. Bassler K, Schulte-Schrepping J, Warnat-Herresthal S et al (2019) The Myeloid Cell Compartment—Cell by Cell. Annu Rev Immunol 37:269–293
85. Park MD, Silvin A, Ginhoux F, Merad M (2022) Macrophages in health and disease. Cell 185:4259–4279
86. Murakami S, Motohashi H (2015) Roles of Nrf2 in cell proliferation and differentiation. Free Radic Biol Med 88:168–178
87. Takubo K, Goda N, Yamada W et al (2010) Regulation of the HIF-1alpha level is essential for hematopoietic stem cells. Cell Stem Cell 7:391–402
88. Owusu-Ansah E, Banerjee U (2009) Reactive oxygen species prime Drosophila haematopoietic progenitors for differentiation. Nature 461:537–541
89. Ludin A, Gur-Cohen S, Golan K et al (2014) Reactive oxygen species regulate hematopoietic

stem cell self-renewal, migration and development, as well as their bone marrow microenvironment. Antioxid Redox Signal 21:1605–1619
90. Merchant AA, Singh A, Matsui W, Biswal S (2011) The redox-sensitive transcription factor Nrf2 regulates murine hematopoietic stem cell survival independently of ROS levels. Blood 118:6572–6579
91. Tsai JJ, Dudakov JA, Takahashi K et al (2013) Nrf2 regulates haematopoietic stem cell function. Nat Cell Biol 15:309–316
92. Andrades A, Peinado P, Alvarez-Perez JC et al (2023) SWI/SNF complexes in hematological malignancies: biological implications and therapeutic opportunities. Mol Cancer 22:39
93. Mas G, Man N, Nakata Y et al (2023) The SWI/SNF chromatin-remodeling subunit DPF2 facilitates NRF2-dependent antiinflammatory and antioxidant gene expression. J Clin Invest 133. https://doi.org/10.1172/JCI158419
94. Zhang J, Ohta T, Maruyama A et al (2006) BRG1 interacts with Nrf2 to selectively mediate HO-1 induction in response to oxidative stress. Mol Cell Biol 26:7942–7952
95. Murakami S, Shimizu R, Romeo P-H et al (2014) Keap1-Nrf2 system regulates cell fate determination of hematopoietic stem cells. Genes Cells 19:239–253
96. Vallelian F, Buzzi RM, Pfefferlé M et al (2022) Heme-stress activated NRF2 skews fate trajectories of bone marrow cells from dendritic cells towards red pulp-like macrophages in hemolytic anemia. Cell Death Differ. https://doi.org/10.1038/s41418-022-00932-1
97. van der Horst D, Carter-Timofte ME, van Grevenynghe J et al (2022) Regulation of innate immunity by Nrf2. Curr Opin Immunol 78:102247
98. Mohan S, Gupta D (2018) Crosstalk of toll-like receptors signaling and Nrf2 pathway for regulation of inflammation. Biomed Pharmacother 108:1866–1878
99. Yin S, Cao W (2015) Toll-Like Receptor Signaling Induces Nrf2 Pathway Activation through p62-Triggered Keap1 Degradation. Mol Cell Biol 35:2673–2683
100. Nadeem A, Siddiqui N, Al-Harbi NO et al (2016) TLR-7 agonist attenuates airway reactivity and inflammation through Nrf2-mediated antioxidant protection in a murine model of allergic asthma. Int J Biochem Cell Biol 73:53–62
101. Tuoheti A, Gu X, Cheng X, Zhang H (2021) Silencing Nrf2 attenuates chronic suppurative otitis media by inhibiting pro-inflammatory cytokine secretion through up-regulating TLR4. Innate Immun 27:70–80
102. Sen R, Baltimore D (1986) Multiple nuclear factors interact with the immunoglobulin enhancer sequences. Cell 46:705–716
103. Taniguchi K, Karin M (2018) NF-κB, inflammation, immunity and cancer: coming of age. Nat Rev Immunol 18:309–324
104. Yu M, Li H, Liu Q et al (2011) Nuclear factor p65 interacts with Keap1 to repress the Nrf2-ARE pathway. Cell Signal 23:883–892
105. Liu G-H, Qu J, Shen X (2008) NF-κB/p65 antagonizes Nrf2-ARE pathway by depriving CBP from Nrf2 and facilitating recruitment of HDAC3 to MafK. Biochimica et Biophysica Acta (BBA) - Molecular. Cell Research 1783:713–727
106. Thimmulappa RK, Lee H, Rangasamy T et al (2016) Nrf2 is a critical regulator of the innate immune response and survival during experimental sepsis. J Clin Invest 116:984–995
107. Morgan MJ, Liu Z-G (2011) Crosstalk of reactive oxygen species and NF-κB signaling. Cell Res 21:103–115
108. Lee D-F, Kuo H-P, Liu M et al (2009) KEAP1 E3 ligase-mediated downregulation of NF-kappaB signaling by targeting IKKbeta. Mol Cell 36:131–140
109. Martinon F, Burns K, Tschopp J (2002) The inflammasome: a molecular platform triggering activation of inflammatory caspases and processing of proIL-beta. Mol Cell 10:417–426
110. Agostini L, Martinon F, Burns K et al (2004) NALP3 forms an IL-1beta-processing inflammasome with increased activity in Muckle-Wells autoinflammatory disorder. Immunity 20:319–325
111. Latz E, Xiao TS, Stutz A (2013) Activation and regulation of the inflammasomes. Nat Rev Immunol 13:397–411
112. Zheng D, Liwinski T, Elinav E (2020) Inflammasome activation and regulation: toward a better understanding of complex mechanisms. Cell Discovery 6:1–22
113. Hennig P, Garstkiewicz M, Grossi S et al (2018) The Crosstalk between Nrf2 and Inflammasomes. Int J Mol Sci 19. https://doi.org/10.3390/ijms19020562
114. Liu X, Zhang X, Ding Y et al (2017) Nuclear Factor E2-Related Factor-2 Negatively Regulates NLRP3 Inflammasome Activity by Inhibiting Reactive Oxygen Species-Induced NLRP3 Priming. Antioxid Redox Signal 26:28–43
115. Garstkiewicz M, Strittmatter GE, Grossi S et al (2017) Opposing effects of Nrf2 and Nrf2-activating compounds on the NLRP3 inflammasome independent of Nrf2-mediated gene expression. Eur J Immunol 47:806–817
116. Zhao C, Gillette DD, Li X et al (2014) Nuclear factor E2-related factor-2 (Nrf2) is required for NLRP3 and AIM2 inflammasome activation. J Biol Chem 289:17020–17029
117. McNab F, Mayer-Barber K, Sher A et al (2015) Type I interferons in infectious disease. Nat Rev Immunol 15:87–103
118. Gunderstofte C, Iversen MB, Peri S et al (2019) Nrf2 Negatively Regulates Type I Interferon Responses and Increases Susceptibility to Herpes Genital Infection in Mice. Front Immunol 10:2101
119. Olagnier D, Lababidi RR, Hadj SB et al (2017) Activation of Nrf2 Signaling Augments Vesicular

Stomatitis Virus Oncolysis via Autophagy-Driven Suppression of Antiviral Immunity. Mol Ther 25:1900–1916
120. Ryan DG, Knatko EV, Casey AM et al (2022) Nrf2 activation reprograms macrophage intermediary metabolism and suppresses the type I interferon response. iScience 25:103827
121. Ablasser A, Chen ZJ (2019) cGAS in action: Expanding roles in immunity and inflammation. Science 363. https://doi.org/10.1126/science.aat8657
122. Hopfner K-P, Hornung V (2020) Molecular mechanisms and cellular functions of cGAS-STING signalling. Nat Rev Mol Cell Biol 21:501–521
123. Olagnier D, Brandtoft AM, Gunderstofte C et al (2018) Nrf2 negatively regulates STING indicating a link between antiviral sensing and metabolic reprogramming. Nat Commun 9:3506
124. Kim M-J, Jeon J-H (2022) Recent Advances in Understanding Nrf2 Agonism and Its Potential Clinical Application to Metabolic and Inflammatory Diseases. Int J Mol Sci 23. https://doi.org/10.3390/ijms23052846
125. Kobayashi EH, Suzuki T, Funayama R et al (2016) Nrf2 suppresses macrophage inflammatory response by blocking proinflammatory cytokine transcription. Nat Commun 7:1–14
126. Wang X, Su S, Zhu Y et al (2023) Metabolic Reprogramming via ACOD1 depletion enhances function of human induced pluripotent stem cell-derived CAR-macrophages in solid tumors. Nat Commun 14:5778
127. Mills EL, Ryan DG, Prag HA et al (2018) Itaconate is an anti-inflammatory metabolite that activates Nrf2 via alkylation of KEAP1. Nature 556:113–117
128. Huot J, Houle F, Marceau F, Landry J (1997) Oxidative stress-induced actin reorganization mediated by the p38 mitogen-activated protein kinase/heat shock protein 27 pathway in vascular endothelial cells. Circ Res 80:383–392
129. Silverstein RL, Febbraio M (2009) CD36, a scavenger receptor involved in immunity, metabolism, angiogenesis, and behavior. Sci Signal 2:re3
130. Maruyama A, Tsukamoto S, Nishikawa K et al (2008) Nrf2 regulates the alternative first exons of CD36 in macrophages through specific antioxidant response elements. Arch Biochem Biophys 477:139–145
131. Olagnier D, Lavergne R-A, Meunier E et al (2011) Nrf2, a PPARγ alternative pathway to promote CD36 expression on inflammatory macrophages: implication for malaria. PLoS Pathog 7:e1002254
132. Mimche PN, Thompson E, Taramelli D, Vivas L (2012) Curcumin enhances non-opsonic phagocytosis of Plasmodium falciparum through up-regulation of CD36 surface expression on monocytes/macrophages. J Antimicrob Chemother 67:1895–1904
133. Wang Z, Sun D, Chen G et al (2017) Tim-3 inhibits macrophage control of Listeria monocytogenes by inhibiting Nrf2. Sci Rep 7:42095
134. Arredouani M, Yang Z, Ning Y et al (2004) The scavenger receptor MARCO is required for lung defense against pneumococcal pneumonia and inhaled particles. J Exp Med 200:267–272
135. Kanno S, Furuyama A, Hirano S (2007) A murine scavenger receptor MARCO recognizes polystyrene nanoparticles. Toxicol Sci 97:398–406
136. Harvey CJ, Thimmulappa RK, Sethi S et al (2011) Targeting Nrf2 signaling improves bacterial clearance by alveolar macrophages in patients with COPD and in a mouse model. Sci Transl Med 3:78ra32
137. Wang Y, Tang B, Li H et al (2023) A small-molecule inhibitor of Keap1-Nrf2 interaction attenuates sepsis by selectively augmenting the antibacterial defence of macrophages at infection sites. EBioMedicine 90:104480
138. Wang Y, Tang B, Long L et al (2021) Improvement of obesity-associated disorders by a small-molecule drug targeting mitochondria of adipose tissue macrophages. Nat Commun 12:102
139. Liang C, Liu L, Bao S et al (2023) Neuroprotection by Nrf2 via modulating microglial phenotype and phagocytosis after intracerebral hemorrhage. Heliyon 9:e13777
140. Wang P, Geng J, Gao J et al (2019) Macrophage achieves self-protection against oxidative stress-induced ageing through the Mst-Nrf2 axis. Nat Commun 10:755
141. Lin X, Bai D, Wei Z et al (2019) Curcumin attenuates oxidative stress in RAW264.7 cells by increasing the activity of antioxidant enzymes and activating the Nrf2-Keap1 pathway. PLoS One 14:e0216711
142. Pfefferlé M, Ingoglia G, Schaer CA et al (2020) Hemolysis transforms liver macrophages into anti-inflammatory erythrophagocytes. J Clin Invest 130:5576–5590
143. Feige JN, Gelman L, Michalik L et al (2006) From molecular action to physiological outputs: peroxisome proliferator-activated receptors are nuclear receptors at the crossroads of key cellular functions. Prog Lipid Res 45:120–159
144. Lee C (2017) Collaborative Power of Nrf2 and PPARγ Activators against Metabolic and Drug-Induced Oxidative Injury. Oxid Med Cell Longev 2017:1378175
145. Christofides A, Konstantinidou E, Jani C, Boussiotis VA (2021) The role of peroxisome proliferator-activated receptors (PPAR) in immune responses. Metabolism 114:154338
146. Huang J, Tabbi-Anneni I, Gunda V, Wang L (2010) Transcription factor Nrf2 regulates SHP and lipogenic gene expression in hepatic lipid metabolism. Am J Physiol Gastrointest Liver Physiol 299:G1211–G1221
147. Polvani S, Tarocchi M, Galli A (2012) PPARγ and Oxidative Stress: Con(β) Catenating NRF2 and FOXO. PPAR Res 2012:641087

148. Kvandová M, Majzúnová M, Dovinová I (2016) The role of PPARgamma in cardiovascular diseases. Physiol Res 65:S343–S363
149. Park EY, Cho IJ, Kim SG (2004) Transactivation of the PPAR-responsive enhancer module in chemopreventive glutathione S-transferase gene by the peroxisome proliferator-activated receptor-gamma and retinoid X receptor heterodimer. Cancer Res 64:3701–3713
150. Desvergne B, Wahli W (1999) Peroxisome proliferator-activated receptors: nuclear control of metabolism. Endocr Rev 20:649–688
151. Varga T, Czimmerer Z, Nagy L (2011) PPARs are a unique set of fatty acid regulated transcription factors controlling both lipid metabolism and inflammation. Biochim Biophys Acta 1812:1007–1022
152. Chawla A, Boisvert WA, Lee CH et al (2001) A PPAR gamma-LXR-ABCA1 pathway in macrophages is involved in cholesterol efflux and atherogenesis. Mol Cell 7:161–171
153. Heming M, Gran S, Jauch S-L et al (2018) Peroxisome Proliferator-Activated Receptor-γ Modulates the Response of Macrophages to Lipopolysaccharide and Glucocorticoids. Front Immunol 9:893
154. Poss KD, Tonegawa S (1997) Heme oxygenase 1 is required for mammalian iron reutilization. Proc Natl Acad Sci U S A 94:10919–10924
155. Buehler PW, Humar R, Schaer DJ (2020) Haptoglobin Therapeutics and Compartmentalization of Cell-Free Hemoglobin Toxicity. Trends Mol Med 26:683–697
156. Vallelian F, Buehler PW, Schaer DJ (2022) Hemolysis, free hemoglobin toxicity, and scavenger protein therapeutics. Blood 140:1837–1844
157. Boyle JJ, Johns M, Lo J et al (2011) Heme induces heme oxygenase 1 via Nrf2: role in the homeostatic macrophage response to intraplaque hemorrhage. Arterioscler Thromb Vasc Biol 31:2685–2691
158. Kasai S, Mimura J, Ozaki T, Itoh K (2018) Emerging Regulatory Role of Nrf2 in Iron, Heme, and Hemoglobin Metabolism in Physiology and Disease. Front Vet Sci 5:242
159. Kerins MJ, Ooi A (2018) The Roles of NRF2 in Modulating Cellular Iron Homeostasis. Antioxid Redox Signal 29:1756–1773
160. Bauer M, Bauer I (2002) Heme oxygenase-1: redox regulation and role in the hepatic response to oxidative stress. Antioxid Redox Signal 4:749–758
161. Kikuchi G, Yoshida T, Noguchi M (2005) Heme oxygenase and heme degradation. Biochem Biophys Res Commun 338:558–567
162. Reichard JF, Motz GT, Puga A (2007) Heme oxygenase-1 induction by NRF2 requires inactivation of the transcriptional repressor BACH1. Nucleic Acids Res 35:7074–7086
163. Pietsch EC, Chan JY, Torti FM, Torti SV (2003) Nrf2 mediates the induction of ferritin H in response to xenobiotics and cancer chemopreventive dithiolethiones. J Biol Chem 278:2361–2369
164. Harada N, Kanayama M, Maruyama A et al (2011) Nrf2 regulates ferroportin 1-mediated iron efflux and counteracts lipopolysaccharide-induced ferroportin 1 mRNA suppression in macrophages. Arch Biochem Biophys 508:101–109
165. Vanden Berghe T, Linkermann A, Jouan-Lanhouet S et al (2014) Regulated necrosis: the expanding network of non-apoptotic cell death pathways. Nat Rev Mol Cell Biol 15:135–147
166. Li J, Cao F, Yin H-L et al (2020) Ferroptosis: past, present and future. Cell Death Dis 11:88
167. Namgaladze D, Fuhrmann DC, Brüne B (2022) Interplay of Nrf2 and BACH1 in inducing ferroportin expression and enhancing resistance of human macrophages towards ferroptosis. Cell Death Discov 8:327
168. He R, Liu B, Xiong R et al (2022) Itaconate inhibits ferroptosis of macrophage via Nrf2 pathways against sepsis-induced acute lung injury. Cell Death Discov 8:43
169. Zhu K, Mrowietz U (2001) Inhibition of dendritic cell differentiation by fumaric acid esters. J Invest Dermatol 116:203–208
170. Litjens NHR, Rademaker M, Ravensbergen B et al (2004) Monomethylfumarate affects polarization of monocyte-derived dendritic cells resulting in down-regulated Th1 lymphocyte responses. Eur J Immunol 34:565–575
171. Peng H, Guerau-de-Arellano M, Mehta VB et al (2012) Dimethyl Fumarate Inhibits Dendritic Cell Maturation via Nuclear Factor κB (NF-κB) and Extracellular Signal-regulated Kinase 1 and 2 (ERK1/2) and Mitogen Stress-activated Kinase 1 (MSK1) Signaling*. J Biol Chem 287:28017–28026
172. Yeang HXA, Hamdam JM, Al-Huseini LMA et al (2012) Loss of transcription factor nuclear factor-erythroid 2 (NF-E2) p45-related factor-2 (Nrf2) leads to dysregulation of immune functions, redox homeostasis, and intracellular signaling in dendritic cells. J Biol Chem 287:10556–10564
173. Hammer A, Waschbisch A, Knippertz I et al (2017) Role of Nuclear Factor (Erythroid-Derived 2)-Like 2 Signaling for Effects of Fumaric Acid Esters on Dendritic Cells. Front Immunol 8:1922
174. Ng LG, Ostuni R, Hidalgo A (2019) Heterogeneity of neutrophils. Nat Rev Immunol 19:255–265
175. Amirova KM, Dimitrova PA, Leseva MN et al (2023) The Triterpenoid Nrf2 Activator, CDDO-Me, Decreases Neutrophil Senescence in a Murine Model of Joint Damage. Int J Mol Sci 24. https://doi.org/10.3390/ijms24108775
176. Zhao Y, Liu Z, Liu G et al (2023) Neutrophils resist ferroptosis and promote breast cancer metastasis through aconitate decarboxylase 1. Cell Metab 35:1688–1703.e10
177. Brinkmann V (2018) Neutrophil Extracellular Traps in the Second Decade. J Innate Immun 10:414–421
178. Urban CF, Nett JE (2019) Neutrophil extracellular traps in fungal infection. Semin Cell Dev Biol 89:47–57
179. Ye S, Li S, Ma Y et al (2021) Curcumin hinders PBDE-47-induced neutrophil extracellular

traps release via Nrf2-associated ROS inhibition. Ecotoxicol Environ Saf 225:112779
180. Thimmulappa RK, Scollick C, Traore K et al (2006) Nrf2-dependent protection from LPS induced inflammatory response and mortality by CDDO-Imidazolide. Biochem Biophys Res Commun 351:883–889
181. Helou DG, Braham S, De Chaisemartin L et al (2019) Nrf2 downregulates zymosan-induced neutrophil activation and modulates migration. PLoS One 14:e0216465
182. Helou DG, Noël B, Gaudin F et al (2019) Cutting Edge: Nrf2 Regulates Neutrophil Recruitment and Accumulation in Skin during Contact Hypersensitivity. J Immunol 202:2189–2194
183. Gomez JC, Dang H, Martin JR, Doerschuk CM (2016) Nrf2 Modulates Host Defense during Streptococcus pneumoniae Pneumonia in Mice. J Immunol 197:2864–2879
184. Singer M, Deutschman CS, Seymour CW et al (2016) The Third International Consensus Definitions for Sepsis and Septic Shock (Sepsis-3). JAMA 315:801–810
185. Huang M, Cai S, Su J (2019) The Pathogenesis of Sepsis and Potential Therapeutic Targets. Int J Mol Sci 20. https://doi.org/10.3390/ijms20215376
186. Wasyluk W, Zwolak A (2021) Metabolic Alterations in Sepsis. J Clin Med Res 10. https://doi.org/10.3390/jcm10112412
187. Singer M (2007) Mitochondrial function in sepsis: acute phase versus multiple organ failure. Crit Care Med 35:S441–S448
188. Thimmulappa RK, Lee H, Rangasamy T et al (2006) Nrf2 is a critical regulator of the innate immune response and survival during experimental sepsis. J Clin Invest 116:984–995
189. Kong X, Thimmulappa R, Kombairaju P, Biswal S (2010) NADPH oxidase-dependent reactive oxygen species mediate amplified TLR4 signaling and sepsis-induced mortality in Nrf2-deficient mice. J Immunol 185:569–577
190. Kong X, Thimmulappa R, Craciun F et al (2011) Enhancing Nrf2 pathway by disruption of Keap1 in myeloid leukocytes protects against sepsis. Am J Respir Crit Care Med 184:928–938
191. Olonisakin TF, Suber T, Gonzalez-Ferrer S et al (2021) Stressed erythrophagocytosis induces immunosuppression during sepsis through heme-mediated STAT1 dysregulation. J Clin Invest 131:e137468
192. Zhao X, Drlica K (2014) Reactive oxygen species and the bacterial response to lethal stress. Curr Opin Microbiol 21:1–6
193. Rothchild AC, Olson GS, Nemeth J et al (2019) Alveolar macrophages generate a noncanonical NRF2-driven transcriptional response to Mycobacterium tuberculosis in vivo. Sci Immunol 4:eaaw6693
194. Cho H-Y, Imani F, Miller-DeGraff L et al (2009) Antiviral activity of Nrf2 in a murine model of respiratory syncytial virus disease. Am J Respir Crit Care Med 179:138–150
195. Ivanciuc T, Sbrana E, Casola A, Garofalo RP (2018) Protective Role of Nuclear Factor Erythroid 2-Related Factor 2 Against Respiratory Syncytial Virus and Human Metapneumovirus Infections. Front Immunol 9:854
196. Kesic MJ, Simmons SO, Bauer R, Jaspers I (2011) Nrf2 expression modifies influenza A entry and replication in nasal epithelial cells. Free Radic Biol Med 51:444–453
197. Kosmider B, Messier EM, Janssen WJ et al (2012) Nrf2 protects human alveolar epithelial cells against injury induced by influenza A virus. Respir Res 13:43
198. Page A, Volchkova VA, Reid SP et al (2014) Marburgvirus hijacks nrf2-dependent pathway by targeting nrf2-negative regulator keap1. Cell Rep 6:1026–1036
199. Edwards MR, Johnson B, Mire CE et al (2014) The Marburg virus VP24 protein interacts with Keap1 to activate the cytoprotective antioxidant response pathway. Cell Rep 6:1017–1025
200. Sugiyama K, Ebinuma H, Nakamoto N et al (2014) Prominent steatosis with hypermetabolism of the cell line permissive for years of infection with hepatitis C virus. PLoS One 9:e94460
201. Olagnier D, Peri S, Steel C et al (2014) Cellular oxidative stress response controls the antiviral and apoptotic programs in dengue virus-infected dendritic cells. PLoS Pathog 10:e1004566
202. Lau A, Villeneuve NF, Sun Z et al (2008) Dual roles of Nrf2 in cancer. Pharmacol Res 58:262–270
203. Menegon S, Columbano A, Giordano S (2016) The Dual Roles of NRF2 in Cancer. Trends Mol Med 22:578–593
204. Pillai R, Hayashi M, Zavitsanou A-M, Papagiannakopoulos T (2022) NRF2: KEAPing Tumors Protected. Cancer Discov 12:625–643
205. Rojo de la Vega M, Chapman E, Zhang DD (2018) NRF2 and the Hallmarks of Cancer. Cancer Cell 34:21–43
206. Liu N, Luo J, Kuang D et al (2019) Lactate inhibits ATP6V0d2 expression in tumor-associated macrophages to promote HIF-2α-mediated tumor progression. J Clin Invest 129:631–646
207. Colegio OR, Chu N-Q, Szabo AL et al (2014) Functional polarization of tumour-associated macrophages by tumour-derived lactic acid. Nature 513:559–563
208. Feng R, Morine Y, Ikemoto T et al (2018) Nrf2 activation drive macrophages polarization and cancer cell epithelial-mesenchymal transition during interaction. Cell Commun Signal 16:54
209. Weiss JM, Davies LC, Karwan M et al (2018) Itaconic acid mediates crosstalk between macrophage metabolism and peritoneal tumors. J Clin Invest 128:3794–3805
210. Zhao H, Teng D, Yang L et al (2022) Myeloid-derived itaconate suppresses cytotoxic CD8+ T cells and promotes tumour growth. Nat Metab 4:1660–1673

211. Humar R, Schaer DJ, Vallelian F (2022) Erythrophagocytes in hemolytic anemia, wound healing, and cancer. Trends Mol Med 28:906–915
212. Schaer DJ, Schulthess-Lutz N, Baselgia L et al (2023) Hemorrhage-activated NRF2 in tumor-associated macrophages drives cancer growth, invasion, and immunotherapy resistance. J Clin Invest. https://doi.org/10.1172/JCI174528
213. Arnold JN, Magiera L, Kraman M, Fearon DT (2014) Tumoral immune suppression by macrophages expressing fibroblast activation protein-α and heme oxygenase-1. Cancer Immunol Res 2:121–126
214. Alaluf E, Vokaer B, Detavernier A et al (2020) Heme oxygenase-1 orchestrates the immunosuppressive program of tumor-associated macrophages. JCI. Insight 5. https://doi.org/10.1172/jci.insight.133929
215. Consonni FM, Bleve A, Totaro MG et al (2021) Heme catabolism by tumor-associated macrophages controls metastasis formation. Nat Immunol 22:595–606
216. Victor P, Sarada D, Ramkumar KM (2020) Pharmacological activation of Nrf2 promotes wound healing. Eur J Pharmacol 886:173395
217. Maassen S, Coenen B, Ioannidis M et al (2023) Itaconate promotes a wound resolving phenotype in pro-inflammatory macrophages. Redox Biol 59:102591
218. Brem H, Tomic-Canic M (2007) Cellular and molecular basis of wound healing in diabetes. J Clin Invest 117:1219–1222
219. Tsourdi E, Barthel A, Rietzsch H et al (2013) Current aspects in the pathophysiology and treatment of chronic wounds in diabetes mellitus. Biomed Res Int 2013:385641
220. Li M, Yu H, Pan H et al (2019) Nrf2 Suppression Delays Diabetic Wound Healing Through Sustained Oxidative Stress and Inflammation. Front Pharmacol 10:1099
221. Abu-Al-Basal MA (2010) Healing potential of Rosmarinus officinalis L. on full-thickness excision cutaneous wounds in alloxan-induced-diabetic BALB/c mice. J Ethnopharmacol 131:443–450
222. Ginwala R, McTish E, Raman C et al (2016) Apigenin, a Natural Flavonoid, Attenuates EAE Severity Through the Modulation of Dendritic Cell and Other Immune Cell Functions. J Neuroimmune Pharmacol 11:36–47
223. Chu C, Deng J, Man Y, Qu Y (2017) Green Tea Extracts Epigallocatechin-3-gallate for Different Treatments. Biomed Res Int 2017:5615647
224. Hozzein WN, Badr G, Badr BM et al (2018) Bee venom improves diabetic wound healing by protecting functional macrophages from apoptosis and enhancing Nrf2, Ang-1 and Tie-2 signaling. Mol Immunol 103:322–335
225. Kamar SS, Abdel-Kader DH, Rashed LA (2019) Beneficial effect of Curcumin Nanoparticles-Hydrogel on excisional skin wound healing in type-I diabetic rat: Histological and immunohistochemical studies. Ann Anat 222:94–102
226. Dick SA, Macklin JA, Nejat S et al (2019) Self-renewing resident cardiac macrophages limit adverse remodeling following myocardial infarction. Nat Immunol 20:29–39
227. Geetika B, Andrea B, Wenjun L et al (2019) Tissue Resident CCR2− and CCR2+ Cardiac Macrophages Differentially Orchestrate Monocyte Recruitment and Fate Specification Following Myocardial Injury. Circ Res 124:263–278
228. Calcagno DM, Ng RP Jr, Toomu A et al (2020) The myeloid type I interferon response to myocardial infarction begins in bone marrow and is regulated by Nrf2-activated macrophages. Sci Immunol 5:eaaz1974
229. King KR, Aguirre AD, Ye Y-X et al (2017) IRF3 and type I interferons fuel a fatal response to myocardial infarction. Nat Med 23:1481–1487
230. Cao DJ, Schiattarella GG, Villalobos E et al (2018) Cytosolic DNA Sensing Promotes Macrophage Transformation and Governs Myocardial Ischemic Injury. Circulation 137:2613–2634
231. Ruotsalainen A-K, Inkala M, Partanen ME et al (2013) The absence of macrophage Nrf2 promotes early atherogenesis. Cardiovasc Res 98:107–115
232. Collins AR, Gupte AA, Ji R et al (2012) Myeloid deletion of nuclear factor erythroid 2-related factor 2 increases atherosclerosis and liver injury. Arterioscler Thromb Vasc Biol 32:2839–2846
233. Abe J-I, Imanishi M, Li S et al (2023) An ERK5-NRF2 Axis Mediates Senescence-Associated Stemness and Atherosclerosis. Circ Res 133:25–44
234. Sussan TE, Jun J, Thimmulappa R et al (2008) Disruption of Nrf2, a key inducer of antioxidant defenses, attenuates ApoE-mediated atherosclerosis in mice. PLoS One 3:e3791
235. Barajas B, Che N, Yin F et al (2011) NF-E2-related factor 2 promotes atherosclerosis by effects on plasma lipoproteins and cholesterol transport that overshadow antioxidant protection. Arterioscler Thromb Vasc Biol 31:58–66
236. Freigang S, Ampenberger F, Spohn G et al (2011) Nrf2 is essential for cholesterol crystal-induced inflammasome activation and exacerbation of atherosclerosis. Eur J Immunol 41:2040–2051
237. Vallelian F, Deuel JW, Opitz L et al (2015) Proteasome inhibition and oxidative reactions disrupt cellular homeostasis during heme stress. Cell Death Differ 22:597–611
238. Vallelian F, Schaer CA, Deuel JW et al (2018) Revisiting the putative role of heme as a trigger of inflammation. Pharmacol Res Perspect 6:e00392
239. Schaer CA, Deuel JW, Schildknecht D et al (2016) Haptoglobin Preserves Vascular Nitric Oxide Signaling during Hemolysis. Am J Respir Crit Care Med 193:1111–1122
240. Akeret K, Buzzi RM, Schaer CA et al (2021) Cerebrospinal fluid hemoglobin drives subarachnoid

hemorrhage-related secondary brain injury. J Cereb Blood Flow Metab 41:3000–3015
241. Schaer DJ, Alayash AI (2010) Clearance and control mechanisms of hemoglobin from cradle to grave. Antioxid Redox Signal 12:181–184
242. Theurl I, Hilgendorf I, Nairz M et al (2016) On-demand erythrocyte disposal and iron recycling requires transient macrophages in the liver. Nat Med 22:945–951
243. Liu Y, Su S, Shayo S et al (2023) Hemolysis dictates monocyte differentiation via two distinct pathways in sickle cell disease vaso-occlusion. J Clin Invest 133:e172087
244. Liu Y, Jing F, Yi W et al (2018) HO-1hi patrolling monocytes protect against vaso-occlusion in sickle cell disease. Blood 131:1600–1610
245. Boyle JJ, Harrington HA, Piper E et al (2009) Coronary intraplaque hemorrhage evokes a novel atheroprotective macrophage phenotype. Am J Pathol 174:1097–1108
246. Ingoglia G, Yalamanoglu A, Pfefferlé M et al (2020) Line-selective macrophage activation with an anti-CD40 antibody drives a hemophagocytic syndrome in mice. Blood Adv 4:2751–2761
247. Pfefferlé M, Ingoglia G, Schaer CA et al (2021) Acute Hemolysis and Heme Suppress Anti-CD40 Antibody-Induced Necro-Inflammatory Liver Disease. Front Immunol 12:680855
248. Pfefferlé M, Dubach IL, Buzzi RM et al (2023) Antibody-induced erythrophagocyte reprogramming of Kupffer cells prevents anti-CD40 cancer immunotherapy-associated liver toxicity. J Immunother Cancer 11:e005718
249. Mizumura K, Maruoka S, Shimizu T, Gon Y (2020) Role of Nrf2 in the pathogenesis of respiratory diseases. Respir Investig 58:28–35
250. Sussan TE, Rangasamy T, Blake DJ et al (2009) Targeting Nrf2 with the triterpenoid CDDO-imidazolide attenuates cigarette smoke-induced emphysema and cardiac dysfunction in mice. Proc Natl Acad Sci U S A 106:250–255
251. Ishii Y, Itoh K, Morishima Y et al (2005) Transcription factor Nrf2 plays a pivotal role in protection against elastase-induced pulmonary inflammation and emphysema. J Immunol 175:6968–6975
252. Wise RA, Holbrook JT, Criner G et al (2016) Lack of Effect of Oral Sulforaphane Administration on Nrf2 Expression in COPD: A Randomized, Double-Blind. Placebo Controlled Trial. PLoS One 11:e0163716
253. Ryan EM, Sadiku P, Coelho P et al (2023) NRF2 Activation Reprograms Defects in Oxidative Metabolism to Restore Macrophage Function in Chronic Obstructive Pulmonary Disease. Am J Respir Crit Care Med 207:998–1011
254. Rangasamy T, Guo J, Mitzner WA et al (2005) Disruption of Nrf2 enhances susceptibility to severe airway inflammation and asthma in mice. J Exp Med 202:47–59
255. Marzec JM, Christie JD, Reddy SP et al (2007) Functional polymorphisms in the transcription factor NRF2 in humans increase the risk of acute lung injury. FASEB J 21:2237–2246
256. Sriram N, Kalayarasan S, Sudhandiran G (2009) Epigallocatechin-3-gallate augments antioxidant activities and inhibits inflammation during bleomycin-induced experimental pulmonary fibrosis through Nrf2-Keap1 signaling. Pulm Pharmacol Ther 22:221–236
257. Tseng K-Y, Stratoulias V, Hu W-F et al (2023) Augmenting hematoma-scavenging capacity of innate immune cells by CDNF reduces brain injury and promotes functional recovery after intracerebral hemorrhage. Cell Death Dis 14:128
258. Pisetsky DS (2023) Pathogenesis of autoimmune disease. Nat Rev Nephrol 19:509–524
259. Ghoreschi K, Brück J, Kellerer C et al (2011) Fumarates improve psoriasis and multiple sclerosis by inducing type II dendritic cells. J Exp Med 208:2291–2303
260. Gold R, Kappos L, Arnold DL et al (2012) Placebo-controlled phase 3 study of oral BG-12 for relapsing multiple sclerosis. N Engl J Med 367:1098–1107
261. Fox RJ, Miller DH, Phillips JT et al (2012) Placebo-controlled phase 3 study of oral BG-12 or glatiramer in multiple sclerosis. N Engl J Med 367:1087–1097
262. Okawa H, Motohashi H, Kobayashi A et al (2006) Hepatocyte-specific deletion of the keap1 gene activates Nrf2 and confers potent resistance against acute drug toxicity. Biochem Biophys Res Commun 339:79–88
263. Taguchi K, Maher JM, Suzuki T et al (2010) Genetic analysis of cytoprotective functions supported by graded expression of Keap1. Mol Cell Biol 30:3016–3026
264. Suzuki T, Murakami S, Biswal SS et al (2017) Systemic Activation of NRF2 Alleviates Lethal Autoimmune Inflammation in Scurfy Mice. Mol Cell Biol 37:e00063–e00017
265. Goess C, Terrillon S, Mayo M et al (2021) NRF2 activator A-1396076 ameliorates inflammation in autoimmune disease models by inhibiting antigen dependent T cell activation. J Transl Autoimmun 4:100079

Part V

Erythroid Cells

Transcriptional Repressor BCL11A in Erythroid Cells

9

Ge Zheng and Stuart H. Orkin

Abstract

BCL11A, a zinc finger repressor, is a stage-specific transcription factor that controls the switch from fetal (HbF, $\alpha_2\gamma_2$) to adult (HbA, $\alpha_2\beta_2$) hemoglobin in erythroid cells. While BCL11A was known as a factor critical for B-lymphoid cell development, its relationship to erythroid cells and HbF arose through genome-wide association studies (GWAS). Subsequent work validated its role as a silencer of γ-globin gene expression in cultured cells and mice. Erythroid-specific loss of BCL11A rescues the phenotype of engineered sickle cell disease (SCD) mice, thereby suggesting that downregulation of BCL11A expression might be beneficial in patients with SCD and β-thalassemia. Common genetic variation in GWAS resides in an erythroid-specific enhancer within the *BCL11A* gene that is required for its own expression. CRISPR/Cas9 gene editing of the enhancer revealed a GATA-binding site that confers a large portion of its regulatory function. Disruption of the GATA site leads to robust HbF reactivation. Advancement of a guide RNA targeting the GATA-binding site in clinical trials has recently led to approval of first-in-man use of *ex vivo* CRISPR editing of hematopoietic stem/progenitor cells (HSPCs) as therapy of SCD and β-thalassemia. Future challenges include expanding access and infrastructure for delivery of genetic therapy to eligible patients, reducing potential toxicity and costs, exploring prospects for *in vivo* targeting of hematopoietic stem cells (HSCs), and developing small molecule drugs that impair function of BCL11A protein as an alternative option.

Keywords

Hemoglobin switch · Sickle cell disease · β-thalassemia · Globin regulation · γ-globin · Gene editing

Abbreviations

ABE	adenine base editor
CBE	cytidine base editor
CHD3/4	chromodomain helicase DNA-binding protein 3/4
ChIP	chromatin immunoprecipitation
CRBN	cereblon
CRISPR	clustered regularly interspaced short palindromic repeats

G. Zheng · S. H. Orkin (✉)
Dana-Farber/Boston Children's Cancer and Blood Disorders Center, Boston, MA, USA

Harvard Medical School and HHMI, Boston, MA, USA
e-mail: stuart_orkin@dfci.harvard.edu

T. Borggrefe, B. D. Giaimo (eds.), *Transcription factors in blood cell development*, Advances in Experimental Medicine and Biology 1459, https://doi.org/10.1007/978-3-031-62731-6_9

CTIP1	COUP-TF-interacting protein 1
DSB	double-stranded break
dTAG	degradation tag
EVI9	ecotropic viral integration site 9
FACS	fluorescence-activated cell sorting
FKBP12	FK506-binding protein 12
GATAD2A/B	GATA zinc finger domain containing 2A/B
GWAS	genome-wide association studies
HbA	adult hemoglobin
HbF	fetal hemoglobin
HbS	sickle hemoglobin
HDAC1/2	histone deacetylase 1/2
HIC2	hypermethylated in cancer 2
HPFH	hereditary persistence of fetal hemoglobin
HRI	heme-regulated inhibitor
HS	hypersensitivity sites
HSC	hematopoietic stem cell
HSPC	hematopoietic stem/progenitor cell
HUDEP	human umbilical cord blood-derived erythroid progenitor
IGF2BP1	insulin-like growth factor 2 mRNA-binding protein 1
KO	knockout
LCR	locus control region
LIN28B	lin-28 homolog B
LRF	leukemia-/lymphoma-related factor
MBD2/3	methyl-CpG-binding domain protein 2/3
MEL	mouse erythroleukemia cells
MHRA	Medicines and Healthcare products Regulatory Agency
MTA1/2/3	metastasis-associated protein 1/2/3
NF-Y	nuclear transcription factor Y
NHEJ	nonhomologous end joining
NuRD	nucleosome remodeling and deacetylase complex
PBM	protein-binding microarray
POK	POZ domain Krüppel-like zinc finger
POKEMON	POZ and Krüppel erythroid myeloid ontogenic factor
PRO-Seq	precision run-on sequencing
PROTAC	proteolysis-targeting chimera
RBBP4/7	retinoblastoma-binding protein 4/7
RNP	ribonucleoprotein
SALL4	spalt-like transcription factor 4
SCD	sickle cell disease
shmiR	shRNA embedded within an microRNA architecture
shRNA	small hairpin RNA
SLAM-Seq	thiol(SH)-linked alkylation for the metabolic sequencing of RNA
SNP	single-nucleotide polymorphism
TALEN	transcription activator-like effector nucleases
TF	transcription factor
TSS	transcription start site
UTR	untranslated region
WHO	World Health Organization
YAC	yeast artificial chromosome
ZBTB7A	zinc finger and BTB domain containing 7A
ZNF	zinc finger
ZNF23	zinc finger 2 and 3
ZNF456	zinc finger 4, 5, and 6

9.1 Background

BCL11A has attracted attention due to its critical role in control of the developmental switch from fetal-to-adult hemoglobin that occurs during normal ontogeny in erythropoiesis.

Hemoglobin, the principal oxygen transport protein in red blood cells, is a hetero-tetramer of two α-like globin peptide chains and two β-like globin peptides, along with a heme prosthetic group and associated iron. The human β-globin locus on chromosome 11 encodes the linked embryonic (ε), fetal (γ), and adult (β) globin genes which are arranged in order and expressed sequentially during development [1–3]. DNase I hypersensitivity (HS) sites in erythroid cell chromatin at the 5′-end of the locus comprises a regulatory element, known as the locus control region (LCR) [4]. The LCR, a super-enhancer, loops to the downstream globin gene promoters and drives high-level transcription in a developmental stage-specific manner [5, 6]. During the first trimester in utero, robust expression of embryonic β-like ε-globin occurs within yolk sac-derived primitive erythrocytes [2]. Thereafter, as the production of enucleated definitive erythrocytes is initiated from stem and progenitor cells in the fetal liver, γ-globin is expressed, constituting the β-like chain of HbF ($\alpha_2\gamma_2$) [3, 7]. The switch from predominant expression of HbF to adult hemoglobin (HbA,$\alpha_2\beta_2$) takes place progressively throughout late fetal life and is completed within the first few months after birth. The change from HbF to HbA is mediated by a transcriptional switch in definitive erythroid progenitors from γ- to β-globin.

Compared to HbA, HbF has a higher affinity for oxygen, which allows the developing fetus to extract oxygen from the maternal circulation. The adult red cells contain residual HbF (~1%), which is largely restricted to a small population of red blood cells termed F cells. The low level of HbF in adults is genetically determined [8, 9]. Reversing the hemoglobin switch by inducing expression of HbF is well-tolerated in adult erythroid cells.

The major hemoglobinopathies, sickle cell disease (SCD) and β-thalassemia, are among the most common inherited disorders [10, 11]. SCD is caused by a single missense mutation which replaces glutamic acid with valine at position 6 of β-globin protein. The sickle hemoglobin tetramer (HbS, $\alpha_2\beta^S{}_2$) is prone to polymerize at low oxygen tension, which leads to the deformation of erythrocytes, the "sickling process." Blockage of microvasculature impairs oxygen delivery to tissues, causing infarction and pain (vaso-occlusive crises). In β-thalassemia, which results from deficient or absent β-globin production, unpaired α-globin chains are unstable and precipitate within erythroid precursors, leading to ineffective erythropoiesis and hemolytic anemia.

Clinical observations revealed that the severity of SCD and β-thalassemia is ameliorated by increased HbF. In 1948 pediatrician Janet Watson astutely noted that children with SCD were asymptomatic in the neonatal period, which she postulated was due to the protective effect of HbF prior to completion of the switch [12]. Later studies showed that specific rare deletions within β-globin gene cluster and single-base substitutions in the γ-globin gene promoters lead to markedly elevated HbF, a condition termed hereditary persistence of fetal hemoglobin (HPFH) [13]. Patients with SCD or β-thalassemia who coinherit an HPFH allele are generally asymptomatic. Natural history studies of patients with SCD or β-thalassemia also have shown that the elevated production of HbF ameliorates clinical severity and improves survival [14–19]. In SCD, increased HbF expression interferes with the process of HbS polymerization (sickling), thereby reducing injury to the red blood cells. Moreover, as the γ-globin and β-globin genes are linked in the locus and individually rely on looping to the LCR, an increase in γ-globin transcription is accompanied by reduced β^S expression. Therefore, increased HbF in SCD is beneficial in two ways. In β-thalassemia, increased γ-globin expression corrects the imbalance of α- to β-globin chains and prevents damaging α-globin precipitation. For these reasons, reactivation of HbF has been considered for decades as an attractive approach to management of the hemoglobin disorders.

9.2 BCL11A as a Critical Regulator in Fetal-to-Adult Hemoglobin Switch

Several transcription factors (TFs), including GATA1, KLF1, and SCL/TAL1, serve critical, but largely stage-independent, roles in globin gene regulation and cell differentiation [20]. Factors with more dedicated roles in HbF control remained elusive until the application of genome-wide association studies (GWAS) [21–23], which assessed the impact of common genetic variation on the level of HbF in several cohorts. GWAS identified associated genomic signals at *BCL11A* on chromosome 2p, the *HBS1L-MYB* on 6q, and the β-globin cluster on 11p. A significant fraction of the genetic contribution to variation in the level of HbF was accounted for by these loci.

BCL11A, a Krüppel zinc finger TF, was first identified as a site of ecotropic viral integration (EVI9) in leukemia studies in mice [24].

BCL11A, which is also known as COUP-TF-interacting protein 1 (CTIP1), is primarily expressed in brain and hematopoietic tissues, including B-cells and early T-cell progenitors, and required for B-lymphoid development [25–29]. The gene has been identified at aberrant t(2;14)(p13;q32.3) chromosomal translocations in human B-cell non-Hodgkin lymphomas [24, 30]. Several isoforms of BCL11A protein (XL, L, S, and XS), which share exons 1 and 2 (Fig. 9.1), are expressed in different tissues. XL and L isoforms are the two major species expressed in adult erythroid cells [31].

In the erythroid lineage, BCL11A is expressed in a developmental stage-specific manner. While it is not expressed during yolk sac (embryonic) erythropoiesis, BCL11A is first detected in definitive erythroid cells in the fetal liver [31]. Experiments in primary human CD34$^+$ hematopoietic stem/progenitor cells (HSPCs) first suggested that BCL11A functions as a transcriptional repressor of γ-globin expression. Small hairpin RNA (shRNA)-mediated knockdown of BCL11A mRNA led to increased γ-globin mRNA expression and HbF [31]. In addition, the failure to silence expression of embryonic β-like globins in the mouse fetal liver in knockout embryos indicated that BCL11A controls the switch during development [32]. Furthermore, inappropriately high level expression of human γ-globin genes in the context of a β-globin cluster yeast artificial chromosome transgene (β-YAC) in BCL11A knockout (KO) embryos provided compelling evidence that BCL11A controls hemoglobin switching *in vivo* [32].

KO mice die in the perinatal period due to requirement(s) for BCL11A outside the hematopoietic system. However, mice with erythroid-specific inactivation of BCL11A develop normally [33]. In β-YAC mice with erythroid-specific knockout of BCL11A, the level of γ-globin is persistently high and >500-fold that in control mice even at 30 weeks after birth [33]. Ostensibly normal erythropoiesis in such mice indicates that BCL11A has few, if any, critical, non-globin gene targets. Erythroid-specific knockout of BCL11A rescues the hematologic and pathologic features of SCD in genetically engineered mouse models due to high-level, pancellular HbF expression [33]. This proof-of-principle study forms the rationale for the downregulation of BCL11A as a therapeutic strategy to reactivate HbF expression in hemoglobinopathies.

9.3 BCL11A as a Quantitative Regulator of HbF Silencing

As sequence changes within the coding regions of BCL11A were not observed, the association of HbF level with the *BCL11A* locus in GWAS implied that common genetic variation affects expression of BCL11A itself. The impact of

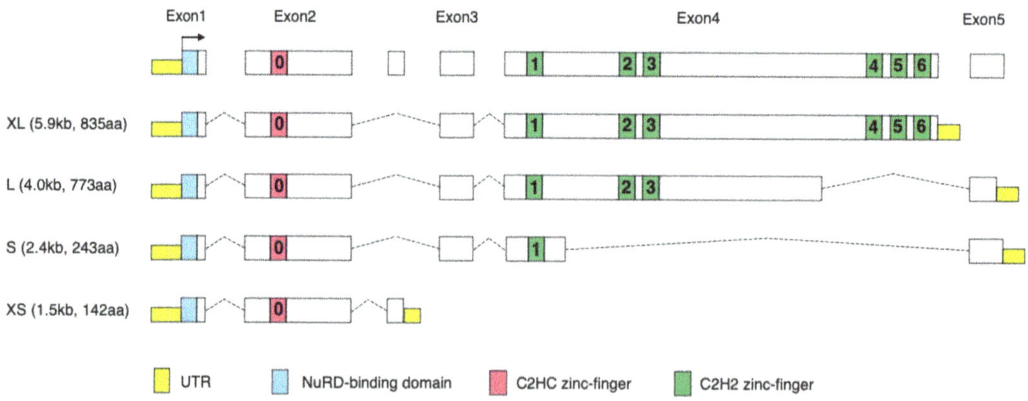

Fig. 9.1 Schematic of human BCL11A isoforms. Several isoforms of BCL11A protein (XL, L, S, and XS) are reported, and they share exons 1 and 2. XL and L isoforms are the two major species expressed in adult erythroid cells, which differs by inclusion of a terminal exon via RNA splicing

common variation, however, leads to exceedingly modest differences in HbF levels (see below). In studies of BCL11A KO mice, repression of endogenous mouse embryonic globin genes or a human γ-globin transgenic in mouse fetal liver is dose-sensitive, as BCL11A$^{+/-}$ heterozygotes display an intermediate phenotype [32]. Moreover, the level of HbF expression in differentiated CD34^{+} cells transduced with lentivirus harboring shRNA targeting BCL11A and differentiated *in vitro* displays a nonlinear, dose relationship with the knockdown of BCL11A [34]. Perhaps most illustrative are findings in rare individuals with haploinsufficiency for BCL11A, either due to microdeletions in the 2p15-p16.1 encompassing the locus or point mutations in coding sequences [35–40]. In the absence of significant regulatory compensation, such individuals express approximately 50% the normal level of BCL11A in all relevant tissues. Haploinsufficiency in the developing brain leads to severe developmental delay in affected children and an autistic-like syndrome [36–39, 41, 42]. Remarkably, although overall hematopoiesis and immune function appear normal in such individuals, the level of HbF is on average in the range of classic HPFH, or ~ 10–15%. Taken together, these findings illustrate the exquisite relationship between the level of BCL11A and HbF in erythroid cells. Such "hard-wiring" of this relationship underlies the attractiveness of BCL11A as a therapeutic target for HbF reactivation.

9.4 The BCL11A Erythroid Enhancer

9.4.1 Genetics and Overall Architecture

With the aim of understanding how common genetic variation at the *BCL11A* locus affects expression of BCL11A, attention was focused on "sentinel" single-nucleotide polymorphisms (SNPs), those most tightly correlated with HbF level [43]. Such SNPs reside in the large intron-2 of the gene, a ~10 kb region distinguished by DNase I HS sites in erythroid cell chromatin, but not in chromatin of B cells, brain tissue, or other sites of BCL11A expression [43]. The DNase I HS sites lie at +55, +58, and +62 kb relative to the transcription start site (TSS) of the *BCL11A* gene. The constellation of these sites, taken together with enrichment of histone marks H3K4me1 and H3K27ac and occupancy by the erythroid TFs GATA1 and SCL/TAL1, strongly suggested that the ~10 kb region might function as a stage-selective, erythroid-specific enhancer. Indeed, a 12.4-kb fragment encompassing the putative enhancer was sufficient to program the expression of a lacZ reporter in fetal livers, but not B cells, of transient transgenic embryos. In addition, TALEN-mediated deletion of the corresponding sequences in mouse erythroleukemia [37] cells virtually ablated BCL11A expression, indicating that the enhancer is not merely capable of programming erythroid-specific transcription but is essential for expression of BCL11A in the native chromosomal context [43]. As anticipated, deletion of the same sequences in mouse pre-B cells failed to alter BCL11A expression. Finally, engineering of mice harboring deletion of the corresponding region demonstrated the strict erythroid specificity of the enhancer [44]. Enhancer-deleted mice were normal in all respects, except for the expected deficit in silencing of endogenous embryonic globin genes and human γ-globin genes in a β-globin cluster YAC transgene. Of special note, hematopoietic stem cell function was normal. The identification of the *BCL11A* enhancer as the site of common genetic variation was an early example of a recurrent theme in human genetics: GWAS "hits" are highly enriched for regulatory, as contrasted with coding, sequences.

Most SNPs associated with a given trait are correlative and not causal. Within the enhancer, one sentinel SNP (rs1427407) in the +62 HS site alters a half E-box/GATA composite motif, a consensus sequence bound by GATA1 and SCL/TAL1 complexes in erythroid cells [43]. Indeed, quantitative chromatin immunoprecipitation (ChIP) revealed reduced occupancy by both TFs on the high-HbF allele and fewer BCL11A RNA transcripts. These findings are consistent with assignment of the SNP as a causal substitution that affects enhancer activity. The difference in BCL11A RNA expression between the low- and high-HbF alleles was 1.7-fold, indicative of the modest effect of this SNP on function.

9.4.2 An Achilles Heel in the Enhancer

The advent of CRISPR/Cas9 gene editing [45, 46] offered an opportunity to dissect the structure-function relationships in the *BCL11A* enhancer with precision. As genetic screens are most readily performed in immortalized cell lines, rather than primary cells, the introduction of transformed human CD34-derived HUDEP cell lines [47] has facilitated research in the field. HUDEP-2 cells, which exhibit an adult-stage phenotype, express predominantly HbA upon differentiation and accurately mimic normal erythroid maturation to a remarkable degree [48]. To confirm a requirement for the human *BCL11A* enhancer, CRIPSR/Cas9 editing was employed to remove the entire enhancer in HUDEP-2 cells. Consistent with prior findings in MEL cells, enhancer-deleted HUDEP-2 cells exhibited near-complete loss of BCL11A RNA and phenocopied BCL11A KO cells with respect to γ-globin expression. To reveal how sequences within the enhancer correlate with function, a tiled pooled editing screen was performed using gRNAs spanning the three HS sites (Fig. 9.2). Since single-nucleotide changes, such as the causal SNP within +62, affect enhancer activity, a goal of the unbiased screen was the identification of sites, if any, that more appreciably impacted function. In practice, a pooled library of gRNAs was introduced into Cas9-expressing HUDEP-2 cells, followed by expansion and differentiation of the population, and sorting for high-HbF-containing cells by intracellular fluorescence-activated cell sorting (FACS). Deconvolution of next-generation DNA sequencing permitted pseudo-mapping of gRNAs and measurement of enrichment scores in sorted cells. In this screen, disruption of

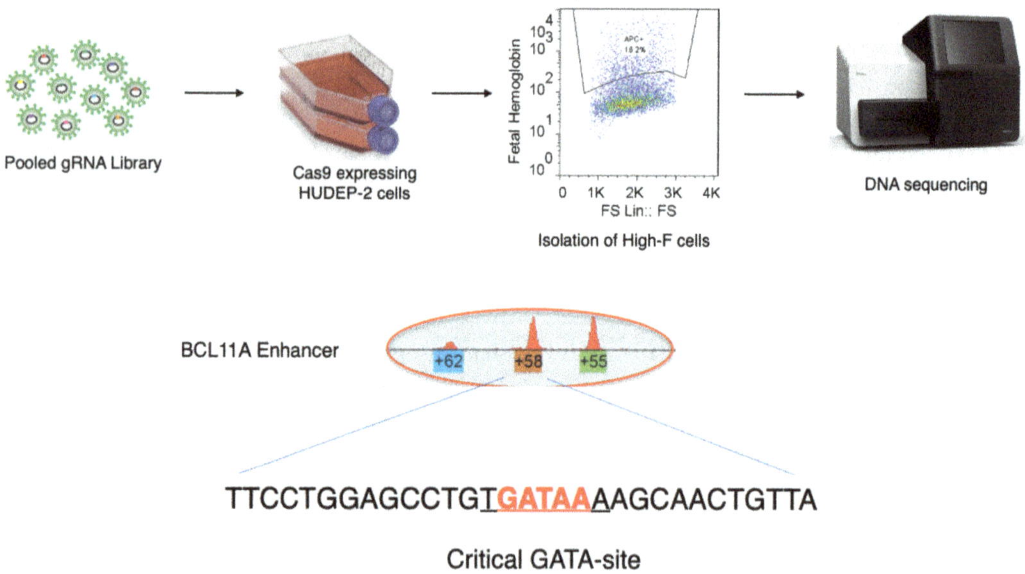

Fig. 9.2 Workflow of tiled pooled CRISPR/Cas9 BCL11A enhancer screen. A pooled library of gRNAs spanning the three HS sites in human BCL11A enhancer was cloned to a lentiviral vector. Cas9-expressing HUDEP-2 cells were transduced with the lentiviral library at low multiplicity, followed by expansion, differentiation, sorting by HbF level, genomic DNA isolation, and deep sequencing of integrated sgRNAs. A cluster of sgRNAs at HS + 58 showed marked enrichment in high-HbF-containing cells and mapped directly onto a GATA1 motif

regions targeted by individual gRNAs is repaired by nonhomologous end joining (NHEJ). Remarkably, the screen identified a discrete region within HS+58 at which disruption of the enhancer was associated with high HbF expression. The gRNAs eliciting the strongest signal interrupted or neighbored a GATA-binding site. Distinct, but smaller signals were also detected in HS+55 and HS+62. A parallel screen performed in CD34⁺ cells using a tiled collection of zinc finger nucleases even more accurately pinpointed the GATA site in HS+58 as the critical sequence vulnerability [49].

Studies of the corresponding mouse enhancer also revealed an unequal contribution of HS sites to overall function. In mouse erythroid cells, HS sites homologous to human HS+55 and +62, but not HS+58, were detected. A similar pooled screen in mouse erythroid cells demonstrated that mouse HS+62 conferred the majority of enhancer activity. Moreover, deletion of this region from the *Bcl11a* locus in engineered mice nearly phenocopied BCL11A KO with respect to impairment of embryonic globin gene silencing [47].

The vulnerability of the ~10-kb *BCL11A* enhancer to disruption of a single GATA site in HS+58 was unanticipated. As all the HS sites contain GATA-binding, the question remains as to why this site is so critical. This finding illustrates how limited are current means to predict enhancer function based on sequence alone. The GATA site in HS+58 represents an "Achilles heel" for the enhancer. Given the exquisite cell specificity of the enhancer and the high efficiency with which it can be edited, the site was proposed as a favorable target for gene editing to reactivate HbF in the β-hemoglobinopathies [47]. Importantly, the marked reduction of BCL11A expression upon genetic manipulation of Achilles heel is restricted to erythroid cells, thereby sparing hematopoietic stem cells (HSCs) or other lineages from any deleterious consequences of BCL11A loss [44]. Indeed, as noted below, a gRNA targeting the HS+58 site, #1617, was advanced to clinical trials by CRISPR Therapeutics and Vertex Pharmaceuticals.

9.5 Mechanism of Action in Globin Repression

Recent studies provide a detailed molecular mechanism of how γ-globin gene repression is controlled by BCL11A. In adult erythroid cells, BCL11A is expressed as two major isoforms, XL and L, which are generated by alternative RNA processing. BCL11A XL, the full-length protein, is more abundant and active for HbF repression, whereas L is inactive by itself. BCL11A XL protein is largely disordered. Seven zinc fingers (ZnFs), numbered 0–6, constitute the only structured domains. ZnF0, located closest to the N-terminus, is a C2HC ZnF, a class that is not well characterized but can mediate protein-protein or protein-nucleic acid interactions. The ZnF0 domain of BCL11B, a close homologue of BCL11A, was reported to mediate homodimerization and serve essential roles for gene expression regulation, induction of cell cycle arrest, and chemoresistance in the human T-cell line [50]. The sequences of the C2HC ZnF0 domains in BCL11A and BCL11B differ by one amino acid (isoleucine in BCL11A replaced by valine in BCL11B). A critical function of BCL11A ZnF0 in globin regulation is implied by pathogenic missense variants (p.Thr47Pro, p. Cys48Phe, and p.His66Gln) reported in individuals with abnormal neurodevelopment accompanied by persistence of HbF [51]. As these missense mutations alter cysteine, histidine, or the neighboring residues that coordinate zinc ions, they are likely to disrupt the integrity of the ZnF. BCL11A carrying these mutations cannot be stably transduced in primary cord blood CD34⁺ cells and is degraded by the proteasomal pathway, suggesting the presence of intact ZnF0 domain is required for BCL11A stability [51].

BCL11A ZnF1 to ZnF6 are C2H2 ZnF domains. The functions of ZnF1 and ZnF23 cluster remain elusive. An *in vitro* protein-binding microarray (PBM) assay suggested that ZnF23 binds GC-rich sequences with low binding affinity [52]. Three closely spaced zinc fingers, ZnF456, located near the C-terminus recognize a novel motif TGACCA in the human γ-globin promoters, as well as all embryonic globin promoters in mouse and human [52]. The crystal structure of BCL11A bound to a 12 bp double-stranded oligo containing the TGACCA motif revealed the details of recognition at atomic resolution [53]. ZnF456 binds to the motif with a K_D value of ~72 nM. Within ZnF456, ZnF45 makes extensive interactions with both DNA strands, whereas ZnF6 has rare contacts to the DNA duplex. Mutation of residues in ZnF4 and ZnF5 strongly decrease binding affinity, suggesting that these fingers are responsible for sequence-specific DNA recognition [53]. In rare individuals with the HFPH syndrome, base substitutions within a TGACCA motif in the γ-globin promoters have been described (e.g., "Greek"-type HPFH, G-A substitution at −117) [54, 55]. At the time these mutations were first described in 1985, it was proposed that they affected factors binding to the ubiquitous CCAAT box within the promoters. The realization that the substitutions, in fact, alter the binding site for BCL11A provided new insight into the mechanism underlying the HPFH phenotype. Specifically, the substitutions disrupt binding of BCL11A to the promoter and relieve repression [52, 56].

Looking more closely at the γ-globin gene promoters, it is notable that the TGACCA motif is duplicated and each motif overlaps a CCAAT box, which is typically bound by the ubiquitous activating factor NF-Y (Fig. 9.3). Indeed, the naturally occurring HPFH substitutions at TGACCA appear restricted to the distal duplicate. An explanation for this observation is provided by high-resolution CUT&RUN analysis which revealed preferential occupancy of BCL11A at the distal TGACCA in adult erythroid chromatin [52]. In cells with low BCL11A and high γ-globin expression, NF-Y selectively binds the proximal CCAAT motif of the promoters and is required for chromosomal looping between the LCR and the γ-globin genes to achieve high-level transcription [57]. Artificial recruitment of dCas9 protein at different positions of γ-globin promoters suggests a competitive binding model between NF-Y and BCL11A in regulation of γ-globin expression: BCL11A initiates γ-globin repression through direct binding at the distal TGACCA sequence, which constitutes a steric barrier to NF-Y binding at the proximal CCAAT motif [57]. BCL11A may also contribute to control of γ-globin expression at a distance, although these have yet to be fully elucidated. As the chromatin conformation of the β-globin locus is altered in absence of BCL11A, it may participate in longer range interactions [58].

Besides its roles in binding the promoter and eviction of the activator NF-Y, BCL11A also recruits the nucleosome remodeling and deacetylase (NuRD) complex though a canonical peptide located at its N-terminus, similar to that also found in TFs FOG1 and SALL4 [59, 60]. Recruitment of the NuRD complex through interaction with its RBBP4 subunit is thought to stabilize γ-globin repression [59, 61]. Indeed, several NuRD complex members including CHD4 are essential for HbF repression, as revealed by dense editing of NuRD genes by CRISPR-Cas9 pooled screening [62]. Additional functional support for a contribution of NuRD to γ-globin repression was provided by the discovery of ZNF410, a pentadactyl DNA-binding protein, which appears to be singularly dedicated to regulation of expression of CHD4, a major component of the complex [63, 64].

Although potent and essential as a regulator of hemoglobin silencing, BCL11A is not the sole factor. ZBTB7A, also known as LRF or POKEMON, exhibits comparable activity in HbF repression, as shown by knockout in HUDEP-2 cells [65]. ZBTB7A recognizes a GC-rich motif that lies distal (~ −200 region) to the BCL11A-binding site in the γ-globin promoters (Fig. 9.3). Base substitutions within the ZBTB7A-binding site have been observed in other rare individuals with the HPFH syndrome [56] which is also the site of rare single-base changes in HPFH. In double BCL11A/ZBTB7A KO HUDEP-2 cells, HbF comprises nearly 100% of the total Hb on differentiation [65]. This observation suggests that these represent the only major, direct repressors of γ-globin transcription. Although the two repressors bind at different regions in the promoters and therefore appear to act independently, both associate

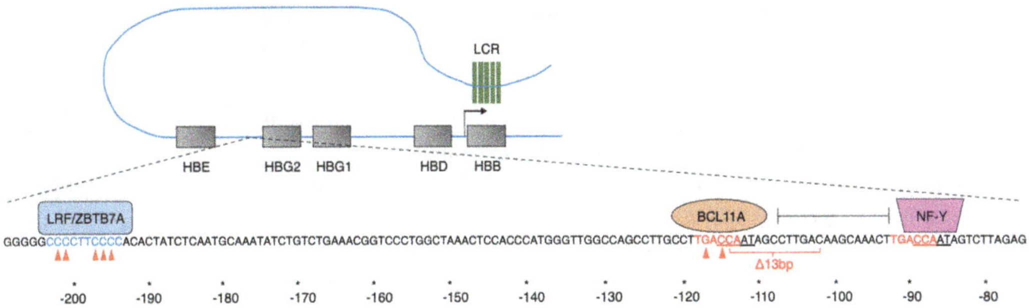

Fig. 9.3 Major transcription factors binding at the γ-globin gene promoter. In adult-stage erythroid cells, BCL11A and LRF/ZBTB7A repress γ-globin expression through binding to the γ-globin promoter in two sites with clustered mutations in HPFH, pointed with red triangles. HPFH-associated mutations disrupt binding of BCL11A or LRF/ZBTB7A, resulting in elevated HbF. The consensus sequences for BCL11A and LRF/ZBTB7A binding are shown in red and blue, respectively. A ubiquitous activator, NF-Y, binds at a CCAAT box (underlined) that overlaps with a proximal BCL11A-binding motif. BCL11A displaces NF-Y by steric hindrance in the γ-globin promoter

with the NuRD complex, suggesting that concerted recruitment of NuRD to the γ-globin promoter is required for stable HbF repression. The derepression of γ-globin transcription with loss of either BCL11A or ZBTB7A is consistent with this relationship.

Apart from the γ-globin promoters, BCL11A also occupies the promoters of embryonic globin genes, including the ζ-gene in the α-globin locus and ε-globin in the human β-globin cluster [52]. BCL11A loss leads to ζ-globin derepression in HUDEP-2 cells [66].

9.6 Few Primary Targets of BCL11A Repression

Most often loss of a critical TF has a profound effect on the proliferation, differentiation, and/or viability of the cells in which it is expressed. The original shRNA knockdown experiments in CD34$^+$ cells [31] and the relative insensitivity of red cell production in mice to loss of BCL11A in the erythroid lineage suggested that BCL11A might control only a limited number of target genes. Typical KO experiments in cells cannot identify primary targets of TFs because they represent static, end-stage analyses, often separated by many hours or days in culture.

To determine the primary (direct) targets of BCL11A in erythroid cells, Mehta et al. [67] used rapid, conditional protein degradation and nascent transcript analysis in HUDEP-2 cells. Rapid degradation of BCL11A in HUDEP-2 cells was achieved with the dTAG platform [68]. The method relies on artificial recruitment of a target protein expressed in fusion with modified FK506-binding protein 12 (FKBP12) by a bifunctional proteolysis-targeting chimera (PROTAC), e.g., dTAG-47, to Cereblon (CRBN), a component of an E3 ubiquitin ligase complex. In this context, the PROTAC binds both FKBP12 and CRBN to bring the target protein in proximity to CRBN for proteolytic degradation. To perform the experiments, Mehta et al. used a high-efficiency editing protocol [69] to engineer HUDEP-2 cells with FKBP12 fused in-frame with BCL11A in both alleles. In cells exposed to dTAG-47, BCL11A protein is degraded to near undetectable levels by 1 h, thereby permitting use of nascent transcription assays (e.g., PRO-Seq or SLAM-Seq) for identification of primary gene targets. Nascent transcription and quantitative proteomic analyses revealed that BCL11A functions solely as a repressor and has only ~31 direct gene targets in erythroid cells. Among this small set of genes, *HBG1/2* genes exhibit the most abundant and progressive changes in transcription and chromatin accessibility upon BCL11A loss. These results demonstrate that the transcriptional effects of BCL11A loss are remarkably specific in erythroid cells. In addition, erythroid cell proliferation and differentiation were minimally perturbed, unlike observations for other TFs [67]. Moreover, BCL11A functions as a continuous regulator of HbF levels. Downregulation of BCL11A expression on addition of dTAG-47 is accompanied by a concomitant increase in HbF expression. Re-expression of BCL11A upon PROTAC washout leads to re-repression of HbF [67]. The paucity of primary targets of BCL11A and dominance of globin regulation in the BCL11A-depleted proteome likely contribute to the ultimate success of downregulation of BCL11A for therapeutic benefit in the hemoglobinopathies (see below).

9.7 Transcriptional Regulation of BCL11A Expression

An understanding of how the expression of BCL11A itself is controlled is incomplete. From the studies of the essential *BCL11A* erythroid enhancer, it is evident that both GATA1 and SCL/TAL1 play important roles in its transcription. Several transcription factors, KLF1, ATF4, and NFIA/X [47, 70], are also implicated in regulation of BCL11A expression (Fig. 9.4). KLF1 was reported to occupy the *BCL11A* promoter and activate its expression. Knockdown of KLF1 in human adult erythroid progenitors reduces BCL11A RNA and protein levels, leading to increased γ-globin expression [70]. Indeed, haploinsufficiency for KLF1 is associated with an HPFH phenotype, which has been attributed to reduced BCL11A expression [71]. ATF4, an HRI (heme-regulated inhibitor)-regulated TF, has been proposed to stimulate transcription of *BCL11A* by binding to HS + 55 and fostering enhancer-promoter contacts [72]. Of possible interest, HRI, an erythroid-specific kinase that controls protein translation, was identified in a protein kinase domain-focused CRISPR/Cas9-based genetic screen for high HbF and suggested to promote transcription of *BCL11A* [73]. NFIA and NFIX, two members of the NFI transcription factor family that are expressed in erythroid cells, occupy HS+55, HS+58, and HS+62 regions in HUDEP-2 and primary erythroid cells. It has been proposed that they may contribute to chromatin opening, thereby facilitating BCL11A expression [74]. The detailed mechanisms employed by ATF4 and NFI TFs remain to be elucidated.

BCL11A mRNA level increases during erythroid cell maturation [31]. A recent study illustrates an additional pathway for control of BCL11A transcription in erythroid cells. HIC2, a pentadactyl zinc finger DNA-binding protein of the POK (POZ domain Krüppel-like zinc finger) family, was identified as a repressor of *BCL11A* transcription in a CRISPR-based genetic screen in HUDEP-2 cells [75]. HIC2 levels are inversely correlated with BCL11A expression and decrease during differentiation [75]. As a potential mechanism, HIC2 has been proposed to compete directly with GATA1 for binding at *BCL11A* HS+55 to repress *BCL11A* transcription at the fetal stage. Downregulation of HIC2 levels in adult cells relieves this repression, allowing for BCL11A expression and HbF silencing [75].

The above factors were identified in genetic screens of HbF-high or HbF-low erythroid cells. Of particular note, they appear to control HbF levels through BCL11A, reinforcing the conclusion that BCL11A is a major, if not the principal physiological, regulator of globin switching.

9.8 Posttranscriptional Regulation of BCL11A

BCL11A is also regulated at the level of mRNA translation during human hematopoietic development. The RNA-binding protein LIN28B, a heterochronic regulator of development, has been shown to suppress BCL11A mRNA translation through direct interaction with ribosomes and BCL11A mRNA [76, 77]. LIN28B is expressed in a developmental stage-specific manner reciprocal to BCL11A expression [77]. At early developmental stages, abundant LIN28B is able to bind BCL11A mRNA and inhibit its translation. The absence of LIN28B expression in adult erythroid cells allows for effective BCL11A protein synthesis, and γ-globin expression is thereby repressed. Independent of LIN28B, experimental

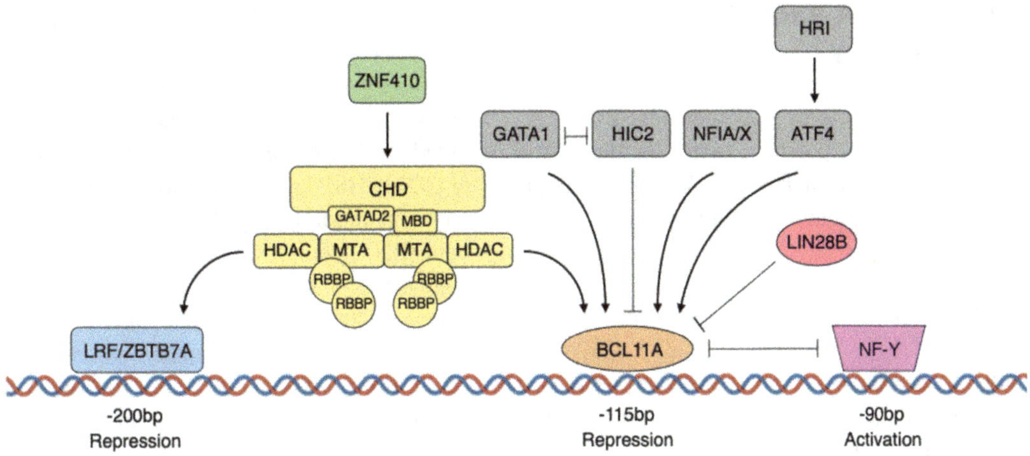

Fig. 9.4 Transcriptional and posttranscriptional regulation of BCL11A. BCL11A and LRF/ZBTB7A independently associate with the NuRD complex to repress γ-globin expression. The NuRD complex consists of six protein subunits: histone deacetylase 1/2 (HDAC1/2), metastasis-associated protein 1/2/3 (MTA1/2/3), GATA zinc finger domain containing 2A/B (GATAD2A/B), retinoblastoma-binding protein 4/7 (RBBP4/7), methyl-CpG-binding domain protein 2/3 (MBD2/3), and chromo-domain helicase DNA-binding protein 3/4 (CHD3/4). NuRD is negatively regulated by ZNF410. ATF4, HRI, and NFIA/X have been proposed to promote BCL11A expression. HIC2 has been proposed to compete with GATA1 for binding at BCL11A enhancer to repress BCL11A transcription. LIN28B has been shown to direct interact with BCL11A mRNA to suppress BCL11A mRNA translation. Many other positive and negative regulators of BCL11A are not shown

suppression of let-7 miRNAs, established targets of LIN28B, also reduced BCL11A expression and significantly increased HbF expression in erythroid cells [76, 78]. Moreover, erythroid-specific overexpression of the insulin-like growth factor 2 mRNA-binding protein 1 (IGF2BP1), a downstream component of the let-7 miRNA pathway, leads to little change in BCL11A mRNA level but significant downregulation of BCL11A protein level [79]. Further RNA immunoprecipitation [19], mRNA stability and polysome studies suggest that IGF2BP1 binds to BCL11A mRNA and mediates posttranscriptional loss of BCL11A [79]. Taken together, these findings suggest that BCL11A expression is post-transcriptionally regulated by the LIN28/let-7/IGF2BP1 axis during the ontogeny of human erythropoiesis. Overexpression of LIN28B, suppression of let-7, or overexpression of IGF2BP1 in adult cells leads to HbF reactivation as a consequence of BCL11A reduction, again demonstrating that BCL11A is a direct and potent regulator of HbF.

In addition to let-7 miRNAs, several miRNAs may regulate BCL11A expression at the translational level in human erythroid cells. Overexpression of miR-30a leads to reduced BCL11A expression and elevated γ-globin expression [80]. MiR-210 inhibits the expression of BCL11A through direct binding to the coding sequence of BCL11A mRNA [81]. Forced expression of miR-210 leads to decrease of BCL11A expression and increase of γ-globin [81]. MiR-138-5p, miR-486-3p, and miR-92a 3p target the 3′UTR of *BCL11A* to restrict its level and induce γ-globin expression in erythroid cells [82–84]. In a more systematic and comprehensive study, expression of RNA and miRNA was compared between fetal and adult human erythroblasts. Several fetal-stage miRNAs were identified and predicted to target BCL11A by computational approaches [85]. However, predicted effects require validation. Overexpression of one or more miRNAs targeting different places in BCL11A mRNA might reduce BCL11A expression to achieve a therapeutic level of HbF. However, the specificity of the miRNAs needs further study if such an approach is considered for clinical trials.

9.9 BCL11A-Based Genetic Therapy

With its central role in HbF silencing, BCL11A quickly became a potential target for the treatment of SCD and β-thalassemia. Two approaches have provided validation in clinical trials.

9.9.1 Lentiviral-Mediated shRNA

Williams and his colleagues modified previously developed lentiviral vectors to express shRNA directed to BCL11A in an erythroid-specific manner [34, 86–88]. Optimized shRNA embedded within an microRNA architecture (shmiR) was transcribed under control of regulatory elements from the β-globin cluster LCR [34, 87]. Preclinical studies showed that transduction of lentivirus into HSCs reactivated HbF to levels sufficient to reverse sickle phenotypes in SCD mice [34]. In a phase I clinical trial of six patients with SCD, transduction of mobilized HSPCs, followed by hematopoietic reconstitution after busulfan preconditioning, led to robust and stable HbF induction in all patients and reduced or absent clinical manifestations of SCD during the follow-up period [88]. Subsequent analysis revealed that the level of HbF in F-expressing cells of these patients exceed the average level of HbF/F-cell in high-F-responsive SCD individuals treated with hydroxyurea [89]. This important observation illustrates the more potent induction of HbF by genetic therapy, as compared with standard-of-care hydroxyurea management. As might be anticipated, a double-shmiR lentivirus targeting BCL11A and ZNF410 elicits even greater HbF induction in erythroid cells [90] but has not yet been advanced to clinical trials.

9.9.2 CRISPR/Cas9 Gene Editing

Applying the findings of Canver et al. [47], CRISPR Therapeutics, in collaboration with Vertex Pharmaceuticals, initiated first-in-man trials of CRISPR/Cas9 in both SCD and β-thalassemia using gRNA#1617 to disrupt the Achilles heel in the *BCL11A* enhancer [91]. In the initial report, single patients with SCD and transfusion-dependent β-thalassemia were reported. Following hematopoietic reconstitution with $CD34^+$ HSPCs subjected to *ex vivo* editing by electroporation of a gRNA-Cas9 ribonucleoprotein (RNP) complex, both individuals exhibited greater than 90% F-cells with HbF/F-cell in the range of 40%. In subsequent patients (>90 in total), the trials met the primary outcomes of freedom from crises in SCD and transfusion-independence in thalassemia for at least 12 months. Following submission of the clinical results to regulatory authorities, this CRISPR-based gene-editing therapy, now called Casgevy, was approved for patients >12 years of age by the Medicines and Healthcare products Regulatory Agency (MHRA) in the UK and the FDA in the USA. Casgevy is the first approved gene editing therapy for a disease. Despite concerns regarding possible off-target editing and chromosomal rearrangements from Cas9 and related nucleases [92–95] due to double-stranded breaks (DSBs) in DNA, no apparent adverse consequences have been observed in the trials for Casgevy. Changes in experimental procedures may further reduce the potential risk of gene editing-related adverse effects. For example, editing without *ex vivo* culture of $CD34^+$ HSPCs avoided genotoxicity [96].

An independent clinical trial for SCD (NCT03653247) targeting the *BCL11A* enhancer was initiated by Sangamo Therapeutics and Sanofi, in which a ZnF nuclease (BIVV003) was employed for editing [49, 97, 98]. Although four patients were treated and appeared to show clinical benefit, the program was halted in early 2022 [97, 98].

An alternative strategy for BCL11A-based gene editing has been brought forward by Editas Medicine. The Editas trials employ Cas12a to disrupt sequences encompassing the BCL11A-binding site in the γ-globin promoters, in effect mimicking naturally occurring HPFH mutations [99]. In principle, prevention of BCL11A binding may elicit even higher levels of HbF than disruption of the HS + 58 GATA1 site. Early clinical results of the RUBY and EdiTHAL trials for SCD and transfusion-dependent β-thalassemia, respectively, are promising [100].

9.9.3 Base Editing

Base editing is a variation on CRISPR/Cas9-based editing that introduces single-point replacements in DNA without inducing DSBs [101, 102]. Two major classes of base editors have been developed: cytidine base editors (CBEs) allowing C > T conversions and adenine base editors (ABEs) allowing A > G conversions [103]. Both CBEs and ABEs have been employed to target core sequences of the HS+58 *BCL11A* enhancer in preclinical experiments [104, 105]. HbF reactivation was similar to that observed with gene editing, suggesting that this approach could be translated to clinical trials.

The recognition sequence for BCL11A, TGACCA, is favorable for use of base editors for reactivation of HbF expression. BEAM-101, a program launched by BEAM Therapeutics, utilizes ABE to introduce point mutations in the *HBG1* and *HBG2* promoters in a clinical trial (BEACON) for SCD. Clinical results are not yet reported.

9.9.4 Multiplex Editing to Increase HbF

Since the degree of HbF reactivation is determined by the extent of reduction in BCL11A expression, various strategies are also being explored to further elevate HbF levels after genetic modifications. For example, multiplex gene editing using CRISPR/Cas9 or base editors to simultaneously edit the *BCL11A* erythroid-specific enhancer and the BCL11A-binding sites in the γ-globin gene promoters results in higher γ-globin induction compared with single editing [105, 106]. Similarly, combined targeting of the HS+58 and other HS sites in the *BCL11A* enhancer is anticipated to increase the level of HbF in edited cells.

9.10 BCL11A Protein as a Therapeutic Target

Besides demonstrating the effectiveness of HbF reactivation in patients with SCD and β-thalassemia, the recent clinical trials thoroughly validate BCL11A as a therapeutic target. The worldwide distribution and high prevalence of SCD and β-thalassemia in medically underserved populations has led the World Health Organization (WHO) to classify the hemoglobinopathies as a global health burden that will increase over time [10, 11]. Current genetic therapies rely on *ex vivo* modification of HSPCs and reinfusion following myeloablative chemotherapeutic preconditioning. As such, they are not well suited to address the true burden of these disorders, particularly in resource-limited regions. Various strategies are being considered to overcome these barriers. In some sense, the simplest approach would be *in vivo* delivery of gene editing components, much like as a vaccine is administered. Early efforts in mice using mRNA encapsulated in lipid nanoparticles to target HSCs have been reported [107, 108]. However, the clinical trials for SCD and β-thalassemia demonstrate that a high proportion of HSCs must be modified to achieve a successful clinical outcome. Whether methods for *in vivo* delivery will be adequate to meet this challenge is uncertain. As a more immediate strategy, efforts are ongoing to reduce preconditioning to lessen potential complications and preserve fertility.

The development of small molecules that impair the function of BCL11A or promote its degradation would permit much wider access of therapy for patients in all regions. Given that BCL11A protein is largely unstructured, the task is daunting. With current advances in protein chemistry and chemical design, new approaches to targeting previously "undruggable" proteins suggest different strategies and offer hope that reaching the goal may be possible. We recently described two sets of nanobodies that bind to BCL11A with high affinity and specificity [109, 110]. The nanobodies were selected against the well-ordered ZnF23 and ZnF456 domains of

BCL11A and further evolved to distinguish BCL11A from its close paralog BCL11B. Expression of nanobody in fusion with an E3 ligase in HUDEP-2 cells or CD34⁺ HSPCs leads to proteolytic degradation of unmodified endogenous BCL11A protein and the anticipated reactivation of HbF [109, 110]. Exposure of primary CD34⁺ HSPCs to a cell-permeant degrader composed of BCL11A-specific nanobody, a genetically encoded cell-penetrating vehicle and an E3 adaptor protein led to sustained loss of BCL11A and significant induction of HbF [109]. In principle, if efficient *in vivo* delivery systems were developed, such nanobodies could be explored as therapeutic tools in the future. Another path to novel therapeutics relies on the development of small molecule ligands that bind BCL11A and either impair its function or elicit degradation. Despite proof-of-principle studies with the dTAG system demonstrating robust HbF reactivation following targeted protein degradation of BCL11A, the feasibility of small molecule targeting of BCL11A remains to be established.

Acknowledgments We thank members of the laboratory for past contributions and their commitment to ongoing studies of BCL11A.

References

1. Fritsch EF, Lawn RM, Maniatis T (1980) Molecular cloning and characterization of the human beta-like globin gene cluster. Cell 19(4):959–972
2. Peschle C et al (1985) Haemoglobin switching in human embryos: asynchrony of zeta—alpha and epsilon—gamma-globin switches in primitive and definite erythropoietic lineage. Nature 313(5999):235–238
3. Sankaran VG, Xu J, Orkin SH (2010) Advances in the understanding of haemoglobin switching. Br J Haematol 149(2):181–194
4. Tuan D et al (1985) The "beta-like-globin" gene domain in human erythroid cells. Proc Natl Acad Sci USA 82(19):6384–6388
5. Grosveld F et al (1987) Position-independent, high-level expression of the human beta-globin gene in transgenic mice. Cell 51(6):975–985
6. Forrester WC et al (1986) A developmentally stable chromatin structure in the human beta-globin gene cluster. Proc Natl Acad Sci USA 83(5):1359–1363
7. McGrath K, Palis J (2008) Ontogeny of erythropoiesis in the mammalian embryo. Curr Top Dev Biol 82:1–22
8. Zago MA et al (1979) Genetic control of F cells in human adults. Blood 53(5):977–986
9. Leonova J et al (1996) Variability in the fetal hemoglobin level of the normal adult. Am J Hematol 53(2):59–65
10. Orkin SH, Higgs DR (2010) Medicine. Sickle cell disease at 100 years. Science 329(5989):291–292
11. Weatherall DJ (2010) The inherited diseases of hemoglobin are an emerging global health burden. Blood 115(22):4331–4336
12. Watson J (1948) The significance of the paucity of sickle cells in newborn Negro infants. Am J Med Sci 215(4):419–423
13. Forget BG (1998) Molecular basis of hereditary persistence of fetal hemoglobin. Ann N Y Acad Sci 850:38–44
14. Castro O et al (1994) The acute chest syndrome in sickle cell disease: Incidence and risk factors. The cooperative study of sickle cell disease. Blood 84(2):643–649
15. Galanello R et al (2009) Amelioration of Sardinian beta thalassemia by genetic modifiers. Blood 114(18):3935–3937
16. Platt OS et al (1994) Mortality in sickle cell disease. Life expectancy and risk factors for early death. N Engl J Med 330(23):1639–1644
17. Platt OS et al (1991) Pain in sickle cell disease. Rates and risk factors. N Engl J Med 325(1):11–16
18. Premawardhena A et al (2005) Haemoglobin E beta thalassaemia in Sri Lanka. Lancet 366(9495):1467–1470
19. Nuinoon M et al (2010) A genome-wide association identified the common genetic variants influence disease severity in beta0-thalassemia/hemoglobin E. Hum Genet 127(3):303–314
20. Cantor AB, Orkin SH (2002) Transcriptional regulation of erythropoiesis: an affair involving multiple partners. Oncogene 21(21):3368–3376
21. Lettre G et al (2008) DNA polymorphisms at the BCL11A, HBS1L-MYB, and beta-globin loci associate with fetal hemoglobin levels and pain crises in sickle cell disease. Proc Natl Acad Sci USA 105(33):11869–11874
22. Menzel S et al (2007) A QTL influencing F cell production maps to a gene encoding a zinc-finger protein on chromosome 2p15. Nat Genet 39(10):1197–1199
23. Uda M et al (2008) Genome-wide association study shows BCL11A associated with persistent fetal hemoglobin and amelioration of the phenotype of beta-thalassemia. Proc Natl Acad Sci USA 105(5):1620–1625
24. Nakamura T et al (2000) Evi9 encodes a novel zinc finger protein that physically interacts with BCL6, a known human B-cell proto-oncogene product. Mol Cell Biol 20(9):3178–3186

25. Saiki Y et al (2000) Human EVI9, a homologue of the mouse myeloid leukemia gene, is expressed in the hematopoietic progenitors and down-regulated during myeloid differentiation of HL60 cells. Genomics 70(3):387–391
26. Luc S et al (2016) Bcl11a deficiency leads to hematopoietic stem cell defects with an aging-like phenotype. Cell Rep 16(12):3181–3194
27. Liu P et al (2003) Bcl11a is essential for normal lymphoid development. Nat Immunol 4(6):525–532
28. Avram D et al (2002) COUP-TF (chicken ovalbumin upstream promoter transcription factor)-interacting protein 1 (CTIP1) is a sequence-specific DNA binding protein. Biochem J 368(Pt 2):555–563
29. Kuo TY, Hsueh YP (2007) Expression of zinc finger transcription factor Bcl11A/Evi9/CTIP1 in rat brain. J Neurosci Res 85(8):1628–1636
30. Satterwhite E et al (2001) The BCL11 gene family: involvement of BCL11A in lymphoid malignancies. Blood 98(12):3413–3420
31. Sankaran VG et al (2008) Human fetal hemoglobin expression is regulated by the developmental stage-specific repressor BCL11A. Science 322(5909):1839–1842
32. Sankaran VG et al (2009) Developmental and species-divergent globin switching are driven by BCL11A. Nature 460(7259):1093–1097
33. Xu J et al (2011) Correction of sickle cell disease in adult mice by interference with fetal hemoglobin silencing. Science 334(6058):993–996
34. Brendel C et al (2016) Lineage-specific BCL11A knockdown circumvents toxicities and reverses sickle phenotype. J Clin Invest 126(10):3868–3878
35. Funnell AP et al (2015) 2p15-p16.1 microdeletions encompassing and proximal to BCL11A are associated with elevated HbF in addition to neurologic impairment. Blood 126(1):89–93
36. Basak A et al (2015) BCL11A deletions result in fetal hemoglobin persistence and neurodevelopmental alterations. J Clin Invest 125(6):2363–2368
37. Dias C et al (2016) BCL11A haploinsufficiency causes an intellectual disability syndrome and dysregulates transcription. Am J Hum Genet 99(2):253–274
38. Cai T et al (2017) Identification of novel mutations in the HbF repressor gene BCL11A in patients with autism and intelligence disabilities. Am J Hematol 92(12):E653–E656
39. Yoshida M et al (2018) Identification of novel BCL11A variants in patients with epileptic encephalopathy: Expanding the phenotypic spectrum. Clin Genet 93(2):368–373
40. Wessels MW et al (2021) Molecular analysis of the erythroid phenotype of a patient with BCL11A haploinsufficiency. Blood Adv 5(9):2339–2349
41. Levy J et al (2017) Molecular and clinical delineation of 2p15p16.1 microdeletion syndrome. Am J Med Genet A 173(8):2081–2087
42. Mimouni-Bloch A et al (2015) A de-novo interstitial microduplication involving 2p16.1-p15 and mirroring 2p16.1-p15 microdeletion syndrome: Clinical and molecular analysis. Eur J Paediatr Neurol 19(6):711–715
43. Bauer DE et al (2013) An erythroid enhancer of BCL11A subject to genetic variation determines fetal hemoglobin level. Science 342(6155):253–257
44. Smith EC et al (2016) Strict in vivo specificity of the Bcl11a erythroid enhancer. Blood 128(19):2338–2342
45. Doudna JA, Charpentier E (2014) Genome editing. The new frontier of genome engineering with CRISPR-Cas9. Science 346(6213):1258096
46. Hsu PD, Lander ES, Zhang F (2014) Development and applications of CRISPR-Cas9 for genome engineering. Cell 157(6):1262–1278
47. Canver MC et al (2015) BCL11A enhancer dissection by Cas9-mediated in situ saturating mutagenesis. Nature 527(7577):192–197
48. Kurita R et al (2013) Establishment of immortalized human erythroid progenitor cell lines able to produce enucleated red blood cells. PLoS One 8(3):e59890
49. Vierstra J et al (2015) Functional footprinting of regulatory DNA. Nat Methods 12(10):927–930
50. Grabarczyk P et al (2018) The N-terminal CCHC zinc finger motif mediates homodimerization of transcription factor BCL11B. Mol Cell Biol 38(5)
51. Shen Y et al (2021) Pathogenic BCL11A variants provide insights into the mechanisms of human fetal hemoglobin silencing. PLoS Genet 17(10):e1009835
52. Liu N et al (2018) Direct promoter repression by BCL11A controls the fetal to adult hemoglobin switch. Cell 173(2):430–442 e17
53. Yang Y et al (2019) Structural insights into the recognition of gamma-globin gene promoter by BCL11A. Cell Res 29(11):960–963
54. Collins FS et al (1985) A point mutation in the A gamma-globin gene promoter in Greek hereditary persistence of fetal haemoglobin. Nature 313(6000):325–326
55. Gelinas R et al (1985) G to A substitution in the distal CCAAT box of the A gamma-globin gene in Greek hereditary persistence of fetal haemoglobin. Nature 313(6000):323–325
56. Martyn GE et al (2018) Natural regulatory mutations elevate the fetal globin gene via disruption of BCL11A or ZBTB7A binding. Nat Genet 50(4):498–503
57. Liu N et al (2021) Transcription factor competition at the gamma-globin promoters controls hemoglobin switching. Nat Genet 53(4):511–520
58. Xu J et al (2010) Transcriptional silencing of gamma-globin by BCL11A involves long-range interactions and cooperation with SOX6. Genes Dev 24(8):783–798
59. Xu J et al (2013) Corepressor-dependent silencing of fetal hemoglobin expression by BCL11A. Proc Natl Acad Sci USA 110(16):6518–6523

60. Moody RR et al (2018) Probing the interaction between the histone methyltransferase/deacetylase subunit RBBP4/7 and the transcription factor BCL11A in epigenetic complexes. J Biol Chem 293(6):2125–2136
61. Shang S et al (2023) MBD2a-NuRD binds to the methylated gamma-globin gene promoter and uniquely forms a complex required for silencing of HbF expression. Proc Natl Acad Sci USA 120(25):e2302254120
62. Sher F et al (2019) Rational targeting of a NuRD subcomplex guided by comprehensive in situ mutagenesis. Nat Genet 51(7):1149–1159
63. Lan X et al (2021) ZNF410 uniquely activates the NuRD component CHD4 to silence fetal hemoglobin expression. Mol Cell 81(2):239–254 e8
64. Vinjamur DS et al (2021) ZNF410 represses fetal globin by singular control of CHD4. Nat Genet 53(5):719–728
65. Masuda T et al (2016) Transcription factors LRF and BCL11A independently repress expression of fetal hemoglobin. Science 351(6270):285–289
66. King AJ et al (2021) Reactivation of a developmentally silenced embryonic globin gene. Nat Commun 12(1):4439
67. Mehta S et al (2022) Temporal resolution of gene derepression and proteome changes upon PROTAC-mediated degradation of BCL11A protein in erythroid cells. Cell Chem Biol 29(8):1273-1287 e8
68. Nabet B et al (2018) The dTAG system for immediate and target-specific protein degradation. Nat Chem Biol 14(5):431–441
69. Mehta S et al (2023) High-efficiency knock-in of degradable tags (dTAG) at endogenous loci in cell lines. Methods Enzymol 681:1–22
70. Zhou D et al (2010) KLF1 regulates BCL11A expression and gamma- to beta-globin gene switching. Nat Genet 42(9):742–744
71. Borg J et al (2010) Haploinsufficiency for the erythroid transcription factor KLF1 causes hereditary persistence of fetal hemoglobin. Nat Genet 42(9):801–805
72. Huang P et al (2020) The HRI-regulated transcription factor ATF4 activates BCL11A transcription to silence fetal hemoglobin expression. Blood 135(24):2121–2132
73. Grevet JD et al (2018) Domain-focused CRISPR screen identifies HRI as a fetal hemoglobin regulator in human erythroid cells. Science 361(6399):285–290
74. Qin K et al (2022) Dual function NFI factors control fetal hemoglobin silencing in adult erythroid cells. Nat Genet 54(6):874–884
75. Huang P et al (2022) HIC2 controls developmental hemoglobin switching by repressing BCL11A transcription. Nat Genet 54(9):1417–1426
76. Lee YT et al (2013) LIN28B-mediated expression of fetal hemoglobin and production of fetal-like erythrocytes from adult human erythroblasts ex vivo. Blood 122(6):1034–1041
77. Basak A et al (2020) Control of human hemoglobin switching by LIN28B-mediated regulation of BCL11A translation. Nat Genet 52(2):138–145
78. de Vasconcellos JF et al (2017) Tough decoy targeting of predominant let-7 miRNA species in adult human hematopoietic cells. J Transl Med 15(1):169
79. de Vasconcellos JF et al (2017) IGF2BP1 overexpression causes fetal-like hemoglobin expression patterns in cultured human adult erythroblasts. Proc Natl Acad Sci USA 114(28):E5664–E5672
80. Gholampour MA et al (2020) miR-30a regulates gamma-globin expression in erythroid precursors of intermedia thalassemia through targeting BCL11A. Mol Biol Rep 47(5):3909–3918
81. Gasparello J et al (2017) BCL11A mRNA targeting by miR-210: A possible network regulating gamma-globin gene expression. Int J Mol Sci 18(12)
82. Sun KT et al (2017) Reciprocal regulation of gamma-globin expression by exo-miRNAs: Relevance to gamma-globin silencing in beta-thalassemia major. Sci Rep 7(1):202
83. Lulli V et al (2013) MicroRNA-486-3p regulates gamma-globin expression in human erythroid cells by directly modulating BCL11A. PLoS One 8(4):e60436
84. Li H et al (2022) MicroRNA-92a-3p-mediated inhibition of BCL11A upregulates gamma-globin expression and inhibits oxidative stress and apoptosis in erythroid precursor cells. Hematology 27(1):1152–1162
85. Lessard S et al (2018) 14q32 and let-7 microRNAs regulate transcriptional networks in fetal and adult human erythroblasts. Hum Mol Genet 27(8):1411–1420
86. Brendel C et al (2020) Preclinical evaluation of a novel lentiviral vector driving lineage-specific BCL11A knockdown for sickle cell gene therapy. Mol Ther Methods Clin Dev 17:589–600
87. Guda S et al (2015) miRNA-embedded shRNAs for lineage-specific BCL11A knockdown and hemoglobin F induction. Mol Ther 23(9):1465–1474
88. Esrick EB et al (2021) Post-transcriptional genetic silencing of BCL11A to treat sickle cell disease. N Engl J Med 384(3):205–215
89. De Souza DC et al (2023) Genetic reversal of the globin switch concurrently modulates both fetal and sickle hemoglobin and reduces red cell sickling. Nat Commun 14(1):5850
90. Liu B et al (2022) Development of a double shmiR lentivirus effectively targeting both BCL11A and ZNF410 for enhanced induction of fetal hemoglobin to treat beta-hemoglobinopathies. Mol Ther 30(8):2693–2708
91. Frangoul H et al (2021) CRISPR-Cas9 gene editing for sickle cell disease and beta-thalassemia. N Engl J Med 384(3):252–260
92. Fu Y et al (2013) High-frequency off-target mutagenesis induced by CRISPR-Cas nucleases in human cells. Nat Biotechnol 31(9):822–826

93. Zuccaro MV et al (2020) Allele-specific chromosome removal after Cas9 cleavage in human embryos. Cell 183(6):1650-1664 e15
94. Leibowitz ML et al (2021) Chromothripsis as an on-target consequence of CRISPR-Cas9 genome editing. Nat Genet 53(6):895–905
95. Cullot G et al (2019) CRISPR-Cas9 genome editing induces megabase-scale chromosomal truncations. Nat Commun 10(1):1136
96. Zeng J et al (2023) Gene editing without ex vivo culture evades genotoxicity in human hematopoietic stem cells. bioRxiv
97. Alavi AK, Abedi M, Galeon I, Reiner D, Smith SE, Wang L, Ramezi A, Rendo P, Walters MC (2021) Preliminary safety and efficacy results from Precizn-1: An ongoing phase 1/2 study on zinc finger nuclease-modified autologous CD34+ HSPCs for Sickle Cell Disease (SCD). Blood:2930
98. Alavi AA, Parikh S, Boismenu R, Chen M, Hsu BL, Cockroft BM, Galeon I, Rendo P, Walters MC (2022) Interim safety and efficacy results from a phase 1/2 study of zinc finger nuclease-modified autologous hematopoietic stem cells for sickle cell disease (PRECIZN-1). Blood 140(Supplement 1): 4907–4909
99. Gilman JG et al (1988) Distal CCAAT box deletion in the A gamma globin gene of two black adolescents with elevated fetal A gamma globin. Nucleic Acids Res 16(22):10635–10642
100. Hanna RF, McKinney C, Pineiro L, Mapara M, Dalal J, Chang K, Jaskolka M, Kim K, Farrington DL, Wally M, Mei B, Lawal A, Afonja OO, Walters MC (2023) AsCas12a gene editing of HBG1/2 promoters with EDIT-301 results in rapid and sustained normalization of hemoglobin and increased fetal hemoglobin in patients with severe sickle cell disease and transfusion-dependent beta-thalassemia. Blood **142**(Supplement 1):4996
101. Komor AC et al (2016) Programmable editing of a target base in genomic DNA without double-stranded DNA cleavage. Nature 533(7603):420–424
102. Gaudelli NM et al (2017) Programmable base editing of A*T to G*C in genomic DNA without DNA cleavage. Nature 551(7681):464–471
103. Anzalone AV, Koblan LW, Liu DR (2020) Genome editing with CRISPR-Cas nucleases, base editors, transposases and prime editors. Nat Biotechnol 38(7):824–844
104. Zeng J et al (2020) Therapeutic base editing of human hematopoietic stem cells. Nat Med 26(4):535–541
105. Liao J et al (2023) Therapeutic adenine base editing of human hematopoietic stem cells. Nat Commun 14(1):207
106. Han Y et al (2022) CRISPR/Cas9-based multiplex genome editing of BCL11A and HBG efficiently induces fetal hemoglobin expression. Eur J Pharmacol 918:174788
107. Breda L et al (2023) In vivo hematopoietic stem cell modification by mRNA delivery. Science 381(6656):436–443
108. Shi D, Toyonaga S, Anderson DG (2023) In vivo RNA delivery to hematopoietic stem and progenitor cells via targeted lipid nanoparticles. Nano Lett 23(7):2938–2944
109. Shen F et al (2022) A cell-permeant nanobody-based degrader that induces fetal hemoglobin. ACS Cent Sci 8(12):1695–1703
110. Yin M et al (2023) Evolution of nanobodies specific for BCL11A. Proc Natl Acad Sci USA 120(3):e2218959120

Erythroid Krüppel-Like Factor (KLF1): A Surprisingly Versatile Regulator of Erythroid Differentiation

10

James J. Bieker and Sjaak Philipsen

Abstract

Erythroid Krüppel-like factor (KLF1), first discovered in 1992, is an erythroid-restricted transcription factor (TF) that is essential for terminal differentiation of erythroid progenitors. At face value, KLF1 is a rather inconspicuous member of the 26-strong SP/KLF TF family. However, 30 years of research have revealed that KLF1 is a jack of all trades in the molecular control of erythropoiesis. Initially described as a one-trick pony required for high-level transcription of the adult *HBB* gene, we now know that it orchestrates the entire erythroid differentiation program. It does so not only as an activator but also as a repressor. In addition, KLF1 was the first TF shown to be directly involved in enhancer/promoter loop formation. *KLF1* variants underlie a wide range of erythroid phenotypes in the human population, varying from very mild conditions such as hereditary persistence of fetal hemoglobin and the In(Lu) blood type in the case of haploinsufficiency, to much more serious nonspherocytic hemolytic anemias in the case of compound heterozygosity, to dominant congenital dyserythropoietic anemia type IV invariably caused by a de novo variant in a highly conserved amino acid in the KLF1 DNA-binding domain. In this chapter, we present an overview of the past and present of KLF1 research and discuss the significance of human *KLF1* variants.

Keywords

EKLF · KLF1 · Transcription factor · Erythropoiesis · Hemoglobin switching · Congenital dyserythropoietic anemia type IV · Nonspherocytic hemolytic anemia

10.1 Initial Discovery and Characterization

Using a subtractive hybridization procedure which enriched for genes differentially expressed between murine erythroleukemia (MEL) cells and the monocyte-macrophage J774 cell line, a cDNA encoding an erythroid-specific transcription factor (TF) was cloned [1]. If translation initiated at the first methionine codon, the cDNA had the capacity to encode a 376 amino acid protein. At amino acid 19, a methionine codon urrounded by a good match to the Kozak

J. J. Bieker (✉)
Department of Cell, Developmental, and Regenerative Biology, Icahn School of Medicine at Mount Sinai, New York, NY, USA
e-mail: james.bieker@mssm.edu

S. Philipsen (✉)
Department of Cell Biology, Erasmus MC, Rotterdam, The Netherlands
e-mail: j.philipsen@erasmusmc.nl

T. Borggrefe, B. D. Giaimo (eds.), *Transcription factors in blood cell development*, Advances in Experimental Medicine and Biology 1459, https://doi.org/10.1007/978-3-031-62731-6_10

consensus sequence for translation initiation (5′-GCCRCC**AUG**G-3′) was observed. It was therefore assumed that the major protein produced would be 358 amino acids in length (37,755 Da). The protein consisted of a proline-rich but otherwise rather featureless N-terminal domain and three Cys2-His2 type zinc fingers (ZFs) at the C-terminus. The ZFs were similar to those first discovered in the general transcription factor (GTF) TFIIIA [2], suggesting that this region was likely to function as a DNA-binding domain (DBD). Further analysis revealed that the ZFs were most closely related to those of mammalian TF SP1 (specificity protein 1), EGR2 (early growth response 2) and WT1 (Wilms tumor transcription factor 1), and of the *Drosophila melanogaster* gap gene Krüppel. On the basis of this strong homology, the newly discovered protein was called erythroid Krüppel-like factor (EKLF). Expression analysis of mouse tissues and a panel of cell lines showed that EKLF mRNA expression is largely restricted to erythroid cells. Even to date, EKLF remains the most erythroid-restricted TF that we know of, although—as we will see later—it is not exclusively erythroid-specific. Based on the arrangement of the three ZFs, which is most akin to that of SP1, a consensus DNA binding site 5′-CCN CNC CCN-3′ was predicted for EKLF. This predicted consensus sequence has been refined by genome-wide analyses (chromatin immunoprecipitation followed by sequencing, ChIP-Seq), which yielded 5′-CCM CRC CCH-3′ (where "M" is A or C; "R" is G or A; "H" is A or T or C) as consensus [3]. This sequence is a common occurrence in the promoters and distal regulatory elements of erythroid-specific genes [4]. For instance, the "CAC" site (5′-CCA CAC CCT-3′) is located at position −90 with respect to the transcription start site (TSS) of the *HBB* (hemoglobin subunit beta) gene. In vitro, "CAC" site variants that give rise to rare forms of β-thalassemia (5′-CCA CA**T** CCT-3′, 5′-CCA CAC **G**CT-3′, or 5′-CCA CAC C**G**T-3′) competed binding of EKLF to the wildtype "CAC" site at an approximately 40- to 100-fold lower level than observed with the wildtype sequence [5]. Furthermore, in transient transfection studies, promoters containing these "CAC" site variants were not activated by EKLF, while a construct containing the wild-type "CAC" site was strongly activated in an EKLF-dependent manner [5, 6]. Collectively, these experiments provided the first evidence for the importance of EKLF as a regulator of transcription in erythroid cells and for its identity as the long-sought factor critical for human fetal to adult β-like globin switching [6]. Given that a number of closely related TFs are co-expressed in erythroid cells, we will first discuss EKLF in the context of the SP/KLF TF family.

10.2 The SP/KLF TF Family: What's in a Name

The similarity of the EKLF DBD to that of SP1 was already noticed in the paper describing the initial cloning and characterization of EKLF [1]. Soon thereafter, an expanding set of related TFs was identified. Firstly, three paralogues of SP1 were identified [7–9]. These are now termed SP2, SP3, and SP4. SP3 was initially also known as SPR-2 (SP1-related protein 2) and SP4 as SPR-1 (SP1-related protein 1) or HF1B (named after a 28-base-pair element in the rat cardiac myosin light-chain-2 gene) [7, 9]. The *SP1–4* genes are closely linked to the genes encoding SP5–9, usually in pairwise combinations. The *SP* genes are linked to the *HOX* clusters, e.g., in humans *SP1/SP7* to *HOXC* (chr12q13.13), *SP2/SP6* to *HOXB* (chr17q21.32), *SP3/SP5/SP9* to *HOXD* (chr2q31.1), and *SP4/SP8* to *HOXA* (chr7p15). Another distinguishing feature of SP factors is the presence of a buttonhead (BTD) box (CXCPXC) 14 to 22 amino acid N-terminal to the ZF domain. The BTD box was discovered in the *Drosophila melanogaster* SP factor Buttonhead [10], it is highly conserved and believed to modulate transcriptional activity [11]. Similarly, an expanding series of factors related to EKLF was described. Many of these were named after the tissue in which they were expressed, e.g., LKLF in the lung [12] and GKLF in the gut [13], or after specific properties, e.g., BKLF due to the abundance of basic amino acid residues present in this protein [14]. However, this unstructured approach led in some cases to multiple names for the same factors and even outright misnomers. A further disadvantage was that the name coined first often didn't reflect major biological functions adequately. For instance, LKLF was found to be required for the quiescent state and viability of single-positive (SP) T cells [15], and GKLF is now best known as one of the four "Yamanaka" factors used to derive induced pluripotent stem cells from somatic cells [16]. Following a proposal from the HUGO Gene Nomenclature Committee, the first nomenclature for KLF factors was introduced by Turner and Crossley [17]. The nomenclature of the entire SP/KLF family was proposed by Suske et al. [18]. The presence of the BTD box was used to define the SP branch of the family (SP1-9). The KLF factors formed the other major branch. Since EKLF was the first KLF factor discovered, it was rebranded KLF1. The mammalian genome contains 17 KLF factors (KLF1-17). KLF18 is a pseudogene located in close proximity to KLF17 (chr1p34.1) [19]; there are 12 additional KLF pseudogenes annotated in the database of the HUGO Gene Nomenclature Committee (March 2023): KLF2P1-4; KLF3P1-2; KLF4P1; KLF7P1; KLF8P1, and KLF17P1-2. Of note, the current approved symbols for LKLF, BKLF, and GKLF are KLF2, KLF3, and KLF4, respectively.

10.3 Genomic Location of Human *KLF1*

Together with *RAD23A*, the human *KLF1* gene is located in a gene-dense area on chr19p13.13 [20], closely linked to *KLF2* (chr19p13.11) and *KLF16* (chr19p13.3). This region is syntenic to human chr9q31.2, which houses *RAD23B* and *KLF4*; this area is linked to *KLF9* (chr9q13). This suggests a common evolutionary origin of the *RAD23-KLF* loci and generation of new *KLF* genes through gene duplication events. Supporting this notion, phylogenetic analyses revealed that *KLF1*, *KLF2*, *KLF4*, and *KLF17* and *KLF9*, *KLF13*, *KLF14*, and *KLF16* form distinct subgroups within the KLF family [21]. Notably, the *RAD23A-KLF1* locus contains several genes with roles in erythropoiesis such as *PRDX2* (peroxiredoxin 2, an antioxidant enzyme that protects red blood cells from oxidative stress [22]), *DNASE2* (mediates the breakdown of DNA from expelled nuclei during erythropoiesis [23]), *NFIX* (a transcriptional regulator of globin expression [24]), and *LYL1* (lymphoblastic leukemia 1, a transcriptional regulator redundant with TAL1 (T-cell acute lymphocytic leukemia 1) in erythropoiesis [25]).

10.4 Human *KLF1* Gene

The human *KLF1* gene contains three exons (Fig. 10.1; [26]). Exon 1 harbors the translation initiation codon and encodes the first 29 amino acids; exon 2 encodes 275 amino acids encompassing the largest part of the N-terminal domain and ZF 1, and exon 3 encodes ZFs 2 and 3. Since there is only a single amino acid C-terminal to the last amino acid of ZF 3, exon 3 encodes 58 amino acids. Unlike mouse KLF1, there is no potential alternative translation initiation codon in exon 1 in human KLF1; the predicted human protein therefore contains 362 amino acid residues. Despite co-expression of at least 12 SP/KLF members in erythroid cells (Table 10.1), KLF1 is a highly specific master regulator of red blood cell development. As yet, it remains poorly understood why these SP/KLF factors, with very similar DNA binding specificities, are unable to compensate for the activity of KLF1 in the erythroid lineage.

Likely, posttranslational modifications (PTMs) in conjunction with erythroid-specific protein-protein interactions have an important role in this process. Early observations indicated closely spaced GATA/CACC boxes as an erythroid module in promoters and distal enhancers [3, 4, 29]. This module is often accompanied by an Ephrussi (E) box 5′-CANNTG-3′ (5′-CAGMTG) which is recognized by a family of heterodimeric TFs. In erythroid cells, TAL1/TCF3 (transcription factor 3) heterodimers and GATA1 form a pentameric complex containing non-DNA binding cofactors LDB1 (LIM domain-

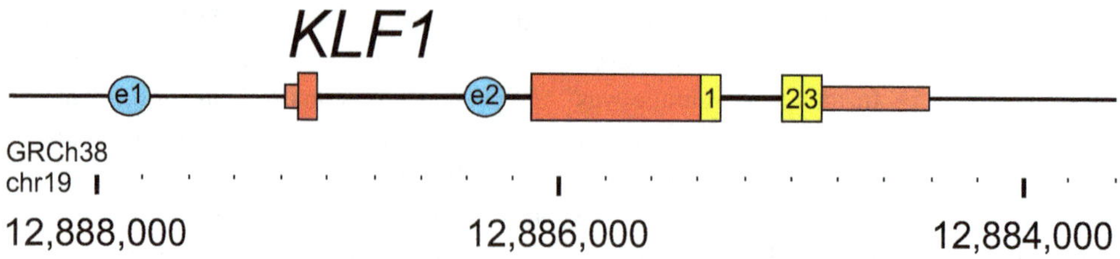

Element	GRCh38 chr19 start	GRCh38 chr19 end	Length (nt)	Description	Domain	Residue
e1	12,887,729	12,887,578	152	upstream enhancer		
exon 1	12,887,201	12,887,141	61	5'UTR		
exon 1	12,887,142	12,887,054	87	ORF		1-29
					TAD1	22-29
intron 1	12,887,053	12,886,143	911			
e2	12,886,464	12,886,246	219	intronic enhancer		
exon 2	12,886,142	12,885,317	826	ORF		30-304
					TAD1	30-47
					TAD2	59-85
					ZF1	264-304
intron 2	12,885,316	12,885,061	256			
exon 3	12,885,060	12,884,888	173	ORF		305-362
					ZF2	305-338
					ZF3	339-361
exon 3	12,884,889	12,8844,22	466	3'UTR		
mRNA			1613			

Fig. 10.1 The human *KLF1* gene. Untranslated regions (UTRs) are in light red; regions coding for the N-terminus are in dark red; regions coding for the zinc fingers are in yellow. Enhancer regions (e) are in blue. *UTR* untranslated region, *ORF* open reading frame, *TAD* transactivation domain, *ZF* zinc finger

Table 10.1 Expression of SP/KLF family members in cultured human erythroid progenitor cells

ENSEMBL ID	Symbol	HEPs Mean ($n = 3$)	stdev	HUDEP1 mean ($n = 3$)	stdev	HUDEP2 mean ($n = 3$)	stdev
ENSG00000185591	SP1	1438	458	3510	592	3409	237
ENSG00000167182	SP2	421	213	732	147	579	28
ENSG00000172845	SP3	741	548	2040	358	2015	17
ENSG00000105866	SP4	130	113	440	84	655	56
ENSG00000105610	KLF1	7592	2776	12,853	2077	7652	1506
ENSG00000109787	KLF3	1864	1331	6023	955	4346	162
ENSG00000067082	KLF6	563	369	846	92	1626	56
ENSG00000119138	KLF9	404	107	667	74	179	12
ENSG00000155090	KLF10	241	234	1249	101	226	15
ENSG00000172059	KLF11	194	90	779	162	707	30
ENSG00000169926	KLF13	4319	1823	7374	1186	3071	217
ENSG00000163884	KLF15	119	46	585	144	74	13
ENSG00000129911	KLF16	841	263	2721	378	1144	163

RNA-Seq data from [27] (primary cells) and [28] (HUDEP cell lines), expressed as reads per kilobase million (RPKM)
HEP human erythroid progenitors

binding protein 1), LMO2 (LIM-domain-only 2), and CBFA2T3 (CBFA2/RUNX1 partner transcriptional corepressor 3) [30]. The observation that the *KLF1* gene is bound by GATA1, TAL1, and LDB1 indicates that it is regulated by this multifactorial erythroid co-regulatory module (Fig. 10.2). KLF1 control elements thus reside within a short (~1 kB) region adjacent to the TSS containing erythroid-specific hypersensitive upstream enhancer elements that exhibit cell-type specificity and are under developmental control during hematopoietic differentiation [33, 34]. These bind general and tissue-restricted regulators and respond to bone morphogenetic protein 4 (BMP4) signaling [35–38]. It should also be noted that an enhancer element in intron 1 plays a critical role in establishing KLF1 maximal expression level [33, 35]. Not surprisingly, the DNA methylation (5-methylcytosine (5mC)) and hydroxymethylation (5-hydromethylcytosine (5hmC)) modification status of these elements are important for KLF1 expression control during hematopoiesis [39–41]. Additional factors have recently been implicated in *KLF1* regulation at these sites (DEK (DEK proto-oncogene), DDX5 (DEAD-box helicase 5), ETV6 (ETS variant transcription factor 6), RIOK2 (RIO kinase 2) [33, 34, 42, 43]), and their co-localization with the erythroid module suggests formation of an enhanceosome, particularly at the upstream enhancer.

10.5 KLF1 Domains

10.5.1 N-Terminal Transactivation Domain

At the N-terminus, KLF1 contains a transcription activation domain (TAD) which has some homology to the TAD of the TP53 (tumor protein p53) TF [44]. The KLF1 TAD can be divided into two subdomains, TAD1 (amino acid 1 to 40) and TAD2 (amino acid 50 to 90). TAD1 contains a phosphorylation site at T23 and a motive interacting with ubiquitin (MIU, amino acid 23–36) which forms non-covalent interactions with ubiquitin [45, 46]. Consistent with these observations,

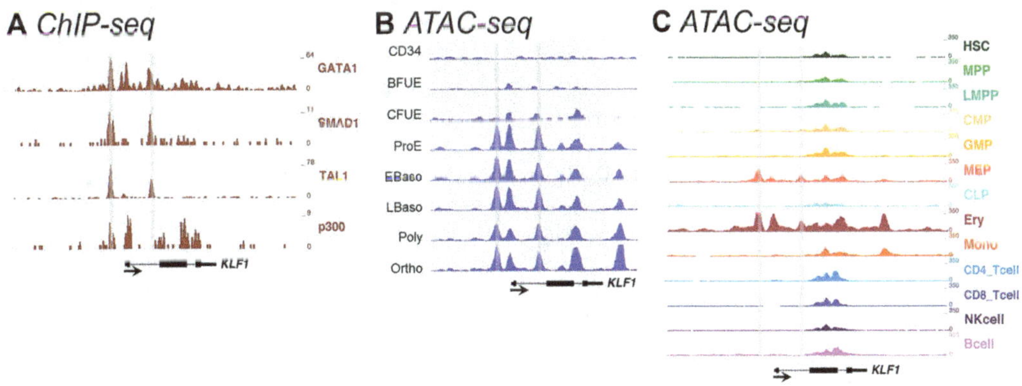

Fig. 10.2 The *KLF1* gene is bound by erythroid transcription factors (TFs) and exhibits a cell- and differentiation-restricted open chromatin domain. (**a**) In vivo GATA1, SMAD1, TAL1, and EP300/KAT3B binding identified from ChIP-Seq analyses of erythroid cells are shown based on genome browser data aligned at the human *KLF1* genomic region. (**b**) Onset of chromatin accessibility as monitored by ATAC-Seq analyses of primary human cells during erythroid differentiation from sorted adult human CD34+ cells [31]. (**c**) Chromatin accessibility as monitored by ATAC-Seq analyses of primary human adult human hematopoietic subpopulations [32]. Shaded bars show the locations of the enhancers upstream of the gene and at intron 1. (Figure modified from [33] via a Creative Commons License http://creativecommons.org/licenses/by/4.0/). BFUE burst-forming unit erythroid, CFUE colony-forming unit erythroid, ProE pro-erythroblast, EBaso early basophilic erythroblast, LBaso late basophilic erythroblast, Poly polychromatic erythroblast, Ortho orthochromatic erythroblast, HSC hematopoietic stem cell, MPP multipotent progenitor, LMPP lympho-myeloid-primed progenitor, CMP common myeloid progenitor, GMP granulocyte/macrophage progenitor, MEP megakaryocyte/erythroid progenitor, CLP common lymphoid progenitor, Ery erythroid progenitor, Mono monocyte; NK natural killer

AlphaFold [47] predicts an alpha-helical structure covering the MIU in TAD1 (Fig. 10.3). TAD2, a conserved feature shared with KLF2, KLF4, KLF5, and KLF15, interacts with the P62 subunit of TFIIH [44] and with the CBP/KAT3A (CREB-binding protein/lysine acetyltransferase 3A) and EP300/KAT3B (E1A-binding protein p300/lysine acetyltransferase 3B) acetyl transferases [48–50]. These interactions are in line with the role of KLF1 as a transcriptional activator and promotor factor. Upon binding to Tfb1PH, the yeast homologue of TFIIH subunit P62, KLF1 TAD2 (amino acids 59–84) adopts an extended form that, although non-helical, is more ordered than in the absence of interaction [44]. Consistent with this, AlphaFold predicts that TAD2 lies within an unstructured domain of KLF1 (Fig. 10.3). This unstructured domain includes the SUMOylation site KLF1 p.K54. SUMO (small ubiquitin-like modifier) modification of this residue invokes interaction with the CHD4 (chromodomain helicase DNA-binding protein 4) subunit of the NuRD (nucleosome remodeling and deacetylase) complex, enabling KLF1 to act as a transcriptional repressor [51], possibly in the context of bipotential lineage decisions within the megakaryocyte/erythroid progenitor (MEP) [52].

10.5.2 The C-Terminal Zinc Finger (ZF) Domain

The DBD of KLF1 consists of three C-terminal ZFs. This ZF array is highly conserved between the members of the SP/KLF transcription family and between species. Deep evolutionary conservation is illustrated by the observation that the 81 amino acid DBD, composed of three Cys-Cys/His-His ZFs spaced by linker sequences of 7 amino acids, is present in SP/KLF homologues of the fruit fly *Drosophila melanogaster* and the roundworm *Caenorhabditis elegans* and is already evident in non-metazoans [21, 53]. The

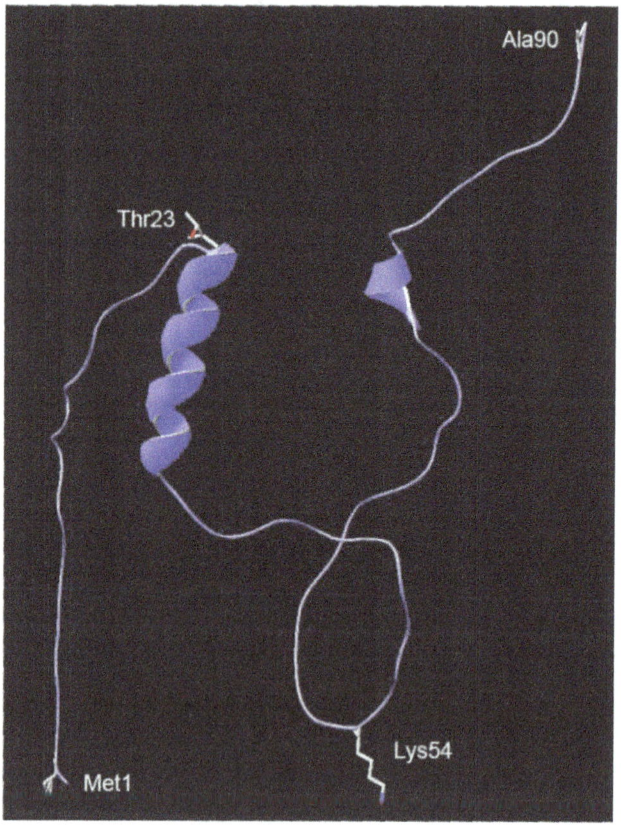

Fig. 10.3 AlphaFold prediction of the structure of the human KLF1 transactivation domain (TAD). Amino acids 1–90 of human KLF1 are displayed

level of conservation is remarkable; not only for the Cys-Cys/His-His motifs but also for the residues directly involved in recognition of the DNA binding site [18]. DNA binding specificity is achieved through base-specific contacts made by three residues which are part of a seven-residue α-helical fold that contacts the DNA. DNA binding occurs in a modular fashion, with each ZF recognizing three base pairs of the nine base pair consensus sequence 5′-CCM CRC CCH-3′. Base-specific interactions are made with the G-rich strand of the binding site, i.e., 5′-DGG GYG KGG-3′ (where "D" is A, G, or T; "Y" is C or T; "K" is G or T) leading to a "lock-and-key" model for binding site specificity (Fig. 10.4). In addition to DNA binding, the ZF domain interacts with an extensive set of proteins that are likely involved in its activation function and include TATA-box-binding protein-associated factor 9 (TAF9) [55], histone H3 [48], the histone H3.3 chaperone HIRA (histone cell cycle regulator homolog A) [56], and the SWI/SNF (switch/sucrose non-fermentable) chromatin remodeler complex [50, 57], the latter likely via SMARCA4 (SWI/SNF-related, matrix-associated, actin-dependent regulator of chromatin, subfamily a, member 4) [58]. The KLF1 DNA-binding domain on its own is capable of DNA binding and SWI/SNF recruitment, leading to chromatin opening as assayed by DNASEI sensitivity. Notably, this is not sufficient for full transcriptional activation for which the presence of the N-terminal domains is required [59].

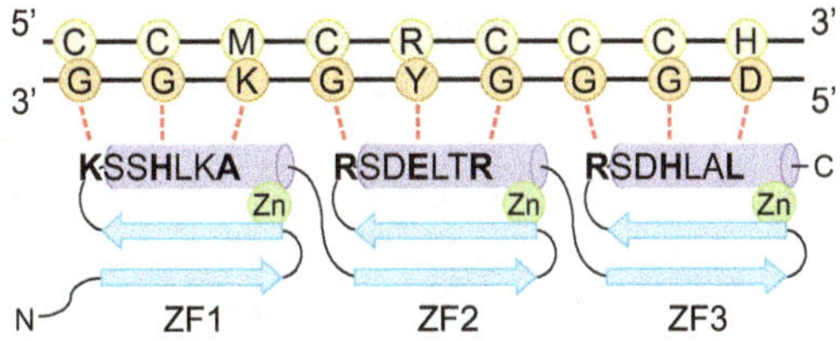

Fig. 10.4 Cartoon of KLF1 zinc finger (ZF) domain interacting with consensus DNA binding site. The modular structure of the KLF1 ZF domain is displayed. Amino acids directly involved in base recognition are indicated in bold. Arrow β-sheet, barrel α-helix, Zn zinc atom coordinated by Cys-Cys/His-His residues, ZF zinc finger. (Modified after [54])

10.6 Cellular Localization

KLF1 activity is required for terminal differentiation of erythroid cells. This involves a series of 4–5 rapid cell divisions during which cell size is reduced, the gene expression program is changed, and nuclear condensation occurs. After the final cell division, the nucleus is expelled and ingested by specialized macrophages. The DNA of the expelled nuclei is degraded by DNASEII which is expressed at high levels by the macrophages. The enucleated cells are called reticulocytes; these cells are released in the circulation where they undergo maturation to fully mature red blood cells. This process, which takes 7 days in humans, involves extensive remodeling of the proteome and degradation of all ribosomes and organelles. KLF1 controls virtually all aspects of terminal erythroid differentiation, including cell cycle regulation; red blood cell antigens; structural proteins; energy, iron, and heme metabolism; and hemoglobinization [27, 60–68]. Enucleation is dependent on KLF1, as erythroid cells stall at the orthochromatic stage in its absence, yielding an asymmetrically located nucleus that is not able to be extruded [69]. Expression of some transcriptional and posttranscriptional regulators such as KLF3 [70], BCL11A (BCL11 transcription factor A) [38, 60], ZBTB7A (zinc finger and BTB domain containing 7A) [71], and PUM1 (pumilio RNA-binding family member 1) [72] is also regulated by KLF1. This wide variety of cellular processes that are required at different stages during terminal differentiation indicates that KLF1 is a highly dynamic regulator of gene expression. Firstly, working in partnership with other factors such as GATA1 and the pentameric complex of GATA1/TAL1/LDB1/LMO2/CBFA2T3 (Fig. 10.2) provides an example of how this dynamism is achieved. Secondly, as we will see later, it is likely that posttranslational modifications are involved in modulation of KLF1 activity. Thirdly, in this context, it is of interest that KLF1 interacts with TET2 (tet methylcytosine dioxygenase 2), connecting KLF1 to DNA epigenetic status [73]. Finally, it has been reported that a substantial proportion of the KLF1 protein pool is located in the cytoplasm of erythroid progenitors [74–76]. A cytoplasmic protein FOE (friend of EKLF), encoded by *ANKRD17* (ankyrin repeat domain 17), was found to sequester KLF1 in the cytoplasm [77]. Transport of KLF1 from the cytoplasm to the nucleus is dependent on nuclear localization signals located adjacent to (aa 275–296 in Mouse; aa 261–282 in Human) and within the ZF DBD (aa 295–375 in Mouse; aa 281–361 in Human) [78, 79]. The nuclear localization signals bind importin-α and importin-β in vitro [78]. In heterokaryon assays, KLF1 is able to shuttle out of the nucleus, but this is quickly followed by rapid reentry [74]. A single region that acts as nuclear export signal (NES) could not be identified; however, the ZF region makes a major contribution. The most common nuclear protein export pathway is mediated by the chromosome region maintenance 1 (CRM1) protein [74]. Inhibition of CRM1 by treatment with leptomycin B inhibits KLF1 nuclear export modestly and only when high concentrations (200 nM) of leptomycin B are used, indicating that this is not a major pathway for KLF1 shuttling. The observation that between 50 and 80% of KLF1 protein resides in the cytoplasm in both embryonic and fetal/adult primary erythroid cells suggests that this functions as a reservoir to maintain KLF1 levels in the nucleus or that KLF1 might have—as yet unidentified—cytoplasmic functions.

10.7 Posttranslational Modifications (PTMs)

The potential role of PTMs in determining KLF1 cellular localization and transcriptional activity has been intensively investigated. This work was inspired by examples of other TFs such as TP53 which is polyubiquitinated by MDM2 (mouse double minute 2), an E3 ubiquitin ligase. Polyubiquitinated TP53 is degraded rapidly via the 26S proteasome system. Cellular stress, for instance, caused by double-strand DNA breaks (DSBs), induces phosphorylation of MDM2 and stabilization of TP53 [80]. Other TFs such as STAT (signal transducer and activator of transcription) proteins are under direct control of phosphorylation switches. In response to cytokine signaling, cytosolic STAT proteins are phosphorylated and undergo dimerization, which triggers active nuclear import [81]. It appears that KLF1 is not subjected to such dramatic on/off switching mechanisms. Rather, PTMs have a modulatory role in KLF1 activity.

10.7.1 SUMOylation

Attachment of a SUMO peptide to p.K74 in mouse KLF1 does not diminish its transcriptional activation potential but creates an interface for binding to CHD4 (also known as Mi2β), the largest subunit of the NuRD repressor complex. This unleashes the capability of KLF1 to repress differentiation of the megakaryocytic lineage from the common MEP [51].

A mouse model of SUMOylation-deficient KLF1 has recently been reported. These mice carry a KLF1 p.K74R variant introduced via homologous recombination in mouse embryonic stem cells (ESCs). Remarkably, homozygous animals ($Klf1^{p.K74R::p.K74R}$) display an expanded lifespan [82]. It was observed that the progress of immune cell-related aging, geriatric disorders, and chronic and malignant conditions is significantly mitigated in $Klf1^{p.K74R::p.K74R}$ mice compared to $Klf1^{WT::WT}$ littermates. Since transfer of $Klf1^{p.K74R::p.K74R}$ hematopoietic stem cells (HSCs) into $Klf1^{WT::WT}$ animals elicits antiaging and antitumor effects, it was concluded that these effects are conveyed via the hematopoietic system. The question remains what the critical target cells of KLF1 p.K74R are. Red blood cells are by far the most abundant cell type in mammals. A detailed molecular and proteomics analysis of the erythro/megakaryocytic system in $Klf1^{p.K74R::p.K74R}$ mice has unfortunately not yet been reported. For human KLF1, the homologous p.K54R variant is present in the database for SNPs (single-nucleotide polymorphisms) in the human genome (dbSNP, rs759212421, searched in July 2023). Given the extremely low frequency of the minor allele, the rare positive individuals carry this SNP most likely in heterozygous state. No phenotype associated with rs759212421 had been reported. Detailed characterization of the phenotype of *Klf1*$^{WT::p.K74R}$ mice could provide pointers to phenotypic consequences of KLF1 p.K54R in humans.

10.7.2 Ubiquitination

Treatment with 26S proteasome inhibitors leads to accumulation of KLF1 protein in MEL cells. KLF1 can be ubiquitinated in vivo; this is not dependent on any particular internal lysine residue. Two PEST (proline (P), glutamic acid (E), serine (S), and threonine (T)) sequences within the N-terminal TADs contribute to its destabilization [45]. It was observed that KLF1 TAD1 can form a non-covalent interaction with ubiquitin. Nuclear magnetic resonance studies demonstrated that TAD1 adopts a 14-residue alpha-helix that forms a recognition interface for ubiquitin [46]. In addition, G-protein pathway suppressor 2 (GPS2), a subunit of the nuclear receptor corepressor/silencing mediator of retinoic acid and thyroid hormone receptor corepressor complex, was found to interact with KLF1 (aa 191–230). This prevents proteasome-mediated degradation of KLF1, thereby increasing its transcriptional activity. Deletion of GPS2 in mice causes impaired erythropoiesis in the fetal liver and leads to severe anemia [83].

10.7.3 Phosphorylation

KLF1 can be phosphorylated at serine and threonine residues within its transactivation regions. Disruption of a protein kinase casein kinase II (CKII) site (p.T41Q42E43D44) located within mouse KLF1 TAD1 and conserved with other mammalian species, abolishes KLF1 transactivation activity. In vitro assays demonstrated that mouse KLF1 TAD1 is phosphorylated by CKII-alpha at p.T41. This residue is not phosphorylated in a CKII-site mutant (p.T41Q42**A**43**G**44). The transactivation capability of KLF1 is augmented by co-transfection of CKII-alpha [84]. PPM1B (protein phosphatase, Mg2+/Mn2+-dependent 1B) is a serine-threonine protein phosphatase that interacts with the PEST1 domain of KLF1. In reporter assays, PPM1B super-activates KLF1 at the *HBB* and *KLF3* promoters, this is dependent on intact PPM1B phosphatase activity. Conversely, depletion of PPM1B in human CD34$^+$ hematopoietic progenitor cells leads to a higher level of endogenous *HBB* gene activation after erythroid differentiation. PPM1B appears to have an indirect role in regulating KLF1 turnover via the KLF1 ZF domain [85]. An independent study showed that nuclear import of mouse KLF1 occurs during the proerythroblast to basophilic erythroblast transition. This involves activation of PKCθ (protein kinase C θ) by phosphorylation of PKCθ p.S676, which then phosphorylates KLF1 at p.S68 followed by SUMOylation at KLF1 p.K74. These KLF1 modifications modulate its interactions with the cytoplasmic ankyrin-repeat-protein FOE (encoded by *ANKRD17*) and importin-β1 (encoded by *KPNB1* (karyopherin subunit beta 1)), respectively [77]. Whether FOE and PKCθ have a similar role in human erythroid cells is currently not known, as the phosphorylation site p.S68 is not conserved in human KLF1 nor in any other mammalian species, which generally have a proline at the orthologous position (p.P50 in human KLF1).

10.7.4 Acetylation

In mouse KLF1, residues acetylated by CBP/KAT3A in vitro map to p.K288 and p.K302 in the KLF1 ZF domain. Although site-specific DNA binding by KLF1 is unaffected by the acetylation status of either of these lysines, K to R mutagenesis of p.K288, but not p.K302, decreases the ability of KLF1 to transactivate the *HBB* promoter in vivo and abolishes super-activation by co-expressed EP300/KAT3B or CBP/KAT3A. Acetylated KLF1 has a higher affinity for the SWI-SNF chromatin-remodeling complex and is a more potent transcriptional activator of chromatin-assembled templates in vitro [50], likely via SMARCA4 interactions [58]. Exogenous expression of KLF1 variants in a mouse KLF1-*null* erythroid line established that acetylation at p.K288 is required for occupancy by KLF1 and recruitment of CBP/KAT3A to the *Hbb* locus, modification of histone H3, opening of the chromatin structure, and transcription of the adult *Hbb* genes [48]. Furthermore, KLF1 interacts with SIN3A (SIN3 transcription regulator family member A) and histone deacetylase 1 (HDAC1) corepressors via its ZF domain [86]. K to R mutagenesis of p.K288 and p.K302 showed that the activation and repression activities of KLF1 can be uncoupled. Remarkably, an intact ZF structure is not required for interactions with SIN3A-HDAC1 and repression, and reversible KLF1-SIN3A interactions have a key role in regulating stage specificity of KLF1-mediated repression. KLF1 p.K302 has been identified as the key residue that is both a substrate for CBP/KAT3A acetylation and required for SIN3A interaction [87]. However, direct in vivo targets of KLF1 repression remain to be identified. Acetylated levels of KLF1 might be altered by its association with the NuRSERY (nuclear remodeling shuttle erythroid) complex that contains HDAC5, GATA1, and ERK/MEK1 (extracellular signal-regulated kinase and mitogen-activated protein kinase 1), which changes during erythroid maturation dependent on ERK phosphorylation as well as HDAC activity [88].

10.8 Molecular Functions

10.8.1 SP/KLF Factor Specificity

The DBD of KLF1 is a modular structure, with each ZF contributing to the recognition of three base pairs of the nine base pair consensus sequence 5′-DGG GYG KGG-3′ (Fig. 10.4). Given that in erythroid cells 12 other SP/KLF family members are co-expressed (Table 10.1) and that—with the exception of SP2—these factors are predicted to recognize very similar DNA motifs, this poses the question by which factor(s) gene expression is regulated in erythroid cells. A pioneering study showed that the function of such closely related proteins can be distinguished in vivo by matching point mutations in an enhancer element (5′HS3 of the *HBB* locus control region) with amino acid changes in the ZFs of SP1 and KLF1. Testing their activity in transgenic mice showed that KLF1, and not SP1, is a direct activator of 5′HS3 in erythroid cells. It was also found that KLF1 is required for hypersensitive site formation over the 5′HS3 element [89]. While this approach was successful, it would be a gargantuan task to extend this to the other 11 SP/KLF factors expressed in erythroid cells and impossible to address specificity for all in vivo binding sites. ChIP-Seq (Fig. 10.5) combined with CRISPR-mediated inactivation of the genes encoding these SP/KLF factors provides a more tractable approach to address these redundancy and specificity issues.

10.8.2 DNA Looping

The three-dimensional organization of a gene locus is important for its regulation, as was first demonstrated for the *Hbb* locus [91, 92]. When actively expressed, the *cis*-regulatory elements of the locus control region and the *Hbb* promoter are in close proximity in nuclear space, forming a compartment termed the active chromatin hub (ACH). Using knockout (KO) and transgenic rescue experiments, it was found that KLF1 is required for the three-dimensional organization of a gene locus [93]. For full ACH formation, KLF1 acts in concert with other factors including the GATA1/TAL1/LDB1/LMO2/CBFA2T3 pentameric complex and ZFPM1 (zinc finger protein, friend of GATA family member 1) [94–96]. Expanding on this idea, additional studies showed that KLF1 has a critical role in recruiting its target genes into nuclear subdomains that provide effective regulatory networks within a small number of transcription factories [75]. The absence of KLF1 leads to the absence of intra- and interchromosomal clustering of its downstream regulated genes at these sites.

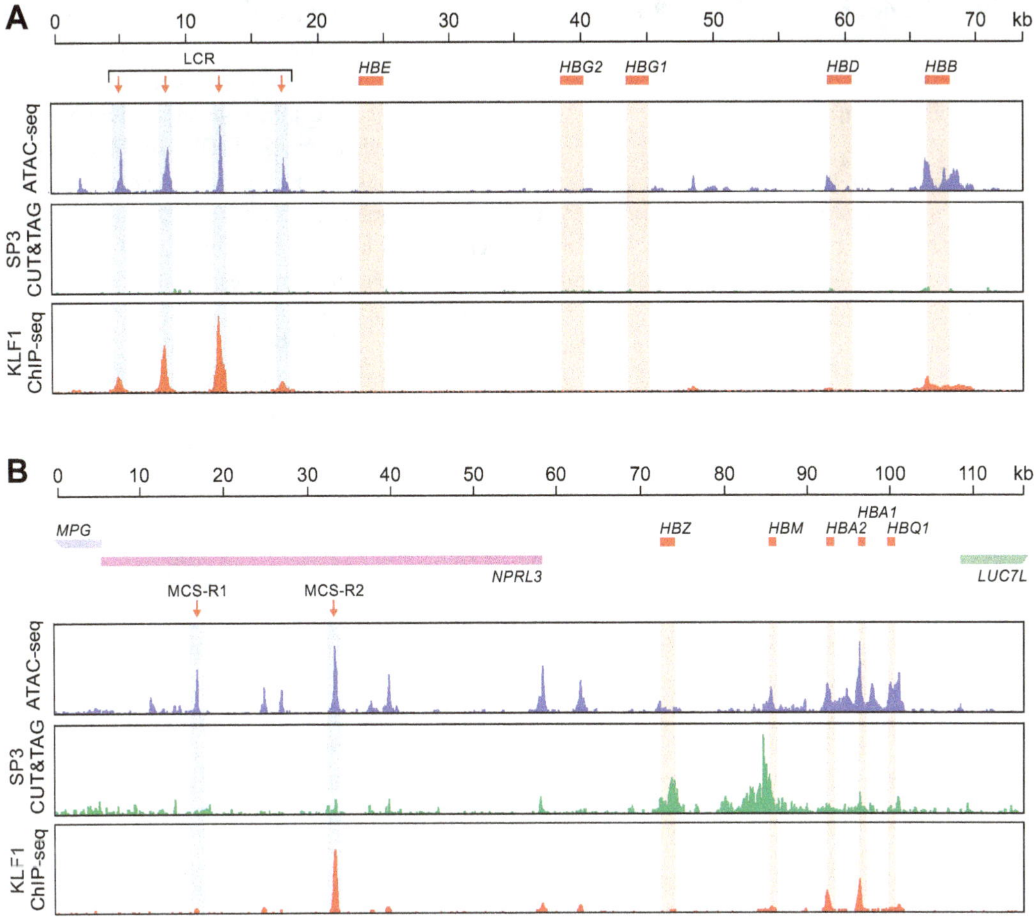

Fig. 10.5 Binding of KLF1 and SP3 to the *HBB* and *HBA* loci. (**a**) KLF1 ChIP-Seq and SP3 CUT&TAG on the *HBB* locus in HUDEP2 cells. (**b**) KLF1 ChIP-Seq and SP3 CUT&TAG on the *HBA* locus in HUDEP2 cells. The ATAC-Seq tracks are included to show areas of high chromatin accessibility. Distal erythroid regulatory elements are indicated by arrows. LCR locus control region, MCS multispecies-conserved sequence. (Data are from Wessels et al. [28] (ATAC-Seq); Cheng et al. [90] (KLF1 ChIP-Seq), and Gillemans and Philipsen, unpublished results (SP3 CUT&TAG))

10.8.3 RNA Polymerase II (RNAPII) Pause Release

KLF1's role in coordinating transcriptional onset of its target genes has been evident from its association with key RNA polymerase-associated players such as TAF9 and TFIIH/P62 and by its ability to integrate signals from its interaction with epigenetic regulators such as SWI/SNF, CBP/KAT3A, and EP300/KAT3B (reviewed in [97]). Recent studies have expanded on mechanisms of KLF1 transcriptional control by showing that KLF1 can act as a pioneer factor, and in this process alters localization of CBP/KAT3A and associated histones such as H3K27ac [98]. These studies were aided by in vivo analyses of erythroid cells from the neonatal anemia (Nan) mutant mouse (described later) and by the availability of a KLF1 antibody that worked for ChIP-Seq. In addition to monitoring RNA polymerase II (RNAPII) phosphorylation status, this combination enabled identification of RNAPII pause-release as a nodal point at which KLF1 fine-tunes the regulation of a subset of downstream targets via its binding to promoters and intergenic/intronic enhancers [98].

10.8.4 Development of Red Blood Cells: KLF1 Activity in Primitive and Definitive Erythropoiesis

The first embryonic red blood cells are made by the yolk sac in a process termed primitive erythropoiesis. During the embryonic/fetal transition, the yolk sac and placenta generate hematopoietic progenitors that migrate to the fetal liver, where they produce the first wave of definitive red blood cells. Then, the aorta/gonads/mesonephros region generates hematopoietic stem cells which migrate to the fetal liver where they produce the second wave of definitive red blood cells. Around birth, the fetal liver hematopoietic stem cells colonize the bone marrow, where they provide lifelong production of definitive red blood cells [99].

Mouse KLF1 mRNA is first expressed at the neural plate stage (embryonic day (E) 7.5) within primitive erythroid cells at the beginning of blood island formation in the yolk sac [100], studies supported by more recent single cell transcriptomic analyses [101, 102]. KLF1 is then expressed by E9 in the hepatic primordia and remains high in the fetal liver, which becomes the sole source of KLF1 mRNA in the E14.5 fetus. Concomitantly with KLF1 mRNA, KLF1 protein is also expressed in primitive erythroid cells and in the fetal liver. KLF1 expression in the adult spleen is localized to the red pulp [100]. Disruption of the *KLF1* gene results in fatal anemia caused by severely reduced expression of the adult *Hbb* genes, while the embryonic εy and βh1 globin genes are expressed normally [103–106]. Thus, KLF1 was thought to be a stage-specific factor acting through the CACC box in the *Hbb* promoter, even though it is already present in embryonic red cells. To investigate whether KLF1 can function as a transcriptional activator in primitive erythropoiesis, expression of erythroid *LacZ* reporter transgenes was investigated. This showed that embryonic expression of reporters driven by 5′HS3 of the human *HBB* LCR requires KLF1 and established that KLF1 can act as a transcriptional activator in primitive erythropoiesis [107]. Subsequent studies revealed that KLF1 *null* primitive red blood cells have membrane defects most likely due to severely reduced expression of erythroid cytoskeleton/membrane proteins including DMTN (dematin), ICAM4 (intercellular adhesion molecule 4, Landsteiner-Wiener blood group), RHD (Rhesus blood group, D antigen), and TFRC (transferrin receptor) [61, 63]. Of note, partially redundant functions between KLF1 and KLF2 in primitive erythroid cells have been firmly established [108–111], providing an explanation for the relatively normal development of these cells in absence of KLF1. In human development, KLF1 is expressed in the blood clusters (likely yolk sac) of gastrulating embryos [112] and at high levels in the fetal liver [113].

10.8.5 Functions in Other Cell Types: Macrophages and Megakaryocytes

KLF1 is the most erythroid-restricted TF known. Yet, KLF1 expression and functions have been reported in other cell types. Regulation of *IL12B* (interleukin 12b) transcription by KLF1 was observed in human macrophages [114]. Erythroblastic islands are composed of a central macrophage surrounded by developing erythroblasts [115]. In mouse fetal liver, DNASE II-alpha is expressed in the central macrophages, where it digests DNA of the extruded nuclei of red blood cells [23]. KLF1 is expressed in central macrophages and binds to the promoter of the *Dnase2a* gene [116, 117]. Note that similar to the human *DNASE2* gene, *Dnase2a* is abutting *Klf1* in the mouse genome. In KLF1 *null* central macrophages, expression of *Dnase2a* is severely downregulated [116]. Consistent with these observations, KLF1 is expressed in the erythromyeloid progenitors, and regulates genes such as *Dnase2a* and *Vcam1* (vascular cell adhesion molecule 1), important for erythroblastic island integrity, within the central macrophages derived from these progenitors [118]. Inducible activation of KLF1 invokes a central macrophage phenotype on macrophages derived from human-induced pluripotent stem cells [119]. A transient population of KLF1-dependent fetal liver central macrophages has been identified using single-cell RNA-Seq analysis [120]. These studies reveal the extensive heterogeneity already present in fetal liver macrophages during development [120] and also define the KLF1-regulated gene signature that identifies potential cell surface markers unique to this subpopulation [117].

KLF1 is highly expressed in the MEPs [52], where its level remains elevated as MEPs differentiate toward the erythroid lineage. However, it is dramatically downregulated as MEPs differentiate toward the megakaryocytic lineage [52]. Gain- and loss-of-function studies show that the KLF1 expression level impacts the bipotential lineage decision made by the MEP [52, 121–123]; specifically, KLF1 inhibits the formation of megakaryocyte colonies while at the same time stimulating erythroid differentiation (reviewed in [124, 125]). As mentioned earlier, the SUMOylation status of KLF1 likely has a role [51], but in addition cross-antagonism between KLF1 and the ETS-family transcription factor FLI1 (friend leukemia integration 1), either directly by protein-protein association or by transcriptional cross-repression, may account for this effect [52, 126]. Support for this idea has come from quantitative single-cell mass cytometry proteomics of differentiating hematopoietic cells, which precisely determined that KLF1/FLI1 proteins are co-expressed in progenitors, but gradually change in stoichiometry, with a KLF1 protein bias as differentiation progresses towards the erythroid trajectory [127], thus altering the gene regulatory network [128].

10.8.6 Lack of Oncogenic Activity

Genomic sequencing of the *KLF1* gene in cell lines, erythroid neoplasms, dysplastic disorders, and leukemias was performed to reveal a potential role in red cell malignancies. In addition, a number of public oncology databases was queried. These analyses yielded only commonly annotated SNPs in the *KLF1* gene. It was concluded that if *KLF1* variants associated with erythroid malignancies exist, they are exceedingly rare [129]. Immunohistological analyses of KLF1 protein in acute myeloid leukemia (AML) samples showed that higher KLF1 levels correlate with cells containing abnormal chromosomes [130]. Surprisingly, high KLF1 expression was not limited to AML subtypes displaying erythroid/megakaryoblastic differentiation, although this may be as a result of aberrant GATA2 expression in B progenitor acute lymphoblastic leukemia (B-ALL) that collaterally increases KLF1 levels [131] as such an effect is proposed to occur normally during development [35]. Expression of KLF1 was associated with poor survival. Thus, while not displaying direct oncogenic activity, KLF1 levels could be a prognostic marker for risk stratification in patients with AML [130].

10.9 Inactivation of the *Klf1* Gene in Mice

The phenotype of *Klf1* gene KO was first described in two independent studies [103, 105]. *Klf1* KO fetuses die in utero around E14.5 due to severe anemia. It was found that expression of the adult *Hbb* globin genes *Hbb-bs* and *Hbb-bt* is severely reduced while expression of the adult *Hba* globin genes *Hba-a1* and *Hba-a2* and embryonic hemoglobins appeared unaffected. Erythroid progenitor cells were present in normal numbers, indicating that KLF1 is not required for erythroid commitment. The phenotype was reminiscent of severe β thalassemia. Unlike humans, mice do not have a fetal β-like globin gene. Attempts were therefore made to rescue *Klf1* KO fetuses by transgenes driving expression of human γ-globin [106, 132]. While this restored the α-like/β-like chain imbalance, the lethal anemia was not rescued indicating that expression of other essential erythroid genes is dependent on KLF1. This was confirmed by genome-wide expression analyses using microarrays and RNA-Seq [61, 63–67, 133–135]. These studies revealed that KLF1 regulates virtually every aspect of terminal erythroid differentiation including the cell cycle, energy metabolism, heme synthesis, membrane proteins, and hemoglobin synthesis. KLF1 regulation of cell cycle inhibitor proteins is critical for enucleation, as in its absence erythroid cells fail to extrude the nucleus; rather, these are stalled at the orthochromatic erythroblast stage [69]. These studies, coupled to those that implicated KLF1 function in promoting early S-phase progression, suggest a dual role at different stages of terminal erythroid differentiation [68]. In addition, it was discovered that KLF1 *null* primitive erythroid cells have membrane and maturation defects [61, 63]. As we will see in the next section, these mouse studies have been instructive for phenotypic analysis of *KLF1* variants that were later discovered in humans.

10.10 Human *KLF1* Variants and Red Blood Cell Disorders/Clinical Significance

Analysis of UK Biobank exomes [136, 137] indicates that rare variants of KLF1 are associated with changes in red blood cell parameters analyzed by complete blood count (Fig. 10.6). Increases in mean corpuscular volume (MCV), red blood cell distribution width (RDW), mean sphered corpuscular volume (MSCV), mean corpuscular hemoglobin (MCH), red blood cell count, and hemoglobin (Hb) concentration are strongly associated with KLF1 variants. By comparing the transcriptomes of erythroblasts derived from individuals with the rare In(Lu) type of Lu(a-b-) blood group to those derived of healthy controls, it was found that many erythroid genes display reduced expression in the In(Lu) samples. This suggested that the primary defect is in an erythroid TF and sequence analysis revealed variants in the promoter or coding sequence of the *KLF1* gene in 21 of 24 individuals with the In(Lu) phenotype [138]. The variants identified included nonsense, frameshift, and missense mutations, the latter affecting conserved residues in the ZF domains. A promoter variant affecting a putative binding site for the erythroid transcription factor GATA1 was also observed. Collectively, it was concluded that the In(Lu) phenotype is caused by reduced levels of KLF1 [138]. This concept was further supported by studies showing that quantitative knockdown of KLF1 transcription is precisely tracked by a similar quantitative reduction in expression of a subset of its erythroid targets [33]. A subsequent study of a large kindred from the island of Malta found that family members with a nonsense variant (p.K288X) in one *KLF1* allele displayed hereditary persistence of fetal hemoglobin (HPFH) [60]. This benign condition, in which the switch from fetal γ-globin to adult β-globin is incomplete, is clinically relevant since reactivation of γ-globin alleviates the symptoms of patients with β-hemoglobinopathies such as sickle cell disease and β-thalassemia. It was discovered that KLF1 is a key activator of the *BCL11A* gene which encodes the major transcriptional repressor of the γ-globin genes *HBG1* and *HBG2* in adult erythroid cells [38, 60]. Currently a large number of *KLF1* variants, associated with a broad variety of human erythroid conditions, has been reported [139]. It was noted that missense variants were largely confined to the ZF domains, highlighting the importance of DNA binding for KLF1 activity. In addition to In(Lu), reduced expression of Kell, Duffy, Kidd, RhD, RhAG (rhesus-associated glycoprotein), Scianna, and LW (Landsteiner-Wiener) blood group antigens has been observed [61, 63, 67, 138, 140, 141]. Increased δ-globin (a minor adult β-like globin) expression [142–150] was also reported; this is clinically important since raised δ-globin is used as a diagnostic marker for detection of carriers of β-thalassemia. Dysregulation of the iron/heme system was suggested by increased zinc protoporphyrin levels; the heme precursor protoporphyrin will bind zinc when insufficient iron is available for binding [151]. Currently (July 2023), over 120 different *KLF1* variants including 60 missense variants affecting 43 KLF1 residues have been associated with human red blood cell phenotypes. Notably, 36 of the 43 variant residues are located in the DNA binding domain [139] indicating the importance of structural integrity of this domain for KLF1 function. In this context, sequencing of the complete *KLF1* transcription unit, including introns and untranslated regions (UTRs) and beyond, can

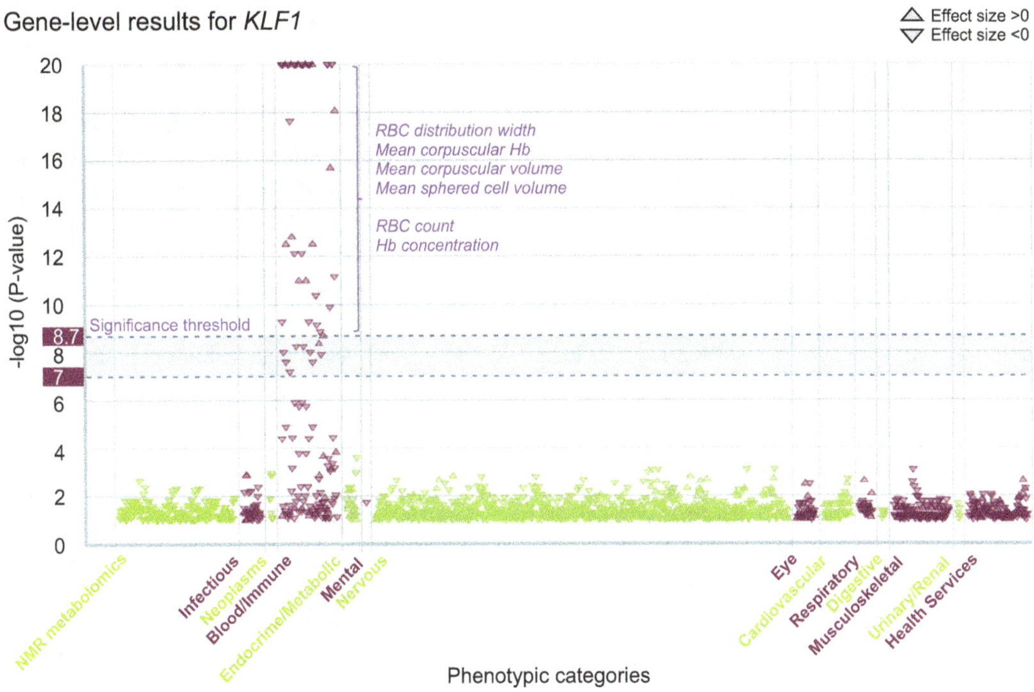

Fig. 10.6 *KLF1* SNPs associated with human traits from the UK Biobank analysis [137]. A gene-based rare variant statistical test was applied to the *KLF1* gene based on analysis of >380,000 individuals (November 2022; https://azphewas.com). Each point represents a gene-phenotype association; the higher the point (y-axis), the more significant the association. The points above the threshold line are considered statistically significant, in this case, red blood cell distribution width, mean corpuscular hemoglobin (Hb), mean corpuscular volume, mean sphered cell volume, red blood cell count, and Hb concentration The phenotypic category is listed along the x-axis; all significant KLF1 variants are within the "blood/immune" set

serve to identify additional genetic variants that alter levels and affect red cell expression [152]. In most populations, *KLF1* variants occur infrequently, e.g., the incidence of In(Lu) blood donors is 1 in 3000 in south-east England [153] and 1 in 5000 in south Wales [154]. Homozygosity or compound heterozygosity for *KLF1* variants would be extremely rare in such populations. In some populations, notably in the isolated island populations of Malta and Sardinia but also in south China, the incidence of *KLF1* variants reaches endemic levels with, for instance, a carrier frequency of 1.25% in south China [142]. In these populations β-thalassemia variants are also found at high frequency. The clinical advantage of *KLF1* heterozygosity can be dramatic, as shown by a case report of β^0/β^0 thalassemic twins with differences in anemia and transfusion dependence from loss of one *KLF1* allele [155]. This was initially thought to indicate that KLF1 acts as disease modifier of β-thalassemia through its role in fetal hemoglobin regulation [142]. Since co-occurrence of β-thalassemia with a *KLF1* variant would still be extremely rare, this notion was reconsidered. Instead, similar to β-thalassemia trait, it was proposed that *KLF1* haploinsufficiency confers protection to cerebral malaria caused by infection with the *Plasmodium falciparum* parasite [139]. Nonetheless, given the critical role of KLF1 in β-like hemoglobin switching, modulation of its levels as a potential therapeutic prerogative for β-hemoglobinopathies has merit, as its expression is tightly restricted, and haploinsufficiency is benign yet leads to the benefit of HPFH. However, attaining this goal has proven to be complex. There is extensive variation in fetal hemoglobin levels in individuals with the same *KLF1* mutation, even in the same family [60, 149]; this has been attributed to micro-variation of *KLF1* expression in the remaining wildtype allele [27, 149]. Attempts to engineer an altered expression level have yielded either too much dysregulation after targeting its coding sequence [156], or minimal effects after in vivo disruption of an intronic enhancer element in primary cells [33]. These latter results have led to the idea that the *KLF1* transcription unit is buffered against effects from localized genetic changes, a notion consistent with its location in the midst of multiple interaction domains [157].

10.10.1 Biallelic KLF1 Variants Are Associated with Strong Erythroid Phenotypes

The presence of compound heterozygosity for *KLF1* variants was first reported for two brothers from the island of Sardinia [151]. They carried *KLF1* p.S270X on the paternal and *KLF1* p.K332Q on the maternal allele. Fetal hemoglobin levels were 32% and 27% of total hemoglobin (normal range 0.1–1.4%). Zinc protoporphyrin was raised to 306 and 251 μg/dl (normal range < 35 μg/dl). In addition, mild hemolysis was observed. Of note, the *KLF1* p.K332Q variant was considered a hypomorphic allele; it was found to be less active than wildtype KLF1 in a transient transactivation assay using the β-globin promoter driving luciferase expression [151]. This highlights the necessity to characterize individual *KLF1* missense variants in detail. Most of these have been reported in the DBD [139] and may therefore differentially affect expression of (subsets of) KLF1 target genes in erythroid cells. This notion was supported by analysis of compound heterozygosity for *KLF1* variants in eight Thai families [158]. The affected individuals had transfusion-dependent anemia defined as congenital non-spherocytic hemolytic anemia (CNSHA). They displayed the In(Lu) phenotype, very high levels of fetal hemoglobin (2.6–54.6% of total hemoglobin), easily detectable levels of embryonic hemoglobins, and reduced activity of red cell pyruvate kinase, an enzyme involved in energy metabolism. Similar to the Sardinian cases, the patients always carried a *KLF1* missense variant in the DBD on one allele. It was proposed that severity of disease correlated with the impact of the *KLF1* missense variant on the function of the KLF1 protein [158]. Only one patient with compound heterozygosity for *KLF1 null* alleles has been reported [140]. This infant was born with severe anemia, jaundice, and fetal distress and was transfusion-dependent from birth. The patient carried a stop-codon variant *KLF1* p.W30X on one allele and a frameshift variant *KLF1* p.R319Efs34X on the other allele. Based on the phenotype of the *Klf1* KO mouse [103, 105], KLF1 *null* status was predicted to lead to prenatal lethal anemia. It was proposed that the child had survived thanks to high fetal Hb expression and possibly presence of other genetic modifiers. This case highlights the importance to be alert of *KLF1* deficiency in areas where *KLF1 null* variants occur relatively frequently. Appropriate measures should be taken as early as possible, in particular to prevent kernicterus (neuronal damage due to high levels of circulating bilirubin, a heme metabolite) [140].

10.10.2 A Rare De Novo KLF1 Variant with Dominant-Negative Phenotype

All *KLF1* variants described above are considered to act as classical loss-of-function alleles. Only one *KLF1* variant, p.E325K, has been reported to display a dominant-negative phenotype [159]. This variant is extremely rare and only occurs as a de novo mutation [159–166]. Patients heterozygous for *KLF1* p.E325K suffer from severe congenital dyserythropoietic anemia (CDA), now termed CDA type IV (OMIM: 613673). Even though the mutation is heterozygous, newborn patients are transfusion-dependent and exhibit a number of morphological defects in their erythroid compartment that lead to osmotic fragility of the red blood cells. Expression of CD44 is decreased and fetal hemoglobin levels are increased, but a more dramatic change is the presence of nucleated blood cells in circulation. The variant changes a highly conserved amino acid in ZF 2 that is directly involved in recognition of the purine in the center of the KLF1 DNA binding site (Fig. 10.4). The change from a negatively charged (E, glutamic acid) to a positively charged residue (K, lysine) (Fig. 10.7) alters the affinity and DNA binding specificity of this KLF1 variant [165, 167]. This results in aberrant gene activation leading to a corrupted transcriptome of the erythroid cells, altering membrane, transport, iron metabolism, and cell cycle proteins that account for much of the defective phenotype of the CDA IV red blood cells and the resultant disease severity [167–169]. CDA IV is a very rare disease, as there are only 12 published cases (Table 10.2). As a result, in addition to a patient-derived iPSC (induced pluripotent stem cell) model [164], a particularly strong source for analysis was the establishment of a cell model of the mutation in the BEL-A erythroid progenitor cell line [168]. Proteomic analyses of these cells complemented that seen from RNA-Seq and demonstrated the wide range of altered intracellular and mitochondrial pathways that illuminate the phenotypic spectrum of red blood cell deficits that follow from the KLF1 p.E325K variant. There have been confusing claims recently of CDA IV occurring in the absence of the *KLF1* p.E325K mutation that are more likely chronic non-spherocytic hemolytic anemia (discussed in [180]). These cases arose from compound heterozygosity [181, 182]; in contrast, the *KLF1* p.E325K mutation always occurs de novo and is a genetically dominant mono-allelic change. The original paper describing CDA IV [159], along with subsequent studies (Table 10.2), provide a framework of expression, morphology, and clinical characteristics for categorization of CDA IV that were mostly unchecked in these two examples. It remains important to pursue multiple avenues of analysis to avoid misdiagnosis of congenital hemolytic anemias with CDA IV (discussed in [161, 177]).

10.10.3 Mouse Klf1 Nan Mutation

The ethylnitrosourea-induced mouse mutant Nan [183] is caused by a mono-allelic change in the residue orthologous to human KLF1 p.E325 [184, 185]. The amino acid change in *Klf1* Nan (p.E339D) is biochemically different from the human CDA IV variant KLF1 p.E325K. In *Klf1*

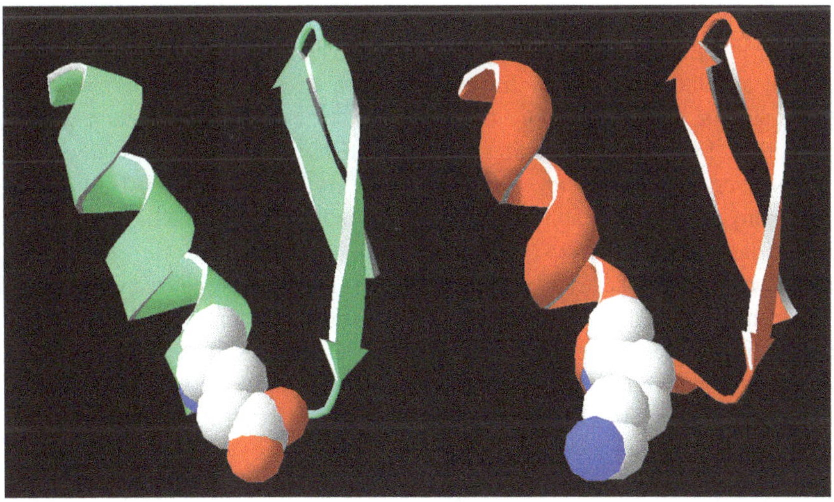

Fig. 10.7 KLF1 zinc finger (ZF) 2. AlphaFold prediction of the structure of human KLF1 zinc finger 2. Left: wildtype zinc finger 2. Right: p.E325K variant of ZF 2 that causes CDA IV. The side chain of KLF1 residue 325 is rendered as a 3D space-filling model, i.e., negatively charged glutamic acid in wildtype KLF1 and positively charged lysine in the p.E325K variant

Table 10.2 CDA IV patients reported in the literature

CDA IV patient KLF1 p.E325K	Reference
Patient 1	Arnaud et al. [159]
Patient 2	Singleton et al. [170] Agre et al. [171] Parsons et al. [172] Tang et al. [173] Wickramasinghe et al. [174]
Patient 3	Jaffray et al. [163]
Patient 4	de-la-Iglesia-Inigo et al. [160]
Patient 5	Ortolano et al. [166]
Patient 6	Ravindranath et al. [175]
Patient 7	Shefer Averbuch et al. [176]
Patient 8	Russo et al. [177]
Patient 9	Kohara et al. [164]
Patient 10	Jamwal et al. [178]
Patient 11	Deguise et al. [161]
Patient 12	Massey et al. [179]

Nan the glutamic acid residue is changed to an aspartic acid residue, both negatively charged amino acids. This is in contrast to KLF1 p.E325K where the glutamic acid residue is replaced by a positively charged lysine residue. The p.E339D change, located within a critical amino acid residue in the second ZF, alters its DNA recognition sequence and leads to hypomorphic and neomorphic expression patterns in the Nan/+ red blood cells [186–188]. There are nevertheless many similarities between the erythroid defects observed in *Klf1* Nan mice and patients with CDA IV. Similar to CDA IV, *Klf1* Nan is genetically dominant, and the mice display life-long spherocytic hemolytic anemia combined with β-thalassemia [184, 185]. Transcriptomic analyses of *Klf1* Nan erythroid cells revealed that expression of many previously established KLF1 target genes is deregulated [186, 187, 189]. In addition, ectopic activation of genes was observed that included misexpressed targets encoding secreted factors that exerted a negative effect on erythropoiesis and iron use [187], likely contributing to the anemia. The nuclear mechanism for these effects not only follows from KLF1 Nan pioneering activity at ectopic sites but also from its remarkable ability to interfere with binding by wildtype KLF1 to a subset of its normal targets [98]. A study investigating the developmental expression patterns of the hemoglobin genes, including the human *HBB* gene cluster present as a transgene, revealed that the switch from primitive, yolk sac-derived erythropoiesis to definitive, fetal liver-derived erythropoiesis was delayed in *Klf1* Nan embryos, and this was reflected in the expression patterns of the hemoglobin genes. In contrast to KLF1 p.E325 in humans, *Klf1* Nan did not lead to persistent high levels of embryonic/fetal hemoglobins in adult erythropoiesis [190]. This lends further support to the notion that KLF1 affects gene expression in a variant-specific manner. Finally, homozygosity for *Klf1* Nan results in early embryonic lethality due to failure of primitive erythropoiesis [185]. This phenotype is more severe than observed for *Klf1* KO fetuses which display relatively normal primitive erythropoiesis [103, 105]. It was proposed that redundancy with KLF2, which rescues primitive erythropoiesis in *Klf1* KO embryos [108–111], is abrogated by the dominant-negative effect of homozygous *Klf1* Nan [185].

10.11 Future Perspectives

Our narrative is a testimony to KLF1 as a master regulator of terminal erythroid differentiation. While much has been learned about its essential role in the erythroid lineage, there is a number of voids in our current knowledge that we hope will be addressed in the near future. It remains unclear how KLF1 activity is regulated during the commitment, expansion, and terminal differentiation of erythroid cells. A conditional KO approach would allow cell-type-specific analysis of KLF1 function, including in macrophages. A "floxed" *Klf1* allele has been created (C57BL/6JSmoc-*Klf1*$^{em1(flox)Smoc}$, NM-CKO-210073); but no studies using this mouse line have been reported yet. How important are PTMs and what properties do they confer to KLF1? In this context, given the high levels of KLF1 protein in the MEP, what is different about it such that the major erythroid targets, most dramatically β-globin, are not expressed? In addition, we know of many protein interaction partners of KLF1, but which ones are important in vivo? Application of BioID (biotin identification), an in vivo proximity-labeling technique [191] might provide important clues to answer this question. In particular, the basis for transcriptional selectivity differences in CDA IV and Nan-KLF1 beyond DNA recognition, perhaps due to changes in protein partners within the KLF1 protein complex, need to be ascertained to come up with a more fully realized mechanistic explanation for their ability to interfere with the function of the normal allele. How is the erythroid metabolism altered in the CDA IV and Nan-KLF1 red blood cells, and how does this exacerbate the associated anemia? It has been notoriously difficult to generate KLF1-specific antibodies. In particular, antibodies suitable for ChIP-Seq have been eluding the red blood cell community, although recently some progress has been made with a reengineered variant [98]. Tagging of the endogenous *KLF1* gene might allow the generation of high quality low background ChIP-Seq and CUT&TAG (cleavage under targets and tagmentation) data comprehensively covering the in vivo KLF1 binding sites in the erythroid genome. In addition, to understand the effect of—in particular—missense KLF1 variants on erythropoiesis, such variants could be introduced in suitable human erythroid cell lines such as BEL-A [168] or HUDEP [192]. Although technically challenging, compound heterozygosity for *KLF1* variants could also be modeled in this system. One perplexing observation is the prevalence of gonadal-related issues in CDA IV patients, with genital dysmorphology and even sex reversal in over half of these patients. The intriguing effect of SUMOylation-deficient KLF1 on aging in mice [82] requires further investigation. Finally, we expect that continued analysis of *KLF1* variants in the population will keep revealing novel aspects of KLF1 structure/function and the role of KLF1 in red blood cell development and human physiology. Indeed, just after completion of this chapter, it was reported that, as an adaptation to hypoxia, chondrocytes express large amounts of hemoglobin and that this is controlled by KLF1 [193]. While these observations have not been independently confirmed, investigating the role of KLF1 in a non-hematopoietic cell type would open up an entirely new research field and may reveal as yet unanticipated roles of KLF1 in chondrocyte biology.

References

1. Miller IJ, Bieker JJ (1993) A novel, erythroid cell-specific murine transcription factor that binds to the CACCC element and is related to the Kruppel family of nuclear proteins. Mol Cell Biol 13(5):2776–2786
2. Miller J, McLachlan AD, Klug A (1985) Repetitive zinc-binding domains in the protein transcription factor IIIA from Xenopus oocytes. EMBO J 4(6):1609–1614
3. Tallack MR, Whitington T, Yuen WS, Wainwright EN, Keys JR, Gardiner BB et al (2010) A global role for KLF1 in erythropoiesis revealed by ChIP-seq in primary erythroid cells. Genome Res 20(8):1052–1063
4. Philipsen S, Talbot D, Fraser P, Grosveld F (1990) The beta-globin dominant control region: hypersensitive site 2. EMBO J 9(7):2159–2167
5. Feng WC, Southwood CM, Bieker JJ (1994) Analyses of beta-thalassemia mutant DNA interactions with erythroid Kruppel-like factor (EKLF), an erythroid cell-specific transcription factor. J Biol Chem 269(2):1493–1500
6. Donze D, Townes TM, Bieker JJ (1995) Role of erythroid Kruppel-like factor in human gamma-to beta-globin gene switching. J Biol Chem 270(4):1955–1959
7. Hagen G, Muller S, Beato M, Suske G (1992) Cloning by recognition site screening of two novel

GT box binding proteins: a family of Sp1 related genes. Nucleic Acids Res 20(21):5519–5525
8. Kingsley C, Winoto A (1992) Cloning of GT box-binding proteins: a novel Sp1 multigene family regulating T-cell receptor gene expression. Mol Cell Biol 12(10):4251–4261
9. Zhu H, Nguyen VT, Brown AB, Pourhosseini A, Garcia AV, van Bilsen M et al (1993) A novel, tissue-restricted zinc finger protein (HF-1b) binds to the cardiac regulatory element (HF-1b/MEF-2) in the rat myosin light-chain 2 gene. Mol Cell Biol 13(7):4432–4444
10. Wimmer EA, Jackle H, Pfeifle C, Cohen SM (1993) A Drosophila homologue of human Sp1 is a head-specific segmentation gene. Nature 366(6456):690–694
11. Blom-Dahl D, Cordoba S, Gabilondo H, Carr-Baena P, Diaz-Benjumea FJ, Estella C (2020) In vivo analysis of the evolutionary conserved BTD-box domain of Sp1 and Btd during Drosophila development. Dev Biol 466(1–2):77–89
12. Anderson KP, Kern CB, Crable SC, Lingrel JB (1995) Isolation of a gene encoding a functional zinc finger protein homologous to erythroid Kruppel-like factor: identification of a new multigene family. Mol Cell Biol 15(11):5957–5965
13. Shields JM, Christy RJ, Yang VW (1996) Identification and characterization of a gene encoding a gut-enriched Kruppel-like factor expressed during growth arrest. J Biol Chem 271(33):20009–20017
14. Crossley M, Whitelaw E, Perkins A, Williams G, Fujiwara Y, Orkin SH (1996) Isolation and characterization of the cDNA encoding BKLF/TEF-2, a major CACCC-box-binding protein in erythroid cells and selected other cells. Mol Cell Biol 16(4):1695–1705
15. Kuo CT, Veselits ML, Leiden JM (1997) LKLF: a transcriptional regulator of single-positive T cell quiescence and survival. Science 277(5334):1986–1990
16. Takahashi K, Yamanaka S (2006) Induction of pluripotent stem cells from mouse embryonic and adult fibroblast cultures by defined factors. Cell 126(4):663–676
17. Turner J, Crossley M (1999) Basic Kruppel-like factor functions within a network of interacting haematopoietic transcription factors. Int J Biochem Cell Biol 31(10):1169–1174
18. Suske G, Bruford E, Philipsen S (2005) Mammalian SP/KLF transcription factors: bring in the family. Genomics 85(5):551–556
19. Pei J, Grishin NV (2013) A new family of predicted Kruppel-like factor genes and pseudogenes in placental mammals. PLoS One 8(11):e81109
20. Jenkins NA, Gilbert DJ, Copeland NG, Gruzglin E, Bieker JJ (1998) Erythroid Kruppel-like transcription factor (Eklf) maps to a region of mouse chromosome 8 syntenic with human chromosome 19. Mamm Genome 9(2):174–176
21. Pei J, Grishin NV (2015) C2H2 zinc finger proteins of the SP/KLF, Wilms tumor, EGR, Huckebein, and Klumpfuss families in metazoans and beyond. Gene 573(1):91–99
22. Johnson RM, Ho YS, Yu DY, Kuypers FA, Ravindranath Y, Goyette GW (2010) The effects of disruption of genes for peroxiredoxin-2, glutathione peroxidase-1, and catalase on erythrocyte oxidative metabolism. Free Radic Biol Med 48(4):519–525
23. Kawane K, Fukuyama H, Kondoh G, Takeda J, Ohsawa Y, Uchiyama Y et al (2001) Requirement of DNase II for definitive erythropoiesis in the mouse fetal liver. Science 292(5521):1546–1549
24. Qin K, Huang P, Feng R, Keller CA, Peslak SA, Khandros E et al (2022) Dual function NFI factors control fetal hemoglobin silencing in adult erythroid cells. Nat Genet 54(6):874–884
25. Chiu SK, Saw J, Huang Y, Sonderegger SE, Wong NC, Powell DR et al (2018) A novel role for Lyl1 in primitive erythropoiesis. Development 145(19):dev162990
26. Bieker JJ (1996) Isolation, genomic structure, and expression of human erythroid Kruppel-like factor (EKLF). DNA Cell Biol 15(5):347–352
27. Heshusius S, Grech L, Gillemans N, Brouwer RWW, den Dekker XT, van Ijcken WFJ et al (2022) Epigenomic analysis of KLF1 haploinsufficiency in primary human erythroblasts. Sci Rep 12(1):336
28. Wessels MW, Cnossen MH, van Dijk TB, Gillemans N, Schmidt KLJ, van Lom K et al (2021) Molecular analysis of the erythroid phenotype of a patient with BCL11A haploinsufficiency. Blood Adv 5(9):2339–2349
29. Philipsen S, Pruzina S, Grosveld F (1993) The minimal requirements for activity in transgenic mice of hypersensitive site 3 of the beta globin locus control region. EMBO J 12(3):1077–1085
30. Wadman IA, Osada H, Grutz GG, Agulnick AD, Westphal H, Forster A et al (1997) The LIM-only protein Lmo2 is a bridging molecule assembling an erythroid, DNA-binding complex which includes the TAL1, E47, GATA-1 and Ldb1/NLI proteins. EMBO J 16(11):3145–3157
31. Schulz VP, Yan H, Lezon-Geyda K, An X, Hale J, Hillyer CD et al (2019) A Unique Epigenomic Landscape Defines Human Erythropoiesis. Cell Rep 28(11):2996–3009 e7
32. Ulirsch JC, Lareau CA, Bao EL, Ludwig LS, Guo MH, Benner C et al (2019) Interrogation of human hematopoiesis at single-cell and single-variant resolution. Nat Genet 51(4):683–693
33. Gnanapragasam MN, Planutis A, Glassberg JA, Bieker JJ (2023) Identification of a genomic DNA sequence that quantitatively modulates KLF1 transcription factor expression in differentiating human hematopoietic cells. Sci Rep 13(1):7589
34. Lohmann F, Dangeti M, Soni S, Chen X, Planutis A, Baron MH et al (2015) The DEK oncoprotein is a critical component of the EKLF/KLF1 enhancer in erythroid cells. Mol Cell Biol 35(21):3726–3738
35. Lohmann F, Bieker JJ (2008) Activation of Eklf expression during hematopoiesis by Gata2 and Smad5 prior to erythroid commitment. Development 135(12):2071–2082

36. Adelman CA, Chattopadhyay S, Bieker JJ (2002) The BMP/BMPR/Smad pathway directs expression of the erythroid-specific EKLF and GATA1 transcription factors during embryoid body differentiation in serum-free media. Development 129(2):539–549
37. Xue L, Chen X, Chang Y, Bieker JJ (2004) Regulatory elements of the EKLF gene that direct erythroid cell-specific expression during mammalian development. Blood 103(11):4078–4083
38. Zhou D, Liu K, Sun CW, Pawlik KM, Townes TM (2010) KLF1 regulates BCL11A expression and gamma- to beta-globin gene switching. Nat Genet 42(9):742–744
39. Fluhr S, Krombholz CF, Meier A, Epting T, Mucke O, Plass C et al (2017) Epigenetic dysregulation of the erythropoietic transcription factor KLF1 and the beta-like globin locus in juvenile myelomonocytic leukemia. Epigenetics 12(8):715–723
40. Madzo J, Liu H, Rodriguez A, Vasanthakumar A, Sundaravel S, Caces DBD et al (2014) Hydroxymethylation at gene regulatory regions directs stem/early progenitor cell commitment during erythropoiesis. Cell Rep 6(1):231–244
41. Zhang X, Su J, Jeong M, Ko M, Huang Y, Park HJ et al (2016) DNMT3A and TET2 compete and cooperate to repress lineage-specific transcription factors in hematopoietic stem cells. Nat Genet 48(9):1014–1023
42. Chen X, Pillay S, Lohmann F, Bieker JJ (2023) Association of Ddx5/p68 protein with the upstream erythroid enhancer element (EHS1) of the Klf1 gene. bioRxiv:743435
43. Ghosh S, Raundhal M, Myers SA, Carr SA, Chen X, Petsko GA et al (2022) Identification of RIOK2 as a master regulator of human blood cell development. Nat Immunol 23(1):109–121
44. Mas C, Lussier-Price M, Soni S, Morse T, Arseneault G, Di Lello P et al (2011) Structural and functional characterization of an atypical activation domain in erythroid Kruppel-like factor (EKLF). Proc Natl Acad Sci USA 108(26):10484–10489
45. Quadrini KJ, Bieker JJ (2006) EKLF/KLF1 is ubiquitinated in vivo and its stability is regulated by activation domain sequences through the 26S proteasome. FEBS Lett 580(9):2285–2293
46. Raiola L, Lussier-Price M, Gagnon D, Lafrance-Vanasse J, Mascle X, Arseneault G et al (2013) Structural characterization of a noncovalent complex between ubiquitin and the transactivation domain of the erythroid-specific factor EKLF. Structure 21(11):2014–2024
47. Jumper J, Evans R, Pritzel A, Green T, Figurnov M, Ronneberger O et al (2021) Highly accurate protein structure prediction with AlphaFold. Nature 596(7873):583–589
48. Sengupta T, Chen K, Milot E, Bieker JJ (2008) Acetylation of EKLF is essential for epigenetic modification and transcriptional activation of the beta-globin locus. Mol Cell Biol 28(20):6160–6170
49. Zhang W, Bieker JJ (1998) Acetylation and modulation of erythroid Kruppel-like factor (EKLF) activity by interaction with histone acetyltransferases. Proc Natl Acad Sci USA 95(17):9855–9860
50. Zhang W, Kadam S, Emerson BM, Bieker JJ (2001) Site-specific acetylation by p300 or CREB binding protein regulates erythroid Kruppel-like factor transcriptional activity via its interaction with the SWI-SNF complex. Mol Cell Biol 21(7):2413–2422
51. Siatecka M, Xue L, Bieker JJ (2007) Sumoylation of EKLF promotes transcriptional repression and is involved in inhibition of megakaryopoiesis. Mol Cell Biol 27(24):8547–8560
52. Frontelo P, Manwani D, Galdass M, Karsunky H, Lohmann F, Gallagher PG et al (2007) Novel role for EKLF in megakaryocyte lineage commitment. Blood 110(12):3871–3880
53. Presnell JS, Schnitzler CE, Browne WE (2015) KLF/SP transcription factor family evolution: expansion, diversification, and innovation in eukaryotes. Genome Biol Evol 7(8):2289–2309
54. Bouwman P, Philipsen S (2002) Regulation of the activity of Sp1-related transcription factors. Mol Cell Endocrinol 195(1–2):27–38
55. Sengupta T, Cohet N, Morle F, Bieker JJ (2009) Distinct modes of gene regulation by a cell-specific transcriptional activator. Proc Natl Acad Sci USA 106(11):4213–4218
56. Soni S, Pchelintsev N, Adams PD, Bieker JJ (2014) Transcription factor EKLF (KLF1) recruitment of the histone chaperone HIRA is essential for beta-globin gene expression. Proc Natl Acad Sci USA 111(37):13337–13342
57. Armstrong JA, Bieker JJ, Emerson BM (1998) A SWI/SNF-related chromatin remodeling complex, E-RC1, is required for tissue-specific transcriptional regulation by EKLF in vitro. Cell 95(1):93–104
58. Kadam S, Emerson BM (2003) Transcriptional specificity of human SWI/SNF BRG1 and BRM chromatin remodeling complexes. Mol Cell 11(2):377–389
59. Kadam S, McAlpine GS, Phelan ML, Kingston RE, Jones KA, Emerson BM (2000) Functional selectivity of recombinant mammalian SWI/SNF subunits. Genes Dev 14(19):2441–2451
60. Borg J, Papadopoulos P, Georgitsi M, Gutierrez L, Grech G, Fanis P et al (2010) Haploinsufficiency for the erythroid transcription factor KLF1 causes hereditary persistence of fetal hemoglobin. Nat Genet 42(9):801–805
61. Drissen R, von Lindern M, Kolbus A, Driegen S, Steinlein P, Beug H et al (2005) The erythroid phenotype of EKLF-null mice: defects in hemoglobin metabolism and membrane stability. Mol Cell Biol 25(12):5205–5214
62. Esteghamat F, Gillemans N, Bilic I, van den Akker E, Cantu I, van Gent T et al (2013) Erythropoiesis and globin switching in compound Klf1::Bcl11a mutant mice. Blood 121(13):2553–2562
63. Hodge D, Coghill E, Keys J, Maguire T, Hartmann B, McDowall A et al (2006) A global role for EKLF

in definitive and primitive erythropoiesis. Blood 107(8):3359–3370
64. Pilon AM, Arcasoy MO, Dressman HK, Vayda SE, Maksimova YD, Sangerman JI et al (2008) Failure of terminal erythroid differentiation in EKLF-deficient mice is associated with cell cycle perturbation and reduced expression of E2F2. Mol Cell Biol 28(24):7394–7401
65. Tallack MR, Keys JR, Humbert PO, Perkins AC (2009) EKLF/KLF1 controls cell cycle entry via direct regulation of E2f2. J Biol Chem 284(31):20966–20974
66. Tallack MR, Magor GW, Dartigues B, Sun L, Huang S, Fittock JM et al (2012) Novel roles for KLF1 in erythropoiesis revealed by mRNA-seq. Genome Res 22(12):2385–2398
67. Tallack MR, Perkins AC (2010) KLF1 directly coordinates almost all aspects of terminal erythroid differentiation. IUBMB Life 62(12):886–890
68. Gnanapragasam MN, Bieker JJ (2017) Orchestration of late events in erythropoiesis by KLF1/EKLF. Curr Opin Hematol 24(3):183–190
69. Gnanapragasam MN, McGrath KE, Catherman S, Xue L, Palis J, Bieker JJ (2016) EKLF/KLF1-regulated cell cycle exit is essential for erythroblast enucleation. Blood 128(12):1631–1641
70. Ilsley MD, Gillinder KR, Magor GW, Huang S, Bailey TL, Crossley M et al (2017) Kruppel-like factors compete for promoters and enhancers to fine-tune transcription. Nucleic Acids Res 45(11):6572–6588
71. Norton LJ, Funnell APW, Burdach J, Wienert B, Kurita R, Nakamura Y et al (2017) KLF1 directly activates expression of the novel fetal globin repressor ZBTB7A/LRF in erythroid cells. Blood Adv 1(11):685–692
72. Elagooz R, Dhara AR, Gott RM, Adams SE, White RA, Ghosh A et al (2022) PUM1 mediates the post-transcriptional regulation of human fetal hemoglobin. Blood Adv 6(23):6016–6022
73. Jeong JJ, Gu X, Nie J, Sundaravel S, Liu H, Kuo WL et al (2019) Cytokine-regulated phosphorylation and activation of TET2 by JAK2 in hematopoiesis. Cancer Discov 9(6):778–795
74. Quadrini KJ, Gruzglin E, Bieker JJ (2008) Non-random subcellular distribution of variant EKLF in erythroid cells. Exp Cell Res 314(7):1595–1604
75. Schoenfelder S, Sexton T, Chakalova L, Cope NF, Horton A, Andrews S et al (2010) Preferential associations between co-regulated genes reveal a transcriptional interactome in erythroid cells. Nat Genet 42(1):53–61
76. Shyu YC, Lee TL, Wen SC, Chen H, Hsiao WY, Chen X et al (2007) Subcellular transport of EKLF and switch-on of murine adult beta maj globin gene transcription. Mol Cell Biol 27(6):2309–2323
77. Shyu YC, Lee TL, Chen X, Hsu PH, Wen SC, Liaw YW et al (2014) Tight regulation of a timed nuclear import wave of EKLF by PKCtheta and FOE during Pro-E to Baso-E transition. Dev Cell 28(4):409–422
78. Quadrini KJ, Bieker JJ (2002) Kruppel-like zinc fingers bind to nuclear import proteins and are required for efficient nuclear localization of erythroid Kruppel-like factor. J Biol Chem 277(35):32243–32252
79. Pandya K, Townes TM (2002) Basic residues within the Kruppel zinc finger DNA binding domains are the critical nuclear localization determinants of EKLF/KLF-1. J Biol Chem 277(18):16304–16312
80. Honda R, Tanaka H, Yasuda H (1997) Oncoprotein MDM2 is a ubiquitin ligase E3 for tumor suppressor p53. FEBS Lett 420(1):25–27
81. Shuai K, Ziemiecki A, Wilks AF, Harpur AG, Sadowski HB, Gilman MZ et al (1993) Polypeptide signalling to the nucleus through tyrosine phosphorylation of Jak and Stat proteins. Nature 366(6455):580–583
82. Shyu YC, Liao PC, Huang TS, Yang CJ, Lu MJ, Huang SM et al (2022) Genetic Disruption of KLF1 K74 SUMOylation in Hematopoietic System Promotes Healthy Longevity in Mice. Adv Sci (Weinh) 9(25):e2201409
83. Ma WB, Wang XH, Li CY, Tian HH, Zhang J, Bi JJ et al (2020) GPS2 promotes erythroid differentiation by control of the stability of EKLF protein. Blood 135(25):2302–2315
84. Ouyang L, Chen X, Bieker JJ (1998) Regulation of erythroid Kruppel-like factor (EKLF) transcriptional activity by phosphorylation of a protein kinase casein kinase II site within its interaction domain. J Biol Chem 273(36):23019–23025
85. Yien YY, Bieker JJ (2012) Functional interactions between erythroid Kruppel-like factor (EKLF/KLF1) and protein phosphatase PPM1B/PP2Cbeta. J Biol Chem 287(19):15193–15204
86. Chen X, Bieker JJ (2001) Unanticipated repression function linked to erythroid Kruppel-like factor. Mol Cell Biol 21(9):3118–3125
87. Chen X, Bieker JJ (2004) Stage-specific repression by the EKLF transcriptional activator. Mol Cell Biol 24(23):10416–10424
88. Varricchio L, Dell'Aversana C, Nebbioso A, Migliaccio G, Altucci L, Mai A et al (2014) Identification of NuRSERY, a new functional HDAC complex composed by HDAC5, GATA1, EKLF and pERK present in human erythroid cells. Int J Biochem Cell Biol 50:112–122
89. Gillemans N, Tewari R, Lindeboom F, Rottier R, de Wit T, Wijgerde M et al (1998) Altered DNA-binding specificity mutants of EKLF and Sp1 show that EKLF is an activator of the beta-globin locus control region in vivo. Genes Dev 12(18):2863–2873
90. Cheng L, Li Y, Qi Q, Xu P, Feng R, Palmer L et al (2021) Single-nucleotide-level mapping of DNA regulatory elements that control fetal hemoglobin expression. Nat Genet 53(6):869–880
91. Carter D, Chakalova L, Osborne CS, Dai YF, Fraser P (2002) Long-range chromatin regulatory interactions in vivo. Nat Genet 32(4):623–626
92. Tolhuis B, Palstra RJ, Splinter E, Grosveld F, de Laat W (2002) Looping and interaction between hyper-

sensitive sites in the active beta-globin locus. Mol Cell 10(6):1453–1465
93. Drissen R, Palstra RJ, Gillemans N, Splinter E, Grosveld F, Philipsen S et al (2004) The active spatial organization of the beta-globin locus requires the transcription factor EKLF. Genes Dev 18(20):2485–2490
94. Li L, Freudenberg J, Cui K, Dale R, Song SH, Dean A et al (2013) Ldb1-nucleated transcription complexes function as primary mediators of global erythroid gene activation. Blood 121(22):4575–4585
95. Song SH, Hou C, Dean A (2007) A positive role for NLI/Ldb1 in long-range beta-globin locus control region function. Mol Cell 28(5):810–822
96. Vakoc CR, Letting DL, Gheldof N, Sawado T, Bender MA, Groudine M et al (2005) Proximity among distant regulatory elements at the beta-globin locus requires GATA-1 and FOG-1. Mol Cell 17(3):453–462
97. Yien YY, Bieker JJ (2013) EKLF/KLF1, a tissue-restricted integrator of transcriptional control, chromatin remodeling, and lineage determination. Mol Cell Biol 33(1):4–13
98. Mukherjee K, Bieker JJ (2022) EKLF/Klf1 regulates erythroid transcription by its pioneering activity and selective control of RNA Pol II pause-release. Cell Rep. 41(12):111830
99. Dzierzak E, Philipsen S (2013) Erythropoiesis: development and differentiation. Cold Spring Harb Perspect Med 3(4):a011601
100. Southwood CM, Downs KM, Bieker JJ (1996) Erythroid Kruppel-like factor exhibits an early and sequentially localized pattern of expression during mammalian erythroid ontogeny. Dev Dyn 206(3):248–259
101. Mittnenzweig M, Mayshar Y, Cheng S, Ben-Yair R, Hadas R, Rais Y et al (2021) A single-embryo, single-cell time-resolved model for mouse gastrulation. Cell 184(11):2825–42 e22
102. Biben C, Weber TS, Potts KS, Choi J, Miles DC, Carmagnac A et al (2023) In vivo clonal tracking reveals evidence of haemangioblast and haematomesoblast contribution to yolk sac haematopoiesis. Nat Commun 14(1):41
103. Nuez B, Michalovich D, Bygrave A, Ploemacher R, Grosveld F (1995) Defective haematopoiesis in fetal liver resulting from inactivation of the EKLF gene. Nature 375(6529):316–318
104. Perkins AC, Gaensler KM, Orkin SH (1996) Silencing of human fetal globin expression is impaired in the absence of the adult beta-globin gene activator protein EKLF. Proc Natl Acad Sci USA 93(22):12267–12271
105. Perkins AC, Sharpe AH, Orkin SH (1995) Lethal beta-thalassaemia in mice lacking the erythroid CACCC-transcription factor EKLF. Nature 375(6529):318–322
106. Wijgerde M, Gribnau J, Trimborn T, Nuez B, Philipsen S, Grosveld F et al (1996) The role of EKLF in human beta-globin gene competition. Genes Dev 10(22):2894–2902
107. Tewari R, Gillemans N, Wijgerde M, Nuez B, von Lindern M, Grosveld F et al (1998) Erythroid Kruppel-like factor (EKLF) is active in primitive and definitive erythroid cells and is required for the function of 5'HS3 of the beta-globin locus control region. EMBO J 17(8):2334–2341
108. Vinjamur DS, Wade KJ, Mohamad SF, Haar JL, Sawyer ST, Lloyd JA (2014) Kruppel-like transcription factors KLF1 and KLF2 have unique and coordinate roles in regulating embryonic erythroid precursor maturation. Haematologica 99(10):1565–1573
109. Alhashem YN, Vinjamur DS, Basu M, Klingmuller U, Gaensler KM, Lloyd JA (2011) Transcription factors KLF1 and KLF2 positively regulate embryonic and fetal beta-globin genes through direct promoter binding. J Biol Chem 286(28):24819–24827
110. Basu P, Lung TK, Lemsaddek W, Sargent TG, Williams DC Jr, Basu M et al (2007) EKLF and KLF2 have compensatory roles in embryonic beta-globin gene expression and primitive erythropoiesis. Blood 110(9):3417–3425
111. Basu P, Morris PE, Haar JL, Wani MA, Lingrel JB, Gaensler KM et al (2005) KLF2 is essential for primitive erythropoiesis and regulates the human and murine embryonic beta-like globin genes in vivo. Blood 106(7):2566–2571
112. Tyser RCV, Mahammadov E, Nakanoh S, Vallier L, Scialdone A, Srinivas S (2021) Single-cell transcriptomic characterization of a gastrulating human embryo. Nature 600(7888):285–289
113. Popescu DM, Botting RA, Stephenson E, Green K, Webb S, Jardine L et al (2019) Decoding human fetal liver haematopoiesis. Nature 574(7778):365–371
114. Luo Q, Ma X, Wahl SM, Bieker JJ, Crossley M, Montaner LJ (2004) Activation and repression of interleukin-12 p40 transcription by erythroid Kruppel-like factor in macrophages. J Biol Chem 279(18):18451–18456
115. Bessis M (1958) Erythroblastic island, functional unity of bone marrow. Rev Hematol 13(1):8–11
116. Porcu S, Manchinu MF, Marongiu MF, Sogos V, Poddie D, Asunis I et al (2011) Klf1 affects DNase II-alpha expression in the central macrophage of a fetal liver erythroblastic island: a non-cell-autonomous role in definitive erythropoiesis. Mol Cell Biol 31(19):4144–4154
117. Mukherjee K, Bieker JJ (2021) Transcriptional control of gene expression and the heterogeneous cellular identity of erythroblastic island macrophages. Front Genet 12:756028
118. Xue L, Galdass M, Gnanapragasam MN, Manwani D, Bieker JJ (2014) Extrinsic and intrinsic control by EKLF (KLF1) within a specialized erythroid niche. Development 141(11):2245–2254
119. Lopez-Yrigoyen M, Yang CT, Fidanza A, Cassetta L, Taylor AH, McCahill A et al (2019) Genetic programming of macrophages generates an in vitro model for the human erythroid island niche. Nat Commun 10(1):881

120. Mukherjee K, Xue L, Planutis A, Gnanapragasam MN, Chess A, Bieker JJ (2021) EKLF/KLF1 expression defines a unique macrophage subset during mouse erythropoiesis. Elife 10:e61070
121. Bouilloux F, Juban G, Cohet N, Buet D, Guyot B, Vainchenker W et al (2008) EKLF restricts megakaryocytic differentiation at the benefit of erythrocytic differentiation. Blood 112(3):576–584
122. Tallack MR, Perkins AC (2010) Megakaryocyte-erythroid lineage promiscuity in EKLF null mouse blood. Haematologica 95(1):144–147
123. Isern J, Fraser ST, He Z, Zhang H, Baron MH (2010) Dose-dependent regulation of primitive erythroid maturation and identity by the transcription factor Eklf. Blood 116(19):3972–3980
124. Siatecka M, Bieker JJ (2011) The multifunctional role of EKLF/KLF1 during erythropoiesis. Blood 118(8):2044–2054
125. Dore LC, Crispino JD (2011) Transcription factor networks in erythroid cell and megakaryocyte development. Blood 118(2):231–239
126. Starck J, Cohet N, Gonnet C, Sarrazin S, Doubeikovskaia Z, Doubeikovski A et al (2003) Functional cross-antagonism between transcription factors FLI-1 and EKLF. Mol Cell Biol 23(4):1390–1402
127. Palii CG, Cheng Q, Gillespie MA, Shannon P, Mazurczyk M, Napolitani G et al (2019) Single-cell proteomics reveal that quantitative changes in co-expressed lineage-specific transcription factors determine cell fate. Cell Stem Cell 24(5):812–20 e5
128. Gillespie MA, Palii CG, Sanchez-Taltavull D, Perkins TJ, Brand M, Ranish JA (2020) Absolute quantification of transcription factors in human erythropoiesis using selected reaction monitoring mass spectrometry. STAR Protoc 1(3):100216
129. Gnanapragasam MN, Crispino JD, Ali AM, Weinberg R, Hoffman R, Raza A et al (2018) Survey and evaluation of mutations in the human KLF1 transcription unit. Sci Rep 8(1):6587
130. Mansoor A, Mansoor MO, Patel JL, Zhao S, Natkunam Y, Bieker JJ (2020) KLF1/EKLF expression in acute leukemia is correlated with chromosomal abnormalities. Blood Cells Mol Dis 83:102434
131. Wang H, Cui B, Sun H, Zhang F, Rao J, Wang R et al (2021) Aberrant GATA2 activation in pediatric B-cell acute lymphoblastic leukemia. Front Pediatr 9:795529
132. Perkins AC, Peterson KR, Stamatoyannopoulos G, Witkowska HE, Orkin SH (2000) Fetal expression of a human Agamma globin transgene rescues globin chain imbalance but not hemolysis in EKLF null mouse embryos. Blood 95(5):1827–1833
133. Coghill E, Eccleston S, Fox V, Cerruti L, Brown C, Cunningham J et al (2001) Erythroid Kruppel-like factor (EKLF) coordinates erythroid cell proliferation and hemoglobinization in cell lines derived from EKLF null mice. Blood 97(6):1861–1868
134. Nilson DG, Sabatino DE, Bodine DM, Gallagher PG (2006) Major erythrocyte membrane protein genes in EKLF-deficient mice. Exp Hematol 34(6):705–712
135. Tallack MR, Keys JR, Perkins AC (2007) Erythroid Kruppel-like factor regulates the G1 cyclin dependent kinase inhibitor p18INK4c. J Mol Biol 369(2):313–321
136. Backman JD, Li AH, Marcketta A, Sun D, Mbatchou J, Kessler MD et al (2021) Exome sequencing and analysis of 454,787 UK Biobank participants. Nature 599(7886):628–634
137. Wang Q, Dhindsa RS, Carss K, Harper AR, Nag A, Tachmazidou I et al (2021) Rare variant contribution to human disease in 281,104 UK Biobank exomes. Nature 597(7877):527–532
138. Singleton BK, Burton NM, Green C, Brady RL, Anstee DJ (2008) Mutations in EKLF/KLF1 form the molecular basis of the rare blood group In(Lu) phenotype. Blood 112(5):2081–2088
139. Perkins A, Xu X, Higgs DR, Patrinos GP, Arnaud L, Bieker JJ et al (2016) Kruppeling erythropoiesis: an unexpected broad spectrum of human red blood cell disorders due to KLF1 variants. Blood 127(15):1856–1862
140. Magor GW, Tallack MR, Gillinder KR, Bell CC, McCallum N, Williams B et al (2015) KLF1-null neonates display hydrops fetalis and a deranged erythroid transcriptome. Blood 125(15):2405–2417
141. Singleton BK, Frayne J, Anstee DJ (2012) Blood group phenotypes resulting from mutations in erythroid transcription factors. Curr Opin Hematol 19(6):486–493
142. Liu D, Zhang X, Yu L, Cai R, Ma X, Zheng C et al (2014) KLF1 mutations are relatively more common in a thalassemia endemic region and ameliorate the severity of beta-thalassemia. Blood 124(5):803–811
143. Yu LH, Liu D, Cai R, Shang X, Zhang XH, Ma XX et al (2015) Changes in hematological parameters in alpha-thalassemia individuals co-inherited with erythroid Kruppel-like factor mutations. Clin Genet 88(1):56–61
144. Paglietti ME, Satta S, Sollaino MC, Barella S, Ventrella A, Desogus MF et al (2016) The Problem of Borderline Hemoglobin A2 Levels in the Screening for beta-Thalassemia Carriers in Sardinia. Acta Haematol 135(4):193–199
145. Satta S, Paglietti ME, Sollaino MC, Barella S, Moi P, Desogus MF et al (2017) Changes in HbA2 and HbF in alpha thalassemia carriers with KLF1 mutation. Blood Cells Mol Dis 64:30–32
146. Borgio JF, AbdulAzeez S, Al-Muslami AM, Naserullah ZA, Al-Jarrash S, Al-Suliman AM et al (2018) KLF1 gene and borderline hemoglobin A(2) in Saudi population. Arch Med Sci 14(1):230–236
147. Fraser NS, Knauth CM, Schoeman EM, Moussa A, Perkins AC, Walsh T et al (2018) Investigation of the variable In(Lu) phenotype caused by KLF1 variants. Transfusion 58(10):2414–2420
148. Tamaddoni A, Khabaz Astaneh S, Tabaripour R, Akhavan-Niaki H (2019) Kruppel-like factor 1 gene

mutations in thalassemia patients from North Iran: report of a new mutation associated with beta-thalassemia intermedia. Hemoglobin 43(1):12–17
149. Eernstman J, Veldhuisen B, Ligthart P, von Lindern M, van der Schoot CE, van den Akker E (2021) Novel variants in Krueppel like factor 1 that cause persistence of fetal hemoglobin in In(Lu) individuals. Sci Rep 11(1):18557
150. Catapano R, Sessa R, Trombetti S, Cesaro E, Russo F, Izzo P et al (2023) Identification and functional analysis of known and new mutations in the transcription factor KLF1 linked with beta-thalassemia-like phenotypes. Biology (Basel) 12(4):510
151. Satta S, Perseu L, Moi P, Asunis I, Cabriolu A, Maccioni L et al (2011) Compound heterozygosity for KLF1 mutations associated with remarkable increase of fetal hemoglobin and red cell protoporphyrin. Haematologica 96(5):767–770
152. Hantaweepant C, Suktitipat B, Pithukpakorn M, Chinthammitr Y, Limwongse C, Tansiri N et al (2023) Whole exome sequencing and rare variant association study to identify genetic modifiers, KLF1 mutations, and a novel double mutation in Thai patients with hemoglobin E/beta-thalassemia. Hematology 28(1):2187155
153. Shaw MA, Leak MR, Daniels GL, Tippett P (1984) The rare Lutheran blood group phenotype Lu(a-b-): a genetic study. Ann Hum Genet 48(3):229–237
154. Rowe GP, Gale SA, Daniels GL, Green CA, Tippett P (1992) A study on Lu-null families in South Wales. Ann Hum Genet 56(3):267–272
155. Xie XM, Liu YN, Li J, Jiang F, Li DZ (2019) A Kruppel-like factor 1 gene mutation ameliorates the severity of beta-thalassemia: a case report. Hemoglobin 43(2):137–139
156. Lamsfus-Calle A, Daniel-Moreno A, Antony JS, Epting T, Heumos L, Baskaran P et al (2020) Comparative targeting analysis of KLF1, BCL11A, and HBG1/2 in CD34(+) HSPCs by CRISPR/Cas9 for the induction of fetal hemoglobin. Sci Rep 10(1):10133
157. Hua P, Badat M, Hanssen LLP, Hentges LD, Crump N, Downes DJ et al (2021) Defining genome architecture at base-pair resolution. Nature 595(7865):125–129
158. Viprakasit V, Ekwattanakit S, Riolueang S, Chalaow N, Fisher C, Lower K et al (2014) Mutations in Kruppel-like factor 1 cause transfusion-dependent hemolytic anemia and persistence of embryonic globin gene expression. Blood 123(10):1586–1595
159. Arnaud L, Saison C, Helias V, Lucien N, Steschenko D, Giarratana MC et al (2010) A dominant mutation in the gene encoding the erythroid transcription factor KLF1 causes a congenital dyserythropoietic anemia. Am J Hum Genet 87(5):721–727
160. de-la-Iglesia-Inigo S, Moreno-Carralero MI, Lemes-Castellano A, Molero-Labarta T, Mendez M, Moran-Jimenez MJ (2017) A case of congenital dyserythropoietic anemia type IV. Clin Case Rep 5(3):248–252
161. Deguise MO, Blain S, Simpson E, Liebman M, Ferretti E (2023) Congenital dyserythropoietic anemia type IV in the genetic era: a rare neonatal case report of rapid identification with a review of the literature. Pediatr Blood Cancer 70(5):e30245
162. Hamada M, Doisaki S, Okuno Y, Muramatsu H, Hama A, Kawashima N et al (2018) Whole-exome analysis to detect congenital hemolytic anemia mimicking congenital dyserythropoietic anemia. Int J Hematol 108(3):306–311
163. Jaffray JA, Mitchell WB, Gnanapragasam MN, Seshan SV, Guo X, Westhoff CM et al (2013) Erythroid transcription factor EKLF/KLF1 mutation causing congenital dyserythropoietic anemia type IV in a patient of Taiwanese origin: review of all reported cases and development of a clinical diagnostic paradigm. Blood Cells Mol Dis 51(2):71–75
164. Kohara H, Utsugisawa T, Sakamoto C, Hirose L, Ogawa Y, Ogura H et al (2019) KLF1 mutation E325K induces cell cycle arrest in erythroid cells differentiated from congenital dyserythropoietic anemia patient-specific induced pluripotent stem cells. Exp Hematol 73(25–37):e8
165. Kulczynska-Figurny K, Bieker JJ, Siatecka M (2020) Severe anemia caused by dominant mutations in Kruppel-like factor 1 (KLF1). Mutat Res Rev Mutat Res 786:108336
166. Ortolano R, Forouhar M, Warwick A, Harper D (2018) A Case of Congenital Dyserythropoeitic Anemia Type IV Caused by E325K Mutation in Erythroid Transcription Factor KLF1. J Pediatr Hematol Oncol 40(6):e389–ee91
167. Ilsley MD, Huang S, Magor GW, Landsberg MJ, Gillinder KR, Perkins AC (2019) Corrupted DNA-binding specificity and ectopic transcription underpin dominant neomorphic mutations in KLF/SP transcription factors. BMC Genomics 20(1):417
168. Ferrer-Vicens I, Ferguson DCJ, Wilson MC, Heesom KJ, Bieker JJ, Frayne J (2023) Novel human cellular model of CDA IV enables comprehensive analysis revealing molecular basis of disease phenotype. Blood 141(25):3039–3054
169. Varricchio L, Planutis A, Manwani D, Jaffray J, Mitchell WB, Migliaccio AR et al (2019) Genetic disarray follows mutant KLF1-E325K expression in a congenital dyserythropoietic anemia patient. Haematologica 104(12):2372–2380
170. Singleton BK, Lau W, Fairweather VS, Burton NM, Wilson MC, Parsons SF et al (2011) Mutations in the second zinc finger of human EKLF reduce promoter affinity but give rise to benign and disease phenotypes. Blood 118(11):3137–3145
171. Agre P, Smith BL, Baumgarten R, Preston GM, Pressman E, Wilson P et al (1994) Human red cell Aquaporin CHIP. II. Expression during normal fetal development and in a novel form of congenital dyserythropoietic anemia. J Clin Invest 94(3):1050–1058
172. Parsons SF, Jones J, Anstee DJ, Judson PA, Gardner B, Wiener E et al (1994) A novel form of congenital

dyserythropoietic anemia associated with deficiency of erythroid CD44 and a unique blood group phenotype [In(a-b-), Co(a-b-)]. Blood 83(3):860–868

173. Tang W, Cai SP, Eng B, Poon MC, Waye JS, Illum N et al (1993) Expression of embryonic zeta-globin and epsilon-globin chains in a 10-year-old girl with congenital anemia. Blood 81(6):1636–1640
174. Wickramasinghe SN, Illum N, Wimberley PD (1991) Congenital dyserythropoietic anaemia with novel intra-erythroblastic and intra-erythrocytic inclusions. Br J Haematol 79(2):322–330
175. Ravindranath Y, Johnson RM, Goyette G, Buck S, Gadgeel M, Gallagher PG (2018) KLF1 E325K-associated congenital dyserythropoietic anemia type IV: insights into the variable clinical severity. J Pediatr Hematol Oncol 40(6):e405–e4e9
176. Shefer Averbuch N, Steinberg-Shemer O, Dgany O, Krasnov T, Noy-Lotan S, Yacobovich J et al (2018) Targeted next generation sequencing for the diagnosis of patients with rare congenital anemias. Eur J Haematol 101(3):297–304
177. Russo R, Andolfo I, Manna F, Gambale A, Marra R, Rosato BE et al (2018) Multi-gene panel testing improves diagnosis and management of patients with hereditary anemias. Am J Hematol 93(5):672–682
178. Jamwal M, Aggarwal A, Sharma P, Bansal D, Das R (2021) Congenital dyserythropoietic anemia type IV with high fetal hemoglobin caused by heterozygous KLF1 p.Glu325Lys: first report in an Indian infant. Ann Hematol 100(1):281–283
179. Massey GV, Chesney A, Gadgeel M, Ravindranath Y (2024) Congenital dyserythropoietic anemia type IV with KLF1 E325K mutation: a new case with dysmorphic male genitalia. Pediatr Blood Cancer 71(6):e30955. https://doi.org/10.22541/au.169053301.19232650/v1
180. Perkins AC, Bieker J (2021) Congenital Anemia Phenotypes Due to KLF1 Mutations. J Pediatr Hematol Oncol 43(1):e148–e1e9
181. Belgemen-Ozer T, Gorukmez O (2020) A very rare congenital dyserythropoietic anemia variant-type IV in a patient with a novel mutation in the KLF1 gene: a case report and review of the literature. J Pediatr Hematol Oncol 42(6):e536–ee40
182. Yang K, Nie W, Huang Q, Liao G, Xiao J, Yin X (2023) Hematopoietic cell transplantation for congenital dyserythropoietic anemia IV caused by compound heterozygous KLF1 mutations. Ann Hematol 102(6):1621–1624
183. Lyon MF, Glenister PH, Loutit JF, Peters J (1983) Dominant haemolytic anemia. Mouse News Letters 68:68
184. Heruth DP, Hawkins T, Logsdon DP, Gibson MI, Sokolovsky IV, Nsumu NN et al (2010) Mutation in erythroid specific transcription factor KLF1 causes Hereditary Spherocytosis in the Nan hemolytic anemia mouse model. Genomics 96(5):303–307
185. Siatecka M, Sahr KE, Andersen SG, Mezei M, Bieker JJ, Peters LL (2010) Severe anemia in the Nan mutant mouse caused by sequence-selective disruption of erythroid Kruppel-like factor. Proc Natl Acad Sci USA 107(34):15151–15156
186. Nebor D, Graber JH, Ciciotte SL, Robledo RF, Papoin J, Hartman E et al (2018) Mutant KLF1 in Adult Anemic Nan Mice Leads to Profound Transcriptome Changes and Disordered Erythropoiesis. Sci Rep 8(1):12793
187. Planutis A, Xue L, Trainor CD, Dangeti M, Gillinder K, Siatecka M et al (2017) Neomorphic effects of the neonatal anemia (Nan-Eklf) mutation contribute to deficits throughout development. Development 144(3):430–440
188. Gillinder KR, Ilsley MD, Nebor D, Sachidanandam R, Lajoie M, Magor GW et al (2017) Promiscuous DNA-binding of a mutant zinc finger protein corrupts the transcriptome and diminishes cell viability. Nucleic Acids Res 45(3):1130–1143
189. Cantu I, van de Werken HJG, Gillemans N, Stadhouders R, Heshusius S, Maas A et al (2019) The mouse KLF1 Nan variant impairs nuclear condensation and erythroid maturation. PLoS One 14(3):e0208659
190. Korporaal A, Gillemans N, Heshusius S, Cantu I, van den Akker E, van Dijk TB et al (2021) Hemoglobin switching in mice carrying the Klf1(Nan) variant. Haematologica 106(2):464–473
191. Roux KJ, Kim DI, Raida M, Burke B (2012) A promiscuous biotin ligase fusion protein identifies proximal and interacting proteins in mammalian cells. J Cell Biol 196(6):801–810
192. Kurita R, Suda N, Sudo K, Miharada K, Hiroyama T, Miyoshi H et al (2013) Establishment of immortalized human erythroid progenitor cell lines able to produce enucleated red blood cells. PLoS One 8(3):e59890
193. Zhang F, Zhang B, Wang Y, Jiang R, Liu J, Wei Y et al (2023) An extra-erythrocyte role of haemoglobin body in chondrocyte hypoxia adaption. Nature 622(7984):834–841

Transcription Factor TAL1 in Erythropoiesis

11

Johannes Fechner and Jörn Lausen

Abstract

Lineage-specific transcription factors (TFs) regulate differentiation of hematopoietic stem cells (HSCs). They are decisive for the establishment and maintenance of lineage-specific gene expression programs during hematopoiesis. For this they create a regulatory network between TFs, epigenetic cofactors, and microRNAs. They activate cell-type specific genes and repress competing gene expression programs. Disturbance of this process leads to impaired lineage fidelity and diseases of the blood system. The TF T-cell acute leukemia 1 (TAL1) is central for erythroid differentiation and contributes to the formation of distinct gene regulatory complexes in progenitor cells and erythroid cells. A TAL1/E47 heterodimer binds to DNA with the TFs GATA-binding factor 1 and 2 (GATA1/2), the cofactors LIM domain only 1 and 2 (LMO1/2), and LIM domain-binding protein 1 (LDB1) to form a core TAL1 complex. Furthermore, cell-type-dependent interactions of TAL1 with other TFs such as with runt-related transcription factor 1 (RUNX1) and Kruppel-like factor 1 (KLF1) are established. Moreover, TAL1 activity is regulated by the formation of TAL1 isoforms, posttranslational modifications (PTMs), and microRNAs. Here, we describe the function of TAL1 in normal hematopoiesis with a focus on erythropoiesis.

Keywords

Transcription factor · TAL1 · Epigenetic regulation · Gene regulation · Erythropoiesis · Hematopoiesis · Posttranslational modification · Protein isoform · MicroRNA

Abbreviations

aa	amino acids
AGM	aorto-gonadal-mesonephric
AKT	AKT serine/threonine kinase
$BCLX_L$	B-cell lymphoma-extra large
BFU-E	burst-forming unit-erythroid
bHLH	basic helix-loop-helix
bp	base pair
BRG1	Brahma-related gene 1
CBP	CREB-binding protein
CD34	cluster of differentiation 34
CDKN1A	cyclin-dependent kinase inhibitor 1A
CFU-E	colony-forming unit-erythroid
CHIP	carboxyl terminus of hsc70-interacting protein

J. Fechner · J. Lausen (✉)
Department of Eukaryotic Genetics, Institute of Biomedical Genetics, University of Stuttgart, Stuttgart, Germany
e-mail: joern.lausen@ibmg.uni-stuttgart.de

T. Borggrefe, B. D. Giaimo (eds.), *Transcription factors in blood cell development*, Advances in Experimental Medicine and Biology 1459, https://doi.org/10.1007/978-3-031-62731-6_11

ChIP-Seq	Chromatin immunoprecipitation followed by deep sequencing
c-MYC	cellular myelocytomatosis
DU.582	Duke University 528
E2a	E2 and E47
E47	E2A immunoglobulin enhancer-binding factors E12/E47 (transcription factor 3, TCF3)
EGF	epidermal growth factor
eIF2	eukaryotic initiation factor 2
eIF4	eukaryotic initiation factor 4
EP300	E1A-binding protein p300
ePBGD	erythroid porphobilinogen deaminase
EPO	erythropoietin
ERG	ETS-related gene
ERK1/MEK1	extracellular signal-regulated kinase 1 and mitogen-activated protein kinase 1
ESCs	embryonic stem cells
ET	embryo transfer
ETO	Eight-twenty-one
FLI1	friend leukemia integration 1
Fl-TAL1	full-length TAL1
FOG	friend of GATA1
GATA1	GATA-binding factor 1
GFI1B	growth factor-independent 1B transcriptional repressor
GPA	glycophorin A
H3	histone 3
HDAC2	histone deacetylase 2
HLH	helix-loop-helix
HSCs	hematopoietic stem cells
hSET1	SET domain containing 1A, histone lysine methyltransferase
HSPCs	hematopoietic stem/progenitor cells
ID1	inhibitor of DNA binding 1
KAT2B	lysine acetyltransferase 2B
KAT3A	lysine acetyltransferase 3A
KAT3B	lysine acetyltransferase 3B
kB	kilobase
kDa	kilodalton
KDM1A	lysine demethylase 1A
KLF1	Kruppel-like factor 1

KMT2F	lysine methyltransferase 2F
KO	knockout
LDB1	LIM domain-binding protein 1
LMO1/2	LIM domain only 1/2
LSD1	lysine-specific histone demethylase 1A
LT-HSC	long-term hematopoietic stem cells
LYL1	lymphoblastic leukemia 1
MEP	megakaryocytic/erythroid progenitor cells
MPP	multipotent progenitor
mSIN3A	SIN3 transcription regulator family member A
MYB	MYB proto-oncogene
NCOR	nuclear receptor corepressor
NF-E2	nuclear factor, erythroid 2
ORFs	open reading frames
P/CAF	EP300/CBP-associated factor
P4.2	human erythrocyte protein 4.2
PI3K	phosphoinositide-3-kinase
pIpC	polyinosinic-polycytidylic acid
PKA	protein kinase A
PKACA	cAMP-dependent protein kinase A
PRMT1	protein arginine methyltransferase 1
PRMT6	protein arginine methyltransferase 6
PU.1/SPI1	hematopoietic transcription factor PU.1/spleen focus-forming virus (SFFV) proviral integration protein
RARS	refractory anemia with ring sideroblasts
RBM15	RNA-binding motif protein 15
RUNX1	runt-related transcription factor 1
SCL	stem cell leukemia
SF3B1	splicing factor 3b subunit 1
shRNA	short hairpin RNA
SKP2	S-phase kinase-associated protein 2
ST-HSC	short-term hematopoietic stem cells
SWI/SNF	switch/sucrose non-fermen table

T-ALL	T-cell acute lymphoblastic leukemia
TAL1	T-cell acute leukemia 1
TCF3	transcription factor 3
TFs	transcription factors
TGFβ	transforming growth factor beta
TIE2	TEK receptor tyrosine kinase
TIE2Cre	Cre recombinase under the control of TIE2 regulatory Elements
TSS	transcription start sites
UCB	umbilical cord blood
UTR	untranslated region
YS	yolk sac

11.1 Introduction

Transcription factors (TFs) play a key role in normal hematopoiesis and in leukemia [1, 2]. In 1989, a translocation between chromosomes 1 and 14 (t(1;14)(p32;q11)) was identified in a leukemia stem cell line (DU.528) [3]. This translocation results in the expression of a fusion transcript that is also present in a subset of patients with T-cell acute lymphoblastic leukemia (T-ALL). The identified gene was named stem cell leukemia (*SCL*). Because of its role in leukemia, the gene is also known as T-cell acute leukemia 1 (*TAL1*). Subsequent studies in the following years confirmed the hypothesis that this gene was important for hematopoiesis. In normal hematopoiesis, TAL1 is required for primitive and definitive hematopoiesis [4–9]. In definitive hematopoiesis, TAL1 is essential for erythropoiesis and megakaryopoiesis [10]. TAL1 also plays an important role in hematopoietic malignancies. The *TAL1* gene is often disrupted in T-cell acute lymphoblastic leukemia, where it's ectopically expressed in T cells due to chromosomal translocations or other genetic events, leading to leukemogenesis [11, 12]. This aberrant expression disrupts the normal program of T-cell differentiation and promotes the proliferation of immature T cells, contributing to the development of leukemia [13].

The *TAL1* gene is located on the human chromosome 1 at position p33 (1p33). It spans approximately 16 kB and consists of three protein-coding and several noncoding exons. Gene expression is regulated by several lineage-specific promoters, which are controlled by partially redundant proximal and distal enhancers. The resulting mRNAs serve as transcripts for two different isoforms of 42 and 22 kDa (Fig. 11.1). The TAL1 protein is a class 2 basic helix-loop-helix (bHLH) transcription factor. bHLH transcription factors can form homo- or heterodimers through their helix-loop-helix (HLH) domain. Class 2 bHLH TFs, such as TAL1, can only form heterodimers with class 1 bHLH proteins. TAL1 primarily heterodimerizes with the class 1 bHLH proteins E2a (E2 and E47), HEB, and E2-2 [14–16]. The heterodimer binds to a palindromic DNA motif known as the E-box (CANNTG) through the basic domain present in both heterodimerization partners [17]. However, full TAL1 activity requires interaction with LMO 1/2 and LDB1 as well as context-dependent additional TFs and cofactors (Fig. 11.1) [4]. Interestingly, TAL1 can play an activating or repressive role depending on the context [18]. In addition, TAL1 can function independently of DNA binding [19]. The different properties of TAL1 will be further discussed in the following sections with a focus on the role of TAL1 in erythropoiesis.

11.2 TAL1 in Embryonic Erythropoiesis

The development of the blood system in vertebrates occurs in several waves. The first two waves occur in the yolk sac (YS) and give rise to transient populations of blood cells such as primitive red blood cells, megakaryocytes, and macrophages [20–22]. The third wave occurs intraembryonically, specifically in the aorto-gonadal-mesonephric (AGM) region. This is where the hematopoietic stem and progenitor cells are generated, which later ensure the continuous production of blood cells throughout life [23]. The molecular program controlling hemato-

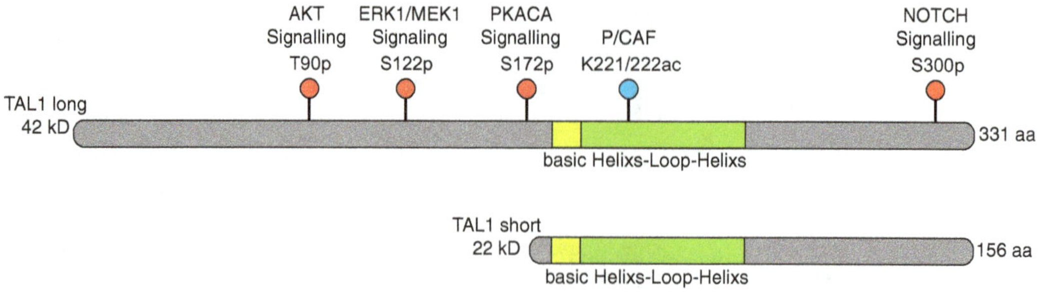

Fig. 11.1 Schematic representation of the TAL1 isoforms. A schematic of the long (42 kD) and the short (22 kD) TAL1 isoforms is represented. The position of the basic helix-loop-helix (bHLH) domain is shown. Furthermore, posttranscriptional modifications (PTMs) of TAL1 are indicated. AKAT AKT serine/threonine kinase, ERK1/MEK1 extracellular-signal regulated kinase 1 and mitogen-activated protein kinase 1, P/CAF EP300/CBP-associated factor, also known as lysine acetyltransferase 2B (KAT2B), PKACA cAMP-dependent protein kinase A, TAL1 T-cell acute leukemia 1

poietic specification is activated shortly after gastrulation in a group of extraembryonic mesodermal cells destined to form blood islands. These blood islands are composed predominantly of primitive erythrocytes and are surrounded by endothelial cells. Beginning at 9.5 embryo transfer (ET), myeloid and lymphoid progenitor cells and hematopoietic stem cells from the YS and AGM region begin to populate the fetal liver. The fetal liver then serves as the primary hematopoietic organ until birth. Hematopoietic stem cells (HSCs) then migrate to the bone marrow [23]. TAL1 plays a very early role in hematopoiesis. The first expression of TAL1 can be detected at 7.5 ET in the mesodermal cells of the YS [24], which form the blood islands [25, 26]. Throughout development, TAL1 is expressed in all hematopoietic tissues. In addition, TAL1 expression can be detected in the developing nervous system, skeletal system, and endothelial cells [24, 27, 28]. Remarkably, several studies have shown that *TAL1*-deficient mouse embryos die very early, between ET 8.5 and 10, due to a complete absence of blood cells [6–9, 29]. Further studies using chimeric mouse embryos demonstrated that *TAL1*$^{-/-}$ embryonic stem cells (ESCs) do not contribute to the formation of the blood system [8]. All other tissues tested showed an involvement of *TAL1*$^{-/-}$ ESCs in their development. Similarly, *TAL1*$^{-/-}$ ESCs do not show any differentiation into blood cells "in vitro." However, an ectopic expression of TAL1 in *TAL1*$^{-/-}$ cells can rescue this differentiation defect [8]. This highlights the critical role of TAL1 in both primitive and definitive hematopoiesis. The same effect has been demonstrated in *Danio rerio* and *Xenopus laevis* embryos. Knockdown of TAL1 resulted in similar phenotypic manifestations as in mouse embryos, demonstrating the conserved mode of action of the TAL1 protein from amphibians over teleost fish to mammals [30–34]. Additionally, it was shown that TAL1 expression is responsible for determining the hematopoietic fate within a very brief developmental window [35]. Using TEK receptor tyrosine kinase-dependent (TIE2-dependent) *TAL1* knockout (KO) mice, it was found that short-term expression of TAL1 prior to the KO controlled by TIE2Cre is sufficient to prevent the early lethality of TAL1-deficient embryos. The TIE2Cre-conditional *TAL1* KO embryos survived the KO until ET 13.5–ET 14.5 and showed circulating primitive erythrocytes and an apparently normal vascular system. Embryonic death occurred due to bleeding and the formation of edema, indicating a defect in erythrocyte maturation [35]. The results also show that TAL1 expression is not essential for the self-renewal and multipotency of long-term hematopoietic stem cells (LT-HSCs) nor for the establishment of the HSCs pool in the fetal liver. However, TAL1 is important for the correct differentiation of erythrocytes and megakaryocytes throughout development. The important role of TAL1 in erythropoiesis was also emphasized in further studies [29]. In this study, TAL1 was expressed under the control of its 3′enhancer in $TAL1^{-/-}$ embryos. The 3′enhancer regulates the expression of TAL1 in fetal and adult HSCs and is therefore also referred to as a stem cell enhancer. The expression of TAL1 by this specific enhancer rescues the formation of early hematopoietic precursors and also leads to an apparently normal vascular development. However, the embryos die at the same stage as the $TAL1^{-/-}$ embryos due to a complete absence of primitive and definitive erythropoiesis [29]. TAL1 is required for the formation of hematopoietic cells during embryonic development. Interestingly, the loss of hematopoietic stem cells in the TAL1 KO can be rescued by a DNA-binding-deficient TAL1 mutant [36]. Furthermore, the DNA-binding domain of TAL1 is not essential for erythroid differentiation of human $CD34^+$ cells. It is possible that TAL1 is recruited to target genes by other TFs, such as GATA1 [37]. In addition to activating the hematopoietic gene expression program during embryogenesis, TAL1 has a repressive effect on the alternative cardiac muscle-specific gene expression program [38]. In this process, TAL1 binds to epigenetically primed enhancers and modulates their activity [39].

11.3 TAL1 in Adult Hematopoiesis and Erythropoiesis

As described in the previous paragraphs, HSCs migrate to the bone marrow toward the end of embryonic development. The subsequent adult or definitive hematopoiesis primarily takes place in the bone marrow. In this process, starting from a self-renewing HSC, over ten specialized hematopoietic cell types are formed [40]. TAL1 plays an essential role at various points during this process. However, some aspects of TAL1's mode of action in adult hematopoiesis have not yet been conclusively clarified. TAL1 is expressed at different levels during blood cell development. In particular, TAL1 is strongly expressed in LT-HSCs but also in short-term hematopoietic stem cells (ST-HSCs) and multipotent progenitor cells (MPPs), albeit somewhat less in comparison to LT-HSCs [41]. In addition, TAL1 can be found in myeloid and lymphoid precursor cells derived from MPPs. In the lymphoid line, TAL1 is downregulated after advanced differentiation, so that TAL1 is not present in B and T cells. However, in the myeloid lineage, TAL1 plays a significant role. Especially the differentiation of megakaryocytes and erythrocytes depends on TAL1 expression. Furthermore, TAL1 can be detected in mast cells and basophilic cells, as well as in smaller amounts in neutrophils, eosinophils, macrophages, and osteoclasts [27, 42–45].

The exact role of TAL1 in HSCs is not yet fully understood. Studying the function of TAL1 during adult hematopoiesis is complicated by the early lethality of *TAL1* KO mice. Initial analyses using conditional *TAL1* KO mice showed no influence of TAL1 on HSCs function. Thus, *TAL1*-deleted HSCs showed no influence on engraftment, self-renewal, or their multipotency [46]. Subsequent studies came to different results; here *TAL1*-deleted HSCs showed a defect in repopulation ability. In competitive repopulation experiments, both *TAL1*-heterozygous cells and, to a greater extent, *TAL1*-deleted HSCs contributed significantly less to the formation of myeloid, B and T cells than wild-type control cells. Since this effect was consistent in serial transplants, researchers concluded that TAL1 is involved in the normal function of ST-HSCs [47]. These results were confirmed in a subsequent study in which TAL1 was reduced in HSCs using shRNAs [48]. The partly differing results could possibly be explained by compensation effects. Lymphoblastic leukemia 1 (LYL1), a paralog of TAL1, is expressed in many cell types during hematopoiesis [49]. Studies showed that LYL1 is essential for normal HSCs function in the absence of TAL1. Cells lacking both *LYL1* and *TAL1* undergo rapid apoptosis. However, if these cells have a copy of *LYL1*, they contribute to the formation of the HSCs pool. TAL1 and LYL1 are thus important for the development and maintenance of HSCs [50]. In experiments where TAL1 was reduced using shRNAs, the normal LYL1 level was not sufficient to compensate for the repopulation defect [48]. Further investigations point to a role of TAL1 in the quiescence and long-term competence of HSCs. LT-HSCs in the G_0 phase show higher TAL1 expression levels compared to cells outside this phase. Downregulation of TAL1 using shRNAs led to a threefold increase in cells in the G_1 phase. It is believed that these effects may be mediated by cyclin-dependent kinase inhibitor 1A (CDKN1A) and inhibitor of DNA binding 1 (ID1). Both proteins are essential for maintaining quiescence and are regulated by TAL1 [41].

In contrast to the role of TAL1 in HSCs, the results about the role of TAL1 in erythro- and megakaryopoiesis are clear: All previous studies confirm an important function of TAL1 in erythro- and megakaryopoiesis. In adult hematopoiesis, erythroid cells and megakaryocytic cells arise from bipotent megakaryocytic/erythroid progenitor cells (MEPs). Erythro-megakaryocytic transcription factors play a role in differentiation into MEPs, and these include GATA1, friend of GATA1 (FOG), and nuclear factor erythroid 2 (NF-E2). Other TFs are restricted to the megakaryocytic lineage, friend leukemia integration 1 (FLI1) or erythroid cells, KLF1. FLI1 and KLF1 regulate each other in MEPs, and their expression is adjusted after lineage decision. In erythroid differentiation, KLF1 plays the role of a master gene. KLF1, GATA1, and TAL1 jointly occupy numerous erythroid genes.

TAL1 is one of the most important factors in the differentiation of MEPs. Compared to other precursor cells, TAL1 is most strongly expressed in MEPs [41, 51]. The deletion of *TAL1* in conditional KO mice shows that TAL1 is necessary for the normal maturation of erythrocytes and megakaryocytes [52]. In this study, the KO of *TAL1* is induced by the injection of polyinosinic-polycytidylic acid (pIpC) into the corresponding transgenic mice. However, pIpC has a nonspecific effect on hematopoiesis: The injection results in a decrease in platelets, red blood cells, and leukocytes. In control animals where *TAL1* was not depleted, wild-type animals recover after a week and show a normal count of platelets, red blood cells, and leukocytes. In the *TAL1*-depleted animals, only the leukocyte count normalizes. This emphasizes the independence of lymphoid cells from TAL1 and the absolute necessity of TAL1 for erythroid and megakaryocytic differentiation [52]. Further studies with conditional KO mice or studies in which TAL1 was downregulated using shRNAs all show a lack of formation of megakaryocytes and erythrocytes from *TAL1*-deleted HSCs [46–48]. This confirms earlier studies that demonstrated that the expression of an antisense mRNA of *TAL1* or the expression of a dominant-negative TAL1 variant blocks the erythroid differentiation of various cell lines. Interestingly, the expression of the antisense mRNA showed a dose-dependent effect: The higher the concentration of the antisense mRNA, the stronger the inhibition of erythroid differentiation [53]. This emphasizes the essential role of TAL1 in erythroid differentiation. It has also been shown that TAL1 can act independently of its DNA binding. After overexpression of TAL1 or a DNA-binding defective TAL1 variant in umbilical cord blood cells (UCB $CD34^+$), enhanced erythroid differentiation is observed. This is characterized by increased β-globin and erythroid porphobilinogen deaminase (ePBGD) expression and an increase in erythroid precursor cells. Furthermore, after overexpression, there is a high proportion of colony-forming unit-erythroid (CFU-E) compared to burst-forming unit-erythroid (BFU-E) in colony assays. These effects occur both after overexpression of TAL1 and after overexpression of the DNA-binding defective TAL1 variant. However, cells after overexpression of the DNA-binding defective TAL1 variant do not show increased production of glycophorin A positive (GPA^+) cells. Accordingly, a TAL1 variant that is unable to bind DNA can take over most of the functions of the TAL1 protein, but for complete erythroid differentiation, the DNA binding of TAL1 is needed [19]. This is consistent with rescue experiments in ESCs: Here, the overexpression of a DNA-binding defective TAL1 variant can rescue primitive erythropoiesis from *TAL1*$^{-/-}$ ES and initiate definitive erythropoiesis, but terminal erythropoiesis remains incomplete [36].

In addition to the previously described qualities, TAL1 exhibits also an anti-apoptotic function in late erythroid precursor cells as well as in early erythrocytes [54]. The formation of erythrocytes is restricted by caspases, which can either cause a reversible halt in growth and differentiation or induce apoptosis of the erythrocytes [55]. Studies have shown that in this context, TAL1 is one of the target proteins of caspases. The activation of caspases in CD34+ precursor cells results in a rapid degradation of TAL1, followed by a degradation of GATA1 and the cessation of cell differentiation or the onset of cell apoptosis. Conversely, the expression of a caspase-resistant TAL1 variant leads to the differentiation of the cells despite the activation of caspases. This is enabled by the absence of degradation of GATA1 and the maintenance of B-cell lymphoma-extra large ($BCLX_L$) protein levels. $BCLX_L$ plays an important role in the survival of immature erythrocytes due to its anti-apoptotic effect [54, 56]. Together this highlights the central role of the TAL1 protein in erythropoiesis.

11.4 The TAL1 Gene Regulatory Complex

TAL1 is a cell-type-specific TF that binds DNA at specific sequences. The consensus TAL1-binding motif is the DNA sequence CANNTG, but it can also be present at other sites. Since TAL1 cannot bind to DNA as a homodimer, it forms heterodimers with ubiquitously expressed bHLH E-proteins, such as E12/E47, HEB, or E2-2 [57]. In addition, the non-DNA-binding proteins LMO1 or LMO2 and the LIM domain-binding protein (LDB1/2) protein are typically found with TAL1 at target genes [58, 59]. The LMO proteins have been described as bridging factors [60], and LDB1 is able to serve as mediator of long-range interactions [61]. The quaternary TAL1 core complex has been described in all TAL1 expressing cells and is extended by other TFs and cofactors, with GATA1/2 being the most prominent (Fig. 11.2a).

In hematopoietic progenitors a "heptad" of TFs with TAL1-, LYL1-, GATA2-, FLI1-, and ETS-related gene (ERG), RUNX1 and LMO2 form regulatory complexes (Fig. 11.2b) [62–67]. The heptad TFs fulfill overlapping and divergent roles during lineage branching. RUNX1, for example, is downregulated during erythroid differentiation but plays a critical role in megakaryocytic gene expression [68, 69]. The association of TAL1 with GATA factors plays a key role in defining TAL1 function. Chromatin immunoprecipitation followed by deep sequencing (ChIP-Seq) revealed that the major determi-

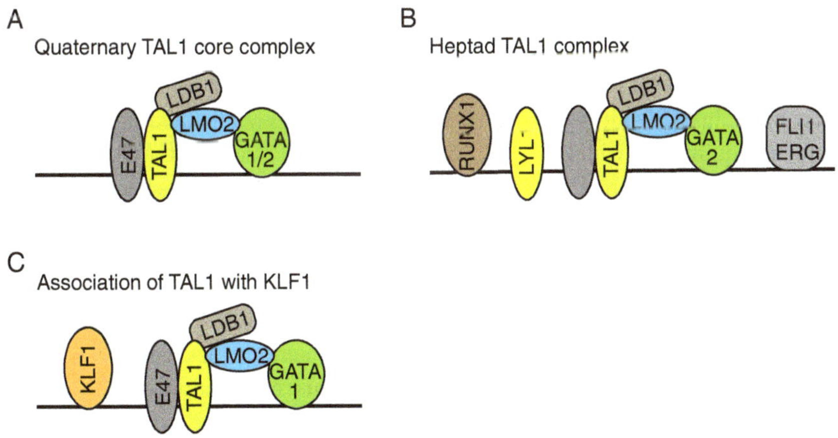

Fig. 11.2 TAL1 complexes in hematopoiesis. (**a**) Quaternary TAL1 core complex, which might be present in all TAL1 expressing cells. (**b**) Heptad TAL1 complex, which plays a role in progenitor cells. **c** Erythroid TAL1 complex with KLF1. This complex activates erythroid specific genes. E47 E2A immunoglobulin enhancer-binding factors E12/E47 (transcription factor 3, TCF3), ERG ETS-related gene, FLI1 friend leukemia integration 1, GATA1/2 GATA-binding factor 1/2, KLF1 Kruppel-like factor 1, LDB1 LIM domain-binding protein 1, LMO2 LIM domain only 2, LYL1 lymphoblastic leukemia 1, RUNX1 runt-related transcription factor 1, TAL1 T-cell acute leukemia 1

nant for TAL1 binding is the presence of a GATA motif, rather than the presence of a perfect CANNTG motif. This argues for the important role of GATA factors as a determinant of TAL1 binding to genomic loci [70, 71]. Indeed, the presence of a perfect E-box motif appears to be less important than the presence of GATA for TAL1 binding [61, 71]. This notion is consistent with the observation that TAL1 function in blood specification is independent of its DNA-binding domain [36]. In progenitor cells TAL1 is foremost associated with GATA2, whereas during erythropoiesis association with GATA1 becomes more prominent. A heptad of TFs including TAL1, LYL1, GATA2, ERG, RUNX1 and LMO2 cooperate in adult blood cell lineage maturation. They act together in hematopoietic stem/progenitor cells (HSPCs) and in erythroid/megakaryocytic differentiation [62, 66]. During early erythropoiesis TAL1/GATA2 also fulfill repressive functions, whereas switching from GATA2 to GATA1 activates erythroid genes [62, 72, 73]. This is accompanied by a shift in TAL1 binding to target genes [62, 70], which is dependent on the presence of GATA1. ChIP-Seq data support the notion that TAL1 is already bound to primed megakaryocytic genes in progenitors [70], whereas erythroid genes are only bound by TAL1 in the erythroid lineage, with varying binding patterns in different subpopulations as maturation progresses. In erythrocytes, TAL1 target genes overlap with KLF1. KLF1 is a central erythroid master gene. In MEPs, it is considered to be in a regulatory equilibrium with FLI1 [74, 75]. When the balance tips toward KLF1, differentiation toward the erythroid lineage takes place. Interestingly TAL1, GATA1, and KLF1 form a core erythroid network as they co-occupy a large number of erythroid genes [76, 77], including the TFs themselves (Fig. 11.2c) [71]. Thus, TAL1 and KLF1 are involved in a positive feedback loop that drives erythroid gene expression [78], such as the human adult β-globin expression [79, 80]. Other TFs such as RUNX1 and SPI1 proto-oncogene (also known as PU.1) counteract the activating function of the TAL1, GATA1, and KLF1 in progenitor cells [69, 76].

The importance of the regulatory TAL1 complex described in the preceding is supported by the fact that an overexpression of the complex components can reprogram non-hematopoietic cells into blood cells. It has been shown that the simultaneous overexpression of ERG, GATA2, LMO2, RUNX1c, and TAL1 results in a robust reprogramming of embryonic and adult fibroblasts into blood progenitor cells [81]. Furthermore, the expression of TAL1 in combination with GATA1, LMO2, and the cellular myelocytomatosis (c-MYC) proto-oncogene can reprogram fibroblasts into erythroid precursor cells that possess an embryonic character. The additional expression of KLF1 or the MYB proto-oncogene (MYB) leads to the formation of adult erythroid progenitors [82].

11.5 Activating and Repressive Functions of TAL1

TAL1 has been shown to act as a transcriptional activator of erythroid genes and megakaryocytic genes. In these cases, association of the TAL1 complex with coactivators, such as CREB-binding protein/lysine acetyltransferase 3A (CBP/KAT3A) and E1A-binding protein P300/lysine acetyltransferase 3B (EP300/KAT3B) and EP300/CBP-associated factor (P/CAF, also known as KAT2B (lysine acetyltransferase 2B)), contributes to gene activation [83, 84]. However, in most cases studied, it is not clear if TAL1 or an associated TF such as GATA1 recruits the coactivators to target genes. Interestingly, TAL1 can also act as a repressor of gene transcription through different mechanisms. Firstly, TAL1/E47 heterodimers are less activating than E47/E47 homodimers, so that the formation of TAL1/E47 heterodimers can act repressive in relative terms [85]. This observation has been linked to the deregulation of gene expression in T-ALL cases where TAL1 is ectopically expressed [86, 87]. On the other hand, TAL1 is involved in active repression and can be associated with transcriptional corepressor proteins. TAL1-mediated repression of alternative cell fates is central to lineage differentiation in development. In particular, TAL1 interacts with subunits of the repressive polycomb complex, PRC1 (Polycomb-repressive complex 1) [88], to repress cardiac genes. TAL1 also binds to cardiac enhancers and represses cardiac-specific TFs [38, 39]. In erythroid cells TAL1 interacts with transcriptional corepressor proteins, such as ETO2 (*CBFA2T3*), nuclear receptor corepressor (NCOR), SIN3 transcription regulator family member A (mSIN3A), and growth factor-independent 1B transcriptional repressor (GFI1B) to prevent developmentally primed genes from being ectopically expressed [89–91]. In addition, TAL1 is associated with the lysine-specific histone demethylase 1A/lysine demethylase 1A (LSD1/KDM1A) on erythroid genes in undifferentiated cells, LSD1 leaves the repressed genes upon onset of erythropoiesis, and the erythroid genes are activated [92]. Similarly, TAL1 is associated with RUNX1 in a repressive complex with protein arginine methyltransferase 6 (PRMT6) on primed genes [69, 93]. TAL1 is also associated with the switch/sucrose non-fermentable (SWI/SNF) protein Brahma-related gene 1 (BRG1). In this context TAL1/BRG1 acts with histone deacetylase 2 (HDAC2) and mSIN3A to repress the erythroid human erythrocyte protein 4.2 (P4.2) gene [94]. In addition, BRG1-induced nucleosome shifting facilitates TAL1 binding [95]. Thus, the interaction of TAL1 with chromatin-remodeling factors may be more complex than gene activation or repression alone. Taken together, the present data suggest that the repressive function of TAL1 contributes to the repression of primed genes of alternate lineages and to the repression of cell-type-dependent genes in progenitors until differentiation is initiated.

11.6 TAL1 Isoforms in Erythropoiesis

The *TAL1* locus consists of three protein-coding exons, several noncoding exons, and a rather long 3′ untranslated region (UTR) of about 3430 bp. Five promoters and several enhancers have been shown to regulate *TAL1* expression [16, 96]. At the protein level, two major isoforms have been described—a long isoform of 331 amino acids with a molecular weight of 42 kDa and a short isoform of 156 amino acids with a molecular weight of 22 kDa (Fig. 11.1). In addition, a third isoform of 39 kDa has been identified, which lacks 22 amino acids at the N-terminus. Various posttranslational modifications (PTMs) lead to other molecular weights for TAL1 being described in the literature [97, 98]. These modifications are discussed in detail in the following section. Our focus in this section is on the formation and function of the short and long isoforms of TAL1. As described above, the expression of the TAL1 transcripts originates from five different promoters, resulting in five transcription start sites (TSS) [16]. Depending on the context, the transcripts originating from TSS 1–4 can lead to the formation of the long or short isoform [16, 99]. The regulation of the formation of the short or long TAL1 isoform occurs in several different ways. The pre-mRNA formed by TSS 1–4 undergoes alternative splicing, which mainly affects the 5′ untranslated exons. However, by skipping the first protein-coding exon, an mRNA encoding the short TAL1 isoform can be produced. Another possibility of regulation is the use of different open reading frames (ORFs). Increased activity of eukaryotic initiation factor 2 (eIF2) or eukaryotic initiation factor 4 (eIF4) at the translational level leads to the generation of the short TAL1 isoform being primarily generated from the long TAL1 mRNA transcripts [10]. TSS 5 is located within the first protein-coding exon and only results in the formation of the short isoform. Expression from TSS 5 is mainly found in T-cell leukemias [100]. It has been shown that this promoter is actively silenced during erythroid differentiation [101]. This is noteworthy, as several studies have demonstrated that the formation of the short TAL1 isoform plays a central role in erythroid differentiation [10, 16, 102]. Overexpression of the short isoform in various cell lines results in preferential differentiation into erythrocytes, whereas the expression of the long isoform leads to megakaryocytic differentiation [10]. It is unclear why the short TAL1 isoform predominantly leads to differentiation into erythrocytes. It is known that the short TAL1 isoform lacks the transactivator domain and the ETO2 interaction domain [102]. ETO2 has a repressive effect on erythroid TAL1 target genes and regulates the expansion of erythroid progenitors [89, 103]. This is most likely an aspect of the predominant differentiation into erythrocytes after overexpression of TAL1 short. It has also been shown that the short TAL1 isoform is a more potent TF than the long isoform. It exhibits a stronger interaction with its binding partners and DNA, leading to increased expression of erythroid target genes such as β-hemoglobin when overexpressed [89]. At the same time, there is evidence that the short isoform is degraded more rapidly. Despite an increase in TAL1 mRNA levels during erythropoiesis, TAL1 protein levels decrease [104]. This suggests that the potency of the short TAL1 isoform, along with its faster degradation, may play a role in erythroid differentiation. There is also evidence that the correct ratio of short to long TAL1 is critical for erythroid differentiation. Disruption of this ratio can lead to diseases. For example, the formation of TAL1 short has been shown to depend on splicing factor 3b subunit 1 (SF3B1) and RNA-binding motif protein 15 (RBM15). Methylation and subsequent degradation of RBM15 by protein arginine methyltransferase 1 (PRMT1) leads to a reduced interaction between SF3B1 and RBM15, favoring the formation of the short TAL1 isoform [102]. Conversely, increased interaction between SF3B1 and RNA-binding motif protein 15 (RBM15) results in the formation of the long TAL1 isoform. The SF3B1K700E mutation enhances the interaction between these binding partners and promotes the formation of the long TAL1 isoform. Approximately 85% of patients with refractory anemia with ring sideroblasts (RARS), a myeloid dysplasia syndrome, have a mutation in SF3B1, primarily affecting K700. This suggests that an abnormal ratio of TAL1 short to TAL1 long may play a role in the development of RARS [102].

The formation of two TAL1 isoforms, one short and one long, appears to be conserved in higher vertebrates. In zebrafish, an N-terminally truncated isoform, produced by an alternative promoter, has been identified [105]. TAL1 β is an N-terminally truncated isoform of 206 amino acids (aa), which is about 50 aa longer than the short isoforms in mouse and human (mouse 154 aa, human 156 aa). All three short isoforms in human, mouse, and zebrafish contain the 100% conserved bHLH domain [106]. In zebrafish, both TAL1 α and β isoforms are redundant for primitive hematopoiesis and have overlapping but distinct expression patterns. TAL1 β is generated earlier and in the dorsal aorta, where it is essential for definitive stem cell development, in contrast to the α form. In the adult kidney, the primary hematopoietic organ in zebrafish, TAL1 α is the dominant form. The long TAL1 α and the shorter TAL1 β isoforms may also have different roles during erythrocyte differentiation in zebrafish [105]. Knockdown of TAL1 β leads to arrest in the first stage of erythroid development (basophilic erythroblasts), whereas knockdown of TAL1 α leads to arrest in the second stage of erythroid development (polychromatophilic erythroblasts) [105]. The early expression of the TAL1 β variant followed by the expression of the TAL1 α variant suggests the possibility of an autoregulatory mechanism of TAL1 β in relation to the TAL1 α variant. The temporal sequence of expression of the short and long TAL1 isoforms in the mouse model has not yet been clarified. However, overexpression experiments suggest that increased expression of TAL1 short leads to increased expression of endogenous TAL1 long, but not to increased formation of TAL1 short [102]. Reporter gene experiments also have shown that TAL1 short prefers to induce TAL1 expression from TSS 1–3, whereas TAL1 long does not show this preference [16]. It is possible that during development, the short TAL1 isoform is generally produced before the long isoform. This would make sense because during early hematopoiesis, primitive erythrocytes are produced first, followed by progenitor cells and finally HSCs [107]. The short isoform plays a critical role, particularly in erythropoiesis, whereas the long isoform is the dominant form in HSCs and during megakaryocyte development.

11.7 Posttranslational Modifications (PTMs) of TAL1

Just as the different isoforms of the TAL1 protein play a role, PTMs are crucial in erythropoiesis. PTMs generally provide a means of temporarily altering the function of a protein, thereby integrating both external and internal signals into process [108]. The TAL1 protein can be modified by several PTMS: These include phosphorylation, acetylation, and ubiquitination. Phosphorylation at four different amino acids are known for TAL1: T90, S122, S172, and S300 (Fig. 11.1) [109–113]. Phosphorylation at T90 is mediated by the AKT serine/threonine kinase (AKT) in a phosphoinositide-3-kinase (PI3K)-dependent manner. This phosphorylation leads to a redistribution of the TAL1 protein in the nucleus, with T90-phosphorylated TAL1 found in the nucleoli. In addition, T90-phosphorylated TAL1 has a reduced repressive effect on E47 homodimer-induced expression in reporter gene assays. Studies also have shown that induced by transforming growth factor beta (TGFβ), phosphorylation at T90 leads to a stronger interaction with carboxyl terminus of hsc70-interacting protein (CHIP), resulting in a faster degradation of the TAL1 protein [112, 114]. Phosphorylation at S122 is one of the most extensively studied PTMS of the TAL1 protein. It is carried out by the extracellular signal-regulated (ERK1) kinase and can be induced by epidermal growth factor (EGF) or erythropoietin (EPO) [109, 111, 115]. The phosphorylation by ERK1 is rapid; in cells stimulated with EGF, increased phosphorylation of the TAL1 protein can be detected as early as 10 min after stimulation. This phosphorylation at S122 increases the transcriptional activity of the TAL1 protein. A TAL1 variant, in which S122 is replaced by alanine, shows only half of the transcriptional activation compared to the wild-type [109, 111]. The fact that the findings on EPO-dependent phosphorylation of TAL1 were obtained in pro-erythroblasts underlines the relevance of phosphorylation in erythropoiesis. In endothelial cells, phosphorylation at S122 accelerates the degradation of the TAL1 protein [116]. Hypoxia in endothelial cells induces phosphorylation at S122, leading to increased ubiquitination of the TAL1 protein and faster degradation. However, this effect was only observed in small vessel endothelial cells [116]. TAL1 is also phosphorylated at S172 by the cAMP-dependent protein kinase A (PKA) [110]. S172 is located close to the DNA-binding basic domain of the TAL1 protein. Phosphorylation at this site selectively reduces DNA binding. As described before, a heterodimer of TAL1 and E-protein binds to the so-called E-box with the consensus sequence CANNTG. But the flanking bases of this sequence are also crucial for recognizing the binding site. The phosphorylation at S172 specifically weakens binding to the binding sites of the μE2 or μE5 immunoglobulin enhancer, whereas phosphorylation does not affect binding to the preferred sequence. Phosphorylation neither affects the localization of the TAL1 protein nor its binding activity to the E-proteins [110]. However, it was found that the binding of TAL1 to LSD1 is destabilized by phosphorylation at S172 [117]. LSD1, a lysine-specific demethylase, can remove methyl groups from histone 3 (H3) [118]. The weakened binding between LSD1 and TAL1 due to phosphorylation at S172 results in hypermethylation at promoters of TAL1 target genes, leading to the activation of these genes. Apart from the reduced LSD1 binding, the attachment of SET domain containing 1A/lysine methyltransferase 2F (hSET1/KMT2F) is increased, which also contributes to hypermethylation. This effect is especially observed during erythropoiesis [117]. It is important to emphasize that all described phosphorylations so far only occur in the long isoform of the TAL1 protein (Fig. 11.1). In addition to the mentioned phosphorylations, TAL1 could also be phosphorylated at S300 [113]. Here, TAL1 is subjected to increased degradation depending on Notch by S-phase kinase-associated protein 2 (SKP2) and CHIP. Mutation of the phosphorylation site at S300 to alanine renders the TAL1 protein resistant to Notch-induced degradation [113]. This degradation may be mediated by a phosphodegron at S300, but direct evidence of phosphorylation at this position is still pending. TAL1 is also acetylated (Fig. 11.1) [83, 84]. P/CAF acetylates a lysine-rich motif in the loop region of the TAL1 protein. This strengthens DNA binding and prevents binding to the corepressor mSIN3A. Acetylation by P/CAF is particularly important during erythropoiesis. Expression of an acetylation-deficient P/CAF variant inhibits erythroid differentiation [83]. At the same time, TAL1 acetylation is augmented after inducing erythroid differentiation. As outlined in the previous sections, TAL1 undergoes proteasomal degradation, which can be initiated by various modifications of the protein [113, 114, 116]. Typically, a protein is marked for proteasomal degradation through polyubiquitination [119]. This was also observed with TAL1, which shows increased ubiquitination during intensified degradation [113, 114]. However, the exact ubiquitination sites have not yet been identified.

11.8 TAL1 3′-UTR

The 3430 bp long 3′-UTR of *TAL1* has numerous potential binding sites for microRNAs, suggesting that TAL1 expression is posttranscriptionally regulated (Fig. 11.3) [120]. A number of microRNAs have been shown to affect TAL1 in context of T-ALL [96, 121, 122]. In megakaryocytic/erythroid differentiation, microRNAs have been predicted to bind TAL1 mRNA [123]. However, detailed analysis of these microRNAs is missing. Recently, it has been shown that microRNAs of the miR-17-92 cluster target TAL1 and form a regulatory loop with this TF, which influences erythroid differentiation. In this case miR-19 of the cluster represses TAL1 expression, while TAL1 binds to the promoter of the microRNA cluster and regulates its expression [124]. Furthermore, TAL1 activates miR144/451, which is essential for erythropoiesis [125, 126]. Interestingly, the 3′-UTR of TAL1 harbors a potential miR-144 binding site (Fig. 11.3) [120]. Target scan analysis for potential microRNA-binding sites at the 3′-UTR of TAL1 [120] revealed six sites for microRNAs, which are broadly conserved among vertebrates. Two of these sites potentially bind several miRNAs (Fig. 11.3). However, the length of the 3′-UTR indicates that many more microRNAs might target TAL1 in distinct biological settings. Taken together, it would be worthwhile to investigate the potential complex regulation between TAL1 and microRNA expression in erythropoiesis.

Acknowledgments We thank Dr. Lei Wang and Michaela Moses-Endewardt for carefully reading the manuscript.

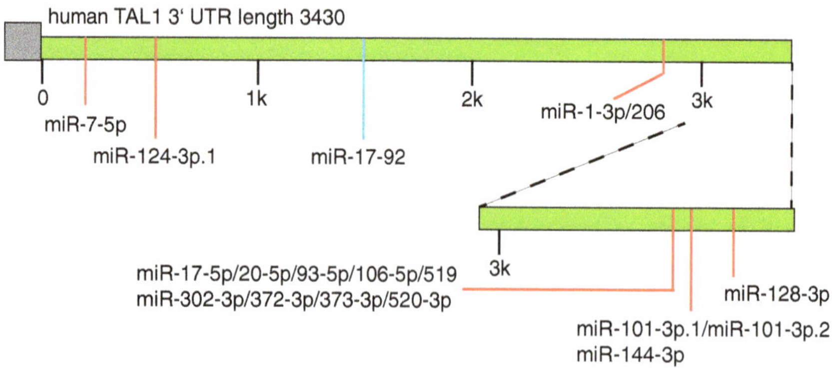

Fig. 11.3 3′-UTR of TAL1 and potential binding sites for microRNAs. The 3430-bp-long 3′UTR of human TAL1 is shown. Conserved sites for miRNA families broadly conserved among vertebrates are indicated according to target scan analysis (red). The position of miR-19 binding of the miR-17-92 cluster in the central region of TAL1 is depicted (blue). In the bottom part, the enlarged 3′-region is displayed. TAL1 T-cell acute leukemia 1

References

1. Orkin SH, Zon LI (2008) Hematopoiesis: an evolving paradigm for stem cell biology. Cell 132(4):631–644
2. Rieger MA, Schroeder T (2012) Hematopoiesis. Cold Spring Harb Perspect Biol 4:12
3. Begley CG et al (1989) Chromosomal translocation in a human leukemic stem-cell line disrupts the T-cell antigen receptor delta-chain diversity region and results in a previously unreported fusion transcript. Proc Natl Acad Sci USA 86(6):2031–2035
4. Porcher C, Chagraoui H, Kristiansen MS (2017) SCL/TAL1: a multifaceted regulator from blood development to disease. Blood 129(15):2051–2060
5. Hoang T, Lambert JA, Martin R (2016) SCL/TAL1 in hematopoiesis and cellular reprogramming. Curr Top Dev Biol 118:163–204
6. Robb L et al (1996) The scl gene product is required for the generation of all hematopoietic lineages in the adult mouse. EMBO J 15(16):4123–4129
7. Robb L et al (1995) Absence of yolk sac hematopoiesis from mice with a targeted disruption of the scl gene. Proc Natl Acad Sci USA 92(15):7075–7079
8. Porcher C et al (1996) The T cell leukemia oncoprotein SCL/tal-1 is essential for development of all hematopoietic lineages. Cell 86(1):47–57
9. Shivdasani RA, Mayer EL, Orkin SH (1995) Absence of blood formation in mice lacking the T-cell leukaemia oncoprotein tal-1/SCL. Nature 373(6513):432–434
10. Calkhoven CF et al (2003) Translational control of SCL-isoform expression in hematopoietic lineage choice. Genes Dev 17(8):959–964
11. Chen Q et al (1990) The tal gene undergoes chromosome translocation in T cell leukemia and potentially encodes a helix-loop-helix protein. EMBO J 9(2):415–424
12. Janssen JW, Ludwig WD, Sterry W, Bartram CR (1993) SIL-TAL1 deletion in T-cell acute lymphoblastic leukemia. Leukemia 7(8):1204–1210
13. Vagapova ER, Spirin PV, Lebedev TD, Prassolov VS (2018) The role of TAL1 in hematopoiesis and leukemogenesis. Acta Nat 10(1):15–23
14. Hsu HL, Wadman I, Baer R (1994) Formation of in vivo complexes between the TAL1 and E2A polypeptides of leukemic T cells. Proc Natl Acad Sci USA 91(8):3181–3185
15. Jones S (2004) An overview of the basic helix-loop-helix proteins. Genome Biol 5(6):226
16. Sharma A et al (2023) Isoforms of the TAL1 transcription factor have different roles in hematopoiesis and cell growth. PLoS Biol 21(6):e3002175
17. Nielsen AL, Norby PL, Pedersen FS, Jorgensen P (1996) E-box sequence and context-dependent TAL1/SCL modulation of basic helix-loop-helix protein-mediated transcriptional activation. J Biol Chem 271(49):31463–31469
18. Hu X et al (2009) Transcriptional regulation by TAL1: a link between epigenetic modifications and erythropoiesis. Epigenetics 4(6):357–361
19. Ravet E et al (2004) Characterization of DNA-binding-dependent and-independent functions of SCL/TAL1 during human erythropoiesis. Blood 103(9):3326–3335
20. Keller G, Lacaud G, Robertson S (1999) Development of the hematopoietic system in the mouse. Exp Hematol 27(5):777–787
21. Ciau-Uitz A, Monteiro R, Kirmizitas A, Patient R (2014) Developmental hematopoiesis: ontogeny, genetic programming and conservation. Exp Hematol 42(8):669–683
22. McGrath KE et al (2015) Distinct sources of hematopoietic progenitors emerge before HSCs and provide functional blood cells in the mammalian embryo. Cell Rep 11(12):1892–1904
23. Baron MH (2003) Embryonic origins of mammalian hematopoiesis. Exp Hematol 31(12):1160–1169
24. Kallianpur AR, Jordan JE, Brandt SJ (1994) The SCL/TAL-1 gene is expressed in progenitors of both the hematopoietic and vascular systems during embryogenesis. Blood 83(5):1200–1208
25. Gering M et al (1998) The SCL gene specifies haemangioblast development from early mesoderm. EMBO J 17(14):4029–4045
26. D'Souza SL, Elefanty AG, Keller G (2005) SCL/Tal-1 is essential for hematopoietic commitment of the hemangioblast but not for its development. Blood 105(10):3862–3870
27. Green AR et al (1992) SCL is coexpressed with GATA-1 in hemopoietic cells but is also expressed in developing brain. Oncogene 7(4):653–660
28. Elefanty AG et al (1999) SCL expression in the mouse embryo detected with a targeted lacZ reporter gene demonstrates its localization to hematopoietic, vascular, and neural tissues. Blood 94(11):3754–3763
29. Sanchez MJ et al (2001) Selective rescue of early haematopoietic progenitors in Scl(-/-) mice by expressing Scl under the control of a stem cell enhancer. Development 128(23):4815–4827
30. Patterson LJ, Gering M, Patient R (2005) Scl is required for dorsal aorta as well as blood formation in zebrafish embryos. Blood 105(9):3502–3511
31. Mead PE et al (1998) SCL specifies hematopoietic mesoderm in Xenopus embryos. Development 125(14):2611–2620
32. Dooley KA, Davidson AJ, Zon LI (2005) Zebrafish scl functions independently in hematopoietic and endothelial development. Dev Biol 277(2):522–536
33. Juarez MA et al (2005) Distinct roles for SCL in erythroid specification and maturation in zebrafish. J Biol Chem 280(50):41636–41644
34. Ciau-Uitz A et al (2013) VEGFA-dependent and -independent pathways synergise to drive Scl expression and initiate programming of the blood stem cell lineage in Xenopus. Development 140(12):2632–2642
35. Schlaeger TM et al (2005) Tie2Cre-mediated gene ablation defines the stem-cell leukemia gene (SCL/tal1)-dependent window during hematopoietic stem-cell development. Blood 105(10):3871–3874

36. Porcher C et al (1999) Specification of hematopoietic and vascular development by the bHLH transcription factor SCL without direct DNA binding. Development 126(20):4603–4615
37. Wilson NK et al (2009) The transcriptional program controlled by the stem cell leukemia gene Scl/Tal1 during early embryonic hematopoietic development. Blood 113(22):5456–5465
38. Van Handel B et al (2012) Scl represses cardiomyogenesis in prospective hemogenic endothelium and endocardium. Cell 150(3):590–605
39. Org T et al (2015) Scl binds to primed enhancers in mesoderm to regulate hematopoietic and cardiac fate divergence. EMBO J 34(6):759–777
40. Belyavsky A, Petinati N, Drize N (2021) Hematopoiesis during ontogenesis, adult life, and aging. Int J Mol Sci 22(17):9231
41. Lacombe J et al (2010) Scl regulates the quiescence and the long-term competence of hematopoietic stem cells. Blood 115(4):792–803
42. Mouthon MA et al (1993) Expression of tal-1 and GATA-binding proteins during human hematopoiesis. Blood 81(3):647–655
43. de Graaf CA et al (2016) Haemopedia: an expression atlas of murine hematopoietic cells. Stem Cell Rep 7(3):571–582
44. Curtis DJ, Salmon JM, Pimanda JE (2012) Concise review: blood relatives: formation and regulation of hematopoietic stem cells by the basic helix-loop-helix transcription factors stem cell leukemia and lymphoblastic leukemia-derived sequence 1. Stem Cells 30(6):1053–1058
45. Courtial N et al (2012) Tal1 regulates osteoclast differentiation through suppression of the master regulator of cell fusion DC-STAMP. FASEB J 26(2):523–532
46. Mikkola HK et al (2003) Haematopoietic stem cells retain long-term repopulating activity and multipotency in the absence of stem-cell leukaemia SCL/tal-1 gene. Nature 421(6922):547–551
47. Curtis DJ et al (2004) SCL is required for normal function of short-term repopulating hematopoietic stem cells. Blood 103(9):3342–3348
48. Brunet de la Grange P et al (2006) Low SCL/TAL1 expression reveals its major role in adult hematopoietic myeloid progenitors and stem cells. Blood 108(9):2998–3004
49. Giroux S et al (2007) lyl-1 and tal-1/scl, two genes encoding closely related bHLH transcription factors, display highly overlapping expression patterns during cardiovascular and hematopoietic ontogeny. Gene Expr Patterns 7(3):215–226
50. Souroullas GP et al (2009) Adult hematopoietic stem and progenitor cells require either Lyl1 or Scl for survival. Cell Stem Cell 4(2):180–186
51. Hoang T et al (1996) Opposing effects of the basic helix-loop-helix transcription factor SCL on erythroid and monocytic differentiation. Blood 87(1):102–111
52. Hall MA et al (2003) The critical regulator of embryonic hematopoiesis, SCL, is vital in the adult for megakaryopoiesis, erythropoiesis, and lineage choice in CFU-S12. Proc Natl Acad Sci USA 100(3):992–997
53. Aplan PD, Nakahara K, Orkin SH, Kirsch IR (1992) The SCL gene product: a positive regulator of erythroid differentiation. EMBO J 11(11):4073–4081
54. Zeuner A et al (2003) Control of erythroid cell production via caspase-mediated cleavage of transcription factor SCL/Tal-1. Cell Death Differ 10(8):905–913
55. De Maria R et al (1999) Negative regulation of erythropoiesis by caspase-mediated cleavage of GATA-1. Nature 401(6752):489–493
56. Motoyama N et al (1995) Massive cell death of immature hematopoietic cells and neurons in Bcl-x-deficient mice. Science 267(5203):1506–1510
57. Massari ME, Murre C (2000) Helix-loop-helix proteins: regulators of transcription in eucaryotic organisms. Mol Cell Biol 20(2):429–440
58. El Omari K et al (2013) Structural basis for LMO2-driven recruitment of the SCL:E47bHLH heterodimer to hematopoietic-specific transcriptional targets. Cell Rep 4(1):135–147
59. El Omari K et al (2011) Structure of the leukemia oncogene LMO2: implications for the assembly of a hematopoietic transcription factor complex. Blood 117(7):2146–2156
60. Wadman IA et al (1997) The LIM-only protein Lmo2 is a bridging molecule assembling an erythroid, DNA-binding complex which includes the TAL1, E47, GATA-1 and Ldb1/NLI proteins. EMBO J 16(11):3145–3157
61. Soler E et al (2010) The genome-wide dynamics of the binding of Ldb1 complexes during erythroid differentiation. Genes Dev 24(3):277–289
62. Pimkin M et al (2014) Divergent functions of hematopoietic transcription factors in lineage priming and differentiation during erythro-megakaryopoiesis. Genome Res 24(12):1932–1944
63. Thoms JAI et al (2021) Disruption of a GATA2-TAL1-ERG regulatory circuit promotes erythroid transition in healthy and leukemic stem cells. Blood 138(16):1441–1455
64. Diffner E et al (2013) Activity of a heptad of transcription factors is associated with stem cell programs and clinical outcome in acute myeloid leukemia. Blood 121(12):2289–2300
65. Wilson NK, Calero-Nieto FJ, Ferreira R, Gottgens B (2011) Transcriptional regulation of haematopoietic transcription factors. Stem Cell Res Ther 2(1):6
66. Wilson NK et al (2010) Combinatorial transcriptional control in blood stem/progenitor cells: genome-wide analysis of ten major transcriptional regulators. Cell Stem Cell 7(4):532–544
67. Pimanda JE et al (2007) Gata2, Fli1, and Scl form a recursively wired gene-regulatory circuit during early hematopoietic development. Proc Natl Acad Sci USA 104(45):17692–17697

68. Tijssen MR et al (2011) Genome-wide analysis of simultaneous GATA1/2, RUNX1, FLI1, and SCL binding in megakaryocytes identifies hematopoietic regulators. Dev Cell 20(5):597–609
69. Kuvardina ON et al (2015) RUNX1 represses the erythroid gene expression program during megakaryocytic differentiation. Blood 125(23):3570–3579
70. Wu W et al (2014) Dynamic shifts in occupancy by TAL1 are guided by GATA factors and drive large-scale reprogramming of gene expression during hematopoiesis. Genome Res 24(12):1945–1962
71. Kassouf MT et al (2010) Genome-wide identification of TAL1's functional targets: insights into its mechanisms of action in primary erythroid cells. Genome Res 20(8):1064–1083
72. Tripic T et al (2009) SCL and associated proteins distinguish active from repressive GATA transcription factor complexes. Blood 113(10):2191–2201
73. Fujiwara T et al (2009) Discovering hematopoietic mechanisms through genome-wide analysis of GATA factor chromatin occupancy. Mol Cell 36(4):667–681
74. Starck J et al (2003) Functional cross-antagonism between transcription factors FLI-1 and EKLF. Mol Cell Biol 23(4):1390–1402
75. Siatecka M, Bieker JJ (2011) The multifunctional role of EKLF/KLF1 during erythropoiesis. Blood 118(8):2044–2054
76. Wontakal SN et al (2012) A core erythroid transcriptional network is repressed by a master regulator of myelo-lymphoid differentiation. Proc Natl Acad Sci USA 109(10):3832–3837
77. Love PE, Warzecha C, Li L (2014) Ldb1 complexes: the new master regulators of erythroid gene transcription. Trends Genet 30(1):1–9
78. Hung CH et al (2021) A positive regulatory feedback loop between EKLF/KLF1 and TAL1/SCL sustaining the erythropoiesis. Int J Mol Sci 22(15):8024
79. Kim YW, Yun WJ, Kim A (2016) Erythroid activator NF-E2, TAL1 and KLF1 play roles in forming the LCR HSs in the human adult beta-globin locus. Int J Biochem Cell Biol 75:45–52
80. Kang Y et al (2015) KLF1 stabilizes GATA-1 and TAL1 occupancy in the human beta-globin locus. Biochim Biophys Acta 1849(3):282–289
81. Batta K, Florkowska M, Kouskoff V, Lacaud G (2014) Direct reprogramming of murine fibroblasts to hematopoietic progenitor cells. Cell Rep 9(5):1871–1884
82. Capellera-Garcia S et al (2016) Defining the minimal factors required for erythropoiesis through direct lineage conversion. Cell Rep 15(11):2550–2562
83. Huang SM et al (2000) P/CAF-mediated acetylation regulates the function of the basic helix-loop-helix transcription factor TAL1/SCL. EMBO J 19(24):6792–6803
84. Huang SM, Qiu Y, Stein RW, Brandt SJ (1999) P300 functions as a transcriptional coactivator for the TAL1/SCL oncoprotein. Oncogene 18(35):4958–4967
85. O'Neil J et al (2004) TAL1/SCL induces leukemia by inhibiting the transcriptional activity of E47/HEB. Cancer Cell 5(6):587–596
86. Parriott G, Kee BL (2022) E protein transcription factors as suppressors of T lymphocyte acute lymphoblastic Leukemia. Front Immunol 13:885144
87. Sanda T, Leong WZ (2017) TAL1 as a master oncogenic transcription factor in T-cell acute lymphoblastic leukemia. Exp Hematol 53:7–15
88. Chagraoui H et al (2018) SCL/TAL1 cooperates with Polycomb RYBP-PRC1 to suppress alternative lineages in blood-fated cells. Nat Commun 9(1):5375
89. Schuh AH et al (2005) ETO-2 associates with SCL in erythroid cells and megakaryocytes and provides repressor functions in erythropoiesis. Mol Cell Biol 25(23):10235–10250
90. Goardon N et al (2006) ETO2 coordinates cellular proliferation and differentiation during erythropoiesis. EMBO J 25(2):357–366
91. Stadhouders R et al (2015) Control of developmentally primed erythroid genes by combinatorial co-repressor actions. Nat Commun 6:8893
92. Hu X et al (2009) LSD1-mediated epigenetic modification is required for TAL1 function and hematopoiesis. Proc Natl Acad Sci USA 106(25):10141–10146
93. Herglotz J et al (2013) Histone arginine methylation keeps RUNX1 target genes in an intermediate state. Oncogene 32(20):2565–2575
94. Xu Z et al (2006) Recruitment of the SWI/SNF protein Brg1 by a multiprotein complex effects transcriptional repression in murine erythroid progenitors. Biochem J 399(2):297–304
95. Hu G et al (2011) Regulation of nucleosome landscape and transcription factor targeting at tissue-specific enhancers by BRG1. Genome Res 21(10):1650–1658
96. Correia NC et al (2016) microRNAs regulate TAL1 expression in T-cell acute lymphoblastic leukemia. Oncotarget 7(7):8268–8281
97. Pulford K et al (1995) Expression of TAL-1 proteins in human tissues. Blood 85(3):675–684
98. Goldfarb AN, Goueli S, Mickelson D, Greenberg JM (1992) T-cell acute lymphoblastic leukemia--the associated gene SCL/tal codes for a 42-Kd nuclear phosphoprotein. Blood 80(11):2858–2866
99. Aplan PD et al (1990) The SCL gene is formed from a transcriptionally complex locus. Mol Cell Biol 10(12):6426–6435
100. Bernard O et al (1992) A third tal-1 promoter is specifically used in human T cell leukemias. J Exp Med 176(4):919–925
101. Courtes C et al (2000) Erythroid-specific inhibition of the tal-1 intragenic promoter is due to binding of a repressor to a novel silencer. J Biol Chem 275(2):949–958
102. Jin S et al (2017) Splicing factor SF3B1K700E mutant dysregulates erythroid differentiation via aberrant alternative splicing of transcription factor TAL1. PLoS One 12(5):e0175523

103. Fujiwara T et al (2013) Role of transcriptional corepressor ETO2 in erythroid cells. Exp Hematol 41(3):303–315. e1
104. Murrell AM et al (1995) Discordant regulation of SCL/TAL-1 mRNA and protein during erythroid differentiation. Oncogene 11(1):131–139
105. Qian F et al (2007) Distinct functions for different scl isoforms in zebrafish primitive and definitive hematopoiesis. PLoS Biol 5(5):e132
106. Barton LM et al (2001) Regulation of the stem cell leukemia (SCL) gene: a tale of two fishes. Proc Natl Acad Sci USA 98(12):6747–6752
107. Baron MH (2013) Concise review: early embryonic erythropoiesis: not so primitive after all. Stem Cells 31(5):849–856
108. Ramazi S, Zahiri J (2021) Post-translational modifications in proteins: resources, tools and prediction methods. Database-Oxford 2021:baab012
109. Cheng JT, Cobb MH, Baer R (1993) Phosphorylation of the TAL1 oncoprotein by the extracellular-signal-regulated protein kinase ERK1. Mol Cell Biol 13(2):801–808
110. Prasad KS, Brandt SJ (1997) Target-dependent effect of phosphorylation on the DNA binding activity of the TAL1/SCL oncoprotein. J Biol Chem 272(17):11457–11462
111. Tang T, Prasad KS, Koury MJ, Brandt SJ (1999) Mitogen-activated protein kinase mediates erythropoietin-induced phosphorylation of the TAL1/SCL transcription factor in murine proerythroblasts. Biochem J 343(Pt 3(Pt 3)):615–620
112. Palamarchuk A et al (2005) Akt phosphorylates Tal1 oncoprotein and inhibits its repressor activity. Cancer Res 65(11):4515–4519
113. Nie LH, Sasaki M, Maki CG (2007) Regulation of p53 nuclear export through sequential changes in conformation and ubiquitination. J Biol Chem 282(19):14616–14625
114. Terme JM, Lhermitte L, Asnafi V, Jalinot P (2009) TGF-beta induces degradation of TAL1/SCL by the ubiquitin-proteasome pathway through AKT-mediated phosphorylation. Blood 113(26):6695–6698
115. Prasad KSS et al (1995) Erythropoietin stimulates transcription of the Tal1/Scl gene and phosphorylation of its protein products. J Biol Chem 270(19):11603–11611
116. Tang T, Arbiser JL, Brandt SJ (2002) Phosphorylation by mitogen-activated protein kinase mediates the hypoxia-induced turnover of the TAL1/SCL transcription factor in endothelial cells. J Biol Chem 277(21):18365–18372
117. Li Y et al (2012) Dynamic interaction between TAL1 oncoprotein and LSD1 regulates TAL1 function in hematopoiesis and leukemogenesis. Oncogene 31(48):5007–5018
118. Perillo B, Tramontano A, Pezone A, Migliaccio A (2020) LSD1: more than demethylation of histone lysine residues. Exp Mol Med 52(12):1936–1947
119. Ciechanover A (2005) Proteolysis: from the lysosome to ubiquitin and the proteasome. Nat Rev Mol Cell Biol 6(1):79–86
120. Agarwal V, Bell GW, Nam JW, Bartel DP (2015) Predicting effective microRNA target sites in mammalian mRNAs. Elife 4:e05005
121. Zhao Y et al (2018) A novel TAL1/miR-149* axis accelerates tumor growth of human T cell acute lymphoblastic leukemia. Int J Clin Exp Pathol 11(8):4026–4034
122. Mansour MR et al (2013) The TAL1 complex targets the FBXW7 tumor suppressor by activating miR-223 in human T cell acute lymphoblastic leukemia. J Exp Med 210(8):1545–1557
123. Garzon R et al (2006) MicroRNA fingerprints during human megakaryocytopoiesis. Proc Natl Acad Sci USA 103(13):5078–5083
124. Meyer A et al (2020) The transcription factor TAL1 and miR-17-92 create a regulatory loop in hematopoiesis. Sci Rep 10(1):21438
125. Dore LC et al (2008) A GATA-1-regulated microRNA locus essential for erythropoiesis. Proc Natl Acad Sci USA 105(9):3333–3338
126. Kohrs N et al (2016) MiR144/451 expression is repressed by RUNX1 during megakaryopoiesis and disturbed by RUNX1/ETO. PLoS Genet 12(3):e1005946

Part VI

Megakaryocytes

GATA1 in Normal and Pathologic Megakaryopoiesis and Platelet Development

12

Kaoru Takasaki and Stella T. Chou

Abstract

GATA1 is a highly conserved hematopoietic transcription factor (TF), essential for normal erythropoiesis and megakaryopoiesis, that encodes a full-length, predominant isoform and an amino (N) terminus-truncated isoform GATA1s. It is consistently expressed throughout megakaryocyte development and interacts with its target genes either independently or in association with binding partners such as FOG1 (friend of GATA1). While the N-terminus and zinc finger have classically been demonstrated to be necessary for the normal regulation of platelet-specific genes, murine models, cell-line studies, and human case reports indicate that the carboxy-terminal activation domain and zinc finger also play key roles in precisely controlling megakaryocyte growth, proliferation, and maturation. Murine models have shown that disruptions to GATA1 increase the proliferation of immature megakaryocytes with abnormal architecture and impaired terminal differentiation into platelets. In humans, germline *GATA1* mutations result in variable cytopenias, including macrothrombocytopenia with abnormal platelet aggregation and excessive bleeding tendencies, while acquired *GATA1s* mutations in individuals with trisomy 21 (T21) result in transient abnormal myelopoiesis (TAM) and myeloid leukemia of Down syndrome (ML-DS) arising from a megakaryocyte-erythroid progenitor (MEP). Taken together, GATA1 plays a key role in regulating megakaryocyte differentiation, maturation, and proliferative capacity. As sequencing and proteomic technologies expand, additional *GATA1* mutations and regulatory mechanisms contributing to human diseases of megakaryocytes and platelets are likely to be revealed.

Keywords

GATA1 · GATA1s · Megakaryocytes · Platelets · Hematopoiesis · Megakaryopoiesis · Thrombocytopenia

K. Takasaki
Department of Pediatrics, Division of Hematology, University of Pennsylvania Perelman School of Medicine, Children's Hospital of Philadelphia, Philadelphia, PA, USA

S. T. Chou (✉)
Department of Pediatrics, Division of Hematology, University of Pennsylvania Perelman School of Medicine, Children's Hospital of Philadelphia, Philadelphia, PA, USA

Department of Pathology and Laboratory Medicine, Division of Transfusion Medicine, University of Pennsylvania Perelman School of Medicine, Children's Hospital of Philadelphia, Philadelphia, PA, USA
e-mail: chous@chop.edu

T. Borggrefe, B. D. Giaimo (eds.), *Transcription factors in blood cell development*, Advances in Experimental Medicine and Biology 1459, https://doi.org/10.1007/978-3-031-62731-6_12

Abbreviations

ADA	Adenosine deaminase
ADP	Adenosine diphosphatase
AML	Acute myeloid leukemia
BM	Bone marrow
C-TAD	Carboxy (C)-terminus transactivation domain
C-ZF	Carboxy (C)-terminus zinc finger
CBF	Core-binding factor
CDK9	Cyclin-dependent kinase 9
CEP	Congenital erythropoietic porphyria
ChIP-Seq	Chromatin immunoprecipitation followed by deep sequencing
CLP	Common lymphoid progenitor
CMP	Common myeloid progenitor
CTBP	Carboxy (C)-terminus binding protein
CTCF	CCCTC-binding factor
DBA	Diamond-Blackfan anemia
DMS	Demarcation membrane system
DNMT1	DNA methyltransferase 1
DS	Down syndrome
E2A	Transcription factor E2 alpha
ESC	Embryonic stem cell
ETO2 (CBFA2T3)	Core binding factor 2 (CBFA2)/Runt-related transcription factor 1 (RUNX1) partner transcriptional co-repressor 3
ETS	Erythroblastosis virus 26 transformation-specific
ETS1	ETS proto-oncogene 1
EZH2	Enhancer of zeste homolog 2
FLI1	Friend leukemia integration 1
FOG1	Friend of GATA1
G1HE	*Gata1* hematopoietic enhancer
G1HRD	*GATA1* hematopoietic regulatory domain
G1ME	*Gata1*⁻ megakaryocyte erythroid
GAPBα	GA-binding protein alpha
GATA1	GATA-binding protein 1
GATA1s	GATA1 short (isoform)
GATA2	GATA-binding protein 2
GFI1B/Gfi1b	Growth factor-independent 1B transcriptional repressor
GMP	Granulocyte/macrophage progenitor
GP1BA/Gp1ba	Glycoprotein 1b subunit alpha
GP1BB/Gp1bb	Glycoprotein 1b subunit beta
GPIIB/Gpiib	Glycoprotein IIb
GPIX/Gpix	Glycoprotein IX
GPV/Gpv	Glycoprotein V
GPVI/Gpvi	Glycoprotein VI
HDAC	Histone deacetylase
Hgb F	Fetal hemoglobin
HS I, II, III	DNase I hypersensitivity region I, II, III
HSC	Hematopoietic stem cell
HSPC	Hematopoietic stem/progenitor cell
IE	Erythroid promoter/first exon
iPSC	Induced pluripotent stem cell
IT	Testis promoter/first exon
ITGA2B/Itga2b	Integrin subunit alpha IIb
Jak2	Janus kinase 2 KIT: KIT proto-oncogene, receptor tyrosine kinase
LDB1	LIN-11, Isl-1 and MEC-3 (LIM) domain-binding 1
LMO2	LIM domain only 2
LMO4	LIM domain only 4
Lu	Lutheran blood group antigen
MBD3	Methyl-CpG-binding domain protein 3
MDS	Myelodysplastic syndrome
MEP	Megakaryocyte-erythroid progenitor
Mi-2β (CHD4)	Chromodomain helicase DNA-binding protein 4
ML-DS	Myeloid leukemia of Down syndrome
c-MPL/c-Mpl	Myeloproliferative leukemia protein
MTA1	Metastasis-associated 1
MTA2	Metastasis-associated 2
Myb	Myeloblastosis viral oncogene homolog
MYC/Myc	Myelocytomatosis viral oncogene homolog

N-TAD	Amino (N)-terminus transactivation domain
N-ZF	Amino (N)-terminus zinc finger
NFE2/Nfe2	Nuclear factor, erythroid 2
NRAS	Neuroblastoma RAS viral oncogene homolog
NuRD	Nucleosome remodeling and deacetylase
P-TEFb	Positive transcription elongation factor b
P66 (GATAD2A)	GATA zinc finger domain containing 2A
PBP/Pbp	Platelet basic protein
PF4/Pf4	Platelet factor 4
PU.1/SPI1	Hematopoietic transcription factor PU.1/spleen focus-forming virus (SFFV) proviral integration protein
RAD21	Double-strand-break repair protein Rad21 homolog
RAS	Rat sarcoma virus
RBAP46 (RBBP7, retinoblastoma-binding protein 7)	Retinoblastoma-binding protein P46
RBAP48 (RBBP4, retinoblastoma-binding protein 4)	Retinoblastoma-binding protein P48
RNAPII	RNA polymerase II
RUNX1	Runt-related transcription factor 1
SCL/TAL1	Stem cell leukemia/T-cell acute leukemia 1
SMC3	Structural maintenance of chromosomes 3
STAG2	Cohesin subunit SA-2
T21	Trisomy 21
TAM	Transient abnormal myelopoiesis
TAD	Transactivation domain
TF	Transcription factor
TMD	Transient myeloproliferative disorder
TP53	Tumor antigen P53
VWF/Vwf	von Willebrand factor
WT	Wild-type
XLTT	X-linked thrombocytopenia with thalassemia
ZBP89	Zinc finger DNA-binding protein 89
ZF	Zinc finger
Znfn1a1	Zinc finger protein, subfamily 1A, 1

12.1 Overview

GATA1 (GATA-binding protein 1) is located on the X chromosome (Xp21-11) and encodes a zinc finger (ZF) transcription factor (TF) essential for normal erythrocyte, megakaryocyte, mast cell, basophil, and eosinophil development, as well as for Sertoli cells in the testis. It is widely conserved between lower vertebrates and mammals, with homology between the human, chicken, mouse, rat, cow, sheep, turtle, and zebrafish forms [1, 2]. In humans, the GATA1 protein contains a transactivation domain (TAD) in each of the amino and carboxy termini (N-TAD and C-TAD, respectively) and two ZFs that are identified by their relative proximity to either terminus (Fig. 12.1). In reporter assays, both the N-TAD and C-TAD function to transcriptionally regulate target genes and have both unique and redundant functions [3, 4]. The N-terminus zinc finger (N-ZF) stabilizes GATA1-DNA-binding interactions, specifically binds to the palindromic (A/T)GATA(A/G):GAT motif [5, 6], and associates with the cofactor FOG1 (friend of GATA1), while the C-terminus zinc finger (C-ZF) facilitates GATA1 binding to its single (A/T)GATA(A/G), palindromic (A/T) GATA(A/G):GAT, and complex inverted or repeated double/tandem (A/T)GATA(A/G) target motifs [7](Fig. 12.2).

Human *GATA1* contains a noncoding first exon and five coding exons. In contrast, the murine *Gata1* ortholog contains two alternative noncoding first exons, the distal testis promoter/first exon (IT) for expression in Sertoli cells and the proximal erythroid promoter/first exon (IE) for expression in hematopoietic cells, followed by five coding exons (Fig. 12.3). *Gata1* also con-

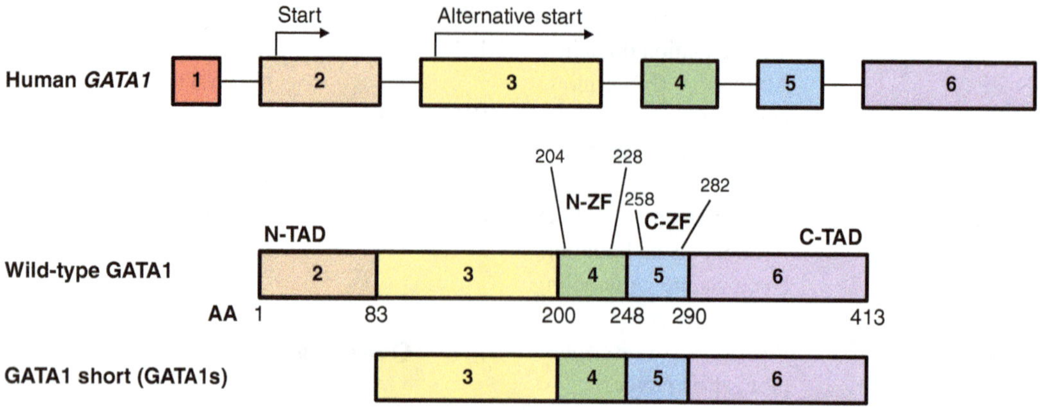

Fig. 12.1 Schematic of human *GATA1* gene and protein isoforms (wild-type GATA1 and GATA1s). N-TAD amino-terminus activation domain, C-TAD carboxy-terminus activation domain, N-ZF amino-terminus zinc finger, C-ZF carboxy-terminus zinc finger

Fig. 12.2 *GATA1*-binding motifs

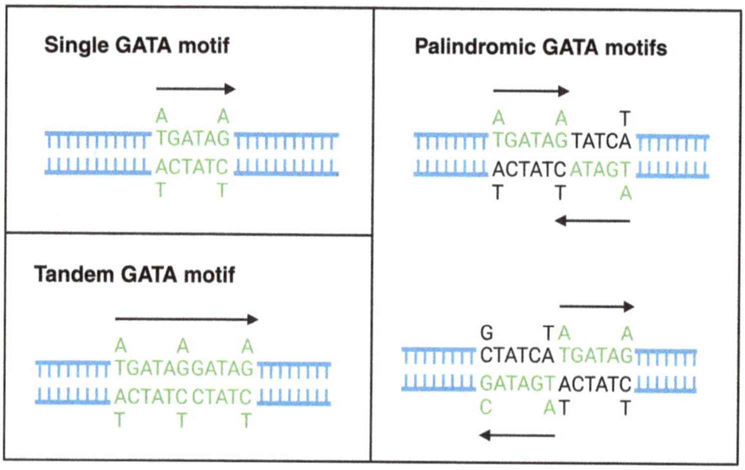

tains 3 DNase I hypersensitivity regions—HS I, HS II, and HS III—that surround IE and are largely conserved in humans [6, 8–12]. HS I, also referred to as G1HE (*Gata1* hematopoietic enhancer), is an enhancer region located 3.9–2.6 kb upstream of the IE promoter. It contains a composite GATA-E-box element and functions in the activation and maintenance of *Gata1* expression in hematopoietic cells [8, 10, 13]. The entire HS I sequence is essential for normal megakaryopoiesis, while only the 5′ region is required for erythropoiesis [12, 14]. HS II encompasses IE and corresponds to the human *GATA1* promoter; it contains a double GATA motif and at least two CACCC elements just proximal to IE [6, 9, 10, 15]. The double GATA motif and at least one of the CACCC elements interacts with the GATA-E-box element of HS I in the regulation of *GATA1* expression [9, 10, 12]. HS III corresponds to intron 1.

There are two isoforms of GATA1: the predominant full-length 42.7-kDa protein is encoded using the canonical start site in exon 2, while the shortened ~40-kDa protein GATA1s (GATA1 short) is missing the first 83 amino acids, including the N-TAD, due to alternative mRNA splicing and use of an alternative start codon in exon 3 (Fig. 12.1). Multiple studies have demonstrated that GATA1s bind similar targets as the full-length isoform but predictably lack transactiva-

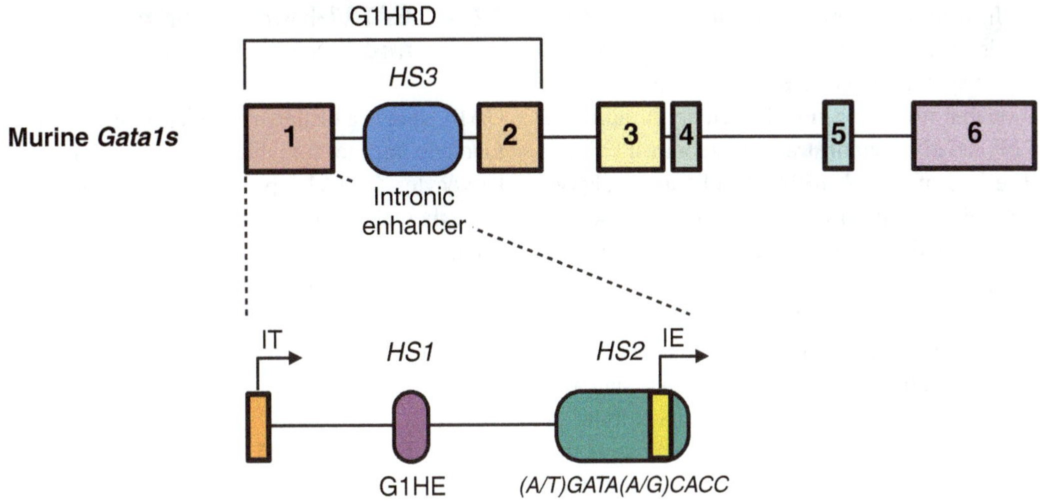

Fig. 12.3 Schematic of murine *Gata1* gene. G1HE *Gata1* hematopoietic enhancer, G1HRD *Gata1* hematopoietic regulatory domain, IT testis promoter/first exon, IE erythroid promoter/first exon, HS hypersensitivity region

tion activity [3, 5, 16, 17]. All humans naturally produce a small amount of GATA1s, but its role in normal hematopoiesis, if any, remains unclear. However, germline mutations resulting in the exclusive production of GATA1s causes thrombocytopenia with variable degrees of anemia and neutropenia [18–21], and somatic *GATA1s* mutations in individuals with Down syndrome (DS), or trisomy 21 (T21), are associated with a leukemia-like condition known as transient abnormal myelopoiesis (TAM) and myeloid leukemia of Down syndrome (ML-DS) [22, 23]. The mechanisms by which GATA1s exerts its phenotypes remain an active area of study.

12.2 GATA1 Regulation in Megakaryopoiesis

GATA1 is expressed in hematopoietic stem cells (HSCs), and its levels gradually increase as differentiation progresses through the common myeloid progenitor (CMP), megakaryocyte-erythroid progenitor (MEP), and then proerythroblasts and megakaryocytes; it is not expressed in the common lymphoid progenitor (CLP) or granulocyte-macrophage progenitor (GMP) [24]. Hematopoiesis therefore requires intricate coordination of transcriptional, posttranslational, and functional regulation of GATA1.

The transcriptional regulation of *GATA1* is primarily driven by the three DNase hypersensitivity regions, or *cis* elements, that are located within an 8.5-kb region known as the *GATA1* hematopoietic regulatory domain (G1HRD) [25, 26]. HS I, the double GATA motif, and a CACCC box are necessary and sufficient for *GATA1* expression in hematopoietic cells, and the relative orientation of these three elements appears critical [13]. The intervening sequences between the three elements also plays a role in *Gata1* expression by directing methylation of the locus by DNA methyltransferase 1 (DNMT1); this methylation is critical to the suppression of aberrant *Gata1* expression in HSCs [27]. It has been suggested that methylation of the region encompassing the G1HRD prevents GATA2 (GATA-binding protein 2) from binding and thereby activating *Gata1* expression in HSCs, and that methylation decreases with the initiation of hematopoietic differentiation [27]. A cell line model similarly found that *GATA1* expression in hematopoietic progenitors is regulated by methylation of the *cis* elements, although this study suggested that GATA2 mediates *GATA1* expression through demethylation of H3K4 [28].

In a murine megakaryocyte model, GATA1 binds to the GATA motif of the HS I GATA-E-box element, suggestive of a positive feedback loop that was first studied in an erythroid model [29] and also demonstrated in zebrafish [30, 31]. The loss of *Gata1* mRNA levels in megakaryocytes derived from mice lacking HS I also suggests that this positive feedback loop regulates the platelet lineage [32]. GATA1 may facilitate hyperacetylation of histone H3 in this region and thus enhance transcription [33]. SCL/TAL1, a TF that also interacts with GATA1 alone and with other binding partners [34], appears to bind the E-box of the HS I element via other proteins such as LMO2, although it is not clear whether these other binding partners are the members of the pentameric complex vs. other proteins [12, 13] and whether SCL/TAL1 is required for *Gata1* transcription [13, 33, 35].

GATA2 has been shown to play a role in *GATA1* regulation in a number of contexts but appears to be nonessential. Cell line studies investigating the megakaryocyte, erythrocyte, and myeloid lineages have diverged on whether *GATA1* expression can be induced by forced expression of *GATA2* [36–40]. GATA1 and GATA2 share redundant functions in *GATA1* regulation, but GATA2's role is more prominent in early hematopoietic development than after lineage commitment [24]. This is consistent with the unique interaction between GATA2 and methylation of the G1HRD in regulating *GATA1* expression in HSCs [27, 28]; although GATA2 regulation of *GATA1* expression has been primarily studied in the context of erythropoiesis, it is likely that differential methylation also marks the transition from HSC to megakaryocyte.

ZBP89 (zinc finger DNA-binding protein 89), a Krüppel-type zinc finger TF that has been primarily studied as a cofactor for the GATA1-FOG1 complex, also regulates *GATA1* expression [27, 33, 41, 42]. It is enriched in the G1HRD in murine cell lines [33, 41], suggesting that it directly controls transcription although its role specifically in megakaryocytes is less clear [33].

12.3 GATA1-Binding Partners and Targets

GATA1 plays a distinct role in each of the hematopoietic lineages for which it is critical. In megakaryocytes, GATA1 expression remains consistent throughout development, and both the N-terminus and the N-ZF are required for normal megakaryopoiesis [32]. GATA1 may interact with its megakaryocyte target genes either independently or as part of a protein complex (Fig. 12.4).

12.3.1 FOG1-GATA1 Complexes

The most well studied of GATA1-binding partners is FOG1 (friend of GATA1), a ZF protein that regulates erythropoiesis and megakaryopoiesis both in association with and independently of GATA1 [43–50]. Depending on the gene and lineage context, FOG1 acts to activate or repress GATA1-mediated gene expression [43, 51]. Megakaryocyte genes that are regulated via GATA1-FOG1 complexes carry tandem GATA- and ETS (erythroblastosis virus 26 transformation-specific)-binding sites in their regulatory regions; ETS domains are characterized by a core GGA(A/T) sequence [52]. The GATA1 N-ZF interacts directly with FOG1, and this complex in turn binds to their target sites via GATA1 [47, 50].

Several studies have investigated mutations affecting the N-ZF, demonstrating that mutations such as V205M, V205G, G208E, and others are sufficient to disrupt the GATA1-FOG1 interaction in megakaryocytes and thus binding to target genes [32, 45–48, 50]. Interestingly, a study in murine G1ME (*Gata1*$^{-}$ megakaryocyte erythroid) cells showed that the N-ZF mutant GATA1^{V205G} binds to chromatin much less selectively than wild-type (WT) GATA1, suggesting that FOG1 regulates GATA1 chromatin occupancy even though FOG1 itself does not bind chromatin directly. Genes relevant to megakaryocyte development and function that have been specifically shown to be dysregulated in the setting of N-ZF mutations include p45 *Nfe2* (nuclear factor, erythroid 2), *Gp1ba* (glycoprotein 1b subunit alpha), *c-Mpl* (myeloproliferative leukemia protein),

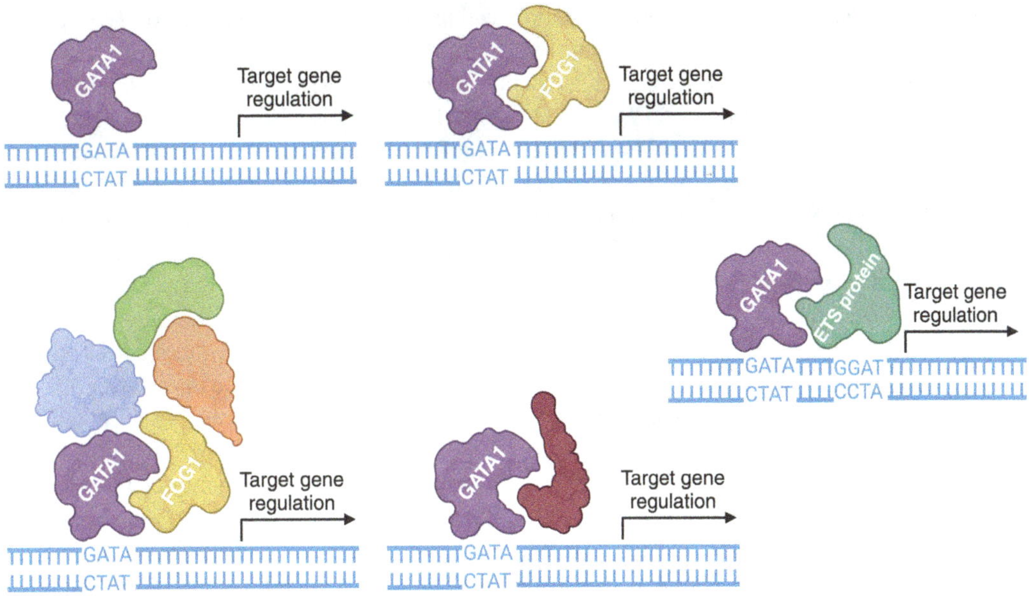

Fig. 12.4 Schematic of different patterns of GATA1 occupancy of its target genes. Unlabeled proteins are representative of a range of binding partners

Gpix (glycoprotein IX), and *Jak2* (Janus kinase 2) [47, 48]. Murine studies have also shown that the GATA1-FOG1 interaction is dispensable for early megakaryocyte development but is required for later megakaryocyte maturation, likely due to partial compensation by GATA2 [46, 53]. It is important to note, however, that in murine models FOG1 itself is required for both megakaryocyte and erythrocyte lineage specification of hematopoietic progenitors, indicating GATA1-dependent and GATA1-independent roles for FOG1 in hematopoiesis [14, 44, 54, 55].

As neither GATA1 nor FOG1 recognizes the ETS sequence, the interaction of the GATA1-FOG1 complex with the ETS site is mediated by an ETS family protein. Multiple platelet-specific genes were in fact identified as containing GATA and ETS sequences before the GATA1-FOG1 interaction with ETS proteins was understood: based on murine and human cell line studies of *GP1BA* [56], *GPIIB* (glycoprotein IIb)/*ITGA2B* (integrin subunit alpha 2b) [57–59], *PF4* (platelet factor 4) [60], *GPIX* [61, 62], *c-MPL* [63, 64], and *GPV* (glycoprotein V) [65], this combined motif was considered a characteristic hallmark of megakaryocytes. Interestingly, the 3′ end of *GATA1* HS1 contains a likely ETS site [12]. At least three ETS proteins—GABPα (GA-binding protein transcription factor subunit alpha) [66], FLI1 (friend leukemia integration 1) [47, 64, 66–68], and ETS1 (ETS proto-oncogene 1) [64, 67]—have since been identified as partners of the GATA1-FOG1 complex in human and murine cell line models.

GABPα is a component of the TF complex GABP and mediates DNA binding. It is active in early megakaryopoiesis and induces the expression of *ITGA2B* and *c-Mpl* [66]. FLI1 is active throughout megakaryopoiesis, although it preferentially binds to genes active in the later stages of this process [66]. While FLI1 is also able to enhance GATA1-FOG1 activation of *ITGA2B*, GAPBα has a stronger affinity to this promoter [47, 66], and instead, FLI plays a more critical role in the expression of *GP1BA* [66, 68], *GPIX* [66–68], *PF4* [66], *c-MPL* [64, 67], and likely *GPIIB* [67]. Despite this importance of FLI1 in GATA1-FOG1 activity, studies have disagreed on the nature and stability of the GATA1-FOG1-FLI1 interaction [47, 50, 69]; some have suggested direct binding between GATA1 and FLI1, while others have demonstrated poor in vitro affinity and instead proposed a more allosteric role for FLI1. Finally, a combination of murine and human cell line studies have shown that ETS1 occupies the promoter regions of *GPIX* [67], *c-MPL* [64, 67], *GPIIB* [67], and likely *PF4* [59].

GATA1-FOG1 also interacts with the multi-protein complex NuRD (nucleosome remodeling and deacetylase) via the N-terminus of FOG1 to mediate both transcriptional repression and activation by GATA1 and maintain hematopoietic lineage fidelity [70–73]. Studies in murine cells have shown that histone deacetylase (HDAC) activity is required for erythrocyte and megakaryocyte lineage commitment. Mice with a FOG1 mutation that disrupts FOG1-NuRD interaction have partial embryonic lethality due to anemia, extramedullary erythropoiesis, and thrombocytopenia due to a block in megakaryocyte maturation; they also show aberrant expression of mast cell and eosinophil genes in megakaryocyte and erythroid lineages and fail to undergo the "GATA switch" that downregulates GATA2 upon terminal differentiation of hematopoietic progenitors into mature erythrocytes [71, 72].

When associated with FOG1, NuRD is a nine-protein complex containing Mi-2β (CHD4, chromodomain helicase DNA-binding protein 4), MTA1 (metastasis-associated 1), MTA2 (metastasis-associated 2), p66 (GATAD2A, GATA zinc finger domain containing 2A), RBAP46 (retinoblastoma-binding protein P46) (RBBP7, retinoblastoma-binding protein 7), RBAP48 (retinoblastoma-binding protein P48) (RBBP4, retinoblastoma-binding protein 4), and MBD3 (methyl-CpG-binding domain protein 3) [70]; in megakaryocytes derived from primary mouse fetal liver, MTA2 and RBAP46 were enriched at GATA1-FOG1 targets including *Itga2b*, *Mpl*, *Gpvi* (glycoprotein VI), *Gpix*, *Gp1ba*, and *Pf4* [70]. The loss of NuRD binding to FOG1 results in significantly decreased activation of *Itga2b*, although the extent of the redundancy between the NuRD- and ETS-mediated GATA1-FOG1 regulation of this gene is unclear as this was demonstrated by a reporter assay in a murine cell line [73]. Similarly, the same study showed that expression levels of other megakaryocyte genes were affected to different extents by loss of the NuRD-FOG1 association [73], indicating alternative mechanisms for gene regulation. Megakaryocytes derived from the fetal livers of mice with *FOG1* mutations preventing NuRD-FOG1 interactions also showed variable changes in mRNA levels of platelet proteins, with significantly decreased levels of *P-selectin* transcript but not *Vwf* (von Willebrand factor), *Pf4*, or *Pbp* (platelet basic protein) [74]. Of note, GATA1-FOG1 also interacts with the CTBP (carboxy-terminus-binding protein) complex, but it has only been shown to drive erythropoiesis with no effect on the megakaryocyte lineage [75].

Finally, in addition to directly occupying the *Gata1* promoter region, ZBP89 appears to associate with GATA1-FOG1 multiprotein complexes in early megakaryocyte development and is required for normal megakaryocyte differentiation of murine embryonic stem cells (ESCs) [41]. Key megakaryocyte genes such as *ITGA2B*, *GPVI*, and *c-MPL* contain GATA and GC-rich binding sites, which are binding motifs recognized by GATA1 and PU.1/SPI1 (hematopoietic transcription factor PU.1/spleen focus-forming virus (SFFV) proviral integration protein); ZBP89 dimerizes with PU.1. Thus, although PU.1 is a hematopoietic TF critical for lymphoid and myeloid but not megakaryocyte or erythroid differentiation [76], it has been hypothesized that GATA1, ZBP89, and PU.1 may together regulate the expression of these genes [41]. PU.1 has separately been shown to regulate the expression of the megakaryocyte gene *PBP* [77], supporting this hypothesis of GATA1, ZBP89, and PU.1 interplay.

12.3.2 Non-FOG1-Binding Partners

In addition to FOG1, GATA1 interacts with a range of protein partners to regulate megakaryopoiesis. These include RUNX1 (runt-related transcription factor 1), FLI1 outside of the context of the GATA1-FOG1 interaction, SCL/TAL1 (stem cell leukemia/T-cell acute lymphoblastic leukemia 1), GATA2, and GATA1 itself; a ChIP-Seq (chromatin immunoprecipitation followed by deep sequencing) study in primary human megakaryocytes suggests that these TFs interact in different combinations, including all five together, to regulate key megakaryocyte genes such as *RUNX1*, *ITGA2B*, and *GP1BB* (glycoprotein 1b platelet subunit beta) [78]. RUNX1 and its cofactor CBFβ, the β subunit of the evolutionarily conserved transcription complex CBF (core-binding factor), are required for definitive hematopoiesis and are upregulated specifically in megakaryocytes but not erythrocytes [79]. The runt domain of RUNX1 binds to the N- and C-TADs of GATA1, although the C-TAD may play a more critical role in this interaction and subsequent transcriptional activity [79, 80]; interaction between GATA1 and RUNX1 has also been shown in *Drosophila melanogaster* and mice. It has been suggested that GATA1 targets RUNX1 for phosphorylation, thereby converting RUNX1 from a repressor to an activator and triggering a megakaryocytic transcriptional program [81].

Several mouse and cell line studies have demonstrated that GATA1 and RUNX1 also interact as part of larger complexes. The GATA1-RUNX1-CBFβ complex synergistically upregulates the *ITGA2B* [79] and *GPIBA* [80] promoters; this activity is mediated in part by P-TEFb (positive transcription elongation factor b), a kinase complex that includes cyclin T1 and CDK9 (cyclin-dependent kinase 9) [82], that functions in various contexts to phosphorylate RNA polymerase II (RNAPII) and promote transcription; chemical inhibition of CDK9 significantly decreased the ability of GATA1-RUNX1-CBFβ to upregulate *ITGA2B* expression [83]. It has also been suggested that GATA1, RUNX1, and CBFβ act within a larger complex with FOG1 and FLI1 to activate *ITGA2B* as well as other key megakaryocyte genes such as *c-MPL* [84].

In addition to complexing with GATA1 alone or in association with GATA2, RUNX1, and/or FLI1 [78], SCL/TAL1 also interacts with GATA1 as part of a pentameric complex with E2A (transcription factor E2 alpha), LDB1 (LIN-11, Isl-1 and MEC-3 (LIM) domain-binding 1), and LMO2 (LIM domain only 2) [7, 34, 85, 86]. This complex binds to a combined E-box (CAGGTG) and GATA motif on its target genes [34]. In megakaryocytes and a murine cell line, GATA1-SCL/TAL1-E2A-LDB1-LMO2 interacts with ETO2 (CBFA2T3, CBFA2/RUNX1 partner transcriptional co-repressor 3) [7, 85, 86], a highly conserved transcriptional corepressor that has also been studied extensively in erythroid differentiation [86, 87]. ETO2 appears to restrain the expression of certain terminal megakaryocyte genes and thus prevent premature megakaryocyte differentiation by binding to the promoter of the key platelet gene *Pf4* [7]. LMO4 (LIM domain only 4) and CDK9 were found to associate with LDB1 in murine cells; both proteins are involved in the regulation of the cell cycle and definitive hematopoiesis [86].

GATA1 binds to itself via a reciprocal interaction between the N-ZF and C-ZF [88, 89]. It has been suggested that dimerization is important for acting on promoters that contain multiple GATA binding sites [89], and a subsequent murine study found that GATA1 homodimerization is required for binding to repeated double/tandem GATA motifs [5]. Overall, investigation into the function of GATA1 homodimers has been limited.

Finally, GFI1B (growth factor-independent 1B transcriptional repressor), a TF originally reported as a GATA1-binding partner in red cells [49], has subsequently been shown in a mouse model to be required for megakaryocyte maturation; fetal liver megakaryocyte colony-forming assays showed smaller colonies and decreased expression of such genes as *Vwf* and *c-Mpl* [90]. *Gfi1b*-mutated mice were also found to have macrothrombocytopenia, megakaryocyte dysplasia, and impaired cytoskeleton organization in megakaryocytes [91]. Based on a human cell line study with constitutive RAS (rat sarcoma virus) activation, GATA1 appears to be required in RAS-mediated differentiation of hematopoietic progenitors to megakaryocytes [92]. However, the mechanism by which GATA1 and the RAS pathway interact remains unclear.

12.4 GATA1 Regulation of Megakaryocyte Development

Through the above interactions, GATA1 induces the expression of numerous megakaryocyte genes that contain GATA-binding sites in their promoters and have been identified in primarily murine-based studies [14, 47, 48, 54, 56–68, 70, 74, 76,

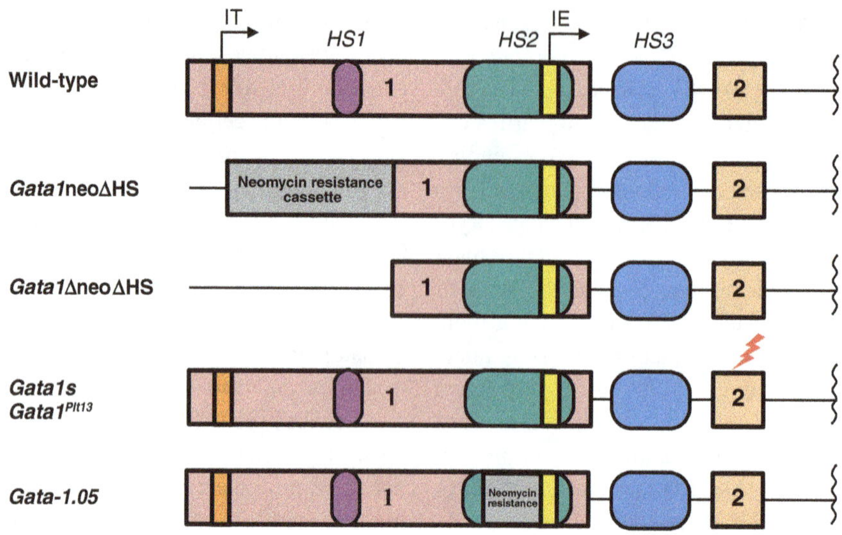

Fig. 12.5 Schematic of gene changes in murine *Gata1* models. IT testis promoter/first exon, IE erythroid promoter/first exon, HS hypersensitivity region

78–80, 83, 84, 90]. GATA1 is essential for normal platelet production, growth, and differentiation; varied GATA1-mutant megakaryocytes are highly proliferative but do not undergo terminal maturation [14, 17, 32, 48, 54, 93, 94]. This has been primarily studied in mice with either the *Gata1*neoΔHS or the *Gata1*ΔneoΔHS construct, both of which lack HS I and the distal promoter [14, 95], with mutations leading to exclusive expression of GATA1s [17, 94], or with the *Gata-1.05* mutation that express 5% of WT levels [93] (Fig. 12.5 and Table 12.1).

Transgenic mice harboring *Gata1*neoΔHS or *Gata1*ΔneoΔHS exhibit increased megakaryocyte progenitors in the fetal liver and markedly increased megakaryocyte number in the adult spleen and bone marrow (BM), consistent with enhanced proliferation [14, 32] (Table 12.1); in vitro experiments with *Gata1*neoΔHS or *Gata1*ΔneoΔHS megakaryocytes reveal a similar proliferative phenotype. The in vivo megakaryocytes are smaller than controls and are dysplastic, with scant granules, disorganization of the demarcation membrane system (DMS), and minimal proplatelet formation [14, 48, 54]; despite the smaller cell size, the megakaryocyte colonies are larger due to proliferation of the immature cells [32], as evidenced by decreased DNA content and ploidy [54]. Transgenic male mice are macrothrombocytopenic, reflecting the impaired proplatelet formation, whereas heterozygous females have a normal platelet count [14, 54]. Microarray [48, 54] and flow cytometry [32] analyses of the in vivo and in vitro megakaryocytes show dysregulation of canonical platelet genes such as *Gp1ba*, *Gp1bb*, *Gpvi*, *Pf4*, *c-Mpl*, and p45 *Nfe2*, as well as those involved in cytokine signaling, calcium signaling, and maintenance of the cytoskeleton. Consistent with this microarray profile and the paucity of intracellular structures, *Gata1*neoΔHS or *Gata1*ΔneoΔHS platelets have decreased activation in response to thrombin or combined ADP (adenosine diphosphatase) and epinephrine [54].

Similarly, Gata1-lox|Pf4-Cre mice with megakaryocyte-specific GATA1 loss exhibit macrothrombocytopenia despite increased megakaryocyte proliferation, aberrant megakaryocyte maturation, and decreased platelet functionality [96]. The GATA1-deficient megakaryocytes appear to have temporally dysregulated maturation, with earlier expression of platelet-specific surface markers such as CD41 or CD61 but lower ploidy when compared to WT. These platelets also have a significantly diminished response to stimulation with aggretin, collagen, convulxin,

Table 12.1 Phenotypes of murine *Gata1* models

	*Gata1*neoΔHS *Gata1*ΔneoΔHS	*Gata1s*	*Gata-1.05*	*Gata1*^{Plt13}
Mutation effect	Four- to five-fold decrease in *Gata1* expression	Translation of GATA1s only	*Gata1* expression 5% of WT	Undetectable *Gata1* expression
Embryonic/fetal phenotype	↑Megakaryocyte progenitors in the liver	↑↑Megakaryocyte progenitors in the liver	Lethal anemia by E12.5 (males) Thrombocytopenia (females)	Lethal anemia (males)
Adult phenotype	↑↑Megakaryocytes in the spleen and bone marrow Macrothrombocytopenia (males)	↑Immature megakaryocytes in the bone marrow Normal to decreased platelet count	Thrombocytopenia, anemia Splenomegaly with megakaryocyte trapping ↑Megakaryocytes in the liver	↑↑Megakaryocytes in the spleen and bone marrow ↑Megakaryocytes in the liver and lung Thrombocytopenia Splenomegaly with megakaryocyte trapping
Platelet phenotype	Small, dysplastic ↑Proliferation ↓Granules and DMS ↓Activation Impaired upregulation of platelet-specific genes	Small ↑Proliferation Moderate upregulation of platelet-specific genes Immature (but more mature than (Δ)neoΔHS)	Morphologically normal Impaired terminal differentiation	Dysplastic ↑Ploidy ↓DMS Immature
References	[8, 14, 32, 48, 54]	[17, 48, 97, 98]	[10, 97]	[101]

DMS demarcation membrane system, *WT* wild-type

and botrocetin attributed to defects in receptors such as integrin β1 or GPVI. Taken together, these findings in distinct mouse models underscore the uncoupling of megakaryocyte proliferation and terminal maturation with GATA1 loss.

The effect of loss of the N-terminus of GATA1 (i.e., GATA1s) appears in mice to be developmental stage-dependent and are distinct from the consequences of *Gata1*neoΔHS and *Gata1*ΔneoΔHS [17]. GATA1s expression in adult mice results in a mild expansion of immature megakaryocytes in the BM [97], but when expressed at the embryonic and fetal stages, there is a significant increase in megakaryocyte progenitors in the yolk sac and fetal liver that are hyperproliferative, immature, and morphologically abnormal, similar to the megakaryocytes observed in *Gata1*neoΔHS and *Gata1*ΔneoΔHS mice (Table 12.1). However, *Gata1s* megakaryocytes appear to be more mature than their neoΔHS and ΔneoΔHS counterparts, with moderate upregulation of megakaryocyte-specific genes but impaired downregulation of non-megakaryocytic genes including five TFs (*Myc* (myelocytomatosis viral oncogene homolog), *Myb* (myeloblastosis viral oncogene homolog), *Gata2*, *Pu.1*, and *Znfn1a1* (zinc finger protein, subfamily 1A, 1)); inadequate repression of these genes that control hematopoietic progenitor proliferation and lineage specification may contribute to the hyperproliferation of *Gata1s* megakaryocytes [17]. Megakaryocytes with exclusive expression of GATA1s form proplatelets at a higher frequency ex vivo compared to *Gata1*ΔneoΔHS but at a lower frequency compared to WT [17, 48], leading to a decreased platelet count overall. A human ex vivo study of hematopoietic stem/progenitor cells (HSPCs) isolated from fetal liver and gene-edited to exclusively express *GATA1s* similarly demonstrated that GATA1s leads to increased cell proliferation with a specific increase in $CD41^+$ cells within all colonies containing megakaryocytes, although this study did not evaluate platelet maturation [98].

In in vitro studies of murine ESC-derived hematopoietic cells, both GATA1 loss [99, 100] and exclusive expression of *Gata1s* [94] lead to increased production of immature, developmentally arrested megakaryocytes. *Gata1s* megakaryocytes generated from ESCs are smaller, more proliferative, and express cell surface CD41 but lack GPIB that is typically expressed on mature megakaryocytes [100] or retain the progenitor cell marker KIT (KIT proto-oncogene, receptor tyrosine kinase) [94]. Not surprisingly, these megakaryocytes express a more immature transcriptional profile when compared to their WT counterparts [94]. In vivo studies of murine yolk sacs [94] and fetal liver [100] yielded similar findings. Meanwhile, GATA1 rescue studies in G1ME cells, a GATA1-deficient murine bipotential MEP line originally derived from differentiation of ESCs, reinforced the role of GATA1 in inhibiting cell division to promote maturation [100].

A more global loss of GATA1 as induced by the *Gata-1.05* mutation induces thrombocytopenia in female heterozygous embryos and adult mice with varying degrees of random X inactivation; male mice die at E12.5 from severe anemia [10, 93] (Table 12.1). The female mice exhibit severe anemia and develop variable degrees of marked splenomegaly in addition to the thrombocytopenia. The megakaryocytes trapped in the spleens appear morphologically normal and express CD41 but fail to undergo terminal differentiation to shed mature platelets [93]. Meanwhile, a mutation in the *Gata1* start codon designated *Plt13*, which is predicted to result in exclusive production of GATA1s but instead results in undetectable levels of either protein isoform, results in moderate thrombocytopenia without anemia in female heterozygous mice, and is embryonic lethal in hemizygous males [101] (Table 12.1). The $Gata1^{Plt13}$ megakaryocytes also accumulate in the spleens of the heterozygous females; they are characterized by hyperchromatic nuclei, disorganized or absent DMS, and increased ploidy when compared to WT. In contrast to the findings in *Gata1*neoΔHS or *Gata1*ΔneoΔHS mice [32, 54], $Gata1^{Plt13}$ megakaryocytes have increased DNA content and form colonies of normal size; of note, increased megakaryocyte colony formation was restricted to $Gata1^{Plt13}$ fetal liver, similar to *Gata1s* mice [17].

Mutations that prevent interactions between GATA1 and its protein partners have also been studied in murine models of hematopoiesis. A near-complete to complete loss of hematopoietic progenitors committed to the megakaryocyte and erythroid lineages has been observed in the absence of FOG1, suggesting that gene regulation through the GATA1-FOG1 and other FOG1 complexes is critical for both erythropoiesis and megakaryopoiesis [44, 48, 55]. The V205G mutation, which eliminates the GATA1-FOG1 interaction, results in large megakaryocytes that are unable to form proplatelets; these megakaryocytes also lack expression of genes involved in terminal maturation [32, 48]. Mice with FOG1 mutations preventing NuRD-FOG1 interaction and thus formation of the GATA1-FOG1-NuRD complex exhibit severe macrothrombocytopenia with intracellular dysplasia and impaired activation, although proliferation and ploidy appear intact [73, 74]. Interestingly, although the α-granules of these platelets appear morphologically abnormal, only P-selectin is significantly decreased among the α-granule proteins [74]. These platelets also lack response to thrombin and convulxin, with absent PF4 release.

Studies in human-derived models are extremely limited and have primarily occurred in the context of DS [102–106]. Although euploid human induced pluripotent stem cells (iPSCs) expressing WT GATA1 vs. exclusively GATA1s generate similar numbers of multipotent hematopoietic progenitors, when cultured in megakaryocyte-specific media, the GATA1s progenitors are more proliferative compared to WT controls [102]. This is accompanied by increased expression of genes associated with megakaryocytes and myeloid cells and decreased expression of those associated with erythrocytes. In human T21 iPSCs, *GATA1s* mutations and GATA1s overexpression both suppress megakaryocyte commitment in hematopoietic progenitors but ultimately result in a population of proliferative, immature CD34$^+$/CD41$^+$ megakaryocytes assayed by cell surface expression markers, cell morphology, and response to thrombin stimulation [103–107].

12.5 GATA1 Mutations in Human Diseases of Platelet Development

12.5.1 Germline GATA1 Mutations

Mutations affecting GATA1 have been recurrently implicated in human hematopoietic disease [108–111]. Germline mutations affecting GATA1 have been reported in association with X-linked congenital cytopenias, a heterogeneous spectrum of hematologic conditions unified by platelet abnormalities. Pathogenic variants can be divided into those that result in loss of the N-TAD, or affecting the C-ZF, N-ZF, or C-TAD (Fig. 12.6 and Table 12.2). While some cases have historically been classified as a variant of Diamond-Blackfan anemia (DBA) [19, 21, 112–116], thalassemia [117–121], or congenital erythropoietic porphyria (CEP) [121–123], GATA1-related cytopenias likely represent a distinct clinical entity. Patients diagnosed with "GATA1-related DBA" were found to have clinical manifestations of DBA, including macrocytic anemia and variable neutropenia, but lack a mutation in one of the ribosomal proteins and instead carry *GATA1* mutations in the N-terminus, resulting in exclusive expression of GATA1s [19, 21, 114, 115, 124], or disrupting the C-TAD [125]. These individuals often have additional hematologic abnormalities, for example, thrombocytopenia, thrombocytosis, or dysmegakaryopoiesis, while lacking extra-hematopoietic manifestations of DBA such as skeletal anomalies. It has been hypothesized that the clinical overlap between germline *GATA1* mutations and DBA is due to the impaired translation of *GATA1* that occurs as a result of the ribosomal insufficiency in DBA [21, 114], but the distinct phenotype of "GATA1-related DBA" suggests that it is more precisely termed GATA-1 related cytopenias.

Similarly, those diagnosed with "X-linked thalassemia" or "X-linked thrombocytopenia with thalassemia" (XLTT) were noted to have ineffective erythropoiesis but also to have incongruous features, such as macrothrombocytopenia, a disproportionately prolonged bleeding time or platelet

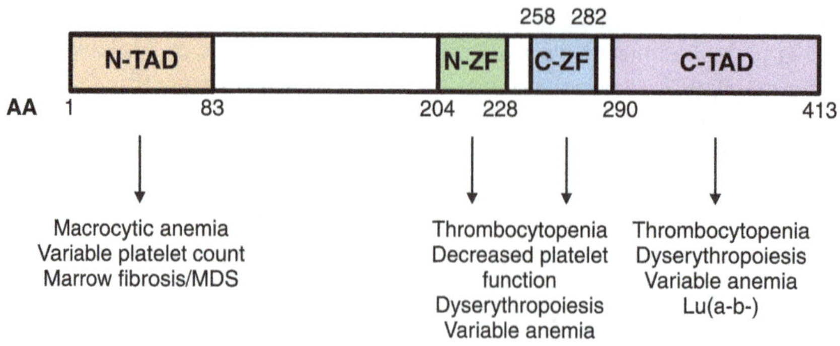

Fig. 12.6 Broad associations between affected region of GATA1 and resulting phenotypes. N-TAD amino-terminus activation domain, C-TAD carboxy-terminus activation domain, N-ZF amino-terminus zinc finger, C-ZF carboxy-terminus zinc finger, Lu Lutheran blood group antigen, MDS myelodysplastic syndrome

dysfunction, and progressive BM fibrosis [117–121, 126]. Instead of a mutation in a globin gene, a *GATA1* mutation affecting the N-ZF and therefore DNA binding, p.Arg216Gln (c.647G.A), has been recurrently identified in these individuals [117–120]. However, it is important to note that the same mutation [127] and other mutations affecting the same residue (e.g., p.Arg216Trp (c.646C > T)) [128, 123] have been described without thalassemia, suggesting that these mutations do not uniformly cause thalassemia and that the interpretation oftheresultingphenotypemaybecontext-orphysician-dependent.

CEP, a group of disorders of heme biosynthesis, is typically caused by autosomal recessive mutations in *UROS* (uroporphyrinogen III synthase) [121, 128]. While the typical hematologic manifestations of CEP are chronic hemolytic anemia and transfusion dependence, those with "X-linked CEP" have been found to also have dyserythropoiesis (described as "thalassemia" in one individual), persistence of Hgb F (fetal hemoglobin), and thrombocytopenia in addition to the characteristic skin findings of porphyria [121, 128]. These individuals have been found to have the mutation p.Arg216Trp (c.646C > T) that affects the N-ZF of *GATA1*. Interestingly, although this mutation has been reported in two probands, the co-inheritance of a *UROS* mutation was reported in one family [128], while this *GATA1* mutation was the only pathogenic variant identified in the other [121]; of note, GATA1 regulates *UROS* expression in developing erythrocytes [129], likely accounting for the clinical overlap. Again, the hematologic features inconsistent with the unifying diagnosis, as well as the overlap in symptoms and terminology with those carrying other diagnoses, implies that "X-linked CEP" is in fact a form of GATA-1-related cytopenias.

Indeed, the wide range of phenotypic manifestations reported in individuals with germline mutations affecting the N-TAD, N-ZF, C-ZF, and C-TAD of *GATA1* suggests that GATA1-related cytopenias are primarily a platelet disorder with variable other hematopoietic and extra-medullary findings that may overlap with those of other conditions (Table 12.2). The heterogeneity is likely due to the differences in the DNA-binding sites and protein-protein interactions that are disrupted by each of the mutations. At the same time, the distinct roles ascribed to each of GATA1's functional domains allow for broad genotype-phenotype associations. The management of GATA1-cytopenias is primarily supportive, although curative HSC transplants have also been reported [20, 21, 53, 112, 121, 128, 130].

Pathogenic variants resulting in exclusive GATA1s expression with loss of the N-TAD occur due to changes in the exon 2 start codon, premature stop codons, or affect the exon 2 donor splice site. They tend to have a more severe red cell phenotype and variable thrombocytopenia despite megakaryocyte dysplasia and may lead to marrow fibrosis or failure including MDS (myelodysplastic syndrome), ML-DS with

Table 12.2 Summary of reported human *GATA1* mutations and associated phenotypes

DNA nucleotide change	Predicted protein change	Megakaryocyte and platelet phenotype	Red cell phenotype	Other hematologic features	Non-hematologic features	Refs.
Pathogenic variants resulting in GATA1s						
c.-21A>G	–	Normal to mild thrombocytosis Dysplastic megakaryocytes	Dyserythropoiesis Mild to severe anemia Macrocytic ↑Hgb F	Occasional neutropenia ML-DS with acquired T21 MDS, AML		[116, 132]
c.2T>C	p.Met1[a]	Normal to mild thrombocytosis Dysplastic megakaryocytes	Dyserythropoiesis Severe anemia Macrocytic "Clinical DBA"	Progression to MDS Megaloblastic changes ML-DS with acquired T21		[21, 114, 115, 132]
c.3G>A	p.Met1[a]	Fluctuating thrombocytopenia	Moderate to severe anemia normocytic "Clinical DBA"		Hypertelorism and flat nasal bridge Hepatomegaly	[124]
c.94delG	p.Val32PhefsTer105	Moderate to severe thrombocytopenia Dysplastic megakaryocytes	Mild to severe anemia Normocytic to macrocytic	TAM with acquired T21 Hypercellular, fibrotic marrow	Reported in heterozygous females only	[131]
c.220+2T>C	–	Mild thrombocytosis Dysplastic megakaryocytes	Dyserythropoiesis Moderate anemia Macrocytic "Clinical DBA"			[21]
c.220G>C	p.Val74Leu	Normal or mild thrombocytopenia Dysplastic megakaryocytes Decreased aggregation	Mild to severe anemia Macrocytic	Neutropenia		[19, 20, 113]
Pathogenic variants affecting N-ZF						
c.613G>A	p.Val205Met	Mild to severe thrombocytopenia Macrocytosis	Moderate to severe anemia Dyserythropoiesis Fetal hydrops		Cryptorchidism	[53, 150]
c.617A>T	p.Asn206Ile	Severe thrombocytopenia Decreased granules	Mild dyserythropoiesis			[146]

(continued)

Table 12.2 (continued)

DNA nucleotide change	Predicted protein change	Megakaryocyte and platelet phenotype	Red cell phenotype	Other hematologic features	Non-hematologic features	Refs.
c.622G>A	p.Gly208Arg	Mild to severe thrombocytopenia Macrocytosis	Moderate to severe anemia Dyserythropoiesis		Cryptorchidism Splenomegaly Hepatomegaly Phenotype may improve with age	[144, 147]
c.622_623delGGinsTC	p.Gly208Ser	Severe thrombocytopenia Hypogranular platelets Macrocytosis Decreased aggregation				[142, 143]
c.647G>A	p.Arg216Gln	Normal to moderate thrombocytopenia Macrocytosis Decreased aggregation, prolonged bleeding time	Mild anemia “β-thalassemia”		Splenomegaly	[117–120, 123, 126–128, 141]
c.646C>T	p.Arg216Trp	Moderate thrombocytopenia	Mild anemia “β-thalassemia”		Splenomegaly “X-linked CEP”	[121–123]
c.653A>G	p.Asp218Gly	Moderate to severe thrombocytopenia Macrocytosis Decreased aggregation	Dyserythropoiesis without anemia			[140, 148, 149]
c.652G>T	p.Asp218Tyr	Severe thrombocytopenia Macrocytosis	Severe anemia	Platelets in heterozygous female expressed only wild-type allele		[139]
c.652G>A	p.Asp218Asn	Severe thrombocytopenia Macrocytosis Decreased aggregation, prolonged bleeding time	Dyserythropoiesis without anemia		Splenomegaly	[137, 145]
Pathogenic variants affecting C-ZF						
c.788C>T	p.Thr263Met	Moderate thrombocytopenia Dysplastic megakaryocytes	Mild dyserythropoiesis Mild anemia	Leukocytosis (neutrophilia) Hypercellular, fibrotic marrow	Reported in heterozygous females only	[138]

c.802C>A	p.Leu268Met	Worsening thrombocytopenia with age Decreased aggregation	Anisocytosis, poikilocytosis Progressive macrocytosis with age ↑Hgb F			[146]
c.865C>T	p.His289Tyr	Normal to mild thrombocytopenia Anisocytosis Decreased aggregation, prolonged bleeding time	Dyserythropoiesis Normal to mild anemia			[137]
Pathogenic variants affecting C-TAD						
c.871-24C>T	–	Normal to moderate thrombocytopenia Macrocytosis Dysplastic megakaryocytes Decreased aggregation	Dyserythropoiesis Hydrops fetalis, severe infantile anemia Worsening anemia with illness ↑Hgb F	Occasional myeloid dysplasia Progression to aplastic anemia	Cryptorchidism Hypospadias	[112, 133]
c.886A>C	p.Thr296Pro	Moderate thrombocytopenia Macrocytosis Decreased aggregation	Mild dyserythropoiesis Mild anemia ↑Hgb F Lu(a-b-)			[135]
c.919C>T	p.Arg307Cys	Mild thrombocytopenia	Mild dyserythropoiesis Hemolysis "Clinical DBA"	ADA overproduction		[125]
c.920G>A	p.Arg307His	Mild thrombocytopenia Macrocytosis	Variable dyserythropoiesis Hyperchromia Macrocytosis Hemolysis ↑Hgb F Hydrops fetalis, severe infantile anemia Lu(a-b-) "Clinical DBA"	ADA overproduction	Low birth weight Hypospadias Splenomegaly	[125, 134]

(continued)

Table 12.2 (continued)

DNA nucleotide change	Predicted protein change	Megakaryocyte and platelet phenotype	Red cell phenotype	Other hematologic features	Non-hematologic features	Refs.
c.1240T>C	p.Ter414Arg	Mild thrombocytopenia Macrocytosis	Lu(a-b-)			[136]
Unclassified pathogenic variants						
c.515T>C	p.Phe172Ser	Severe thrombocytopenia	Severe anemia	Leukopenia Bone marrow hypoplasia	Mild splenomegaly	[130]

AML acute myeloid leukemia, *ADA* adenosine deaminase, *CEP* congenital erythropoietic porphyria, *C-TAD* carboxy-terminus transactivation domain, *C-ZF* carboxy-terminus zinc finger, *DBA* Diamond-Blackfan anemia, *Hgb F* fetal hemoglobin, *Lu* Lutheran blood group antigen, *MDS* myelodysplastic syndrome, *ML-DS* myeloid leukemia of Down syndrome, *N-ZF* amino-terminus zinc finger, *T21* trisomy 21, *TAM* transient abnormal myelopoiesis

[a]Single nucleotide mutation that disrupts start codon; an alternative start codon results in GATA1s expression

GATA1s expressed in the blast cells only (*see* GATA1 mutations in Down syndrome (DS) *below*), or ML-DS-like AML (acute myeloid leukemia) with acquired T21 [19–21, 113–116, 124, 131, 132]. Of note, only one variant (p.Val74Leu) is reported to result in decreased platelet function [20], although platelet aggregation testing was not pursued in most cases likely as bleeding symptoms were minor or consistent with the degree of thrombocytopenia.

Clinical phenotypes of mutations affecting the C-TAD are generally similar to those affecting the N-TAD, but with overall milder thrombocytopenia and more severe dyserythropoiesis and/or anemia. While most of these mutations are single-amino acid changes, one is intronic and hypothesized to affect splicing [112, 133]. There is a tendency for the anemia to be most severe in infancy, with multiple reports of hydrops fetalis [112, 133] as well as severe neonatal anemia requiring multiple transfusions [112, 133, 134], but there are also reports of possible progression to aplastic anemia later in life [112]. These C-TAD mutations are uniquely associated with Lu(a-b-) erythrocytes [135, 136], adenosine deaminase (ADA) overproduction and hemolysis [125], and an elevated Hgb F fraction [112, 133–135]. In comparison to germline *GATA1s* mutations, C-TAD variants appear more likely to cause abnormal platelet function [112, 133] or a bleeding diathesis, with easy bruising and petechiae [135], epistaxis [133, 135], gastrointestinal bleeding [133], and/or excessive bleeding with trauma or surgery [133].

Meanwhile, variants affecting either the N-ZF or C-ZF are associated with more severe macrothrombocytopenia, decreased platelet function and prolonged bleeding time, and dyserythropoiesis with variable anemia. This group of variants includes p.Val205Met, which disrupts GATA1-FOG1 binding and has been extensively studied in primarily murine-derived models [46, 47, 50, 53, 70, 72, 123]. Interestingly, even carrier females may have mild cytopenias or bleeding tendencies [53, 128, 137, 138], and both female heterozygotes [127, 139, 140] and male hemizygotes [119, 141] have rarely been reported to have two platelet populations on peripheral smear. The reported bleeding in individuals with *GATA1* mutations affecting either ZF range from none [138, 142] to severe, regardless of tested platelet function, and include easy bruising and petechiae [126, 127, 137, 143–147], recurrent epistaxis [126, 130, 137, 143, 145, 147], other mucocutaneous bleeding [119, 145–148], excessive bleeding with trauma or surgery [119, 126, 137, 143, 148, 149], gastrointestinal hemorrhage [143, 149], menorrhagia [137], fetal cerebral hemorrhage [150], and other non-specified bleeding requiring medical evaluation with or without platelet transfusions or other intervention [53, 118, 120, 127, 139, 144]. Of note, two related individuals with the p.Leu258Met mutation had a worsening phenotype with age [146], in contrast to the stable cytopenias seen with the majority of the variants. Another mutation, p.Thr263Met, has been reported only in two related heterozygous females [138].

12.5.2 GATA1 Mutations in Down Syndrome (DS)

GATA1s uniquely interacts with DS, or T21, to result in TAM and ML-DS [16, 22, 151, 152]. Formerly known as transient myeloproliferative disorder (TMD), TAM is a preleukemic or leukemia-like condition that occurs exclusively in ~10% of infants with DS in the first 3 months of life [153–155]. In the majority of cases, the clonal proliferation of blast cells remits spontaneously, and no chemotherapy is required unless the blast burden causes end-organ dysfunction, most commonly massive hepatomegaly that may lead to cardiopulmonary compromise. Despite this, however, the 5-year overall survival (OS) for TAM is 80–90%, and ~25% of those who survive subsequently develop ML-DS, a true leukemia that requires standard chemotherapy, within 4 years [153–155]. The blasts of both TAM and ML-DS carry a *GATA1s* mutation while ML-DS blasts harbor at least one additional mutation, often in epigenetic regulators or members of the cohesin complex, suggesting that ML-DS arises from a residual TAM subclone that has acquired an additional oncogenic "hit" [23, 156–158] (Fig. 12.7). Both conditions are thought to arise

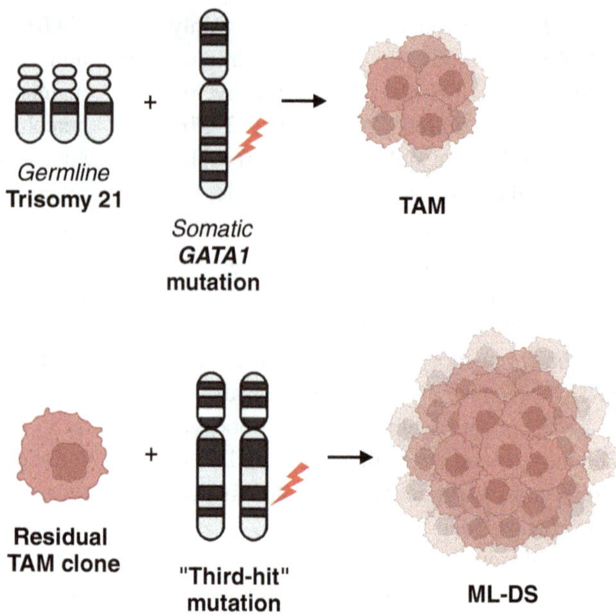

Fig. 12.7 Progression from germline trisomy 21 state to TAM with acquisition of *GATA1s* mutation, then to ML-DS with acquisition of "third-hit" mutation. TAM transient abnormal myelopoiesis, ML-DS myeloid leukemia of Down syndrome

from an aberrant MEP clone, and the blasts typically carry a megakaryocyte-like phenotype by flow cytometry.

T21 itself perturbs fetal liver hematopoiesis [159–161], and the acquisition of the *GATA1s* mutation, while somatic, is early and often during fetal life [162, 163]. Given that TAM can present on the day of birth or even cause in utero complications such as hydrops fetalis, it is conceptualized as a disorder of fetal hematopoiesis [154, 164], and its self-limiting nature is therefore likely due to the postnatal transition from fetal to adult hematopoiesis. Interestingly, the frequency of *GATA1s* mutations in the DS population may be as high as 30% when assessed by targeted next-generation sequencing instead of standard Sanger sequencing, indicating that ~20% of infants with DS have "silent TAM" [22]. Whether this is because some infants with T21 and a *GATA1s* mutation truly do not develop clinical TAM or because the disease process peaked and spontaneously remitted in fetal life remains unclear.

ML-DS is both genetically and clinically distinct from myeloid leukemias not associated with DS [165]. In addition to the hallmark *GATA1s* mutation, there are recurrent oncogenic mutations that are more common in ML-DS or do not carry the same prognostic significance when compared to non-DS leukemias. ML-DS blasts are exquisitely chemosensitive, especially when compared to other myeloid leukemias, leading to a >90% 5-year OS although the survival rate for relapsed cases remains <20% [166, 167]. Individuals without T21 but with germline *GATA1* mutations do not have a leukemic predisposition. Yet there also exist case reports of patients without T21 but with germline *GATA1s* mutations who developed TAM [131] or ML-DS-like leukemia [132] and were found to have acquired T21 in the blast cells only. Furthermore, there are case reports of individuals without either germline T21 or *GATA1* mutations who developed TAM and/or ML-DS-like leukemia and were found to have acquired T21 and GATA1s in the blast cells only [168]. The characteristics of these patients mirrored what would be expected for those with DS with regards to age of onset, response to chemotherapy, and associated genetic changes in the blasts. Given that in vivo and in vitro murine studies have demonstrated that GATA1s increase megakaryocyte proliferation at the expense of terminal maturation [14, 32, 48, 54, 93, 94, 96, 99, 100], it is not clear why a leukemic risk is increased only in the setting of T21. These cases therefore emphasize the unique interaction between chromosome 21 gene dosage imbalance and GATA1s, regardless of the order in which the genetic events occur.

Studies in human fetal liver and neonatal peripheral blood samples have demonstrated differential methylation patterns induced by both T21 and GATA1s [169, 170]. While the addition of the third copy of chromosome 21 appears to trigger hypomethylation of genes globally implicated in the broad phenotype of DS, the addition of a *GATA1s* mutation is associated with aberrant hypermethylation specifically of genes involved in hematopoiesis and cell cycle regulation, including the GATA1 target and key platelet gene *NFE2* [169]. Interestingly, there appear to be limited differences between the epigenetic signature and transcriptome of TAM and ML-DS blasts.

As demonstrated in whole-exome and whole-genome sequencing studies, the additional mutation(s) found in ML-DS blasts when compared to those of TAM are typically those that affect chromatin organization and gene expression, with *CTCF* (CCCTC-binding factor), *EZH2* (enhancer of zeste homolog 2), *NRAS* (neuroblastoma RAS viral oncogene homolog), *RAD21* (double-strand-break repair protein Rad21 homolog), *STAG2* (cohesin subunit SA-2), and *TP53* (tumor antigen P53) among the recurrently mutated genes [23, 156]. A study of human fetal liver xenografts in mice demonstrated that the addition of a *STAG2* mutation to a *GATA1s* background significantly increases engraftment and leukemic progression regardless of chromosome 21 status; in the T21 context, GATA1s alone is sufficient to drive blast production, but the addition of a *STAG2* mutation amplifies this effect [98]. Another study of human iPSCs with T21, a *GATA1s* mutation, and haploinsufficiency of the cohesin protein SMC3 (structural maintenance of chromosomes 3) found that the combination of all three genetic alterations enhanced megakaryocyte colony-forming potential, colony size, and proliferation [106]. SMC3 haploinsufficiency appears to exacerbate the megakaryocyte maturation arrest at the megakaryoblast stage, giving rise to a leukemic phenotype.

ML-DS blasts demonstrate increased expression of *GATA1* and of genes such as *GATA2* and *MYC* that are typically downregulated by GATA1 [171]. The human iPSC model with T21, GATA1s, and SMC3 haploinsufficiency found that *NFE2*, which was found to be hypermethylated in TAM fetal liver samples [169], is also downregulated in this context and that forced *NFE2* expression partially restores the maturation defect in megakaryocytes [106]. These examples of gene dysregulation imply that without the N-TAD, GATA1s is functionally unable to regulate these targets. Taken together, these findings suggest that altered chromatin structure and accessibility induced by "third-hit" mutations exacerbate the impaired regulation of hematopoietic genes by GATA1s to give rise to ML-DS.

In summary, GATA1 plays a key role in regulating megakaryocyte differentiation, maturation, and proliferative capacity. Varied murine and cellular models have elucidated the functions of the N- and C-TAD and ZFs. Human *GATA1* mutations in these regions also provide mechanistic insight, and as sequencing technologies expand, additional mutations will likely be identified that contribute to *GATA1*-related cytopenias, including thrombocytopenia. With single-cell-based technology, future studies in human primary cells and/or iPSCs harboring distinct *GATA1* mutations may improve our understanding of GATA1 regulatory mechanisms and how their disruptions lead to human diseases of megakaryocytes and platelets.

Acknowledgments This work was supported by NIH grants R01 HL151260 and U01 HL134696 (to STC), the Doris Duke Charitable Foundation Physician Scientist Fellowship (to KT), and the Ken Mizuno Fund in Hematology (to KT). Figures were generated on Biorender.com.

References

1. Gillis WQ, St John J, Bowerman B et al (2009) Whole genome duplications and expansion of the vertebrate GATA transcription factor gene family. BMC Evol Biol 9:207
2. Tang Y, Wei Y, He W et al (2014) GATA transcription factors in vertebrates: evolutionary, structural and functional interplay. Mol Gen Genomics 289:203–214
3. Martin DI, Orkin SH (1990) Transcriptional activation and DNA binding by the erythroid factor GF-1/NF-E1/Eryf 1. Genes Dev 4:1886–1898

4. Kaneko H, Kobayashi E, Yamamoto M et al (2012) N- and C-terminal transactivation domains of GATA1 protein coordinate hematopoietic program. J Biol Chem 287:21439–21449
5. Hasegawa A, Kaneko H, Ishihara D et al (2016) GATA1 binding kinetics on conformation-specific binding sites elicit differential transcriptional regulation. Mol Cell Biol 36:2151–2167
6. Trainor CD, Omichinski JG, Vandergon TL et al (1996) A palindromic regulatory site within vertebrate GATA-1 promoters requires both zinc fingers of the GATA-1 DNA-binding domain for high-affinity interaction. Mol Cell Biol 16:2238–2247
7. Hamlett I, Draper J, Strouboulis J et al (2008) Characterization of megakaryocyte GATA1-interacting proteins: the corepressor ETO2 and GATA1 interact to regulate terminal megakaryocyte maturation. Blood 112:2738–2749
8. McDevitt MA, Fujiwara Y, Shivdasani RA et al (1997) An upstream, DNase I hypersensitive region of the hematopoietic-expressed transcription factor GATA-1 gene confers developmental specificity in transgenic mice. Proc Natl Acad Sci USA 94:7976–7981
9. Moriguchi T, Suzuki M, Yu L et al (2015) Progenitor stage-specific activity of a *cis* -acting double GATA Motif for *Gata1* gene expression. Mol Cell Biol 35:805–815
10. Ohneda K, Shimizu R, Nishimura S et al (2002) A minigene containing four discrete *cis* elements recapitulates GATA-1 gene expression *in vivo*: A GATA-1 minigene. Genes Cells 7:1243–1254
11. Valverde-Garduno V, Guyot B, Anguita E et al (2004) Differences in the chromatin structure and cis-element organization of the human and mouse GATA1 loci: implications for cis-element identification. Blood 104:3106–3116
12. Vyas P, McDevitt MA, Cantor AB et al (1999) Different sequence requirements for expression in erythroid and megakaryocytic cells within a regulatory element upstream of the *GATA-1* gene. Development 126:2799–2811
13. Nishimura S, Takahashi S, Kuroha T et al (2000) A GATA box in the *GATA-1* gene hematopoietic enhancer is a critical element in the network of GATA factors and sites that regulate this gene. Mol Cell Biol 20:713–723
14. Shivdasani RA, Fujiwara Y, McDevitt MA et al (1997) A lineage-selective knockout establishes the critical role of transcription factor GATA-1 in megakaryocyte growth and platelet development. EMBO J 16:3965–3973
15. Gutiérrez L, Caballero N, Fernández-Calleja L et al (2020) Regulation of GATA1 levels in erythropoiesis. IUBMB Life 72:89–105
16. Wechsler J, Greene M, McDevitt MA et al (2002) Acquired mutations in GATA1 in the megakaryoblastic leukemia of Down syndrome. Nat Genet 32:148–152
17. Li Z, Godinho FJ, Klusmann J-H et al (2005) Developmental stage–selective effect of somatically mutated leukemogenic transcription factor GATA1. Nat Genet 37:613–619
18. Crispino JD, Weiss MJ (2014) Erythro-megakaryocytic transcription factors associated with hereditary anemia. Blood 123:3080–3088
19. Sankaran VG, Ghazvinian R, Do R et al (2012) Exome sequencing identifies GATA1 mutations resulting in Diamond-Blackfan anemia. J Clin Invest 122:2439–2443
20. Hollanda LM, Lima CSP, Cunha AF et al (2006) An inherited mutation leading to production of only the short isoform of GATA-1 is associated with impaired erythropoiesis. Nat Genet 38:807–812
21. Van Dooijeweert B, Kia SK, Dahl N et al (2022) GATA-1 defects in Diamond–Blackfan anemia: phenotypic characterization points to a specific subset of disease. Genes 13:447
22. Roberts I, Alford K, Hall G et al (2013) GATA1-mutant clones are frequent and often unsuspected in babies with Down syndrome: identification of a population at risk of leukemia. Blood 122:3908–3917
23. Yoshida K, Toki T, Okuno Y et al (2013) The landscape of somatic mutations in Down syndrome-related myeloid disorders. Nat Genet 45:1293–1299
24. Kobayashi M, Yamamoto M (2007) Regulation of GATA1 Gene Expression. J Biochem 142:1–10
25. Onodera K, Takahashi S, Nishimura S et al (1997) GATA-1 transcription is controlled by distinct regulatory mechanisms during primitive and definitive erythropoiesis. Proc Natl Acad Sci USA 94:4487–4492
26. Motohashi H, Katsuoka F, Shavit JA et al (2000) Positive or negative MARE-dependent transcriptional regulation is determined by the abundance of small Maf proteins. Cell 103:865–875
27. Takai J, Moriguchi T, Suzuki M et al (2013) The Gata1 5′ region harbors distinct cis-regulatory modules that direct gene activation in erythroid cells and gene inactivation in HSCs. Blood 122:3450–3460
28. Guo Y, Fu X, Huo B et al (2016) GATA2 regulates GATA1 expression through LSD1-mediated histone modification. Am J Transl Res 8:2265–2274
29. Tsai SF, Strauss E, Orkin SH (1991) Functional analysis and in vivo footprinting implicate the erythroid transcription factor GATA-1 as a positive regulator of its own promoter. Genes Dev 5:919–931
30. Kobayashi M, Nishikawa K, Yamamoto M (2001) Hematopoietic regulatory domain of *gata1* gene is positively regulated by GATA1 protein in zebrafish embryos. Development 128:2341–2350
31. Nishikawa K, Kobayashi M, Masumi A et al (2003) Self-association of Gata1 enhances transcriptional activity in vivo in zebra fish embryos. Mol Cell Biol 23:8295–8305
32. Kuhl C, Atzberger A, Iborra F et al (2005) GATA1-mediated megakaryocyte differentiation and growth control can be uncoupled and mapped to different domains in GATA1. Mol Cell Biol 25:8592–8606

33. Guyot B, Murai K, Fujiwara Y et al (2006) Characterization of a megakaryocyte-specific enhancer of the key hemopoietic transcription factor GATA1. J Biol Chem 281:13733–13742
34. Wadman IA, Osada H, Grütz GG et al (1997) The LIM-only protein Lmo2 is a bridging molecule assembling an erythroid, DNA-binding complex which includes the TAL1, E47, GATA-1 and Ldb1/NLI proteins. EMBO J 16:3145–3157
35. Wu W, Morrissey CS, Keller CA et al (2014) Dynamic shifts in occupancy by TAL1 are guided by GATA factors and drive large-scale reprogramming of gene expression during hematopoiesis. Genome Res 24:1945–1962
36. Visvader J, Adams J (1993) Megakaryocytic differentiation induced in 416B myeloid cells by GATA-2 and GATA-3 transgenes or 5-azacytidine is tightly coupled to GATA-1 expression. Blood 82:1493–1501
37. Ikonomi P, Noguchi CT, Miller W et al (2000) Levels of GATA-1/GATA-2 transcription factors modulate expression of embryonic and fetal hemoglobins. Gene 261:277–287
38. Lugus JJ, Chung YS, Mills JC et al (2007) GATA2 functions at multiple steps in hemangioblast development and differentiation. Development 134:393–405
39. Harigae H, Okitsu Y, Yokoyama H et al (2006) Induction of erythroid-specific genes by overexpression of GATA-2 in K562 cells. Int J Hematol 84:38–42
40. Huang Z, Dore LC, Li Z et al (2009) GATA-2 reinforces megakaryocyte development in the absence of GATA-1. Mol Cell Biol 29:5168–5180
41. Woo AJ, Moran TB, Schindler YL et al (2008) Identification of ZBP-89 as a novel GATA-1-associated transcription factor involved in megakaryocytic and erythroid development. Mol Cell Biol 28:2675–2689
42. Ohneda K, Ohmori S, Ishijima Y et al (2009) Characterization of a functional ZBP-89 binding site that mediates Gata1 gene expression during hematopoietic development. J Biol Chem 284:30187–30199
43. Tsang AP, Visvader JE, Turner CA et al (1997) FOG, a multitype zinc finger protein, acts as a cofactor for transcription factor GATA-1 in erythroid and megakaryocytic differentiation. Cell 90:109–119
44. Tsang AP, Fujiwara Y, Hom DB et al (1998) Failure of megakaryopoiesis and arrested erythropoiesis in mice lacking the GATA-1 transcriptional cofactor FOG. Genes Dev 12:1176–1188
45. Crispino JD, Lodish MB, MacKay JP et al (1999) Use of altered specificity mutants to probe a specific protein–protein interaction in differentiation. Mol Cell 3:219–228
46. Chang AN, Cantor AB, Fujiwara Y et al (2002) GATA-factor dependence of the multitype zinc-finger protein FOG-1 for its essential role in megakaryopoiesis. Proc Natl Acad Sci USA 99:9237–9242
47. Wang X, Crispino JD, Letting DL et al (2002) Control of megakaryocyte-specific gene expression by GATA-1 and FOG-1: role of Ets transcription factors. EMBO J 21:5225–5234
48. Muntean AG, Crispino JD (2005) Differential requirements for the activation domain and FOG-interaction surface of GATA-1 in megakaryocyte gene expression and development. Blood 106:1223–1231
49. Rodriguez P, Bonte E, Krijgsveld J et al (2005) GATA-1 forms distinct activating and repressive complexes in erythroid cells. EMBO J 24:2354–2366
50. Chlon TM, Doré LC, Crispino JD (2012) Cofactor-mediated restriction of GATA-1 chromatin occupancy coordinates lineage-specific gene expression. Mol Cell 47:608–621
51. Fox AH, Liew C, Holmes M et al (1999) Transcriptional cofactors of the FOG family interact with GATA proteins by means of multiple zinc fingers. EMBO J 18:2812–2822
52. Hollenhorst PC, McIntosh LP, Graves BJ (2011) Genomic and biochemical insights into the specificity of ETS transcription factors. Annu Rev Biochem 80:437–471
53. Nichols KE, Crispino JD, Poncz M et al (2000) Familial dyserythropoietic anaemia and thrombocytopenia due to an inherited mutation in GATA1. Nat Genet 24:266–270
54. Vyas P, Ault K, Jackson CW et al (1999) Consequences of GATA-1 deficiency in megakaryocytes and platelets. Blood 93:2867–2875
55. Mancini E, Sanjuan-Pla A, Luciani L et al (2012) FOG-1 and GATA-1 act sequentially to specify definitive megakaryocytic and erythroid progenitors. EMBO J 31:351–365
56. Hashimoto Y, Ware J (1995) Identification of essential GATA and Ets binding motifs within the promoter of the platelet glycoprotein Ibα gene. J Biol Chem 270:24532–24539
57. Lemarchandel V, Ghysdael J, Mignotte V et al (1993) GATA and Ets *cis* -acting sequences mediate megakaryocyte-specific expression. Mol Cell Biol 13:668–676
58. Martin F, Prandini MH, Thevenon D et al (1993) The transcription factor GATA-1 regulates the promoter activity of the platelet glycoprotein IIb gene. J Biol Chem 268:21606–21612
59. Block K, Shou Y, Poncz M (1996) An Ets/Sp1 interaction in the 5′-flanking region of the megakaryocyte- specific alpha IIb gene appears to stabilize Sp1 binding and is essential for expression of this TATA-less gene. Blood 88:2071–2080
60. Minami T, Tachibana K, Imanishi T et al (1998) Both Ets-1 and GATA-1 are essential for positive regulation of platelet factor 4 gene expression. Eur J Biochem 258:879–889
61. Hickey MJ, Roth GJ (1993) Characterization of the gene encoding human platelet glycoprotein IX. J Biol Chem 268:3438–3443

62. Bastian LS, Yagi M, Chan C et al (1996) Analysis of the megakaryocyte glycoprotein IX promoter identifies positive and negative regulatory domains and functional GATA and Ets sites. J Biol Chem 271:18554–18560
63. Mignotte V, Vigon I, Boucher de Crèvecoeur E et al (1994) Structure and transcription of the human c-mpl gene (MPL). Genomics 20:5–12
64. Deveaux S, Filipe A, Lemarchandel V et al (1996) Analysis of the thrombopoietin receptor (MPL) promoter implicates GATA and Ets proteins in the coregulation of megakaryocyte-specific genes. Blood 87:4678–4685
65. Lanza F, Morales M, La Salle C de et al (1993) Cloning and characterization of the gene encoding the human platelet glycoprotein V. A member of the leucine-rich glycoprotein family cleaved during thrombin-induced platelet activation. J Biol Chem 268:20801–20807
66. Pang L, Xue H-H, Szalai G et al (2006) Maturation stage-specific regulation of megakaryopoiesis by pointed-domain Ets proteins. Blood 108:2198–2206
67. Jackers P, Szalai G, Moussa O et al (2004) Ets-dependent regulation of target gene expression during megakaryopoiesis. J Biol Chem 279:52183–52190
68. Eisbacher M, Holmes ML, Newton A et al (2003) Protein-protein interaction between Fli-1 and GATA-1 mediates synergistic expression of megakaryocyte-specific genes through cooperative DNA binding. Mol Cell Biol 23:3427–3441
69. Starck J, Cohet N, Gonnet C et al (2003) Functional cross-antagonism between transcription factors FLI-1 and EKLF. Mol Cell Biol 23:1390–1402
70. Hong W, Nakazawa M, Chen Y-Y et al (2005) FOG-1 recruits the NuRD repressor complex to mediate transcriptional repression by GATA-1. EMBO J 24:2367–2378
71. Gao Z, Huang Z, Olivey HE et al (2010) FOG-1-mediated recruitment of NuRD is required for cell lineage re-enforcement during haematopoiesis. EMBO J 29:457–468
72. Gregory GD, Miccio A, Bersenev A et al (2010) FOG1 requires NuRD to promote hematopoiesis and maintain lineage fidelity within the megakaryocytic-erythroid compartment. Blood 115:2156–2166
73. Miccio A, Wang Y, Hong W et al (2010) NuRD mediates activating and repressive functions of GATA-1 and FOG-1 during blood development. EMBO J 29:442–456
74. Wang Y, Meng R, Hayes V et al (2011) Pleiotropic platelet defects in mice with disrupted FOG1-NuRD interaction. Blood 118:6183–6191
75. Katz SG, Cantor AB, Orkin SH (2002) Interaction between FOG-1 and the corepressor C-terminal binding protein is dispensable for normal erythropoiesis in vivo. Mol Cell Biol 22:3121–3128
76. Chou ST, Khandros E, Bailey LC et al (2009) Graded repression of PU.1/Sfpi1 gene transcription by GATA factors regulates hematopoietic cell fate. Blood 114:983–994
77. Zhang C, Gadue P, Scott E et al (1997) Activation of the megakaryocyte-specific gene platelet basic protein (PBP) by the Ets family factor PU.1. J Biol Chem 272:26236–26246
78. Tijssen MR, Cvejic A, Joshi A et al (2011) Genome-wide analysis of simultaneous GATA1/2, RUNX1, FLI1, and SCL binding in megakaryocytes identifies hematopoietic regulators. Dev Cell 20:597–609
79. Elagib KE, Racke FK, Mogass M et al (2003) RUNX1 and GATA-1 coexpression and cooperation in megakaryocytic differentiation. Blood 101:4333–4341
80. Xu G, Kanezaki R, Toki T et al (2006) Physical association of the patient-specific GATA1 mutants with RUNX1 in acute megakaryoblastic leukemia accompanying Down syndrome. Leukemia 20:1002–1008
81. Elagib KE, Goldfarb AN (2007) Regulation of RUNX1 transcriptional function by GATA-1. Crit Rev Eukaryot Gene Expr 17:271–280
82. Waltzer L, Ferjoux G, Bataillé L et al (2003) Cooperation between the GATA and RUNX factors Serpent and Lozenge during Drosophila hematopoiesis. EMBO J 22:6516–6525
83. Elagib KE, Mihaylov IS, Delehanty LL et al (2008) Cross-talk of GATA-1 and P-TEFb in megakaryocyte differentiation. Blood 112:4884–4894
84. Huang H, Yu M, Akie TE et al (2009) Differentiation-dependent interactions between RUNX-1 and FLI-1 during megakaryocyte development. Mol Cell Biol 29:4103–4115
85. Schuh AH, Tipping AJ, Clark AJ et al (2005) ETO-2 associates with SCL in erythroid cells and megakaryocytes and provides repressor functions in erythropoiesis. Mol Cell Biol 25:10235–10250
86. Meier N, Krpic S, Rodriguez P et al (2006) Novel binding partners of Ldb1 are required for haematopoietic development. Development 133:4913–4923
87. Fujiwara T, Alqadi YW, Okitsu Y et al (2013) Role of transcriptional corepressor ETO2 in erythroid cells. Exp Hematol 41:303–315.e1
88. Crossley M, Merika M, Orkin SH (1995) Self-Association of the Erythroid Transcription Factor GATA-1 Mediated by Its Zinc Finger Domains. Mol Cell Biol 15:2448–2456
89. Mackay JP, Kowalski K, Fox AH et al (1998) Involvement of the N-finger in the self-association of GATA-1. J Biol Chem 273:30560–30567
90. Saleque S, Cameron S, Orkin SH (2002) The zinc-finger proto-oncogene *Gfi-1b* is essential for development of the erythroid and megakaryocytic lineages. Genes Dev 16:301–306
91. Beauchemin H, Shooshtarizadeh P, Vadnais C et al (2017) Gfi1b controls integrin signaling-dependent cytoskeleton dynamics and organization in megakaryocytes. Haematologica 102:484–497
92. Matsumura I, Kawasaki A, Tanaka H et al (2000) Biologic significance of GATA-1 activities in Ras-mediated megakaryocytic differentiation of hematopoietic cell lines. Blood 96:2440–2450

93. Takahashi S, Komeno T, Suwabe N et al (1998) Role of GATA-1 in proliferation and differentiation of definitive erythroid and megakaryocytic cells in vivo. Blood 92:434–442
94. Juban G, Sakakini N, Chagraoui H et al (2020) Oncogenic Gata1 causes stage-specific megakaryocyte differentiation delay. Haematologica 106:1106–1119
95. McDevitt MA, Shivdasani RA, Fujiwara Y et al (1997) A "knockdown" mutation created by cis-element gene targeting reveals the dependence of erythroid cell maturation on the level of transcription factor GATA-1. Proc Natl Acad Sci USA 94:6781–6785
96. Meinders M, Hoogenboezem M, Scheenstra MR et al (2016) Repercussion of megakaryocyte-specific Gata1 loss on megakaryopoiesis and the hematopoietic precursor compartment. PLoS One 11:e0154342
97. Ling T, Zhang K, Yang J et al (2023) Gata1s mutant mice display persistent defects in the erythroid lineage. Blood Adv 7:3253–3264
98. Wagenblast E, Araújo J, Gan OI et al (2021) Mapping the cellular origin and early evolution of leukemia in down syndrome. Science 373:eabf6202
99. Noh J-Y, Gandre-Babbe S, Wang Y et al (2015) Inducible Gata1 suppression expands megakaryocyte-erythroid progenitors from embryonic stem cells. J Clin Invest 125:2369–2374
100. Stachura DL, Chou ST, Weiss MJ (2006) Early block to erythromegakaryocytic development conferred by loss of transcription factor GATA-1. Blood 107:87–97
101. Majewski IJ, Metcalf D, Mielke LA et al (2006) A mutation in the translation initiation codon of *Gata-1* disrupts megakaryocyte maturation and causes thrombocytopenia. Proc Natl Acad Sci USA 103:14146–14151
102. Byrska-Bishop M, VanDorn D, Campbell AE et al (2015) Pluripotent stem cells reveal erythroid-specific activities of the GATA1 N-terminus. J Clin Invest 125:993–1005
103. Barwe SP, Sidhu I, Kolb EA et al (2020) Modeling transient abnormal myelopoiesis using induced pluripotent stem cells and CRISPR/Cas9 technology. Mol Ther Methods Clin Dev 19:201–209
104. Matsuo S, Nishinaka-Arai Y, Kazuki Y et al (2021) Pluripotent stem cell model of early hematopoiesis in down syndrome reveals quantitative effects of short-form GATA1 protein on lineage specification. PLoS One 16:1–18
105. Barwe SP, Sebastian A, Sidhu I et al (2022) Modeling Down syndrome myeloid leukemia by sequential introduction of GATA1 and STAG2 mutations in induced pluripotent stem cells with trisomy 21. Cells 11:628
106. Arkoun B, Robert E, Boudia F et al (2022) Stepwise GATA1 and SMC3 mutations alter megakaryocyte differentiation in a Down syndrome leukemia model. J Clin Invest 132:e156290
107. Sit YT, Takasaki K, An HH et al (2023) Synergistic roles of DYRK1A and GATA1 in trisomy 21 megakaryopoiesis. JCI Insight 8(23):e172851
108. Cines DB, Bussel JB, McMillan RB et al (2004) Congenital and acquired thrombocytopenia. Hematology 2004:390–406
109. Ciovacco WA, Raskind WH, Kacena MA (2008) Human phenotypes associated with GATA-1 mutations. Gene 427:1–6
110. Crispino JD, Horwitz MS (2017) GATA factor mutations in hematologic disease. Blood 129:2103–2110
111. Takasaki K, Kacena MA, Raskind WH et al (2023) GATA1-related cytopenia. In: GeneReviews. University of Washington, Seattle, WA
112. Kobayashi A, Ohtaka R, Toki T et al (2022) Dyserythropoietic anaemia with an intronic *GATA1* splicing mutation in patients suspected to have Diamond-Blackfan anaemia. eJHaem 3:163–167
113. Klar J, Khalfallah A, Arzoo PS et al (2014) Recurrent GATA1 mutations in Diamond-Blackfan anaemia. Br J Haematol 166:949–951
114. Ludwig LS, Gazda HT, Eng JC et al (2014) Altered translation of GATA1 in Diamond-Blackfan anemia. Nat Med 20:748–753
115. Parrella S, Aspesi A, Quarello P et al (2014) Loss of GATA-1 full length as a cause of Diamond-Blackfan anemia phenotype. Pediatr Blood Cancer 61:1319–1321
116. Zucker J, Temm C, Czader M et al (2016) A child with dyserythropoietic anemia and megakaryocyte dysplasia due to a novel 5'UTR GATA1s splice mutation. Pediatr Blood Cancer 63:917–921
117. Raskind WH, Niakan KK, Wolff J et al (2000) Mapping of a syndrome of X-linked thrombocytopenia with Thalassemia to band Xp11-12: further evidence of genetic heterogeneity of X-linked thrombocytopenia. Blood 95:2262–2268
118. Yu C, Niakan KK, Matsushita M et al (2002) X-linked thrombocytopenia with thalassemia from a mutation in the amino finger of GATA-1 affecting DNA binding rather than FOG-1 interaction. Blood 100:2040–2045
119. Balduini CL, Pecci A, Loffredo G et al (2004) Effects of the R216Q mutation of GATA-1 on erythropoiesis and megakaryocytopoiesis. Thromb Haemost 91:129–140
120. Åström M, Hahn-Strömberg V, Zetterberg E et al (2015) X-linked thrombocytopenia with thalassemia displays bone marrow reticulin fibrosis and enhanced angiogenesis: comparisons with primary myelofibrosis. Am J Hematol 90:E44–E48
121. Phillips JD, Steensma DP, Pulsipher MA et al (2007) Congenital erythropoietic porphyria due to a mutation in GATA1: the first trans-acting mutation causative for a human porphyria. Blood 109:2618–2621
122. Hindmarsh JT (1986) The porphyrias: recent advances. Clin Chem 32:1255–1263

Part VII

Leukemia/Lymphoma

The Diverse Roles of *ETV6* Alterations in B-Lymphoblastic Leukemia and Other Hematopoietic Cancers

13

Alexander C. Monovich, Aishwarya Gurumurthy, and Russell J. H. Ryan

Abstract

Genetic alterations of the repressive ETS family transcription factor gene *ETV6* are recurrent in several categories of hematopoietic malignancy, including subsets of B-cell and T-cell acute lymphoblastic leukemias (B-ALL and T-ALL), myeloid neoplasms, and mature B-cell lymphomas. *ETV6* is essential for adult hematopoietic stem cells (HSCs), contributes to specific functions of some mature immune cells, and plays a key role in thrombopoiesis as demonstrated by familial *ETV6* mutations associated with thrombocytopenia and predisposition to hematopoietic cancers, particularly B-ALL. *ETV6* appears to have a tumor suppressor role in several hematopoietic lineages, as demonstrated by recurrent somatic loss-of-function (LoF) and putative dominant-negative alterations in leukemias and lymphomas. *ETV6* rearrangements contribute to recurrent fusion oncogenes such as the B-ALL-associated transcription factor (TF) fusions *ETV6::RUNX1* and *PAX5::ETV6*, rare drivers such as *ETV6::NCOA6*, and a spectrum of tyrosine kinase gene fusions encoding hyperactive signaling proteins that self-associate via the ETV6 N-terminal pointed domain. Another subset of recurrent rearrangements involving the *ETV6* gene locus appear to function primarily to drive overexpression of the partner gene. This review surveys what is known about the biochemical and genome regulatory properties of ETV6 as well as our current understanding of how alterations in these functions contribute to hematopoietic and nonhematopoietic cancers.

Keywords

ETV6 · ETS · RUNX1 · Leukemia · Lymphoma · B-ALL · T-ALL · AML · Tumor suppressor · Thrombocytopenia

Abbreviations

ABC-DLBCL	diffuse large B-cell lymphoma and activated B-cell subtype
ABL1	Abelson leukemia virus proto-oncogene 1
AID/AICDA	activation-induced cytidine deaminase
AML	acute myeloid leukemia
AML1	acute myeloid leukemia 1 (former name of RUNX1)
Aop	anterior open (ETV6/ETV7 homolog)

A. C. Monovich · A. Gurumurthy · R. J. H. Ryan (✉)
Department of Pathology, University of Michigan Medical School, Ann Arbor, MI, USA
e-mail: rjhryan@med.umich.edu

T. Borggrefe, B. D. Giaimo (eds.), *Transcription factors in blood cell development*, Advances in Experimental Medicine and Biology 1459, https://doi.org/10.1007/978-3-031-62731-6_13

APOBEC	apolipoprotein B mRNA-editing enzyme, catalytic polypeptide
B-ALL	B-cell acute lymphoblastic leukemia
$BCLX_L$	B-cell lymphoma—extra large
BCR	Breakpoint cluster region activator of RhoGEF and GTPase
CDL	Central disordered/linker region of ETV6
cFos	FOS protooncogene, AP-1 transcription factor subunit
ChIP	chromatin immunoprecipitation
CID	C-terminal inhibitory domain of ETV6
CMML	chronic myelomonocytic leukemia
CRISPR	clustered regularly interspaced short palindromic repeats
CSF2	colony-stimulating factor 2
CUT&RUN	cleavage under targets and release using nuclease
DBD	DNA binding domain
DUX4	double homeobox four gene
ER	ETV6::RUNX1
EICE	ETS/IRF composite element
ENCODE	ENCyclopedia Of DNA Elements consortium
EPOR	erythropoietin receptor
ERG	ETS-related gene
ERK1	extracellular signal-regulated kinase
ETP-ALL	early T precursor acute lymphoblastic leukemia
ETS	erythroblastosis virus 26 transformation-specific
ETV6/7	ETS variant transcription factor 6/factor 7
EWS	Ewing sarcoma RNA binding protein 1 (encoded by EWSR1 gene)
FBL6	F-box and leucine-rich repeat protein 6 (former name of FBXL6)
FBXL6	F-box and leucine-rich repeat protein 6
FISH	fluorescence in situ hybridization
FLI1	Friend leukemia integration 1
Fos	Fos protooncogene, AP-1 transcription factor subunit
FOXO3	Fork head box 03
GATA1/2	GATA binding factor 1/factor 2
Gfi1b	growth factor-independent 1B
GSX2/GSH2	GS homeobox 2
H3K27ac	histone H3 lysine 27 acetylation
HER2	human epidermal growth factor receptor 2
HDAC3	histone deacetylase 3
HOXA9	homeobox A9
HSC	hematopoietic stem cell
HSPC	hematopoietic stem/progenitor cell
H-L(3)MBT	human lethal(3)malignant brain tumor-like
iAMP21	intrachromosomal amplification of chromosome 21 (B-ALL subtype)
IFNγ	interferon gamma
IGF2BP1	insulin-like growth factor 2 mRNA binding protein 1
IgH	immunoglobulin heavy chain
IKZF1	Ikaros family zinc finger 1
IL3/7	interleukin 3/interleukin 7
ISRE	interferon-stimulated response elements
HLXB9	homeobox HB9 (former name of MNX1)
IRF4/8	interferon regulatory factor 4/factor 8
ISRE	interferon-stimulated response element
JAK	Janus kinase
KAT6A	lysine acetyltransferase 6A
Kdm5C	lysine demethylase 5C
KLF4	Kruppel-like transcription factor 4
L3MBTL1	L3MBTL histone methyl-lysine binding protein 1 (formerly H-L(3)MBT)
Lck	lymphoid-specific protein-tyrosine kinase
LoF	loss of function

MAPK	mitogen-activated protein kinase
MCD	MYD88-CD79B mutation cluster
MEIS1	myeloid ecotropic insertion site 1
MEK	mitogen-activated protein kinase
MEP	megakaryocytic/erythroid progenitor
MDS	myelodysplastic syndrome
MMP3	matrix metallopeptidase 3
MN1	meningioma-1 protooncogene, transcriptional regulator
MNX1	motor neuron and pancreas homeobox 1
MPP4	multipotent progenitor 4
mSin3A	SIN3 transcription regulator family member A
M/LN-eo-TK	myeloid and lymphoid neoplasms with eosinophilia and tyrosine kinase fusion genes
NCOA6	nuclear receptor coactivator 6
NCoR	nuclear receptor corepressor
NCOR1	nuclear receptor corepressor 1 (formerly N-CoR)
NCOR2	nuclear receptor corepressor 2 (aka, SMRT)
ND	N-terminal disordered region of ETV6
NTRK3	neurotrophic receptor tyrosine kinase 3
PAX5	paired box 5
PAX5alt	PAX5-altered (B-ALL subtype)
PBX1	pre-B-cell leukemia homeobox 1
PDGFRB	platelet-derived growth factor receptor beta
PIAS3	protein inhibitor of activated STAT 3
PIK3C3	phosphatidylinositol 3-kinase catalytic subunit type 3
PNT	pointed (domain)
PTM	posttranslational modification
PU.1	purine-rich box 1 (encoded by SPI1)
RAG1/2	recombination-activating gene 1/ gene 2
Ras	Rat sarcoma virus GTPase
RAR	retinoic acid receptor
ROSE	Ranking Of Super Enhancer
RUNX1	Runt-related transcription factor 1
RXR	retinoid X receptor
Sca1	stem cell antigen 1 (encoded by Ly6a)
SIN3A	SIN3 transcription regulator family member A (formerly mSin3A)
SMRT	silencing mediator for retinoid or thyroid-hormone receptors (aka NCOR2)
SOX11	SRY-box transcription factor 11
SPI1	SPI1 proto-oncogene (encodes PU.1)
SPIB	Spi-B transcription factor
STAT	signal transducer and activator of transcription
SUMO-1	small ubiquitin-like modifier 1
T-ALL	T-cell acute lymphoblastic leukemia
TCF3	transcription factor 3 (encodes E2A)
TEL	translocation ETS leukemia (former name of ETV6)
TEL2	translocation ETS leukemia 2 (former name of ETV7)
TF	transcription factor
TGFβ	transforming growth factor beta
TK	tyrosine kinase
Th9	T-helper 9
Th17	T-helper 17
THC5	thrombocytopenia 5
TNF	tumor necrosis factor
TRK	tropomyosin receptor kinase
UBC9	ubiquitin carrier protein 9
VEGFA	vascular endothelial growth factor A
V_H	immunoglobulin heavy chain variable gene segment
XPO1	exportin 1
Yan	former name of Aop
ZNF384	zinc finger protein 384

13.1 Introduction

The *ETV6* gene (ETS variant transcription factor 6), formerly *TEL* (translocation, ETS, leukemia), was first identified 30 years ago as a target of recurrent somatic alterations in B-cell acute lymphoblastic leukemia (B-ALL) and acute myeloid leukemia (AML). Since that time, a diverse spectrum of recurrent somatic and germline alterations of *ETV6* have been associated with hematopoietic cancers, including B-ALL, myeloid and T-lymphoblastic leukemias, and mature B-cell lymphomas. Here, we review what is known about the molecular functions of ETV6, the phenotypic consequences of its disruption in hematopoietic development, and the current evidence for mechanisms by which its alteration contributes to hematopoietic cancers.

13.2 Overview of ETV6 Molecular Features

ETV6 is a sequence-specific DNA binding transcription factor (TF), and one of 28 human TFs containing the ETS domain, which binds to a core recognition sequence of GGAA/T. *ETV6* is broadly expressed in human tissues but shows increased expression in a subset of cell types, including endothelial cells, hematopoietic stem/progenitor cells (HSPCs), and a subset of mature lymphoid and myeloid cell populations [1, 2]. The structure of ETV6 resembles that of many other ETS factors, with a DNA binding ETS domain located in the C-terminal half of the protein, while additional domains associated with protein-protein interactions are found in the N-terminal portion (Fig. 13.1 and 13.2a). ETV6 is one of eight human class II ETS factors as determined by systematic determination of in vitro DNA binding sequence preferences [3]. Class I and II ETS factors (together representing 23 of 27 human ETS factors) recognize a canonical motif consisting of a core recognition sequence of "GGAA" (motif positions +1 to +4) and a preference for cytosines in the preceding −1 and−2 base positions. The latter characteristic distinguishes class I/class II factors from the purine-rich upstream sequences preferred by class III ETS factors like the key hematopoietic lineage regulator PU.1/SPI1 [3, 4], while motif differences between the class I/ class II group are more subtle based on preferences at the −3 position. Thus, the intrinsic DNA binding preferences of ETV6 seem to be highly redundant with many other ETS factors expressed in blood cells.

A number of key biochemical properties distinguish ETV6 from most other members of the ETS family with the exception of ETV7 (formerly TEL2), a close ETV6 homolog that shows a more restricted expression pattern in human tissues [5, 6] and is absent in mice and many other rodents. First, while many ETS factors act primarily as transcriptional activators, ETV6 acts as a transcriptional repressor of reporter constructs [7–10] and endogenous promoters [11, 12] containing canonical ETS motifs. This property has been attributed to binding of chromatin corepressors including NCOR2/SMRT, NCOR1 (formerly N-CoR), SIN3A (formerly mSin3A), HDAC3, and L3MBTL1 (formerly H-L(3)MBT) to specific domains of ETV6 (Fig. 13.1), including the N-terminal PNT domain and central disordered/linker (CDL) region [7, 9, 10, 13, 14], resulting in corepressor recruitment to ETV6 binding sites. Second, the full-length ETV6 protein has a low intrinsic affinity for DNA due to the presence of autoinhibitory domain in the central linker region and C-terminus [15]. The C-terminal inhibitory domain (CID) has been structurally well characterized and was shown to dynamically bind and sterically inhibit the ETV6 ETS domain by a mechanism distinct from other ETS factor autoinhibitory domains [16, 17]. Third, the N-terminal PNT domain of ETV6 is able to undergo homo-oligomerization via a head-to-tail structure [18, 19], a property not shared by PNT domains in other ETS factors (with the likely exception of ETV7 [5]). Together, the autoinhibitory and homo-oligomerization properties of ETV6 appear to reduce its affinity relative to other ETS factors for binding to single canonical ETS motifs and increase its selectivity for binding sites containing multiple ETS motifs [15], although these effects have been described

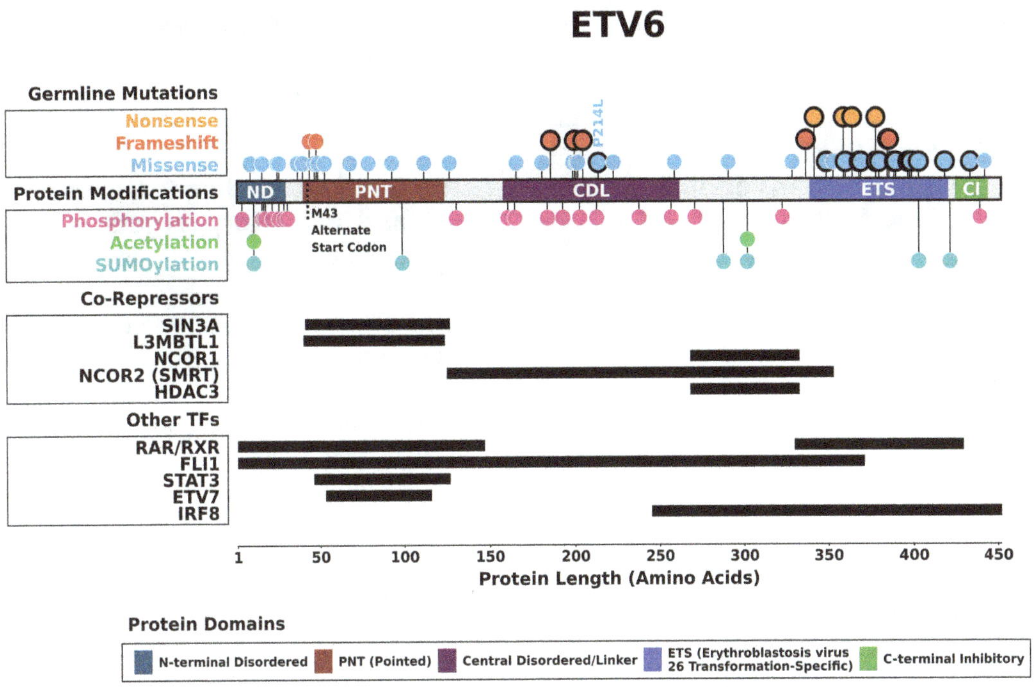

Fig. 13.1 Detailed molecular structure, mutations, modifications, and interactions of the ETV6 transcriptional repressor. Top: Diagram of ETV6 protein domains, known posttranslational modifications [24, 26, 226], and reported germline mutations [31, 60, 61, 66, 227–229]. Mutations shown experimentally to damage specific ETV6 transcriptional repressor functions [31, 60, 61, 66, 229] are outlined in black. The ETV6 missense mutation P214L highlighted in the main text is uniquely labeled. Bottom: ETV6 regions reported to be important for facilitating interactions with indicated corepressor or transcription factor (TF) [5, 7, 9, 13, 14, 35, 43–45]

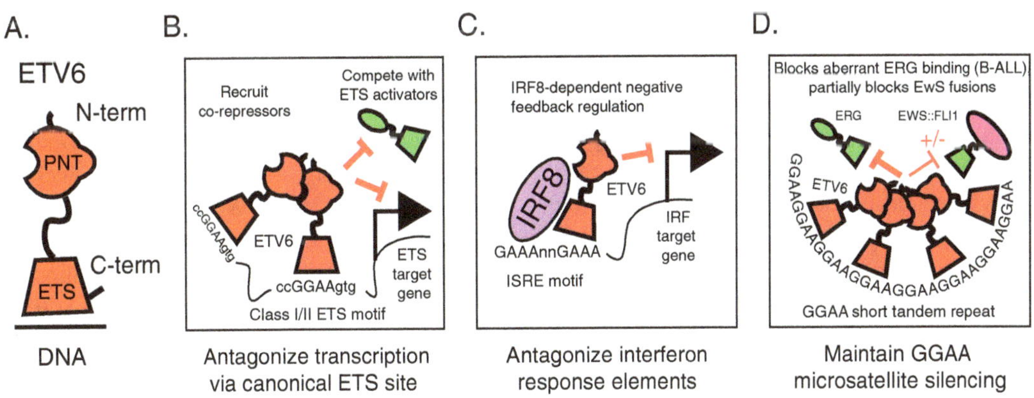

Fig. 13.2 Reported functions of the ETV6 transcriptional repressor. (**a**) Simplified representation of ETV6 and critical domains used in Figs. 13.2, 13.3 and 13.4. (**b**) Competitive antagonism of ETS activators by the ETV6 repressor via canonical ETS binding motifs. (**c**) ETV6 antagonism of interferon-activated genes via co-binding with IRF8. (**d**) ETV6 cooperative binding at GGAA tandem repeat elements

primarily with reporter constructs rather than in the genomic context.

13.3 Signaling and Posttranslational Modifications

Signaling-dependent posttranslational modification (PTM) has been implicated as key regulatory mechanism for ETV6, as it is for several other ETS family factors [16]. Numerous ETV6 PTMs have been identified in high-throughput proteomic studies (Fig. 13.1) although only a minority have been functionally characterized. Gene repression by the *Drosophila melanogaster* ETV6 homolog Aop (formerly Yan) is relieved by phosphorylation downstream of RAS/MAPK signaling [20, 21], a key regulatory step in differentiation of neurons and other tissues. ETV6 was shown to be phosphorylated by two MAPK pathway effectors, the stress-activated p38 kinase (phosphorylates ETV6 Serine 257) [22] and RAS/MEK downstream effector ERK1 (phosphorylates ETV6 Serine 213 and 257) [23], which result in decreased transcriptional repressive activity. ETV6 is also negatively regulated by conjugation to the small ubiquitin-like modifier protein SUMO1 at lysine 11, a highly conserved residue in the N-terminal disordered region (ND), via the E3 SUMO-protein ligase PIAS3 [24]. SUMOylation at this site interferes with ETV6 binding to DNA and promotes its proteasomal degradation [24]. An alternative ETV6 protein isoform initiated from an internal methionine (Met 43) lacks this SUMOylation site, which appears to explain the greater repressive activity of this isoform. A earlier-described SUMOylation site associated with SUMO-conjugating enzyme UBC9 within the PNT domain at lysine 99 was shown to promote nuclear export in some cellular contexts [25–27]. The PNT domain of ETV6 is additionally reported to participate in the recruitment of the E3 ubiquitin ligase FBXL6 (formerly FBL6) to promote ETV6 ubiquitination [28]. Studies with oligomerization-deficient mutants of ETV6 suggested that monomeric ETV6 is more readily SUMOylated and ubiquitinated, resulting in increased proteasomal degradation of the monomeric forms [24, 28].

13.4 Genomic Binding Patterns and Regulatory Targets of ETV6

There have been longstanding efforts to identify the direct genomic regulatory targets of ETV6 relevant to its function in the hematopoietic system and in blood cancers. One paradigm of ETV6 genomic function involves competition between ETV6 and activating ETS factors at gene promoters (Fig. 13.2b). The first identified direct ETV6 target gene, MMP3 (stromolysin-1) [11], was revealed via studies in mouse fibroblasts and belongs to a family of genes previously identified as activation targets of oncogenic RAS and MAPK-regulated ETS activators. The *MMP3* promoter contains multiple ETS binding sites, is strongly repressed by *ETV6* transgene overexpression in RAS-transformed mouse fibroblasts, and has been widely used in plasmid-based luciferase reporter for ETV6 repressive activity in biochemical studies. However, MMP3 is not expressed in hematopoietic cells or leukemias, illustrating that the regulatory roles of ETV6 are likely to be tissue- and regulatory state-specific. Other proposed direct ETV6 repression targets that similarly contain multiple promoter ETS motifs and have been shown to be transcriptionally upregulated by activating ETS TFs include *BCL2L1*/BCLX$_L$ (fibroblasts) [12, 29] and *PF4* (platelet factor 4, megakaryocytes) [30, 31]. Recent genome-wide mapping of ETV6 binding sites by CUT&RUN in primary mouse HSPCs showed a majority of ETV6 peaks at gene promoter-proximal sites, with strong enrichment for canonical ETS binding motifs [32], supporting the relevance of this regulatory model in early hematopoiesis. ETV6 ChIP-Seq datasets generated by the ENCODE project in mature B-cell (GM12878) and myeloid/erythroid leukemia (K562) cell lines also showed canonical ETS motifs as the most significantly enriched sequence in ETV6 binding peaks [33, 34] although the majority of peaks in these datasets were distal to gene promoters, suggesting that ETV6 may also antagonize other ETS factors at distal enhancers.

Transcriptional repressive effects of ETV6 that are independent of canonical ETS motif binding have also been described. The best characterized of these occurs in a subset of mature immune cells that express the hematopoietic lineage-specific IRF family TF IRF8 (Fig. 13.2c). Interferon gamma (IFNγ) stimulates transcription of interferon-induced genes via IRF1 binding to interferon-stimulated response elements (ISRE) in their promoters. In a murine macrophage-like cell line, this early activation was blunted by a negative feedback loop consisting of upregulation of IRF8 and subsequent co-binding of IRF8 and ETV6 to those same ISRE elements [35]. Interactions between IRF8 and ETV6 require the ETV6 ETS domain and c-terminal loop, rather than the PNT domain that confers ETV6 self-association and involve ETV6-dependent recruitment of HDAC3 (Fig. 13.1). Binding site mutagenesis studies suggested that IRF8 and ETV6 bind cooperatively to different parts of the ISRE motif (GAAAnnGAAA), despite the lack of a core GGAA sequence typical of ETS binding sites. The apparent repressive function of IRF8-ETV6 co-binding at ISRE elements contrasts with the function of the better-characterized cooperativity between IRF8 or its homolog IRF4 and the class III ETS factors PU.1/SPI1 and SPIB at ETS/IRF composite elements (EICE) [36]. EICEs contain a more conventional ETS core motif and are associated with gene activation in interferon-treated macrophages and certain immune cell developmental contexts. IRF8 is differentially expressed in many myeloid and lymphoid cell subsets, and evidence for cooperativity of IRF8 and ETV6 in repression of specific developmentally repressed gene targets was shown for the Th9 subset of $CD4^+$ T cells [37] and dendritic cells [38, 39]. Interestingly, single-cell RNA-Seq profiles of peripheral blood cells from patients with germline *ETV6* mutations showed increased expression of interferon-activated genes in monocytes and dendritic cells compared to control patients [40]. Wild-type IRF4, a close relative of IRF8, has not been shown to bind ETV6. However, a recently described germline variant, IRF4 F359L, was intriguingly reported to gain neomorphic ETV6 binding [41]. Patients with *IRF4* F359L mutations exhibit hypogammaglobulinemia, reduced plasma cell frequency, and altered T-cell subsets. While wild-type IRF4 binds ISRE elements as a dimer and activates transcription, IRF4 F359L mediates ETV6-dependent repression of an ISRE-linked reporter similar to the effects of wild-type IRF8. *ETV6* knockdown in patient-derived B-lymphoblastoid cell lines restored aberrantly low expression of genes associated with IRF4-driven plasma cell differentiation, suggesting that the neomorphic IRF4-ETV6 interaction might be a key driver of these patients' phenotype. Another recent report found that transcription of the ETV6 homolog ETV7 is strongly induced by interferon treatment, and ETV7 was also shown to antagonize the transcriptional upregulation of many interferon-activated genes [42]. Thus, several independent lines of evidence support an immune cell-specific role for ETV6/ETV7 in modulation of gene regulation by IRF family factors.

Other studies provided evidence that ETV6 can negatively regulate transcriptional activation by JAK-STAT signaling [43] and retinoic acid signaling [44] through direct binding to STAT3 and RAR/RXR, respectively. ETV6-dependent repression of transcriptional activation from STAT3 response elements was dependent on binding to STAT3 via the ETV6 PNT domain but did not require the ETV6 ETS (DNA-binding) domain, while both the ETV6 PNT and ETS domains were required for RAR/RXR binding and transcriptional repression (Fig. 13.1). ETV6 was also reported to directly bind FLI1, an activating ETS factor with a key role in megakaryopoiesis [45]. More data is needed to determine whether these regulatory modes involve binding of ETV6 and its partner factor to chromatin and to identify physiological contexts in which they play a significant regulatory role.

Another distinctive pattern of genomic binding by ETV6 was recently reported by the Vakoc and Stegmeier labs in Ewing sarcoma [46, 47] and by our laboratory in B-lymphoblastic leukemia [48]. In these diseases, a large proportion of ETV6 binding peaks detected by ChIP-Seq are at short tandem repeats of the sequence GGAA (Fig. 13.2d). GGAA repeats were previously known to be important binding targets of the Ewing sarcoma fusion oncoproteins EWS::FLI1 (E::F) and EWS::ERG, which extensively self-associate through their N-terminal EWS domains, resulting in cooperative binding to many low-affinity "GGAA" motifs that lack the flanking bases typically seen in canonical class I/class II ETS motifs. Binding of E::F or homologous fusions to GGAA repeats recruits activating chromatin regulatory proteins that convert these repeats to an active enhancer-like state, resulting in activation of nearby genes that are essential for the oncogenic state of Ewing sarcoma [49, 50]. While systematic CRISPR knockout (KO) screens have shown that ETV6 is dispensable for most cancer types, it is a highly essential gene in Ewing sarcoma. This dependency was shown to result from the ability of ETV6 to antagonize E::F activity at a subset of moderate-length repeats, which became active enhancers upon experimental depletion of ETV6 protein, resulting in increased expression and tumor suppressor activity of SOX11. The homo-oligomerization property of ETV6 was critical for its ability to antagonize E::F at these repeats, which could be disrupted by expression of a dominant-negative peptide consisting of a free ETV6 PNT domain [47]. Our work in B-ALL (detailed in a latter section) also found that ETV6 repressed enhancer-like activity of GGAA repeats and that ETV6 deficiency facilitated repeat enhancer-dependent activation of leukemia subtype-specific genes. Thus, ETV6 seems to have a specific role in antagonizing aberrant activity of ETS activators at GGAA repeats in certain malignant cell contexts. Whether this property is biologically significant in any normal developmental context remains to be seen.

Although ETV6 has been primarily implicated in repression of chromatin state and gene expression, some reports have indicated the possibility of a transcriptional activating function of ETV6 binding in specific contexts [51, 52]. Further studies are required to characterize biochemical mechanisms underlying putative transcriptional activating functions of ETV6.

13.5 ETV6 Function in Hematolymphoid Development

Initial mouse modeling studies in the Orkin lab indicated that KO of *Etv6/Tel* in mice leads to embryonic lethality at embryonic day 11 (E11), which was attributed to defective angiogenesis in the yolk sac and other structures [53]. Interestingly, hematopoiesis in the embryonic yolk sac and fetal liver are preserved in *Etv6* KO mice [53, 54], and some later *Etv6* KO strains showed occasional survival to E18, with death attributed to the near-complete failure of bone marrow hematopoiesis [55]. Elegant mouse chimera studies showed that *Etv6−/−* cells are maintained in most adult tissues and contribute to multilineage hematopoiesis and lymphopoiesis from the yolk sac and fetal liver but fail to sustain ongoing production of any hematopoietic or lymphoid lineages from the adult bone marrow [54]. Accordingly, conditional KO studies demonstrated that loss of *Etv6* in the mouse leads to failure of adult hematopoietic stem cell (HSC) maintenance, while allowing differentiation of most hematopoietic lineages to proceed until the short-term stem/progenitor pool is exhausted [55]. Thus, *Etv6* was the first gene shown to be specifically required for adult, but not fetal, HSC function.

Further evidence for the importance of *Etv6* in establishing the HSC gene regulatory program came from the finding that transgene-mediated expression of *Etv6* along with three other TFs, *Gata2*, *Gfi1b*, and *Fos*, is sufficient for reprogramming of mouse fibroblasts to a HSC state [56]. Although the other three factors alone were sufficient for derivation of small numbers of HSCs, addition of an *Etv6* transgene significantly increased the efficiency of the process. In addition to cell-autonomous effects in HSCs, studies in *Xenopus laevis* models demonstrated that Etv6 plays a key role in HSC development from dorsal lateral plate mesoderm as loss of Etv6 prevents activation of VEGFA in hemangioblasts [57]. Two mechanisms were recently proposed for ETV6-dependent activation of VEGFA: ETV6-dependent repression of the VEGFA repressor FOXO3 and paradoxical direct activation of the VEGFA activator KLF4 by ETV6 [52].

Gross effects of *Etv6* KO on differentiation of committed hematopoietic and lymphoid progenitors were largely limited to the megakaryocyte lineage [55]. *Etv6* knockout in murine megakaryocytic/erythrocyte progenitors (MEPs) with GATA1-Cre led to a 50% decrease in platelets counts, which likely represented a specific requirement for *Etv6* in terminal platelet production, since megakaryocyte progenitor colony formation increased by fivefold. This contrasted with an absence of clear effects of *Etv6* KO in other lineage-committed progenitors, which maintained normal production of erythrocytes (GATA1-Cre), B-cells (CD19-Cre), and T-cells (Lck-Cre). As discussed earlier in the context of ETV6 interactions with IRF family members, ETV6 does have significant effects on the development of specific subsets of T-cells and dendritic cells in mice. In addition to the subsets discussed earlier, knockdown and overexpression experiments have implicated ETV6 in promoting differentiation of Th17 cells [58], a T-cell subset with a key role in several autoimmune disorders.

13.6 Germline ETV6 Mutations

Interest in the hematopoietic effects of ETV6 was reinvigorated by the discovery that human germline *ETV6* mutations underlie a characteristic syndrome of thrombocytopenia and leukemia predisposition [31, 59–62], now referred to as thrombocytopenia 5 (THC5, OMIM #616216). Affected patients have a variable history of increased bleeding and show mild to moderate thrombocytopenia with defects in platelet spreading [62, 63], possibly related to altered expression of cytoskeleton regulatory genes [40, 63]. Erythrocyte mean corpuscular volume (MCV) in THC5 patients is significantly higher than the population average, although true macrocytosis above the reference range is only seen in a minority [64]. Interestingly, patients have increased numbers of circulating CD34^{+} stem/progenitor cells [40, 63]. THC5 shows autosomal dominant inheritance, and *ETV6* variants that co-segregate with thrombocytopenia in affected families include truncating (frameshift/nonsense) mutations that eliminate the ETS domain, ETS domain missense mutations that abrogate DNA binding, and a missense mutation in the central domain, ETV6 P214L, that has been reported in several apparently unrelated families. The rate of hematological malignancy for *ETV6* mutation carriers in the first 20 reported families across six studies was 25% [65], with a predominance of B-ALL and less frequent myeloid malignancies (AML, myelodysplastic syndrome [MDS], myeloproliferative disorders, and mixed-phenotype acute leukemia).

When expressed in model systems, proteins encoded by germline *ETV6* variants showed a number of altered properties, including loss of DNA binding and repression of ETS motif-bearing transcriptional reporters, cytoplasmic mislocalization, and dominant-negative interference with reporter repression by wild-type ETV6, likely mediated by interactions between the PNT domains of mutant and wild-type ETV6 [31, 60, 61, 66]. Cytoplasmic mislocalization of ETV6 and its binding partner HDAC3 were also noted in THC5 patient peripheral blood mononuclear cells, indicating that this phenomenon is not an artifact of transgene overexpression [40]. The central domain mutant ETV6 P214L shows LoF and apparent dominant-negative activities similar to those of the ETS domain mutants, possibly related to the known role of the central domain in regulating DNA binding autoinhibition [15] or gain of ETV6 P214L interactions with the nuclear export protein XPO1 [67]. It should be noted that truncated ETV6 proteins predicted to be encoded by frameshift or nonsense mutations have not been detectable in primary patient cells [60, 66] or in an *Etv6* truncation mutation mouse model [32], suggesting instability at the transcript or protein level. On the other hand, a few germline N-terminal *ETV6* truncating alterations that eliminate the PNT domain have been associated with leukemia development but not thrombocytopenia [68, 69], suggesting that PNT domain-mediated dominant-negative effects of mutant ETV6 proteins may be important for the platelet phenotype.

THC5-related ETV6 mutants, but not wild-type ETV6, reduced proliferation of human CD34+ HSPCs [31]. In vitro megakaryocyte differentiation assays conducted in *ETV6* transgene-expressing human CD34+ cells showed increased numbers of small platelet progenitors and decreased numbers of more mature large megakaryocytes and proplatelets in cells expressing THC5-related ETV6 mutants but not wild-type ETV6 [40, 60, 63]. Immunofluorescence showed almost complete relocalization of ETV6 from the nucleus to the cytoplasm of cultured megakaryocytes expressing ETV6 mutants, suggesting that mutant ETV6 was able to bind and mislocalize wild-type ETV6 [60]. Failure of proplatelet production was also noted in induced pluripotent stem cell differentiation model of ETV6-deficient megakaryocyte differentiation [70], although increased numbers of early megakaryocyte progenitors were only seen in the setting of homozygous but not heterozygous *ETV6* knockout.

The direct mechanism by which ETV6 variants alter platelet development remains unclear. One study [40] found increased expression of certain interferon-regulated gene transcripts in the platelets of THC5 patients and in in vitro differentiated megakaryocytes derived from *ETV6*-knockout human CD34+ cells. The authors proposed that mutant ETV6-dependent mislocalization of HDAC3 might derepress those targets in megakaryocytes. The authors also noted decreased expression of cytoskeletal genes in *ETV6* KO megakaryocyte differentiation, which may relate to platelet function defects. A more recent single-cell analysis of in vitro megakaryopoiesis from THC5 patient-derived CD34+ cells [71] noted increased expression of genes with promoter ETS motifs, including upregulation of RPS6 and other ribosomal genes in THC5 patient megakaryocytes. Since ETV6 was reported to bind and antagonize the ETS activator FLI1, which is critical for megakaryocyte development and is affected by hemizygous gene deletions in a different hereditary thrombocytopenia syndrome, Paris-Trousseau [72], it is possible that altered regulation of FLI1 targets may contribute to THC5, but direct evidence for this is currently lacking.

Two recent reports describe mouse models of THC5-related ETV6 mutations: *Etv6*^{R355X} which models the human truncating mutation *ETV6*^{R359X} [32] and Etv6$^{h.P214L}$ which models the human mutation *ETV6* P214L [73]. Unlike human THC5, both models lacked thrombocytopenia in the heterozygous state, although mice homozygous for the R355X allele showed subtle platelet function abnormalities and increased proportions of B-cells. The Etv6$^{h.P214L}$ model showed subtly altered expression of inflammatory genes in the multipotent progenitor 4 (MPP4) population [73]. Genome-wide ETV6 mapping by CUT&RUN in mouse and human hematopoietic progenitors highlighted the locus containing TNF and several other inflammatory genes notable for multiple strong ETV6 binding sites [32]. While HSPCs in mice heterozygous for the Etv6$^{R355X/+}$ allele showed no significant difference in inflammatory gene expression under baseline conditions, TNF and other inflammatory genes were significantly overexpressed in Etv6$^{R355X/+}$ mice after bone marrow transplantation [32]. HSCs from both the Etv6$^{R355X/+}$ and Etv6$^{h.P214L}$ models showed a disadvantage versus wild-type stem cells in competitive transplant experiments, suggesting deficiencies in stress hematopoiesis that could be affected by altered inflammatory signaling. These findings further underscore a connection between ETV6 and inflammatory gene regulation in the hematopoietic system but also underscore the investigative challenges posed by apparent phenotypic differences between mice and humans with *ETV6* gene alterations.

13.7 Patterns of *ETV6* Alterations in Leukemia

ETV6 alterations identified in hematologic cancers are highly diverse. The most common patterns, seen in B- and T-lymphoblastic leukemias, myeloid neoplasms, and mature B-cell lymphomas, are deletions, rearrangements, nonsense mutations, and missense mutations in key structural domains that lead to loss of ETV6 function (Fig. 13.3a), supporting *ETV6* as an important tumor suppressor gene. Due to the ability of ETV6 to homo-oligomerize through the PNT domain, at least some truncated or DNA binding-defective ETV6 mutants likely act as dominant-negative inhibitors of wild-type ETV6 in heterozygous cells. Another pattern involves *ETV6* fusion with another TF gene (e.g., *RUNX1* or *PAX5* in B-ALL), resulting in fusion proteins that include the DNA binding domain (DBD) of the partner TF and the oligomerization and corepressor recruitment domains of ETV6, thus repressing targets normally activated by the partner TF (Fig. 13.3b). In contrast, *ETV6* fusions

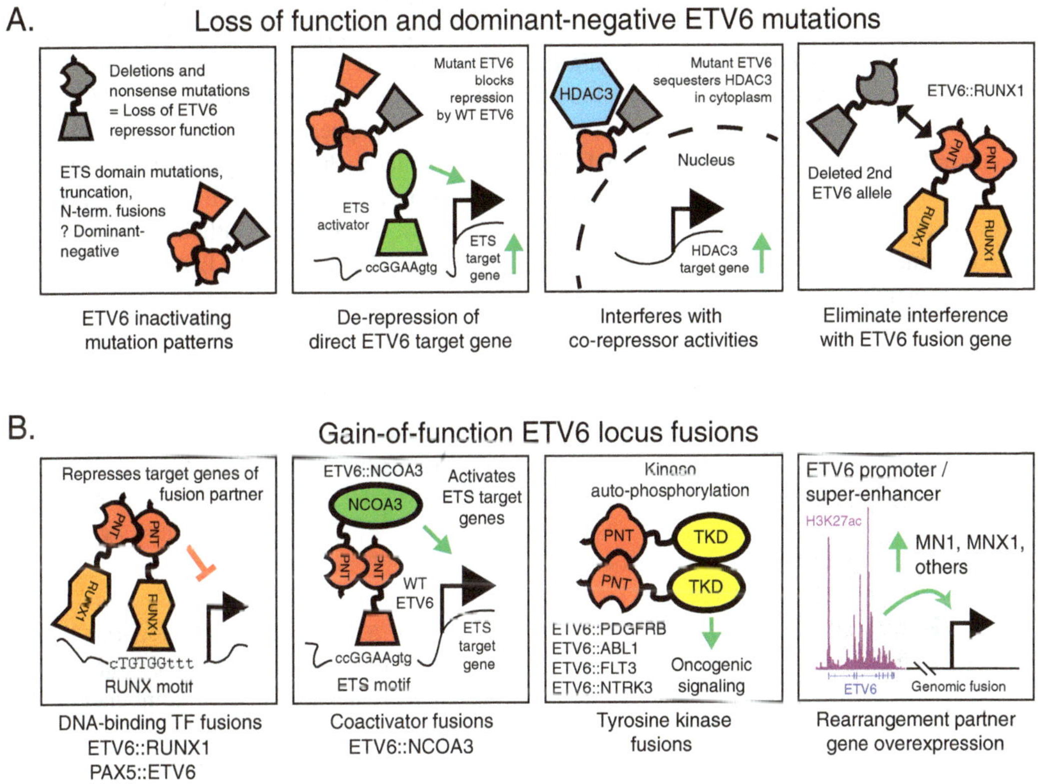

Fig. 13.3 Proposed molecular activities of cancer-associated ETV6 alterations. (**a**) ETV6 loss-of-function (LoF) mutations and other genomic alterations (first panel) may eliminate the protein entirely or may produce a mutant protein lacking DNA-binding function but retaining an intact PNT domain. The latter category of mutations may show dominant-negative activity by binding and interfering with the repressor function of wild-type ETV6 (second panel) or by binding to corepressors and sequestering them in the cytoplasm (third panel). A special situation arises in the setting of PNT domain-containing fusion proteins such as ETV6::RUNX1, where loss of the second ETV6 allele might eliminate PNT domain-mediated interactions with the fusion protein (fourth panel). (**b**) Fusion proteins that include the oligomerization and corepressor recruitment domains of ETV6 and the DNA binding domains (DBDs) of other TFs such as RUNX1 and PAX5 (first panel) may result in repression of genes normally regulated by the partner TF. A fusion protein that links the oligomerization domain of ETV6 to the NCOA3 coactivator may activate genes normally repressed by ETV6 (second panel). Fusion proteins containing the ETV6 oligomerization domain and tyrosine kinase (TK) domains (third panel) may result in kinase autophosphorylation and aberrant activation of downstream signaling pathways. *ETV6* locus fusions may also result in overexpression of the partner gene, with or without production of an ETV6 fusion protein (fourth panel), due to strong transcriptional activating elements present in the *ETV6* locus

with the coactivator gene *NCOA6* appear to drive gene activation. Another important class is *ETV6* fusions with tyrosine kinase (TK) genes that facilitate homo-oligomerization and autoactivation of oncogenic tyrosine kinases. Finally, the *ETV6* locus contains powerful gene regulatory elements that drive strong expression of the wild-type gene in normal hematopoietic progenitors and can be coopted to overexpress oncogenic partner genes via genomic rearrangements, including via putative "enhancer hijacking" rearrangements that lack a detectable gene fusion [74–77].

It should be noted that LoF and gain-of-function (GoF) *ETV6* alterations are not mutually exclusive, and in fact frequently coexist, since formation of a neomorphic *ETV6* gene fusion also destroys one wild-type *ETV6* allele. At least some N-terminal ETV6 fusion proteins also appear to interfere with wild-type ETV6 function in a dominant-negative fashion. The potential multifunctionality of *ETV6* locus rearrangements can make the biological function of rare or novel *ETV6* alterations difficult to predict, but attention to the principles of recurrent and well-studied *ETV6* alterations may allow for an informed hypothesis.

13.8 Diagnostic and Clinical Features of *ETV6::RUNX1*+ B-ALL

The single most common leukemia-associated *ETV6* gene fusion is the t(12;21)(p13;q22) rearrangement leading to the *ETV6::RUNX1* (*E::R*) fusion oncogene (originally known as *TEL::AML1*), which defines a common genetic subtype of B-lymphoblastic leukemia representing 18% of pediatric B-ALL in a recent large series [78]. Distinctive features of *E::R*$^+$ B-ALL pathogenesis and phenotype are listed in Fig. 13.4. The presence of the *E::R* fusion is mutually exclusive with the primary genetic lesions that define other subtypes of B-ALL, such as high-hyperdiploid karyotype, *KMT2A*-rearrangement, *BCR::ABL1*, *TCF3::PBX1*, *DUX4*-rearrangement, and others, indicating that each of the lesions marks a distinct leukemogenic pathway. Despite its high prevalence, the t(12;21) rearrangement was described relatively late, as it is cryptic to conventional karyotypic analysis [79] and in clinical practice is typically detected by fluorescence in situ hybridization (FISH) or RNA-based techniques. The encoded fusion protein contains the oligomerizing N-terminal PNT domain of ETV6 and nearly the full length of RUNX1 [80–82], a TF that plays a key role in myeloid and lymphoid gene regulatory programs. Recently, rare cryptic fusion genes resulting from focal insertion of exons 3–5 of *ETV6* into *RUNX1* intron 2 were described in a few cases of B-ALL; these events appear to generate fusion oncoproteins similar to those expressed in common *E::R* fusions but are not detectable by standard clinical FISH assays [83, 84]. Although the primitive neoplastic blasts of *E::R*+ B-ALL are morphologically indistinguishable from those seen in other forms of acute lymphoblastic leukemia, they demonstrate a characteristic CD27$^+$/CD44$^{low\text{-}neg}$ immunophenotype by flow cytometry [85] and show a distinctive pattern of altered gene expression that readily distinguishes *E::R*$^+$ B-ALL from other driver-defined subtypes by gene expression profiling [78, 86], although this signature is shared with another rare subtype, *E::R*-like B-ALL, as described below.

E::R$^+$ B-ALL is considered a low-risk B-ALL subtype with excellent prognosis for patients treated with standard-of-care combination chemotherapy [87]. The high rates of complete remission and cure in this subtype have been attributed to higher sensitivity of *E::R*$^+$ B-ALL blasts to several agents including L-asparaginase and corticosteroids compared to other B-ALL subtypes [88–90], although the biological basis of this difference is poorly understood. Despite these excellent outcomes, leukemia therapy still imposes a great burden on the lives of children with *E::R*$^+$ B-ALL, and there is hope that improved biological understanding of low-risk B-ALL may allow for further therapeutic de-escalation [91, 92] or early disease detection and intervention [93].

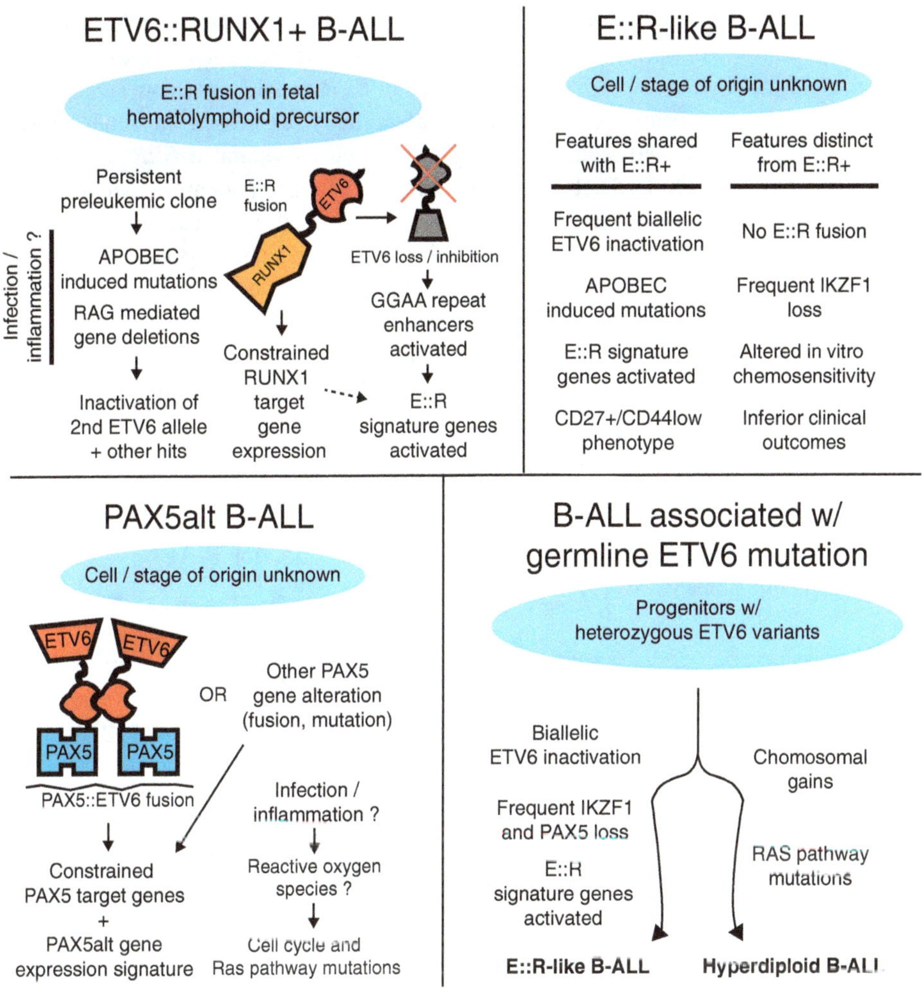

Fig. 13.4 B-ALL subtypes associated with recurrent ETV6 alterations. Comparison of the proposed roles of ETV6 fusion proteins and inactivating ETV6 mutations across four categories of B-ALL, along with known or hypothesized leukemogenic mechanisms. Note that recurrent inactivating *ETV6* mutations are seen at lower frequency across many other B-ALL subtypes and *ETV6*::tyrosine kinase (TK) fusions can occur in *BCR::ABL1*-like B-ALL

13.9 Molecular Function of the ETV6::RUNX1 Fusion Protein

The *E::R* fusion gene encodes nearly the full length of *RUNX1*, a critical regulator of hematopoiesis at all developmental stages [94]. Like *ETV6*, germline mutations in *RUNX1* cause a familial platelet disorder and predisposition to leukemia [95], albeit with a bias toward myeloid malignancies rather than B-ALL. *RUNX1* is required for initial establishment of an open chromatin state at the loci of other key hematopoietic regulators [96] and is directed by the Runt DBD binding to sequence-specific response elements. Regulatory effects of *RUNX1*-encoded proteins are complex and varied at different stages of normal hematopoietic differentiation, being affected by cell state-dependent splice variants that encode different protein isoforms, differences in proteins binding partners, and other differences in the chromatin regulatory landscape. In B-cell progenitors, RUNX1 appears to have transcription-enhancing activity at most of its target sites [97].

The E::R fusion protein bears the N-terminal PNT oligomerization domain and central domain of ETV6, but lacks an ETS DBD, binding to chromatin instead through the Runt domain of the RUNX1 portion of the protein. E::R appears to suppress RUNX1-dependent transactivation at many of its target sites [97–99], likely by competing with wild-type RUNX1 for binding and recruiting corepressors such as mSin3A that normally interact with ETV6 [9]. Antagonism of normal RUNX1 function by E::R appears to be incomplete, as the second *RUNX1* allele is never lost in *E::R*+ B-ALL (unlike the second copy of *ETV6*), and wild-type RUNX1 appears to be essential for leukemia homeostasis [97].

13.10 Fetal Origin of Human E::R+ B-ALL

E::R$^+$ B-ALL is almost exclusively a pediatric disease, with a median onset at 4 years of age [78]. Fascinatingly, formation of the *E::R* fusion gene, believed to be the first step in *E::R*$^+$ leukemogenesis, has been shown to occur prenatally [93]. Pairs of monozygotic twins that share a common placental circulation in utero have been shown to develop clonally related *E::R*$^+$ B-ALL [100–102]. Furthermore, *E::R* fusion transcripts are detectable in the neonatal blood spots of many infants who later developed *E::R*+ B-ALL as children [103, 104]. Similar "backtracking" studies were able to retrospectively detect B-ALL clone-specific immunoglobulin gene rearrangements in neonatal blood spots of patients with other common pediatric B-ALL subtypes [105–107]. Unfortunately, prospective detection of an *E::R* fusion at birth is not currently a useful diagnostic approach, because such transcripts are detectable in up to 5% of all infants, the vast majority of whom will not go on to develop leukemia [108–110]. This suggests that several additional pathogenic events, likely occurring over many years, are required for leukemia development. There is some evidence that epigenetic preservation of fetal hematopoiesis-specific gene expression programs may contribute to the biology of *E::R*$^+$ B-ALL, as a core signature of genes differentially expressed in fetal versus adult hematopoiesis separated *E::R*$^+$ B-ALL (particularly in younger patients) from other subtypes [111]. One fetal core gene, *IGF2BP1*, is among the most upregulated genes in the *E::R*$^+$/like B-ALL gene signature and appears to be a bona fide oncogenic driver in B-ALL, as activating *IGH::IGF2BP1* rearrangements are recurrent in a subset of Down syndrome-associated B-ALL [112]. Retrovirally delivered *IGF2BP1* transgene expression appears to synergize with *E::R* in expansion of HSC and lymphoid progenitor populations in a mouse model [113], although these mice did not develop B-ALL.

13.11 Lessons and Limitations of E::R$^+$ B-ALL Leukemogenesis Models

Biologically faithful models of *E::R*$^+$ B-ALL leukemogenesis have proven difficult to develop, but a variety of complementary mouse models and human cell in vitro and xenograft-based systems have provided key insights into the "first-hit" biology of *E::R* acquisition in fetal progenitors, as well as potential cooperating genetic and non-genetic second-hit events. Given the long delay between acquisition of the *E::R* fusion human fetal hematolymphoid progenitor clones and the eventual development of disease, it is not surprising that *E::R* does not show strong transforming properties in short-term experimental systems.

Unlike *BCR::ABL1* or *ETV6*::TK gene fusions (discussed below), expression of an *E::R* transgene does not confer growth factor independence to murine lymphoid and hematopoietic progenitor cell lines in vitro [114, 115] nor is *E::R* expression alone sufficient to drive B-ALL in engineered mouse models when *E::R* is expressed from B-cell-specific promoters [116, 117] or the endogenous *ETV6* locus [118, 119]. Fetal liver hematopoietic progenitors transduced with *E::R*, but not control vector, were reported to give rise to B-cell precursor clones with enhanced expansion in growth factor-dependent culture [114], a finding that was recapitulated in fetal progenitors from a transgenic mouse model of *E::R* expressed from the endogenous *Etv6* promoter [118], although cells from this model could not be propagated indefinitely and did not give rise to leukemia in mice. Truncation analysis showed that expansion of fetal B-cell precursors required an E::R fusion protein that contained the PNT and central domains of ETV6 as well as the Runt DBD of RUNX1 [120]. Studies using different E::R-expressing lentiviral designs suggested that effects on murine fetal pro-B-cells are highly dependent on E::R expression levels, with low transgenic E::R protein expression causing increased colony formation and replating capacity relative to control vectors, while stronger *E::R* transgene expression had the opposite effect [121]. This observation highlights the challenges in comparing studies across labs that used different vector designs. Fetal mouse pro-B-cells transduced with the low-expressing *E::R* vector engrafted in recipient mice but did not cause leukemia on their own; however cotransduction of *E::R* with an oncogenic RAS G12D transgene was reported to drive leukemia more rapidly than the RAS G12D transgene alone.

In contrast to the preceding fetal progenitor studies, studies involving *E::R* retrovirus transduction of adult mouse bone marrow showed reduced expansion of B-cell progenitors in IL7-dependent culture [122, 123]. Another study provided evidence that the ability of *E::R* to inhibit or expand B-cell progenitor clones may be dependent on a specific cytokine context, as *E::R*-transduced murine B-cell progenitor cell lines, B-cell progenitors from mice bearing *E::R* under control of an *IGH* enhancer, and $CD34^+CD19^+CD38^-$ pre-leukemia-like cells derived from *E::R*-transduced human $CD34^+$ cells were resistant to the growth-inhibiting effects of TGFβ, possibly because E::R blocked TGFβ-dependent induction of the cell cycle inhibitor *CDKN1B* [124].

Some evidence has suggested that HSCs, rather than committed B-cell progenitors, might be a more likely cell of origin for the first hit of $E::R^+$ B-ALL. Conditional activation in adult mice of *E::R* expressed from the *Etv6* locus in HSCs leads to expanded numbers of HSPCs but depletion of lymphoid progenitors and mature B-cells [118]. B-cell depletion was also observed when $E::R^+$ fetal liver cells were transplanted into adult recipients. The ability of *E::R* to expand the HSC population in mice raised the possibility that *E::R*-expressing HSCs could represent a preleukemic population, but chemical mutagenesis with N-ethyl-N-nitrosourea in this model resulted only in T-cell neoplasms [118]. Other studies involving *E::R* transgene transduction of sorted cord blood-derived human cell populations suggested that only *E::R*-transduced HSCs but not adult multipotent progenitors or pro-B-cells could engraft in mice [125].

A subsequent transgenic mouse model that combined *E::R* expression from the *ETV6* locus with sleeping beauty transposon-based random genetic perturbations did give rise to B-ALL but with lower frequency than T-ALL or AML [119]. Interestingly, B-ALL development in this model was associated with transposon insertions that resulted in overexpression of the erythropoietin receptor gene *Epor*, which is ubiquitously overexpressed (but not genetically altered) in human $E::R^+$ B-ALL, and inactivation of the B-cell developmental TF gene *Ikzf1*, a common genetic event in human *E::R*-like (but not $E::R^+$) B-ALL. However, loss of the second *Etv6* allele, the most common secondary genetic event in human $E::R^+$ B-ALL, was not seen in this model.

Several independent lines of evidence suggest that an abnormal state of immune activation may promote progression of $E::R^+$ B-ALL. Sequencing of $E::R^+$ B-ALL showed that genomic rearrangement and deletion breakpoints in $E::R^+$ B-ALL, including those resulting in alteration of tumor suppressor genes such as *ETV6*, *BTG1*, *TBL1XR1*, and *CDKN2A/CDKN2B*, are enriched for sequence motifs resembling those recognized by the recombination-activating enzymes RAG1 and RAG2 [126]. Importantly, the breakpoints of the putative first-hit *E::R* rearrangement itself did not show this signature. Furthermore, frequent V_H replacement in $E::R^+$ B-ALL subclones indicated ongoing RAG activity during leukemia evolution [127], suggesting that aberrant activity of immunoglobulin gene rearrangement mechanisms may contribute to leukemogenesis. Unlike other defined-driver subtypes of B-ALL, genome-wide mutation signatures in *E::R*+ B-ALL show frequent enrichment for mutations in a context consistent with generation by APOBEC family enzymes [78]. These signatures (SBS2 and SBS13) occur in diverse other cancer types and may reflect episodic activation of an APOBEC3A/B-mediated response against viruses and retrotransposons [128]. It has been suggested that another APOBEC family enzyme, AID, which physiologically drives somatic hypermutation of immunoglobulin genes in mature germinal center B-cells, might contribute to B-ALL, as *Aicda* reportedly shows detectable levels of expression and evidence of activity in some pre-B-cell subsets [127]. Mouse pre-B-cells transduced with *E::R* and cultured in the presence of IL7 were subjected to repeat in vitro cycles of IL7 withdrawal and immune activation with lipopolysaccharide (LPS), after which recipient immunodeficient mice developed rapid and highly penetrant leukemia, which was abrogated in mice with deficiency of either *Aicda* or *Rag1* [127]. The role of AID in inflammation-associated leukemogenesis in vivo has been disputed, since *Aicda* is dispensable for B-ALL development in an infection-triggered mouse models of *PAX5*-altered B-ALL [129]. Furthermore, AID-associated mutation signatures have not been reported as enriched in B-ALL, in contrast to normal post-germinal center mature B-cells and B-cell neoplasms with a geminal center, memory B-cell, or plasma cell phenotype [130, 131].

A mouse model in which *E::R* was expressed from a Sca1 promoter in the HSC, but not committed B-cell compartment, was used to further investigate the role of infection in leukemogenesis in vivo [132]. Unlike models where *E::R* was expressed from the native *ETV6* locus, these mice lacked a block in B-cell development. Sca1-*E::R*+ mice developed B-ALL at high latency (>15 months) and low penetrance (~10%) when exposed to a common infectious environment but not when kept in specific pathogen-free conditions. Supporting a role for RAG gene hyperactivity in this process, expression of RAG1 and RAG2 was markedly higher in infection-exposed pro-B-cells/pre-B-cells from Sca1-*E::R*+ versus control mice. B-ALL development in these mice was not associated with *ETV6* gene inactivation, and it was unclear if these leukemias showed other distinctive features of human *E::R*+ B-ALL, such as *E::R*+/like signature gene expression or APOBEC signature mutations. A similar model of Sca1-Cre-dependent *E::R* expression from the native *ETV6* locus developed a low rate of B-ALL when exposed to common pathogens but much more frequently developed T-ALL [133]. B-ALL was also seen in models where Sca1-driven *E::R* was combined with heterozygosity for *Pax5* (recurrently deleted in E::R+ B-ALL) or *Kdm5C* (infrequently altered in E::R+ B-ALL), even in the absence of infection exposure [133].

To address concerns that mouse-human differences may underlie the shortcomings of *E::R*+ B-ALL leukemogenesis models, efforts have also been made to model *E::R*+ B-ALL in human cells. Human CD34+ HPCs transduced with *E::R* gave rise to an engrafted B-cell progenitor population with an abnormal CD34+CD19+CD38− phenotype when injected into immunodeficient mice [134]. This resembled the phenotype of leukemic and preleukemic *E::R*+ cells identified in a pair of monozygotic twins discordant for leukemia development. However, the vector-transduced cells did not expand or give rise to frank leukemia. Stromal coculture and cytokine-dependent directed differentiation of human pluripotent stem cells was shown to recapitulate a fetal-specific CD19-IL7R+ B-cell progenitor population, and engineered expression of *E::R* from the native human *ETV6* locus in this model blocked differentiation of CD19-IL7R+ progenitors to pro-B-cells and later stages of differentiation [135], suggesting a role for *E::R* in B-cell maturation arrest of preleukemic clones. However, these cells did not engraft or cause leukemia in recipient immunodeficient mice.

13.12 Secondary *ETV6* Inactivation in *E::R*+ B-ALL

In addition to gain-of-function properties of the E::R fusion protein, it has long been clear that loss of wild-type ETV6 function plays a key role in E::R+ B-ALL. In the same year that the *E::R* fusion was first described, several groups identified loss of heterozygosity of the *ETV6* locus as a common event in B-ALL [136–139] usually but not always coinciding with the presence of the *E::R*+ fusion. This phenomenon is driven primarily by deletion of the *ETV6* allele not involved in the *E::R*+ fusion, resulting in complete absence of wild-type *ETV6* transcripts in about 40% of *E::R*+ B-ALL [78, 140, 141]. This observation suggests that *ETV6* functions as a dose-dependent tumor suppressor gene, with all *E::R*+ leukemias and preleukemic clones possessing low ETV6 activity due to loss of one *ETV6* allele and dominant-negative effects of the E::R protein on wild-type ETV6 [142], but with ongoing selection for loss of the second allele in many patients. Another possibility is that PNT domain-mediated interactions between wild-type ETV6 and E::R may partially interfere with the oncogenic effects of E::R, driving selection for loss of the second *ETV6* allele to relieve that interference (*see* Fig. 13.3a), but direct evidence for this is lacking.

Our laboratory recently identified a surprising mechanism by which ETV6 deficiency facilitated overexpression of previously identified *E::R*+ B-ALL signature genes. We found that such genes were enriched for proximity to short tandem repeats (microsatellites) of the sequence "GGAA," which took on an active enhancer-like chromatin state marked by high levels of chromatin accessibility and the H3K27ac modification in *E::R*+ B-ALL [48] a finding that is absent from other genetic driver-defined B-ALL subtypes [143]. Active repeat enhancers were bound by the ETS activator ERG in E::R+ B-ALL cells, while these same repeats were bound in a repressed state by the ETV6 in other B-ALL subtypes. Knockdown of ERG, reexpression of ETV6, or CRISPR interference-mediated repression of repeat enhancers led to downregulation of the repeat-associated *E::R*+ B-ALL signature genes, including candidate drivers of leukemia fitness such as *EPOR* [144, 145] and *PIK3C3* [146]. Most repeat enhancers and target genes showed no evidence of direct binding by the E::R fusion protein, indicating that this aspect of the *E::R*+ B-ALL phenotype is related to ETV6 deficiency rather than being direct consequences of the E::R fusion. It remains unclear why ETV6 binds disproportionately to GGAA repeats in B-ALL [and in Ewing sarcoma [46, 47]], and further investigation is needed to determine whether ETV6-dependent regulation of GGAA repeats plays a role in normal development or is purely an emergent property of certain malignant cells. The relatively low conservation of GGAA repeat positions across species may pose a challenge to such investigations, meriting careful comparison of appropriate human and mouse models.

13.13 Somatic *ETV6* Alterations in Other B-ALL Subtypes

ETV6 is one of the most commonly mutated genes in B-lymphoblastic leukemia, with genetic alterations, most often focal deletions and other inactivating events, occurring in about 17% of pediatric B-ALL [78]. Excluding *E::R*+ B-ALL, somatic *ETV6* mutations have been reported as most frequent in the *ETV6::RUNX1*-like, iAMP21, *ZNF384*, and PAX5alt subtypes.

ETV6::RUNX1-like B-ALL (*E::R*-like) is a poorly understood subtype of B-ALL defined by a signature of aberrant gene expression that closely resembles that of *E::R*+ B-ALL but that lacks a detectable *E::R* fusion gene [147–149]. Like *E::R*+ B-ALL, *E::R*-like B-ALL most often arises in young children [median onset 2.8 years of age in a large pediatric series [78]]. Other features that unite *E::R*+ and *E::R*-like B-ALL and distinguish them from other B-ALL subtypes (Fig. 13.4) include the CD27^{pos}/CD44$^{low\text{-}neg}$ immunophenotype [148] and enrichment for the

APOBEC mutation signatures [78, 150]. The latter finding suggests that these two subtypes do not merely converge on a common final phenotype but likely share common mechanisms of genomic alteration during leukemogenesis. Nearly all *E::R*-like B-ALL cases show *ETV6* gene alterations [147–150]. Many cases show either biallelic *ETV6* LoF alterations (rare in other subtypes besides E::R+ B-ALL) or *ETV6* gene fusions with non-*RUNX1* partners. *E::R*-like B-ALL also shows high rates of *PAX5* and *IKZF1* LoF alterations; the latter finding is uncommon in *E::R*+ B-ALL but is common in high-risk B-ALL subtypes such as *BCR::ABL1*+ and *BCR::ABL1*-like B-ALL. *E::R*-like B-ALL patients appear to have worse event-free survival than E::R+ B-ALL with modern risk-adapted therapy [87], and unlike *E::R*+ B-ALL [88], they do not appear to show high sensitivity to L-asparaginase, a key component of B-ALL induction chemotherapy regimens [90]. Further investigation into the biological basis of the similarities and differences between these subtypes may therefore be of clinical importance.

Another recurrent *ETV6* gene fusion, *PAX5::ETV6*, is associated with the "*PAX5*-altered" (PAX5alt) subtype of B-ALL [78, 149], which is recognized by a distinctive gene expression signature and a spectrum of mutations and fusions affecting the B-lineage-defining TF gene *PAX5*. Dysregulation of PAX5 target genes is thought to contribute to B-ALL via arrest of B-cell progenitor differentiation, and rare germline *PAX5* mutations are associated with B-ALL susceptibility [151, 152]. *PAX5::ETV6* is the most common gene fusion in PAX5alt B-ALL, reported in about 13% of cases in one large series [149]. Although ETV6 is the distal (C-terminal) partner in the PAX5::ETV6 fusion protein, contrasting with the arrangement seen in E::R, it still contains the PNT oligomerization and corepressor recruitment domains of ETV6, fused to the N-terminal paired-box DBD of *PAX5* [149, 153, 154]. Analogous to E::R, PAX5::ETV6 has been show to bind to physiologic PAX5 target sites and repress gene expression [155–158], possibly in part through recruitment of corepressors that normally bind ETV6 [156]. Interestingly, several other *PAX5* fusion gene partners associated with PAX5alt B-ALL also contain oligomerization domains and are capable of repressing transcription via PAX5 binding sites, suggesting that oligomerization of PAX5 fusion proteins may be a unifying mechanism of gene dysregulation for this class of fusions [157]. Although the ETS DBD of ETV6 is present in the PAX5::ETV6 fusion protein (unlike in E::R), it is uncertain whether this contributes to its oncogenic function, given that other mutant forms of PAX5 seen in this B-ALL subtype lack an ETS domain. Like *E::R*+ B-ALL, PAX5alt B-ALL tends to occur at an early age, and infection/immune system activation has been linked to its pathogenesis [132, 159], although PAX5alt B-ALL lacks the APOBEC signature seen in *E::R*+ and *E::R*-like B-ALL, and instead frequently shows a signature linked to reactive oxygen species [78].

A final B-ALL subtype that can show a unique pattern of *ETV6* alteration is *BCR::ABL1*-like B-ALL, which is defined by a broad spectrum of tyrosine kinase-activating fusion genes [160]. Some fusions seen in this subtype link the proximal portion of the *ETV6* gene to tyrosine kinase genes including *ABL1*, *JAK2*, *NTRK3*, and *PDGFRB* [78, 160]. The resulting fusion proteins presumably undergo ETV6 PNT domain-dependent kinase self-association, autophosphorylation, and constitutive downstream signaling. Detection of such fusions is clinically important, since addition of small molecule inhibitors against the involved kinase may improve therapeutic response compared to standard chemotherapy alone. These de novo *BCR::ABL1*-like B-ALLs need to be distinguished clinically from the blast phase of myeloid/lymphoid neoplasms with eosinophilia, which are discussed below.

13.14 B-ALL in the Setting of Germline *ETV6* Mutations

As mentioned earlier, germline *ETV6* alterations have been identified as a strong B-ALL risk factor with 25–30% of carriers developing leukemia [65, 66]. Most germline *ETV6* mutations linked to B-ALL development either alter conserved residues within the ETS DBD or resulted in a truncating protein that lacked the ETS domain entirely. Transgenic expression of such variants typically showed loss of transcriptional repressor activity and dominant-negative interference with wild-type *ETV6* in reporter assays, impaired DNA binding, and localization to the cytoplasm rather than nucleus. Fascinatingly, B-ALL developing in patients with damaging *ETV6* germline mutations was nearly always one of two subtypes, high-hyperdiploid (70%) or *E::R*-like [66], suggesting that early ETV6 deficiency can contribute to two distinct leukemogenic pathways. Hyperdiploid B-ALL, both sporadic and in germline *ETV6* mutant patients, is enriched for RAS/MAPK pathway alterations such as *NRAS* and *KRAS* mutations [66, 78]. Wild-type but not mutant *ETV6* transgene expression was found to inhibit the growth of mutant RAS-transformed BaF3 mouse B-cells [66], consistent with long-standing (albeit circumstantial) evidence implicating ETV6 as an antagonist of RAS/MAPK pathway target genes in mammalian cells. Most nonhyperdiploid B-ALL arising in patients with germline *ETV6* mutations show characteristic features of *E::R*-like B-ALL, including the defining *E::R*+/-like gene expression signature (but not an *E::R* fusion gene) frequent inactivating alterations of the second *ETV6* allele, mutations of *PAX5* and *IKZF1*, and infrequent RAS pathway mutations. Given the relatively high penetrance of B-ALL in patients with damaging germline *ETV6* alterations, it is notable that mouse models of THC5-associated *ETV6* mutations have not yet been reported to develop B-ALL.

13.15 *ETV6*:: TK Fusions in Myeloid and Lymphoma Neoplasms with Eosinophilia

Fusions between the N-terminal PNT oligomerization domain of *ETV6* and various TK genes result in autophosphorylated and constitutively active oncogenic tyrosine kinases that are seen in rare subsets of hematopoietic cancers known as myeloid/lymphoid neoplasms with eosinophilia and TK fusion genes [M/LN-eo-TK; [161, 162]]. The first of these to be described was the *ETV6::PDGFRB* fusion in a case of apparent chronic myelomonocytic leukemia ([CMML; [163]]. M/LN-eo with *PDGFRB* rearrangement (which usually but not always bears *ETV6* as the N-terminal fusion partner) is notable for frequent eosinophilia and presentation as either a "chronic phase" myeloproliferative/myelodysplastic overlap neoplasm such as CMML or a "blast phase" acute myeloid or B-lymphoblastic leukemia. Other categories of M/LN-eo-TK include *ETV6::ABL1* [164] and *ETV6::FLT3* fusions. *ETV6::FLT3* and other *FLT3* fusions are more likely to present with a T-lymphoblastic component. In addition to conferring self-association, the N-terminal portion of the ETV6 protein shares features with proteins encoded by other N-terminal kinase fusion partners like *BCR* (breakpoint cluster region activator of RhoGEF and GTPase) that seem to contribute to fusion kinase activity, such as a phosphorylation-dependent binding site (tyrosine 314 on ETV6) for the adaptor protein GRB2 [165]. It is important to note, however, that not all N-terminal partner genes confer the same phenotype in leukemogenic kinase fusions. *ABL1* fusions are a notable example of this, as the common *BCR::ABL1* fusions are associated with a different spectrum of disease (chronic myeloid leukemia and acute myeloid and lymphoblastic leukemias) than the M/LN-eo seen with *ETV6::ABL1*. As in other leukemias with *ETV6*::TK fusion genes, appropriate tyrosine kinase inhibitors may play a significant role in treatment of M/LN-eo-TK.

13.16 ETV6 Alterations in Acute Myeloid Leukemia and MDS

Inactivating *ETV6* mutations similar to those seen in B-ALL are recurrent events in myelodysplastic neoplasms and AML at relatively low frequency (1–3%), indicating that *ETV6* also plays a tumor suppressor role in myeloid neoplasia [166–170]. Most such events appear to be sporadic, although AML and MDS have been reported in patients with germline *ETV6* mutations, albeit at a lower frequency than B-ALL [166, 169]. Although *ETV6* mutations are relatively uncommon in MDS, they are significantly associated with poor clinical outcomes [167], and *ETV6* is one of 16 prognostic genes included in the recently developed Molecular International Prognostic Scoring System for MDS [170]. As in other diseases, *ETV6* alterations in AML are associated with loss of detectable protein expression in many instances, suggesting that many mutations confer protein instability [166]. *ETV6* fusions with a number of TK genes have been reported in AML, including *ABL1*, *ABL2*, *JAK2*, *FGFR3*, *SYK*, *FRK*, *LYN*, *NTRK3*, *PDGFRA*, *PDGFRB*, and *FLT3* [reviewed in [171]].

Genomic rearrangements of *ETV6* occurring in pediatric AML merit specific discussion, due to several recurrent events that exemplify the ability of *ETV6* locus regulatory elements to drive overexpression of partner genes (e.g., Fig. 13.3b). A recent study used ChIP-Seq to map the distribution of the active enhancer histone mark H3K27ac in a cohort of pediatric AML. Applying the Ranking Of Super-Enhancers (ROSE) algorithm to this data, they found that the *ETV6* locus contained the most highly ranked super-enhancer in the genome for these leukemias [172]. Strong tissue-specific superenhancers have often been implicated in so-called enhancer-hijacking rearrangements in which juxtaposition of an oncogene to a super-enhancer bearing locus results in aberrant oncogene overexpression. Indeed, transcriptional overexpression in the setting of *ETV6* locus rearrangements has been observed for a number of partner genes, including *GSX2* (*GSH2*), *IL3*, and *CSF2* [74]. Similar *ETV6* enhancer hijacking appears to occur rarely in B-ALL [173].

Recurrent rearrangements in pediatric AML between *ETV6* and the homeobox TF gene *MNX1* (previously *HLXB9*) are structurally diverse, with some, but not all leading to *ETV6::MNX1* fusion genes. However, a common feature in all such rearrangements appears to be *MNX1* overexpression [75, 76], likely attributable to the activity of *ETV6* locus regulatory elements and/or topological repositioning of the *MNX1* gene to an active rather than heterochromatic nuclear compartment [174]. *MNX1*-driven leukemia has been difficult to model, but a recent induced pluripotent stem cell iPSC model in which CRISPR/Cas9 was used to generate fusion of the *MNX1* and *ETV6* loci resulted in overexpression of *MNX1* upon hematopoietic-directed differentiation, with evidence of reduced erythroid and megakaryocyte maturation [175].

The similarly named gene *MN1* (Meningioma-1) also undergoes a mix of gene fusion and nonfusion rearrangements with the *ETV6* locus (and occasionally other partners) that lead to *MN1* overexpression [77]. Such rearrangements are likely multifunctional in that they likely inactivate the wild-type function of the involved *ETV6* allele and in some cases may result in expression of a dominant-negative N-terminal ETV6 fusion protein. *MN1* overexpression alone is sufficient for transformation of mouse hematopoietic progenitors [176], suggesting that this may be the main driver function of these rearrangements, but additional oncogenic hits appear to be required in human leukemogenesis models [177]. Interestingly, *MN1*-overexpressing AML also overexpresses the leukemogenic early myeloid progenitor transcription factors *HOXA9* and *MEIS1*, which also require MLL1/KMT2A for activation [178–180]. Preclinical models of MN1 overexpressing AML are sensitive to small molecule inhibitors of the MLL1/KMT2A-menin interaction [181], a class of drugs that has shown great promise for other HOXA9-overexpressing subsets of AML [182, 183], thus highlighting the importance of better understanding the functional significance of *ETV6::MN1* rearrangements.

13.17 *ETV6* Alterations in T-Lymphoblastic and T/Myeloid Leukemia

Genetic alteration of *ETV6* also seems to play a key role in a subset of T-cell precursor leukemias and lymphomas. *ETV6* mutations are enriched in early T-precursor acute lymphoblastic leukemia (ETP-ALL), which is distinguished from classic

T-lymphoblastic leukemia/lymphoma by co-expression of myeloid and hematopoietic stem cell genes along with early T lineage markers such as cytoplasmic CD3 and CD7, absence of markers associated with later stages of T-cell progenitor differentiation, and enrichment for gene mutations also associated with myeloid leukemias [184]. ETP-ALL has been generally associated with worse clinical outcomes and decreased sensitivity to many chemotherapeutic agents used for T-ALL therapy [90, 184]. *ETV6* alterations have been reported in 24–33% of ETP-ALL [versus 0–10% of other T-ALL [185, 186]] and include a spectrum of deletions, gene fusions, and point mutations. Mutations in ETP-ALL included predicted C-terminal and N-terminal (alternative translation start site) truncation mutants resembling those seen in AML and capable of dominant-negative interference with normal *ETV6* repressive function, although expression of truncated proteins in ETP-ALL blasts was not confirmed [185]. *ETV6*-mutated T-ALL has been noted to have a characteristic gene expression signature [185], and a subset of these genes, including *CD33*, *HOXA13*, *PTEN*, and *PRDM16*, could be activated by *ETV6* knockdown in an ETP-ALL-like cell line [185].

A rare, recurrent transcriptional regulator fusion, *ETV6::NCOA2*, has been reported in mixed-phenotype T/myeloid leukemia [187–189]. This fusion incorporates the N-terminal portions of *ETV6*, but not the ETS domain; however, unlike *E::R*, the partner gene lacks a DNA binding domain of its own. *NCOA2* encodes a transcriptional coactivator and also undergoes rare fusions with the acetyltransferase gene *MOZ/KAT6A* in AML. A recent functional study compared these two oncogenic fusions [189] and found that while *MOZ/KAT6A::NCOA2* strongly transformed mouse hematopoietic stem cells to generate myeloid leukemia, *ETV6::NCOA2* only weakly extended the colony-forming abilities of hematopoietic stem cells in vitro. Instead, a lentiviral *ETV6::NCOA2* transgene transduced into mouse hematopoietic stem cells led to an early block in T-cell development. *ETV6::NCOA2* collaborated with NOTCH mutations to generate T/myeloid leukemias from both mouse and human HSCs. Although ETV6::NCOA2 lacks an ETS DBD, it retains the ability to oligomerize with wild-type ETV6, and it was proposed that this might lead to transcriptional activation of wild-type ETV6 binding targets via EP300/KAT3B acetyltransferase recruitment by the NCOA2 AD1 and AD2 domains, an interpretation supported by ETV6/ETV6::NCOA2 coexpression in reporter assays and significant overlap between genes upregulated by *ETV6::NCOA2* transduction and in *ETV6*-mutated ETP-ALL. Nonetheless, the direct regulatory targets of ETV6 in normal T-cell development and T-lineage leukemias remain poorly defined, as are the specific targets though which disruption of ETV6 promotes ETP and T/myeloid leukemias.

13.18 *ETV6* Alterations in Mature B-Cell Lymphomas

Predicted LoF mutations in *ETV6* are also recurrent in mature B-cell lymphomas, notably diffuse large B-cell lymphoma (DLBCL). DLBCL is a genetically and biologically heterogeneous disease, and as with B-ALL and T-ALL, *ETV6* mutations in DLBCL are not evenly distributed across subtypes but are most often seen in a genetically defined subtype of DLBCL, "MYD88-CD79A" (MCD), with an activated B-cell-like gene expression program (ABC-DLBCL) and frequent cooccurrence of mutations in *MYD88*, *CD79B*, *PIM1*, *PRDM1*, *TBL1XR1*, and *BTG1*, among other genes [190, 191]. Interestingly, *TBL1XR1* and *BTG1* deletions, like *ETV6* deletions, also frequently co-occur in *E::R*+ B-ALL [78, 155], hinting at possible common mechanisms between these otherwise very different B-cell neoplasms. As in other hematological cancers, the oncogenic mechanism of *ETV6* loss of function in MCD DLBCL remains poorly understood, although ongoing efforts to model this subtype in genetically engineered mice [192–194] may provide good opportunities for future experimental investigation. In addition to representing a substantial proportion of systemic DLBCL, the MCD spectrum of gene mutations is the predominant pattern for DLBCL arising in certain anatomic sites, including primary central nervous system and vitreoretinal, testicular, and intravascular DLBCL. *ETV6* alterations are very frequent for DLBCL originating in these sites [195–199]. Surprisingly, ETV6 missense mutations occurring in vitreoretinal lymphoma most often occurred in the first exon, which encodes a portion of *ETV6* N-terminal to the SAM domain and whose function is poorly understood [197].

13.19 *ETV6* Aberrations in Nonhematopoietic Cancers

ETV6 alterations in solid tumors have been recently reviewed [200] and will only be briefly discussed here. A recurrent fusion between *ETV6* and the TK gene *NTRK3* (*ETV6::NTRK3*; *E::N*) can occur in myeloid and lymphoid leukemias but also occurs in a distinctive array of solid tumors including congenital fibrosarcoma [201–203], congenital mesoblastic nephroma [204, 205], pediatric glioma [206, 207], thyroid carcinoma [208–210], and secretory carcinomas of the breast [211], salivary gland [212], and skin [213]. The resulting fusion protein is structurally similar to other ETV6::TK fusions, undergoing ETV6 PNT domain-mediated oligomerization, kinase domain autophosphorylation, and constitutive activation of downstream signaling pathways. E::N expression leads to the constitutive activation of two major cell survival pathways, namely, the RAS-MAPK mitogenic pathway and the phosphatidylinositol-3-kinase (PI3K)-AKT pathway [211, 214]. *E::N* is considered a disease-defining driver oncogene in congenital fibrosarcoma, occurring in essentially all cases, while in other mesenchymal tumor types, *E::N* may occur as an alternate driver, as in inflammatory myofibroblastic tumor where it can occur in place of more common ALK fusions [215, 216]. Among epithelial tumors, the *E::N* fusion gene is strongly associated with secretory carcinomas, rare cancers with a characteristic histological appearance that can arise in the breast, salivary glands, or cutaneous adnexae. It seems likely that the *E::N* fusion interacts with shared biological features of these three developmentally related tissues to drive a similar malignancy in each site. Notably, *ETV6* transcript expression was higher in human salivary gland than in any other nonhematopoietic bulk human tissue profiled by two independent consortia [217, 218]. Importantly, many *NTRK3* fusion-dependent solid tumors are clinically sensitive to recently developed selective TRK kinase inhibitors [219, 220]. Other reported *ETV6*::TK fusions in solid tumors include *ETV6::RET* [221, 222] and *ETV6::MET* [223] in secretory carcinomas and *ETV6::ALK* in epithelioid fibrous histiocytoma [224] and may also be targetable with selective inhibitors of the involved kinase.

ETV6 fusions with nontyrosine kinase genes are less well described in solid tumors. Recurrent fusions between *ETV6* and an adjacent gene, *BCL2L14* [225], were recently described in triple-negative breast cancer (TNBC), an aggressive subtype whose survival is independent of targetable hormone receptor and HER2 proteins. Expression of *BCL2L14::ETV6* fusion transgenes appeared to alter cell motility, invasiveness, and gene expression in vitro, but the mechanistic contribution of *ETV6* is unclear, since only a subset of evaluated *BCL2L14::ETV6* fusion transcripts encoded in-frame portions of the *ETV6* coding sequence.

13.20 Summary and Future Questions

Despite longstanding interest in the biology of ETV6 in the hematopoietic system, many key questions remain unanswered regarding its role in blood cancers. Models for the contribution of ETV6 to leukemia have proven difficult to develop, with transgenic mouse models of *E::R*+ B-ALL and thrombocytopenia-associated germline *ETV6* mutations recapitulating only limited features of the human disease to date. Important controversies remain about the cell of origin of *E::R*+ and other *ETV6* mutation-associated B-ALL subtypes, including the requirement for a fetal-specific epigenetic state and whether the initial hit occurs in a committed B-cell progenitor or multipotent stem cell. Answering these questions, as well as better defining the factors associated with *ETV6*-altered B-ALL progression, could pay important dividends for early intervention in young patients at high risk for developing these diseases. Our understanding of the contribution of *ETV6* LoF mutations in other hematopoietic tumors such as T-ALL, AML, and mature B-cell lymphoma is still rudimentary, and further work could provide critical information about the biology of high-risk forms of these diseases. Linking information about the lineage- and state-specific biochemistry of ETV6 and its interactors, direct genomic regulatory targets, and phenotypic consequences of ETV6 disruption in primary human samples and appropriate model systems will be essential for further advances.

Acknowledgments The authors would like to thank Dr. Mark Chiang for the helpful feedback on this review.

References

1. Karlsson M, Zhang C, Méar L et al (2021) A single-cell type transcriptomics map of human tissues. Sci Adv 7:eabh2169
2. Yoshida H, Lareau CA, Ramirez RN et al (2019) The cis-regulatory atlas of the mouse immune system. Cell 176:897–912.e20
3. Wei GH, Badis G, Berger MF et al (2010) Genome-wide analysis of ETS-family in vitro and in vivo. EMBO J 29:2147–2160
4. Terrell JR, Taylor SJ, Schneider AL et al (2023) DNA selection by the master transcription factor PU.1. Cell Rep 42:112671
5. Potter MD, Buijs A, Kreider B et al (2000) Identification and characterization of a new human ETS-family transcription factor, TEL2, that is expressed in hematopoietic tissues and can associate with TEL1/ETV6. Blood 95:3341–3348
6. Gu X, Shin B-H, Akbarali Y et al (2001) Tel-2 is a novel transcriptional repressor related to the Ets factor Tel/ETV-6*. J Biol Chem 276:9421–9436
7. Chakrabarti SR, Nucifora G (1999) The leukemia-associated gene TEL encodes a transcription repressor which associates with SMRT and mSin3A. BBRC 264:871–877
8. Lopez RG, Carron C, Oury C et al (1999) TEL is a sequence-specific transcriptional repressor. J Biol Chem 274:30132–30138
9. Fenrick R, Amann JM, Lutterbach B et al (1999) Both TEL and AML-1 contribute repression domains to the t(12;21) fusion protein. Mol Cell Biol 19:6566–6574
10. Guidez F, Petrie K, Ford AM et al (2000) Recruitment of the nuclear receptor corepressor N-CoR by the TEL moiety of the childhood leukemia-associated TEL-AML1 oncoprotein. Blood 96:2557–2561
11. Fenrick R, Wang L, Nip J et al (2000) TEL, a putative tumor suppressor, modulates cell growth and cell morphology of Ras-transformed cells while repressing the transcription of stromelysin-1. Mol Cell Biol 20:5828–5839
12. Irvin BJ, Wood LD, Wang L et al (2003) TEL, a putative tumor suppressor, induces apoptosis and represses transcription of Bcl-XL*. J Biol Chem 278:46378–46386
13. Wang L, Hiebert SW (2001) TEL contacts multiple co-repressors and specifically associates with histone deacetylase-3. Oncogene 20:3716–3725
14. Boccuni P, MacGrogan D, Scandura JM et al (2003) The human L(3)MBT polycomb group protein is a transcriptional repressor and interacts physically and functionally with TEL (ETV6). J Biol Chem 278:15412–15420
15. Green SM, Coyne HJ, McIntosh LP et al (2010) DNA binding by the ETS protein TEL (ETV6) is regulated by autoinhibition and self-association. J Biol Chem 285:18496–18504
16. Hollenhorst PC, McIntosh LP, Graves BJ (2011) Genomic and biochemical insights into the specificity of ETS transcription factors. Annu Rev Biochem 80:437–471
17. Coyne HJ, De S, Okon M et al (2012) Autoinhibition of ETV6 (TEL) DNA binding: appended helices sterically block the ETS domain. J Mol Biol 421:67–84
18. Tran HH, Kim CA, Faham S et al (2002) Native interface of the SAM domain polymer of TEL. BMC Struct Biol 2:5
19. Kim CA, Phillips ML, Kim W et al (2001) Polymerization of the SAM domain of TEL in leukemogenesis and transcriptional repression. EMBO J 20:4173–4182
20. Rebay I, Rubin GM (1995) Yan functions as a general inhibitor of differentiation and is negatively regulated by activation of the Ras1/MAPK pathway. Cell 81:857–866
21. Qiao F, Song H, Kim CA et al (2004) Derepression by depolymerization: structural insights into the regulation of Yan by Mae. Cell 118:163–173
22. Arai H, Maki K, Waga K et al (2002) Functional regulation of TEL by p38-induced phosphorylation. BBRC 299:116–125
23. Maki K, Arai H, Waga K et al (2004) Leukemia-related transcription factor TEL is negatively regulated through extracellular signal-regulated kinase-induced phosphorylation. Mol Cell Biol 24:3227–3237
24. Roukens MG, Alloul-Ramdhani M, Vertegaal ACO et al (2008) Identification of a new site of sumoylation on Tel (ETV6) uncovers a PIAS-

dependent mode of regulating Tel function. Mol Cell Biol 28:2342–2357
25. Wood LD, Irvin BJ, Nucifora G et al (2003) Small ubiquitin-like modifier conjugation regulates nuclear export of TEL, a putative tumor suppressor. Proc Natl Acad Sci USA 100:3257–3262
26. Chakrabarti SR, Sood R, Nandi S et al (2000) Posttranslational modification of TEL and TEL/AML1 by SUMO-1 and cell-cycle-dependent assembly into nuclear bodies. Proc Natl Acad Sci USA 97:13281–13285
27. Chakrabarti SR, Sood R, Ganguly S et al (1999) Modulation of TEL transcription activity by interaction with the ubiquitin-conjugating enzyme UBC9. Proc Natl Acad Sci USA 96:7467–7472
28. Roukens MG, Alloul-Ramdhani M, Moghadasi S et al (2008) Downregulation of vertebrate Tel (ETV6) and drosophila Yan is facilitated by an evolutionarily conserved mechanism of F-box-mediated ubiquitination. Mol Cell Biol 28:4394–4406
29. Sevilla L, Aperlo C, Dulic V et al (1999) The Ets2 transcription factor inhibits apoptosis induced by colony-stimulating factor 1 deprivation of macrophages through a Bcl-xL-dependent mechanism. Mol Cell Biol 19:2624–2634
30. Lemarchandel V, Ghysdael J, Mignotte V et al (1993) GATA and Ets cis-acting sequences mediate megakaryocyte-specific expression. Mol Cell Biol 13:668–676
31. Zhang MY, Churpek JE, Keel SB et al (2015) Germline ETV6 mutations in familial thrombocytopenia and hematologic malignancy. Nat Genet 47:180–185
32. Bloom M, Oak N, Baskin-Doerfler R et al (2023) ETV6 represses inflammatory response genes and regulates HSPC function during stress hematopoiesis in mice. Blood Adv 7:5608. https://doi.org/10.1182/bloodadvances.2022009313
33. Gerstein MB, Kundaje A, Hariharan M et al (2012) Architecture of the human regulatory network derived from ENCODE data. Nature 489:91–100
34. Neveu B, Caron M, Lagacé K et al (2018) Genome wide mapping of ETV6 binding sites in pre-B leukemic cells. Sci Rep 8:1–9
35. Kuwata T, Gongora C, Kanno Y et al (2002) Gamma interferon triggers interaction between ICSBP (IRF-8) and TEL, recruiting the histone deacetylase HDAC3 to the interferon-responsive element. Mol Cell Biol 22:7439–7448
36. Kanno Y, Levi B-Z, Tamura T et al (2005) Immune cell-specific amplification of interferon signaling by the IRF-4/8-PU.1 complex. J Interf Cytokine Res 25:770–779
37. Humblin E, Thibaudin M, Chalmin F et al (2017) IRF8-dependent molecular complexes control the Th9 transcriptional program. Nat Commun 8:2085
38. Villar J, Cros A, De Juan A et al (2022) ETV3 and ETV6 enable monocyte differentiation into dendritic cells by repressing macrophage fate commitment. Nat Immunol 24:84–95
39. Lau CM, Tiniakou I, Perez OA et al (2018) Transcription factor Etv6 regulates functional differentiation of cross-presenting classical dendritic cells. J Exp Med 215:2265–2278
40. Fisher MH, Kirkpatrick GD, Stevens B et al (2020) ETV6 germline mutations cause HDAC3/ NCOR2 mislocalization and upregulation of interferon response genes. JCI Insight 5:e140332
41. Thouenon R, Chentout L, Moreno-Corona N et al (2023) A neomorphic mutation in the interferon activation domain of IRF4 causes a dominant primary immunodeficiency. J Exp Med 220:e20221292
42. Froggatt HM, Harding AT, Chaparian RR et al (2021) ETV7 limits antiviral gene expression and control of influenza viruses. Sci Signal 14:eabe1194
43. Schick N, Oakeley EJ, Hynes NE et al (2004) TEL/ETV6 is a signal transducer and activator of transcription 3 (Stat3)-induced repressor of Stat3 activity *. J Biol Chem 279:38787–38796
44. Meester-Smoor MA, Janssen MJFW, ter Haar WM et al (2011) The ETS family member TEL binds to nuclear receptors RAR and RXR and represses gene activation. PLoS One 6:e23620
45. Kwiatkowski BA, Bastian LS, Bauer TR et al (1998) The ets family member Tel binds to the Fli-1 oncoprotein and inhibits its transcriptional activity*. J Biol Chem 273:17525–17530
46. Lu DY, Ellegast JM, Ross KN et al (2023) The ETS transcription factor ETV6 constrains the transcriptional activity of EWS–FLI to promote Ewing sarcoma. Nat Cell Biol 25:285–297
47. Gao Y, He X-Y, Wu XS et al (2023) ETV6 dependency in Ewing sarcoma by antagonism of EWS-FLI1-mediated enhancer activation. Nat Cell Biol 25:298–308
48. Kodgule R, Goldman JW, Monovich AC et al (2023) ETV6 deficiency unlocks ERG-dependent microsatellite enhancers to drive aberrant gene activation in B-lymphoblastic leukemia. Blood Cancer Discov 4:34–53
49. Gangwal K, Sankar S, Hollenhorst PC et al (2008) Microsatellites as EWS/FLI response elements in Ewing's sarcoma. Proc Natl Acad Sci USA 105:10149–10154
50. Riggi N, Knoechel B, Gillespie SM et al (2014) EWS-FLI1 utilizes divergent chromatin remodeling mechanisms to directly activate or repress enhancer elements in Ewing sarcoma. Cancer Cell 26:668–681
51. Unnikrishnan A, Guan YF, Huang Y et al (2016) A quantitative proteomics approach identifies ETV6 and IKZF1 as new regulators of an ERG-driven transcriptional network. Nucleic Acids Res 44:10644–10661
52. Li L, Rispoli R, Patient R et al (2019) Etv6 activates VEGFA expression through positive and negative transcriptional regulatory networks in Xenopus embryos. Nat Commun 10:1083
53. Wang LC, Kuo F, Fujiwara Y et al (1997) Yolk sac angiogenic defect and intra-embryonic apoptosis in

mice lacking the Ets-related factor TEL. EMBO J 16:4374–4383
54. Wang LC, Swat W, Fujiwara Y et al (1998) The TEL/ETV6 gene is required specifically for hematopoiesis in the bone marrow. Genes Dev 12:2392–2402
55. Hock H, Meade E, Medeiros S et al (2004) Tel/Etv6 is an essential and selective regulator of adult hematopoietic stem cell survival. Genes Dev 18:2336–2341
56. Pereira C-F, Chang B, Qiu J et al (2013) Induction of a hemogenic program in mouse fibroblasts. Cell Stem Cell 13:205–218
57. Ciau-Uitz A, Pinheiro P, Gupta R et al (2010) Tel1/ETV6 specifies blood stem cells through the agency of VEGF signaling. Dev Cell 18:569–578
58. Ciofani M, Madar A, Galan C et al (2012) A validated regulatory network for Th17 cell specification. Cell 151:289–303
59. Moriyama T, Metzger ML, Wu G et al (2015) Germline genetic variation in ETV6 and risk of childhood acute lymphoblastic leukaemia: a systematic genetic study. Lancet Oncol 16:1659–1666
60. Noetzli L, Lo RW, Lee-Sherick AB et al (2015) Germline mutations in ETV6 are associated with thrombocytopenia, red cell macrocytosis and predisposition to lymphoblastic leukemia. Nat Genet 47:535–538
61. Topka S, Vijai J, Walsh MF et al (2015) Germline ETV6 mutations confer susceptibility to acute lymphoblastic leukemia and thrombocytopenia. PLoS Genet 11:e1005262
62. Melazzini F, Palombo F, Balduini A et al (2016) Clinical and pathogenic features of ETV6-related thrombocytopenia with predisposition to acute lymphoblastic leukemia. Haematologica 101:1333–1342
63. Poggi M, Canault M, Favier M et al (2017) Germline variants in ETV6 underlie reduced platelet formation, platelet dysfunction and increased levels of circulating CD34+ progenitors. Haematologica 102:282–294
64. Di Paola J, Porter CC (2019) ETV6-related thrombocytopenia and leukemia predisposition. Blood 134:663–667
65. Hock H, Shimamura A (2017) ETV6 in hematopoiesis and leukemia predisposition. Semin Hematol 54:98–104
66. Nishii R, Baskin-Doerfler R, Yang W et al (2021) Molecular basis of ETV6-mediated predisposition to childhood acute lymphoblastic leukemia. Blood 137:364–373
67. McConville M, Thomas T, Valadez C et al (2023) Familial thrombocytopenia-associated germline ETV6 P214L mutation results in XPO1-mediated nuclear export. Blood [65th ASH Annual Meeting Abstracts] 142:1298
68. Dirse V, Norvilas R, Gineikiene E et al (2018) ETV6 and NOTCH1 germline variants in adult acute leukemia. Leuk Lymphoma 59:1022–1024
69. Järviaho T, Bang B, Zachariadis V et al (2019) Predisposition to childhood acute lymphoblastic leukemia caused by a constitutional translocation disrupting ETV6. Blood Adv 3:2722–2731
70. Borst S, Nations CC, Klein JG et al (2021) Study of inherited thrombocytopenia resulting from mutations in ETV6 or RUNX1 using a human pluripotent stem cell model. Stem Cell Rep 16:1458–1467
71. Bigot T, Gabinaud E, Hannouche L et al (2023) Single-cell analysis of megakaryopoiesis in peripheral CD34+ cells: insights into ETV6-related thrombocytopenia. J Thromb Haemost 21:2528–2544
72. Hart A, Melet F, Grossfeld P et al (2000) Fli-1 is required for murine vascular and megakaryocytic development and is hemizygously deleted in patients with thrombocytopenia. Immunity 13:167–177
73. Zhou C, Uluisik R, Rowley JW et al (2022) Germline ETV6 mutation promotes inflammation and disrupts lymphoid development of early hematopoietic progenitors. Exp Hematol 112–113:24–34
74. Cools J, Mentens N, Odero MD et al (2002) Evidence for position effects as a variantETV6-mediated leukemogenic mechanism in myeloid leukemias with a t(4;12)(q11-q12;p13) or t(5;12)(q31;p13). Blood 99:1776–1784
75. Tosi S, Mostafa Kamel Y, Owoka T et al (2015) Paediatric acute myeloid leukaemia with the t(7;12)(q36;p13) rearrangement: a review of the biological and clinical management aspects. Biomark Res 3:21
76. von Bergh ARM, van Drunen E, van Wering ER et al (2006) High incidence of t(7;12)(q36;p13) in infant AML but not in infant ALL, with a dismal outcome and ectopic expression ofHLXB9. Genes Chromosomes Cancer 45:731–739
77. Wang T, Chen X, Hui S et al (2020) Ectopia associated MN1 fusions and aberrant activation in myeloid neoplasms with t(12;22)(p13;q12). Cancer Gene Ther 27:810–818
78. Brady SW, Roberts KG, Gu Z et al (2022) The genomic landscape of pediatric acute lymphoblastic leukemia. Nat Genet 54:1376–1389
79. Romana SP, Coniat ML, Berger R (1994) t(12;21): A new recurrent translocation in acute lymphoblastic leukemia. Genes Chromosomes Cancer 9:186–191
80. Romana S, Poirel H, Leconiat M et al (1995) High frequency of t(12;21) in childhood B-lineage acute lymphoblastic leukemia. Blood 86:4263–4269
81. Golub TR, Barker GF, Bohlander SK et al (1995) Fusion of the TEL gene on 12p13 to the AML1 gene on 21q22 in acute lymphoblastic leukemia. Proc Natl Acad Sci USA 92:4917–4921
82. Shurtleff SA, Buijs A, Behm FG et al (1995) TEL/AML1 fusion resulting from a cryptic t(12;21) is the most common genetic lesion in pediatric ALL and defines a subgroup of patients with an excellent prognosis. Leukemia 9:1985–1989
83. Hahm C, Han SH, Mun YC et al (2014) ETV6/RUNX1 rearrangement identified by RT-PCR without evidence on FISH. Acta Haematol 132:122–124
84. Mueller SB, Pikman Y, Tasian SK et al (2023) *ETV6* fusions from insertions of exons 3-5 in

pediatric hematologic malignancies. Haematologica 108:3471–3476
85. Vaskova M, Mejstrikova E, Kalina T et al (2005) Transfer of genomics information to flow cytometry: expression of CD27 and CD44 discriminates subtypes of acute lymphoblastic leukemia. Leukemia 19:876–878
86. Ross ME, Zhou X, Song G et al (2003) Classification of pediatric acute lymphoblastic leukemia by gene expression profiling. Blood 102:2951–2959
87. Jeha S, Choi J, Roberts KG et al (2021) Clinical significance of novel subtypes of acute lymphoblastic leukemia in the context of minimal residual disease–directed therapy. Blood Cancer Discov 2:326–337
88. Woerden NLR, Pieters R, Loonen AH et al (2000) TEL/AML1 gene fusion is related to in vitro drug sensitivity for L-asparaginase in childhood acute lymphoblastic leukemia. Blood 96:1094–1099
89. Frost B-M, Forestier E, Gustafsson G et al (2004) Translocation t(12;21) is related to in vitro cellular drug sensitivity to doxorubicin and etoposide in childhood acute lymphoblastic leukemia. Blood 104:2452–2457
90. Lee SHR, Yang W, Gocho Y et al (2023) Pharmacotypes across the genomic landscape of pediatric acute lymphoblastic leukemia and impact on treatment response. Nat Med 29:170–179
91. Raetz EA, Bhojwani D, Devidas M et al (2023) Children's Oncology Group blueprint for research: acute lymphoblastic leukemia. Pediatr Blood Cancer 70:e30585
92. Pieters R, de Groot-Kruseman H, Van der Velden V et al (2016) Successful therapy reduction and intensification for childhood acute lymphoblastic leukemia based on minimal residual disease monitoring: study ALL10 from the Dutch Childhood Oncology Group. J Clin Oncol 34:2591–2601
93. Marcotte EL, Spector LG, Mendes-de-Almeida DP et al (2021) The prenatal origin of childhood leukemia: potential applications for epidemiology and newborn screening. Front Pediatr 9:639479
94. de Bruijn M, Dzierzak E (2017) Runx transcription factors in the development and function of the definitive hematopoietic system. Blood 129:2061–2069
95. Song W-J, Sullivan MG, Legare RD et al (1999) Haploinsufficiency of CBFA2 causes familial thrombocytopenia with propensity to develop acute myelogenous leukaemia. Nat Genet 23:166–175
96. Lichtinger M, Hoogenkamp M, Krysinska H et al (2010) Chromatin regulation by RUNX1. Blood Cells Mol Dis 44:287–290
97. Wray JP, Deltcheva EM, Boiers C et al (2022) Regulome analysis in B-acute lymphoblastic leukemia exposes Core Binding Factor addiction as a therapeutic vulnerability. Nat Commun 13:7124
98. Teppo S, Laukkanen S, Liuksiala T et al (2016) Genome-wide repression of eRNA and target gene loci by the ETV6-RUNX1 fusion in acute leukemia. Genome Res 26:1468–1477
99. Hiebert SW, Sun W, Davis JN et al (1996) The t(12;21) translocation converts AML-1B from an activator to a repressor of transcription. Mol Cell Biol 16:1349–1355
100. Wiemels JL, Ford AM, Van Wering ER et al (1999) Protracted and variable latency of acute lymphoblastic leukemia after TEL-AML1 gene fusion in utero. Blood 94:1057–1062
101. Maia AT, Ford AM, Jalali GR et al (2001) Molecular tracking of leukemogenesis in a triplet pregnancy. Blood 98:478–482
102. Ford AM, Bennett CA, Price CM et al (1998) Fetal origins of the TEL-AML1 fusion gene in identical twins with leukemia. Proc Natl Acad Sci USA 95:4584–4588
103. Wiemels JL, Cazzaniga G, Daniotti M et al (1999) Prenatal origin of acute lymphoblastic leukaemia in children. Lancet 354:1499–1503
104. McHale CM, Wiemels JL, Zhang L et al (2003) Prenatal origin of TEL-AML1–positive acute lymphoblastic leukemia in children born in California. Genes Chromosomes Cancer 37:36–43
105. Taub JW, Konrad MA, Ge Y et al (2002) High frequency of leukemic clones in newborn screening blood samples of children with B-precursor acute lymphoblastic leukemia. Blood 99:2992–2996
106. Maia AT, Tussiwand R, Cazzaniga G et al (2004) Identification of preleukemic precursors of hyperdiploid acute lymphoblastic leukemia in cord blood. Genes Chromosomes Cancer 40:38–43
107. Gruhn B, Taub JW, Ge Y et al (2008) Prenatal origin of childhood acute lymphoblastic leukemia, association with birth weight and hyperdiploidy. Leukemia 22:1692–1697
108. Eguchi-Ishimae M, Eguchi M, Ishii E et al (2001) Breakage and fusion of the TEL (ETV6) gene in immature B lymphocytes induced by apoptogenic signals. Blood 97:737–743
109. Mori H, Colman SM, Xiao Z et al (2002) Chromosome translocations and covert leukemic clones are generated during normal fetal development. Proc Natl Acad Sci USA 99:8242–8247
110. Schäfer D, Olsen M, Lähnemann D et al (2018) Five percent of healthy newborns have an ETV6-RUNX1 fusion as revealed by DNA-based GIPFEL screening. Blood 131:821–826
111. Sommarin MNE, Olofzon R, Palo S et al (2023) Single-cell multiomics of human fetal hematopoiesis define a developmental-specific population and a fetal signature. Blood Adv 7:5325–5340
112. Li Z, Chang T-C, Junco JJ et al (2023) Genomic landscape of Down syndrome–associated acute lymphoblastic leukemia. Blood 142:172–184
113. Sharma G, Tran TM, Bansal I et al (2023) RNA binding protein IGF2BP1 synergizes with ETV6-RUNX1 to drive oncogenic signaling in B-cell Acute Lymphoblastic Leukemia. J Exp Clin Cancer Res 42:231
114. Morrow M, Horton S, Kioussis D et al (2004) TEL-AML1 promotes development of specific hemato-

poietic lineages consistent with preleukemic activity. Blood 103:3890–3896
115. Andreasson P, Schwaller J, Anastasiadou E et al (2001) The expression of ETV6/CBFA2 (TEL/AML1) is not sufficient for the transformation of hematopoietic cell lines in vitro or the induction of hematologic disease in vivo. Cancer Genet Cytogenet 130:93–104
116. Andersson A, Olofsson T, Lindgren D et al (2005) Molecular signatures in childhood acute leukemia and their correlations to expression patterns in normal hematopoietic subpopulations. Proc Natl Acad Sci USA 102:19069–19074
117. Kantner H-P, Warsch W, Delogu A et al (2013) ETV6/RUNX1 induces reactive oxygen species and drives the accumulation of DNA damage in B cells. Neoplasia 15:1292–1300
118. Schindler JW, Van Buren D, Foudi A et al (2009) TEL-AML1 corrupts hematopoietic stem cells to persist in the bone marrow and initiate leukemia. Cell Stem Cell 5:43–53
119. van der Weyden L, Giotopoulos G, Rust AG et al (2011) Modeling the evolution of ETV6-RUNX1–induced B-cell precursor acute lymphoblastic leukemia in mice. Blood 118:1041–1051
120. Morrow M, Samanta A, Kioussis D et al (2007) TEL-AML1 preleukemic activity requires the DNA binding domain of AML1 and the dimerization and corepressor binding domains of TEL. Oncogene 26:4404–4414
121. Tsuzuki S, Seto M (2013) TEL (ETV6)-AML1 (RUNX1) initiates self-renewing fetal pro-B cells in association with a transcriptional program shared with embryonic stem cells in mice. Stem Cells 31:236–247
122. Fischer M, Schwieger M, Horn S et al (2005) Defining the oncogenic function of the TEL/AML1 (ETV6/RUNX1) fusion protein in a mouse model. Oncogene 24:7579–7591
123. Tsuzuki S, Seto M, Greaves M et al (2004) Modeling first-hit functions of the t(12;21) TEL-AML1 translocation in mice. Proc Natl Acad Sci USA 101:8443–8448
124. Ford AM, Palmi C, Bueno C et al (2009) The TEL-AML1 leukemia fusion gene dysregulates the TGF-β pathway in early B lineage progenitor cells. J Clin Invest 119:826–836
125. Fan D, Zhou X, Li Z et al (2015) Stem cell programs are retained in human leukemic lymphoblasts. Oncogene 34:2083–2093
126. Papaemmanuil E, Rapado I, Li Y et al (2014) RAG-mediated recombination is the predominant driver of oncogenic rearrangement in ETV6-RUNX1 acute lymphoblastic leukemia. Nat Genet 46:116–125
127. Swaminathan S, Klemm L, Park E et al (2015) Mechanisms of clonal evolution in childhood acute lymphoblastic leukemia. Nat Immunol 16:766–774
128. Mertz TM, Collins CD, Dennis M et al (2022) APOBEC-induced mutagenesis in cancer. Annu Rev Genet 56:229–252
129. Rodríguez-Hernández G, Opitz FV, Delgado P et al (2019) Infectious stimuli promote malignant B-cell acute lymphoblastic leukemia in the absence of AID. Nat Commun 10:5563
130. Kasar S, Kim J, Improgo R et al (2015) Whole-genome sequencing reveals activation-induced cytidine deaminase signatures during indolent chronic lymphocytic leukaemia evolution. Nat Commun 6:8866
131. Machado HE, Mitchell E, Øbro NF et al (2022) Diverse mutational landscapes in human lymphocytes. Nature 608:724–732
132. Rodríguez-Hernández G, Hauer J, Martín-Lorenzo A et al (2017) Infection exposure promotes ETV6-RUNX1 precursor B-cell leukemia via impaired H3K4 demethylases. Cancer Res 77:4365–4377
133. Rodríguez-Hernández G, Casado-García A, Isidro-Hernández M et al (2021) The second oncogenic hit determines the cell fate of ETV6-RUNX1 positive. Leukemia 9:704591
134. Hong D, Gupta R, Ancliff P et al (2008) Initiating and cancer-propagating cells in TEL-AML1-associated childhood leukemia. Science 319:336–339
135. Böiers C, Richardson SE, Laycock E et al (2018) A human IPS model implicates embryonic B-myeloid fate restriction as developmental susceptibility to B acute lymphoblastic leukemia-associated ETV6-RUNX1. Dev Cell 44:362–377.e7
136. Takeuchi S, Bartram CR, Wada M et al (1995) Allelotype analysis of childhood acute lymphoblastic leukemia. Cancer Res 55:5377–5382
137. Sato Y, Suto Y, Pietenpol J et al (1995) TEL and KIP1 define the smallest region of deletions on 12pl3 in hematopoietic malignancies. Blood 86:1525–1533
138. Cave H, Gerard B, Martin E et al (1995) Loss of heterozygosity in the chromosomal region 12p12-13 is very common in childhood acute lymphoblastic leukemia and permits the precise localization of a tumor-suppressor gene distinct from p27KIP1. Blood 86:3869–3875
139. Stegmaier K, Pendse S, Barker GF et al (1995) Frequent loss of heterozygosity at the TEL gene locus in acute lymphoblastic leukemia of childhood. Blood 86:38–44
140. Kim DH, Moldwin RL, Vignon C et al (1996) TEL-AML1 translocations with TEL and CDKN2 inactivation in acute lymphoblastic leukemia cell lines. Blood 88:785–794
141. Raynaud S, Cave H, Baens M et al (1996) The 12;21 translocation involving TEL and deletion of the other TEL allele: two frequently associated alterations found in childhood acute lymphoblastic leukemia. Blood 87:2891–2899
142. Gunji H, Waga K, Nakamura F et al (2004) TEL/AML1 shows dominant-negative effects over TEL as well as AML1. BBRC 322:623–630
143. Barnett KR, Mobley RJ, Diedrich JD et al (2023) Epigenomic mapping reveals distinct B cell acute lymphoblastic leukemia chromatin architectures and regulators. Cell Genom 3.100442

144. Torrano V, Procter J, Cardus P et al (2011) ETV6-RUNX1 promotes survival of early B lineage progenitor cells via a dysregulated erythropoietin receptor. Blood 118:4910–4918
145. Iacobucci I, Li Y, Roberts KG et al (2016) Truncating erythropoietin receptor rearrangements in acute lymphoblastic leukemia. Cancer Cell 29:186–200
146. Polak R, Bierings MB, van der Leije CS et al (2019) Autophagy inhibition as a potential future targeted therapy for ETV6-RUNX1-driven B-cell precursor acute lymphoblastic leukemia. Haematologica 104:738–748
147. Lilljebjörn H, Henningsson R, Hyrenius-Wittsten A et al (2016) Identification of ETV6-RUNX1-like and DUX4-rearranged subtypes in paediatric B-cell precursor acute lymphoblastic leukaemia. Nat Commun 7:11790
148. Zaliova M, Kotrova M, Bresolin S et al (2017) ETV6/RUNX1-like acute lymphoblastic leukemia: a novel B-cell precursor leukemia subtype associated with the CD27/CD44 immunophenotype. Genes Chromosomes Cancer 56:608–616
149. Gu Z, Churchman ML, Roberts KG et al (2019) PAX5-driven subtypes of B-progenitor acute lymphoblastic leukemia. Nat Genet 51:296–307
150. Ryan SL, Peden JF, Kingsbury Z et al (2023) Whole genome sequencing provides comprehensive genetic testing in childhood B-cell acute lymphoblastic leukaemia. Leukemia 37:518–528
151. Shah S, Schrader KA, Waanders E et al (2013) A recurrent germline PAX5 mutation confers susceptibility to pre-B cell acute lymphoblastic leukemia. Nat Genet 45:1226–1231
152. Auer F, Rüschendorf F, Gombert M et al (2014) Inherited susceptibility to pre B-ALL caused by germline transmission of PAX5 c.547G>A. Leukemia 28:1136–1138
153. Cazzaniga G, Daniotti M, Tosi S et al (2001) The paired box domain gene PAX5 is fused to ETV6/TEL in an acute lymphoblastic leukemia case. Cancer Res 61:4666–4670
154. Strehl S, König M, Dworzak MN et al (2003) PAX5/ETV6 fusion defines cytogenetic entity dic(9;12)(p13;p13). Leukemia 17:1121–1123
155. Mullighan CG, Goorha S, Radtke I et al (2007) Genome-wide analysis of genetic alterations in acute lymphoblastic leukaemia. Nature 446:758–764
156. Fazio G, Palmi C, Rolink A et al (2008) PAX5/TEL acts as a transcriptional repressor causing down-modulation of CD19, enhances migration to CXCL12, and confers survival advantage in pre-BI cells. Cancer Res 68:181–189
157. Kawamata N, Pennella MA, Woo JL et al (2012) Dominant-negative mechanism of leukemogenic PAX5 fusions. Oncogene 31:966–977
158. Fazio G, Cazzaniga V, Palmi C et al (2013) PAX5/ETV6 alters the gene expression profile of precursor B cells with opposite dominant effect on endogenous PAX5. Leukemia 27:992–995
159. Vicente-Dueñas C, Janssen S, Oldenburg M et al (2020) An intact gut microbiome protects genetically predisposed mice against leukemia. Blood 136:2003–2017
160. Roberts KG, Li Y, Payne-Turner D et al (2014) Targetable kinase-activating lesions in Ph-like acute lymphoblastic leukemia. N Engl J Med 371:1005–1015
161. Tzankov A, Reichard KK, Hasserjian RP et al (2023) Updates on eosinophilic disorders. Virchows Arch 482:85–97
162. Saft L, Kvasnicka HM, Boudova L et al (2023) Myeloid/lymphoid neoplasms with eosinophilia and tyrosine kinase fusion genes: a workshop report with focus on novel entities and a literature review including paediatric cases. Histopathology 83:829–849
163. Golub TR, Barker GF, Lovett M et al (1994) Fusion of PDGF receptor β to a novel ets-like gene, tel, in chronic myelomonocytic leukemia with t(5;12) chromosomal translocation. Cell 77:307–316
164. Golub TR, Goga A, Barker GF et al (1996) Oligomerization of the ABL tyrosine kinase by the Ets protein TEL in human leukemia. Mol Cell Biol 16:4107–4116
165. Million RP, Harakawa N, Roumiantsev S et al (2004) A direct binding site for Grb2 contributes to transformation and leukemogenesis by the Tel-Abl (ETV6-Abl) tyrosine kinase. Mol Cell Biol 24:4685–4695
166. Barjesteh van Waalwijk van Doorn-Khosrovani S, Spensberger D, de Knegt Y et al (2005) Somatic heterozygous mutations in ETV6 (TEL) and frequent absence of ETV6 protein in acute myeloid leukemia. Oncogene 24:4129–4137
167. Bejar R, Stevenson K, Abdel-Wahab O et al (2011) Clinical effect of point mutations in myelodysplastic syndromes. N Engl J Med 364:2496–2506
168. Wang Q, Dong S, Yao H et al (2014) ETV6 mutation in a cohort of 970 patients with hematologic malignancies. Haematologica 99:e176–e178
169. Gurney M, Chekkaf I, Baranwal A et al (2023) The clinical and molecular spectrum of ETV6 mutated myeloid neoplasms. Br J Haematol 202(2):279–283
170. Bernard E, Tuechler H, Greenberg PL et al (2022) Molecular international prognostic scoring system for myelodysplastic syndromes. NEJM Evid 1:EVIDoa2200008
171. Biswas A, Rajesh Y, Mitra P et al (2020) ETV6 gene aberrations in non-haematological malignancies: a review highlighting ETV6 associated fusion genes in solid tumors. Biochim Biophys Acta 1874:188389
172. Perez MW, Sias-Garcia O, Daramola A et al (2021) Defining the transcriptional control of pediatric AML highlights RARA as a superenhancer-regulated druggable dependency. Blood Adv 5:4864–4876
173. Xu W, Tian F, Tai X, et al (2024) *ETV6::ACSL6* translocation-driven super-enhancer activation leads to eosinophilia in acute lymphoblastic leukemia through IL-3 overexpression. Online ahead of print
174. Ballabio E, Cantarella CD, Federico C et al (2009) Ectopic expression of the HLXB9 gene is associated

with an altered nuclear position in t(7;12) leukaemias. Leukemia 23:1179–1182

175. Nilsson T, Waraky A, Östlund A et al (2022) An induced pluripotent stem cell t(7;12)(q36;p13) acute myeloid leukemia model shows high expression of MNX1 and a block in differentiation of the erythroid and megakaryocytic lineages. Int J Cancer 151:770–778
176. Heuser M, Argiropoulos B, Kuchenbauer F et al (2007) MN1 overexpression induces acute myeloid leukemia in mice and predicts ATRA resistance in patients with AML. Blood 110:1639–1647
177. Imren S, Heuser M, Gasparetto M et al (2014) Modeling de novo leukemogenesis from human cord blood with MN1 and NUP98HOXD13. Blood 124:3608–3612
178. Riedel SS, Lu C, Xie HM et al (2021) Intrinsically disordered Meningioma-1 stabilizes the BAF complex to cause AML. Mol Cell 81:2332–2348.e9
179. Heuser M, Yun H, Berg T et al (2011) Cell of origin in AML: susceptibility to MN1-induced transformation is regulated by the MEIS1/AbdB-like HOX protein complex. Cancer Cell 20:39–52
180. Riedel SS, Haladyna JN, Bezzant M et al (2016) MLL1 and DOT1L cooperate with meningioma-1 to induce acute myeloid leukemia. J Clin Invest 126:1438–1450
181. Libbrecht C, Xie HM, Kingsley MC et al (2021) Menin is necessary for long term maintenance of meningioma-1 driven leukemia. Leukemia 35:1405–1417
182. Klossowski S, Miao H, Kempinska K et al (2020) Menin inhibitor MI-3454 induces remission in *MLL1*-rearranged and *NPM1*-mutated models of leukemia. J Clin Invest 130:981–997
183. Issa GC, Aldoss I, DiPersio J et al (2023) The menin inhibitor revumenib in KMT2A-rearranged or NPM1-mutant leukaemia. Nature 615:920–924
184. Coustan-Smith E, Mullighan CG, Onciu M et al (2009) Early T-cell precursor leukaemia: a subtype of very high-risk acute lymphoblastic leukaemia. Lancet Oncol 10:147–156
185. Van Vlierberghe P, Ambesi-Impiombato A, Perez-Garcia A et al (2011) ETV6 mutations in early immature human T cell leukemias. J Exp Med 208:2571–2579
186. Zhang J, Ding L, Holmfeldt L et al (2012) The genetic basis of early T-cell precursor acute lymphoblastic leukaemia. Nature 481:157–163
187. Homminga I, Pieters R, Langerak AW et al (2011) Integrated transcript and genome analyses reveal NKX2-1 and MEF2C as potential oncogenes in T cell acute lymphoblastic leukemia. Cancer Cell 19:484–497
188. Strehl S, Nebral K, König M et al (2008) ETV6-NCOA2: a novel fusion gene in acute leukemia associated with coexpression of T-lymphoid and myeloid markers and frequent NOTCH1 mutations. Clin Cancer Res 14:977–983
189. Fishman H, Madiwale S, Geron I et al (2022) ETV6-NCOA2 fusion induces T/myeloid mixed-phenotype leukemia through transformation of nonthymic hematopoietic progenitor cells. Blood 139:399–412
190. Chapuy B, Stewart C, Dunford AJ et al (2018) Molecular subtypes of diffuse large B cell lymphoma are associated with distinct pathogenic mechanisms and outcomes. Nat Med 24:679–690
191. Schmitz R, Wright GW, Huang DW et al (2018) Genetics and pathogenesis of diffuse large B-cell lymphoma. N Engl J Med 378:1396–1407
192. Venturutti L, Rivas MA, Pelzer BW et al (2023) An aged/autoimmune B-cell program defines the early transformation of extranodal lymphomas. Cancer Discov 13:216–243
193. Flümann R, Rehkämper T, Nieper P et al (2021) An autochthonous mouse model of Myd88- and BCL2-driven diffuse large B-cell lymphoma reveals actionable molecular vulnerabilities. Blood Cancer Discov 2:70–91
194. Flümann R, Hansen J, Pelzer BW et al (2023) Distinct genetically determined origins of Myd88/BCL2-driven aggressive lymphoma rationalize targeted therapeutic intervention strategies. Blood Cancer Discov 4:78–97
195. Radke J, Ishaque N, Koll R et al (2022) The genomic and transcriptional landscape of primary central nervous system lymphoma. Nat Commun 13:2558
196. Chapuy B, Roemer MGM, Stewart C et al (2016) Targetable genetic features of primary testicular and primary central nervous system lymphomas. Blood 127:869–881
197. Bonzheim I, Sander P, Salmerón-Villalobos J et al (2022) The molecular hallmarks of primary and secondary vitreoretinal lymphoma. Blood Adv 6:1598–1607
198. Kodgule R, Chen J, Khonde P et al (2022) Recurrent switch 2 domain RAC2 mutations in intravascular large B-cell lymphoma. Blood Adv 6:6051–6055
199. Gonzalez-Farre B, Ramis-Zaldivar JE, Castrejón de Anta N et al (2023) Intravascular large B-cell lymphoma genomic profile is characterized by alterations in genes regulating NF-κB and immune checkpoints. Am J Surg Pathol 47:202
200. Biswas A, Rajesh Y, Mitra P et al (2020) ETV6 gene aberrations in non-haematological malignancies: a review highlighting ETV6 associated fusion genes in solid tumors. Biochim Biophys Acta Rev Cancer 1874:188389
201. Knezevich SR, McFadden DE, Tao W et al (1998) A novel ETV6-NTRK3 gene fusion in congenital fibrosarcoma. Nat Genet 18:184–187
202. Bourgeois JM, Knezevich SR, Mathers JA et al (2000) Molecular detection of the ETV6-NTRK3 gene fusion differentiates congenital Fibrosarcoma from other childhood spindle cell tumors. Am J Surg Pathol 24:937
203. Sheng WQ, Hisaoka M, Okamoto S et al (2001) Congenital-infantile fibrosarcoma. A clinicopathologic study of 10 cases and molecular detection of

the ETV6-NTRK3 fusion transcripts using paraffin-embedded tissues. Am J Clin Pathol 115:348–355
204. Vokuhl C, Nourkami-Tutdibi N, Furtwängler R et al (2018) ETV6-NTRK3 in congenital mesoblastic nephroma: a report of the SIOP/GPOH nephroblastoma study. Pediatr Blood Cancer 65:e26925
205. Rubin BP, Chen CJ, Morgan TW et al (1998) Congenital mesoblastic nephroma t(12;15) is associated with ETV6-NTRK3 gene fusion: cytogenetic and molecular relationship to congenital (infantile) fibrosarcoma. Am J Pathol 153:1451–1458
206. Wu G, Diaz AK, Paugh BS et al (2014) The genomic landscape of diffuse intrinsic pontine glioma and pediatric non-brainstem high-grade glioma. Nat Genet 46:444–450
207. Xu T, Wang H, Huang X et al (2018) Gene fusion in malignant glioma: an emerging target for next-generation personalized treatment. Transl Oncol 11:609–618
208. Bastos AU, de Jesus AC, Cerutti JM (2018) ETV6-NTRK3 and STRN-ALK kinase fusions are recurrent events in papillary thyroid cancer of adult population. Eur J Endocrinol 178:83–91
209. Seethala RR, Chiosea SI, Liu CZ et al (2017) Clinical and morphologic features of ETV6-NTRK3 translocated papillary thyroid carcinoma in an adult population without radiation exposure. Am J Surg Pathol 41:446–457
210. Leeman-Neill RJ, Kelly LM, Liu P et al (2014) ETV6-NTRK3 is a common chromosomal rearrangement in radiation-associated thyroid cancer. Cancer 120:799–807
211. Tognon C, Knezevich SR, Huntsman D et al (2002) Expression of the ETV6-NTRK3 gene fusion as a primary event in human secretory breast carcinoma. Cancer Cell 2:367–376
212. Guilmette J, Dias-Santagata D, Nosé V et al (2019) Novel gene fusions in secretory carcinoma of the salivary glands: enlarging the ETV6 family. Hum Pathol 83:50–58
213. Bishop JA, Taube JM, Su A et al (2017) Secretory carcinoma of the skin harboring ETV6 gene fusions: a cutaneous analogue to secretory carcinomas of the breast and salivary glands. Am J Surg Pathol 41:62
214. Wai DH, Knezevich SR, Lucas T et al (2000) The ETV6-NTRK3 gene fusion encodes a chimeric protein tyrosine kinase that transforms NIH3T3 cells. Oncogene 19:906–915
215. Alassiri AH, Ali RH, Shen Y et al (2016) ETV6-NTRK3 is expressed in a subset of ALK-negative inflammatory myofibroblastic tumors. Am J Surg Pathol 40:1051
216. Yamamoto H, Yoshida A, Taguchi K et al (2016) ALK, ROS1 and NTRK3 gene rearrangements in inflammatory myofibroblastic tumours. Histopathology 69:72–83
217. Lonsdale J, Thomas J, Salvatore M et al (2013) The genotype-tissue expression (GTEx) project. Nat Genet 45:580–585
218. Uhlén M, Fagerberg L, Hallström BM et al (2015) Tissue-based map of the human proteome. Science 347:1260419
219. Hong DS, DuBois SG, Kummar S et al (2020) Larotrectinib in patients with TRK fusion-positive solid tumours: a pooled analysis of three phase 1/2 clinical trials. Lancet Oncol 21:531–540
220. Doebele RC, Drilon A, Paz-Ares L et al (2020) Entrectinib in patients with advanced or metastatic NTRK fusion-positive solid tumours: integrated analysis of three phase 1–2 trials. Lancet Oncol 21:271–282
221. Skalova A, Vanecek T, Martinek P et al (2018) Molecular profiling of mammary analog secretory carcinoma revealed a subset of tumors harboring a novel ETV6-RET translocation: report of 10 cases. Am J Surg Pathol 42:234–246
222. Black M, Liu CZ, Onozato M et al (2020) Concurrent identification of novel EGFR-SEPT14 fusion and ETV6-RET fusion in secretory carcinoma of the salivary gland. Head Neck Pathol 14:817–821
223. Rooper LM, Karantanos T, Ning Y et al (2018) Salivary secretory carcinoma with a novel ETV6-MET fusion: expanding the molecular Spectrum of a recently described entity. Am J Surg Pathol 42:1121
224. Dickson BC, Swanson D, Charames GS et al (2018) Epithelioid fibrous histiocytoma: molecular characterization of ALK fusion partners in 23 cases. Mod Pathol 31:753–762
225. Lee S, Hu Y, Loo SK et al (2020) Landscape analysis of adjacent gene rearrangements reveals BCL2L14–ETV6 gene fusions in more aggressive triple-negative breast cancer. Proc Natl Acad Sci USA 117:9912–9921
226. The UniProt Consortium (2021) UniProt: the universal protein knowledgebase in 2021. Nucleic Acids Res 49:D480–D489
227. Duployez N, Abou Chahla W, Lejeune S et al (2018) Detection of a new heterozygous germline ETV6 mutation in a case with hyperdiploid acute lymphoblastic leukemia. Eur J Haematol 100:104–107
228. Junk SV, Klein N, Schreek S et al (2019) TP53, ETV6 and RUNX1 germline variants in a case series of patients developing secondary neoplasms after treatment for childhood acute lymphoblastic leukemia. Haematologica 104:e402–e405
229. Kozubik KS, Radova L, Reblova K et al (2021) Functional analysis of germline ETV6 W380R mutation causing inherited thrombocytopenia and secondary acute lymphoblastic leukemia or essential thrombocythemia. Platelets 32:838–841

Function of PML-RARA in Acute Promyelocytic Leukemia

14

Ugo Testa and Elvira Pelosi

Abstract

The transformation of acute promyelocytic leukemia (APL) from the most fatal to the most curable subtype of acute myeloid leukemia (AML), with long-term survival exceeding 90%, has represented one of the most exciting successes in hematology and in oncology. APL is a paradigm for oncoprotein-targeted cure.

APL is caused by a 15/17 chromosomal translocation which generates the PML-RARA fusion protein and can be cured by the chemotherapy-free approach based on the combination of two therapies targeting PML-RARA: retinoic acid (RA) and arsenic. PML-RARA is the key driver of APL and acts by deregulating transcriptional control, particularly RAR targets involved in self-renewal or myeloid differentiation, also disrupting PML nuclear bodies. PML-RARA mainly acts as a modulator of the expression of specific target genes: genes whose regulatory elements recruit PML-RARA are not uniformly repressed but also may be upregulated or remain unchanged. RA and arsenic trioxide directly target PML-RARA-mediated transcriptional deregulation and protein stability, removing the differentiation block at promyelocytic stage and inducing clinical remission of APL patients.

Keywords

Acute promyelocytic leukemia · Retinoic acid · All-trans-retinoic acid (ATRA) · Arsenic trioxide (ATO) · Gene regulation · Transcription · Differentiation

Abbreviations

AF-1	activation function 1
AF-2	activation function 2
AML	acute myeloid leukemia
APL	acute promyelocytic leukemia
ATO	arsenic trioxide
ATRA	all-trans-retinoic acid
cAMP	cyclic adenyl monophosphate
CRABP1	cellular retinoic acid binding protein 1
CREs	Cis-regulatory regions
CSFR	colony-stimulating factor receptor
DBD	DNA binding domain
DR	DNA repeat
EYA2	eyes absent homolog 2
EZH2	enhancer of Zeste homolog 2 (EZH2)
FISH	fluorescent in situ hybridization

U. Testa (✉) · E. Pelosi
Department of Oncology, Istituto Superiore di Sanità, Rome, Italy
e-mail: ugo.testa@iss.it

T. Borggrefe, B. D. Giaimo (eds.), *Transcription factors in blood cell development*, Advances in Experimental Medicine and Biology 1459, https://doi.org/10.1007/978-3-031-62731-6_14

GO	gemtuzumab ozigomicin
HDAC	histone deacetylase
HPC	hematopoietic progenitor cell
HSC	hematopoietic stem cell
HSCT	hematopoietic stem cell transplantation
HSPCs	hematopoietic stem/progenitor cells
IDA	idarubicin
IFR	interferon response factor
IFR2BP	interferon response 2 binding protein
JNK	c-Jun N-terminal kinase
LBD	ligand binding domain
NBs	nuclear bodies
NCoR	nuclear receptor corepressor
NES	nuclear export signal
NFkB	nuclear factor kappa chain enhancer of activated B-cells
NGS	next-generation sequencing
NLS	nuclear localization signal
PKCA	protein kinase-C A
PLZF	promyelocytic leukemia zinc finger
PML	promyelocytic leukemia protein
PRC1	polycomb repressive complex 1
RA	retinoic acid
RARA	retinoic acid receptor alfa
RARB	retinoic acid receptor beta
RARG	retinoic acid receptor gamma
RBCC	RINGB-box-coiled coil
RT-PCR	reverse transcriptase-polymerase chain reaction
RXR	retinoid X receptor
SIM	SUMO-interacting motif
SMRT	silencing mediator for retinoid thyroid hormone receptor
STAT	signal transducer and activator of transcription
TRADD	tumor necrosis factor 1-associated via death domain
VCP	valosin-containing protein
VPA	valproic acid

14.1 Introduction

Acute promyelocytic leukemia (APL) is a distinct and peculiar subtype of acute myeloid leukemia (AML), characterized by a myeloid differentiation block at the level of myeloid precursors, with consequent accumulation in bone marrow of leukemic promyelocytes. This leukemia is defined by a single pathognomonic genetic alteration consisting in the chromosomal translocation 15;17, with consequent formation of the oncogenic fusion gene *PML-RARA* encoding the fusion protein PML-RARA (promyelocytic leukemia-retinoic acid receptor alpha). In a minority of cases (<5%), APL may harbor fusion of *RARA* gene with other partners, such as *PLZF*, *NPM1*, *NUMA1*, and *STAT5B*.

The discovery of the molecular alteration involved in APL has completely revolutionized the treatment of this AML subtype. Three key discoveries have contributed to unravel the molecular genetics of APL: (i) typical chromosomal translocation t(15;17) present in most APL patients [1], (ii) clinical study showing a peculiar sensitivity to retinoic acid (RA) of an AML subtype characterized by the accumulation of promyelocytes [2] and the translocation t(15;17) at molecular level through analysis of the *RARA* gene with identification of the fusion of the *RARA* gene with a novel gene [3, 4].

Four evidences supported a key pathogenetic role of PML-RARA fusion protein in the genesis of APL:

(i) The t(15;17) is very frequently the only cytogenetic abnormality observed in APL.
(ii) The introduction of *PML-RARA* (transgenic mice-expressing PML-RARA) is able to initiate a leukemic process in mouse models, generating a leukemic condition recapitulating the main clinical and morphological features of the human disease [5].
(iii) The expression of PML-RARA fusion protein inhibits differentiation and promotes survival of myeloid precursors [6].
(iv) Overexpression of PML-RARA fusion in normal human progenitor/stem cells induces rapid differentiation to the promyelocytic stage followed by maturation arrest which is abolished by retinoic acid [7].

In the in vivo model of transgenic PML-RARA mice, the onset of APL occurs after a long latency and with low penetrance, suggesting that APL development is a multistep process requiring other cooperating genetic alterations. In line with this interpretation, during the preleukemic phase of the disease, PML-RARA expression induced DNA damage in hematopoietic stem cells (HSCs)/hematopoietic progenitor cells (HPCs) and activated DNA repair [8].

All these studies provided clear evidence that the PML-RARA fusion protein is the master driver of APL leukemogenesis.

Before the introduction of the current therapy based on the administration of all-trans-retinoic acid (ATRA) and arsenic trioxide (ATO), APL was the most dangerous AML subtype due to the high frequency of fatal bleeding occurring during chemotherapy induction [9]. With the introduction of the differentiation induction therapy with ATRA and ATO, the outcome of APL patients has strongly improved with a long-term survival rate around 95% [10]. A therapeutic algorithm was now established for the treatment of APL patients, worldwide accepted (Fig. 14.1).

In 98% of cases (typical APL), APLs are related to the t(15;17) translocation with generation of the *PML-RARA* fusion gene. In about 1% of cases, APLs are related to the t(11;17) translocation, generating the *PLZF-RARA* fusion gene (atypical APL). In the remaining 1% of cases (atypical APL), the *RARA* gene or the *RARB* or *RARG* genes are fused to other genetic partners (Table 14.1).

14.2 RARA and PML: The Two Constituents of the PML-RARA Fusion Protein

RARA gene is located on chromosome 21 (17q21) and is composed by 10 exons encoding two isoforms RARA1 and RARA2, differing between them in the N-terminal region. RARA, together with two other receptors retinoic acid receptor beta (RARB) and retinoic acid receptor gamma (RARG), are the physiological nuclear receptors of vitamin A. The RARA protein is a member of nuclear receptor superfamily and exhibits a high homology (90%) with RARB and RARG. RARs exert their biological activity as obligate heterodimers with retinoid X receptor (RXR): both RARs and RXRs exhibit a conserved modular structure with a N-terminal ligand-independent activation function (AF-1), a central conserved DNA-binding domain (DBD) and a C-terminal ligand-binding domain (LBD). The multifunctional LBD is required for ligand binding and dimerization and contains a ligand-dependent activation function (AF-2) which contains a motif that can be modulated by ligand (RA) [11] (Fig. 14.2). RARs are obligate partners with RXRs which form heterodimers. In contrast, RXR is not an obligate RAR partner in that it can form heterodimers with other nuclear receptors.

The biological activity of the RAR-RXR heterodimer is modulated by the ligand RA that, after binding to the receptor, induces a conformational change. The activated RAR-RXR acts as a modulator of the transcriptional activity of target genes characterized by the presence of a DNA motif (AGGTCA core motif) known as retinoic acid response elements (RAREs) [12]. Activated RARs can either activate or repress transcription of target genes. When RARs are not activated by ligand, the transcriptional activity of target genes is repressed by corepressor complexes bound at the level of promoter regions. The corepressors that most frequently interact with RAR-RXR heterodimers are represented by the nuclear receptor corepressor (NCoR) and the silencing mediator for retinoid and thyroid hormone receptors (SMRT). These corepressors are able to recruit histone deacetylases (HDACs) [13].

RARs play an important role in embryogenesis and in physiology of many tissues. Concerning the hematopoietic system, RARA is required for normal myeloid differentiation [14]. Studies in mouse have shown that the emergence of mouse HSCs from hemogenic endothelium requires RAR signaling [15], and in adult mouse, RA signaling is required for HSC quiescence [16].

PML gene is located on chromosome 15 (15q24), contains nine exons, and generates several alternative spliced transcripts. The various PML isoforms have a common N-terminal region containing a RINGB-box-coiled-coil (RBCC)

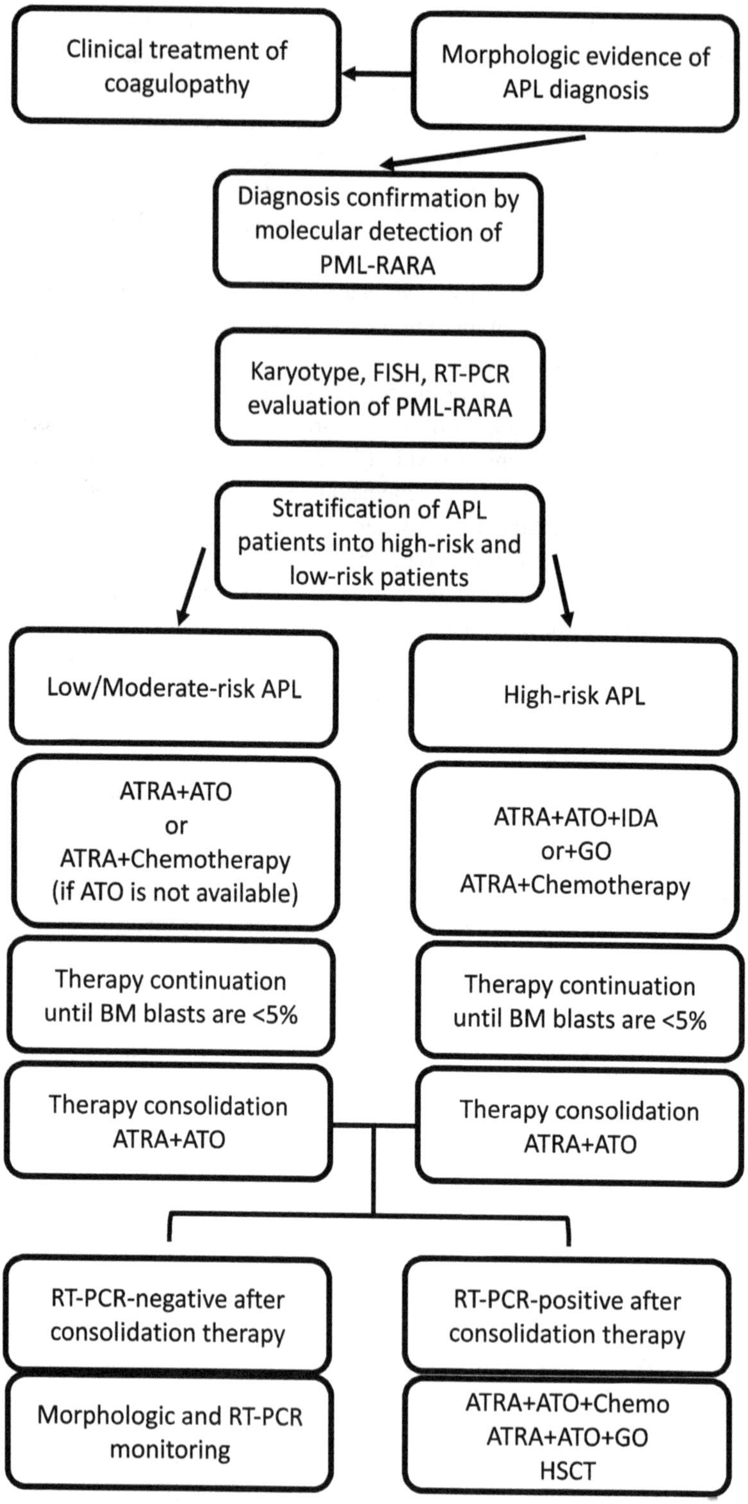

Fig. 14.1 APL current treatment algorithms. A patient with suspect APL immediately undergoes clinical treatment for coagulopathy and is evaluated through molecular test for confirmation of APL diagnosis. Patients positive for PML-RARA fusion protein are then stratified into two groups: a low-moderate-risk group undergoing treatment with ATRA+ATO or ATRA + chemotherapy and a high-risk group treated with ATRA+ATO+Idarubicin (IDA) or gemtuzumab ozogamicin (GO). The patients are treated until their BM count decreases to >5%. After a consolidation therapy based on a ATRA+ATO regimen, the patients are evaluated by RT-PCR for residual disease. According to the negativity or positivity to the RT-PCR test, the positive patients are treated with additional ATRA-based regimens up to hematopoietic stem cell transplantation (HSCT), while the negative patients are just monitored by blood morphological analyses and RT-PCR detection of PML-RARA

Table 14.1 Genetic and clinical features of the most recurrent typical and atypical APL

Diagnosis	Karyotype	Fusion genes	Frequency (%)	Drug sensitivity				Prognosis
				ATRA	ATO	Chemo	Combi	
Typical APL	t(15;17)(q22;q21)	PML-RARA	98	S	S	S	S	Good
Atypical APL	t(11;17)(11q23;q21)	PLZF-RARA	1	R	R	S	S	Poor
Atypical APL	t(17;17)(q21;q21)	STAT5B-RARA	Rare	R	R	S	S	Poor
Atypical APL	t(5;17)(5q35;q21)	NPM1-RARA	Rare	R	R	S	S	Poor
Atypical APL	t(1;17)(q42;q21)	IRF2BP2-RARA	Rare	S	U	S	U	Poor
Atypical APL	t(3;3)(q24;q26)/inv.(3)	TBLR1-RARB	Rare	R	U	U	S	Poor
Atypical APL	t(12;12)(q13;q15)	CPSF6-RARG	Rare	R	R	S	S	Poor
Atypical APL	t(11;12)(p15;q13)	NUP98-RARG	Rare	R	R	S	U	Poor

S sensitive, *R* resistant, *U* undetermined

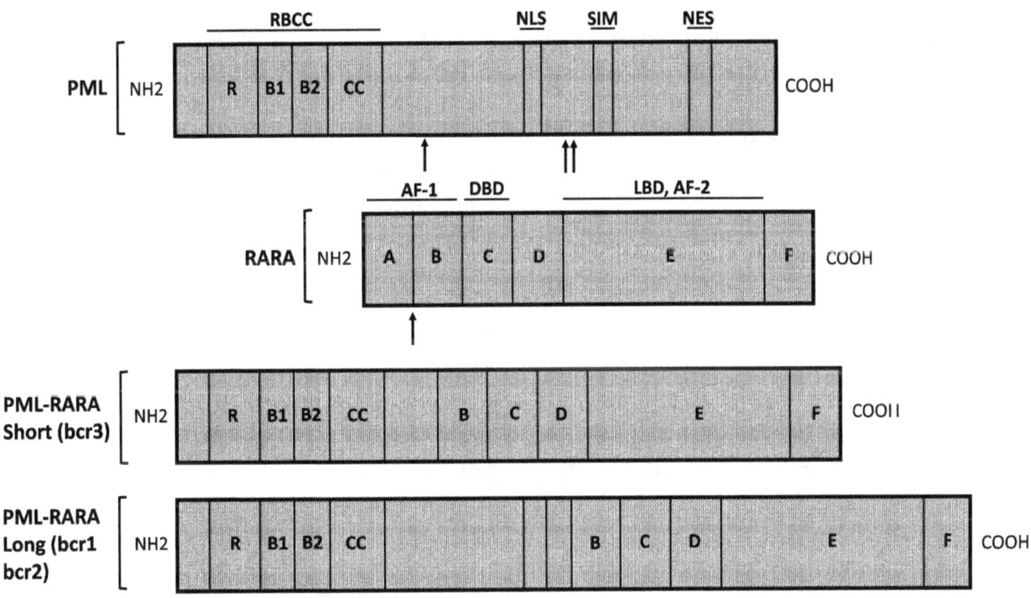

Fig. 14.2 Structure of PML, RARA, and PML-RARA fusion protein. Arrows indicate the breakpoint regions in RARA (at the level of amino acid residue 60) and in PML (at the level of amino acid residues 394 or 552, generating the bcr3, bcr1, and bcr2 fusion proteins). In PML, protein structural domains are represented by RING finger (R), B boxes (B1 and B2), coiled-coil domain (CC) (all together forming the RBCC motif), nuclear localization signal (NLS), SUMO-interacting motif (SIM), and nuclear export signal (NES). In RARA, structural elements are represented by N-terminal domain (A, B), forming the activation function-1 (AF-1) motif, DNA-binding domain (DBD, C), the ligand binding domain (LBD, E), and other regulatory domains (D and F)

motif (Fig. 14.2) but differ in the central and C-terminal regions [17].

PML protein is localized at the level of nuclear bodies (NBs) and spherical structures localized in the nucleus and involved in the regulation of complex biological processes, such as stem cell quiescence and self-renewal and senescence [18]. PML is capable of interacting with more than 170 different proteins: these interactions are usually mediated by the RBCC domain, which induces multimerization of PML molecules and organization at the level of NBs [18]. A posttranslational modification controls the recruitment in NBs of several proteins. The posttranslational modification of these PML partners determines their sequestration in NBs or their activation [19].

14.3 PML-RARA Isoforms

The t(15;17) translocation generates a fusion protein composed of a NH2 terminal region corresponding to a part of the *PML* gene containing all the RBCC domain and a COOH terminal region corresponding to a part of the *RARA* gene containing the DBD, the LBD, and the AF-2 (Fig. 14.2). Different points of the chromosome breakpoints of the t(15;17) generate different PML-RARA isoforms. Three different typical PML-RARA isoforms were reported, generated by three different breakpoints within the *PML* locus: in intron 6, generating the bcr1 isoform (L, long isoform); in exon 6, generating the bcr2 isoform (V, variant isoform); and in intron 3, generating the bcr3 isoform (S, short isoform) [20]. Only one breakpoint within intron 2 was identified in the *RARA* gene.

In addition to the typical PML-RARA isoforms, sporadic atypical isoforms were observed in a minority of APL patients, generated by atypical breakpoints [20].

The possible consequences of the different PML-RARA isoforms in APL biology and their possible prognostic implications are unclear. Therefore, the latest guidelines on APL treatment do not recommend any therapeutic change based on PML-RARA isoforms [21].

The t(15;17) generates two chimeric genes *PML-RARA* and *RARA-PML*. Therefore, it is not surprising that, while 1005 patients harboring the t(15;17) express a PML-RARA isoform, the reciprocal RARA-PML fusion is observed in 70–80% of APL patients; therefore, the corresponding RARA-PML protein does not seem to be required for APL development [22].

14.4 Mutational Landscape of APLs

Although PML-RARA represents the main genetic driver of APL development, additional mutations (co-mutations) are observed in APL samples that could cooperate with PML-RARA in driving leukemia development [23, 24]. Primary APLs harbor an average of eight nonsilent somatic mutations per exome [24]. The most recurrently mutated genes in APL are represented by *FLT3*, with 27% of cases with *FLT3-ITD* (internal tandem duplication) and 16% with *FLT3* point mutations (in a minority of cases, *FLT3-ITD* and *FLT3* point mutations coexist in the same APL samples). Other genes recurrently mutated in APL were *WT1* (14%), *NRAS*(10%), *ARID1A* (5%), *KRAS* (4%), *ETV6* (1%), and *EZH2* (1%). Mutations in genes of the RAS pathway predominantly occurred in cases with WT-*FLT3*. *FLT3-ITD* and *FLT3* point mutations rarely overlap with other mutations. About 40% of APL patients display an alteration of *FLT3* [24]. The mutational profile of APLs was significantly different from that observed in other AML subtypes in that (i) *FLT3-ITD* and *FLT3* point mutations are more frequent in APLs than in non-APL AMLs and (ii) mutations in well-known leukemia-related genes, such as *DNMT3A*, *NPM1*, *TET2*, *IDH1*, *IDH2*, and *ASXL1*, frequently mutated in non-APL AMLs, are absent or rarely mutated in APLs [24]. Analysis of APL patients relapsing after ATRA or ATO therapy showed that some mutations, such as *PML*, *RARA*, *RUNX1*, and *ARID1B*, are more frequent at relapse than in primary APLs [24].

In line with these findings in primary APLs, studies in PML-RARA transgenic mice have supported a functional cooperation between *PML-RARA* and *FLT3-ITD* genes in promoting APL development in mouse. Retroviral transduction of FLT3-ITD into bone marrow cells derived from PML-RARA transgenic mice resulted in the development with short latency of an APL-like disease with complete penetrance [25].

A recent study reported in APL patients the occurrence of frequent mutations in the noncoding genome at the level of cis-regulatory regions

(CREs) [26]. Mutations at the level of CREs occur preferentially in active enhancers bound by some master transcription factors (TFs) in APL, such as MYB, PD1, IRF1, GFI1, and RUNX1 [26]. A recurrently mutated region is located within the third intron of *WT1*, a gene playing a key role in the control of normal and malignant hematopoiesis. The *WT1* gene encodes for a TF that plays an important role in the control of cell growth, differentiation, and apoptosis. WT1 protein acts regulating the activity of other genes by binding to specific DNA regions. The WT1 protein is highly expressed in leukemia. In 169 APL patients, somatic *WT1* CRE mutations were observed in 3.6% of cases and germline *WT1* CRE mutations in 4.7% of cases [26]. Biallelic *WT1* inactivation was frequently observed in APL patients with noncoding *WT1* variants [26]. Germline *WT1* CRE mutations suggest the occurrence of a germline polymorphism associated with an increased risk of APL development [26]. At the level of mechanism, noncoding *WT1* variants disrupt the binding of MYB on chromatin and through this mechanism determine a decrease of *WT1* expression [26].

14.5 Molecular Mechanism of the PML-RARA Oncoprotein

14.5.1 PML-RARA as a Transcriptional Repressor

The fundamental biological effects induced by PML-RARA fusion protein consist in a block of cell differentiation and in a stimulation of cell proliferation. A consistent number of studies in suitable experimental models have elucidated the molecular mechanisms mediating these biological activities of PML-RARA. PML-RARA exerts its activity acting as an altered TF that represses the expression of these target genes [27]. This transcriptional repressive activity is largely dependent on the capacity of PML-RARA to interact with other protein forming a molecular complex that includes as binding partners RXR and PML [27]. This complex promotes the recruitment of corepressors and, through this mechanism, represses the transcription of its target genes [27] (Fig. 14.3). These repressors are mainly represented by HDACs [28]. In line with this finding, pharmacological inhibition of HDACs with inhibitors such as valproic acid (VPA) induces remission in transgenic models of therapy-resistant APL [29, 30]. On the other hand, pharmacological doses of retinoic acid promote significant changes of histone 3 acetylation, thus inducing a modulation of the epigenetic landscape of APL cells [31]. On the other side, VPA increases the sensitivity of APL cells to retinoic acid [32]. It is important to note that VPA, in contrast to ATRA and ATO, does not induce degradation of PML-RARA and activates in APL cells a program of cell death [33].

PLZF-RARA fusions, at variance with *PML-RARA*, are not sensitive to ATRA, and this different responsiveness seems to be related to different interactions with corepressors: in fact, PLZF-RARA, but not PML-RARA, forms via its PLZF moiety corepressor complexes which are insensitive to ATRA [34].

A recent study showed that HDAC3 is highly expressed in APL cells and at protein level associates with PML-RARA; HDAC3 induces a deacetylation of PML-RARA at the level of lysine 394, reducing its sumoylation and subsequent ubiquitylation. Particularly, the reduced PIAS1-mediated sumoylation of PML-RARA determines a reduced ubiquitylation of this protein by RNF4. HDAC3 inhibitors promote PML-RARA ubiquitylation and degradation and reduce the expression of PML-RARA in both wild-type and ATRA- or ATO-resistant APL cells [35]. These observations suggest that HDAC3 is not only a repressor of transcription but also a positive regulator of PML-RARA oncoprotein through its deacetylation and inhibition of its degradation.

Interestingly, a recent study suggested a differential effect of different HDAC inhibitors on APL cells: in fact, in mouse models of APL, it was shown that the class I inhibitor VPA in association with ATRA targets the bulk leukemic population, while the pan-HDAC inhibitor Vorinostat targets the population of leukemia-

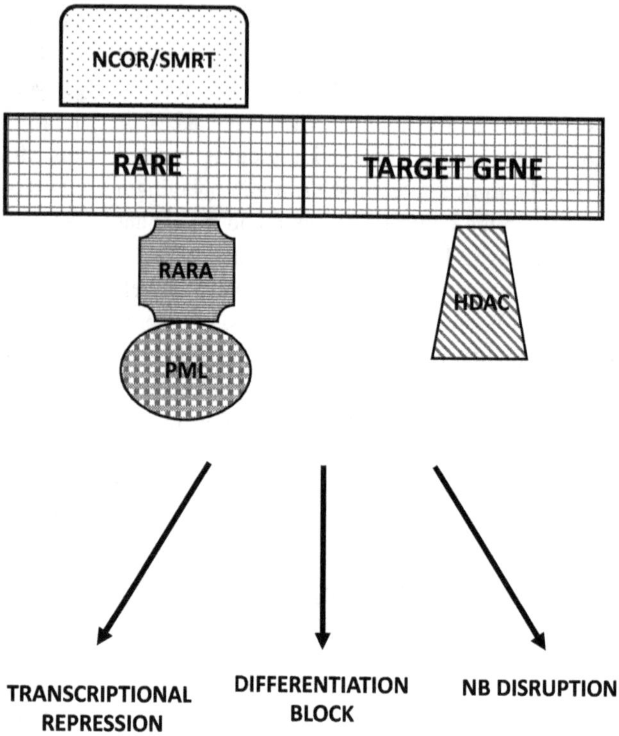

Fig. 14.3 Molecular mechanism of PML-RARA-mediated transcriptional repression. The binding PML-RARA to RAREs present in RA-responsive genes interferes with the transcription of these genes and disrupts nuclear bodies and blocks granulocytic differentiation. The PML-RARA protein induces the recruitment of corepressors (NCOR/SMART) to orchestrate the silencing of the transcription of genes required for granulocytic differentiation

initiating cells [36]. The triplet ATRA+VPA+Vorinostat provided the optimal therapeutic benefit [36].

14.5.2 PML-RARA Sumoylation

PML-RARA sumoylation is essential for leukemic transforming capacity, as directly supported by experiments in transgenic mice generated using a PML-RARA mutated at the level of the sumoylation site [37].

PML as well as PML-RARA sumoylation requires the activity of the SUMO E3-ligase inhibitor PIAS1; PIAS1 suppression abrogates the ability of arsenic trioxide to induce the degradation of PML-RARA and the apoptosis of APL cells [38].

The PML-RARA sumoylation is required to induce the degradation of PML-RARA protein by arsenic trioxide: in fact, arsenic trioxide induces PML sumoylation which in turn triggers its Lys48-linked polyubiquitination and proteasome-dependent degradation [39]. This process requires the recruitment of sumoylated PML of two proteins: RING finger protein 4 (RNF4), a member of the family of SUMO targeted ubiquitin ligase, containing multiple SUMO interacting motifs and a RING domain [39], and p97/valosin-containing protein (VCP) segregase which is required for the extraction of the ubiquitinated proteins from PML bodies prior to ubiquitin-mediated proteolysis [40].

14.5.3 PML-RARA as a Direct Regulator of Gene Transcription

PML-RARA fusion protein possesses other important properties related to the capacity to interact with RXR. In fact, in APL cells, PML-RARA is constantly bound to RXR, and this heterodimerization with RXR is essential for PML-RARA-mediated transcriptional control and for its in vivo leukemogenic activity [41, 42].

The RXR binding capacity is essential also to confer to PML-RARA the ability to directly modulate gene expression through binding to specific DNA motifs (direct repeat, DR). In fact, Martens et al. identified genome-wide association of RXR with PML-RARA at the level of the PML-RARA response elements: in addition to response elements, such as DR2 and DR5, PML-RARA binds also at the level of DR1, DR3, and DR5 motifs [31]. This extended DNA binding capacity of PML-RARA compared to RARA implies a gain of DNA binding capacity as an essential feature of PML-RARA-mediated transformation [31]. ATRA treatment induces a loss of most of the PML-RARA/RXR binding sites through changes in histone acetylation. These observations suggest that PML-RARA-RXR complexes modulate gene expression through binding to specific DNA regions, producing local changes of histone H3 acetylation [31].

Other studies supported a direct modulatory effect of PML-RARA through its binding at the level of specific DNA regions. Thus, Wang et al. explored genome-wide binding of PML-RARA binding using chromatin immunoprecipitation followed by deep sequencing (ChIP-Seq) in a PML-RARA-inducible cell model (U937 cells transduced with *PML-RARA* gene. Using this approach, they showed that PML-RARA predominantly targets gene promoters regulated by the TF PU.1 through both protein-protein interactions and DNA binding through RARE (DAR) half sites forming complex binding motifs [43]. These promoters regulate genes with a key role in hematopoiesis. Most of these PML-RARA targets are genes whose expression is inhibited by PML-RARA and reactivated following ATRA treatment [43]. PML-RARA binding sites were observed also at the level of other key regulators of hematopoiesis, including *AP1*, *GATA2*, *RUNX1*, *RUNX3*, and *GFI1* genes [31].

Tan et al. have further characterized the DNA targets of PML-RARA binding and defined two types of PML-RARA targets that diverged in function during leukemogenesis: repressed and activated genes [44]. Most of the genes repressed by PML-RARA are uniquely expressed. In the mature myeloid progeny, genes activated by PML-RARA mainly correspond to genes uniquely expressed in promyelocytes and GM-progenitors. Repressive and activating functions of PML-RARA were attributed to distinct HDAC1 and E1A binding protein p300/lysine acetyltransferase 3B (EP300/KAT3B) binding patterns: HDAC1. And EP300/KAT3B are counterplayers that can control histone acetylation levels, determining either transcriptional repression or activation. Activated genes are related to the binding of PML-RARA in association with the HDAC1 corepressor and the EP300/KAT3B coactivator and by the formation of superenhancers (303 superenhancers activated by PML-RARA were identified), while repressed genes are characterized by the binding of PML-RARA associated only with HDAC1 [44] (Fig. 14.4). Using this approach, *GFI1* was identified as a gene essential for maintenance of APL phenotype and whose expression is activated by PML-RARA through binding at the level of a superenhancer element present in the *GFI1*gene [44]. Knockdown of GFI1 expression in APL cells induced granulocytic differentiation and a decrease in the binding of PML-RARA at the level of activated genes [44].

MYB is another gene whose expression is activated by PML-RARA through a transcriptional regulation related to the binding of PML-RARA at the level of *MYB* gene promoter with consequent increase of the acetylation of the promoter region of this gene [45]. Increased MYB expression in APL cells stimulates cell proliferation and inhibits cell differentiation.

Recent studies provided evidence that the DNA binding capacity of PML-RARA plays an essential role in mediating its leukemogenic activity. In fact, the ability of PML-RARA to promote aberrant self-renewal of leukemic cells is dependent upon its ability to bind its DNA targets, in that PML-RARA mutated at the level of its DNA binding domain of RARA was unable to promote leukemic self-renewal [46]. This observation supported the view that PML-RARA could promote the initiation of APL formation through its binding at the level of specific DNA regions, thus dysregulating the expression of genes that are essential for self-renewal control [46].

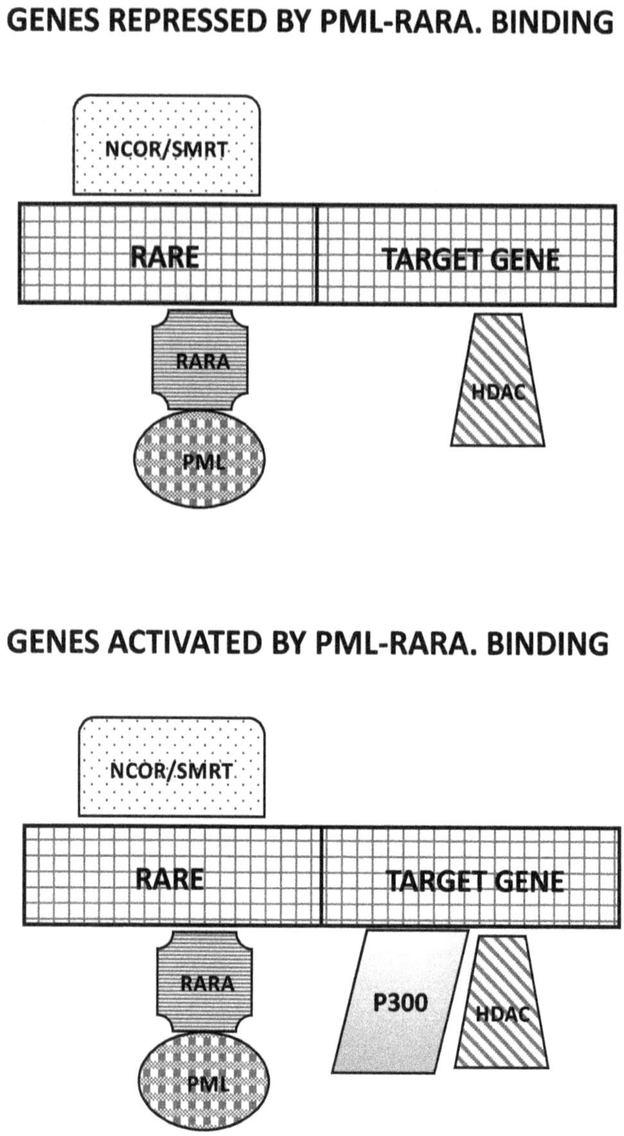

Fig. 14.4 Differential gene regulation exerted by PML-RARA on target genes is dictated by the binding of either HDAC1 alone resulting in gene repression or of HDAC1 and EP300/KAT3B resulting in gene activation

A recent study adopted a new strategy for obtaining unequivocal evidence on the identification of the genes bound by PML-RARA, based on the generation of an epitope-tagged PML-RARA retroviral construct. Using this approach, it was possible to identify, within genomes of primary murine hematopoietic progenitor cells, 67 binding sites associated with concurrent changes in DNA accessibility and expression by single-cell RNA sequencing, thus supporting the definition as direct target genes [47]. The extension of these studies to purified normal human CD34+ HSPCs led to the identification of 14 genes bound and coordinately dysregulated by PML-RARA. These PML-RARA-regulated genes involve some genes highly relevant for hematopoiesis such as *GATA2*, *RUNX1*, *SPI1*, and *SOCS2* [47]. Importantly, through those gene expression analyses, it was also possible to identify 224 genes whose expression is dysregulated by PML-RARA (127 upregulated and 97 down-regulated) in human hematopoietic stem/progenitor cells (HSPCs). These genes were not directly bound by PML-RARA, thus suggesting that their expression is influenced by PML-RARA through indirect mechanisms [47]. These studies have elucidated the mechanisms through which PML-RARA induces a reprogramming of gene expression in hematopoietic stem progenitor cells (HSPCs) as a fundamental step in the leukemogenic activity of this fusion protein.

Villiers et al. have explored the potential mechanisms for the differential regulation of gene expression induced by PML-RARA binding at the level of RAREs. Thus, genes whose regulatory elements recruit PML-RARA can be repressed, upregulated, or not modified [48]. The mechanisms regulating this differential gene expression regulation by PML-RARA are complex and can be analyzed using multiomics and a machine learning approach, called regulatory element behavior extraction learning, able to deconvolute the complex local mechanisms at the level of transcription machinery regulating gene expression at PML-RARA bound positions [48].

14.5.4 Nuclear Bodies and Their Destruction by PML-RARA

In APL cells, PML-RARA fusion protein alters the nuclear structure of NBs determining their disruption into nuclear microspeckles. NBs are membraneless insoluble structures present in the nucleus and recruiting and assembling many partner proteins. As above discussed, PML protein is localized in NBs and represents a crucial component of these subnuclear structures. In fact, PML is involved in NB biogenesis, maintenance of its integrity, and the recruitment and localization of about 100 proteins in this structure [49, 50]. The posttranslational modifications of PML and, particularly, its sumoylation are key steps in the dynamic formation of NBs and support the fundamental role of this protein in their generation [19, 51]. PML RING domains play an essential role in NB formation. These domains induce a tetramerization of PML molecules which is required for NB assembly and PML sumoylation. In fact, the tetramerized PML acquires the property to bind UBC9, the SUMO E2 enzyme [52]. RING tetramerization and PML-RARA sumoylation are required for leukemogenetic activity in PML-RARA-driven transgenic mice.

PML-RARA expression in APL cells determines an important impact on nuclear architecture, inducing the disruption of NBs and their transformation into nuclear microspeckles, a feature considered as a diagnostic hallmark of APL [53]. This change in nuclear architecture was considered as a key step in APL development [53].

Voisset et al. have evaluated in an experimental model the potential leukemogenic effect of NB disruption [54]. To this end, they have generated a knock-in mouse model with NB disruption induced by two-point mutations at the level of the RING domain (C62A/C65A, $PML^{C62A/C65A}$) of PML: those mice developed no leukemia. However, the expression in those mice also of a concomitant RARA linked to a dimerization domain (p50-RARA) displayed a doubling in the rate of leukemia formation, with a reduced latency period [54]. The analysis of primary $PML^{C62A/C65A}$ cells generated into transgenic mice displayed mutagenic properties, as evidenced by increased sister-chromatid exchange and chromosome abnormalities. Furthermore, these cells exhibited impaired homologous recombination and nonhomologous end-joining repair pathways [54]. These observations supported the view that NB disruption contributes to APL pathogenesis.

14.6 Molecular Mechanisms of ATRA and ATO Resistance: Role of PML-RARA Mutations

As above discussed, the dramatic progresses achieved in the last 30 years in the molecular pathogenesis of APL have contributed to completely revolutionize its treatment moving to molecular targeted therapies. Two key steps of this process have been represented by the demonstration that ATRA induces granulocytic differentiation of APL cells and ATO promotes PML-RARA fusion protein degradation. These two drugs synergize in their antileukemic effects. This new therapeutic approach was widely shown to induce complete remission in more than 90%

of patients, with a five-year disease-free survival of about 80%. However, about 10–20% of APL patients are either refractory or relapse after the two standard APL treatments (ATO+ATRA or ATRA+ chemotherapy), and growing evidence showed that a consistent number of these patients display point mutations in the PML or RARA moieties of the PML-RARA fusion protein.

PML mutations are more frequently observed in relapsed APL patients receiving ATO treatment. ATO binds directly to cysteine residues present in the RBCC domain of PML-RARA. Initial studies have shown A216V and L218P mutations in the PML moiety of PML-RARA fusion molecule, observed in two APL patients developing resistance to ATO [55]. A subsequent study reported other mutations in addition to A216V, including S214L, A216T, L217F, and S220G in APL patients with resistance to ATO [56]. All these mutations occurred in the PML-B2 domain of PML-RARA preventing ATO binding and inducing resistance to this drug [56]. B1 and B2 PML boxes are cysteine-/histidine-rich protein motifs, able to bind Zn^{2+}, required for NB formation and protein dimerization. This observation suggests the existence of a *PML* mutational hot spot domain (C212-S220) in the B2 PML domain, involved in arsenic binding to PML-RARA. These mutations were observed in a group of APL patients receiving ATO treatment in monotherapy as induction treatment and the in vitro analysis of these PML-RARA mutants showed that A216V, S214L, and A216T mutations inhibit the response of PML-RARA to ATO, while L217F and S220G mutations display only a weak inhibitory effect [57]. In vitro ATO resistance could be bypassed using higher ATO concentrations or combining ATO with ATRA [57].

Iaccarino et al. reported an extensive molecular characterization of 30 relapsing APL patients: in 15 patients relapsing after ATRA+chemotherapy treatment, no PML mutations were observed and *RARA* mutations (located in LBD) were observed in 2/15 patients. In 15 APL patients relapsing after ATO+ATRA treatment, *PML* mutations were observed in 6/15 patients and *RARA* mutations in 3/15 patients [58]. The *PML* mutations were observed in three patients only in PML-RARA, in one patient in the un-rearranged *PML* allele, and in one case PML and RARA moieties of PML-RARA [58, 59].

A recent study elucidated the molecular mechanism of resistance to ATO of two *PML* mutants: The A216T PML was almost completely resistant to ATO and did not recruit the SUMO targeted ubiquitin E3 ligase RNF4. The L217F was modified by SUMO and recruits RNF4 but was unable to develop the ubiquitin conjugate required for the recruitment of the segregase p97, essential for ATO-induced PML-RARA degradation [60].

Mutations of the *RARA* moiety of PML-RARA are frequently observed in relapsed APL patients treated with ATRA+chemotherapy or ATRA+ATO. An initial study reported two patients acquiring ATRA resistance after treatment with ATRA: both these patients displayed point mutations at the level of the LBD of the RARA moiety of PML-RARA [61]. Gallagher et al. reported in about one-third of ATRA-treated, relapsing APL patients PML-RARA mutations at the level of the RARA LBD: in 4/6 of these patients, *RARA* mutations emerged off-ATRA treatment, suggesting an active role of these mutations in ATRA resistance [62]. In a subsequent study, the same authors reported PML-RARA mutations at the level of the LBD of RARA moiety in 40% of relapsing APL patients treated with ATRA+ chemotherapy, and 39% of these mutations emerged off-ATRA selection pressure [63].

Lehmann-Che et al. have analyzed by next-generation sequencing (NGS) 64 APL patients relapsing after treatment with ATRA+ chemotherapy and showed that these patients can be subdivided into two subgroups: One subgroup displays at diagnosis a high incidence of comutations affecting epigenetic or transcriptional regulators such as *WT1* or activating mitogen-activated protein kinase (MAPK) signaling and at relapse retain these mutations and acquire mutations inhibiting therapy response, such as *RARA* and Cytosolyc 5′ nucleotidase (*NT5C2*), encoding a 5′ nucleotidase enzyme that is responsible for the inactivation of nucleoside analog chemotherapy drugs. The other subgroup primarily shows *FLT3*-activating mutations at diagnosis that are lost at relapse, thus suggesting that these relapses originate from ancestral preleukemic PML-RARA-expressing cells that survived ATRA+ chemotherapy [64].

14.7 Variant APL: Different Partners of RARA May Modify the Sensitivity to ATRA and ATO

About 2% of APLs are associated with atypical rearrangements, where the *RARA* gene is fused to partners other than *PML* or where *RARB* or *RARG* genes are fused with different partners [65] (Table 14.1). Unfortunately, the gene fusions involved in variant APLs frequently generate fusion proteins that are not sensitive to ATRA nor to ATO and poorly respond to standard treatments.

Many APL variants involve the fusion of the *RARA* gene with different partners. APL variant with PLFZ-RARA fusion gene is the most recurrent APL variant representing about 1% of APLs. PLZF is a zinc finger TF playing an important role in the control of HSCs' self-renewal and differentiation. *PLZF* gene is located at the level of chromosome 11 (11q23), and the translocation t(11;17)(q23;q21) generates the two fusion proteins PLZF-RARA and RARA-PLZF. Two isoforms of PLZF-RARA were identified [66]. Similarly to what observed for PML-RARA, PLZF-RARA blocks myeloid differentiation and promotes leukemia development [67]. However, at variance with PML-RARA which increases retinoic acid sensitivity of hematopoietic precursors, PLZF-RARA exerts an inhibitory effect on ATRA-induced myeloid differentiation and inhibits RARA- and RARG-dependent genes [68]. A 120 amino acid N-terminal motif, called the POZ domain-mediating protein-protein interactions, is essential for the inhibitory activity of PLZF-RARA on retinoic acid-dependent myeloid differentiation [68]. PLZF-RARA forms homodimers or heterodimers with RXRA; PLZF-RARA-RXR heterodimers are able to bind RARE elements present at the level of target genes, where they recruit various types of corepressors, resulting in an inhibitory effect on the expression of RARA and RARG-dependent genes [69, 70].

PLZF-RARA variant APLs are resistant to ATRA treatment, and this property seems to be related to the capacity of the PLZF-RARA fusion protein to form, via its PLZF moiety, corepressor complexes which are unsensitive to ATRA [34]. Rego et al. have explored the response to ATRA or to ATO in transgenic mice models of APL induced by PML-RARA or by PLZF-RARA: PML-RARA-driven leukemias were responsive to ATRA or to ATO or to both agents in association, while PLZF-RARA-driven leukemias were resistant. Surprisingly, in vitro and in vivo studies showed that ATO+ ATRA induced the degradation of both PML-RARA and PLZF-RARA fusion oncoproteins [71]. These observations supported the conclusion that (i) the type of fusion RARA oncoprotein is a key dominant of the response to ATRA and ATO and its role in the maintenance of the leukemic phenotype is different for PML-RARA and for PLZF-RARA; ATRA+ ATO is a very efficacious treatment strategy for classical PML-RARA APL but not for variant PLZF-RARA APL [71].

Interestingly, experimental studies have shown two potential strategies to restore an ATRA sensitivity in PLZF-RARA-positive APL cells: (i) an HDAC inhibitor, inducing accumulation of acetylated histones, inhibited the transcriptional repression elicited by PLZF-RARA [72]; (ii) B-CPT-cAMP, an activator of PKCA inducing the phosphorylation of PLZF-RARA molecule at the level of Ser 765, inhibits the association between PLZF-RARA and corepressors [73].

In addition to PLZF-RARA also RARA-PLZF, the reciprocal translocation product significantly contributed to the generation of the APL variant leukemic phenotype. In fact, studies in transgenic mice showed that PLZF-RARA transgenic mice develop leukemia that lacks the differentiation inhibition at the promyelocytic stage [74]. However, PLZF-RARA/RARA-PLZF double transgenic mice develop leukemia with classical features of APL [75].

At transcriptional level, RARA-PLZF interacted with C/EBPA gene and recruited the HDAC1 corepressor to induce histone H3 deacetylation and then decreased expression of C/EBPA target genes required for myeloid differentiation [76]. RARA-PLZF, but not PLZF-RARA, overcomes PLZF-mediated repression of CRABP1 (cellular retinoic acid binding protein I) through binding to this gene, recruitment of EP300/KAT3B, induction of promoter hypomethylation, and CRABLI gene upregulation [77]. This mechanism contributes to the development of retinoid resistance observed in PLZF-RARA APL variant [77].

Given the rarity of PLZF-RARA-positive APLs, only one study reported the analysis of the mutational spectrum observed in seven patients [78]. Concerning the myeloid-specific genes, PLZF-RARA APLs displayed frequent *TET2* (4/7), *RUNX1*, and *CSF3R* (2/7 patients, each); only 1/7 patients displayed *FLT3-TKD* mutations, while *FLT3-ITD* mutations were not observed [78]. Interestingly, an extended mutational analysis showed particularly frequent (5/7) *ARID1A* mutations. One of these patients displayed mutations in three genes belonging to the SWI/SNF chromatin remodeling complexes, *ARID1A*, *ARID2*, and *SMARCA4* [78]. Importantly, studies in mice showed that ARID1A is essential for maintaining the number and the function of HSCs and its loss affects the differentiation of both myeloid and lymphoid lineages [79]. ARID1A silencing affected ATRA-induced differentiation of APL NB4 cells, supporting its role in granulocytic differentiation [79].

As it was observed for PML-RARA, also PLZF-RARA induces the selective upregulation of the expression of some genes playing an important role in the leukemogenic activity of this fusion protein. One of these genes is the transcription coactivator Eyes Absent Homolog 2 (EYA2), a target of PLZF-RARA as well as of activated PLZF [80]. The EYA2 gene is one of four EYA family genes that encode vertebrate homolog proteins of the *Drosophila melanogaster* protein eyes. Absent involved in fly eye development. EYA possesses tyrosine and threonine phosphatase activities and transactivation activity and is involved in various biological functions such as damage DNA repair, innate immunity, and cellular proliferation. The upregulation of the expression of EYA2 is involved in promoting the self-renewal capacity on hematopoietic. Stem/progenitor cells and the proliferation of leukemic cells but not their differentiation block [80].

The inappropriate recruitment of functional epigenetic repressors, such as the polycomb repressive complex 1 (PRC1), determines epigenetic imbalance at some specific genomic loci in an ATRA-insensitive manner [69]. In line with this finding, the observed ATRA-induced degradation of PLZF-RARA under ATRA treatment [71] suggests the existence of a leukemogenic event persistent even in the absence of the PLZF-RARA fusion protein. Through single-cell multiomic studies, it was identified a subset of PLZF-RARA-positive leukemic cells ATRA-resistant, with proliferation, DNA replication, and repair signatures dependent on an E2F transcriptional network targeting the epigenetic regulator EZH2 (enhancer of zeste homolog 2) [81]. PLZF induced a change in histone marks at specific enhancer genes, underlying the development of RA relapse-initiating cells. Particularly, the shift in histone marks at the level of enhancers underlined a shift in H3K27me3 enrichment from enhancers resulting developmental processes to those regulating kinase signaling and bloodstream systems [81]. In experimental models of PLZF-RARA-driven leukemia, targeting of canonical (catalytic activities of EZH2 based on histone transferase) and noncanonical (noncatalytic EZH2 oncogenic activity) EZH2 activities induces the elimination of leukemia-initiating cells [81]. Functional studies have supported the essential role of EZH2 in mediating ATRA resistance of PLZF-RARA-positive leukemic cells [81].

PLZF-RARA variant APLs are resistant to treatment with ATRA or ATO. However, a large proportion of these patients achieve CR when treated with a combination therapy based on ATRA+chemotherapy or ATO+ATRA+chemotherapy [82].

The t(17;17)(p21;p21) generates the fusion gene *STAT5B-RARA* related to the development of a rare APL variant. The two genes involved in this translocation are located both at 17p21. This translocation is generated by two cooperating events, the inversion of 17p211.2p11 and 17q21.2 interstitial microdeletion. Three different transcripts of this fusion gene were identified, generated by three different breakpoints in the *STAT5B* gene (at the level of exon 14, 15, or 16) and one breakpoint in the RARA gene (exon 3) [83].

Experimental studies have shown the capacity of STAT5B-RARA fusion protein to inhibit myeloid differentiation through a mechanism involving binding at the level of RARE present in target genes as a STAT5B-RARA homodimer or a heterodimer with RXR, resulting in binding of SMRT corepressor and subsequent inhibition of RARA-mediated transcriptional activity [84, 85]. The STAT5B-RARA-mediated repression of gene transcription requires the coiled-coil domain of STAT5 which, through its oligomerization, acquires the capacity of stably binding the corepressor SMRT, independent of ATRA stimulation [84].

APLs with STAT5B-RARA are resistant to both ATRA and ATO [86]. These APL variants can be treated with chemotherapy or alternatively with hypomethylating agents in combination with ATRA [87].

The t(5;17)(q32;q21) generates the NPM1-RARA fusion gene observed in some rare cases of variant APL [88]. The breakpoint of *NPM1* gene in NPM1-RARA is variable and most frequently involves *NPM1* at exon 4 and *RARA* at exon 3 [89]. At cytological level, NPM1-RARA APLs display a morphology different from typical APL [89], a phenomenon confirmed also in the study of NPM1-RARA transgenic mice [74]. NPM1-RARA binds to RARE present at the level of target genes as homodimers or as heterodimers with RXR and then recruits corepressors at the level of these sites, resulting in a predominantly negative regulation of RARA-regulated gene expression [90]. Retinoic acid promotes the partial dissociation of these corepressors from NPM1-RARA and induces the association of the coactivator protein receptor-associated coactivator 3 (RAC3), a member of the Rac subfamily of the Rho family of small G proteins, with consequent activation of gene transcription [90]. The site of interaction between NPM-RARA and RAC3 remained undetermined and could imply the RARA E-domain [90]. NPM1-RARA, as well as PML-RARA, promotes the block of myeloid differentiation, and for this biological activity, the interaction with RXR is required [91]. Furthermore, NPM1-RARA exerted an inhibitory effect on extrinsic apoptosis through a mechanism involving binding to TNFR1-associated death domain (TRADD), with consequent inhibition of caspase activation and activation of NF-kB and JNK signaling pathway [92, 93]. TRADD is a death domain protein containing adaptor molecule that interacts with TNFR1 and mediates programmed cell death signaling and NF-kappa B activation.

IRFBP2-RARA is a rare fusion event observed in an APL variant originated by the t(1;17)(q42.3;q21.2) with variable breakpoints generating five different types of transcripts [94–97]. The IRF2BP2-RARA fusion protein displays biological properties similar to PML-RARA: through homodimerization, this fusion protein interacts with DNA RARE elements and induces a repression of gene expression, rescued by ATRA. The IRF2BP2-RARA fusion protein transforms normal HSCs through a process that can be abrogated by ATRA [98]. At clinical level, IRF2BP2-RARA APLs are sensitive to treatment with ATRA+chemotherapy [97].

Some rare patients with variant APL display rearrangements of the RARB gene with a unique partner represented by the *TBL1XR1* gene, generating the fusion gene *TBL1XR1-RARB* [99]. At molecular level, TBL1XR1-RARB forms homodimers and exerts a dominant negative effect against RARB as well as RARA, resulting in a block of cell differentiation and stimulation of proliferation [99]. RARB-rearranged APLs are resistant to ATRA as well as to ATO [99].

Another group of rare variant APLs is represented by APL-like AMLs with a rearrangement of *RARG* gene. These patients display clinical presentation and cytomorphological and immunophenotypic features similar to those observed in typical APL patients [100]. Three more frequent partner genes of RARG were identified, including *CPSF6*, *NUP98*, and *NHRNP*$_c$ [100]. *WT1* and *NRAS/KRAS* mutations are common comutations in RARG-rearranged APLs. However, *FLT3-ITD* and *FLT3-TKD* mutations, frequent in typical APLs, are absent in RARG-rearranged APLs [100]. These patients are not responsive to ATRA or ATO and their prognosis is poor [100]. Interestingly, RARG-rearranged APLs display a peculiar profile of RNA gene expression, thus suggesting that these AMLs could represent a new AML molecular subtype [100].

14.8 Conclusion

In the last four decades, there was a complete revolution in the treatment of APL transforming this AML subtype from the most lethal to the more curable. These advances were largely due to cellular and molecular studies that have contributed to unveil the complex mechanisms through which the fusion protein PML-RARA drives the leukemic transformation. The molecular targeting of the fusion PML-RARA protein using two different drugs ATRA and ATO allowed to obtain a durable remission of most of APL patients, in large part cured of their leukemic disease without the need of bone marrow transplantation. Thus, APL is a paradigm for oncoprotein-targeted cure.

Acknowledgments This study did not receive external fundings.

References

1. Rowley JD, Golomb HM, Dougherty C (1977) 15/17 translocation, a consistent chromosomal change in acute promyelocytic leukaemia. Lancet 1:549–550
2. Huang M, Ye Y, Chen R et al (1988) Use of all trans retinoic acid in the treatment of acute promyelocytic leukemia. Blood 72:567–572
3. Borrow J, Goddart A, Sheer D et al (1990) Molecular analysis of acute promyelocytic leukemia breakpoint cluster region on chromosome 17. Science 249:1577–1580
4. De Thé H, Chomienne C, Lanotte M et al (1990) The t(15 ;17) translocation of acute promyelocytic leukemia fuses the retinoic acid receptor alpha gene to a novel transcribed locus. Nature 347:558–561
5. Grisolano JL, Wesselschmidt RL, Pelicci PG et al (1997) Altered myeloid development and acute leukemia in transgenic mice expressing PML-RAR alpha under control of cathepsin G regulatory sequences. Blood 89:376–387
6. Grignani F, Ferrucci PF, Testa U et al (1993) The acute promyelocytic leukemia-specific PML-RAR alpha fusion protein inhibits differentiation and promotes survival of myeloid precursor cells. Cell 74:423–431
7. Grignani F, Valtieri M, Gabbianelli M et al (2000) PML/RARαfusion protein expression in normal human hematopoietic progenitors dictates myeloid commitment and promyelocytic phenotype. Blood 96:1531–1537
8. Viale A, De Franco F, Orleth A et al (2009) Cell-cycle restriction limits DNA damage and maintains self-renewal and leukaemia stem cells. Nature 457:51–56
9. Wang ZY, Chen Z (2008) Acute promyelocytic leukemia: from highly fatal to highly curable. Blood 111:2505–2515
10. Lo Coco F, Avvisati G, Vignetti M et al (2013) Retinoic acid and arsenic trioxide for acute promyelocytic leukemia. N Engl J Med 369:111–121
11. Gronemeyer H, Gustafsson JA, Laudet V (2004) Principles for modulation of the nuclear receptor superfamily. Nat Rev Drug Discov 3:950–964
12. De Thé H, Vivanco-Ruiz MM, Tiollais P et al (1990) Identification of a retinoic acid receptor beta gene. Nature 343:177–180
13. Das BC, Thapa P, Karki R et al (2014) Retinoic acid signaling pathways in development and diseases. Bioorg Med Chem 22:673–683
14. Kastner P, Lawrence HJ, Waltzinger C et al (2001) Positive and negative regulation of granulopoiesis by endogenous RARalpha. Blood 97:1314–1320
15. Chanda B, Ditadi A, Iscove N et al (2013) Retinoic acid signaling is essential for embryonic hematopoietic stem cell development. Cell 155:215–227
16. Cabezas-Wallscheid N, Buettner F, Sommerkamp P et al (2017) Vitamin A-retinoic acid signaling regulates hematopoietic stem cell dormancy. Cell 167:807–823
17. Jensen K, Shiele C, Freemont PS (2001) PML protein isoforms and the RBCC/TRIM motif. Oncogene 20:7223–7233
18. Lallemand-Breittenbach V, de Thé H (2010) PML nuclear bodies. Cold Spring Harb Perspect Biol 2:a000661
19. Zhong S, Muller S, Ronchetti S et al (2000) Role of SUMO-1-modiifed PML in nuclear body formation. Blood 95:2748–2752
20. Liquori A, Ibanez M, Sargas C et al (2020) Acute promyelocytic leukemia: a constellation of molecular events around a single PML-RARA fusion gene. Cancers 12:624
21. Sanz MA, Fenaux P, Tallman S et al (2019) Management of acute promyelocytic leukemia:

updated recommendations from an expert panel of the European Leukemia Net. Blood 95:41–52
22. Alcalay M, Zangrilli D, Fagioli M et al (1992) Expression pattern of the RARalpha-PML fusion gene in acute promyelocytic leukemia. Proc Natl Acad Sci USA 89:4840–4844
23. Riva L, Ronchini C, Bodini M et al (2013) Acute promyelocytic leukemias share cooperative mutations with other myeloid-leukemia subgroups. Blood Cancer J 3:e147
24. Madan V, Shyamsunder P, Han L et al (2016) Comprehensive mutational analysis of primary and relapse acute promyelocytic leukemia. Leukemia 30:1672–1681
25. Kelly LM, Kutok JL, Williams IR et al (2002) PML/RARα and FLT3-ITD induce an APL-like disease in a mouse model. Proc Natl Acad Sci USA 99:8283–8288
26. Song H, Liu Y, Tan Y et al (2022) Recurrent noncoding somatic and germline WT1 variants converge to disrupt MYB binding in acute promyelocytic leukemia. Blood 140:1132–1141
27. De Thé H, Lavau C, Marchio A et al (1991) The PML-RAR alpha fusion mRNA generated by the t(15;17) translocation in acute promyelocytic leukemia encodes a functionally altered RAR. Cell 66:675–684
28. Lin RJ, Nagy L, Inoue S et al (1998) Role of the histone deacetylase complex in acute promyelocytic leukemia. Nature 391:811–814
29. He LZ, Tolentino T, Grayson P et al (2001) Histone deacetylase inhibitors induce remission in transgenic models of therapy-resistant acute promyelocytic leukemia. J Clin Invest 08:1321–1330
30. Warrell RP, He LZ, Richon V et al (1998) Therapeutic targeting of transcription in acute promyelocytic leukemia by use of an inhibitor of histone deacetylase. J Natl Cancer Inst 90:1621–1625
31. Martens JH, Brinkman AB, Simmer F et al (2010) PML-RARalpha/RXR alters the epigenetic landscape in acute promyelocytic leukemia. Cancer Cell 17:173–185
32. Trus MR, Yang L, Suarez Saiz F et al (2005) The histone deacetylase inhibitor valproic acid alters sensitivity towards all trans retinoic acid in acute myeloblastic leukemia cells. Leukemia 19:1161–1168
33. Leiva M, Moretti S, Soilihi H et al (2012) Valproic acid induces differentiation and transient tumor regression, but spares leukemia-initiating activity in mouse models of APL. Leukemia 26:1630–1637
34. He LZ, Guidez F, Tribioli C et al (1998) Distinct interactions of PML-RARalpha and PLZF-RARalpha with co-repressors determine differential responses to RA in APL. Nat Genet 18:126–135
35. Dai B, Wang F, Wang Y et al (2023) Targeting HDAC3 to overcome the resistance to ATRA of arsenic in acute promyelocytic leukemia through ubiquitination and degradation of PML-RARα. Cell Death Differ 30(5):1320–1333
36. Moretti S, Abdel-Aziz AK, Ceccacci E et al (2022) Co-targeting leukemia-initiating cells and leukemia bulk leads to disease eradication. Leukemia 36:1306–1312
37. Zhu J, Zhou J, Peres L et al (2005) A sumoylation site in PML/RARA is essential for leukemic transformation. Cancer Cell 7:143–153
38. Rabellino A, Carter B, Konstantidinou G et al (2012) The SUMO E3-ligase PIAS1 regulates the tumor suppressor PML and its oncogenic counterpart PML-RARA. Cancer Res 72:2275–2284
39. Lallemand-Breitenbach V, Jeanne M, Benhenda S et al (2008) Arsenic degrades PML or PML-RARalpha through a SUMO-triggered RNF4/ubiquitin-mediated pathway. Nat Cell Biol 10:547–555
40. Jaffray EG, Tathman MH, Mojsa B et al (2023) The p97/VCP segregase is essential for arsenic-induced degradation of PML and PML-RARA. Cell Biol 222:e202201027
41. Zeisig BB, Kwok C, Zelent A et al (2007) Recruitment of XR homotetrameric RARalpha fusion proteins is essential for transformation. Cancer Cell 12:36–51
42. Zhu J, Nasr R, Riaucoux-Lormiere F et al (2007) RXR is an essential component of the oncogenic PML/RAR complex in vivo. Cancer Cell 12:23–35
43. Wang K, Wang P, Shi J et al (2010) PML/RARα targets promoter regions containing PU.1 consensus and RARE hal-sites in acute promyelocytic leukemia. Cancer Cell 17:186–197
44. Tan Y, Wang X, Song H et al (2021) A PML/RARα direct target atlas redefines transcriptional deregulation in acute promyelocytic cells. Blood 137:1503–1516
45. Wang G, Tian Y, Hu Q et al (2019) PML/RARa blocks the differentiation and promotes the proliferation of acute promyelocytic leukemia through activating MYB expression by transcriptional regulation mechanisms. J Cell Biochem 120:1210–1220
46. Katerndahl C, Rogers O, Day R et al (2021) Tumor suppressor function of Gata2 in acute promyelocytic leukemia. Blood 138:1148–1157
47. Katerndahl C, Rogers O, Ramakrishnan SM et al (2022) A comprehensive reassessment of the direct DNA binding sites of PML-RARA within the genomes of primary hematopoietic progenitor cells, and their effects on chromatin structure and gene expression. Blood 140(suppl. 1):5858–5859
48. Villiers W, Kelly A, He X et al (2023) Multi-omics and machine learning reveal context-specific gene regulatory activities of PML::RARA in acute promyelocytic leukemia. Nat Commun 14:724
49. Ishov AM, Sotnikov AG, Negorev D et al (1999) PML is critical for ND10 formation and recruits the PML-interacting protein Daxx to this nuclear structure when modified by SUMO-1. J Cell Biol 147:221–234
50. Zhong S, Hu P, Tz Y et al (1999) A role for PML and the nuclear body in genomic stability. Oncogene 18:7941–7947

51. Chen X, Kao HY (2013) Post-translational modifications of PML: consequences and implications. Front Oncol 2:210
52. Wang P, Benhanda S, Wu H et al (2018) RING tetramerization is required for nuclear body biogenesis and PML sumoylation. Nat Commun 9:1277
53. Dick JA, Maul GG, Miller WH et al (1994) A novel macromolecular structure is a target of the promyelocyte-retinoic acid receptor oncoprotein. Cell 76:333–343
54. Voisset E, Moravcsik E, Stratford EW et al (2018) Pml nuclear body disruption cooperates in APL pathogenesis and impairs DNA damage repair pathways in mice. Blood 131:636–648
55. Goto E, Tomita A, Hayakawa F et al (2011) Missense mutations in PML-RARA are critical for the lack of responsiveness to arsenic trioxide treatment. Blood 118:1600–1609
56. Zhu HH, Qin YZ, Huang XJ et al (2014) Resistance to arsenic therapy in acute promyelocytic leukemia. N Engl J Med 370:1864–1866
57. Liu J, Zhu HH, Jiong H et al (2016) Varying responses of PML-RARA with different genetic mutations to arsenic trioxide. Blood 127:243–250
58. Iaccarino L, Ottone T, Alfonso V et al (2019) Mutational landscape of patients with acute promyelocytic leukemia at diagnosis and relapse. Am J Hematol 94:1091–1097
59. Iaccarino L, Ottone T, Divona M et al (2016) Mutations affecting both the rearranged and the unrearranged PML alleles in refractory acute promyelocytic leukemia. Br J Haematol 172:909–913
60. Jaffray EG, Tatham MH, Rojas-Fernandez A, et al. (2023) PML mutants resistant to arsenic induced degradation fail to generate the appropriate SUMO and ubiquitin signals required for RNF4 and p97 recruitment. BioRxiv, in press. https://doi.org/10.1101/2023.091.15524136
61. Marasca R, Zucchini P, Galimberti S et al (1999) Missense mutations in the PML/RARalpha ligand binding domain in ATRA-resistant As_2O_3 sensitive relapsed acute promyelocytic leukemia. Haematologica 84:963–968
62. GalLagher RE, Scachter-Tokarz EL, Zhou DC et al (2006) Relapse of acute promyelocytic leukemia with PML-RA alpha mutant subclones independent of proximate all-trans retinoic acid selection pressure. Leukemia 20:556–562
63. Gallagher RE, Moser BK, Racevskis J et al (2012) Treatment-influenced associations of PML-RARα mutations, FLT3 mutations, and additional chromosome abnormalities in relapsed acute promyelocytic leukemia. Blood 120:2098–2108
64. Lehmann-Che J, Bally C, Letouzé E et al (2018) Dual origin of relapses in retinoic-acid resistant acute promyelocytic leukemia. Nat Commun 9:1–8
65. Zhang X, Sun J, Yu W et al (2021) Current views on the genetic landscape and management of variant acute promyelocytic leukemia. Biomark Res 9:33
66. Licht JD, Chomienne C, Goy A et al (1995) Clinical and molecular characterization of a rare syndrome of acute promyelocytic leukemia associated with translocation (11;17). Blood 85:1083–1094
67. Chen Z, Guidez F, Rousselot P et al (1994) PLZF-RAR alpha fusion proteins generated from the variant t(11;17)(q23;q21) translocation in acute promyelocytic leukemia inhibit ligand-dependent transactivation of wild-type retinoic acid receptors. Proc Natl Acad Sci USA 91:1178–1182
68. Ruthard M, Testa U, Nervi C et al (1997) Opposite effects of the acute promyelocytic leukemia PML-retinoic acid receptor alpha (RARalpha) and PLZF-RARA fusion proteins on retinoic acid signalling. Mol Cell Biol 17:4859–4869
69. Boukarabila H, Saurin AJ, Batsche E et al (2009) The PCR1 Polycomb group complex interacts with PLZF/RARA to mediate leukemic transformation. Genes Dev 23:1195–1206
70. Hong SH, David G, Wong CW et al (1997) SMRT corepressor interacts with PLZF and with PLZF-RARalpha oncoproteins associated with acute promyelocytic leukemia. Proc Natl Acad Sci USA 94:9028–9033
71. Rego EM, He LZ, Warrell RP et al (2000) Retinoic acid (RA) and As_2O_3treatment in transgenic models of acute promyelocytic leukemia (APL) unravel the distinct nature of the leukemogenic process induced the PML-RARalpha and PLZF-RARalpha oncoproteins. Proc Natl Acad Sci USA 97:10173–10178
72. He LZ, Tolentino T, Grayson P et al (2001) Histone deacetylase inhibitors induce remission in transgenic models of therapy-resistant acute promyelocytic leukemia. J Clin Invest 108:1321–1330
73. Jiao B, Ren ZH, Liu P et al (2013) 8-CPT-cAML/all-trans retinoic acid targets t(11;17) acute promyelocytic leukemia through enhanced cell differentiation and PLZF/RARA degradation. Proc Natl Acad Sci USA 110:3495–3500
74. Rego EM, Ruggero D, Tribioli C et al (2006) Leukemia with distinct phenotypes in transgenic mice expressing PML/RARA, PLZF/RARA alpha or NPM/RAR alpha. Oncogene 25:1974–1979
75. He LZ, Bhaumik M, Tribioli C et al (2000) Two critical hits for promyelocytic leukemia. Mol Cell 6:1131–1141
76. Girard N, Tremblay M, Humbert M et al (2013) RARalpha-PLZF oncogene inhibits C/EBPalpha function in myeloid cells. Proc Natl Acad Sci USA 110:13522–13527
77. Guidez F, Parks S, Wong H et al (2007) RARA-PLZF overcomes PLZF-mediated repression of CRABPI, contributing to retinoid resistance in t(11;17) acute promyelocytic leukemia. Proc Natl Acad Sci USA 104:18694–18699
78. Fabiani E, Cicconi L, Nardozza AM et al (2021) Mutational profile of ZBTB16-RARA-positive acute myeloid leukemia. Cancer Med 10;3839–3847
79. Han L, Madan V, Mayakonda A et al (2019) Chromatin remodeling mediated by ARID1A is

indispensable for normal hematopoiesis in mice. Leukemia 33:2291–2305

80. Ono R, Masuya M, Ishii S et al (2017) Eya2, a target activated by Plzf, is critical for PLZF-RARA-induced leukemogenesis. Mol Cell Biol 37:e00585–e00516
81. Poplineau M, Platet N, Mazuel A et al (2022) Noncanonical EZH2 drives retinoic acid resistance of variant acute promyelocytic leukemia. Blood 140:2358–2370
82. Cicconi L, Testi AM, Montesinos P et al (2021) Characteristics and outcome of acute myeloid leukemia with uncommon retinoic acid receptor alpha (RARA) fusion variants. Blood Cancer J 11:167
83. Pessina C, Basilico C, Genoni A et al (2017) A new acute myeloid leukemia case with STAT5B-RARA gene fusion due to 17q21.2 interstitial deletion. Leuk Lymphoma 58:1977–1980
84. Mauer AB, Wichmann C, Gross A et al (2002) Interactions of STAT5b-RARalpha fusion protein represses transcription and differentiation through interaction with a corepressor complex. Blood 99:2647–2652
85. Dong S, Tweardy DJ (2002) Interactions of STAT5b-RARalpha, a novel acute promyelocytic leukemia fusion protein, with retinoic acid receptor and STAT3 signaling pathways. Blood 99:2637–2646
86. Strehl S, Konig M, Boztug H et al (2013) All-trans retinoic acid and arsenic trioxide resistance of acute promyelocytic leukemia with the variant STAT5B-RARA fusion gene. Leukemia 27:1606–1610
87. Crangola G, Gurnari C, Paterno G et al (2019) Stat5b-RARa-positive acute myeloid leukemia: diagnostic and therapeutic challenges of a rare AML subtype. Leuk Res 78:21–23
88. Redner RL, Rush EA, Faas S et al (1996) The t(15;17) variant of acute promyelocytic leukemia expresses a nucleophosmin-retinoic acid receptor fusion. Blood 87:882–886
89. Kikuma T, Nakamachi Y, Nogichi Y et al (2015) A new transcriptional variant and small azurophilic granules in an acute promyelocytic leukemia case with NPM1/RARA fusion gene. Int J Hematol 102:713–718
90. Redner RL, Chen JD, Rush EA et al (2000) The t(5;17) acute promyelocytic leukemia fusion protein NPM1-RARA interacts with co-repressor and co-activator proteins and exhibits both positive and negative transcriptional properties. Blood 95:2683–2690
91. Rush EA, Pollock SL, Abecassis I et al (2013) Interaction with RXR is necessary for NPM-RAR-induced myeloid differentiation blockade. Leuk Res 37:1704–1710
92. Chattopahyay A, Abecassis I, Redner RL (2015) NPM-RAR binding to TRADD selectively inhibits caspase activation, while allowing activation of NF-kB and JNK. Leuk Lymphoma 56:3401–3406
93. Chattopahyay A, Hood BL, Conrads TP et al (2014) Extrinsic apoptosis impeded by direct binding of the APL fusion protein NPM-RAR to TRADD. Mol Cancer Res 1:1283–1291
94. Shimomura Y, Pitsui H, Yamashita Y et al (2016) New variant of acute promyelocytic leukemia with IRF2BP2-RARA fusión. Cancer Sci 107:1165–1168
95. Yin CC, Jain N, Mehrotra M et al (2015) Identification of a novel fusion gene IRF2BP2-RARA in acute promyelocytic leukemia. J Natl Compr Canc Netw 13:19–22
96. Mazharuddin S, Chattopahyay A, Levy MY et al (2018) IRF2BP2-RARA t(1;17)(q42;q21.2) APL blasts differentiate in response to all-trans retinoic acid. Leuk Lymphoma 59:2246–2249
97. Liu Y, Xu F, Wen J et al (2019) A rare case of acute promyelocytic leukemia with IRF2BP2-RARA fusion; and literature review. Onco Targets Ther 12;6157–6163
98. Jovanovic JV, Chillon MC, Vincent-Faber C et al (2017) The cryptic IRF2BP2-RARA fusion transforms hematopoietic stem/progenitor cells and induces retinoid-sensitive acute promyelocytic leukemia. Leukemia 31(3):747–751
99. Osumi T, Tsujimoto S, Tanura M et al (2018) Recurrent RARB translocations in acute promyelocytic leukemia lacking RARA translocation. Cancer Res 78:4452–4458
100. Zhu HH, Qin YZ, Zhang ZL et al (2023) A global study for acute myeloid leukemia with RARG rearrangement. Blood Adv 7:2972–2982

15 MYB as a Critical Transcription Factor and Potential Therapeutic Target in AML

Mary Louise Clarke, Odd Stokke Gabrielsen, and Jon Frampton

Abstract

Myb was identified over four decades ago as the transforming component of acute leukemia viruses in chickens. Since then it has become increasingly apparent that dysregulated MYB activity characterizes many blood cancers, including acute myeloid leukemia, and that it represents the most "addictive" oncoprotein in many, if not all, such diseases. As a consequence of this tumor-specific dependency for MYB, it has become a major focus of efforts to develop specific antileukemia drugs. Much attention is being given to ways to interrupt the interaction between MYB and cooperating factors, in particular EP300/KAT3B and CBP/KAT3A. Aside from candidates identified through screening of small molecules, the most exciting prospect for novel drugs seems to be the design of peptide mimetics that interfere directly at the interface between MYB and its cofactors. Such peptides combine a high degree of target specificity with good efficacy including minimal effects on normal hematopoietic cells.

Keywords

MYB · Leukemia · AML · Oncogene addiction · Small molecule inhibitors · Peptide mimetics

Abbreviations

AKT	AKT serine/threonine kinase
AML	acute myeloid leukemia
AMV	avian myeloblastosis virus
BRD2, BRD3, BRD4, and BRDT	bromodomain-containing protein 2, 3, 4, and T
CBX5	chromobox protein homolog 5 of the heterochromatin family of proteins
C/EBPα and β	CCAAT/enhancer-binding proteins alpha and beta

M. L. Clarke
Department of Biomedical Sciences, College of Medicine & Health, University of Birmingham, Edgbaston, Birmingham, UK

Department of Cancer & Genomic Sciences, College of Medicine & Health, University of Birmingham, Edgbaston, Birmingham, UK

O. S. Gabrielsen
Department of Biosciences, University of Oslo, Oslo, Norway

J. Frampton (✉)
Department of Cancer & Genomic Sciences, College of Medicine & Health, University of Birmingham, Edgbaston, Birmingham, UK
e-mail: j.frampton@bham.ac.uk

T. Borggrefe, B. D. Giaimo (eds.), *Transcription factors in blood cell development*, Advances in Experimental Medicine and Biology 1459, https://doi.org/10.1007/978-3-031-62731-6_15

CBP/KAT3A	CREB binding protein (CBP), lysine acetyltransferase 3A
CDK9	cyclin-dependent kinase 9
CRISPR	clustered regularly interspaced short palindromic repeats
EP300/KAT3B	E1A binding protein p300, lysine acetyltransferase 3B
FLI1	Friend leukemia integration 1
ERG	ETS-related gene
ERK	extracellular signal-regulated kinases, also known as mitogen-activated protein kinases
FKBP12	FK506-binding protein 12
FOXP1	forkhead box P1 transcription factor
GLI	glioma-associated oncogene homolog
HBS1L	HBS1-like translational GTPase
HMGB3	high-mobility group box 3
HOXA9	homeobox A9
KIX	kinase-inducible interacting
IRX5	iroquois-class homeodomain protein 5
MEIS1	myeloid ecotropic viral integration site homolog 1
MYC	MYC protooncogene
PU.1/SPI1	hematopoietic transcription factor PU.1/spleen focus-forming virus (SFFV) proviral integration protein
SAGA	Spt-Ada-Gcn5-acetyltransferase complex
PBX1 and PBX2	pre-B-cell leukemia homeoboxes 1 and 2
SAGA	Spt-Ada-Gcn5-acetyltransferase complex, SAGA coactivator complex
TAD	transactivation domain
TAF4 and TAF12	TATA-box binding protein-associated factors 4 and 12
T-ALL	T-cell acute lymphoblastic leukemia
TAL1	T-cell acute leukemia 1
TAT	cationic, cell-penetrating peptide sequence
TFIID	transcription factor II D
TSHZ1	teashirt zinc finger homeobox 1
Vav-iCre	VAV gene-driven Cre recombinase
VAV	Vav guanine nucleotide exchange factor

15.1 Introduction

Acute myeloid leukemia (AML) is the most common adult leukemia that, despite intensive research, remains a real therapeutic challenge, especially in the elderly. Poor survival rates following treatment in the over 65s as well as the heterogenous nature of the disease have stimulated the search for more targeted therapies, including inhibition of aberrant gain-of-function (GoF) mutations affecting signal transduction and gene regulation as well as focusing on leukemia stem cell-specific characteristics including immune system and niche interactions and the application of personalized CAR-T (chimeric antigen receptor T) cell approaches [1, 2].

Alterations in a variety of transcription factors (TFs) underlie the origins and evolution of many types of AML [3] and, although previously considered to be "undruggable," could be attractive targets for intervention now that their mechanisms of action are much better understood. One such TF is MYB, which is critical throughout the hematopoietic hierarchy and has been linked in a number of ways to the behavior of many forms of leukemia, in particular AML. In this chapter, we will discuss the nature of the critical involvement of MYB in leukemia, with a focus on AML, and how knowledge of disease-specific aspects of its function are providing new insight into potential strategies for antileukemia therapy, which could have both AML-specific efficacy and possibly broader applicability across other myeloid and lymphoid leukemia subtypes.

15.2 AML

AML encompasses a variety of diseases with the common feature of uncontrolled proliferation and perturbed differentiation of myeloid lineage progenitors with concomitant suppression of normal hematopoietic cells, leading to anemia, leukocytopenia, and thrombocytopenia. Although generally considered as an uncommon cancer, AML is the most common acute leukemia in adults, with a particularly high frequency in individuals over the age of 65 and represents around 1% of all newly diagnosed tumors [4].

Combination chemotherapy with cytarabine and anthracycline has remained the standard of care for AML for over 40 years, and the current complete remission ratio of patients with newly diagnosed AML is 70–80%. However, the 5-year survival rate remains at 30–40%. Moreover, this chemotherapy regimen is poorly tolerated by older and less fit patients, and any improvements seen in overall survival have largely been limited to those patients under the age of 65 and those without adverse-risk cytogenetics. Furthermore, the only potentially curative strategy for those with intermediate- or adverse-risk disease is allogeneic hematopoietic stem cell (HSC) transplantation, which is not an option for many patients, again because of age, frailty, and medical comorbidities. With these limitations in mind, and with the global trend in demographics being toward a much older population, it is clear that there is an urgent need for novel approaches in AML therapy. The growing understanding of the way in which the normal regulation of gene expression is perturbed in AML, particularly in establishing and maintaining the leukemia stem cell component of the tumor, offers new opportunities to tackle this still relatively intractable disease.

15.3 MYB as an Initiator and Maintenance Factor in AML

The identification and initial understanding of the role of MYB in hematopoietic cells were largely the result of studies on animal models of hematological tumors, with clear links to human leukemia only becoming apparent much later. The role and mechanisms of action of MYB in normal hematopoiesis are covered extensively in Chap. 1 of this book, so here we will concentrate on the involvement of MYB in leukemogenesis, with an emphasis on mechanisms that apply to AML, although learnings from other leukemias are likely to be relevant in the identification of potential new targeted therapies for AML.

15.3.1 The Discovery of MYB as a Leukemia-Inducing Oncoprotein

Acute leukemia-inducing retroviruses isolated from chickens were instrumental in the characterization of a number of oncogenes in the laboratory of Thomas Graf during the 1970s, not least of which was *Myb* [5]. Two acute leukemia retroviruses were identified that had captured cDNA sequences representing transcripts from the *Myb* gene, namely, acute myeloblastosis virus (AMV) and E26 [6]. These "transduced" *Myb* sequences contained a number of point mutations and truncations affecting the coding sequence, while in the case of the E26 *Myb*, these had also become fused to sequences deriving from the chicken *Ets1* gene [7]. Rapidly following the discovery of the avian *Myb* viral sequences, genomic homologues were identified in birds, mice, and humans (reviewed in [8]), and further examples of transduced oncogenic cDNA sequences were discovered in murine leukemia retroviruses [9].

15.3.2 Leukemia-Associated Features Involving Mutation to the MYB Gene or Its Target DNA Motifs

Although early experiments demonstrated that the sequence differences between genomic *Myb* and the avian leukemia transforming viral versions influenced the oncogenic potential of the retroviruses [10, 11], no such mutations were immediately apparent in the *MYB* gene in patient leukemia samples, thereby probably limiting interest and progress in understanding the role of MYB in human disease.

Heightened interest in the leukemia-inducing potential of MYB resulted when duplications of the *MYB* gene or translocations of the *MYB* gene to the T-cell receptor gene locus were recognized as a recurrent feature of T-cell acute lymphoblastic leukemia (T-ALL) [12, 13], and subsequently other translocations were found in a variety of leukemias, including some AMLs [14, 15]. Such translocations are assumed to be leukemogenic due to dysregulation of the normal pattern of *MYB* expression. Similarly, and providing very compelling evidence for the role that MYB can play in leukemogenesis, an acquired mutation was identified in some childhood T-ALL in a noncoding intergenic region near the *SCL* (*TAL1*) gene creating a "super enhancer" element that results in altered expression of *SCL*, which is itself linked to leukemia induction [16].

Inherited variations in noncoding sequences in the vicinity of the *MYB* gene can also play an important role in affecting its expression in ways that have consequences for an individual's hematological profile and disease predisposition. Extensive analysis linking genetic variations, especially single nucleotide polymorphisms (SNPs), to phenotypic traits has revealed that the intergenic region upstream of the *MYB* gene, spanning over 100kbp up to the adjacent *HBS1L* gene, contains a number of inherited variants that influence hematological parameters, such as red blood cell count and hemoglobin levels [17], which in some instances have been linked to effects on the binding of TFs to presumed enhancer elements that act upon *MYB* gene expression [18, 19]. Some SNPs have an influence on myeloid disease occurrence, one notable example being rs9376092 76kbp upstream of the transcription start site. The variant allele of this SNP predisposes to myeloproliferative neoplasm (MPN) and appears to be associated with a corresponding lower level of *MYB* RNA expression in bone marrow (BM) progenitors [20].

Modeling a lower constitutive level of *Myb* in genetically altered mice revealed oncogenic consequences that differed depending on the degree of reduction. An approximately 80% reduction of *Myb* expression in hematopoietic cells, elicited by an insertion in intron 6 [21], produced a condition resembling the MPN disease essential thrombocythemia. This essential thrombocythemia-like phenotype was notably characterized by the acquisition of stem cell properties in committed myeloid progenitors, possibly reflecting enhanced responsiveness to cytokine stimulation, including the IL3 receptor [22]. In contrast, a 50% reduction, as seen in haploinsufficient animals containing one *Myb* knockout (KO) allele [23], leads to the age-related development of a range of myeloid preleukemic conditions, as well as AML, seemingly due to a progressive accumulation of defects in proteostasis [24, 25].

The hematooncological diseases exhibited in these models of lower *Myb* expression illustrate the importance of maintaining precise MYB levels and also that MYB loss of function (LoF) can be as important a proleukemogenic factor as GoF through overexpression, presumably linked to the fact that target genes can be either positively and negatively regulated by MYB depending upon the particular cell type and the associating cofactors in each context.

15.3.3 Leukemia-Associated Features Involving Indirect Effects on MYB Activity

Probably the most frequent way in which the activity of MYB is influenced to affect the initiation or maintenance of leukemia is through indirect routes that impinge on its transcription, translation, or protein modification and stability. Numerous studies in the 1980s pointed to higher *MYB* expression in cells derived from various leukemia types, including AML [reviewed in [14, 15]]. Experimental modeling of *Myb* overexpression to determine a causal link with leukemogenesis for a long time proved difficult to achieve in animal models. However, recently Pieters et al. [26] were able to achieve overexpression in vivo limited to the hematopoietic system using mice engineered conditionally to express *Myb*. Combining an *R26-Myb* knock-in containing a floxed stop cassette with a *Vav-iCre* transgene, they were able to achieve a tenfold increase of *Myb* transcripts and a threefold upregulation of MYB protein. For the first time, this made it possible to demonstrate that hematopoietic-specific overexpression of MYB is sufficient to drive B-cell neoplasms and myeloid malignancies in mice.

Genetic routes leading to perturbed MYB activity, including overexpression, could be either inherited or acquired. That inherited genetic differences can impact on MYB in this way is illustrated by the recent description of a germline mutation that results in a glycine to serine substitution (G3131S) in the MLL1/KMT2A ("mixed lineage leukemia 1") protein [27]. This alteration confers susceptibility to familial MPN and was proposed to have its effect through a subnormal level of MYB based on the observation that introduction of the G3131S variant into the K562 leukemia cell lines resulted in a reduced level of MYB expression.

Although other such indirect inherited genetic linkages are undoubtedly yet to be identified, several aspects of the regulation of MYB expression can clearly be influenced by acquired mutations in upstream signal transduction components. A recurrent player in this respect is the aforementioned lysine methyl transferase MLL1/KMT2A, the gene for which is frequently involved in chromosomal translocations characterizing acute leukemia. At least 50 different translocations have been reported, each generating a distinct chimeric MLL protein that is functional and can result in dysregulated transcription. In respect to the link between MLL, MYB, and leukemia, experiments in the laboratory of Robert Slany were key [28]. They utilized a conditionally active version of the MLL-ENL protein to transform mouse hematopoietic progenitors and then to determine how the expression of genes was affected upon its subsequent inactivation. Myb and the homeodomain (HD) containing proteins HOXA9 and MEIS1 stood out as being highly dependent on the presence of active MLL-ENL. Further, Hess et al. [28] showed that coexpression of *Hoxa9* and *Meis1* during inactivation of MLL-ENL was able to maintain the transformed phenotype and restore the expression of *Myb*. In the context of MLL-ENL transformation, suppression of *Myb* activity using siRNA or a dominant negative version reversed the transformed phenotype, although overexpression of *Myb* was unable to rescue the effects of loss of active MLL-ENL, suggesting that Myb is essential but not sufficient for the maintenance of the leukemia state in this situation. Employing a similar system, Michael Cleary's group examined the gene expression signature of various MLL fusion protein model AMLs, concluding that a stem cell-like profile characterizes maintenance of the leukemia in particular featuring Myb but also Hmgb3 and Cbx5 as key factors [29]. As in the experiments by Hess et al. [28], knockdown of *myb* or either of *Hmgb3* and *Cbx5* revealed a critical dependency and resulted in differentiation of the leukemic cells. Interestingly, they also observed a failure of each of the three factors individually to elicit leukemia; however, their coexpression was found to be leukemogenic.

A number of other studies have also demonstrated links between MLL and MLL fusion proteins and MYB. Hence, MYB was shown to bind to MLL1/KMT2A through contact with MENIN in human leukemia cells, including both cells with or without an MLL fusion [30]. Silencing *MYB* in human leukemic cell lines and primary patient material caused a global decrease in H3K4 methylation, a decrease in *HOXA9* and *MEIS1* gene expression, and decreased MLL and MENIN occupancy on the *HOXA9* gene locus. Putting this together with the earlier results that *Myb* expression is regulated by *Hoxa9* and *Meis1* [28] suggested the existence of an autoregulatory feedback loop. Continuing with this theme that MLL fusion proteins can both indirectly and directly operate via effects on *Myb* gene expression, Tom Gonda's group showed that the MLL-AF9 protein can directly act upon *Myb* promoter elements recruiting Cdk9 to the first intron of the gene [31].

Most recently, another connection between MLL1 and *MYB* expression has emerged that involves the participation of a long noncoding RNA, *Myrlin*. This RNA, which is expressed from an intergenic promoter 81kbp upstream of the *Myb* promoter, is coexpressed with *Myb* in erythroid progenitors (and presumably some or all other hematopoietic progenitors). Interestingly, reduction of *Myrlin* by CRISPRi compromised MLL1/KMT2A occupancy in the *Myb* locus and decreased CDK9 and RNA polymerase II (RNAPII) binding, resulting in transcriptional pausing in the *Myb* first intron [32].

It has been known for some time that *Myb* RNA can be transcribed from two alternative promoters: one upstream of exon 1 and another at the end of intron 1 adjacent to the beginning of exon 2 [33]. Importantly, the balance in expression from the two promoters is emerging as another potential way in which *MYB* expression can be perturbed in the leukemic context. It is important to mention that aside from the fact that transcripts deriving from the two promoters actually encode proteins that differ at their N-terminus by 6–20 amino acids, which have been shown to influence the profile of genes controlled by MYB [34], starting transcription from the second promoter effectively avoids the possibility of transcriptional pausing in the first intron. This pausing, which has in some genes such as *Myc* been referred to as transcriptional "attenuation," was recognized early on as a feature of the *Myb* gene [35, 36]. By avoiding such pausing, transcription from the second promoter could, especially if enhanced through acquired upstream influences, create a situation of dysregulated *Myb* expression and thereby potentially a leukemogenic stimulus. Studies from our laboratory suggested that such an imbalance distinguishes myeloid leukemia cells from normal hematopoietic progenitors. Moreover, the difference in usage of the two promoters correlated with a difference in the occupancy of HD-containing proteins across the region [37]. Comparing normal hematopoietic progenitors and transformed myelomonocytic blasts, we identified a difference in promoter usage such that proportionately more transcription initiation occurs from the downstream promoter in the leukemia context. This difference was correlated with a distinct distribution of occupancy by the HD-containing proteins HOXA9, MEIS1, PBX1, and PBX2. Most notably, PBX2 binding around the dual promoter region exhibits a shift from upstream of the exon 1 promoter in normal hematopoietic cells to the downstream promoter at the intron 1/exon 2 boundary in leukemia cells. Although a functional association is yet to be proven for these correlations, it is very interesting that a recent study has shown that relapse in pediatric T-ALL is associated with a relative increase in the use of the downstream *MYB* promoter [38].

From the studies reviewed until this point, it is clear that the MLL-HD protein-MYB axis plays a prominent role in normal hematopoietic gene regulation and that perturbations at multiple points on this axis can influence MYB expression and consequently be leukemogenic. Other mechanisms undoubtedly must impinge on MYB and be susceptible to changes that can disrupt normal *MYB* expression and have leukemogenic consequences. However, even when not immediately obvious, leukemia-driving mutations can be linked to the MLL-HD protein-MYB axis. This is illustrated by *SETBP1*, missense mutations which have been frequently identified in myeloid neoplasms, including AML [39]. SETBP1 binds the SET nuclear oncoprotein, which inhibits acetylation of nucleosomes, especially on histone H4, and is involved in DNA replication among other cell processes. Overexpression of wild-type or mutant versions of Setbp1 in mouse bone marrow cells can induce an AML-like phenotype, which can be reversed by *Myb* knockdown [40]. Interestingly, *Myb* has been shown to be a direct target for transcriptional activation involving Setbp1, which involves interaction with and recruitment of MLL1/KMT2A to the *Myb* promoter [41].

15.3.4 Leukemia-Specific Dependency on MYB

Given that MYB is widely expressed throughout the hematopoietic hierarchy and has critical roles in normal stem cells, progenitors, and differentiated cells (*see* Chap. 1), it is no surprise that it also features widely the regulation of gene expression in leukemia. Although direct mutation to the coding sequence of *Myb* has been observed in animal models of AML, it seems that something equivalent is not seen in human leukemia. However, from the above discussion of factors that influence *MYB* gene expression and how many of these are components of leukemogenesis, it should be obvious that perturbations in MYB expression, either in level or timing in particular cellular contexts, can be an important and perhaps essential part of the leukemic transformation or leukemia maintenance mechanisms. A number of studies support this view, showing that in various leukemia scenarios, MYB serves to regulate a program of gene expression that is compatible with stem cell self-renewal and furthermore that the cancer cells are "addicted" [42] in that they are more sensitive to reductions in MYB activity compared to normal hematopoietic cells.

Building on the earlier work by Somervaille and colleagues [29], Scott Lowe's group addressed the issue of the critical regulated target genes downstream of a MLL fusion protein and how these related to the leukemia stem cell (LSC) signature of a model leukemia [43]. The authors used a retroviral vector system to provide coexpression of MLL-AF9 and oncogenic RAS ($NRAS^{G12D}$) with the additional feature that the MLL-AF9 component was controlled by a "Tet-off" regulator also encoded by the retrovirus. Mouse hematopoietic progenitors transformed ex vivo were transplanted to establish in vivo leukemia, and subsequently MLL-AF9 expression was inactivated by application of doxycycline. The inactivation of MLL-AF9 led to disease regression, and 6 days after administration of doxycycline, microarray analysis was performed to assess changes in gene expression. This revealed a strong inverse correlation between global gene expression changes and loss of the LSC-associated signature, although this is not in itself unexpected given the long timeframe during which cells would have differentiated significantly. Perhaps the more meaningful data in this study derived from using chromatin immunoprecipitation followed by deep sequencing (ChIP-Seq) to look at MLL-AF9 binding to gene promoter regions of those TFs in the LSC signature that were most affected by the loss of MLL-AF9. This approach identified *Hoxa9* and *Meis1*, as expected, as well as *Myb*, *Myc*, *Tshz1*, *Foxp1*, and *Irx5*, with *Myb* being by far the most affected. Following on from these observations, the authors then used a controllable shRNA-mediated knockdown strategy ("Tet-on") to enable reduction of *Myb* levels in MLL-AF9/$NRAS^{G12D}$-transformed cells, assessing the impact on the leukemia by competitive proliferation assay. Loss of cells expressing the *Myb* shRNA from the mixed population clearly demonstrated the need for the persistent presence of high levels of *Myb*. Using the same competitive assay, the effects of MYB inhibition were also assessed in 10 human leukemia cell lines; this revealed impaired proliferation of all four MLL fusion-expressing lines tested as well as two other AML lines without MLL fusion proteins, although four lines exhibited only minimal or no effects. Furthermore, analysis of the RNA transcriptome following *MYB* knockdown in this system revealed that the changes were very similar to those observed after MLL-AF9 withdrawal, with the exception that MLL target genes such as *HOXA9* and *MEIS1* were not in the MYB-specific signatures and suggest that these HD-containing proteins may be less important than MYB for maintaining aberrant self-renewal. Supporting this conclusion, similar knockdowns combined with the competitive proliferation assay did not reveal any effect of loss of HOXA9 and MEIS1 on MLL-AF9/$NRAS^{G12D}$-transformed AML. Importantly, and again utilizing the same assay, Zuber and colleagues [43] showed that *Myb* knockdown in normal hematopoietic stem and progenitor cells (HSPCs) did not compromise their ability to reconstitute lethally irradiated mouse recipients. Collectively, this landmark study strongly supported the idea that a wide range of AMLs are addicted to the activity of MYB and that this addiction could be the "Achilles heel" that could be targeted therapeutically.

Sensitivity of leukemia to perturbations in mechanisms linked to MYB has also been observed through studies on the effects of knockdown of individual components of the TFIID and SAGA coactivator complexes. Hence, using a retroviral library of shRNAs to target 16 genes encoding the subunits of TFIID, the laboratory of Christopher Vakoc showed that the MLL-AF9/NRAS^{G12D} AML cell line is hypersensitive to *Taf12* knockdown [44]. This dependence on TAF12 appears to be due to a direct interaction between the TAF12/TAF4 histone-fold heterodimer and the transactivation domain of Myb, ectopic expression of the Taf4 histone-fold fragment being shown to efficiently squelch TAF12 in cells, suppress Myb, and cause regression of AML in the mouse model.

Taking a more general view of genes that are selectively required for the growth and survival of leukemia as compared to other cancer types, Takao et al. [45] recently queried 688 human cancer cell lines tested as part of the DepMap Cancer Dependency Map developed at the Broad Institute (https://depmap.org/portal/) and found that MYB is the most significantly required human gene in 37 leukemia cell lines, including 20 of diverse molecular subtypes derived from AML patients (Fig. 15.1).

15.4 MYB as a Therapeutic Target in Treating AML

Studies such as those described above indicating that MYB oncogene addiction may characterize many AMLs; in particular those transformed through mechanisms involving MLL form a theoretical basis for the design of novel strategies for the selective targeting of leukemia cells with minimal or no adverse consequences for normal hematopoietic cells. Indeed, a significant number of studies over the last decade or so have attempted, with varying degrees of success, to tackle MYB addiction in AML, often paying particular attention to knowledge gained over the last 40 years about the cooperating factors that are necessary for MYB activity. As will be seen below, there is a sense that the field is reaching a turning point as investigators focus on specific features of MYB protein-protein interactions (PPIs) and appreciate that the long-held view that transcription factors may not be druggable is actually shown not to be the case for MYB and other regulators of gene expression [46, 47].

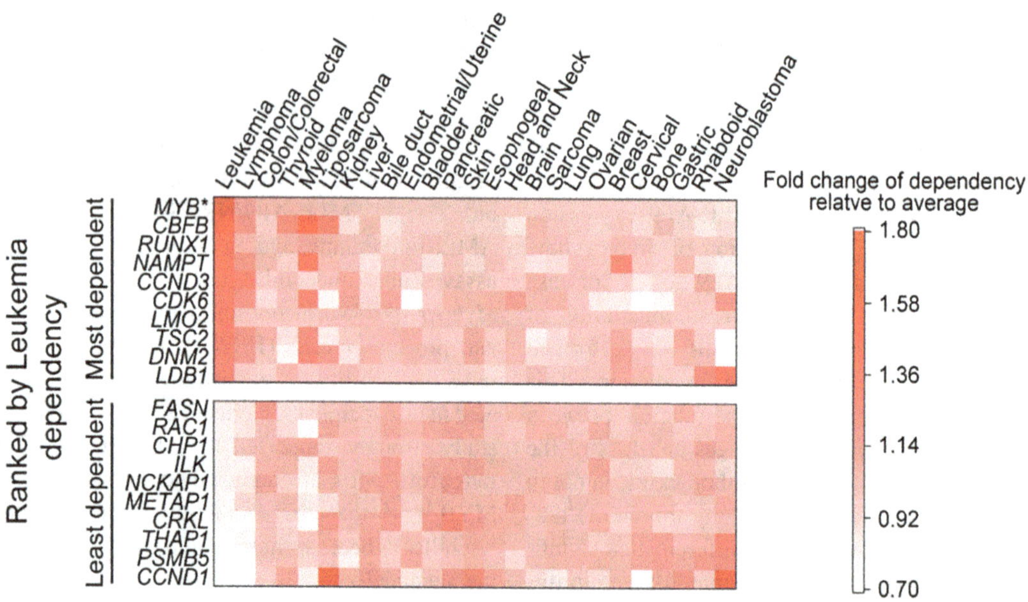

Fig. 15.1 Analysis of MYB using the DepMap Cancer Dependency Map. Heatmap showing the top and bottom 10 gene dependencies for survival and proliferation of 652 cancer cell lines in the DepMap Cancer Dependency Map Project, ranked by the greatest dependency for 37 leukemia lines, including 20 AML cell lines. The fold dependency is indicated by the red shading; *p = 1.1e−15, t-test of leukemia versus other tumor types. (Taken from [45])

15.4.1 Indirect Targeting of MYB Oncogene Addiction

Chromatin regulators, like TFs, are frequently found to be mutated in tumors, and themselves have been a focus of interest as potential therapeutic targets for small molecule inhibitors. Bromodomain-containing proteins, which are able to bind acetyl lysine residues that when on histones are usually associated with gene activation, have received a great deal of attention as possible drug targets in recent years. The first examples of bromodomain inhibitors selective for the bromodomain and extraterminal (BET) class [48] showed enticing anticancer activities, and several compounds have since advanced to human clinical trials [49]. The BET family (BRD2, BRD3, BRD4, and BRDT) features two amino-terminal bromodomains, with the JQ1 inhibitor targeted against BRD4 emerging as a lead antileukemia candidate.

The laboratory of Christopher Vakoc has been prominent in defining the importance of BRD4 in AML, which was approached initially by screening a library of shRNA molecules targeting known chromatin regulators against an MLL-AF9/NRAS^{G12D}-transformed AML mouse model. The predominant "hit" from this screen was Brd4, the inactivation of which by either shRNA or use of the JQ1 inhibitor led to antileukemia effects both in vitro and in vivo [50]. Subsequently, the group showed that in the AML model the location of Brd4 binding to chromatin correlated closely with hematopoietic transcription factors PU.1, FLI1, ERG, C/EBPα, C/EBPβ, and MYB at nucleosome-depleted enhancer and promoter regions in conjunction with the lysine acetyltransferase activity of EP300/KAT3B and CBP/KAT3A [51]. This recruitment leads to interaction with the "Mediator" coactivator complex, which is constituted of about 30 subunits and participates in the recruitment and activation of RnapII [52]. BET inhibition causes a rapid release of Mediator from a subset of *cis*-regulatory elements in the genome of AML cells that are enriched for targets of MYB and for functions related to leukemogenesis. This result implies that MYB is more suppressed by JQ1 than other hematopoietic TFs, which might be a contributor to the overall sensitivity of leukemia cells to JQ1, linking with the previous observations that AML cells are more sensitive to MYB suppression than normal myeloid cells [53, 54].

15.4.2 Small Molecule Inhibition of Essential MYB-Cofactor Interactions

Over the last decade, a number of screens have been performed to identify candidate small molecules that have the potential selectively to inhibit MYB activity in AML. The laboratory of Karl-Heinz Klempnauer has been prominent in this regard, particularly focusing on compounds that appear to act either directly on MYB or on the cooperating factors C/EBPβ, EP300/KAT3B, and CBP/KAT3A, which are both crucial in the induction of AML by MYB [55, 56]. The group developed cell-based assays incorporating GFP- or luciferase-expressing transgenic reporters derived from a MYB-dependent myeloid-specific gene to screen small molecule libraries, including collections of compounds already approved in other therapeutic contexts. The nature of the reporter gene promoter used in the screen for inhibitors means that molecules could be identified that influence MYB either directly or indirectly through their action on cooperating factors.

Several sesquiterpene lactones (STL) were identified using the MYB-dependent reporter assay and appeared to act through the MYB cooperation partner C/EBPβ. These STL include mexicanin-I [57], helenalin acetate [58], and atriplicolide tiglate [59], which affect the expression of a large number of MYB-regulated genes, although the comprehensiveness of the impact on the overall program of MYB transcriptional targets in AML and the extent of off-target effects was not fully explored.

In addition to the STL, the Klempnauer group also identified naphthol AS-E phosphate as an inhibitor of MYB function in vitro and in vivo [60]. This compound binds to the "KIX" domain of EP300/KAT3B and CBP/KAT3A, interfering with its known interaction with MYB [61, 62] and suppressing the expression of MYB target genes in mouse model AML cells and human leukemia cell lines. The same group also identified the triterpenoid celastrol as a compound capable of interrupting the MYB-EP300/KAT3B and MYB-CBP/KAT3A interaction, with selective antiproliferative effects on AML cells in vitro and some indication of a survival enhancing influence in an in vivo mouse model of MLL fusion AML, although its unfavorable pharmacological properties excluded further development [63]. Lastly, another compound identified by the Klempnauer group, the naphthoquinone plumbagin, also seemed to disrupt the MYB-EP300/KAT3B and MYB-CBP/KAT3A cooperation, in this case by directly binding to the MYB transactivation domain, but this too has failed to progress beyond the initial stage of identification [64].

Using the same or similar screening strategies, the Klempnauer group also identified a number of compounds that do not act through the C/EBP-MYB-EP300/KAT3B and C/EBP-MYB-CBP/KAT3A axis but nevertheless impact on MYB activity, although there is little indication that there is any selectivity in the effects and probably many cellular processes are impacted. Hence, the polyether ionophore monensin seemed to cause downregulation of MYB expression in AML, although no mechanism was deduced [65], while the proteasome inhibitor oprozomib [66] and protein kinase inhibitors such as bosutinib [67] were also found to limit MYB activity and to exert antiproliferative effects on human AML cells in vitro.

Others have adopted a similar approach in screening novel inhibitors that affect MYB activity. One candidate, which is already approved as a treatment against parasitic worms, is mebendazole. The group of Owen Williams used a MYB gene expression signature from MLL-rearranged AML to probe the Connectivity Map (CMAP) database [68], thereby identifying mebendazole as a potential MYB targeting drug [69]. Transient exposure to mebendazole was found to be sufficient to inhibit colony formation by AML cells, but not normal cord blood-derived progenitor cells, and was effective at impairing AML progression in vivo in a mouse xenotransplantation model. Mechanistically, mebendazole induces degradation via the proteasome, but it is by no means selective for MYB, and indeed anti-AML effects of the drug have also been attributed to inhibition of activation of AKT and ERK [70] and to degradation of GLI TFs [71]. The Williams group identified another hit from their CMAP screen, namely, the steroidal lactone withaferin A (WFA), which likewise induced rapid ablation of MYB protein with consequences for leukemia cell viability and model disease progression [72]. Although WFA has been reported to have multiple anticancer effects, its anti-AML activity could at least be partially reversed using a stabilized MYB mutant.

In assessing the effects of the agents emerging as potential inhibitors of MYB in the context of AML, there will be many factors to weigh up in deciding whether any are suitable for progression in preclinical studies. Aside from pharmacokinetic and toxicity considerations, it is important to have a picture of the balance between on- and off-target effects, which for many of the examples described here could be highly unfavorable. A recent study has sought to assess the efficacy of MYB inhibitors in terms of the extent to which they influence the expression of MYB target genes defined as those that respond rapidly following loss of MYB protein using a degron-mediated strategy [73]. The latter approach entailed the knock-in of the FKBP12^{F36V}-mScarlet-coding DNA sequence into both alleles of the MYB gene in the MV411 AML cell line, enabling rapid degradation of the hybrid MYB protein following treatment with dTAG^{V}-1, a highly specific VHL-engaging "PROTAC" (proteolysis targeting chimera) [74]. The immediate effects of degron-mediated loss of MYB compared to gene expression changes elicited by MYB inhibitors or the BET inhibitor JQ1 applied to MV411 cells were assessed by quantifying nascent transcripts using thiol (SH)-linked alkylation metabolic sequencing of RNA (SLAM-seq [75]). It was concluded that MYB inhibitors influenced a relatively minor portion of the direct MYB program (1–34% of the 450 targets defined by induced degradation of MYB), compared with almost half of the MYB program being affected by JQ1. As well as not affecting the entire MYB transcriptional program, significant portions of the RNAs altered by the inhibitors appeared to be off-target, although in the case of the KIX-binding inhibitors some at least of these may in fact represent on-target modulation of EP300/KAT3B and CBP/KAT3A in a MYB-independent manner. These findings clearly shed some light on MYB inhibitor specificity; however, there is a possible limitation in the study in terms of the setting of a threshold for judging whether a particular gene is significantly affected when comparing total loss of MYB protein function through degradation with partial loss of activity due to interference with a cooperating factor interaction as in the case of the KIX-binding inhibitors. The MYB-degron system may also have an inherent limitation as well in the sense that it is possible that the peptide sequences fused to MYB could lead to a hypomorphic phenotype, thereby potentially skewing the overall profile of MYB-regulated genes.

15.4.3 Peptide Mimetics Targeted at MYB-Cofactor Interactions

Although the search for small molecules that might target MYB addiction in AML has yielded a number of candidates, several of which are approved drugs that can potentially be repurposed from use in other clinical indications; the general sense is that none of them are likely to progress beyond the stage of laboratory investigation, not least because of lack of specificity for MYB and the risk of off-target effects. Perhaps anticipating such an outcome, there has been a growing interest in the idea of generating peptides that are able to engage with and thereby disrupt PPI surfaces that are unique to the cooperation of MYB with crucial cofactors.

Unsurprisingly, the focus of attention in designing peptide mimetics to inhibit MYB activity has been the MYB-KIX domain involved in its interaction with EP300/KAT3B and CBP/KAT3A. Based on the high-resolution structure of the MYB:EP300/KAT3B and MYB:CBP/KAT3A complex and using molecular mechanic simulations to ensure preservation of key protein-protein contacts, the group of Alex Kentsis generated a peptide mimetic designed to disrupt the interaction [76]. This peptide, which was called MYBMIM, comprised the helical MYB transactivation domain (TAD, residues 293–310) that binds to the EP300/KAT3B and CBP/KAT3A KIX domain, fused to the cationic cell-penetrant TAT peptide. Additionally, the MYBMIM peptide was designed in the *retro-inverso* orientation and contained D-amino acids to confer proteolytic stability (Fig. 15.2). MYBMIM-induced sustained reduction of growth of AML cell lines, either with or without MLL rearrangements, when compared to untreated control, while no effect was seen on normal cells. Furthermore, MYBMIM treatment significantly delayed leukemia progression and extended survival in xenografts of patient-derived MLL-rearranged human leukemia.

Treatment of the MOLM13 human AML cell line with MYBMIM resulted in downregulation of 1730 genes as well as upregulation of 2232 genes, consistent with other reports of MYB-induced gene repression. Notably, the genes affected by MYBMIM treatment exhibited significant enrichment for direct MYB target genes. Furthermore, ChIP-Seq to analyze genome-wide distribution of MYB protein complexes in cells treated with MYBMIM demonstrated the elimination of 2690 MYB complexes bound to promoters and enhancers.

The MYB-degron study already mentioned also compared MYB protein depletion effects on gene expression with the results obtained using MYBMIM. The fractional overlap was found not to be extensive, but again this may reflect to some extent the fact that by interrupting a cofactor interaction, the peptide mimetic only influences a subset of MYB target genes. The same study also allowed comparison of gene expression changes caused by MYBMIM and those small molecule inhibitors supposedly acting upon the MYB:EP300/KAT3B and MYB:CBP/KAT3A interaction, but this likewise only showed a partial overlap.

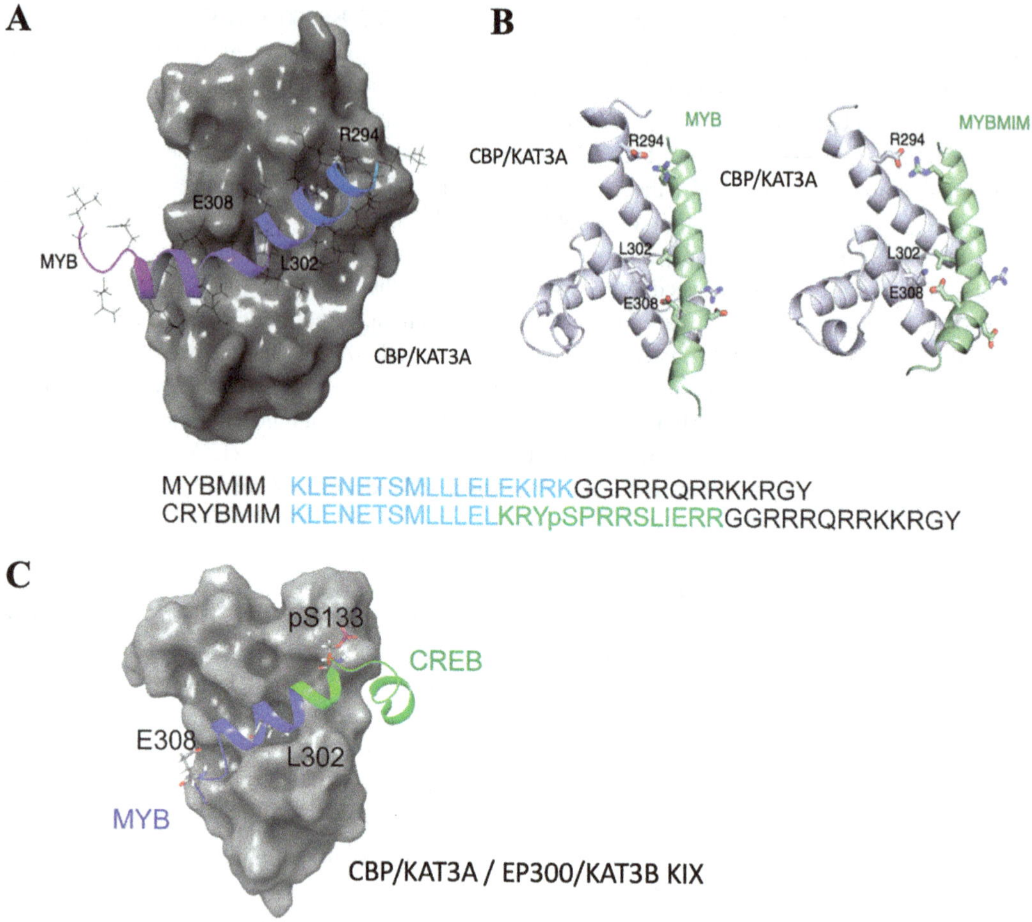

Fig. 15.2 Peptide mimetics targeting the MYB-EP300/KAT3B and CBP/KAT3A interface. (**a**) Molecular structure of the complex of the transactivation domain of MYB (blue) with the KIX domain of CBP/KAT3A (gray). MYB residues making contacts with CBP/KAT3A are labeled as indicated. (**b**) Ribbon diagrams illustrating the molecular structures of the transactivation domain (green) of MYB (left) and MYBMIM (right) in complex with the KIX domain of CBP/KAT3A (gray). Both MYB and MYBMIM retain E308 and R294 salt bridge and L302 hydrophobic interactions, as marked by the sidechain representations. (**c**) Molecular model of the CRYBMIM:KIX complex. Residues in contact with the KIX domain are labeled, with portions derived from MYB and CREB marked in blue and green, respectively. The peptide sequences of MYBMIM and CRYBMIM are shown in the middle of the figure. (Adapted from [76] and [45])

Before the recent comparative analysis of MYB inhibitor effects on MYB target gene expression, the Kentsis group had already sought to improve upon the efficacy of MYBMIM [45]. Since the MYBMIM peptide binds to recombinant CBP/KAT3A KIX domain with a dissociation constant (determined by using microscale thermophoresis) of about 20 μM, as compared to near 4 μM in the case of the native MYB peptide, the Kentsis group reasoned that a peptide mimetic inhibitor with higher affinity to EP300/KAT3B CBP/KAT3A would be more effective. They used molecular modeling to extend MYBMIM into the adjoining site on CBP/KAT3A that binds CREB (cAMP response element-binding protein), adding CREB residues 124–147, and making other adjustments to maintain a favorable backbone geometry (Fig. 15.2). Termed CRYBMIM, this peptide was shown to have a dissociation constant for binding to CBP/KAT3A-KIX of 5.7 μM, a considerable improvement on MYBMIM and close to that of MYB itself. Importantly, CRYBMIM exhibited greater potency in terms of disruption on the MYB:EP300/KAT3B and MYB:CBP/KAT3A interaction in AML cells and a much higher potency against AML cell line growth in culture. Moreover, the range of potency extended to almost all AML cell lines, while retaining relatively minimal effects on the clonogenic growth of normal progenitor cells.

A similar approach to improving upon the selectivity of a MYB sequence peptide targeted at the KIX domain took the approach of combining the transactivation of MYB with that of the MLL protein, which itself engages with a distinct part of the CBP/KAT3A protein [77]. As for other peptide mimetics, the peptide created, MYBLL-tide, additionally incorporated cell-penetrating moieties (Fig. 15.3). Highly selective binding to the EP300/KAT3B and CBP/KAT3A KIX domain was demonstrated, and some preliminary antileukemia cell line findings suggest efficacy in influencing AML, but no direct comparison has yet been made to determine how MYBLL-tide performs against MYBMIM or CRYBMIM.

Several laboratories continue to search for improvements in the specificity and affinity of inhibitors of the MYB:EP300/KAT3B and MYB:CBP/KAT3A interaction [78, 79]. Worth noting perhaps, the study by Jones et al. provides a clear indication that rationale, structural chemistry-based considerations can go a long way to creating peptide mimetics with even more favorable properties. Hence, they showed that by mutating only two MYB residues that bind in or near a hotspot at the surface of EP300/KAT3B KIX, it is possible to obtain single-digit nanomolar inhibitors of the MYB:KIX interaction that bind 400-fold tighter than MYB itself [79]. It seems very likely therefore that over the course of the next few years, highly selective MYB-based peptide mimetics will progress to preclinical studies and ultimately clinical trials.

Interestingly, transcriptome analysis of the effects of CRYBMIM on an AML cell line revealed that in addition to the expected suppression of MYB target genes, there is a significant induction of myeloid and monocyte differentiation programs, which was linked to remodeling of chromatin-bound transcription factor complexes upon disruption of the MYB:EP300/KAT3B and CBP/KAT3A interaction [45]. This could represent an important advantage to the therapeutic use of the MYB peptide mimetics in the context of the observations that relapsed AML exhibits stable nongenetic resistance that is due at least in part to the use of alternative enhancers to sustain aberrant gene expression [80]. Therapeutic remodeling of complexes of master regulators such as MYB may therefore constitute an effective strategy to reprogram gene

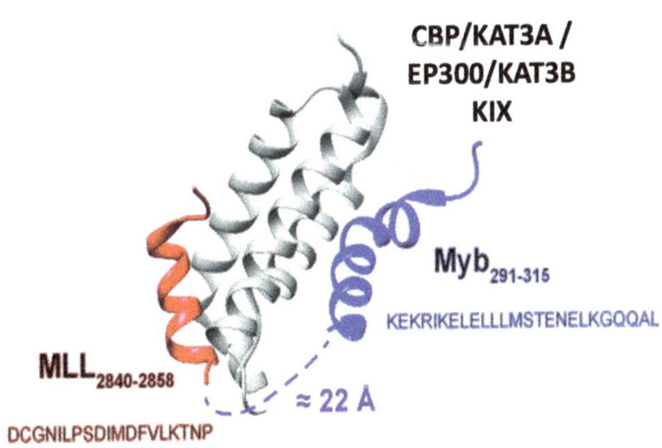

Fig. 15.3 Design of MYBLL-tide peptide mimetic fusing MYB and MLL sequences to selectively target the EP300/KAT3B and CBP/KAT3A KIX domain. Ribbon diagram illustrating the structure of the EP300/KAT3B and CBP/KAT3A KIX domain (gray) and its interaction with MYBLL-tide through the MLL (red) and MYB (blue) transactivation domain peptides. The distance between the N-terminus of MYB and the C-terminus of MLL is indicated by the purple dotted line. (Adapted from [77])

expression in leukemia that may prevent or overcome such resistance.

15.5 Conclusion

After over four decades of investigation of animal model systems and human AML cells, the transcription factor MYB has emerged as one of a group of oncoproteins to which AML is "addicted" and that therefore represents a good candidate for the development of novel therapeutic agents to be used either alone or in combination with existing drugs. Indeed, MYB is perhaps the most addictive oncoprotein in AML and is proving to be a paradigm for the generation of highly targeted peptide mimetics that work by interfering with PPI between MYB and its critical cofactors. The efficacy and selectivity being exhibited by the most recently developed peptide mimetics against MYB bode well for their eventual progression to clinical trials. Furthermore, the wider role of MYB in human tumors, including breast and colorectal, suggests that such peptide mimetics could eventually become relevant in the treatment of some of the most common cancers.

References

1. Roloff GW, Odenike O, Bajel A, Wei AH, Foley N, Uy GL (2022) Contemporary approach to acute myeloid leukemia therapy in 2022. Am Soc Clin Oncol Educ Book 42:1–16. https://doi.org/10.1200/EDBK_349605
2. Hofmann S, Schubert M-L, Wang L, He B, Neuber B, Dreger P, Müller-Tidow C, Schmitt M (2019) Chimeric antigen receptor (CAR) T cell therapy in acute myeloid leukemia (AML). J Clin Med 8:200. https://doi.org/10.3390/jcm8020200
3. Assi SA, Imperato MR, Coleman DJL, Pickin A, Potluri S, Ptasinska A, Chin PS, Blair H, Cauchy P, James SR, Zacarias-Cabeza J, Gilding LN, Beggs A, Clokie S, Loke JC, Jenkin P, Uddin A, Delwel R, Richards SJ, Raghavan M, Griffiths MJ, Heidenreich O, Cockerill PN, Bonifer C (2019) Subtype-specific regulatory network rewiring in acute myeloid leukemia. Nat Genet 51:151–162. https://doi.org/10.1038/s41588-018-0270-1
4. De Kouchkovsky I, Abdul-Hay M (2016) Acute myeloid leukemia: a comprehensive review and 2016 update. Blood Cancer J 6:e441. https://doi.org/10.1038/bcj.2016.50
5. Beug H, von Kirchbach A, Döderlein G, Conscience JF, Graf T (1979) Chicken hematopoietic cells transformed by seven strains of defective avian leukemia viruses display three distinct phenotypes of differentiation. Cell 18:375–390. https://doi.org/10.1016/0092-8674(79)90057-6
6. Roussel M, Saule S, Lagrou C, Rommens C, Beug H, Graf T, Stehelin D (1979) Three new types of viral oncogene of cellular origin specific for haematopoietic cell transformation. Nature 281:452–455. https://doi.org/10.1038/281452a0
7. Nunn M, Weiher H, Bullock P, Duesberg P (1984) Avian erythroblastosis virus E26: nucleotide sequence of the tripartite onc gene and of the LTR, and analysis of the cellular prototype of the viral ets sequence. Virology 139:330–339. https://doi.org/10.1016/0042-6822(84)90378-7
8. Lipsick JS, Manak J, Mitiku N, Chen CK, Fogarty P, Guthrie E (2001) Functional evolution of the Myb oncogene family. Blood Cells Mol Dis 27:456–458. https://doi.org/10.1006/bcmd.2001.0404
9. Shen-Ong GL, Wolff L (1987) Moloney murine leukemia virus-induced myeloid tumors in adult BALB/c mice: requirement of c-myb activation but lack of v-abl involvement. J Virol 61:3721–3725. https://doi.org/10.1128/jvi.61.12.3721-3725.1987
10. Introna M, Golay J, Frampton J, Nakano T, Ness SA, Graf T (1990) Mutations in v-myb alter the differentiation of myelomonocytic cells transformed by the oncogene. Cell 63:1289–1297. https://doi.org/10.1016/0092-8674(90)90424-d
11. Metz T, Graf T (1991) Fusion of the nuclear oncoproteins v-Myb and v-Ets is required for the leukemogenicity of E26 virus. Cell 66:95–105. https://doi.org/10.1016/0092-8674(91)90142-l
12. Lahortiga I, De Keersmaecker K, Van Vlierberghe P, Graux C, Cauwelier B, Lambert F, Mentens N, Beverloo HB, Pieters R, Speleman F, Odero MD, Bauters M, Froyen G, Marynen P, Vandenberghe P, Wlodarska I, Meijerink JPP, Cools J (2007) Duplication of the MYB oncogene in T cell acute lymphoblastic leukemia. Nat Genet 39:593–595. https://doi.org/10.1038/ng2025
13. Clappier E, Cuccuini W, Kalota A, Crinquette A, Cayuela J-M, Dik WA, Langerak AW, Montpellier B, Nadel B, Walrafen P, Delattre O, Aurias A, Leblanc T, Dombret H, Gewirtz AM, Baruchel A, Sigaux F, Soulier J (2007) The C-MYB locus is involved in chromosomal translocation and genomic duplications in human T-cell acute leukemia (T-ALL), the translocation defining a new T-ALL subtype in very young children. Blood 110:1251–1261. https://doi.org/10.1182/blood-2006-12-064683
14. Pattabiraman DR, Gonda TJ (2013) Role and potential for therapeutic targeting of MYB in leukemia. Leukemia 27:269–277. https://doi.org/10.1038/leu.2012.225

15. Wang X, Angelis N, Thein SL (2018) MYB—a regulatory factor in hematopoiesis. Gene 665:6–17. https://doi.org/10.1016/j.gene.2018.04.065
16. Mansour MR, Abraham BJ, Anders L, Berezovskaya A, Gutierrez A, Durbin AD, Etchin J, Lawton L, Sallan SE, Silverman LB, Loh ML, Hunger SP, Sanda T, Young RA, Look AT (2014) An oncogenic super-enhancer formed through somatic mutation of a non-coding intergenic element. Science 346:1373–1377. https://doi.org/10.1126/science.1259037
17. Thein SL, Menzel S, Peng X, Best S, Jiang J, Close J, Silver N, Gerovasilli A, Ping C, Yamaguchi M, Wahlberg K, Ulug P, Spector TD, Garner C, Matsuda F, Farrall M, Lathrop M (2007) Intergenic variants of HBS1L-MYB are responsible for a major quantitative trait locus on chromosome 6q23 influencing fetal hemoglobin levels in adults. Proc Natl Acad Sci USA 104:11346–11351. https://doi.org/10.1073/pnas.0611393104
18. Stadhouders R, Thongjuea S, Andrieu-Soler C, Palstra R-J, Bryne JC, van den Heuvel A, Stevens M, de Boer E, Kockx C, van der Sloot A, van den Hout M, van Ijcken W, Eick D, Lenhard B, Grosveld F, Soler E (2012) Dynamic long-range chromatin interactions control Myb proto-oncogene transcription during erythroid development. EMBO J 31:986–999. https://doi.org/10.1038/emboj.2011.450
19. Stadhouders R, Aktuna S, Thongjuea S, Aghajanirefah S, Pourfarzad F, van IJcken W, Lenhard B, Rooks H, Best S, Menzel S, Grosveld F, Thein SL, Soler E. (2014) HBS1L-MYB intergenic variants modulate fetal hemoglobin via long-range MYB enhancers. J Clin Invest 124:1699–1710. https://doi.org/10.1172/JCI71520
20. Tapper W, Jones AV, Kralovics R, Harutyunyan AS, Zoi K, Leung W, Godfrey AL, Guglielmelli P, Callaway A, Ward D, Aranaz P, White HE, Waghorn K, Lin F, Chase A, Baxter EJ, Maclean C, Nangalia J, Chen E, Evans P, Short M, Jack J, Wallis L, Oscier D, Duncombe AS, Schuh A, Mead AJ, Griffiths M, Ewing J, Gale RE, Schnittger S, Haferlach T, Stegelmann F, Döhner K, Grallert H, Strauch K, Tanaka T, Bandinelli S, Giannopoulos A, Pieri L, Mannarelli C, Gisslinger H, Barosi G, Cazzola M, Reiter A, Harrison C, Campbell P, Green AR, Vannucchi A, Cross NCP (2015) Genetic variation at MECOM, TERT, JAK2 and HBS1L-MYB predisposes to myeloproliferative neoplasms. Nat Commun 6:6691. https://doi.org/10.1038/ncomms7691
21. Emambokus N, Vegiopoulos A, Harman B, Jenkinson E, Anderson G, Frampton J (2003) Progression through key stages of haemopoiesis is dependent on distinct threshold levels of c-Myb. EMBO J 22:4478–4488. https://doi.org/10.1093/emboj/cdg434
22. Clarke M, Volpe G, Sheriff L, Walton D, Ward C, Wei W, Dumon S, García P, Frampton J (2017) Transcriptional regulation of SPROUTY2 by MYB influences myeloid cell proliferation and stem cell properties by enhancing responsiveness to IL-3. Leukemia 31:957–966. https://doi.org/10.1038/leu.2016.289
23. Mucenski ML, McLain K, Kier AB, Swerdlow SH, Schreiner CM, Miller TA, Pietryga DW, Scott WJ Jr, Potter SS (1991) A functional c-myb gene is required for normal murine fetal hepatic hematopoiesis. Cell 65:677–689. https://doi.org/10.1016/0092-8674(91)90099-k
24. Clarke ML, Lemma RB, Walton DS, Volpe G, Noyvert B, Gabrielsen OS, Frampton J (2023) MYB insufficiency disrupts proteostasis in hematopoietic stem cells leading to age-related neoplasia. Blood 141:1858–1870. https://doi.org/10.1182/blood.2022019138
25. Gonda TJ (2023) MYB deficiency leads to myeloid neoplasia too. Blood 141:1786–1787. https://doi.org/10.1182/blood.2022019533
26. Pieters T, Almeida A, T'Sas S, Lemeire K, Hochepied T, Berx G, Kentsis A, Goossens S, Van Vlierberghe P (2022) Myb drives B-cell neoplasms and myeloid malignancies in vivo. Blood Adv 6:2987–2991. https://doi.org/10.1182/bloodadvances.2021005955
27. Yin L, Xie S, Chen Y, Li W, Jiang X, Li H, Li J, Wu Z, Xiao X, Zhang G, Cheng Z, Peng H (2021) Novel germline mutation KMT2A G3131S confers genetic susceptibility to familial myeloproliferative neoplasms. Ann Hematol 100:2229–2240. https://doi.org/10.1007/s00277-021-04562-4
28. Hess JL, Bittner CB, Zeisig DT, Bach C, Fuchs U, Borkhardt A, Frampton J, Slany RK (2006) c-Myb is an essential downstream target for homeobox-mediated transformation of hematopoietic cells. Blood 108:297–304. https://doi.org/10.1182/blood-2005-12-5014
29. Somervaille TC, Matheny CJ, Spencer GJ, Iwasaki M, Rinn JL, Witten DM, Chang HY, Shurtleff SA, Downing JR, Cleary ML (2009) Hierarchical maintenance of MLL myeloid leukemia stem cells employs a transcriptional program shared with embryonic rather than adult stem cells. Cell Stem Cell 4:129–140. https://doi.org/10.1016/j.stem.2008.11.015
30. Jin S, Zhao H, Yi Y, Nakata Y, Kalota A, Gewirtz AM (2010) c-Myb binds MLL through menin in human leukemia cells and is an important driver of MLL-associated leukemogenesis. J Clin Invest 120:593–606. https://doi.org/10.1172/jci38030
31. Cao L, Mitra P, Gonda TJ (2019) The mechanism of MYB transcriptional regulation by MLL-AF9 oncoprotein. Sci Rep 9:20084. https://doi.org/10.1038/s41598-019-56426-7
32. Kim J, Diaz LF, Miller MJ, Leadem B, Krivega I, Dean A (2023) An enhancer RNA recruits MLL1 to regulate transcription of Myb. https://doi.org/10.1101/2023.09.26.559528
33. Jacobs SM, Gorse KM, Westin EH (1994) Identification of a second promoter in the human c-myb proto-oncogene. Oncogene 9:227–235
34. Frerich CA, Sedam HN, Kang H, Mitani Y, El-Naggar AK, Ness SA (2019) N-terminal truncated Myb with new transcriptional activity produced through use

of an alternative MYB promoter in salivary gland adenoid cystic carcinoma. Cancers 12:45. https://doi.org/10.3390/cancers12010045

35. Bender TP, Kuehl WM (1986) Murine myb protooncogene mRNA: cDNA sequence and evidence for 5′ heterogeneity. Proc Natl Acad Sci USA 83:3204–3208. https://doi.org/10.1073/pnas.83.10.3204
36. Watson RJ, Dyson PJ, McMahon J (1987) Multiple c-myb transcript cap sites are variously utilized in cells of mouse haemopoietic origin. EMBO J 6:1643–1651. https://doi.org/10.1002/j.1460-2075.1987.tb02413.x
37. Dassé E, Volpe G, Walton DS, Wilson N, Del Pozzo W, O'Neill LP, Slany RK, Frampton J, Dumon S (2012) Distinct regulation of c-myb gene expression by HoxA9, Meis1 and Pbx proteins in normal hematopoietic progenitors and transformed myeloid cells. Blood Cancer J 2:e76. https://doi.org/10.1038/bcj.2012.20
38. Fehr A, Arvidsson G, Nordlund J, Lönnerholm G, Stenman G, Andersson MK (2023) Increased MYB alternative promoter usage is associated with relapse in acute lymphoblastic leukemia. Genes Chromosomes Cancer 62:597–606. https://doi.org/10.1002/gcc.23151
39. Makishima H, Yoshida K, Nguyen N, Przychodzen B, Sanada M, Okuno Y, Ng KP, Gudmundsson KO, Vishwakarma BA, Jerez A, Gomez-Segui I, Takahashi M, Shiraishi Y, Nagata Y, Guinta K, Mori H, Sekeres MA, Chiba K, Tanaka H, Muramatsu H, Sakaguchi H, Paquette RL, McDevitt MA, Kojima S, Saunthararajah Y, Miyano S, Shih LY, Du Y, Ogawa S, Maciejewski JP (2013) Somatic SETBP1 mutations in myeloid malignancies. Nat Genet 45:942–946. https://doi.org/10.1038/ng.2696
40. Nguyen N, Vishwakarma BA, Oakley K, Han Y, Przychodzen B, Maciejewski JP, Du Y (2016) Myb expression is critical for myeloid leukemia development induced by Setbp1 activation. Oncotarget 7:86300–86312. https://doi.org/10.18632/oncotarget.13383
41. Nguyen N, Gudmundsson KO, Soltis AR, Oakley K, Roy KR, Han Y, Gurnari C, Maciejewski JP, Crouch G, Ernst P, Dalgard CL, Du Y (2021) Recruitment of MLL1 complex is essential for SETBP1 to induce myeloid transformation. iScience 25:103679. https://doi.org/10.1016/j.isci.2021.103679
42. Weinstein IB, Joe A (2008) Oncogene addiction. Cancer Res 68:3077–3080. https://doi.org/10.1158/0008-5472.CAN-07-3293
43. Zuber J, Rappaport AR, Luo W, Wang E, Chen C, Vaseva AV, Shi J, Weissmueller S, Fellmann C, Fellman C, Taylor MJ, Weissenboeck M, Graeber TG, Kogan SC, Vakoc CR, Lowe SW (2011) An integrated approach to dissecting oncogene addiction implicates a Myb-coordinated self-renewal program as essential for leukemia maintenance. Gene Dev 25:1628–1640. https://doi.org/10.1101/gad.17269211
44. Xu Y, Milazzo JP, Somerville TDD, Tarumoto Y, Huang Y-H, Ostrander EL, Wilkinson JE, Challen GA, Vakoc CR (2018) A TFIID-SAGA perturbation that targets MYB and suppresses acute myeloid leukemia. Cancer Cell 33:13–28.e8. https://doi.org/10.1016/j.ccell.2017.12.002
45. Takao S, Forbes L, Uni M, Cheng S, Pineda JMB, Tarumoto Y, Cifani P, Minuesa G, Chen C, Kharas MG, Bradley RK, Vakoc CR, Koche RP, Kentsis A (2021) Convergent organization of aberrant MYB complex controls oncogenic gene expression in acute myeloid leukemia. Elife 10:e65905. https://doi.org/10.7554/elife.65905
46. Bushweller JH (2019) Targeting transcription factors in cancer—from undruggable to reality. Nat Rev Cancer 19:611–624. https://doi.org/10.1038/s41568-019-0196-7
47. Henley MJ, Koehler AN (2021) Advances in targeting 'undruggable' transcription factors with small molecules. Nat Rev Drug Discov 20:669–688. https://doi.org/10.1038/s41573-021-00199-0
48. Filippakopoulos P, Qi J, Picaud S, Shen Y, Smith WB, Fedorov O, Morse EM, Keates T, Hickman TT, Felletar I, Philpott M, Munro S, McKeown MR, Wang Y, Christie AL, West N, Cameron MJ, Schwartz B, Heightman TD, La Thangue N, French CA, Wiest O, Kung AL, Knapp S, Bradner JE (2010) Selective inhibition of BET bromodomains. Nature 468:1067–1073. https://doi.org/10.1038/nature09504
49. Cochran AG, Conery AR, Sims RJ 3rd. (2019) Bromodomains: a new target class for drug development. Nat Rev Drug Discov 18:609–628. https://doi.org/10.1038/s41573-019-0030-7
50. Zuber J, Shi J, Wang E, Rappaport AR, Herrmann H, Sison EA, Magoon D, Qi J, Blatt K, Wunderlich M, Taylor MJ, Johns C, Chicas A, Mulloy JC, Kogan SC, Brown P, Valent P, Bradner JE, Lowe SW, Vakoc CR (2011) RNAi screen identifies Brd4 as a therapeutic target in acute myeloid leukaemia. Nature 478:524–528. https://doi.org/10.1038/nature10334
51. Roe J-S, Mercan F, Rivera K, Pappin DJ, Vakoc CR (2015) BET bromodomain inhibition suppresses the function of hematopoietic transcription factors in acute myeloid leukemia. Mol Cell 58:1028–1039. https://doi.org/10.1016/j.molcel.2015.04.011
52. Allen BL, Taatjes DJ (2015) The mediator complex: a central integrator of transcription. Nat Rev Mol Cell Biol 16:155–166. https://doi.org/10.1038/nrm3951
53. Bhagwat AS, Roe J-S, Mok BYL, Hohmann AF, Shi J, Vakoc CR (2016) BET bromodomain inhibition releases the mediator complex from select cis-regulatory elements. Cell Rep 15:519–530. https://doi.org/10.1016/j.celrep.2016.03.054
54. Roe JS, Vakoc CR (2016) The essential transcriptional function of BRD4 in acute myeloid leukemia. Cold Spring Harb Symp Quant Biol 81:61–66. https://doi.org/10.1101/sqb.2016.81.031039
55. Pattabiraman DR, McGirr C, Shakhbazov K, Barbier V, Krishnan K, Mukhopadhyay P, Hawthorne P, Trezise A, Ding J, Grimmond SM, Papathanasiou P, Alexander WS, Perkins AC, Levesque J-P, Winkler IG, Gonda TJ (2014) Interaction of c-Myb with P300 is required for the induction of acute myeloid

leukemia (AML) by human AML oncogenes. Blood 123:2682–2690. https://doi.org/10.1182/blood-2012-02-413187
56. Klempnauer K-H (2023) C/EBPβ cooperates with MYB to maintain the oncogenic program of AML cells. Oncotarget 14:174–177. https://doi.org/10.18632/oncotarget.28377
57. Bujnicki T, Wilczek C, Schomburg C, Feldmann F, Schlenke P, Müller-Tidow C, Schmidt TJ, Klempnauer K-H (2012) Inhibition of Myb-dependent gene expression by the sesquiterpene lactone mexicanin-I. Leukemia 26:615–622. https://doi.org/10.1038/leu.2011.275
58. Jakobs A, Steinmann S, Henrich SM, Schmidt TJ, Klempnauer K-H (2016) Helenalin acetate, a natural sesquiterpene lactone with anti-inflammatory and anti-cancer activity, disrupts the cooperation of CCAAT box/enhancer-binding protein β (C/EBPβ) and co-activator p300. J Biol Chem 291:26098–26108. https://doi.org/10.1074/jbc.M116.748129
59. Yusenko MV, Trentmann A, Casolari DA, Ghani LA, Lenz M, Horn M, Dörner W, Klempnauer S, Mootz HD, Arteaga MF, Mikesch J-H, D'Andrea RJ, Gonda TJ, Müller-Tidow C, Schmidt TJ, Klempnauer K-H (2012) C/EBPβ is a MYB- and p300-cooperating pro-leukemogenic factor and promising drug target in acute myeloid leukemia. Oncogene 40:4746–4758. https://doi.org/10.1038/s41388-021-01800-x
60. Uttarkar S, Dukare S, Bopp B, Goblirsch M, Jose J, Klempnauer K-H (2015) Naphthol AS-E phosphate inhibits the activity of the transcription factor Myb by blocking the interaction with the KIX domain of the coactivator P300. Mol Cancer Ther 14:1276–1285. https://doi.org/10.1158/1535-7163.mct-14-0662
61. Kasper LH, Boussouar F, Ney PA, Jackson CW, Rehg J, van Deursen JM, Brindle PK (2002) A transcription-factor-binding surface of coactivator p300 is required for haematopoiesis. Nature 419:738–743. https://doi.org/10.1038/nature01062
62. Zor T, De Guzman RN, Dyson HJ, Wright PE (2004) Solution structure of the KIX domain of CBP bound to the transactivation domain of c-Myb. J Mol Biol 337:521–534. https://doi.org/10.1016/j.jmb.2004.01.038
63. Uttarkar S, Dassé E, Coulibaly A, Steinmann S, Jakobs A, Schomburg C, Trentmann A, Jose J, Schlenke P, Berdel WE, Schmidt TJ, Müller-Tidow C, Frampton J, Klempnauer K-H (2016) Targeting acute myeloid leukemia with a small molecule inhibitor of the Myb/P300 interaction. Blood 127:1173–1182. https://doi.org/10.1182/blood-2015-09-668632
64. Uttarkar S, Piontek T, Dukare S, Schomburg C, Schlenke P, Berdel WE, Müller-Tidow C, Schmidt TJ, Klempnauer K-H (2016) Small-molecule disruption of the Myb/P300 cooperation targets acute myeloid leukemia cells. Mol Cancer Ther 15:2905–2915. https://doi.org/10.1158/1535-7163.mct-16-0185
65. Yusenko MV, Trentmann A, Andersson MK, Ghani LA, Jakobs A, Arteaga Paz MF, Mikesch JH, Peter von Kries J, Stenman G, Klempnauer K-H (2020) Monensin, a novel potent MYB inhibitor, suppresses proliferation of acute myeloid leukemia and adenoid cystic carcinoma cells. Cancer Lett 479:61–70. https://doi.org/10.1016/j.canlet.2020.01.039
66. Yusenko MV, Biyanee A, Andersson MK, Radetzki S, von Kries JP, Stenman G, Klempnauer K-H (2021) Proteasome inhibitors suppress MYB oncogenic activity in a p300-dependent manner. Cancer Lett 520:132–142. https://doi.org/10.1016/j.canlet.2021.07.010
67. Biyanee A, Yusenko MV, Klempnauer K-H (2022) Src-family protein kinase inhibitors suppress MYB activity in a p300-dependent manner. Cells 11:1162. https://doi.org/10.3390/cells11071162
68. Lamb J, Crawford ED, Peck D, Modell JW, Blat IC, Wrobel MJ, Lerner J, Brunet JP, Subramanian A, Ross KN, Reich M, Hieronymus H, Wei G, Armstrong SA, Haggarty SJ, Clemons PA, Wei R, Carr SA, Lander ES, Golub TR (2006) The connectivity map: using gene-expression signatures to connect small molecules, genes, and disease. Science 313:1929–1935. https://doi.org/10.1126/science.1132939
69. Walf-Vorderwülbecke V, Pearce K, Brooks T, Hubank M, Heuvel-Eibrink MM, van den Zwaan CM, Adams S, Edwards D, Bartram J, Samarasinghe S, Ancliff P, Khwaja A, Goulden N, Williams G, de Boer J, Williams O (2018) Targeting acute myeloid leukemia by drug-induced c-MYB degradation. Leukemia 32:882–889. 10.1038/leu.2017.317
70. He L, Shi L, Du Z, Huang H, Gong R, Ma L, Chen L, Gao S, Lyu J, Gu H (2018) Mebendazole exhibits potent anti-leukemia activity on acute myeloid leukemia. Exp Cell Res 369:61–68. https://doi.org/10.1016/j.yexcr.2018.05.006
71. Freisleben F, Modemann F, Muschhammer J, Stamm H, Brauneck F, Krispien A, Bokemeyer C, Kirschner KN, Wellbrock J, Fiedler W (2021) Mebendazole mediates proteasomal degradation of GLI transcription factors in acute myeloid leukemia. Int J Mol Sci 22:10670. https://doi.org/10.3390/ijms221910670
72. Clesham K, Walf-Vorderwülbecke V, Gasparoli L, Virely C, Cantilena S, Tsakaneli A, Inglott S, Adams S, Samarasinghe S, Bartram J, Williams G, de Boer J, Williams O (2022) Identification of a c-MYB-directed therapeutic for acute myeloid leukemia. Leukemia 36:1541–1549. https://doi.org/10.1038/s41375-022-01554-9
73. Harada T, Perez MW, Kalfon J, Braes FD, Batley R, Eagle K, Nabet B, Leifer B, Kruell J, Paralkar VR, Stegmaier K, Koehler AN, Orkin SH, Pimkin M (2023) Rapid-kinetics degron benchmarking reveals off-target activities and mixed agonism-antagonism of MYB inhibitors. https://doi.org/10.1101/2023.04.07.536032
74. Nabet B, Ferguson FM, Seong BKA, Kuljanin M, Leggett AL, Mohardt ML, Robichaud A, Conway AS, Buckley DL, Mancias JD, Bradner JE, Stegmaier K, Gray NS (2020) Rapid and direct control of target protein levels with VHL-recruiting dTAG molecules. Nat Commun 11:4687–4688. https://doi.org/10.1038/s41467-020-18377-w

75. Herzog VA, Reichholf B, Neumann T, Rescheneder P, Bhat P, Burkard TR, Wlotzka W, von Haeseler A, Zuber J, Ameres SL (2017) Thiol-linked alkylation of RNA to assess expression dynamics. Nat Methods 14:1198–1204. https://doi.org/10.1038/nmeth.4435
76. Ramaswamy K, Forbes L, Minuesa G, Gindin T, Brown F, Kharas MG, Krivtsov AV, Armstrong SA, Still E, de Stanchina E, Knoechel B, Koche R, Kentsis A (2018) Peptidomimetic blockade of MYB in acute myeloid leukemia. Nat Commun 9:110. https://doi.org/10.1038/s41467-017-02618-6
77. Joy ST, Henley MJ, De Salle SN, Beyersdorf MS, Vock IW, Huldin AJL, Mapp AK (2021) A dual-site inhibitor of CBP/p300 KIX is a selective and effective modulator of Myb. J Am Chem Soc 143:15056–15062. https://doi.org/10.1021/jacs.1c04432
78. Suetaka S, Oka Y, Kunihara T, Hayashi Y, Arai M (2022) Rational design of a helical peptide inhibitor targeting c-Myb–KIX interaction. Sci Rep-uk 12:816. https://doi.org/10.1038/s41598-021-04497-w
79. Jones M, Grosche P, Floersheimer A, André J, Gattlen R, Oser D, Tinchant J, Wille R, Chie-Leon B, Gerspacher M, Ertl P, Ostermann N, Altmann E, Manchado E, Vorherr T, Chène P (2023) Design and biochemical characterization of peptidic inhibitors of the Myb/P300 interaction. Biochemistry-us. https://doi.org/10.1021/acs.biochem.2c00690
80. Bell CC, Fennell KA, Chan Y-C, Rambow F, Yeung MM, Vassiliadis D, Lara L, Yeh P, Martelotto LG, Rogiers A, Kremer BE, Barbash O, Mohammad HP, Johanson TM, Burr ML, Dhar A, Karpinich N, Tian L, Tyler DS, MacPherson L, Shi J, Pinnawala N, Fong CY, Papenfuss AT, Grimmond SM, Dawson S-J, Allan RS, Kruger RG, Vakoc CR, Goode DL, Naik SH, Gilan O, Lam EYN, Marine J-C, Prinjha RK, Dawson MA (2019) Targeting enhancer switching overcomes non-genetic drug resistance in acute myeloid leukaemia. Nat Commun 10:2723. https://doi.org/10.1038/s41467-019-10652-9

ETS1 Function in Leukemia and Lymphoma

16

Rebecca A. Luchtel

Abstract

ETS proto-oncogene 1 (ETS1) is a transcription factor (TF) critically involved in lymphoid cell development and function. ETS1 expression is tightly regulated throughout differentiation and activation in T-cells, natural killer (NK) cells, and B-cells. It has also been described as an oncogene in a range of solid and hematologic cancer types. Among hematologic malignancies, its role has been best studied in T-cell acute lymphoblastic leukemia (T-ALL), adult T-cell leukemia/lymphoma (ATLL), and diffuse large B-cell lymphoma (DLBCL). Aberrant expression of *ETS1* in these malignancies is driven primarily by chromosomal amplification and enhancer-driven transcriptional regulation, promoting the ETS1 transcriptional program. ETS1 also facilitates aberrantly expressed or activated transcriptional complexes to drive oncogenic pathways. Collectively, ETS1 functions to regulate cell growth, differentiation, signaling, response to stimuli, and viral interactions in these malignancies. A tumor suppressor role has also been indicated for ETS1 in select lymphoma types, emphasizing the importance of cellular context in ETS1 function. Research is ongoing to further characterize the clinical implications of ETS1 dysregulation in hematologic malignancies, to further resolve binding complexes and transcriptional targets, and to identify effective therapeutic targeting approaches.

Keywords

ETS1 · ETS family · Hematologic malignancies · Leukemia · Lymphoma · Transcription factor · T-cell · B-cell

Abbreviations

AP1	activator protein 1
ATAC-Seq	assay for transposase-accessible chromatin sequencing
ATLL	adult T-cell leukemia/lymphoma
BATF3	basic leucine zipper ATF-like transcription factor 3
bHLH	basic helix-loop-helix
BRD4	bromodomain containing 4
CaMKII	calcium/calmodulin-dependent protein kinase II
CBFβ	RUNX1-core binding factor subunit β
CBP	CREB binding protein
CCR4	C-C motif chemokine receptor 4

R. A. Luchtel (✉)
Division of Hematology and Oncology, Department of Medicine, Northwestern University, Chicago, IL, USA
e-mail: rebecca.luchtel@northwestern.edu

T. Borggrefe, B. D. Giaimo (eds.), *Transcription factors in blood cell development*, Advances in Experimental Medicine and Biology 1459, https://doi.org/10.1007/978-3-031-62731-6_16

CDC73	cell division cycle 73
ChIP-Seq	chromatin immunoprecipitation followed by deep sequencing
cHL	classical Hodgkin's lymphoma
COP1	COP1 E3 ubiquitin ligase
CXCR5	C-X-C motif chemokine receptor 5
DBD	DNA binding domain
DLBCL	diffuse large B-cell lymphoma
DN	double negative
DP	double positive
E2A/TCF3	transcription factor 3
E2F	E2F transcription factor
EBV	Epstein-Barr virus
ELK1	ETS transcription factor ELK1
EP300	E1A binding protein p300
ETS1	ETS protooncogene 1
FLI1	Friend leukemia integration 1
FOXP3	Forkhead box P3
GABPA	GA binding protein transcription factor subunit alpha
H3K27ac	histone 3 lysine 27 acetylation
HBZ	HTLV-1 basic leucine zipper
HEB/TCF12	transcription factor 12
HTLV-1	human T-cell leukemia virus type 1
IL2	interleukin 2
IL7R	interleukin 7 receptor
KO	knockout
LTR	long terminal repeat
MAPK	mitogen-activated protein kinase
MHC	major histocompatibility complex
mTORC1	mechanistic target of rapamycin kinase complex 1
NFAT	nuclear factor of activated T-cells
NK cells	natural killer
PAX5	paired box 5
PRDM1	PR/SET domain 1
PTMs	posttranslational modifications
RAG1	recombination-activating gene 1
SP1	specificity protein 1
SPI1/PU.1	spleen-focus forming virus (SFFV) proviral integration protein/hematopoietic transcription factor PU.1
STAT5	signal transducer and activator of transcription 5A
SUMO-2	small ubiquitin like modifier 2
T-ALL	T-cell acute lymphoblastic leukemia
TAL1	T-cell acute leukemia 1
TAX	transactivator protein/p40
TCR	T-cell receptor
TF	transcription factor
Th17	T-helper 17
TLX1	T-cell leukemia homeobox 1
Treg	regulatory T-cell
TSS	transcription start site
USP9X	ubiquitin-specific peptidase 9 X-linked
VEGF	vascular endothelial growth factor

16.1 ETS1 Function

ETS proto-oncogene 1 (ETS1, encoded by *ETS1*) is a highly conserved transcription factor (TF) with critical roles in immune cell development and function. In addition to these functions in the normal cellular context, ETS1 has been described as a proto-oncogene in a range of solid and hematologic malignancies. This chapter will summarize the normal and oncogenic functions of ETS1 with focus on the role of ETS1 in leukemias and lymphomas.

ETS1 was first discovered as part of a fusion gene with *v-myb* in the E26 avian leukemia retrovirus [1]. ETS1 is the founding member of the ETS family of transcription factors, composed of 28 family members in humans (27 in mice). The ETS family shares an 85-amino acid winged helix-turn-helix DNA binding domain known as the ETS domain. The ETS domain recognizes the 5′-GGA(A/T)-3′ DNA motif. While all ETS family members share the ETS domain, they differ in the presence of other protein domains as well as in their expression and function across cell types and contexts.

The ETS1 amino acid sequence is highly conserved across a wide range of species, suggesting an essential and nonredundant function. Of the ETS family members, ETS1 shares the greatest homology and similarity in domain structure with ETS2. However, despite the similarity in sequence, ETS1 and ETS2 have unique as well as redundant functions, likely mediated in part by varied interacting proteins as well as cell type-specific expression.

16.1.1 Expression and Isoforms

Under normal conditions, ETS1 is expressed predominantly in lymphoid tissues where ETS1 plays a critical role in normal lymphoid cell development. The expression of ETS1 varies across immune cell subsets and in response to differentiation stage and activation with strong expression in T-cell subsets, natural killer (NK) cells, and naïve and memory B-cells. ETS1 expression generally decreases upon activation, and as such, *ETS1* expression and polymorphisms have been linked to inflammatory and autoimmune diseases [2]. Aberrant expression is also observed in lymphoid cells and nonlymphoid tissues in various cancer types.

Multiple isoforms of *ETS1* have been characterized, with three isoforms primarily expressed in humans. The longest *ETS1* isoform (p68) containing a novel N-terminus consisting of exons Ia, I, and II is not expressed in humans. The dominant isoform in humans is p51 (also known as p54) (Fig. 16.1a). p51 is characterized by an alternate start site and retention of exons III–VI, encoding pointed and acidic domains; exon VII, encoding the first of two autoinhibitory domains; and exons VIII and IX, encoding the ETS and second inhibitory domains (Fig. 16.1a–b).

p42 is the second most abundant *ETS1* isoform in humans, albeit at a significantly lower expression level than p51. The p42 isoform differs from p51 by deletion of exon VII, resulting in loss of the N-terminal autoinhibitory domain and subsequent increased DNA binding activity (Fig. 16.1a) [3]. Another instance in which loss of an autoinhibitory domain exists is in the E26 avian leukemia retrovirus, wherein the C-terminal autoinhibitory domain of ETS1 is deleted, resulting in oncogenic activity [4].

The third *ETS1* isoform found in humans, p27, lacks exons III–VI corresponding to pointed and acidic/transactivation domains (Fig. 16.1a). This isoform has ETS and autoinhibitory domains, exerting dominant negative effects on p51/p54 [5]. It has been suggested that the p27 isoform binds DNA, preventing p51/54 transactivation as well as causing translocation of p51/p54 from the nucleus to cytoplasm [5].

16.1.2 Protein Domains and Posttranslational Modifications

Central to all ETS family members is the ETS DNA binding domain (DBD) that recognizes the 5′-GGA(A/T)-3′ ETS motif. In ETS1, this domain resides at the C-terminus (Fig. 16.1b).

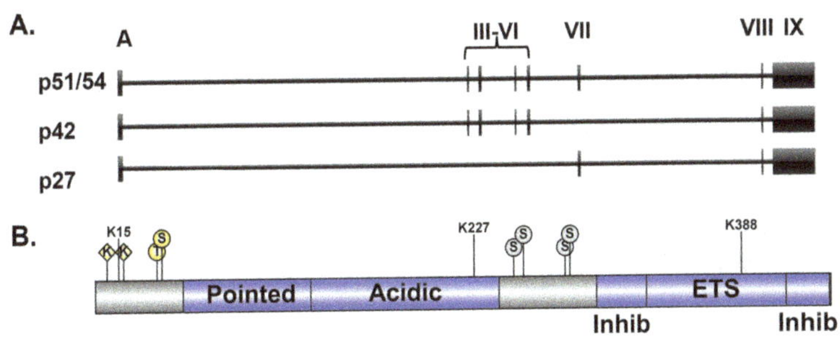

Fig. 16.1 Gene and protein models of ETS1. (**a**) Gene models for p51/p54, p42, and p27 isoforms of human *ETS1*. Vertical black bars depict exons. A, alternate start site. (**b**) Protein domains and key residues of ETS1 p51/p54. Inhib, autoinhibitory domains. Yellow, activating; gray, inhibitory. Circle, phosphorylation (T38, S41, S251, S257, S282, S285); diamond, acetylation (K8, K18). Sumoylation (K15, K227) and ubiquitination (K388) sites marked. Models were constructed using Group-Based Prediction System IBS 1.0 software [102]

In the dominant isoform of ETS1, p51/p54, the ETS domain is flanked by autoinhibitory domains composed of a hydrophobic core surrounded by four α-helices (HI-1, HI-2, H4, and H5). The autoinhibitory domains of ETS1 are regulated by phosphorylation as well as interaction with binding factors [6]. The pointed domain, found in about one-third of ETS family members, is located in the N-terminus of ETS1. This domain serves as a mitogen-activated protein kinase (MAPK) docking site [7], enhancing phosphorylation [8], which in turn is proposed to recruit transcriptional machinery. The acidic or transactivation domain is a group of acidic residues required for transcriptional activation [9] and CREB binding protein/lysine acetyltransferase 3A (CBP)/KAT3A and E1A binding protein p300/lysine acetyltransferase 3B (EP300/KAT3B, encoded by *EP300*) coactivator binding [10].

Protein modifications such as phosphorylation, acetylation, sumoylation, and ubiquitination facilitate or counteract domain functions as well as control protein activity, stability, and turnover (Table 16.1). As such, some of these modifications and interactions have also been explored as therapeutic targets for ETS1 inhibition.

Phosphorylation Phosphorylation can result in either inhibition or activation of ETS1 function, which is dependent on the protein domain and specific residues. The autoinhibitory domains flanking the DNA-binding ETS domain are stabilized by phosphorylation, resulting in reduced affinity for DNA. This phosphorylation is carried out by calcium-/calmodulin-dependent protein kinase II (CaMKII) calcium signaling, resulting in phosphorylation at S251, S257, S282, and S285 [11, 12]. Phosphorylation of these serine residues reinforces the autoinhibitory domain and results in progressive weakening of DNA binding [13]. In contrast, phosphorylation of T38 and S41 function to enhance the transcriptional activity of ETS1. T38 and S41 have been shown to be phosphorylated in response to a number of signaling factors, including vascular endothelial growth factor (VEGF) [14] and Ras/ERK/MAPK [8, 15].

Acetylation Lysine acetylation of K8 and K18 has been shown to promote transcriptional activity of ETS1. ETS1 is thought to be acetylated by complexing with CBP/KAT3A and EP300/KAT3B. Phosphorylation of T38 and S41 resulting from various signaling pathways (e.g., VEGF and ERK/MAPK) promotes binding of CBP/KAT3A and EP300/KAT3B and therefore increases acetylation [14]. ETS1 acetylation may then enhance interaction with bromodomain containing 4 (BRD4) to promote transcription of target genes [14].

Sumoylation Sumoylation at K15 and K227 is an inhibitory posttranslational modification to ETS1. Small ubiquitin-like modifier 2 (SUMO-2) sumoylation of ETS1 was stimulated in response to H_2O_2 treatment, indicating a response to cellular stress [16]. Sumoylation does not have a major effect on ubiquitination, protein stability, or nuclear localization of the protein. Rather, SUMO-2 was shown to repress transcriptional activity of ETS1.

Table 16.1 Posttranslational modifications (PTMs) of ETS1

Modification	Residues	Factors involved	Implication	References
Phosphorylation (activation)	T38, S41	RAS/MEK/ERK and VEGF	Induces transcriptional activity and enhances acetylation	[8, 14, 103]
Phosphorylation (repression)	S251, S257, S282, S285	CaMKII calcium signaling	Inhibits DNA binding activity through decreased affinity	[11, 13]
Sumoylation	K15, K227	SUMO-2	Inhibits transcriptional activity	[16]
Ubiquitination	K388	USP9X (deubiquitinase) and COP1 (ubiquitin ligase)	Proteasomal degradation	[16–18]
Acetylation	K8, K18	CBP/KAT3A EP300/KAT3B	Increases BRD4 binding	[14]

Ubiquitination Ubiquitination of ETS1 induces degradation of ETS1 through the proteasome pathway, as demonstrated by increased abundance of ETS1 in the presence of 26S proteasome inhibitor [16]. Ubiquitin-specific peptidase 9 X-linked (USP9X) is a deubiquitinating enzyme involved in processing ubiquitinated proteins as well as ubiquitin precursors. Knockdown of USP9X resulted in decreased ETS1 protein expression and increased ETS1 ubiquitination at K388, a putative ubiquitination site located in the ETS domain [17]. In addition to ETS1, several other ETS family proteins have also been shown to interact with and be targeted by USP9X. The CaMKII serine phosphorylation sites S276 and S282 have also been shown to serve as binding sites for COP1 E3 ubiquitin ligase (COP1), a part of the ubiquitin ligase complex that promotes polyubiquitylation of ETS1 [18]. Phosphorylation of these serine residues promotes COP1 binding while phosphorylation of Y283 was shown to reduce the affinity of COP1 for ETS1 [18].

16.1.3 Role in Blood Cell Development and Transcriptional Target Genes

ETS1 is predominantly expressed in lymphoid cells and is essential for lymphoid cell development and function. This has been demonstrated using *Ets1* knockout (KO) mouse models in which *Ets1* loss has a marked effect on T-cells, NK cells, and B-cells. *Ets1*-deficient mice display marked growth retardation and reduced survival within the first 3–4 weeks of life [19, 20]. Therefore, most mouse models of *Ets1* deficiency have focused on cell-type-specific *Ets1* loss.

T-cells The essential functions of ETS1 in T-cells are wide ranging, including differentiation, development, activation, and survival. In a *Rag2* KO mouse model introduced with *Ets1*-deficient embryonic stem cells, thymocytes and peripheral T-cells were reduced in number in vivo, and T-cells were highly susceptible to apoptosis in vitro [21, 22].

More extensive profiling of T-cell development in a similar model revealed that ETS1 is required for the development of double-positive (DP) thymocytes in the αβ lineage, with cell death occurring in the double-negative subsets [19]. Development of γδ thymocytes was not similarly affected and matured normally. This is likely due to a role of ETS1 in pre-T-cell receptor (TCR) signaling, which is required for allelic exclusion of the *TCRβ* locus in αβ thymocytes [19]. In a Vav-Cre-mediated conditional mouse model of *Ets1* deficiency, a similar phenotype was observed, with profoundly reduced total thymocyte number and number of cells at double-negative (DN) stages, as well as hampered DN to DP transition [23].

Lineage differentiation differences have also been demonstrated in *Ets1* KO thymocytes. ETS1 has been shown to contribute to the differentiation of major histocompatibility complex (MHC) class 1 thymocytes and expression of CD8 [24]. It has been proposed that ETS1 plays a role in promoting the CD8 lineage through regulating gene expression of *Cd8a* [25] as well as *Runx3*, a TF responsible for *Cd4* silencing [26]. ETS1 has also been shown to be required for regulatory T-cell (Treg) development via a proposed role in regulating Forkhead box P3 (*Foxp3*) expression, in which ETS1 interacts with *Foxp3* enhancer and is required for DNA demethylation of this locus [27]. In contrast, ETS1 has been shown to be a negative regulator of T-helper 17 (Th17) differentiation, with enhanced Th17 differentiation in *Ets1*-deficient Th cells to Th17 cells, possibly due in part to resistance to interleukin 2 (IL2) [28].

It was also demonstrated that ETS1 is required for activation, with *Ets1*-deficient T-cells having reduced proliferation in response to anti-CD3 stimulation [21]. It has been proposed that ETS1 may play a central role, together with nuclear factor of activated T-cells (NFAT) 2, in the stimulation and restimulation transcriptional response of memory T-cells [29].

NK Cells Similar to its role in T-cells, ETS1 is also required for development and activity of NK cells. *Ets1* KO mice (generated via deletion of exons III and IV of the *Ets1* gene) have severely reduced number of mature NK cells and reduced antitumor/cytolytic activity [30, 31]. The reduced cytolytic activity is proposed to be a function of deficiency in degranulation or expression of inhibitory NK cell receptors [31]. These roles for ETS1 were replicated in human embryonic stem cells models of *ETS1* knockdown and overexpression of the dominant-negative *ETS1* isoform, p27 [32]. Genome-wide transcriptional profiling and ETS1 binding experiments have identified a number of key NK cell regulatory genes to be transcriptionally driven by ETS1 [31, 32]. Of note, NK T-cells, like NK cells, depend on ETS1 for development and are similarly reduced in number in *Ets1* KO mice [33].

B-Cells Unlike T- and NK cells, the total number of B-cells has not been shown to be affected by ETS1 status. However, mouse models of B-cell *Ets1* deficiency revealed skewed B-cell subsets with a marked accumulation of IgM plasma cells and reduced proportion of marginal zone B-cells [22, 34, 35]. This was accompanied by an increased differentiation to memory B-cells [35], secretion of autoantibodies [34, 35], and loss of self-tolerance [36]. In support of this, chromatin immunoprecipitation followed by deep sequencing (ChIP-Seq) analysis of genome-wide murine ETS1 DNA binding and transcriptional response in B-cells identified direct transcriptional targets of ETS1, which included genes involved in autoimmunity [37].

16.1.4 DNA Binding and Interacting Factors

Protein-protein interactions play a critical role in ETS1 DNA binding through influence on ETS1 activity, enhancer and promoter localization, transcriptional targets, and transactivation or repressive function. The sequence, organization, and proximity of DNA motifs are important to this regulation, and the specific ETS1 DNA binding sequence and flanking region as well as nature of interaction can vary by genomic region and binding partner context. The transcriptional targets of ETS1 encompass both unique and redundant roles for ETS1. Genome-wide analysis of ETS1 occupancy has demonstrated distinct consensus sequences for housekeeper and nonhousekeeper genes as well as proximity to the transcription start site (TSS) [38]. However, specificity of ETS1 for nonhousekeeper genes as well as the role of sequence and/or binding partners on the relative activity of ETS1 are not yet comprehensively defined [39].

The best characterized structural relationships representing ETS1 heterodimeric DNA binding are ETS1-Runt-related transcription factor 1 (RUNX1) and murine ETS1-paired box 5 (PAX5) in T- and B-cells, respectively (Fig. 16.2). Homodimerization of ETS1 binding to tandem head-to-head ETS motifs has also been characterized [40, 41]. In these interactions, the autoinhibition that is reinforced by serine phosphorylation is counteracted by interactions with partner TFs. This is a key unifying feature of ETS1 DNA binding, although the specific allosteric mechanisms contributing to antagonization of autoinhibition may vary based on the partner TF.

RUNX1 is a TF involved in lymphoid development (among other cell types). ETS1 interacts with the RUNX1-core binding factor subunit β (CBFβ) heterodimer at *TCRα/TCRβ* loci, leading to reciprocal activation of DNA binding/transactivation [42]. Structural analyses of this relationship have been reported by several groups (Fig. 16.2a) [6, 43, 44]. Comparison of this binding interaction to both murine ETS1-PAX5 (*Mb-1* locus) and human ETS1 homodimer (stromelysin-1 locus, encoded by *MMP3*) led to the proposal of an allosteric regulation unique to ETS1-RUNX1 wherein the stability of the complex is refractory to the inhibitory effect of ETS1 serine phosphorylation [43]. Genome-wide analysis of ETS1 occupancy in the Jurkat T-cell line revealed a preferential ETS1 composite binding motif associated with RUNX1 co-occupancy (CAGGA[A/T]GTGG) which is similar to *TCRα/β* enhancer binding sequences (Fig. 16.2a) [38]. Of note, a high degree of overlap was observed between RUNX-bound

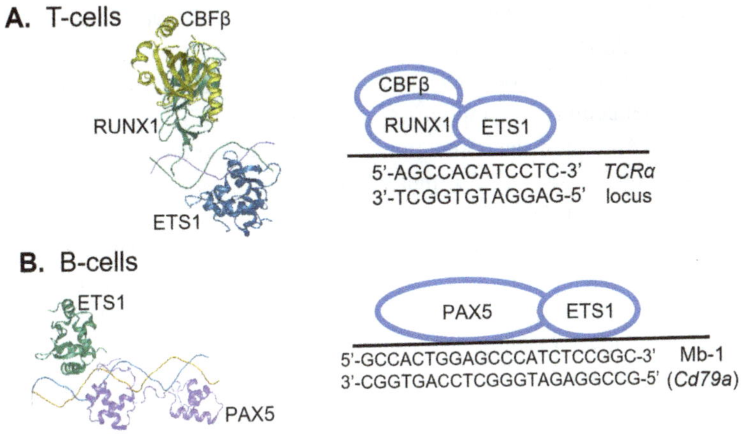

Fig. 16.2 ETS1 interacting partner examples in T- and B-cells. (**a**) Interaction between ETS1 and the RUNX1-CBFβ dimer at the *TCRα* enhancer [43]. (**b**) Interaction between ETS1 and PAX5 at the murine *Mb-1* (*Cd79a*) locus [50]. Protein structures and DNA sequence are shown for each interaction. Protein structures were visualized using Protean 3D software (DNASTAR, Inc)

regions and cooccupancy with ETS1, and many of these regions were localized >500 bp from the transcriptional start site, indicating a role in T-cell enhancer regulation [38]. Adding to our understanding of ETS1-RUNX1 interactions is the recent report of a role for signal transducer and activator of transcription 5A (STAT5) in the recruitment of and interaction with these factors at T-cell-specific loci [45]. STAT5-ETS1 interactions have been reported upon IL2 stimulation in T-cells [46]. Cooperative interactions have also been reported between ETS1 and other protein regulators of the T- and NK cell transcriptional program such as NFAT, activator protein 1 (AP1), and T-box transcription factor 21 (T-BET, encoded by *TBX21*) [47, 48].

In B-cells, the interaction between murine PAX5, a key regulator of B-cell lineage commitment, and ETS1 has been structurally characterized (Fig. 16.2b) [49, 50]. Cooperative binding of PAX5 functions to counteract autoinhibition of ETS1 through ternary complex assembly. This complex alters the conformation of ETS1 to increase DNA binding affinity, promoting binding of ETS1 to the *Mb-1* (*Cd79a*) promoter and transcriptional activator function [50–52].

In addition to cooperation with TFs involved in lymphoid development, ETS1 has been shown to interact with epigenetic and viral factors and other proteins involved in specific malignancies (detailed in the next section). Phosphorylation-enhanced binding of ETS1 to CBP/KAT3A and EP300/KAT3B histone acetyltransferases has been described by multiple groups to have important roles in ETS1 activation and transcriptional regulation [8, 10, 53]. Binding of ETS1 to CBP/KAT3A may also function to enhance acetylation of TFs through complex formation [54]. In the angiogenesis response, acetylated ETS1 has been shown to bind to BRD4, recruiting a transcriptional complex to promote transcription [14]. There is also suggestion that ETS1 interacts with histone deacetylase 1 (HDAC1) [55]. The interaction between ETS1 and viral proteins, as well as the presence of ETS motifs in viral genes, represents another potentially important role of ETS1 that requires further study as it relates to viral-associated cancers as well as other diseases [56].

16.2 Role of ETS1 in Hematologic Malignancies

Given the importance of ETS1 in regulating differentiation and development and response to stimuli in lymphoid cells, it follows that ETS1 would also serve as a target for malignant transformation and growth in these same cell types. In line with this, an oncogenic role for ETS1 has been described in both leukemia and lymphomas. However, its varied expression and function across immune cell types and subsets also make the impact of ETS1 context-dependent, with tumor suppressor functions also proposed in select hematologic malignancies.

In contrast to some other well-down oncogenic driver genes in leukemias/lymphomas, *ETS1* has a relatively low mutation rate with no highly recurrent mutation sites identified to date. While chromosomal translocations involving the 11q23-24 region have been reported, the majority of these do not involve *ETS1*; as such, translocation does not appear to be a major mechanism of ETS1 dysregulation. Instead, ETS1 has been shown to be overexpressed or dysregulated in leukemias and lymphomas through a range of alternate mechanisms such as copy number alterations, enhancer-driven expression, and protein-protein interactions (Table 16.2, Fig. 16.3). Aberrant expression and/or regulation of ETS1 then promotes a range of prooncogenic functions affecting cell growth, differentiation, signaling, response to stimuli, and viral interactions (Fig. 16.4). Specific associations between overexpression of ETS1 and clinical parameters such as outcome/survival or prognosis are yet to be determined for most hematologic malignancies. Importantly, tumor suppressor function has also been suggested for ETS1 in some types of lymphoma such as large B-cell lymphoma with 11q aberration and classical Hodgkin's lymphoma (cHL).

Although KO mouse models have shown ETS1 to be particularly important in normal lymphoid development and function, oncogenic roles for ETS1 have also been well characterized in a range of solid tumor types [57] where aberrant ETS1 expression has been associated with increased invasion and modulation of DNA repair factors, metabolic regulation, and the tumor microenvironment. Many of the emerging functions of ETS1 in hematologic malignancies mirror what has been shown in solid tumors. In the remainder of this chapter, the role of ETS1 in leukemias and lymphomas that have been reported to date is summarized.

Table 16.2 Genomic alterations in *ETS1* in leukemias and lymphomas

Disease	Prevalence	Functional significance	Reference
CNA—Gain			
Diffuse large B-cell lymphoma (DLBCL)	23–24%	Increased expression of *ETS1* and *FLI1*	[84, 85]
Acute myeloid leukemia (AML)	29–40%	Complex karyotype associated with poor prognosis; unclear whether *ETS1* is a functional target of gain	[104, 105]
Gastrointestinal marginal zone B-cell lymphoma	> 20%	Found in large cell variants	[94]
Nonsynonymous mutations			
Diffuse large B-cell lymphoma (DLBCL)	~5%	May be associated with poor OS	[87, 88]
Large B-cell lymphoma with 11q aberration	~25%		[90, 92]
Burkitt lymphoma	~30%	Higher incidence in EBV+ cases	[95]
CNA—Loss			
Large B-cell lymphoma with 11q aberration	100%	Downregulated *ETS1* expression	[90, 91]
Classical Hodgkin's lymphoma (cHL)	>60%	Downregulated *ETS1* expression	[93]

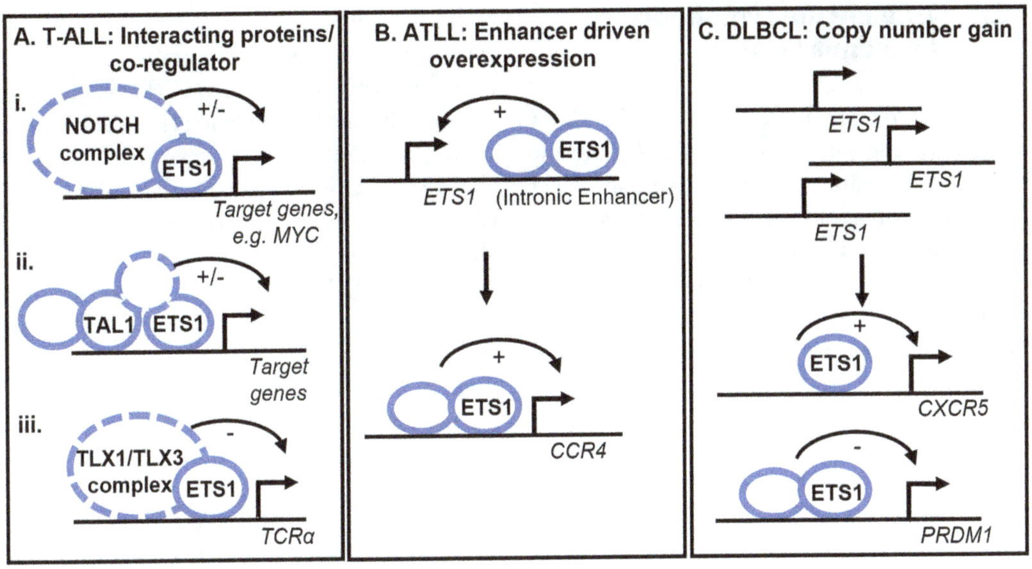

Fig. 16.3 Mechanisms of ETS1 oncogenic dysregulation and aberrant expression in hematologic malignancies. (**a**) In T-ALL, ETS1 interacts with oncogenic protein complexes to facilitate transcriptional regulation, resulting in activation or repression of target genes. (**i**) ETS1 interacts with the NOTCH transcriptional complex and promotes H3K27ac, coregulating the NOTCH transcriptional program [23]. (**ii**) ETS1 interacts with the TAL1 transcriptional complex, regulating T-cell-specific and apoptotic target genes [58, 66]. (**iii**) ETS1 interacts with the TLX1/TLX3 complex to repress *TCRα* transcription and mediate T-cell maturation arrest [68, 69]. (**b**) In ATLL, ETS1 promotes its overexpression through intronic enhancer driven transcriptional activation. ETS1 also regulates transcription of ETS1 target genes, such as *CCR4* [73]. (**c**) In DLBCL, the *ETS1* locus is amplified, resulting in overexpression of *ETS1* and transcriptional activation and inhibition of ETS1 target genes, *CXCR5* and *PRDM1*, respectively [85]. Dashed lines indicate multiprotein complexes

Fig. 16.4 Proposed cellular functions of ETS1 in hematologic malignancies

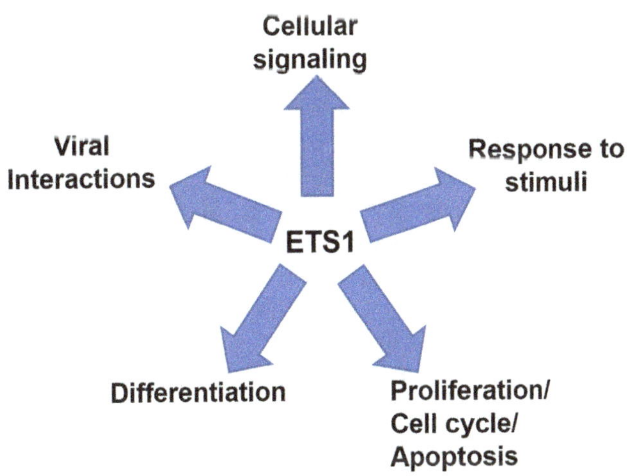

16.3 T-Cell Acute Lymphoblastic Leukemia (T-ALL)

T-cell acute lymphoblastic leukemia (T-ALL) is an aggressive leukemia characterized by arrest in normal thymic differentiation at various stages and transformation of T-cell progenitors. Leukemogenesis is driven by a combination of genetic events and aberrant expression of transcriptional regulators. ETS1 is highly expressed in T-ALL and has been associated with hyperleukocytosis but not survival [23]. Knockdown of *ETS1* resulted in reduced proliferation/cell cycle progression and increased apoptosis in T-ALL cell lines and reduced disease progression in a patient-derived xenograft (PDX) mouse model [23, 58]. ETS1 has been implicated in T-ALL biology through facilitation of binding of key oncogenic drivers such as NOTCH, T-cell acute leukemia 1 (TAL1), and T-cell leukemia homeobox 1 (TLX1/TLX3) (Fig. 16.3a). In each of these examples, ETS1 is hijacked by genetically driven oncogenic transcriptional regulators that are aberrantly expressed or activated to promote oncogenic transcriptional programs.

Approximately 60% of T-ALLs are NOTCH-driven through activating NOTCH receptor 1 (*NOTCH1)* mutations. NOTCH proteins are master regulatory factors involved in cell- and tissue-specific development and are activated in a number of tumor types. NOTCH1 mutations drive expression of oncogenic target genes and have been a mechanistic and therapeutic focus in T-ALL. ChIP-Seq revealed significant overlap in DNA binding between NOTCH1 and the T-cell transcriptional regulators, ETS1, RUNX1, and GA binding protein transcription factor subunit alpha (GABPA, an ETS family member) [59]. Building on this, the role of ETS1 in NOTCH signaling was studied in T-ALL [23]. Using a transgenic mouse model of activated *Notch1*-induced T-ALL, *Ets1* deletion was found to significantly reduce leukemic cells and extend survival [23]. Overlap was observed between genes regulated by ETS1 and NOTCH1 in leukemia cells, which included the c-MYC pathway. Further analyses demonstrated cooccupancy and physical interaction of ETS1 and NOTCH1 at NOTCH binding sites [23]. *ETS1* knockdown experiments further demonstrated a role for ETS1 in recruitment of NOTCH complex and histone 3 lysine 27 acetylation (H3K27ac) deposition [23]. Cell division cycle 73 (CDC73), a NOTCH cofactor, was later found to interact with ETS1 and to at least in part require ETS1 for NOTCH binding to enhancers of oncogenic target genes [60]. Taken together, it was proposed that ETS1 facilitates recruitment of NOTCH complex to specific NOTCH targets in T-ALL (Fig. 16.3ai). Knockdown/loss of *ETS1* in vitro and in a mouse model of NOTCH-induced T-ALL resulted in enhanced response to NOTCH inhibition [23].

TAL1, an essential regulator of hematopoiesis, is overexpressed in a subset of T-ALLs with chromosomal rearrangement driving a minority of these cases [61, 62]. TAL1 overexpression has been implicated in T-ALL leukemogenesis [63–65]. As a basic helix-loop-helix (bHLH) TF, TAL1 heterodimerizes with bHLH E-proteins, transcription factor 3 (E2A, encoded by *TCF3*), and transcription factor 12 (HEB, encoded by *TCF12*) and along with other transcriptional regulators forms a core transcriptional complex. In T-ALL cell lines, ETS1 and RUNX1 were both directly regulated by TAL1 and were shown to also interact with TAL1 through immunoprecipitation experiments [58]. Significant overlap was observed between ETS1, RUNX1, and TAL1 DNA binding sites, and loss of either ETS1 or RUNX1/RUNX3 resulted in reduced TAL1 binding [58, 66]. Target genes of cooccupancy sites include T-cell-specific and apoptotic genes, suggesting a possible role for ETS1, in concert with RUNX and TAL1, in leukemogenesis. LIM domain-only 1 (LMO1) and LMO2 function together with TAL1 as part of this regulatory transcriptional complex [66]. Of note, predicted ETS1 DNA binding sites have also been shown to be gained through insertion and deletion events (indels) in the *LMO2* promoter in T-ALL, although ETS1 binding to these gained consensus sequences is yet to be demonstrated [67].

TLX1 and TLX3 (also known as HOX11 and HOX11L2) are aberrantly expressed in a subset of T-ALL owing to chromosomal translocations.

TLX1 is an oncogenic transcriptional repressor that has been implicated in leukemogenesis through interactions with pathways involved in growth, differentiation, and signaling. ChIP-Seq analysis identified ETS1 and RUNX1 as the most enriched motifs in TLX1 binding sites and showed overlap between DNA binding of TLX1, ETS1, and RUNX1 at TLX1-repressed genes [68]. ETS1 was shown to physically interact with TLX1/TLX3 and to be required for TLX1-/TLX3-mediated transcriptional repression of *TCRα* [69]. ETS1/RUNX1-associated TLX1 transcriptional repression of T-cell recombination genes has also been shown for *TCRβ*, recombination-activating gene 1 (*RAG1*), and interleukin 7 receptor (*IL7R*) genes, together demonstrating a role for ETS1 as a mediator of TLX1/TLX3 in T-cell maturation arrest [68, 69]. Notably, overlap has also been observed between both NOTCH1 targets and tumor suppressor genes and genes repressed by TLX1 [68]. It has been proposed that antagonism of NOTCH signaling early in pathogenesis may serve as genetic pressure for acquisition of NOTCH1 mutations [68]. Cobinding of TLX1 and ETS1 was supported in another study using transgenic mice expressing *TLX1* and the *NUP214-ABL1* fusion gene [70]. In this study, $TLX1^+$ leukemic cells derived from this mouse model compared to $CD4^+CD8^+$ wild-type cells showed enrichment of ETS1, STAT5, and RUNX1 motifs in accessible chromatin regions [70]. This study further showed that TLX1 and STAT5 cobound enhancer regions and that the ETS motif was enriched in genes negatively regulated by STAT5 [70].

16.4 Adult T-Cell Leukemia/Lymphoma (ATLL)

Adult T-cell leukemia/lymphoma (ATLL) is an aggressive type of peripheral T-cell lymphoma driven by the human T-cell leukemia virus-type 1 (HTLV-1) retrovirus. HTLV-1 is endemic in many parts of the world, including Japan, Caribbean, Africa, and Latin America, among other regions. ATLL develops in approximately 2–5% of people infected with HTLV-1 virus, often decades after the initial infection.

ATLL cells often exhibit a Treg-like immunophenotype with characteristic positivity for CD4, CD25, and C-C motif chemokine receptor 4 (CCR4) and partial positivity for CD30 and FOXP3; however, variability exists in the immunophenotype [71]. Interestingly, the somatic variant profile has been shown to vary based on geographic region of descent, indicating possible roles of host or virus genetics, among other potential environmental factors, in disease pathogenesis [72]. While nonsynonymous variants have been identified in *ETS1* with low frequency, no recurrent somatic mutations or other genetic events have been reported to date for *ETS1* in this malignancy. p51 represents the major *ETS1* isoform in ATLL cells and is highly expressed, particularly for ATLL of North American descent, with an oncogenic role recently reported [73].

Assay for transposase-accessible chromatin sequencing (ATAC-Seq) of ATLL patient-derived cell lines identified the ETS motif as a highly enriched DNA binding motif in open chromatin, suggesting a regulatory role for the ETS family in ATLL [73]. Transcriptional analyses predicted *ETS1* as a primary ETS family member overexpressed in ATLL. ATAC-Seq of ATLL cells revealed open chromatin in an intronic enhancer for *ETS1*, and ChIP-Seq demonstrated ETS1 binding to this region, together providing evidence that ETS1 is driving its own expression (Fig. 16.3b).

Functional studies showed that ETS1 drove cell growth and *CCR4* expression in ATLL cells of North American descent [73]. *CCR4* gain-of-function mutations are recurrent in ATLL, and elevated expression of *CCR4* is associated with poor prognosis [74]. *CCR4* is a transcriptional target of ETS1 in normal Treg cells as well as ATLL. In a small cohort of ATLL patients of North American descent, *ETS1* expression was higher in cases without gain-of-function *CCR4* mutations, indicating that elevated ETS1 may be an alternate mechanism to increase *CCR4* expression or activity. While this study focused on

ETS1, there is also the possibility that other ETS family members expressed in ATLL may also play a role in disease pathogenesis.

A pathogenic role for ETS1 in ATLL is supported by additional lines of evidence, potentially indicating multiple regulatory mechanisms, including interaction with HTLV-1. Of note, *ETS1* expression was higher in ATLL cases with higher HTLV-1 proviral load [73]. Because ATLL is driven by the HTLV-1 retrovirus, much research in ATLL has rightly focused on the role of HTLV-1 viral proteins in disease pathogenesis. Of the HTLV-1 proteins, transactivator protein/p40 (TAX) and HTLV-1 basic leucine zipper (HBZ) are the best characterized in disease pathogenesis. The majority of ATLL cells express HBZ, which is thought to be involved in ATLL maintenance, whereas expression of TAX is more transient owing to its higher immunogenicity.

Characterization of HBZ-interacting factors has suggested ETS1 as an interacting host transcriptional regulator. Analysis of mutation burden at viral and host TF sites identified HBZ binding sites containing ETS or AP1 recognition sequences as having the highest mutation rates [75]. Among host transcription factors, ETS1 binding sites also had elevated mutation burden. In addition, $CD4^+$ T-cells transduced with HBZ RNA showed increased activation of ETS1 target sequences (along with basic leucine zipper ATF-like transcription factor 3 (BATF3) and RUNX1) [76]. This is in alignment with ATAC-Seq of ATLL cells in which ETS1, BATF3, and RUNX1 were identified as the dominant motifs in accessible chromatin [73]. Together, these data support a consistent trend in which both AP1 (BATF3) and ETS1 are key host transcriptional regulators interacting with HTLV-1 HBZ.

Potentially relevant to ATLL are reports of direct interactions between ETS1 and TAX. It was initially reported that ETS1 is capable of binding to the HTLV-1 long terminal repeat (LTR) regulatory region, enhancing transcriptional activity [77, 78]. Follow-up studies demonstrated cooperation between TAX and ETS1 in activation of the HTLV-1 LTR [79, 80]. Interaction between ETS1 and TAX has also been shown in another genomic context in which TAX cooperates with ETS1 in the formation of the ternary complex composed of TAX-ETS1-specificity protein 1 (SP1) [81]. This complex was shown to interact with DNA to transactivate the parathyroid hormone-related protein (PTHrP encoded by *PTHLH*) promoter. Similar relationships have been described between TAX and other ETS family members including SPI1 protooncogene (SPI1) and ETS transcription factor ELK1 (ELK1) [82, 83]. While these interactions have been demonstrated in T-cells (among other cell types), primarily through the use of reporter assays, they are yet to be demonstrated in the context of ATLL.

Further work is required to comprehensively understand the mechanisms by which ETS1 is supporting oncogenic activity in ATLL as well as ways in which it may be targeted therapeutically.

16.5 Diffuse Large B-Cell Lymphoma (DLBCL)

Diffuse large B-cell lymphoma (DLBCL) is the most common type of lymphoma and is characterized by distinct molecular subgroups, each with defining genomic and transcriptomic features. Copy number gains of the 11q24.3 locus involving *ETS1* have been reported in approximately 25% of DLBCLs (Table 16.2) [84, 85]. Friend leukemia integration 1 (*FLI1*), an ETS family member located adjacent to *ETS1* in a head-to-head fashion, is similarly amplified. Copy number gains were associated with increased expression of both *ETS1* and *FLI1* transcripts [85]. Subgroup-specific differences in *ETS1* expression have been shown, with *ETS1* higher in activated B-cell-like (ABC) DLBCL compared to germinal center B-cell-like (GCB) DLBCL [85, 86]. Analysis of ETS1 T38 phosphorylation, a marker of activation, also showed increased nuclear T38 phosphorylation in ABC-DLBCL compared to GCB-DLBCL tumors [15]. Importantly, T38 phosphorylation was shown to be regulated by MEK/ERK, a target of B-cell activation [15].

Functional characterization of both ETS1 and FLI1 in DLBCL demonstrated that each independently drives proliferation in a DLBCL cell line [85]. Loss of ETS1 and FLI1 was also shown to promote apoptosis in a cooperative manner. ETS1 T38 phosphorylation was required for growth promotion in DLBCL cell lines [15]. ETS1 but not FLI1 was shown to have a direct, repressive effect on PR/SET domain 1 (*PRDM1*), a key regulator of plasma cell differentiation. ChIP confirmed binding of ETS1 to the *PRDM1* promoter [85]. ETS1 was also shown to positively regulate the chemokine receptor C-X-C motif chemokine receptor 5(*CXCR5*), whereas chemokine receptors *CXCR4* and *CXCR7* were positively regulated by FLI1 [85]. This is reminiscent to the promotion of *CCR4* expression in ATLL and suggests a common cellular function for ETS1 (and ETS family members) in lymphoid malignancies of influencing cellular signaling pathways through regulating chemokine receptor expression. Transcriptomic analyses also identified cell cycle pathways as well as E2F transcription factor (E2F), mechanistic target of rapamycin kinase complex 1 (mTORC1), and MYC targets to be associated with ETS1 T38 phosphorylation in DLBCL [15].

In addition to copy number alterations, somatic mutations have also been reported for ETS1 in DLBCL affecting approximately 5–8% of cases [87, 88]. These mutations are not recurrent in terms of site position and are characterized by nonsynonymous, splicing, and frameshift mutations. One study noted that *ETS1* mutations were observed exclusively in relapsed and refractory patients and as such were associated with shorter overall survival [88]. Functional characterization of somatic *ETS1* mutations in DLBCL have not yet been reported.

16.6 Proposed Tumor Suppressor Role in B-Cell Malignancies

It is important to note that a tumor suppressor role has also been proposed for ETS1 in specific B-cell malignancies: large B-cell lymphoma with 11q aberration and cHL. Both malignancies are characterized by copy number losses affecting the *ETS1* locus in addition to other potential mechanisms of inactivation. This is in contrast to the copy number gains observed in DLBCL and highlights the importance of cellular context to the functions of ETS1. To date, functional analyses of ETS1 have not been reported in these malignancies; therefore, a tumor suppressor role has not yet been determined.

Large B-cell lymphoma with 11q aberration is currently considered a provisional entity by the World Health Organization [89]. *ETS1* has been proposed as a target gene affected by focal homozygous deletion in 11q24 in this entity [90, 91]. Moreover, *ETS1* is mutated in about 25–30% of investigated cases in small cohorts [90, 92]. Mutations have been mapped to the DBD among other regions across *ETS1* but have not been functionally characterized.

Similarly, deletion of the *ETS1* locus was observed in over 60% of cHLs [93]. In addition to this, DNA methylation analysis suggested hypermethylation of regions containing ETS1 binding motif [93]. The *ETS1* locus itself is not hypermethylated; however, *ETS1* expression is significantly lower in cHL cell lines than normal B-cells.

16.7 Other B-Cell Lymphoma Types

There are other B-cell lymphoma types for which genomic aberrations have been reported for *ETS1*, but the function of *ETS1* has not yet been studied. Copy number gains were reported for gastrointestinal marginal zone B-cell lymphoma, occurring in >20% of cases [94]. Gains affecting the *ETS1* locus were found exclusively in cases with large cell cytology. In Burkitt lymphoma, *ETS1* mutations were identified in ~30% of cases [95]. These mutations were localized to intronic, exonic, splicing, and UTR regions and may be associated with Epstein-Barr virus (EBV) positivity. *ETS1*, among other ETS family members, was also identified as a gene downregulated by EBV in Burkitt lymphoma cells [96].

16.8 Summary of the Role of ETS1 in Leukemias and Lymphomas

Of the three main types of leukemia or lymphoma in which ETS1 has been studied, T-ALL, ATLL, and DLBCL, *ETS1* expression/dysregulation is driven by different factors (Fig. 16.3) but acts to mediate similar functional consequences. In these malignancies, canonical ETS1 is overexpressed and functions to either drive known target genes or interact with genetically driven TFs to promote an oncogenic transcriptional program. ETS1 binding is often localized to enhancer regions and characterized by H3K27ac. ETS1 binds in concert with other transcriptional regulators and larger transcriptional complexes, and these regulatory factors are critical in mediating the transcriptional activity and target genes of ETS1.

In all three malignancies, knockdown of *ETS1* resulted in reduced cell proliferation or cell cycle arrest. Protection from apoptosis was an observed function of *ETS1* in T-ALL and DLBCL. Similarities were also seen in the classes of target genes and interacting proteins, primarily falling into cell signaling and differentiation categories. Response to stimuli or cell signaling target genes included *CCR4* in ATLL, *CXCR5* in DLBCL, and NOTCH signaling in T-ALL. Effects on cellular development and differentiation included TAL1 and TLX1/TLX3 complex interactions in T-ALL and *PRDM1* repression in DLBCL. There is also intriguing evidence suggesting a role for ETS1 in viral-mediated pathogenesis for virus-associated malignancies such as ATLL.

In contrast, ETS1 has been implicated as a tumor suppressor in large B-cell lymphoma with 11q aberration and cHL. Additional work is needed to mechanistically define and understand the specific function of ETS1 in these malignancies. These opposing oncogenic and tumor suppressor roles proposed for ETS1 highlight the importance of cellular context in dictating ETS1 function.

16.9 Targetability

Given the prooncogenic role of ETS1 demonstrated in a variety of malignancies including leukemia and lymphoma, as well as its relative specificity in expression for lymphoid and tumor cell types, some efforts have been directed toward therapeutic inhibition/targeting of ETS1. While a specific inhibitor has not yet been developed, several strategies have been put forward for inhibition of ETS1 targeting protein turnover, activation, epigenetic coactivators, and DNA and protein interactions.

ETS1 protein turnover has been targeted by two main routes: deubiquitinase and kinase inhibition. The USP9X deubiquitinating enzyme has been shown to prevent ETS1 proteasomal destruction in melanoma [17], and USP9X expression is elevated in a number of tumor types. An inhibitor of USP9X, USP24, and USP5 deubiquitinases has been shown to target ETS1 for degradation in melanoma [17].

Phosphorylation of ETS1 at Y283 has been proposed to reverse CAMKII-dependent serine phosphorylation and subsequent COP1-mediated ubiquitination [18]. SRC family kinases, specifically SRC or YES1, have been shown to facilitate Y283 phosphorylation in breast cancer cells, thereby promoting ETS stability. As such, SRC family inhibitors have been proposed as a possible mechanism of ETS1 targeting. While SRC expression is not high in many types of lymphoma, it is unclear what role other SRC-family kinases may play in regulating COP1-ETS1 interactions. ETS1 T38 phosphorylation has also been proposed as a target for inhibiting ETS1 activation. T38 phosphorylation is a marker for ETS1 activation that was shown to be regulated by MEK/ERK signaling in DLBCL and decreased upon MEK inhibition [15].

Epigenetic drugs have shown some promise in ETS1 targeting. Acetylation of ETS1 by EP300/KAT3B promotes interaction with BRD4. Evidence suggests that ETS1 interacts with BRD4 to promote transcription of target genes

[14], and enrichment of ETS1 motifs has been overserved to be enriched at BRD4 binding sites, including DLBCL and T-ALL cells [97]. ETS1 specific binding, like BRD4, is often localized to enhancer regions. Therefore, EP300/KAT3B and/or BRD4 inhibition has been suggested as an ETS1 targeting strategy. In support of this, DLBCL cell lines have also shown sensitivity to a BRD4 inhibitor [98, 99], which has been suggested to be mediated by inhibition of PAX5, an ETS1 binding partner [99]. ATLL and T-ALL cell lines are also sensitive to BRD4 inhibition, either as monoagent or in combination [70, 100, 101].

As ETS1 protein complexes continue to be resolved, a role for small molecule inhibitors aimed at disrupting specific protein-protein interactions may be of promise. Inhibiting the ETS1-DNA interface is another theoretical approach, although complicated by the high degree of homology between ETS1 and other ETS family members. ETS1 targeting is also being proposed to complement other therapeutic approaches such as NOTCH inhibition [23]. In addition to ETS1 targeting mechanisms, defining specific tumor types and/or specific molecular biomarkers that confer sensitivity to ETS1 targeting is also needed to advance the potential therapeutic application of ETS1 inhibition.

16.10 Future Directions

While there is strong evidence for an oncogenic role of ETS1 in leukemias and lymphomas, additional research is needed to more comprehensively characterize target genes and biological functions of ETS1 in specific cellular contexts, including its presumed role as a tumor suppressor in specific lymphoma types. Increased understanding of associations between ETS1 expression and clinical parameters and outcomes is also needed across leukemias and lymphomas. Moreover, additional work is needed to define ETS1 protein interactions and more fully resolve larger protein binding complexes. This work will be essential to developing effective and specific therapeutic targeting strategies.

Acknowledgments RL is supported by an American Society of Hematology (ASH) Research Scholar Award.

References

1. Leprince D, Gegonne A, Coll J, de Taisne C, Schneeberger A, Lagrou C et al (1983) A putative second cell-derived oncogene of the avian leukaemia retrovirus E26. Nature 306(5941):395–397. https://doi.org/10.1038/306395a0
2. Garrett-Sinha LA (2023) An update on the roles of transcription factor Ets1 in autoimmune diseases. WIREs Mech Dis 15:e1627. https://doi.org/10.1002/wsbm.1627
3. Lionneton F, Lelievre E, Baillat D, Stehelin D, Soncin F (2003) Characterization and functional analysis of the p42Ets-1 variant of the mouse Ets-1 transcription factor. Oncogene 22(57):9156–9164. https://doi.org/10.1038/sj.onc.1207241
4. Lim F, Kraut N, Framptom J, Graf T (1992) DNA binding by c-Ets-1, but not v-Ets, is repressed by an intramolecular mechanism. EMBO J 11(2):643–652. https://doi.org/10.1002/j.1460-2075.1992.tb05096.x
5. Laitem C, Leprivier G, Choul-Li S, Begue A, Monte D, Larsimont D et al (2009) Ets-1 p27: a novel Ets-1 isoform with dominant-negative effects on the transcriptional properties and the subcellular localization of Ets-1 p51. Oncogene 28(20):2087–2099. https://doi.org/10.1038/onc.2009.72
6. Garvie CW, Pufall MA, Graves BJ, Wolberger C (2002) Structural analysis of the autoinhibition of Ets-1 and its role in protein partnerships. J Biol Chem 277(47):45529–45536. https://doi.org/10.1074/jbc.M206327200
7. Seidel JJ, Graves BJ (2002) An ERK2 docking site in the Pointed domain distinguishes a subset of ETS transcription factors. Genes Dev 16(1):127–137. https://doi.org/10.1101/gad.950902
8. Foulds CE, Nelson ML, Blaszczak AG, Graves BJ (2004) Ras/mitogen-activated protein kinase signaling activates Ets-1 and Ets-2 by CBP/p300 recruitment. Mol Cell Biol 24(24):10954–10964. https://doi.org/10.1128/MCB.24.24.10954-10964.2004
9. Gegonne A, Punyammalee B, Rabault B, Bosselut R, Seneca S, Crabeel M et al (1992) Analysis of the DNA binding and transcriptional activation properties of the Ets1 oncoprotein. New Biol 4(5):512–519
10. Yang C, Shapiro LH, Rivera M, Kumar A, Brindle PK (1998) A role for CREB binding protein and p300 transcriptional coactivators in Ets-1 transactivation functions. Mol Cell Biol 18(4):2218–2229. https://doi.org/10.1128/MCB.18.4.2218
11. Rabault B, Ghysdael J (1994) Calcium-induced phosphorylation of ETS1 inhibits its specific DNA binding activity. J Biol Chem 269(45):28143–28151
12. Cowley DO, Graves BJ (2000) Phosphorylation represses Ets-1 DNA binding by reinforcing autoinhibition. Genes Dev 14(3):366–376

13. Pufall MA, Lee GM, Nelson ML, Kang HS, Velyvis A, Kay LE et al (2005) Variable control of Ets-1 DNA binding by multiple phosphates in an unstructured region. Science 309(5731):142–145. https://doi.org/10.1126/science.1111915
14. Chen J, Fu Y, Day DS, Sun Y, Wang S, Liang X et al (2017) VEGF amplifies transcription through ETS1 acetylation to enable angiogenesis. Nat Commun 8(1):383. https://doi.org/10.1038/s41467-017-00405-x
15. Chung EYL, Sartori G, Ponzoni M, Cascione L, Priebe V, Xu-Monette ZY et al (2023) ETS1 phosphorylation at threonine 38 is associated with the cell of origin of diffuse large B cell lymphoma and sustains the growth of tumour cells. Br J Haematol 203:244. https://doi.org/10.1111/bjh.19018
16. Ji Z, Degerny C, Vintonenko N, Deheuninck J, Foveau B, Leroy C et al (2007) Regulation of the Ets-1 transcription factor by sumoylation and ubiquitinylation. Oncogene 26(3):395–406. https://doi.org/10.1038/sj.onc.1209789
17. Potu H, Peterson LF, Kandarpa M, Pal A, Sun H, Durham A et al (2017) Usp9x regulates Ets-1 ubiquitination and stability to control NRAS expression and tumorigenicity in melanoma. Nat Commun 8:14449. https://doi.org/10.1038/ncomms14449
18. Lu G, Zhang Q, Huang Y, Song J, Tomaino R, Ehrenberger T et al (2014) Phosphorylation of ETS1 by Src family kinases prevents its recognition by the COP1 tumor suppressor. Cancer Cell 26(2):222–234. https://doi.org/10.1016/j.ccr.2014.06.026
19. Eyquem S, Chemin K, Fasseu M, Bories JC (2004) The Ets-1 transcription factor is required for complete pre-T cell receptor function and allelic exclusion at the T cell receptor beta locus. Proc Natl Acad Sci USA 101(44):15712–15717. https://doi.org/10.1073/pnas.0405546101
20. Eyquem S, Chemin K, Fasseu M, Chopin M, Sigaux F, Cumano A et al (2004) The development of early and mature B cells is impaired in mice deficient for the Ets-1 transcription factor. Eur J Immunol 34(11):3187–3196. https://doi.org/10.1002/eji.200425352
21. Muthusamy N, Barton K, Leiden JM (1995) Defective activation and survival of T cells lacking the Ets-1 transcription factor. Nature 377(6550):639–642. https://doi.org/10.1038/377639a0
22. Bories JC, Willerford DM, Grevin D, Davidson L, Camus A, Martin P et al (1995) Increased T-cell apoptosis and terminal B-cell differentiation induced by inactivation of the Ets-1 proto-oncogene. Nature 377(6550):635–638. https://doi.org/10.1038/377635a0
23. McCarter AC, Della Gatta G, Melnick A, Kim E, Sha C, Wang Q et al (2020) Combinatorial ETS1-dependent control of oncogenic NOTCH1 enhancers in T-cell leukemia. Blood Cancer Discov 1(2):178–197. https://doi.org/10.1158/2643-3230.BCD-20-0026
24. Clements JL, John SA, Garrett-Sinha LA (2006) Impaired generation of CD8+ thymocytes in Ets-1-deficient mice. J Immunol 177(2):905–912. https://doi.org/10.4049/jimmunol.177.2.905
25. Hambor JE, Mennone J, Coon ME, Hanke JH, Kavathas P (1993) Identification and characterization of an Alu-containing, T-cell-specific enhancer located in the last intron of the human CD8 alpha gene. Mol Cell Biol 13(11):7056–7070. https://doi.org/10.1128/mcb.13.11.7056-7070.1993
26. Zamisch M, Tian L, Grenningloh R, Xiong Y, Wildt KF, Ehlers M et al (2009) The transcription factor Ets1 is important for CD4 repression and Runx3 up-regulation during CD8 T cell differentiation in the thymus. J Exp Med 206(12):2685–2699. https://doi.org/10.1084/jem.20092024
27. Mouly E, Chemin K, Nguyen HV, Chopin M, Mesnard L, Leite-de-Moraes M et al (2010) The Ets-1 transcription factor controls the development and function of natural regulatory T cells. J Exp Med 207(10):2113–2125. https://doi.org/10.1084/jem.20092153
28. Moisan J, Grenningloh R, Bettelli E, Oukka M, Ho IC (2007) Ets-1 is a negative regulator of Th17 differentiation. J Exp Med 204(12):2825–2835. https://doi.org/10.1084/jem.20070994
29. Tu WJ, Hardy K, Sutton CR, McCuaig R, Li J, Dunn J et al (2017) Priming of transcriptional memory responses via the chromatin accessibility landscape in T cells. Sci Rep 7:44825. https://doi.org/10.1038/srep44825
30. Barton K, Muthusamy N, Fischer C, Ting CN, Walunas TL, Lanier LL et al (1998) The Ets-1 transcription factor is required for the development of natural killer cells in mice. Immunity 9(4):555–563. https://doi.org/10.1016/s1074-7613(00)80638-x
31. Ramirez K, Chandler KJ, Spaulding C, Zandi S, Sigvardsson M, Graves BJ et al (2012) Gene deregulation and chronic activation in natural killer cells deficient in the transcription factor ETS1. Immunity 36(6):921–932. https://doi.org/10.1016/j.immuni.2012.04.006
32. Taveirne S, Wahlen S, Van Loocke W, Kiekens L, Persyn E, Van Ammel E et al (2020) The transcription factor ETS1 is an important regulator of human NK cell development and terminal differentiation. Blood 136(3):288–298. https://doi.org/10.1182/blood.2020005204
33. Walunas TL, Wang B, Wang CR, Leiden JM (2000) Cutting edge: the Ets1 transcription factor is required for the development of NK T cells in mice. J Immunol 164(6):2857–2860. https://doi.org/10.4049/jimmunol.164.6.2857
34. Wang D, John SA, Clements JL, Percy DH, Barton KP, Garrett-Sinha LA (2005) Ets-1 deficiency leads to altered B cell differentiation, hyperresponsiveness to TLR9 and autoimmune disease. Int Immunol 17(9):1179–1191. https://doi.org/10.1093/intimm/dxh295

35. Sunshine A, Goich D, Stith A, Sortino K, Dalton J, Metcalfe S et al (2019) Ets1 controls the development of B cell autoimmune responses in a cell-intrinsic manner. Immunohorizons 3(7):331–340. https://doi.org/10.4049/immunohorizons.1900033
36. Russell L, John S, Cullen J, Luo W, Shlomchik MJ, Garrett-Sinha LA (2015) Requirement for transcription factor Ets1 in B cell tolerance to self-antigens. J Immunol 195(8):3574–3583. https://doi.org/10.4049/jimmunol.1500776
37. Saelee P, Kearly A, Nutt SL, Garrett-Sinha LA (2017) Genome-wide identification of target genes for the key B cell transcription factor Ets1. Front Immunol 8:383. https://doi.org/10.3389/fimmu.2017.00383
38. Hollenhorst PC, Chandler KJ, Poulsen RL, Johnson WE, Speck NA, Graves BJ (2009) DNA specificity determinants associate with distinct transcription factor functions. PLoS Genet 5(12):e1000778. https://doi.org/10.1371/journal.pgen.1000778
39. Hollenhorst PC, McIntosh LP, Graves BJ (2011) Genomic and biochemical insights into the specificity of ETS transcription factors. Annu Rev Biochem 80:437–471. https://doi.org/10.1146/annurev.biochem.79.081507.103945
40. Baillat D, Begue A, Stehelin D, Aumercier M (2002) ETS-1 transcription factor binds cooperatively to the palindromic head to head ETS-binding sites of the stromelysin-1 promoter by counteracting autoinhibition. J Biol Chem 277(33):29386–29398. https://doi.org/10.1074/jbc.M200088200
41. Lamber EP, Vanhille L, Textor LC, Kachalova GS, Sieweke MH, Wilmanns M (2008) Regulation of the transcription factor Ets-1 by DNA-mediated homo-dimerization. EMBO J 27(14):2006–2017. https://doi.org/10.1038/emboj.2008.117
42. Kim WY, Sieweke M, Ogawa E, Wee HJ, Englmeier U, Graf T et al (1999) Mutual activation of Ets-1 and AML1 DNA binding by direct interaction of their autoinhibitory domains. EMBO J 18(6):1609–1620. https://doi.org/10.1093/emboj/18.6.1609
43. Shiina M, Hamada K, Inoue Bungo T, Shimamura M, Uchiyama A, Baba S et al (2015) A novel allosteric mechanism on protein-DNA interactions underlying the phosphorylation-dependent regulation of Ets1 target gene expressions. J Mol Biol 427(8):1655–1669. https://doi.org/10.1016/j.jmb.2014.07.020
44. Shrivastava T, Mino K, Babayeva ND, Baranovskaya OI, Rizzino A, Tahirov TH (2014) Structural basis of Ets1 activation by Runx1. Leukemia 28(10):2040–2048. https://doi.org/10.1038/leu.2014.111
45. Pham D, Silberger DJ, Nguyen KN, Gao M, Weaver CT, Hatton RD (2023) Batf stabilizes Th17 cell development via impaired Stat5 recruitment of Ets1-Runx1 complexes. EMBO J 42(8):e109803. https://doi.org/10.15252/embj.2021109803
46. Rameil P, Lecine P, Ghysdael J, Gouilleux F, Kahn-Perles B, Imbert J (2000) IL-2 and long-term T cell activation induce physical and functional interaction between STAT5 and ETS transcription factors in human T cells. Oncogene 19(17):2086–2097. https://doi.org/10.1038/sj.onc.1203542
47. Tsao HW, Tai TS, Tseng W, Chang HH, Grenningloh R, Miaw SC et al (2013) Ets-1 facilitates nuclear entry of NFAT proteins and their recruitment to the IL-2 promoter. Proc Natl Acad Sci USA 110(39):15776–15781. https://doi.org/10.1073/pnas.1304343110
48. Pham D, Moseley CE, Gao M, Savic D, Winstead CJ, Sun M et al (2019) Batf pioneers the reorganization of chromatin in developing effector T cells via Ets1-dependent recruitment of Ctcf. Cell Rep 29(5):1203–20 e7. https://doi.org/10.1016/j.celrep.2019.09.064
49. Wheat W, Fitzsimmons D, Lennox H, Krautkramer SR, Gentile LN, McIntosh LP et al (1999) The highly conserved beta-hairpin of the paired DNA-binding domain is required for assembly of Pax-Ets ternary complexes. Mol Cell Biol 19(3):2231–2241. https://doi.org/10.1128/MCB.19.3.2231
50. Garvie CW, Hagman J, Wolberger C (2001) Structural studies of Ets-1/Pax5 complex formation on DNA. Mol Cell 8(6):1267–1276. https://doi.org/10.1016/s1097-2765(01)00410-5
51. Fitzsimmons D, Lukin K, Lutz R, Garvie CW, Wolberger C, Hagman J (2009) Highly cooperative recruitment of Ets-1 and release of autoinhibition by Pax5. J Mol Biol 392(2):452–464. https://doi.org/10.1016/j.jmb.2009.07.028
52. Fitzsimmons D, Lutz R, Wheat W, Chamberlin HM, Hagman J (2001) Highly conserved amino acids in Pax and Ets proteins are required for DNA binding and ternary complex assembly. Nucleic Acids Res 29(20):4154–4165. https://doi.org/10.1093/nar/29.20.4154
53. Jayaraman G, Srinivas R, Duggan C, Ferreira E, Swaminathan S, Somasundaram K et al (1999) p300/cAMP-responsive element-binding protein interactions with ets-1 and ets-2 in the transcriptional activation of the human stromelysin promoter. J Biol Chem 274(24):17342–17352. https://doi.org/10.1074/jbc.274.24.17342
54. Li K, Qiu C, Sun P, Liu DC, Wu TJ, Wang K et al (2019) Ets1-mediated acetylation of FoxO1 is critical for gluconeogenesis regulation during feed-fast cycles. Cell Rep 26(11):2998–3010 e5. https://doi.org/10.1016/j.celrep.2019.02.035
55. Lee CG, Kwon HK, Sahoo A, Hwang W, So JS, Hwang JS et al (2012) Interaction of Ets-1 with HDAC1 represses IL-10 expression in Th1 cells. J Immunol 188(5):2244–2253. https://doi.org/10.4049/jimmunol.1101614
56. Sementchenko VI, Watson DK (2000) Ets target genes: past, present and future. Oncogene 19(55):6533–6548. https://doi.org/10.1038/sj.onc.1204034
57. Sizemore GM, Pitarresi JR, Balakrishnan S, Ostrowski MC (2017) The ETS family of oncogenic transcription factors in solid tumours. Nat

Rev Cancer 17(6):337–351. https://doi.org/10.1038/nrc.2017.20
58. Palii CG, Perez-Iratxeta C, Yao Z, Cao Y, Dai F, Davison J et al (2011) Differential genomic targeting of the transcription factor TAL1 in alternate haematopoietic lineages. EMBO J 30(3):494–509. https://doi.org/10.1038/emboj.2010.342
59. Wang H, Zang C, Taing L, Arnett KL, Wong YJ, Pear WS et al (2014) NOTCH1-RBPJ complexes drive target gene expression through dynamic interactions with superenhancers. Proc Natl Acad Sci USA 111(2):705–710. https://doi.org/10.1073/pnas.1315023111
60. Melnick AF, Mullin C, Lin K, McCarter AC, Liang S, Liu YE et al (2023) Cdc73 protects Notch-induced T-cell leukemia cells from DNA damage and mitochondrial stress. Blood 142:2159. https://doi.org/10.1182/blood.2023020144
61. Carroll AJ, Crist WM, Link MP, Amylon MD, Pullen DJ, Ragab AH et al (1990) The t(1;14)(p34;q11) is nonrandom and restricted to T-cell acute lymphoblastic leukemia: a Pediatric Oncology Group study. Blood 76(6):1220–1224
62. Sacchi N, de Klein A, Showalter SD, Bigi G, Papas TS (1988) High expression of ets-1 gene in human thymocytes and immature T leukemic cells. Leukemia 2(1):12–18
63. Ferrando AA, Neuberg DS, Staunton J, Loh ML, Huard C, Raimondi SC et al (2002) Gene expression signatures define novel oncogenic pathways in T cell acute lymphoblastic leukemia. Cancer Cell 1(1):75–87. https://doi.org/10.1016/s1535-6108(02)00018-1
64. O'Neil J, Shank J, Cusson N, Murre C, Kelliher M (2004) TAL1/SCL induces leukemia by inhibiting the transcriptional activity of E47/HEB. Cancer Cell 5(6):587–596. https://doi.org/10.1016/j.ccr.2004.05.023
65. Palomero T, Odom DT, O'Neil J, Ferrando AA, Margolin A, Neuberg DS et al (2006) Transcriptional regulatory networks downstream of TAL1/SCL in T-cell acute lymphoblastic leukemia. Blood 108(3):986–992. https://doi.org/10.1182/blood-2005-08-3482
66. Sanda T, Lawton LN, Barrasa MI, Fan ZP, Kohlhammer H, Gutierrez A et al (2012) Core transcriptional regulatory circuit controlled by the TAL1 complex in human T cell acute lymphoblastic leukemia. Cancer Cell 22(2):209–221. https://doi.org/10.1016/j.ccr.2012.06.007
67. Rahman S, Magnussen M, Leon TE, Farah N, Li Z, Abraham BJ et al (2017) Activation of the LMO2 oncogene through a somatically acquired neomorphic promoter in T-cell acute lymphoblastic leukemia. Blood 129(24):3221–3226. https://doi.org/10.1182/blood-2016-09-742148
68. Durinck K, Van Loocke W, Van der Meulen J, Van de Walle I, Ongenaert M, Rondou P et al (2015) Characterization of the genome-wide TLX1 binding profile in T-cell acute lymphoblastic leukemia. Leukemia 29(12):2317–2327. https://doi.org/10.1038/leu.2015.162
69. Dadi S, Le Noir S, Payet-Bornet D, Lhermitte L, Zacarias-Cabeza J, Bergeron J et al (2012) TLX homeodomain oncogenes mediate T cell maturation arrest in T-ALL via interaction with ETS1 and suppression of TCRalpha gene expression. Cancer Cell 21(4):563–576. https://doi.org/10.1016/j.ccr.2012.02.013
70. Vanden Bempt M, Demeyer S, Broux M, De Bie J, Bornschein S, Mentens N et al (2018) Cooperative enhancer activation by TLX1 and STAT5 drives development of NUP214-ABL1/TLX1-positive T cell acute lymphoblastic leukemia. Cancer Cell 34(2):271–85 e7. https://doi.org/10.1016/j.ccell.2018.07.007
71. Tamaki T, Karube K, Sakihama S, Tsuruta Y, Awazawa R, Hayashi M et al (2023) A comprehensive study of the immunophenotype and its clinicopathologic significance in adult T-cell leukemia/lymphoma. Mod Pathol 36(8):100169. https://doi.org/10.1016/j.modpat.2023.100169
72. Shah UA, Chung EY, Giricz O, Pradhan K, Kataoka K, Gordon-Mitchell S et al (2018) North American ATLL has a distinct mutational and transcriptional profile and responds to epigenetic therapies. Blood 132(14):1507–1518. https://doi.org/10.1182/blood-2018-01-824607
73. Luchtel RA, Zhao Y, Aggarwal RK, Pradhan K, Maqbool SB (2022) ETS1 is a novel transcriptional regulator of adult T-cell leukemia/lymphoma of North American descent. Blood Adv 6(20):5613–5624. https://doi.org/10.1182/bloodadvances.2022007725
74. Ishida T, Utsunomiya A, Iida S, Inagaki H, Takatsuka Y, Kusumoto S et al (2003) Clinical significance of CCR4 expression in adult T-cell leukemia/lymphoma: its close association with skin involvement and unfavorable outcome. Clin Cancer Res 9(10 Pt 1):3625–3634
75. Kogure Y, Kameda T, Koya J, Yoshimitsu M, Nosaka K, Yasunaga JI et al (2022) Whole-genome landscape of adult T-cell leukemia/lymphoma. Blood 139(7):967–982. https://doi.org/10.1182/blood.2021013568
76. Toyoda K, Yasunaga JI, Shichijo T, Arima Y, Tsujita K, Tanaka A et al (2023) HTLV-1 bZIP factor-induced reprogramming of lactate metabolism and epigenetic status promote leukemic cell expansion. Blood Cancer Discov 4:374. https://doi.org/10.1158/2643-3230.BCD-22-0139
77. Bosselut R, Duvall JF, Gegonne A, Bailly M, Hemar A, Brady J et al (1990) The product of the c-ets-1 proto-oncogene and the related Ets2 protein act as transcriptional activators of the long terminal repeat of human T cell leukemia virus HTLV-1. EMBO J 9(10):3137–3144. https://doi.org/10.1002/j.1460-2075.1990.tb07511.x
78. Gitlin SD, Bosselut R, Gegonne A, Ghysdael J, Brady JN (1991) Sequence-specific interaction of the Ets1 protein with the long terminal

repeat of the human T-lymphotropic virus type I. J Virol 65(10):5513–5523. https://doi.org/10.1128/JVI.65.10.5513-5523.1991
79. Gitlin SD, Dittmer J, Shin RC, Brady JN (1993) Transcriptional activation of the human T-lymphotropic virus type I long terminal repeat by functional interaction of Tax1 and Ets1. J Virol 67(12):7307–7316. https://doi.org/10.1128/JVI.67.12.7307-7316.1993
80. Seeler JS, Muchardt C, Podar M, Gaynor RB (1993) Regulatory elements involved in tax-mediated transactivation of the HTLV-I LTR. Virology 196(2):442–450. https://doi.org/10.1006/viro.1993.1500
81. Dittmer J, Pise-Masison CA, Clemens KE, Choi KS, Brady JN (1997) Interaction of human T-cell lymphotropic virus type I Tax, Ets1, and Sp1 in transactivation of the PTHrP P2 promoter. J Biol Chem 272(8):4953–4958. https://doi.org/10.1074/jbc.272.8.4953
82. Schaefer M, Hopkins RG, Failla ML, Gitlin JD (1999) Hepatocyte-specific localization and copper-dependent trafficking of the Wilson's disease protein in the liver. Am J Phys 276(3):G639–G646. https://doi.org/10.1152/ajpgi.1999.276.3.G639
83. Tsukada J, Misago M, Serino Y, Ogawa R, Murakami S, Nakanishi M et al (1997) Human T-cell leukemia virus type I Tax transactivates the promoter of human prointerleukin-1beta gene through association with two transcription factors, nuclear factor-interleukin-6 and Spi-1. Blood 90(8):3142–3153
84. Pasqualucci L, Trifonov V, Fabbri G, Ma J, Rossi D, Chiarenza A et al (2011) Analysis of the coding genome of diffuse large B-cell lymphoma. Nat Genet 43(9):830–837. https://doi.org/10.1038/ng.892
85. Bonetti P, Testoni M, Scandurra M, Ponzoni M, Piva R, Mensah AA et al (2013) Deregulation of ETS1 and FLI1 contributes to the pathogenesis of diffuse large B-cell lymphoma. Blood 122(13):2233–2241. https://doi.org/10.1182/blood-2013-01-475772
86. Priebe V, Sartori G, Napoli S, Chung EYL, Cascione L, Kwee I et al (2020) Role of ETS1 in the transcriptional network of diffuse large B cell lymphoma of the activated B cell-like type. Cancers (Basel) 12(7). https://doi.org/10.3390/cancers12071912
87. Morin RD, Mendez-Lago M, Mungall AJ, Goya R, Mungall KL, Corbett RD et al (2011) Frequent mutation of histone-modifying genes in non-Hodgkin lymphoma. Nature 476(7360):298–303. https://doi.org/10.1038/nature10351
88. Pedrosa L, Fernandez-Miranda I, Perez-Callejo D, Quero C, Rodriguez M, Martin-Acosta P et al (2021) Proposal and validation of a method to classify genetic subtypes of diffuse large B cell lymphoma. Sci Rep 11(1):1886. https://doi.org/10.1038/s41598-020-80376-0
89. Campo E, Jaffe ES, Cook JR, Quintanilla-Martinez L, Swerdlow SH, Anderson KC et al (2022) The international consensus classification of mature lymphoid neoplasms: a report from the Clinical Advisory Committee. Blood 140(11):1229–1253. https://doi.org/10.1182/blood.2022015851
90. Gonzalez-Farre B, Ramis-Zaldivar JE, Salmeron-Villalobos J, Balague O, Celis V, Verdu-Amoros J et al (2019) Burkitt-like lymphoma with 11q aberration: a germinal center-derived lymphoma genetically unrelated to Burkitt lymphoma. Haematologica 104(9):1822–1829. https://doi.org/10.3324/haematol.2018.207928
91. Ferreiro JF, Morscio J, Dierickx D, Marcelis L, Verhoef G, Vandenberghe P et al (2015) Post-transplant molecularly defined Burkitt lymphomas are frequently MYC-negative and characterized by the 11q-gain/loss pattern. Haematologica 100(7):e275–e279. https://doi.org/10.3324/haematol.2015.124305
92. Salaverria I, Martin-Guerrero I, Wagener R, Kreuz M, Kohler CW, Richter J et al (2014) A recurrent 11q aberration pattern characterizes a subset of MYC-negative high-grade B-cell lymphomas resembling Burkitt lymphoma. Blood 123(8):1187–1198. https://doi.org/10.1182/blood-2013-06-507996
93. Overbeck BM, Martin-Subero JI, Ammerpohl O, Klapper W, Siebert R, Giefing M (2012) ETS1 encoding a transcription factor involved in B-cell differentiation is recurrently deleted and down-regulated in classical Hodgkin's lymphoma. Haematologica 97(10):1612–1614. https://doi.org/10.3324/haematol.2012.061770
94. Flossbach L, Holzmann K, Mattfeldt T, Buck M, Lanz K, Held M et al (2013) High-resolution genomic profiling reveals clonal evolution and competition in gastrointestinal marginal zone B-cell lymphoma and its large cell variant. Int J Cancer 132(3):E116–E127. https://doi.org/10.1002/ijc.27774
95. Panea RI, Love CL, Shingleton JR, Reddy A, Bailey JA, Moormann AM et al (2019) The whole-genome landscape of Burkitt lymphoma subtypes. Blood 134(19):1598–1607. https://doi.org/10.1182/blood.2019001880
96. Baran-Marszak F, Fagard R, Girard B, Camilleri-Broet S, Zeng F, Lenoir GM et al (2002) Gene array identification of Epstein Barr virus-regulated cellular genes in EBV-converted Burkitt lymphoma cell lines. Lab Investig 82(11):1463–1479. https://doi.org/10.1097/01.lab.0000035025.51772.2b
97. Jiang G, Deng W, Liu Y, Wang C (2020) General mechanism of JQ1 in inhibiting various types of cancer. Mol Med Rep 21(3):1021–1034. https://doi.org/10.3892/mmr.2020.10927
98. Trabucco SE, Gerstein RM, Evens AM, Bradner JE, Shultz LD, Greiner DL et al (2015) Inhibition of bromodomain proteins for the treatment of human diffuse large B-cell lymphoma. Clin Cancer Res 21(1):113–122. https://doi.org/10.1158/1078-0432.CCR-13-3346
99. Oien DB, Sharma S, Hattersley MM, DuPont MM, Criscione SW, Prickett LB et al (2023) BET inhibition targets ABC-DLBCL constitu-

tive B-cell receptor signaling through PAX5. Blood Adv 7:5108. https://doi.org/10.1182/bloodadvances.2022009257
100. Nakagawa M, Shaffer AL 3rd, Ceribelli M, Zhang M, Wright G, Huang DW et al (2018) Targeting the HTLV-I-regulated BATF3/IRF4 transcriptional network in adult T cell leukemia/lymphoma. Cancer Cell 34(2):286–97 e10. https://doi.org/10.1016/j.ccell.2018.06.014
101. Knoechel B, Roderick JE, Williamson KE, Zhu J, Lohr JG, Cotton MJ et al (2014) An epigenetic mechanism of resistance to targeted therapy in T cell acute lymphoblastic leukemia. Nat Genet 46(4):364–370. https://doi.org/10.1038/ng.2913
102. Liu W, Xie Y, Ma J, Luo X, Nie P, Zuo Z et al (2015) IBS: an illustrator for the presentation and visualization of biological sequences. Bioinformatics 31(20):3359–3361. https://doi.org/10.1093/bioinformatics/btv362
103. Nelson ML, Kang HS, Lee GM, Blaszczak AG, Lau DK, McIntosh LP et al (2010) Ras signaling requires dynamic properties of Ets1 for phosphorylation-enhanced binding to coactivator CBP. Proc Natl Acad Sci USA 107(22):10026–10031. https://doi.org/10.1073/pnas.0915137107
104. Rucker FG, Bullinger L, Schwaenen C, Lipka DB, Wessendorf S, Frohling S et al (2006) Disclosure of candidate genes in acute myeloid leukemia with complex karyotypes using microarray-based molecular characterization. J Clin Oncol 24(24):3887–3894. https://doi.org/10.1200/JCO.2005.04.5450
105. Poppe B, Vandesompele J, Schoch C, Lindvall C, Mrozek K, Bloomfield CD et al (2004) Expression analyses identify MLL as a prominent target of 11q23 amplification and support an etiologic role for MLL gain of function in myeloid malignancies. Blood 103(1):229–235. https://doi.org/10.1182/blood-2003-06-2163

17 The Molecular and Biological Function of MEF2D in Leukemia

Pengcheng Zhang and Rui Lu

Abstract

Myocyte enhancer factor 2 (MEF2) is a key transcription factor (TF) in skeletal, cardiac, and neural tissue development and includes four isoforms: MEF2A, MEF2B, MEF2C, and MEF2D. These isoforms significantly affect embryonic development, nervous system regulation, muscle cell differentiation, B- and T-cell development, thymocyte selection, and effects on tumorigenesis and leukemia. This chapter describes the multifaceted roles of MEF2 family proteins, covering embryonic development, nervous system regulation, and muscle cell differentiation. It further elucidates the contribution of MEF2 to various blood and immune cell functions. Specifically, in B-cell precursor acute lymphoblastic leukemia (BCP-ALL), MEF2D is aberrantly expressed and forms a fusion protein with BCL9, CSF1R, DAZAP1, HNRNPUL1, and SS18. These fusion proteins are closely related to the pathogenesis of leukemia. In addition, it specifically introduces the regulatory effect of MEF2D fusion protein on the proliferation and growth of B-cell acute lymphoblastic leukemia (B-ALL) cells. Finally, we detail the positive feedback loop between MEF2D and IRF8 that significantly promotes the progression of acute myeloid leukemia (AML) and the importance of the ZMYND8-BRD4 interaction in regulating the IRF8 and MYC transcriptional programs. The MEF2D-CEBPE axis is highlighted as a key transcriptional mechanism controlling the block of leukemic cell self-renewal and differentiation in AML. This chapter starts with the structure and function of MEF2 family proteins, specifically summarizing and analyzing the role of MEF2D in B-ALL and AML, mediating the complex molecular mechanisms of transcriptional regulation and exploring their implications for human health and disease.

Keywords

MEF2 family proteins · Transcription factor · MEF2D · MEF2D fusion proteins · B-ALL · AML

P. Zhang · R. Lu (✉)
Department of Medicine, Division of Hematology/Oncology, University of Alabama at Birmingham Heersink School of Medicine, Birmingham, AL, USA

O'Neal Comprehensive Cancer Center, University of Alabama at Birmingham Heersink School of Medicine, Birmingham, AL, USA
e-mail: ruilu1@uabmc.edu

T. Borggrefe, B. D. Giaimo (eds.), *Transcription factors in blood cell development*, Advances in Experimental Medicine and Biology 1459, https://doi.org/10.1007/978-3-031-62731-6_17

Abbreviations

6-OHDA	6-hydroxydopamine
ABL1	Abelson tyrosine kinase 1
ALS	amyotrophic lateral sclerosis
AML	acute myeloid leukemia
ARID1B	AT-rich interaction domain 1B
ASH2L	ASH2 like, histone lysine methyltransferase complex subunit
ATF	activating transcription factor
B-ALL	B-cell acute lymphoblastic leukemia
BCL2/c-MYC	B-cell lymphoma 2/cellular myelocytomatosis
BCL6	B-cell lymphoma 6
BCL9	B-cell lymphoma 9
BCP-ALL	B-cell precursor acute lymphoblastic leukemia
BCR	B-cell receptor
Blimip1	B-lymphocyte-induced maturation protein 1
BRD4	bromodomain-containing protein 4
BRD7	bromodomain-containing protein 7
CABIN1	calcineurin binding protein 1
CaN	calcineurin
CEBPA	CCAAT/enhancer binding protein alpha
CEBPE	CCAAT/enhancer binding protein epsilon
CLP	common lymphoid progenitor
CLTC	clathrin heavy chain
CM	cardiac myxoma
CMA	chaperone-mediated autophagy
CREB	cAMP response element-binding protein
CREBBP	CREB binding protein
CSF1R	colony-stimulating factor 1 receptor
CTLA4	cytotoxic T-lymphocyte antigen 4
DA	dopaminergic
DAZAP1	DAZ-associated protein 1
DBD	DNA binding domain
DOT1L	DOT1-like histone lysine methyltransferase
DUX4	double homeobox 4
DYRK1A	dual-specificity tyrosine-phosphorylation-regulated kinase 1A
E2A	transcription factor 3
EBNA-1	Epstein-Barr nuclear antigen 1
EBV	Epstein-Barr virus
EP300/KAT3B	E1A binding protein p300/lysine acetyltransferase 3B
ETV6-RUNX1	ETS variant transcription factor 6-Runt-related transcription factor 1 fusion gene
EWSR1	Ewing sarcoma RNA binding protein 1
Flt3	Fms-like tyrosine kinase 3
Fos-B	FBJ murine osteosarcoma viral oncogene homolog B
FOXJ2	Forkhead box J2
FOXO1	Forkhead box O1
GATA2	GATA binding factor 2
GSK3β	glycogen synthase kinase-3 β
HATs	histone acetyltransferases
HDAC4	histone deacetylase 4
HDAC5	histone deacetylase 5
HDACs	histone deacetylases
HIV-1	human immunodeficiency virus 1
HNRNPH1	heterogeneous nuclear ribonucleoprotein H1
HNRNPM	heterogeneous nuclear ribonucleoprotein M
HNRNPUL1	heterogeneous nuclear ribonucleoprotein U-like 1
HOX-A	HOX cluster A
HOXA9	homeobox A9
HTLV-1	human T-cell leukemia virus type 1
ICOS	Inducible T-cell costimulator
IEGs	immediate early genes
IFNγ	interferon-gamma
IGF	insulin-like growth factor
IKAROS	IKAROS family zinc finger

IKZF1	Ikaros family zinc finger protein 1
IL10	interleukin 10
IL2	interleukin 2
IRF4	interferon regulatory factor 4
IRF8	interferon regulatory factor 8
KAT3A	lysine acetyltransferase 3A
KAT3B	lysine acetyltransferase 3B
KLF2	Kruppel-like factor 2
KLF6	Kruppel-like factor 6
KMT1C	lysine methyltransferase 1C
KMT2A	lysine methyltransferase 2A
lncRNA	noncoding RNA
LoF	loss of function
LSD1/KDM1A	lysine-specific histone demethylase 1A/lysine demethylase 1A
MADS	MCM-agamous-definiens-serum
MAPK	mitogen-activated protein kinase
MB	MEF2D-BCL9
MBNL3	muscleblind-like splicing regulator 3
MEF2A	myocyte-specific enhancer factor 2A
MEF2B	myocyte-specific enhancer factor 2B
MEF2C	myocyte-specific enhancer factor 2C
MEF2D	myocyte-specific enhancer factor 2D
MH	MEF2D-HNRNPUL1
MLLr	MLL rearrangement
MLL1	mixed-lineage leukemia 1
MMP9	matrix metallopeptidase 9
MPTP	1-methl-4-phenyl-1,2,3,6-tetrahydropyridine
MRF4	myogenic regulatory Factor 4
mtDNA	mitochondrial DNA
MYF5	myogenic factor 5
MYF6	myogenic factor 6
MYOD	myogenic differentiation 1
MYOG	myogenin
ND6	NADH dehydrogenase subunit 6
NGS	Next-generation sequefcing
NIPBL	nipped-B-like protein
NK	natural killer
NLS	nuclear localization signal
NRAS	neuroblastoma RAS viral oncogene homolog
NRF2	nuclear factor erythroid 2-related factor 2
Nurr77	nuclear receptor 4A1
NUTM1	nuclear protein in testis midline carcinoma 1
OGD/R-injured neurons	oxygen-glucose deprivation/reperfusion-injured neurons
P13K/AKT	phosphoinositide 3-kinase/AKT serine/threonine kinase
PAK4	P21 (RAC1)-activated kinase 4
PAX5	paired box 5
PBX1	pre-B-cell leukemia homeobox 1
PD	Parkinson's disease
PETN	phosphatase and tensin homolog
PKA	protein kinase A
PP2A-B55δ	protein phosphatase 2A regulatory subunit B55 delta
PP2B	protein phosphatase 2B
PPP3	protein phosphatase 3
Pre-BCR	pre-B-cell receptor
PTMs	posttranslational modifications
PU.1/MEIS1	hematopoietic transcription factor PU.1/myeloid ecotropic insertion site 1
RAG1	recombination-activating gene 1
RBFOX2	RNA binding fox-1 homolog 2
ROS	reactive oxygen species
RUNX1	Runt-related transcription factor 1
SCN	suprachiasmatic nucleus
SENP3	SUMO-specific peptidase 3
siRNAs	small interfering RNAs
SLUG	snail family transcriptional repressor 2
SMARCA2	SWI-/SNF-related, matrix-associated, actin-dependent regulator of chromatin, subfamily A, member 2

SMARCA4	SWI/SNF-related, matrix-associated, actin-dependent regulator of chromatin, subfamily A, member 4
SOD3	superoxide dismutase 3
SRF	serum response factor
SS18	synovial sarcoma translocation, chromosome 18
STAT6	signal transducer and activator of transcription 6
SUMO	sumoylation
SYNRG	synergin gamma
TAD	transcription activation domain
TAF15	TATA-box binding protein-associated factor 15
TCF3-PBX1	TCF3-PBX1 fusion gene
TFs	transcription factors
TLR	Toll-like receptor
TP53	tumor protein 53
TPA	12-O-tetradecanoylphorbol-13-acetate
Tpms	transcripts per million
VCAM-1	vascular cell adhesion molecule-1
ZEB1	zinc finger E-box binding homeobox 1
ZEB2	zinc finger E-box binding homeobox 2
ZMYND8	zinc finger MYND-type containing 8
ZNF384/ZNF362	zinc finger protein 384/zinc finger protein 362

17.1 The MEF2 Family Transcription Factors

17.1.1 The Introduction of MEF2 Family Proteins

The human myocyte enhancer factor 2 (MEF2) transcription factor (TF) family is comprised of four proteins: MEF2A, MEF2B, MEF2C, and MEF2D. MEF2A and MEF2C share similar sequences, likely due to vertebrate evolution duplications, while MEF2B originated from a distinct gene [1]. Model organisms like *Saccharomyces cerevisiae*, *Caenorhabditis elegans*, and *Drosophila melanogaster* have only one MEF2 gene.

MEF2 proteins contain three domains: the N-terminal DNA binding MADS (MCM-agamous-definiens serum) domain, the central MEF2 domain, and the C-terminal transactivation domain (TAD) (Fig. 17.1a). The MADS and MEF2 domains are highly conserved within the family, with 91% and 68% amino acid identity, respectively, between MEF2A and the most divergent MEF2B. However, the TAD shows lower conservation, with only 6% amino acid identity between MEF2A and MEF2B. MEF2 transcription factors play crucial roles in various tissues, influencing cell differentiation, proliferation, apoptosis, migration, morphology, and metabolism in muscles, heart, bone, blood vessels, nerves, blood, and immune system cells. Altered MEF2 proteins have been linked to developmental and neurological disorders and have recently been implicated in cancer development, particularly leukemia [1–7].

This subsection focuses on recent discoveries in the disease-related functions of MEF2 family proteins, focusing on their involvement in cancer development. It discusses the proteins' roles in specific cell types, such as neurons, muscles, and hematopoietic cells.

17.1.2 The MEF2 Proteins' Structure and Function

The four MEF2 family members share nearly identical N-termini (amino acids 1–86) but have different C-termini (Fig. 17.1). At the N-terminus, there is a highly conserved MADS domain (approximately 55 amino acids) followed by the MEF2 domain (29 amino acids), which functions as a DNA binding domain (DBD), interaction site for other proteins, and dimerization site. The MADS and MEF2 domains faithfully enable high-affinity binding to the DNA sequence CTA(A/T)4TAG/A and interactions with cofactors. The C-terminus plays a vital role in the MEF2 family proteins' transcriptional activation and nuclear localization [8, 9]. Different

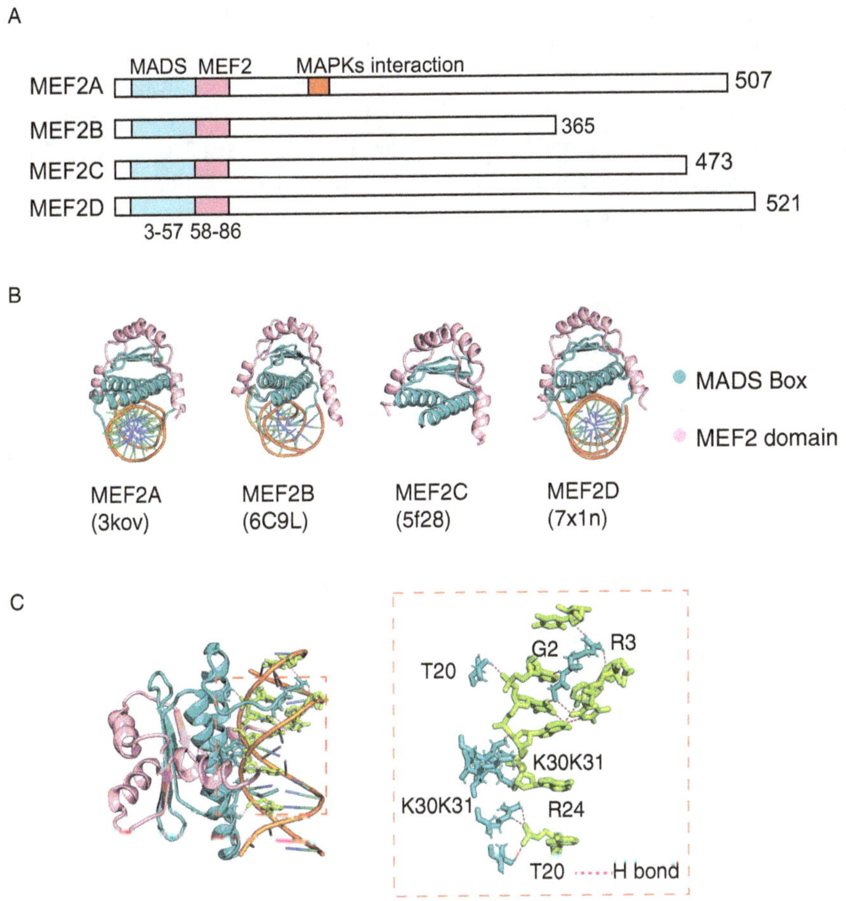

Fig. 17.1 Domain and structure of MEF2 family proteins. (**a**) Schematic diagram of MEF2 family proteins. Human MEF2 proteins (MEF2A, MEF2B, MEF2C, and MEF2D). All four MEF2 proteins share a similar N-terminal catalytic domain but have different C-termini. At the N-terminus, there is a highly conserved MADS domain (approximately 55 amino acids) followed by the MEF2 domain (29 amino acids), which functions as a DNA binding domain, interaction site for other proteins, and dimerization site. Both the MADS and MEF2 domains enable high-affinity binding to the DNA sequence CTA(A/T)4TAG/A and interactions with cofactors. The C-terminus plays a vital role in the MEF2 protein's transcriptional activation and nuclear localization. (**b**) The protein structure of the MEF2 family N-terminal with DNA or without DNA. MEF2A(3kov), MEF2B(6C9L), MEF2C(5f28), and MEF2D(7x1n). (**c**) The hydrogen bond of MEF2D with DNA

C-terminal amino acid sequences among MEF2 members likely determine their specific activities for transcriptional activation. The C-terminal region (amino acids 472–507) contains the crucial nuclear localization signal (NLS), and the area between amino acids 321–472 is believed to be the TAD [9]. The TAD further comprises four subdomains: TAD1 (amino acids 289–296), TAD2 (amino acids 312–322), TAD3 (amino acids 338–348), and TAD4 (amino acids 364–373). These subdomains are conserved in MEF2A, MEF2C, and MEF2D and are essential for transcriptional activation. The MEF2 domain serves as a docking site for various coactivators [e.g., EP300/KAT3B (E1A binding protein p300/lysine acetyltransferase 3B), CREB (cAMP response element-binding protein), etc.], corepressors [including CABIN1 (calcineurin binding protein 1), class IIa HDACs (histone deacetylases)], kinases [e.g., GSK3β (*glycogen synthase kinase-3β*)], and other selective regulators [e.g., protein kinase A (PKA)] [2, 3, 10, 11].

The transcriptional program of MEF2 involves interactions with several coactivators such as EP300/KAT3B, CBP/KAT3A (CREB-binding protein/lysine acetyltransferase 3A), cFOS, cJUN, and ATF (activating transcription factor) [2, 3, 10–21]. MEF2 family transcription factors are crucial for the cardiovascular system's development, morphogenesis, and homeostasis (Table 17.1). Four human MEF2 TFs have been identified and delineated as MEF2A, MEF2B, MEF2C, and MEF2D, each with homologs in other vertebrates. MEF2A and MEF2D are particularly interesting to adult cardiovascular homeostasis, having been identified as the predominant isoforms in the adult heart and exhibiting partial endothelial redundancy [6, 22–26].

MEF2 has distinct spatial and temporal expression during development and is involved in B-cell development, T-cell differentiation, and thymocyte selection. Different MEF2 members exhibit overlapping patterns when compared across embryogenic and adult tissues [27]. Next-generation sequencing (NGS) of human tissues revealed that MEF2A mRNA is most highly expressed in the heart, followed by skin and brain tissues [28]. MEF2C mRNA expression was highest in the brain, almost three times higher than in lymph nodes (next highest), and six times higher than in the heart, suggesting a role in brain development. MEF2A, MEF2C, and MEF2D have redundant functions during skeletal muscle regeneration [29]; their unique expression profiles suggest specific roles in distinct regeneration processes. For instance, MEF2A is essential for myoblast differentiation, while MEF2B, MEF2C, and MEF2D play minor roles in this process. Different MEF2 members can selectively target other genes within the muscle genome. For example, MEF2A and MEF2D are essential for cardiomyocyte survival, while MEF2C primarily contributes to cardiac morphogenesis [30, 31]. Conditional knockout (KO) of *Mef2c* in mouse neuronal populations leads to severe changes in behavioral and synaptic phenotypes of the mice, and *Mef2c* haploinsufficiency due to microdeletion of chromosome 5q14.3 is associated with neurodevelopmental disorders [32].

Mef2a KO mice and mice with loss-of-function (LoF) mutations in *Mef2a* still express

Table 17.1 MEF2 family proteins and their activators, repressors, and posttranslational modifications

Protein	Cofactors/Modification	Interaction architecture	Interaction domain	Biology functions	References
MEF2A	Phosphorylation and SUMO	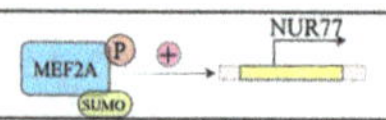	TAD	Promoting synaptic differentiation.	[13, 66]
MEF2C	Myogenin, E2A	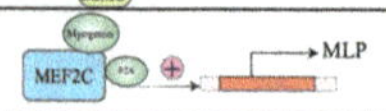	MADS, MEF2	Muscle differentiation	[2, 49]
MEF2C	MyoD, P300, E2A, MLP	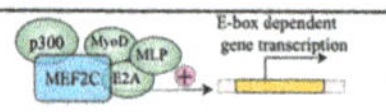	MEF2	Muscle cell fate maintenance	[2, 18]
MEF2D	FOXP3	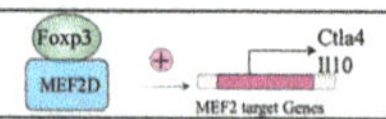	Unknow	Effector Treg phenotypes	[70]
MEF2D	FAK	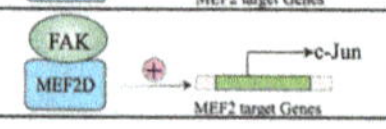	MEF2	Cell proiferation or with FOSB induced cell death of T cells	[50]
MEF2D	No	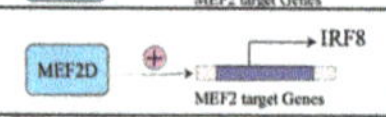	No	AML maintenance	[142–144, 150]
MEF2D	G9a, LSD1	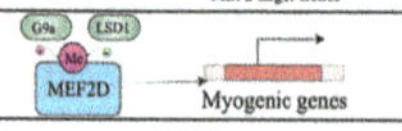	TAD	Skeletal muscle differentiation.	[39, 40, 41]
MEF2A/C/D	HDAC, mSin3, Cabin	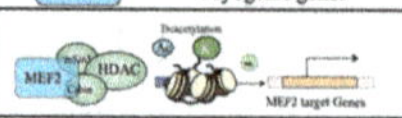	MADS, MEF2	Differentiation in myocytes transcription repression	[17, 21, 42, 62]
MEF2A/C/D	P300, NFAT,	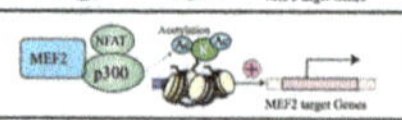	MADS, MEF2	Transcription activation	[12, 21, 41]
MEF2D	P300	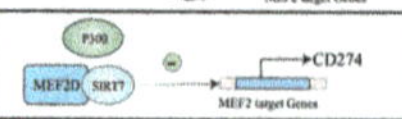	MADS, MEF2	PD-L1 expression UP, Cancer immune surveillance DOWN	[80]
MEF2D	SIRT7	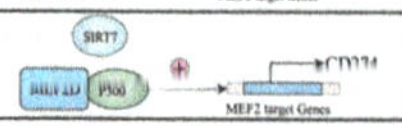	MADS, MEF2	PD-L1 expression DOWN, Cancer immune surveillance UP	[80]

disease phenotypes despite the normal function of other *Mef2* members, highlighting the irreplaceable role of *Mef2a* [33]. Studies have revealed redundancy among MEF2 proteins wherein one paralog may replace another following loss. Notably, MEF2A exhibits vigorous compensatory binding activity at sites typically associated with the significant genome occupant MEF2D. For instance, MEF2A and MEF2D play similar roles in repressing cell cycle genes and activating a subset of sarcomere markers in cardiomyocytes. Additionally, both MEF2A and MEF2D may be redundant for the survival of neonatal cardiomyocytes, as overexpressing either factor reduces programmed cell death in the presence of MEF2A deficiency. While the DBDs of MEF2A and MEF2D share >95% amino acid identity, a few nonconsensus amino acids may contribute to their differential binding. Interestingly, granule neurons occasionally coexpress high levels of both MEF2A and MEF2D, although MEF2 protein concentrations vary across different neuronal cell types. Moreover, the crystal structure of MEF2A homodimers suggests that regions outside the DBD may interact with the genome, possibly conferring distinct binding activities to various MEF2 family members. Besides potential DNA interactions, the TAD of each MEF2 member may undergo unique posttranslational modifications (PTMs) or bind different cofactors, stabilizing the binding of one paralog over the other [6, 31, 34–36].

MEF2 is a scaffolding protein that interacts with HDACs or histone acetyltransferases (HATs) and regulates gene expression. MEF2A and MEF2D mediate 12-O-tetradecanoylphorbol-13-acetate (TPA)-triggered superoxide dismutase 3 (SOD3) expression by interacting with HDACs or EP300/KAT3B. SOD3 is an isoenzyme that plays a crucial role in extracellular redox homeostasis [10]. Multisite phosphorylation is known to control the transcriptional activity of MEF2 proteins, but it is unclear whether other modifications are involved. It is reported that human MEF2D and MEF2C are modified by SUMO-2 and SUMO-3 at a motif highly conserved among MEF2 proteins from diverse organisms. This motif is located within the C-terminal TAD, and its sumoylation inhibits transcription. As a transcriptional corepressor of MEF2, histone deacetylase 4 (HDAC4) potentiates sumoylation. This potentiation is dependent on the N-terminal region but not the C-terminal deacetylase domain of HDAC4 and is inhibited by the sumoylation of HDAC4 itself. Moreover, HDAC5, HDAC7, and an HDAC9 isoform also stimulate sumoylation of MEF2. Opposing the action of class IIa HDACs, the SUMO protease SUMO-specific peptidase 3 (SENP3) reverses the sumoylation to augment the transcriptional and myogenic activities of MEF2.

Similarly, the calcium-/calmodulin-dependent kinases and extracellular signal-regulated kinase 5 signaling pathways negatively regulate the sumoylation. These results thus identify sumoylation as a novel regulatory mechanism for MEF2 and suggest that this modification interplays with phosphorylation to promote intramolecular signaling for coordinated regulation in vivo [37]. In intact cells, the interaction of EP300/KAT3B with its protein targets is highly selective, and MEF2D is the preferred channel for EP300/KAT3B-mediated transcriptional control in the heart [38].

Epigenetic modifiers are thought to regulate MEF2-dependent transcription by editing histones and MEF2. Reversible lysine methylation of MEF2D by G9a/KMT1C (lysine methyltransferase 1C) and LSD1/KDM1A (lysine-specific histone demethylase 1A/lysine demethylase 1A) serves as a regulatory mechanism for MEF2D activity and skeletal muscle differentiation, with G9a/KMT1C methylating lysine 267 to repress its transcriptional activity, while LSD1/KDM1A counteracts this effect [39–41]. The MEF2 TF is a crucial regulator of developmental and adult phenotypes in various tissues, including skeletal and cardiac muscle. Multiple PTMs control MEF2D and are effectors of the Ca2-/calmodulin-dependent protein phosphatases calcineurin (CaN), PP2B (protein phosphatase 2B), and PPP3 (protein phosphatase 3). CaN-catalyzed dephosphorylation promotes desumoylation and acetylation of MEF2D, increasing its transcriptional activity. Both MEF2D and CaN bind the scaffold protein myo-A kinase-anchored protein (mAKAP) located at the nuclear membrane, so targeting the mAKAP signalosome inhibits C2C12 skeletal myoblast differentiation and neonatal rat ventricular myocyte hypertrophy. Formation of the mAKAP signalosome is required for MEF2D dephosphorylation and acetylation in C2C12 cells. Reduced phosphorylation of MEF2D is associated with a shift from type IIa HDACs binding to EP300/KAT3B histone acetylase binding, which is associated with increased MEF2D-dependent gene expression and ventricular myocyte hypertrophy [41].

17.2 MEF2D in the Development

17.2.1 The Role of MEF2D in Nonhematopoietic Development

MEF2 plays a vital role in the proliferation, differentiation, and development of various cell types in a tissue-specific manner. Four isoforms of MEF2 (A-D) are differentially involved in controlling cell fate during developmental stages of the cardiac, muscular, vascular, immune, and skeletal systems. Through association with various cytokines, MEF2 isoforms can trigger changes in complex protein networks and regulate multiple stages of cell differentiation, proliferation, survival, and apoptosis. The role of the MEF2 family of transcription factors in development has been studied in various cell types, and shifting changes in this family of TFs contributes to a wide range of disease phenotypes, from cancer to infection.

17.2.2 The Role of MEF2D in Neuronal Cells

Like all the members of the MEF2 family, MEF2D also plays a crucial role in neuronal cells, muscle cells, and other developmental pathways. It is regulated by various class II HDACs, specifically HDAC4 and HDAC5. MEF2D also undergoes alternative splicing, which yields multiple transcript variants. The MEF2D gene is located on chromosome 1q12–q23. It is ubiquitously expressed and exhibits a high and uniform distribution of protein expression in most cells and tissues. However, elevated mRNA transcripts are found in the cerebral cortex, bone marrow, and skeletal muscle with 50.9, 49.8, and 55.5 tpms (transcripts per millions), respectively [17]. Genetic compensatory mechanisms coordinated within class IIa HDACs by a reactive backup circuit of JUN and MEF2D ensure that Schwann cells differentiate in response to cAMP and efficiently produce myelin at developmental stages and after neural injury [42]. Highly conserved DBD, including the MADS-box, is critical for expressing multiple muscle genes with homologous target MEF2 loci in *cis*. MEF2D contains multiple alternatively spliced products of a single gene, one of which is the human homolog of the *Xenopus laevis* SRF-associated factor SL-1 (serum response factor-associated factor SL1). Like its relatives, cloned MEF2D can recapitulate endogenous tissue-specific MEF2 activity by activating transcription through sequence-specific binding to MEF2 loci. Alternative splicing may be necessary in this process: two alternative MEF2D domains, at least one of which is explicitly involved during myogenic differentiation, are also precisely associated with endogenous MEF2 activity [43].

MEF2D was shown by Northern blot hybridization to be strongly expressed in the head portion of mouse embryos at later stages of ontogenesis, in the cerebellum and the cerebrum of adult mice, in cultured cell lines of neuronal origin, and in skeletal and cardiac muscles. During ontogenesis, MEF2D transcripts were detected by in situ hybridization in the olfactory bulb, entorhinal cortex, pyriform cortex, and hippocampus, in Purkinje and granule cells, and in large neurons in both the ventral and dorsal horns of the spinal cord. Adult mice continued to express MEF2D in these limited areas of the central nervous system. Thus, MEF2D seems to be involved in either the differentiation process or the function of these neurons [44].

Accumulating evidence indicates that MEF2D is a prosurvival factor for neurons. Upregulation of MEF2D significantly rescues the decreased viability of OGD/R-injured neurons (oxygen-glucose deprivation/reperfusion-injured neuron) and suppresses OGD/R-induced apoptosis and reactive oxygen species (ROS) production. MEF2D overexpression increases nuclear factor erythroid 2-related factor 2 (NRF2) expression and enhances activation of NRF2 antioxidant signaling. MEF2D may serve as a neuroprotective target with potential applications in treating cerebral ischemia/reperfusion injury [45]. MEF2D regulates neuronal differentiation and development. DYRK1A (dual specificity tyrosine-phosphorylation-regulated kinase 1A) is crucial for cell proliferation and brain development. MEF2D activates DYRK1A isoform 5 transcriptions, suggesting potential cross-interaction during neurodevelopment. MEF2D could upregulate DYRK1A gene expression through specific activation of DYRK1A isoform 5 gene transcription [46]. Although MEF2 family proteins have long been known to be involved in mitochondrial biogenesis, the localization of MEF2D in mitochondria was recently confirmed. MEF2D binds to a consensus sequence in the light chain of mitochondrial DNA that encodes a complex I subunit called NADH dehydrogenase subunit 6(ND6). Disruption of MEF2D resulted in decreased complex I activity, increased ROS production, and decreased ATP levels [47, 48]. MEF2D is linked to neuronal survival and differentiation. cAMP/PKA signaling negatively regulates MEF2D function in myogenic cells. In hippocampal neurons, Krueppel-like factor 6 (KLF6) is a critical transcriptional target of MEF2. Knockdown of KLF6 through small interfering RNAs (siRNAs) leads to neuronal cell death and counteracts the prosurvival effects of MEF2D [49]. MEF2D in rodent neurons' mitochondria regulate mitochondrial DNA (mtDNA)-encoded gene expression. MEF2D is directed to mitochondria through the N-terminal motif and mitochondrial heat shock protein 70 (mtHsp70). Blocking MEF2D in mitochondria reduces complex I activity, increases H_2O_2 levels, reduces ATP production, and sensitizes neurons to stress-induced death. Direct regulation of complex I by mitochondrial MEF2D underlies its neuroprotective effects, possibly linked to Parkinson's disease (PD) [47]. The hypothalamus' suprachiasmatic nucleus (SCN) coordinates peripheral circadian clocks. In mice, neuronal MEF2D is necessary for regular circadian and sleep behavior [50].

17.2.3 The Role of MEF2D in the Muscle and Heart

MEF2 family (MEF2A, MEF2B, MEF2C, and MEF2D) genes and the myogenesis regulator (MRF) family (myogenic differentiation 1 (MYOD), myogenin (MYOG), myogenic factor 5 (MYF5), and myogenic regulatory factor 4 (MRF4), also known as myogenic factor 6 (MYF6)) play critical roles in the myogenesis process, synergistically stimulating and initiating the differentiation process of myoblasts and by targeting them induce multiple waves of muscle-specific gene expression [51, 52]. MEF2D is a critical transcriptional mediator of pathological remodeling in the adult heart downstream of various stress-promoted injuries. The upregulation of many positive cell cycle regulators characterizes the cell cycle's marked impairment in MEF2D-depleted neonatal cardiomyocytes. The expression of phosphatase and Tensin Homolog (PTEN), a central negative regulator of phosphoinositide-3-kinase/AKT serine/threonine kinase (PI3K/AKT), was significantly reduced in MEF2D-deficient cardiomyocytes and found to be a direct target gene of MEF2D. MEF2D plays a crucial role in cell cycle regulation of postmitotic neonatal cardiomyocytes in vitro [53]. MEF2 is a vital transcriptional regulator in embryonic and postnatal muscle. The four *MEF2* genes undergo alternate splicing with unknown specific isoform functions [6, 54–57]. RNA binding Fox-1 homolog 2 (RBFOX2) directly regulates the muscle-specific splicing mode of MEF2D and, when RBFOX2 depletion during differentiation, hinders the production of differentiated MEF2A and MEF2D isoforms and increases the expression of undifferentiated isoforms, potentially inhibiting myoblast fusion. Overexpression of undifferentiated MEF2D isoforms has been shown to inhibit myoblast differentiation. MEF2D isoform partially rescues the fusion defect caused by RBFOX2 depletion, demonstrating its role in myoblast fusion [58]. The MEF2 family of TFs plays a vital role in muscle differentiation during embryogenesis. Deleting a single MEF2 gene does not affect muscle regeneration induced by cardiotoxin, but regeneration is blocked when MEF2A, MEF2C, and MEF2D genes are collectively deleted. MEF2A, MEF2C, and MEF2D loss in satellite cells leads to abnormal expression of numerous protein-coding and long noncoding RNA (lncRNA) genes [29]. The MEF2 TF regulates cell differentiation through tissue-specific splicing to produce muscle-specific isoforms. Ubiquitous (MEF2Dα1) and muscle-specific (MEF2Dα2) isoforms are expressed in muscle and antagonistically regulate differentiation. While *Mef2d* isoforms share a common gene binding set, only MEF2Dα2 activates late muscle transcription. PKA phosphorylation of MEF2Dα1 triggers corepressor association, while exon switching in MEF2Dα2 prevents this repression, enabling recruitment of ASH2-like histone lysine methyltransferase complex subunit (ASH2L) for muscle gene transactivation [59]. Muscleblind-like splicing regulator 3 (MBNL3) promotes the exclusion of alternatively spliced β-exons in *Mef2*, a muscle TF. The more transcriptionally active (+) β-isoform of *Mef2d* overcomes MBNL3's inhibitory effect on muscle differentiation. MBNL3 directly binds to intron 7 downstream of alternatively spliced exons in pre-mRNAs to regulate *Mef2d* splicing [60].

Expression was detected for all four MEF2 transcripts (MEF2A, MEF2B, MEF2C, and MEF2D) at all developmental stages in the human heart, while the *MEF2B* transcript was downregulated in mouse heart development. Although transcripts of none of the MEF2 genes other than mouse *Mef2b* showed significant quantitative changes, qualitative changes in MEF2 transcripts were found during heart development. In particular, the MEF2D transcript exhibited substantial changes during the perinatal period due to alternative splicing [61].

Muscle formation relies on key myogenic regulators controlling gene expression and signaling pathways. Protein phosphatase 2A regulatory subunit B55 delta (PP2A-B55δ) primarily dephosphorylates HDAC4 post-myocyte differentiation to suppress MEF2D-dependent gene expression during myogenic fusion [62].

MEF2D in the adult heart regulates stress-induced cardiac growth and gene expression reprogramming. *Mef2d*-deficient mice resist pressure overload and β-chronic adrenergic stimulation-induced cardiac hypertrophy, fetal gene activation, and fibrosis [63].

17.2.4 The Role of MEF2D in Hematopoietic Development

The role of the MEF2 TF is well documented, especially in muscle, cardiovascular, skeletal, and neuronal development, with many reports also indicating that the MEF2 TF plays a crucial role in developing the hematopoietic system. Some evidence of this mechanism is that loss of MEF2C function early in hematopoietic development leads to blockade of the common lymphoid progenitor (CLP) stage, resulting in significantly reduced numbers of T-, B-, and natural killer (NK) cells but increased numbers of myeloid cells [64]. Deletion of MEF2C in B-cells using Cd19-Cre did not impair B-cell development, suggesting that MEF2C is not required for B-cell development once past the CLP stage. However, B-cell receptor (BCR)-induced B-cell proliferation and T-cell-dependent antibody responses do require MEF2C [64].

In contrast to BCR-mediated responses, Toll-like receptor (TLR)- or CD40-induced B-cell increase does not require MEF2C [64]. Compensation between MEF2C and MEF2D occurs in B-cell development, as the loss of both isoforms results in a block in B-cell development. Loss of MEF2D did not have a significant effect on the macrophage, T-, or B-cell development. However, macrophages from MEF2D KO mice showed reduced production of proinflammatory cytokines. This is due to increased production of interleukin 10 (IL10) [64, 65]. MEF2A and

MEF2D are moderately expressed in primary human monocytes and highly in monocyte-derived macrophages. In macrophages, MEF2A/MEF2D dimers strongly interact with HDAC1 and, to a lesser extent, HDAC7, whereas low levels of MEF2A/D-HDAC1 complexes are found in undifferentiated cells or monocytes. The MEF2A/D-EP300/KAT3B activator complex activates c-Jun expression in macrophages [21]. The transcription factor MEF2D is essential for regulating differentiation and adaptive responses in many cell types and is known to be highly expressed in T-cells [66]. By limiting T-cell function, autoimmunity can be suppressed, as well as anticancer immunity. In the thymus, MEF2D regulates the adverse selection of double-positive T-cells by promoting the transcription of nuclear receptor 4A1(Nurr77) in an HDAC7-dependent manner [67]. In activated T-cells, MEF2D is released from class IIa HDACs and calcineurin binding protein 1 (CABIN1) repression, thereby maintaining the production of interleukin 2 (IL2), IL4, and interferon-gamma (IFNγ) [68–70]. MEF2D/ZEB1 (zinc finger E-box binding homeobox 1) and snail family transcriptional repressor 2 (SLUG)-mediated mechanism by which P21 (RAC1)-activated kinase 4 (PAK4) reprograms the endothelial cell transcriptome and downregulates CLAUDIN-14 and vascular cell adhesion molecule-1 (VCAM-1) expression enhance vessel permeability and reduce T-cell adhesion to the endothelium [71].

In Foxp3$^+$ regulatory T-cells (Treg), MEF2D interacts with the TF Foxp3 and is released from canonical partners such as histone/protein deacetylases. Although MEF2D is not required for Treg generation and maintenance, it is necessary to express IL10, cytotoxic T-lymphocyte antigen 4 (CTLA4), and inducible T-cell costimulator (Icos) and acquire effector Treg phenotypes. MEF2D acts in synergy and additively at these locations with Foxp3 and downstream of B-lymphocyte-induced maturation protein 1 (Blimp1). Furthermore, depletion of MEF2D in Treg cells suppresses the Foxp3 program, especially under supraphysiological conditions that stimulate inflammation, such as transplantation and cancer. Moreover, inhibition or knockdown of the MEF2 repressor HDAC9 increases the suppressive properties of Treg cells [70, 72, 73]. MEF2 plays a role in calcium-dependent transcription of the IL2 gene in T-lymphocytes, and its downregulation results in repression of IL2 transcription in primary human T-cells. MEF2 subtype dysregulation is found in T-cell acute lymphoblastic leukemia (T-ALL) with nonrearranged T-cell receptors in disease fields [74, 75]. A series of developmental processes of B-cells are regulated by homeopathic transcription, primarily the regulation of intercellular signals and specific TFs. For example, the generation of the earliest B-cell progenitors depends on Fms-like tyrosine kinase 3 (Flt3) ligand and IL7-mediated signaling, and transcription factor 3 (E2A), Forkhead Box O1 (FOXO1), IKAROS family zinc finger (IKAROS), and Runt-related transcription factor 1 (RUNX1) that initiate the B-cell program activate the B-cell gene expression program and suppression of lineage inappropriate genes.

MEF2 proteins belong to the MADS-box family, which includes four paralogues (MEF2A-D) in vertebrates with a highly conserved DNA binding domain. These factors are critical for muscle and neuronal development and are highly sensitive to extracellular signaling pathways. MEF2C and MEF2D are highly expressed throughout B-cell differentiation [76, 77]. MEF2C was previously identified as a regulator of B-cell homeostasis and antigen activation [78, 79]. MEF2C and MEF2D are critical integrators of pre-B-cell receptor (pre-BCR) signaling, translating mitogen-activated protein kinase (MAPK) cascade-triggered phosphorylation events into transcriptional regulators that coordinate progression to the immature B-cell stage through positive and negative feed-forward loops. In addition to the well-characterized immediate early genes (IEGs) that initiate cell proliferation and differentiation programs, other essential regulatory pre-B-cell genes [e.g., Krüppel-like factor 2 (KLF2), histone deacetylase 5 (HDAC5), and interferon regulatory factor 4 (IRF4)] may interact through combinational interactions with MEF2C/2D [80].

17.2.5 The Role of MEF2D in the Disease

MEF2 family proteins were initially discovered in cardiac muscle cells and were named for their role in regulating the transcription of cardiac-specific genes. Subsequent studies revealed that MEF2 family proteins are highly expressed in various tissues and cell types and regulate biological processes in cardiovascular, nervous, immune, muscle, and skeletal systems. MEF2 proteins exert their effects by interacting with cotranscription factors, histone-modifying enzymes, and other signaling pathway proteins, thereby modulating the transcriptional levels of target genes and influencing cell function and fate [81].

While MEF2 family proteins are essential for normal physiological processes, mounting evidence suggests their crucial regulatory functions in the pathogenesis and development of diseases. In cardiovascular diseases, MEF2 family proteins are closely associated with cardiac hypertrophy, myocardial infarction, and arrhythmias. In neurological disorders, MEF2 proteins are involved in the maturation of nerve cells and synaptic plasticity and are linked to neurodegenerative diseases such as Alzheimer's and Parkinson's diseases. Furthermore, MEF2 family proteins are also implicated in the pathogenesis of motor neuron diseases like amyotrophic lateral sclerosis (ALS) and musculoskeletal disorders such as muscle atrophy and skeletal developmental anomalies [27, 29, 82, 83].

MEF2D null mice experience suppressed cardiac hypertrophy, decreased fibrosis, and an inability to undergo fetal activation. DYRK1A is encoded in chromosome 21, a critical Down syndrome region. MEF2D is involved in the upregulation of this kinase and activates isoform-5 of DYRK1A, and the interaction of these genes promotes neurodevelopment [46]. MEF2D acts as a neuronal survival factor in maintaining dopaminergic (DA) neurons, the progressive death of which leads to Parkinson's disease (PD). PD-associated neurotoxins destabilize MEF2D by PTMs. MPP^+, a toxic metabolite, decreases the half-life of MEF2D in dopamine neuronal cell lines by MEF2D protein degradation and destabilizing MEF2D mRNA [84]. It has been reported that modulation of transcription factor Nur77 via MEF2D results in DA neuron loss in response to another neurotoxin, 1-Methl-4-phenyl-1,2,3,6-tetrahydropyridine (MPTP). The neurotoxin affects the binding of MEF2D and Nur77 under basal conditions and significantly decreases the expression of both TFs. This results in severe DA loss and an increase in the postsynaptic FBJ murine osteosarcoma viral oncogene homolog B (Fos-B) activity indicative of nigrostriatal damage in mice. Dysfunctions in chaperone-mediated autophagy (CMA) via α-synuclein are also associated with PD. Rotenone, a mitochondrial complex-1 inhibitor, has been used to study CMA substrates. The inhibitor induces the accumulation of α-synuclein and MEF2D in the cytoplasm due to impaired lysosome-mediated degradation. It affects the protective role of MEF2D, resulting in mitochondrial dysfunction and oxidative stress. The accumulation of α-synuclein in neurons of the nigral pathway reduces the protective activity of MEF2D in idiopathic and experimental PD.

The expression of MEF2D in microglia modulates neuroinflammation. MEF2D binds to a MEF2 binding site in the promoter region of the IL10 gene and transcribes I-10; the silencing of MEF2D decreases IL10 induction and increases the levels of tumor necrosis factor alpha (TNF-α) expression, resulting in neuroinflammation. MEF2D also acts as a potential oncogene in the tumorigenesis of malignant glioma and is highly upregulated in grade 3 and 4 malignant glioma patients [85–88]. MEF2D represses skeletal myogenesis upon phosphorylation by PKA and HDAC4 activity in the nucleus. Recently, MEF2D has been implicated in protumorigenic effects, causing cardiac myxoma (CM), the most common cardiac tumor. CM tissue has upregulated expression of MEF2D, which correlates with increased tumor size. MEF2D regulates insulin-like growth factor (IGF)-induced proliferation and control of apoptosis in the pathophysiology of CM, and knockdown of MEF2D reduced the activity of IGF-1 and matrix metallopeptidase 9 (MMP9) and associated tumorigenesis. Similar studies reveal that MEF2D expression positively correlates with the proliferation of CM cells; MEF2D expression can be suppressed by miR-218, a tumor suppressor that is downregulated in myxoma cells and serves as an important target for the treatment of cardiac myxoma [63, 89–91].

MEF2's impact is not limited to development, regulation, and carcinogenesis; it also has connections with chronic viral infections such as Epstein-Barr virus (EBV) [92, 93], human immunodeficiency virus 1 (HIV-1), and human T-cell leukemia virus type 1 (HTLV-1). MEF2A regulates viral gene expression in retroviral HTLV-1 pathogenesis, and binding of Epstein-Barr nuclear antigen 1(EBNA-1) to B-cell-specific transcription factors inhibits MEF2B, leading to inhibited growth and viability of latently infected EBV cells [94, 95]. MEF2D dysfunction underlies DA neuron loss in PD. 6-Hydroxydopamine (6-OHDA) exposure in DA neuronal cell lines causes direct oxidative modification of MEF2D. Oxidized MEF2D binds more strongly to the critical regulator of CMA, heat shock homolog 70 kDa. This oxidation inhibits MEF2D's function, contributing to oxidative stress-induced neurotoxicity and potentially the pathogenesis of PD [96].

17.3 The Biology Function of MEF2D in B-ALL

17.3.1 B-Cell Acute Lymphoblastic Leukemia (B-ALL)

Acute lymphoblastic leukemia (ALL) is a hematologic malignancy characterized by the malignant multiplication of lymphoid progenitor cells in the bone marrow, peripheral blood, and extramedullary sites. According to clinical statistics, leukemia accounts for 4% and 3% of oncological diseases in males and females, respectively, with 4% and 3% mortality rates, respectively [97, 98]. B-cell ALL (B-ALL) incidence accounts for 85% of adult and 75% of pediatric ALL cases [99]. Recently, frontline regimens for ALL treatment have resulted in a five-year overall survival (OS) rate of 90% in pediatric patients but less than 45% in adults. In older patients (>60 years), the 5-year OS rate is under 20%. In addition, approximately one-third of patients with standard-risk ALL will experience relapse, with the likelihood doubling for those patients found to be high risk [100–105]. Hematopoietic stem cell transplantation remains the most potent therapeutic approach for B-ALL, but some patients are ineligible for the procedure, and transplant-related complications and posttransplantation relapse are still clinical obstacles. Therefore, there is an urgent need to explore novel therapeutic targets and new therapeutic strategies to combat B-ALL.

17.3.2 MEF2D Fusions in B-ALL

17.3.2.1 MEF2D Fusion Proteins Appearing in B-ALL and the Clinical Features

ALL is a complex and life-threatening group of hematological malignancies resulting from the clonal proliferation of lymphoid progenitor cells. It poses a significant global health challenge, affecting people of all ages [106]. ALL can be categorized into two primary subtypes based on the immunophenotype: B-cell precursor ALL (BCP-ALL) and T-cell ALL. BCP-ALL accounts for approximately 80% of ALL cases and is characterized by its association with various genetic lesions, including oncogenic fusions arising from chromosomal translocations. These genetic aberrations have been shown to disrupt multiple cellular events critical for normal function, such as lymphoid development, cytokine receptors, RAS signaling, tumor suppression, and epigenetic control [107]. Recent advancements in RNA followed by deep sequencing (RNA-Seq) technology have identified recurrent fusions between MEF2D and different partner genes in BCP-ALL patients, both in adult and pediatric populations. The frequencies of these fusions are approximately 4.2% in adults and 3.2% in pediatric cases [108]. Remarkably, individuals with MEF2D fusions form a distinct subgroup exhibiting a pre-B-cell immunophenotype, demonstrating similar gene expression profiles and experiencing poorer clinical outcomes.

Additionally, MEF2D fusions often cooccur with hotspot point mutations in the neuroblastoma RAS viral oncogene homolog (NRAS) gene, suggesting potential cooperative effects driving the disease's progression [109, 110]. Researchers have reported nine different types of MEF2D fusions in BCP-ALL, with MEF2D-BCL9 (BCL9 transcription coactivator) and MEF2D-HNRNPUL1 (heterogeneous nuclear ribonucleoprotein U like 1) being the most frequently observed in patients. These fusion variants share a conserved N-terminal region composed of MEF and MADS-box domains (residues 1–86 from the wild-type MEF2D protein), indicating the crucial role of this region in leukemogenesis. The wild-type MEF2D protein belongs to the MEF2 transcription factor family, which is essential in regulating cell differentiation, survival, proliferation, and adaptive responses. Of note, MEF2C and MEF2D have been recognized as critical components of transcriptional complexes coordinating B-cell development [80].

BCP-ALL is characterized by genetic alterations with high heterogeneity. Precise subtypes with distinct genomic and gene expression patterns have been recently revealed using high-throughput sequencing technology. Most of these profiles are associated with recurrent nonoverlapping rearrangements or hotspot point mutations that are analogous to the established subtypes, such as rearrangements to/in double homeobox 4 (DUX4), MEF2D, nuclear protein in testis midline carcinoma 1 (NUTM1), B-cell lymphoma 6 (BCL6), B-cell lymphoma 2/cellular myelocytomatosis (BCL2/c-MYC) and zinc finger protein 384/zinc finger protein 362 (ZNF384/ZNF362) as well as ETS variant transcription factor 6-runt-related transcription factor 1 fusion gene (TV6-RUNX1)-like gene expression, PAX5alt [diverse paired box 5 (PAX5) alterations, including rearrangements, intragenic amplifications, or mutations], and hotspot mutations PAX5 with biallelic PAX5 alterations, IKAROS family zinc finger 1 (IKZF1), and zinc finger E-box binding homeobox 2 (ZEB2).

Three new subtypes of adult and childhood BCP-ALL have recently been described [94, 95, 97–101]. The subtypes are involved in the rearrangements of DNA binding factors, including DUX4, MEF2D, and ZNF384, which separately account for approximately 4–7%, 2–4%, and 3–5% of childhood BCP-ALL and 4–7%, 2–7%, and 3–8% of adult BCP-ALL. In most patients, DUX4 overexpression is the consequence of DUX4 fusion, typically IGH, but rarely ERG [108, 111–119]. By contrast, the partner genes of MEF2D and ZNF384 are complex and diverse: a total of 9 genes (BCL9, SS18, FOXJ2, CSF1R, DAZAP1, STAT6, HNRNPUL1, HNRNPH1, and HNRNPM) and 11 genes (EP300/KAT3B, TCF3, TAF15, CREBBP, EWSR1, ARID1B, SMARCA2, SMARCA4, SYNRG, NIPBL, and CLTC) have been reported as fusion partners of MEF2D and ZNF384 [8, 15–18, 20, 22, 66–68], respectively. MEF2D is a TF that can specifically bind to the MEF2 element 5′-YTA[AT](4)TAR-3′. It belongs to the MEF2 gene family that contributes to the differentiation of muscle and neural cells, cardiac morphogenesis, formation of blood vessels, growth factor responsiveness, survival of neuronal cells, and acute leukemia. As a member of the MEF2 gene family, MEF2C is an activated oncogene in early T-cell precursor ALL, a subtype of high-risk T- lineage ALL [120–126]. In 2005, MEF2D-DAZAP1 was reported in the TS-2 cell line, as established from a three-year-old girl with ALL; it was found to contain t(1;19)(q23;p13.3) but was lacking the TCF3-PBX1 (pre-B-cell leukemia homeobox 1) fusion. These studies determined that MEF2D fusions can retain the MADS-box domain required to mediate DNA binding with enhanced MEF2D transcriptional activity. The heterogeneous nuclear ribonucleoproteins, including HNRNPUL1, heterogeneous nuclear ribonucleoprotein H1(HNRNPH1), and heterogeneous nuclear ribonucleoprotein HM(HNRNPM), are involved in MEF2D-r ALL [108, 116, 118]. These proteins can bind RNA and are associated with pre-mRNA processing in the nucleus. A target gene of MEF2D, HDAC9, is significantly upregulated in

MEF2D-r patients and provides an optional therapeutic strategy by using histone deacetylase inhibitors, such as Panobinostat [107, 109, 110]. B-ALL is characterized by multiple fusion genes resulting from chromosomal translocations, and MEF2D-associated fusions have recently been identified in a relatively poor surviving subtype of B-ALL. B-ALL is characterized by frequent and severe chromosomal changes, such as aneuploidy and rearrangements, which disrupt oncogenes and create fusion proteins with altered functions. Fusion proteins like ETV6-RUNX1 fusion gene (ETV6-RUNX1), TCF3-PBX1 fusion gene (TCF3-PBX1), PAX5, and lysine methyltransferase 2A [KMT2A (MLL)] and tyrosine kinases such as BCR-ABL1 (Abelson tyrosine kinase 1) are often acquired early in the development of leukemia. They disrupt cellular pathways involved in hematopoietic development and tumor formation through inhibition, kinase signaling, and chromatin remodeling. Identifying these rearrangements is crucial for effective treatment of ALL, as they can impact treatment outcomes and serve as potential targets for therapy.

17.3.2.2 MEF2D Fusion Proteins Regulate ALL Occurrences and their Role in Proliferation

In ALL cases, specific rearrangements occur between MEF2D and five genes (BCL9, CSF1R, DAZAP1, HNRNPUL1, and SS18) (Fig. 17.2), each with distinct gene expression profiles. MEF2D-rearranged ALL has four unique features: a distinct immunophenotype, altered DNA copy number at the rearrangement site, older age at diagnosis, and poor prognosis. These rearrangements boost MEF2D's transcriptional activity, trigger lymphoid transformation, activate HDAC9 expression, and increase histone deacetylase inhibitor treatment sensitivity. Consequently, MEF2D-rearranged ALL represents a distinct and high-risk form of leukemia,

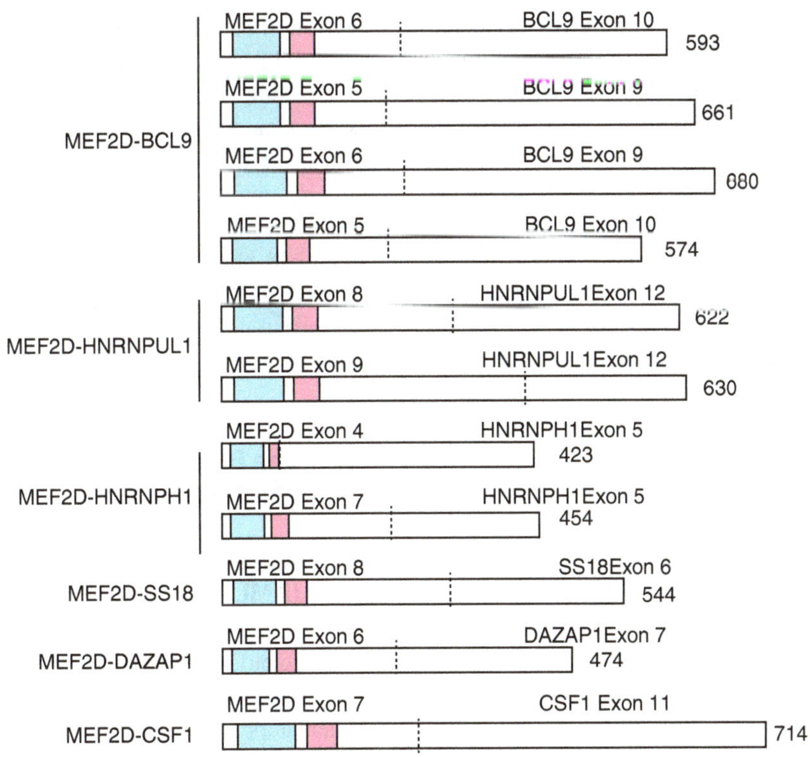

Fig. 17.2 Schematic diagrams of MEF2D fusion proteins in B-ALL

warranting significant attention [110]. MEF2D was aberrantly overexpressed in B-ALL, and MEF2D overexpression was remarkably correlated to disease progression in ALL patients [110, 127–129]. Also, MEF2D knockdown in B-ALL cells significantly restrained cell viability, induced cell apoptosis, arrested cell cycle progression, and enhanced drug resistance in vitro, as well as downregulating the expression of the PI3K/AKT signaling pathway in B-ALL. The B-ALL tumor load markedly reduced in the B-ALL xenograft mouse model following MEF2D knockdown. Thus, MEF2D has become a potential therapeutic target in B-ALL [53]. In REH cells overexpressing MEF2D-SS18, genes related to B-cell development were downregulated, whereas genes related to drug sensitivity and B-cell survival were upregulated. In vitro B-cell differentiation experiments showed that MEF2D-SS18 significantly inhibited the differentiation of mouse standard lymphoid progenitor cells into CD19-positive B-cells [130].

The fusion partner of MEF2D, HNRNPUL1, encodes a nuclear RNA-binding protein belonging to the heterogeneous nuclear ribonucleoprotein family. HNRNPUL1 involves several crucial cellular processes, including messenger RNA processing and trafficking, DNA double-strand breaks and repair, and histone gene transcription. This gene has a high mutation frequency in certain cancers like uterine endometrial carcinoma and gastric adenocarcinoma, suggesting its involvement in cancer-related pathways [80, 131, 132]. In the context of MEF2D fusions, HNRNPUL1 retains domains that interact with tumor protein 53 (TP53) and bromodomain-containing protein 7 (BRD7), suggesting a regulatory role in transcription and RNA binding. Moreover, previous studies have shown that nuclear expression of MEF2D-BCL9 (MB) and MEF2D-HNRNPUL1 (MH) may lead to dysregulation of HDAC9 and recombination-activating gene 1 (RAG1), resulting in the partial developmental arrest of B-cells in vitro. Aberrant HDAC9 expression has also been associated with developing lymphoproliferative disorders and lymphomas in experimental models [133]. Current MEF2D fusions in BCP-ALL patients have been linked to a poor prognosis. Experimental studies involving MH knockin mice have demonstrated progressive disease, with advancement from impaired pre-B-cell development to preleukemia within 10–12 months. In NRASG12D BCP-ALL cells, MH has been shown to trigger a more aggressive phenotype of BCP-ALL compared to NRASG12D-induced leukemia. Notably, MH exhibits an enhanced chromatin binding capacity primarily through MEF2D response element (MRE) motifs in target genes. Disrupting MH-DNA interactions has been found to alleviate aberrant target gene expression and arrest B-cell differentiation. The C-terminal portion of MH (HNRNPUL1 portion) is involved in counter-regulatory activity, cofactor recruitment, and homodimerization of fusion proteins [110]. Promisingly, therapeutic interventions targeting MH-driven HDAC family transactivation using the HDAC inhibitor Panobinostat in combination with traditional chemotherapy have improved overall survival in experimental MH/NRASG12D BCP-ALL models [134].

In conclusion, MEF2D fusions represent a significant molecular subtype of BCP-ALL with distinct clinical and biological characteristics. Understanding the mechanisms underlying the leukemogenic potential of MEF2D fusions is essential for developing targeted therapies and improving outcomes for patients affected by this aggressive form of leukemia. Ongoing research in this field holds great promise for advancing our knowledge of ALL pathogenesis and refining treatment strategies to combat this challenging disease.

17.4 The Biology Function of MEF2D in AML

17.4.1 Acute Myeloid Leukemia (AML)

Acute myeloid leukemia (AML) is a fatal disease characterized by the clonal proliferation of primitive hematopoietic stem or progenitor cells. It manifests as abnormal differentiation of bone marrow cells, leading to higher levels of immature malignant cells and fewer differentiated red blood cells, platelets, and white blood cells. The disease occurs in all age groups, with a higher prevalence in the elderly (>60 years). The genomic landscape of AML is well defined by the disease's uncommon genetic instability and a relatively low number of driver mutations. The mutation of genes involved in epigenetic regulation is a common and early event in leukemogenesis. Chromosomal translocations involving the MLL gene at chromosome 11q23 cause aggressive leukemias, responsible for approximately 10% of acute leukemias in humans [135, 136]. MLL rearrangements generate a chimeric gene encoding an oncogenic fusion protein consisting of the N-terminus of MLL fused to the C-terminus of one of more than 80 known fusion partners [137]. For example, MLL rearrangement (MLLr) produces a chimeric protein that recruits the transcription elongation complex containing the DOT1-like histone lysine methyltransferase/lysine methyltransferase 4 (DOT1L/KMT4) to selected oncogenic transcription factors, including the HOX cluster A (HOX-A) cluster genes [138]. MLL fusion proteins bind DNA/chromatin and induce leukemic transformation of hematopoietic stem and progenitor cells through transcriptional deregulation of fusion protein target genes [139].

17.4.2 MEF2D Molecular and Function in AML

Human cancer initiation and progression are driven by complex genetic alterations that influence transcriptional and chromatin dysregulation [140, 141]. AML is an aggressive, heterogeneous hematopoietic malignancy characterized by abnormal self-renewal and impaired differentiation of myeloid progenitor cells. Recurrent oncogenic changes can occur in AML due to cancer's characteristic overproduction of immature myeloid precursor cells, TF, and epigenetic regulators such as CCAAT/enhancer binding protein alpha (CEBPA), Runt-related transcription factor 1 (RUNX1), GATA binding factor 2 (GATA2), and lysine methyltransferase 2A (KMT2A). Mixed-lineage leukemia-1 (MLL1) is critical for stem cell renewal and myeloid differentiation [142]. AML often relies on regulatory proteins such as TFs or chromatin regulators to control the expression of genes critical for maintaining the malignant state. TFs contain one or more direct DBDs, representing the most important class of proteins in interpreting eukaryotic genomes. An estimated 1600 TFs in the human genome potentially bind DNA in a sequence-specific manner, many of which are biochemically and functionally ambiguous [143, 144]. Master TFs often form interconnected signaling regulatory loops that control each other's expression and their downstream transcriptional programs by coordinating the regulation of enhancer activity. Extensive molecular studies of aggressive, heterogeneous hematopoietic stem cell malignancies such as AML have identified transcriptional and dysregulation of TF levels and transcription as key hallmarks of the disease [141, 145]. One of the leading causes of leukemia is transcriptional dysregulation, and transcriptional dysregulation is prevalent in leukemia. However, little is known about the regulatory role of the transcriptional regulator MEF2D in AML. A recent study employing an elegant CRISPR-/Cas9-based genetic screen to reveal novel transcriptional addiction pathways in AML systematically showed that a direct physical association between the chromatin readers zinc finger MYND-type containing 8 (ZMYND8) and bromodomain-containing protein 4 (BRD4) is essential for the regulation of interferon regulatory factor 8 (IRF8) and c-MYC transcriptional programs [146–148]. IRF8, a hematopoietic lineage-specific TF, is considered an essential myeloid cell and B lymphocyte fate determinant [149, 150]. MEF2A-D proteins are evolutionarily conserved MADS TFs implicated in normal muscle, neuronal, and blood system cell development as well as human disease [23, 72]. MEF2D has been reported to be upregulated in some types of cancer, and MEF2D may be involved in leukemogenesis [23, 72, 98, 127, 146, 151, 152]. Interestingly, acute IRF8 depletion led to downregulation of MEF2D, which was also identified as an essential AML gene in a CRISPR/Cas9 genetic screen, and

MEF2D depletion led to downregulation of IRF8, suggesting a positive feedback loop between these two TFs. This positive feedback loop could sustain the hematopoietic transcription factor PU.1/myeloid ecotropic insertion site 1 (PU.1/MEIS1)-coregulated transcriptional output by stabilizing PU.1's chromatin occupancy [146–148]. Among them, PU.1 and MEIS1 play an essential role in maintaining the proliferation of AML cells. Interestingly, ZMYND8 is an AML-specific dependency. Ablation of ZMYND8 resulted in the abrogation of AML cell proliferation in vitro and in vivo and significantly increased AML heterogeneity. Mechanistically, ZMYND8 promotes maintenance of the leukemic state through parallel activation of IRF8 and c-MYC transcriptional programs. The PHD-BD-PWWP (plant homeodomain-bromodomain-Pro Trp Trp Pro) reader cassette of ZMYND8 directly interacts with the extra-terminal domain of BRD4. The ZMYND8-BRD4 interaction is critical for the regulation of both IRF8 and MYC, as disruption of the ZMYND8-BRD4 interaction by mutating the PHD-BD-PWWP domain of ZMYND8 resulted in the termination of AML cell proliferation in vitro and in vivo. Most importantly, genetic deletion of ZMYND8 in normal hematopoietic stem cells (HSCs) did not affect myeloid differentiation in vitro. In addition, MEF2D plays a vital regulatory role in the proliferation and survival of MLLr AML. MEF2D binds introns and enhancers, and it has high chromatin accessibility in some patient-associated genetic alterations, such as KMT2A rearrangements. The TF MEF2D is a super-enhancer-associated, highly expressed gene in MLLr AML. Knockdown of MEF2D severely impaired leukemic growth, induced myeloid differentiation, and delayed oncogenic progression in vivo. Mechanistically, loss of MEF2D leads to robust activation of the CCAAT/enhancer binding protein epsilon (CEBPE)-centric myeloid differentiation program in AML cells. Chromatin analysis revealed that MEF2D binds and inhibits chromatin accessibility to CEBPE cis-regulatory regions. Homeobox A9 (HOXA9) positively regulates MEF2D, and the MEF2D-CEBPE axis is a critical transcriptional mechanism regulating leukemia cell self-renewal and differentiation blockage (Fig. 17.3). In previous studies, IRF8 was also reported in myelomonocytic differentiation and AML pathogenesis [153]. Cancer genome sequencing projects in AML revealed frequent genetic alterations in regulators controlling gene transcription, chromatin state, and DNA covalent modification [154, 155]. While a small number of these driver gene alterations have become drug targets in AML, most are not viable targets for

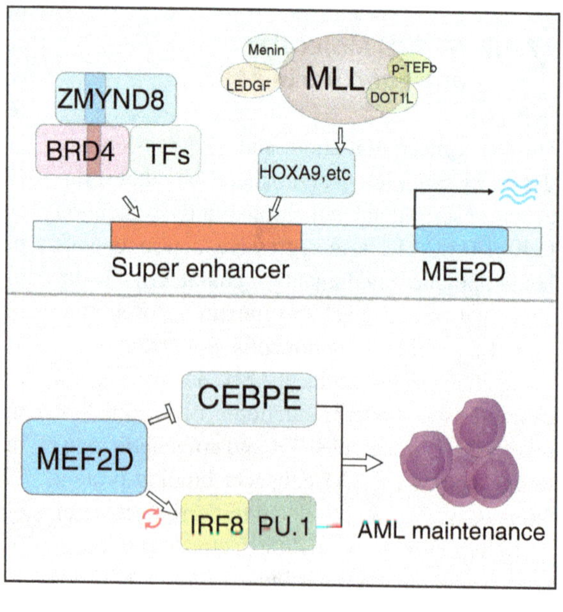

Fig. 17.3 Model of regulation and function of MEF2D in AML

small molecule inhibitors but the positive feedback regulatory loop of ZMYND8-BRD4 interaction, MEF2D, IRF8, and MEF2D -CEBPE axis, both as achievable possible targets for small molecule inhibitors.

17.5 Conclusion

This chapter explores human MEF2 transcription factors (MEF2A, MEF2B, MEF2C, and MEF2D) and their roles in various physiological processes. It also discusses their significance in development, neurological disorders, and leukemia. MEF2D stands out in muscle and neuronal development.

In leukemia, MEF2D forms fusion proteins with BCL9, CSF1R, DAZAP1, HNRNPUL1, and SS18, contributing to disease pathogenesis. This chapter uncovers a positive feedback loop between MEF2D and IRF8, crucial in AML progression. Additionally, it highlights the MEF2D-CEBPE axis as a vital mechanism in leukemia cell regulation. These findings offer potential therapeutic avenues in leukemia treatment.

In conclusion, this study offers a comprehensive overview of the human MEF2 TFs' diverse functions in development and disease. The findings underscore the existence of complex regulatory networks involving MEF2 proteins and their potential as therapeutic targets for various pathological conditions, including leukemia. Future investigations in this field are warranted to unravel the intricate molecular mechanisms of MEF2-mediated transcriptional regulation and explore their impact on human health and disease.

Acknowledgments The authors thank the Lu Lab members Virginia Vasey, Xinyue Zhou, and Sajesan Aryal at the University of Alabama at Birmingham (UAB) for helping with critical feedbacks. This work was supported in part by National Institutes of Health grants R01 CA259480. RL is an American Society of Hematology scholar in basic science and a research scholar of the American Cancer Society (RSG-22-036-01-DMC). We apologize to researchers whose work could not be cited owing to space limitations.

References

1. Canté-Barrett K, Pieters R, Meijerink JP (2014) Myocyte enhancer factor 2C in hematopoiesis and leukemia. Oncogene 33(4):403–410
2. Rashid AJ, Cole CJ, Josselyn SA (2014) Emerging roles for MEF2 transcription factors in memory. Genes Brain Behav 13(1):118–125
3. Dietrich JB (2013) The MEF2 family and the brain: from molecules to memory. Cell Tissue Res 352(2):179–190
4. Yang Q, Mao Z (2010) Dysregulation of autophagy and Parkinson's disease: the MEF2D link. Apoptosis 15(11):1410–1414
5. Aziz A, Liu QC, Dilworth FJ (2010) Regulating a master regulator: establishing tissue-specific gene expression in skeletal muscle. Epigenetics 5(8):691–695
6. Potthoff MJ, Olson EN (2007) MEF2: a central regulator of diverse developmental programs. Development 134(23):4131–4140
7. McGee SL (2007) Exercise and MEF2-HDAC interactions. Appl Physiol Nutr Metab 32(5):852–856
8. Yu YT (1996) Distinct domains of myocyte enhancer binding factor-2A determining nuclear localization and cell type-specific transcriptional activity. J Biol Chem 271(40):24675–24683
9. Borghi S, Molinari S, Razzini G, Parise F, Battini R, Ferrari S (2001) The nuclear localization domain of the MEF2 family of transcription factors shows member-specific features and mediates the nuclear import of histone deacetylase 4. J Cell Sci 114(Pt 24):4477–4483
10. Ichihara M, Kamiya T, Hara H, Adachi T (2018) The MEF2A and MEF2D function as scaffold proteins that interact with HDAC1 or p300 in SOD3 expression in THP-1 cells. Free Radic Res 52(7):799–807
11. Di Giorgio E, Franforte E, Cefalù S, Rossi S, Dei Tos AP, Brenca M et al (2017) The co-existence of transcriptional activator and transcriptional repressor MEF2 complexes influences tumor aggressiveness. PLoS Genet 13(4):e1006752
12. Zheng M, Zhu J, Lu T, Liu L, Sun H, Liu Z et al (2013) p300-mediated histone acetylation is essential for the regulation of GATA4 and MEF2C by BMP2 in H9c2 cells. Cardiovasc Toxicol 13(4):316–322
13. Youn HD, Liu JO (2000) Cabin1 represses MEF2-dependent Nur77 expression and T cell apoptosis by controlling association of histone deacetylases and acetylases with MEF2. Immunity 13(1):85–94
14. Youn HD, Chatila TA, Liu JO (2000) Integration of calcineurin and MEF2 signals by the coactivator p300 during T-cell apoptosis. EMBO J 19(16):4323–4331
15. Shin HM, Han TH (1999) CD28-mediated regulation of the c-jun promoter involves the MEF2 transcription factor in Jurkat T cells. Mol Immunol 36(3):197–203

16. Sharma-Walia N, Krishnan HH, Naranatt PP, Zeng L, Smith MS, Chandran B (2005) ERK1/2 and MEK1/2 induced by Kaposi's sarcoma-associated herpesvirus (human herpesvirus 8) early during infection of target cells are essential for expression of viral genes and for establishment of infection. J Virol 79(16):10308–10329
17. Madugula K, Mulherkar R, Khan ZK, Chigbu DI, Patel D, Harhaj EW et al (2019) MEF-2 isoforms' (A-D) roles in development and tumorigenesis. Oncotarget 10(28):2755–2787
18. Ma K, Chan JK, Zhu G, Wu Z (2005) Myocyte enhancer factor 2 acetylation by p300 enhances its DNA binding activity, transcriptional activity, and myogenic differentiation. Mol Cell Biol 25(9):3575–3582
19. Katsuyama M, Ozgur Cevik M, Arakawa N, Kakehi T, Nishinaka T, Iwata K et al (2007) Myocyte enhancer factor 2B is involved in the inducible expression of NOX1/NADPH oxidase, a vascular superoxide-producing enzyme. FEBS J 274(19):5128–5136
20. He J, Ye J, Cai Y, Riquelme C, Liu JO, Liu X et al (2011) Structure of p300 bound to MEF2 on DNA reveals a mechanism of enhanceosome assembly. Nucleic Acids Res 39(10):4464–4474
21. Aude-Garcia C, Collin-Faure V, Bausinger H, Hanau D, Rabilloud T, Lemercier C (2010) Dual roles for MEF2A and MEF2D during human macrophage terminal differentiation and c-Jun expression. Biochem J 430(2):237–244
22. Samak M, Kues A, Kaltenborn D, Klösener L, Mietsch M, Germena G et al (2023) Dysregulation of Krüppel-like factor 2 and myocyte enhancer factor 2D drive cardiac microvascular inflammation and dysfunction in diabetes. Int J Mol Sci 24(3):2482
23. Pon JR, Marra MA (2016) MEF2 transcription factors: developmental regulators and emerging cancer genes. Oncotarget 7(3):2297–2312
24. Lu YW, Martino N, Gerlach BD, Lamar JM, Vincent PA, Adam AP et al (2021) MEF2 (myocyte enhancer factor 2) is essential for endothelial homeostasis and the atheroprotective gene expression program. Arterioscler Thromb Vasc Biol 41(3):1105–1123
25. Cornwell JD, McDermott JC (2023) MEF2 in cardiac hypertrophy in response to hypertension. Trends Cardiovasc Med 33(4):204–212
26. Chen X, Gao B, Ponnusamy M, Lin Z, Liu J (2017) MEF2 signaling and human diseases. Oncotarget 8(67):112152–112165
27. McKinsey TA, Zhang CL, Olson EN (2002) MEF2: a calcium-dependent regulator of cell division, differentiation and death. Trends Biochem Sci 27(1):40–47
28. Fagerberg L, Hallström BM, Oksvold P, Kampf C, Djureinovic D, Odeberg J et al (2014) Analysis of the human tissue-specific expression by genome-wide integration of transcriptomics and antibody-based proteomics. Mol Cell Proteomics 13(2):397–406
29. Liu N, Nelson BR, Bezprozvannaya S, Shelton JM, Richardson JA, Bassel-Duby R et al (2014) Requirement of MEF2A, C, and D for skeletal muscle regeneration. Proc Natl Acad Sci USA 111(11):4109–4114
30. Estrella NL, Desjardins CA, Nocco SE, Clark AL, Maksimenko Y, Naya FJ (2015) MEF2 transcription factors regulate distinct gene programs in mammalian skeletal muscle differentiation. J Biol Chem 290(2):1256–1268
31. Desjardins CA, Naya FJ (2017) Antagonistic regulation of cell-cycle and differentiation gene programs in neonatal cardiomyocytes by homologous MEF2 transcription factors. J Biol Chem 292(25):10613–10629
32. Harrington AJ, Bridges CM, Berto S, Blankenship K, Cho JY, Assali A et al (2020) MEF2C hypofunction in neuronal and neuroimmune populations produces MEF2C haploinsufficiency syndrome-like behaviors in mice. Biol Psychiatry 88(6):488–499
33. Qiao Q, Zhao CM, Yang CX, Gu JN, Guo YH, Zhang M et al (2021) Detection and functional characterization of a novel MEF2A variation responsible for familial dilated cardiomyopathy. Clin Chem Lab Med 59(5):955–963
34. Shalizi AK, Bonni A (2005) Brawn for brains: the role of MEF2 proteins in the developing nervous system. Curr Top Dev Biol 69:239–266
35. Majidi SP, Reddy NC, Moore MJ, Chen H, Yamada T, Andzelm MM et al (2019) Chromatin environment and cellular context specify compensatory activity of paralogous MEF2 transcription factors. Cell Rep 29(7):2001–15.e5
36. Macneil LT, Walhout AJ (2011) Gene regulatory networks and the role of robustness and stochasticity in the control of gene expression. Genome Res 21(5):645–657
37. Grégoire S, Yang XJ (2005) Association with class IIa histone deacetylases upregulates the sumoylation of MEF2 transcription factors. Mol Cell Biol 25(6):2273–2287
38. Slepak TI, Webster KA, Zang J, Prentice H, O'Dowd A, Hicks MN et al (2001) Control of cardiac-specific transcription by p300 through myocyte enhancer factor-2D. J Biol Chem 276(10):7575–7585
39. Chopra A, Cho WC, Willmore WG, Biggar KK (2020) Hypoxia-inducible lysine methyltransferases: G9a and GLP hypoxic regulation, non-histone substrate modification, and pathological relevance. Front Genet 11:579636
40. Majello B, Gorini F, Saccà CD, Amente S (2019) Expanding the role of the histone lysine-specific demethylase LSD1 in cancer. Cancers (Basel) 11(3):324
41. Li J, Aponte Paris S, Thakur H, Kapiloff MS, Dodge-Kafka KL (2019) Muscle A-kinase-anchoring protein-β-bound calcineurin toggles active and repressive transcriptional complexes of myocyte enhancer factor 2D. J Biol Chem 294(7):2543–2554
42. Velasco-Aviles S, Patel N, Casillas-Bajo A, Frutos-Rincón L, Velasco E, Gallar J et al (2022) A genetic compensatory mechanism regulated by Jun and

Mef2d modulates the expression of distinct class IIa Hdacs to ensure peripheral nerve myelination and repair. elife 11:e72917
43. Breitbart RE, Liang CS, Smoot LB, Laheru DA, Mahdavi V, Nadal-Ginard B (1993) A fourth human MEF2 transcription factor, hMEF2D, is an early marker of the myogenic lineage. Development 118(4):1095–1106
44. Ikeshima H, Imai S, Shimoda K, Hata J, Takano T (1995) Expression of a MADS box gene, MEF2D, in neurons of the mouse central nervous system: implication of its binary function in myogenic and neurogenic cell lineages. Neurosci Lett 200(2):117–120
45. Wang N, Yang W, Li L, Tian M (2020) MEF2D upregulation protects neurons from oxygen-glucose deprivation/re-oxygenation-induced injury by enhancing Nrf2 activation. Brain Res 1741:146878
46. Wang P, Wang L, Chen L, Sun X (2017) Dual-specificity tyrosine-phosphorylation regulated kinase 1A gene transcription is regulated by myocyte enhancer factor 2D. Sci Rep 7(1):7240
47. She H, Yang Q, Shepherd K, Smith Y, Miller G, Testa C et al (2011) Direct regulation of complex I by mitochondrial MEF2D is disrupted in a mouse model of Parkinson disease and in human patients. J Clin Invest 121(3):930–940
48. Sepuri NBV, Tammineni P, Mohammed F, Paripati A (2017) Nuclear transcription factors in the mitochondria: a new paradigm in fine-tuning mitochondrial metabolism. Handb Exp Pharmacol 240:3–20
49. Salma J, McDermott JC (2012) Suppression of a MEF2-KLF6 survival pathway by PKA signaling promotes apoptosis in embryonic hippocampal neurons. J Neurosci 32(8):2790–2803
50. Mohawk JA, Cox KH, Sato M, Yoo SH, Yanagisawa M, Olson EN et al (2019) Neuronal myocyte-specific enhancer factor 2D (MEF2D) is required for normal circadian and sleep behavior in mice. J Neurosci 39(40):7958–7967
51. Zhang S, Zhang Y, Chen C, Hu Q, Fu Y, Xu L et al (2022) Identification of robust and key differentially expressed genes during C2C12 cell myogenesis based on multiomics data. Int J Mol Sci 23(11):6002
52. Cardoso AC, Pereira AHM, Ambrosio ALB, Consonni SR, Rocha de Oliveira R, Bajgelman MC et al (2016) FAK forms a complex with MEF2 to couple biomechanical signaling to transcription in cardiomyocytes. Structure 24(8):1301–1310
53. Estrella NL, Clark AL, Desjardins CA, Nocco SE, Naya FJ (2015) MEF2D deficiency in neonatal cardiomyocytes triggers cell cycle re-entry and programmed cell death in vitro. J Biol Chem 290(40):24367–24380
54. Yu YT, Breitbart RE, Smoot LB, Lee Y, Mahdavi V, Nadal-Ginard B (1992) Human myocyte-specific enhancer factor 2 comprises a group of tissue-restricted MADS box transcription factors. Genes Dev 6(9):1783–1798
55. Ohkawa Y, Marfella CG, Imbalzano AN (2006) Skeletal muscle specification by myogenin and Mef2D via the SWI/SNF ATPase Brg1. EMBO J 25(3):490–501
56. Martin JF, Miano JM, Hustad CM, Copeland NG, Jenkins NA, Olson EN (1994) A Mef2 gene that generates a muscle-specific isoform via alternative mRNA splicing. Mol Cell Biol 14(3):1647–1656
57. Black BL, Olson EN (1998) Transcriptional control of muscle development by myocyte enhancer factor-2 (MEF2) proteins. Annu Rev Cell Dev Biol 14:167–196
58. Singh RK, Xia Z, Bland CS, Kalsotra A, Scavuzzo MA, Curk T et al (2014) Rbfox2-coordinated alternative splicing of Mef2d and Rock2 controls myoblast fusion during myogenesis. Mol Cell 55(4):592–603
59. Sebastian S, Faralli H, Yao Z, Rakopoulos P, Palii C, Cao Y et al (2013) Tissue-specific splicing of a ubiquitously expressed transcription factor is essential for muscle differentiation. Genes Dev 27(11):1247–1259
60. Lee KS, Cao Y, Witwicka HE, Tom S, Tapscott SJ, Wang EH (2010) RNA-binding protein Muscleblind-like 3 (MBNL3) disrupts myocyte enhancer factor 2 (Mef2) {beta}-exon splicing. J Biol Chem 285(44):33779–33787
61. Iida K, Hidaka K, Takeuchi M, Nakayama M, Yutani C, Mukai T et al (1999) Expression of MEF2 genes during human cardiac development. Tohoku J Exp Med 187(1):15–23
62. Veloso A, Martin M, Bruyr J, O'Grady T, Deroanne C, Mottet D et al (2019) Dephosphorylation of HDAC4 by PP2A-Bδ unravels a new role for the HDAC4/MEF2 axis in myoblast fusion. Cell Death Dis 10(7):512
63. Kim Y, Phan D, van Rooij E, Wang DZ, McAnally J, Qi X et al (2008) The MEF2D transcription factor mediates stress-dependent cardiac remodeling in mice. J Clin Invest 118(1):124–132
64. Stehling-Sun S, Dade J, Nutt SL, DeKoter RP, Camargo FD (2009) Regulation of lymphoid versus myeloid fate 'choice' by the transcription factor Mef2c. Nat Immunol 10(3):289–296
65. Pattison MJ, Naik RJ, Reyskens K, Arthur JSC (2020) Loss of Mef2D function enhances TLR induced IL-10 production in macrophages. Biosci Rep 40(8):BSR20201859
66. Swanson BJ, Jäck HM, Lyons GE (1998) Characterization of myocyte enhancer factor 2 (MEF2) expression in B and T cells: MEF2C is a B cell-restricted transcription factor in lymphocytes. Mol Immunol 35(8):445–458
67. Dequiedt F, Kasler H, Fischle W, Kiermer V, Weinstein M, Herndier BG et al (2003) HDAC7, a thymus-specific class II histone deacetylase, regulates Nur77 transcription and TCR-mediated apoptosis. Immunity 18(5):687–698

68. Matsuoka H, Fujimura T, Hayashi M, Matsuda K, Ishii Y, Aramori I et al (2007) Disruption of HDAC4/N-CoR complex by histone deacetylase inhibitors leads to inhibition of IL-2 gene expression. Biochem Pharmacol 74(3):465–476
69. Pan F, Ye Z, Cheng L, Liu JO (2004) Myocyte enhancer factor 2 mediates calcium-dependent transcription of the interleukin-2 gene in T lymphocytes: a calcium signaling module that is distinct from but collaborates with the nuclear factor of activated T-cells (NFAT). J Biol Chem 279(15):14477–14480
70. Esau C, Boes M, Youn HD, Tatterson L, Liu JO, Chen J (2001) Deletion of calcineurin and myocyte enhancer factor 2 (MEF2) binding domain of Cabin1 results in enhanced cytokine gene expression in T cells. J Exp Med 194(10):1449–1459
71. Ma W, Wang Y, Zhang R, Yang F, Zhang D, Huang M et al (2021) Targeting PAK4 to reprogram the vascular microenvironment and improve CAR-T immunotherapy for glioblastoma. Nat Cancer 2(1):83–97
72. Di Giorgio E, Wang L, Xiong Y, Akimova T, Christensen LM, Han R et al (2020) MEF2D sustains activation of effector Foxp3+ Tregs during transplant survival and anticancer immunity. J Clin Invest 130(12):6242–6260
73. Tao R, de Zoeten EF, Ozkaynak E, Chen C, Wang L, Porrett PM et al (2007) Deacetylase inhibition promotes the generation and function of regulatory T cells. Nat Med 13(11):1299–1307
74. Zuurbier L, Gutierrez A, Mullighan CG, Canté-Barrett K, Gevaert AO, de Rooi J et al (2014) Immature MEF2C-dysregulated T-cell leukemia patients have an early T-cell precursor acute lymphoblastic leukemia gene signature and typically have non-rearranged T-cell receptors. Haematologica 99(1):94–102
75. Nagel S, Venturini L, Meyer C, Kaufmann M, Scherr M, Drexler HG et al (2011) Transcriptional deregulation of oncogenic myocyte enhancer factor 2C in T-cell acute lymphoblastic leukemia. Leuk Lymphoma 52(2):290–297
76. Seita J, Sahoo D, Rossi DJ, Bhattacharya D, Serwold T, Inlay MA et al (2012) Gene expression commons: an open platform for absolute gene expression profiling. PLoS One 7(7):e40321
77. Heng TS, Painter MW (2008) The Immunological Genome Project: networks of gene expression in immune cells. Nat Immunol 9(10):1091–1094
78. Gekas C, Rhodes KE, Gereige LM, Helgadottir H, Ferrari R, Kurdistani SK et al (2009) Mef2C is a lineage-restricted target of Scl/Tal1 and regulates megakaryopoiesis and B-cell homeostasis. Blood 113(15):3461–3471
79. Wilker PR, Kohyama M, Sandau MM, Albring JC, Nakagawa O, Schwarz JJ et al (2008) Transcription factor Mef2c is required for B cell proliferation and survival after antigen receptor stimulation. Nat Immunol 9(6):603–612
80. Herglotz J, Unrau L, Hauschildt F, Fischer M, Kriebitzsch N, Alawi M et al (2016) Essential control of early B-cell development by Mef2 transcription factors. Blood 127(5):572–581
81. Xiang J, Zhang N, Sun H, Su L, Zhang C, Xu H et al (2020) Disruption of SIRT7 increases the efficacy of checkpoint inhibitor via MEF2D regulation of programmed cell death 1 ligand 1 in hepatocellular carcinoma cells. Gastroenterology 158(3):664–78.e24
82. Small EM, Olson EN (2011) Pervasive roles of microRNAs in cardiovascular biology. Nature 469(7330):336–342
83. Mao Z, Bonni A, Xia F, Nadal-Vicens M, Greenberg ME (1999) Neuronal activity-dependent cell survival mediated by transcription factor MEF2. Science 286(5440):785–790
84. Zeng W, Zhang W, Lu F, Gao L, Gao G (2017) Resveratrol attenuates MPP(+)-induced mitochondrial dysfunction and cell apoptosis via AKT/GSK-3β pathway in SN4741 cells. Neurosci Lett 637:50–56
85. Zhao Y, Li Y, Ma Y, Wang S, Cheng J, Yang T et al (2016) Myocyte enhancer factor 2D promotes tumorigenicity in malignant glioma cells. Tumour Biol 37(1):601–610
86. Yang S, Gao L, Lu F, Wang B, Gao F, Zhu G et al (2015) Transcription factor myocyte enhancer factor 2D regulates interleukin-10 production in microglia to protect neuronal cells from inflammation-induced death. J Neuroinflammation 12:33
87. Sala G, Arosio A, Stefanoni G, Melchionda L, Riva C, Marinig D et al (2013) Rotenone upregulates alpha-synuclein and myocyte enhancer factor 2D independently from lysosomal degradation inhibition. Biomed Res Int 2013:846725
88. Chu Y, Mickiewicz AL, Kordower JH (2011) α-synuclein aggregation reduces nigral myocyte enhancer factor-2D in idiopathic and experimental Parkinson's disease. Neurobiol Dis 41(1):71–82
89. Huo Y, Zhao Q, Wang C, Zhao F, Du Y, Sun W (2016) The involvement of myocyte enhancer factor 2D in regulating tumor biology of cardiac myxoma. Tumour Biol 37(4):5405–5411
90. Du M, Perry RL, Nowacki NB, Gordon JW, Salma J, Zhao J et al (2008) Protein kinase A represses skeletal myogenesis by targeting myocyte enhancer factor 2D. Mol Cell Biol 28(9):2952–2970
91. Cao Q, Dong P, Wang Y, Zhang J, Shi X, Wang Y (2015) miR-218 suppresses cardiac myxoma proliferation by targeting myocyte enhancer factor 2D. Oncol Rep 33(5):2606–2612
92. Tempera I, De Leo A, Kossenkov AV, Cesaroni M, Song H, Dawany N et al (2016) Identification of MEF2B, EBF1, and IL6R as direct gene targets of Epstein-Barr virus (EBV) nuclear antigen 1 critical for EBV-infected B-lymphocyte survival. J Virol 90(1):345–355
93. Murata T, Narita Y, Sugimoto A, Kawashima D, Kanda T, Tsurumi T (2013) Contribution of myocyte enhancer factor 2 family transcription factors to BZLF1 expression in Epstein-Barr virus reactivation from latency. J Virol 87(18):10148–10162

94. Yndart A, Kaushik A, Agudelo M, Raymond A, Atluri VS, Saxena SK et al (2015) Investigation of neuropathogenesis in HIV-1 clade B and C infection associated with IL-33 and ST2 regulation. ACS Chem Neurosci 6(9):1600–1612
95. Jain P, Lavorgna A, Sehgal M, Gao L, Ginwala R, Sagar D et al (2015) Myocyte enhancer factor (MEF)-2 plays essential roles in T-cell transformation associated with HTLV-1 infection by stabilizing complex between Tax and CREB. Retrovirology 12:23
96. Gao L, She H, Li W, Zeng J, Zhu J, Jones DP et al (2014) Oxidation of survival factor MEF2D in neuronal death and Parkinson's disease. Antioxid Redox Signal 20(18):2936–2948
97. Siegel RL, Miller KD, Fuchs HE, Jemal A (2021) Cancer statistics, 2021. CA Cancer J Clin 71(1):7–33
98. Chang N, Feng J, Liao P, Hu Y, Li M, He Y et al (2023) Knockdown of MEF2D inhibits the development and progression of B-cell acute lymphoblastic leukemia. Transl Cancer Res 12(2):287–300
99. Kline KAF, Kallen ME, Duong VH, Law JY (2021) Acute lymphoblastic leukemia and acute lymphoblastic lymphoma: same disease Spectrum but two distinct diagnoses. Curr Hematol Malig Rep 16(5):384–393
100. Sasaki K, Jabbour E, Short NJ, Jain N, Ravandi F, Pui CH et al (2021) Acute lymphoblastic leukemia: a population-based study of outcome in the United States based on the surveillance, epidemiology, and end results (SEER) database, 1980-2017. Am J Hematol 96(6):650–658
101. Neaga A, Jimbu L, Mesaros O, Bota M, Lazar D, Cainap S et al (2021) Why do children with acute lymphoblastic leukemia fare better than adults? Cancers (Basel) 13(15):3886
102. Jen EY, Xu Q, Schetter A, Przepiorka D, Shen YL, Roscoe D et al (2019) FDA approval: blinatumomab for patients with B-cell precursor acute lymphoblastic leukemia in morphologic remission with minimal residual disease. Clin Cancer Res 25(2):473–477
103. Inaba H, Mullighan CG (2020) Pediatric acute lymphoblastic leukemia. Haematologica 105(11):2524–2539
104. Brown PA, Shah B, Advani A, Aoun P, Boyer MW, Burke PW et al (2021) Acute lymphoblastic leukemia, version 2.2021, NCCN clinical practice guidelines in oncology. J Natl Compr Cancer Netw 19(9):1079–1109
105. Advani AS (2022) Novel strategies in the treatment of acute lymphoblastic leukaemia. Lancet Haematol 9(4):e240–e2e1
106. Malard F, Mohty M (2020) Acute lymphoblastic leukaemia. Lancet 395(10230):1146–1162
107. Iacobucci I, Mullighan CG (2017) Genetic basis of acute lymphoblastic leukemia. J Clin Oncol 35(9):975–983
108. Li JF, Dai YT, Lilljebjörn H, Shen SH, Cui BW, Bai L et al (2018) Transcriptional landscape of B cell precursor acute lymphoblastic leukemia based on an international study of 1,223 cases. Proc Natl Acad Sci USA 115(50):E11711–E11e20
109. Liu YF, Wang BY, Zhang WN, Huang JY, Li BS, Zhang M et al (2016) Genomic profiling of adult and pediatric B-cell acute lymphoblastic leukemia. EBioMedicine 8:173–183
110. Gu Z, Churchman M, Roberts K, Li Y, Liu Y, Harvey RC et al (2016) Genomic analyses identify recurrent MEF2D fusions in acute lymphoblastic leukaemia. Nat Commun 7:13331
111. Zhang J, McCastlain K, Yoshihara H, Xu B, Chang Y, Churchman ML et al (2016) Deregulation of DUX4 and ERG in acute lymphoblastic leukemia. Nat Genet 48(12):1481–1489
112. Yasuda T, Tsuzuki S, Kawazu M, Hayakawa F, Kojima S, Ueno T et al (2016) Recurrent DUX4 fusions in B cell acute lymphoblastic leukemia of adolescents and young adults. Nat Genet 48(5):569–574
113. Suzuki K, Okuno Y, Kawashima N, Muramatsu H, Okuno T, Wang X et al (2016) MEF2D-BCL9 fusion gene is associated with high-risk acute B-cell precursor lymphoblastic leukemia in adolescents. J Clin Oncol 34(28):3451–3459
114. Qian M, Zhang H, Kham SK, Liu S, Jiang C, Zhao X et al (2017) Whole-transcriptome sequencing identifies a distinct subtype of acute lymphoblastic leukemia with predominant genomic abnormalities of EP300 and CREBBP. Genome Res 27(2):185–195
115. Ping N, Qiu H, Wang Q, Dai H, Ruan C, Ehrentraut S et al (2015) Establishment and genetic characterization of a novel mixed-phenotype acute leukemia cell line with EP300 ZNF384 fusion. J Hematol Oncol 8:100
116. Ohki K, Kiyokawa N, Saito Y, Hirabayashi S, Nakabayashi K, Ichikawa H et al (2019) Clinical and molecular characteristics of MEF2D fusion-positive B-cell precursor acute lymphoblastic leukemia in childhood, including a novel translocation resulting in MEF2D-HNRNPH1 gene fusion. Haematologica 104(1):128–137
117. McClure BJ, Heatley SL, Kok CH, Sadras T, An J, Hughes TP et al (2018) Pre-B acute lymphoblastic leukaemia recurrent fusion, EP300-ZNF384, is associated with a distinct gene expression. Br J Cancer 118(7):1000–1004
118. Gu Z, Churchman ML, Roberts KG, Moore I, Zhou X, Nakitandwe J et al (2019) PAX5-driven subtypes of B-progenitor acute lymphoblastic leukemia. Nat Genet 51(2):296–307
119. Bastian L, Schroeder MP, Eckert C, Schlee C, Tanchez JO, Kämpf S et al (2019) PAX5 biallelic genomic alterations define a novel subgroup of B-cell precursor acute lymphoblastic leukemia. Leukemia 33(8):1895–1909

120. Zhang J, Ding L, Holmfeldt L, Wu G, Heatley SL, Payne-Turner D et al (2012) The genetic basis of early T-cell precursor acute lymphoblastic leukaemia. Nature 481(7380):157–163
121. Teachey DT, Pui CH (2019) Comparative features and outcomes between paediatric T-cell and B-cell acute lymphoblastic leukaemia. Lancet Oncol 20(3):e142–ee54
122. Seki M, Kimura S, Isobe T, Yoshida K, Ueno H, Nakajima-Takagi Y et al (2017) Recurrent SPI1 (PU.1) fusions in high-risk pediatric T cell acute lymphoblastic leukemia. Nat Genet 49(8):1274–1281
123. Ma M, Wang X, Tang J, Xue H, Chen J, Pan C et al (2012) Early T-cell precursor leukemia: a subtype of high risk childhood acute lymphoblastic leukemia. Front Med 6(4):416–420
124. Liu Y, Easton J, Shao Y, Maciaszek J, Wang Z, Wilkinson MR et al (2017) The genomic landscape of pediatric and young adult T-lineage acute lymphoblastic leukemia. Nat Genet 49(8):1211–1218
125. Homminga I, Pieters R, Langerak AW, de Rooi JJ, Stubbs A, Verstegen M et al (2011) Integrated transcript and genome analyses reveal NKX2-1 and MEF2C as potential oncogenes in T cell acute lymphoblastic leukemia. Cancer Cell 19(4):484–497
126. Chen B, Jiang L, Zhong ML, Li JF, Li BS, Peng LJ et al (2018) Identification of fusion genes and characterization of transcriptome features in T-cell acute lymphoblastic leukemia. Proc Natl Acad Sci USA 115(2):373–378
127. Tsuzuki S, Yasuda T, Kojima S, Kawazu M, Akahane K, Inukai T et al (2020) Targeting MEF2D-fusion oncogenic transcriptional circuitries in B-cell precursor acute lymphoblastic leukemia. Blood Cancer Discov 1(1):82–95
128. Sadras T, Müschen M (2020) MEF2D fusions drive oncogenic pre-BCR signaling in B-ALL. Blood Cancer Discov. 1(1):18–20
129. Mullighan CG (2019) How advanced are we in targeting novel subtypes of ALL? Best Pract Res Clin Haematol 32(4):101095
130. Zhang M, Mao D, Zhang W (2018) The pathogenic role of MEF2D-SS18 fusion gene in B-cell acute lymphoblastic leukemia. Biochem Biophys Res Commun 496(4):1331–1336
131. Li H, Liu J, Shen S, Dai D, Cheng S, Dong X et al (2020) Pan-cancer analysis of alternative splicing regulator heterogeneous nuclear ribonucleoproteins (hnRNPs) family and their prognostic potential. J Cell Mol Med 24(19):11111–11119
132. Ideue T, Adachi S, Naganuma T, Tanigawa A, Natsume T, Hirose T (2012) U7 small nuclear ribonucleoprotein represses histone gene transcription in cell cycle-arrested cells. Proc Natl Acad Sci USA 109(15):5693–5698
133. Gil VS, Bhagat G, Howell L, Zhang J, Kim CH, Stengel S et al (2016) Deregulated expression of HDAC9 in B cells promotes development of lymphoproliferative disease and lymphoma in mice. Dis Model Mech 9(12):1483–1495
134. Zhang M, Zhang H, Li Z, Bai L, Wang Q, Li J et al (2022) Functional, structural, and molecular characterizations of the leukemogenic driver MEF2D-HNRNPUL1 fusion. Blood 140(12):1390–1407
135. Charles NJ, Boyer DF (2017) Mixed-phenotype acute leukemia: diagnostic criteria and pitfalls. Arch Pathol Lab Med 141(11):1462–1468
136. Marschalek R (2015) MLL leukemia and future treatment strategies. Arch Pharm (Weinheim) 348(4):221–228
137. Meyer C, Burmeister T, Gröger D, Tsaur G, Fechina L, Renneville A et al (2018) The MLL recombinome of acute leukemias in 2017. Leukemia 32(2):273–284
138. Nguyen AT, Taranova O, He J, Zhang Y (2011) DOT1L, the H3K79 methyltransferase, is required for MLL-AF9-mediated leukemogenesis. Blood 117(25):6912–6922
139. Xu J, Li L, Xiong J, denDekker A, Ye A, Karatas H et al (2016) MLL1 and MLL1 fusion proteins have distinct functions in regulating leukemic transcription program. Cell Discov 2:16008
140. Dawson MA (2017) The cancer epigenome: concepts, challenges, and therapeutic opportunities. Science 355(6330):1147–1152
141. Bradner JE, Hnisz D, Young RA (2017) Transcriptional addiction in cancer. Cell 168(4):629–643
142. Khwaja A, Bjorkholm M, Gale RE, Levine RL, Jordan CT, Ehninger G et al (2016) Acute myeloid leukaemia. Nat Rev Dis Primers 2:16010
143. Lambert SA, Jolma A, Campitelli LF, Das PK, Yin Y, Albu M et al (2018) The human transcription factors. Cell 172(4):650–665
144. Vaquerizas JM, Kummerfeld SK, Teichmann SA, Luscombe NM (2009) A census of human transcription factors: function, expression and evolution. Nat Rev Genet 10(4):252–263
145. Tenen DG (2003) Disruption of differentiation in human cancer: AML shows the way. Nat Rev Cancer 3(2):89–101
146. Pingul BY, Huang H, Chen Q, Alikarami F, Zhang Z, Qi J et al (2022) Dissection of the MEF2D-IRF8 transcriptional circuit dependency in acute myeloid leukemia. iScience 25(10):105139
147. Taylor SJ, Sundaravel S, Steidl U (2021) Exploiting a key transcriptional dependency: ZMYND8 and IRF8 in AML. Mol Cell 81(17):3445–3446
148. Cao Z, Budinich KA, Huang H, Ren D, Lu B, Zhang Z et al (2021) ZMYND8-regulated IRF8 transcription axis is an acute myeloid leukemia dependency. Mol Cell 81(17):3604–22.e10
149. Tamura T, Kurotaki D, Koizumi S (2015) Regulation of myelopoiesis by the transcription factor IRF8. Int J Hematol 101(4):342–351

150. Carotta S, Willis SN, Hasbold J, Inouye M, Pang SH, Emslie D et al (2014) The transcription factors IRF8 and PU.1 negatively regulate plasma cell differentiation. J Exp Med 211(11):2169–2181
151. Zhao L, Zhang P, Galbo PM, Zhou X, Aryal S, Qiu S et al (2021) Transcription factor MEF2D is required for the maintenance of MLL-rearranged acute myeloid leukemia. Blood Adv 5(22):4727–4740
152. Li X, Zhang Y, Chai X, Zhou S, Zhang H, He J et al (2019) Overexpression of MEF2D contributes to oncogenic malignancy and chemotherapeutic resistance in ovarian carcinoma. Am J Cancer Res 9(5):887–905
153. Moorman HR, Reategui Y, Poschel DB, Liu K (2022) IRF8: mechanism of action and health implications. Cells 11(17):2630
154. Desai P, Mencia-Trinchant N, Savenkov O, Simon MS, Cheang G, Lee S et al (2018) Somatic mutations precede acute myeloid leukemia years before diagnosis. Nat Med 24(7):1015–1023
155. Papaemmanuil E, Gerstung M, Bullinger L, Gaidzik VI, Paschka P, Roberts ND et al (2016) Genomic classification and prognosis in acute myeloid leukemia. N Engl J Med 374(23):2209–2221

18 HOXA9 Regulome and Pharmacological Interventions in Leukemia

Sajesan Aryal and Rui Lu

Abstract

HOXA9, an important transcription factor (TF) in hematopoiesis, is aberrantly expressed in numerous cases of both acute myeloid leukemia (AML) and acute lymphoblastic leukemia (ALL) and is a strong indicator of poor prognosis in patients. HOXA9 is a proto-oncogene which is both sufficient and necessary for leukemia transformation. HOXA9 expression in leukemia correlates with patient survival outcomes and response to therapy. Chromosomal transformations (such as NUP98-HOXA9), mutations, epigenetic dysregulation (e.g., MLL- MENIN -LEDGF complex or DOT1L/KMT4), transcription factors (such as USF1/USF2), and noncoding RNA (such as HOTTIP and HOTAIR) regulate HOXA9 mRNA and protein during leukemia. HOXA9 regulates survival, self-renewal, and progenitor cell cycle through several of its downstream target TFs including LMO2, anti-apoptotic BCL2, SOX4, and receptor tyrosine kinase FLT3 and STAT5. This dynamic and multilayered HOXA9 regulome provides new therapeutic opportunities, including inhibitors targeting DOT1L/KMT4, MENIN, NPM1, and ENL proteins. Recent findings also suggest that HOXA9 maintains leukemia by actively repressing myeloid differentiation genes. This chapter summarizes the recent advances understanding biochemical mechanisms underlying HOXA9-mediated leukemogenesis, the clinical significance of its abnormal expression, and pharmacological approaches to treat HOXA9-driven leukemia.

Keywords

Homeobox A9 (HOXA9) · Leukemia · Hematopoiesis · Epigenetics · Transcription factors · Gene regulation · Therapeutic strategies · Prognostic implications · Biomarkers

Abbreviations

5′ UTR	5′ untranslated region
AFF	ALF transcription elongation factor
AL	acute leukemia
ALL	acute lymphoblastic leukemia
AML	acute myeloid leukemia

S. Aryal · R. Lu (✉)
Department of Medicine, Division of Hematology/Oncology, University of Alabama at Birmingham Heersink School of Medicine, Birmingham, AL, USA

O'Neal Comprehensive Cancer Center, University of Alabama at Birmingham Heersink School of Medicine, Birmingham, AL, USA
e-mail: ruilu1@uabmc.edu

T. Borggrefe, B. D. Giaimo (eds.), *Transcription factors in blood cell development*, Advances in Experimental Medicine and Biology 1459, https://doi.org/10.1007/978-3-031-62731-6_18

Brg1	Brahma-related gene 1
CA-AML	cytogenetically abnormal acute myeloid leukemia
CBP/KAT3A	CREB binding protein/lysine acetyltransferase 3A
CDX2	caudal-type homeobox 2
CDX4	caudal-type homeobox 4
CEBPA	CCAAT/enhancer binding protein alpha
CLP	common lymphoid progenitor
CMP	common myeloid progenitor
CN-AML	cytogenetically normal AML
DFS	disease-free survival
DNMT3A	DNA methyltransferase 3 alpha
DOTCom	DOT1L/KMT4-containing complex
EAF	ELL-associated factor
ELL	elongation factor for RNA polymerase
EP300/KAT3B	E1A binding protein p300/ lysine acetyltransferase 3B
ERG	ETS-related gene
ETV6	ETS variant transcription factor 6
EZH2	enhancer of zeste homolog 2
FAB	French American-British
FLT3	Fms-like tyrosine kinase 3
FLT3-ITD	FMS-like tyrosine kinase 3-internal tandem duplication
G9A/KMT1C	histone methyltransferase G9A/lysine methyltransferase 1C
H3K27ac	histone 3 lysine 27 acetylation
H3K27me3	histone 3 lysine 27 trimethylation
H3K4me3	histone 3 lysine 4 trimethylation
H3K79	histone 3 lysine 79
H3K79me2	histone 3 lysine 79 dimethylation
H3K9ac	histone 3 lysine 9 acetylation
HBO1	H4-specific histone acetylase
HD	homeodomain
HE	hemogenic endothelial
hESCs	human embryonic stem cells
HOTAIR	HOX transcript antisense RNA
HOTTIP	HOXA transcript at distal tip
HOX	homeobox
HSCs	hematopoietic stem cells
HSPC	hematopoietic stem/progenitor cell
HX	HOX-specific hexapeptide
HXR9	HXR9 peptide
IDH1/2	isocitrate dehydrogenase 1/dehydrogenase 2
KMT2A	lysine methyltransferase 2A (formerly known as MLL)
KMT2B	lysine methyltransferase 2B (or MLL2)
KO	knockout
LEDGF	lens epithelium-derived growth factor (also referred to as PSIP1)
LMO2	LIM domain-only 2
lncRNAs	long noncoding RNAs
LYN	LYN proto-oncogene
MEIS	myeloid ecotropic viral integration site homolog
MEIS1	myeloid ecotropic viral integration site homolog 1
Meis1-3	myeloid ecotropic viral integration site homologs 1–3
miRNAs	microRNAs
MLL	mixed-lineage leukemia
MLL1/KMT2A	mixed-lineage leukemia 1/ lysine methyltransferase 2A
MLL3/KMT2C	mixed-lineage leukemia 3/ lysine methyltransferase 2C
MLL4/KMT2D	mixed-lineage leukemia 4/ lysine methyltransferase 2D
MLLr	MLL rearrangement
MYB	MYB proto-oncogene
NK	natural killer
NPM1	nucleophosmin
NPM1c	NPM1 gene mutation variant
NUP98	nucleoporin 98
NUP98-HOXA9	fusion of the NUP98 gene with HOXA9
OS	overall survival
PBX	pre-B-cell leukemia homeobox

Pbx1-4	pre-B-cell leukemia homeoboxes 1–4
PBX3	pre-B-cell leukemia homeobox 3
PRC1	polycomb repressive complex 1
PRC2	polycomb repressive complex 2
Prep1-2	PKNOX family transcription factors 1–2
PSIP1	PC4 and SRSF1-interacting protein 1
PTD	partial tandem duplication
P-TEFb	positive transcription elongation factor b
PU.1/SPI1	hematopoietic transcription factor PU.1/spleen focus-forming virus (SFFV) proviral integration protein
SEC	super elongation complex
shRNA	short hairpin RNA
SOX4	SRY-box transcription factor 4
SUZ12	suppressor of zeste 12 protein homolog
SWI/SNF	SWItch/sucrose nonfermentable
TALE	three-amino-acid-loop-extension
TET2	Tet methylcytosine dioxygenase 2
TFs	transcription factors
TIP60/KAT5	TAT-interacting protein 60 kDa/lysine acetyltransferase 5
Ubx	ultrabithorax
USF1	upstream stimulation factor 1
USF2	upstream stimulation factor 2
XPO1	exportin 1, a nuclear protein export receptor

18.1 Overview of the Hematopoietic System and HOX Genes

The hematopoietic system is crucial to the production of blood cells and plasma. This process, known as hematopoiesis, starts with the emergence of hematopoietic stem cells (HSCs) [1, 2]. The maturation of blood cell lineages from HSCs involves two significant steps: the production of primitive erythroid cells, followed by an expanding population of definitive erythroid cells [3]. During hematopoietic development, a complex network of intrinsic factors and environmental cues govern the continuous process of lineage commitment through the activation and repression of distinct signaling events [4]. HSCs are recognized by their capability to self-renew and replenish the entire blood system when transplanted into immunodeficient mice [5]. HSCs can self-renew, generating more stem cells, or differentiate into specialized progenitor cells, leading to specific blood cell types [6]. As they differentiate, HSCs give rise to a series of progenitor cell intermediates that gradually adopt the identity of a mature blood cell [7]. In adults, hematopoiesis primarily occurs in the bone marrow, whereas in infants and children, it also continues in the spleen and liver [8]. HSCs differentiate into common myeloid progenitor (CMP) cells and common lymphoid progenitor (CLP) cells. The former gives rise to red blood cells, platelets, granulocytes, and monocytes. The latter leads to lymphocytes, which include B-cells, T-cells, and natural killer (NK) cells [9]. The differentiation of these progenitor cells into specific blood cell types is governed by a complex network of cytokines and growth factors [10]. Once produced, these cells are released into the bloodstream and circulate throughout the body. Each type has a different lifespan: red blood cells last about 120 days, white blood cells range from days to years, and platelets have a lifespan of about 10 days [11]. Moreover, HSCs serve as a compelling model for understanding tissue stem cells and their role in ageing and oncogenesis.

Homeobox (HOX) genes play a crucial role in the development and regulation of the hematopoietic system [12]. They are characterized by a unique 183 nucleotide-long DNA sequence known as the homeobox. This highly conserved region binds DNA on one side and cofactors on the other during transcription, thereby facilitating precise gene expression regulation. Comprising 39 genes organized into 4 clusters (HOXA, HOXB, HOXC, and HOXD),

HOX genes regulate cell differentiation, proliferation, and survival [13]. These genes are crucial for the proper formation of the hematopoietic system, with HOXA9 being especially important for hematopoiesis [14]. The HOXA-C clusters play a pivotal role during hematopoiesis, modulating the equilibrium between cellular proliferation and differentiation. HOXA genes exhibit predominant expression in myeloid cells, HOXB in erythroid cells, and HOXC in lymphoid cells [15]. Mammalian HOX genes typically feature two exons and one intron. The homeobox, situated in the second exon, translates into a 60–61 amino acid homeodomain that interacts with DNA utilizing a helix-turn-helix DNA-binding motif [16]. Upstream of the homeodomain (HD) is a three-amino-acid-loop-extension (TALE) motif binding site, facilitating interaction with TALE proteins, specifically members of the myeloid ecotropic viral integration site homolog (MEIS) and pre-B-cell leukemia transcription factor (PBX) families, which serve as key HOX cofactors. The DNA-binding efficiency of HOX proteins is dependent on their interaction with these cofactors, enhancing their specificity. HOX genes encode DNA-binding TFs that perform a pivotal role in cellular differentiation during embryogenesis, establishing themselves as primary regulators of hematopoiesis [17].

HD proteins, including HOX proteins, are evolutionarily conserved TFs possessing diverse developmental functions. The HD exhibits a high degree of conservation and consists of three helical regions folded into a tight globular structure that binds to a 5′-TAAT-3′ core motif [18]. In *Drosophila melanogaster*, this TAATNA motif occurs more than 100,000 times throughout the genome, which does not correspond to the distinct sets of target genes for each HOX protein. However, the presence of this recognition sequence appears to be critically significant for DNA binding [19]. For instance, chromatin immunoprecipitation followed by deep sequencing (ChIP-Seq) study investigating genome-wide HOXA9 binding sites in transformed myeloblasts found that over 98% of sites contain a HOX motif. Moreover, HD comprises three α-helices, with the third helix known as the "recognition helix," providing DNA-binding specificity, a feature that has been evolutionarily conserved across the HD protein family [20]. The underlying molecular mechanisms of how HDs recognize target DNA sequences have been a focal point of extensive research. For HOX proteins, their HDs carry identical DNA base-contacting residues and exhibit highly similar DNA sequence specificity. The recognition helix within the HD binds to the DNA major groove, while the amino-terminal tail contacts the DNA minor groove [21]. It is hypothesized that a divergent N-terminus that interacts with variable cofactors and cooperative DNA binding with other cofactors may enhance the transcriptional specificity of HOX proteins [22].

The genetic regulation and the cell-specific binding activity of HD proteins, such as HOX, rely not only on their affinity toward specific motifs but also on factors such as chromatin accessibility and the presence of specific cofactors and collaborators within the cellular context. Research on the *Drosophila* HOX protein ultrabithorax (Ubx) demonstrated that its binding activity across various developmental stages is strongly dictated by chromatin accessibility [23]. In the hematopoietic system, early factors like hematopoietic transcription factor PU.1/spleen focus-forming virus (SFFV) proviral integration protein (PU.1/SPI1) and CCAAT/enhancer binding protein alpha (CEBPA) have been identified as key agents in establishing areas of relaxed chromatin. This creates an environment conducive to the recruitment of various TFs in response to cellular signals [24]. An intriguing overlap has been identified between CEBPA, SWItch/sucrose nonfermentable (SWI/SNF) factor Brahma-related gene 1 (BRG1), and HOXA9. These proteins colocalize at numerous promoter distal regulatory regions throughout the genome in myeloblasts transformed by HOXA9/MEIS1, implying that chromatin accessibility plays a significant role in directing HOXA9 to specific genomic loci. Moreover, this targeting is further refined through unique protein-protein interactions with cofactors and collaborators that are coexpressed with HOXA9 in a particular cellular milieu [25]. Although the HD proteins are evolutionarily conserved, they exhibit specificity in their roles across various tissues and diseases.

For example, while HOXA9 plays a pivotal role in hematopoiesis and leukemogenesis, other HOX genes are vital in the development and maintenance of different tissues, contributing to the pathogenesis of other diseases [26].

18.2 HOX Genes in Normal Hematopoiesis

During embryogenesis, Hox genes are pivotal in somitogenesis, playing an important role in the formation of somites and shaping the body plan [27]. Notably, their sequences and functions have been highly conserved through evolution, from flies to humans [28]. These genes are located within a specific cluster known as the HOX cluster. HOX genes exhibit spatiotemporal activity, indicating their key functions in hierarchical commitment of cell fate in embryonic development and early hematopoiesis [29–31]. While *HOXC* genes play a role in erythropoiesis, *HOXD* genes are not involved in hematopoietic development [32]. Of all clusters, *HOXA* and *HOXB* genes are the most dominant in hematopoiesis, with *HOXA* genes mainly expressed in myeloid lineages. Expression of *HOXA* is high in human HSPCs and downregulated upon differentiation and maturation [19, 33]. Overexpression and knockout (KO) studies have also demonstrated that HOXA9 is the most abundant HOX gene in HSCs and has the most significant effect on hematopoietic development [34–36]. *Hoxa*-deficient mice exhibit impaired HSPC activity, which is partially rescued by *Hoxa9* alone [37]. Functional cooperation between HOXA9 and HSC-specific TFs is observed. In particular, HOXA9 interacts with enhancers of proto-oncogenes like ETS-related gene (ERG), Fms-like tyrosine kinase 3 (FLT3), LIM domain-only 2 (LMO2), MYB proto-oncogene (MYB), and SRY-box transcription factor 4 (SOX4) [35]. HOXA9 also works together with HOXB4 in HSCs to maintain potency [38]. Hematopoietic differentiation of human embryonic stem cells (hESCs) is also enhanced by HOXA9 overexpression, as it commits hemogenic endothelial (HE) precursor cells into primitive blood cells [39]. Other studies have also demonstrated that HOXA9 particularly is a crucial factor in obtaining long-term repopulating HSCs [34–36]. On the contrary, differentiation of HSCs to mature blood cells is associated with downregulation of HOXA9 [40]. Indeed, the progressive loss of HOX gene expression is a vital signal for HSC proliferation and the start of lymphoid and myeloid differentiation [26]. Regulation of HOXA9 expression in HSCs and progenitor cells is intricate and multilayered. Several mechanisms—transcriptional, posttranscriptional, and epigenetic—are involved in its complex regulation. One crucial factor in the transcriptional regulation is the polycomb repressive complex 2 (PRC2), which suppresses HOXA9 expression in HSCs and is critical for maintaining stem cell identity. In general, two master regulators of HOXA9 expression, the mixed-lineage leukemia (MLL) proteins and the polycomb group histone methyltransferases, activate and repress *HOXA9* transcription, respectively. Mixed-lineage leukemia 1/lysine methyltransferase 2A (MLL1/KMT2A) positively regulates HOXA9 expression through H3K4me3 at its promoter [41]. In contrast, the HOXA9 transcription is repressed by the sequential activity of polycomb repressive complexes PRC1 and PRC2, responsible for H3K27me3 [42]. In addition, HOXA9 expression is highly correlated with H3K79me2 status and DOT1-like histone lysine methyltransferase/lysine methyltransferase 4 (DOT1L/KMT4), an H3K79 methyltransferase required for sustaining HOXA9 expression in HSCs [43]. Overall, manipulation of HOXA expression leads to lineage bias through a combination of altered lineage commitment and blocked differentiation. Therefore, deregulation of HOXA genes is a common feature of malignant hematopoiesis, primarily leukemia.

18.3 HOX Genes in Malignant Hematopoiesis

HOX genes are established key TFs regulating normal hematopoiesis, and their aberrant expression is strongly associated with leukemia [44] (Fig. 18.1). Particularly in AML, HOXA and HOXB gene families were highly expressed,

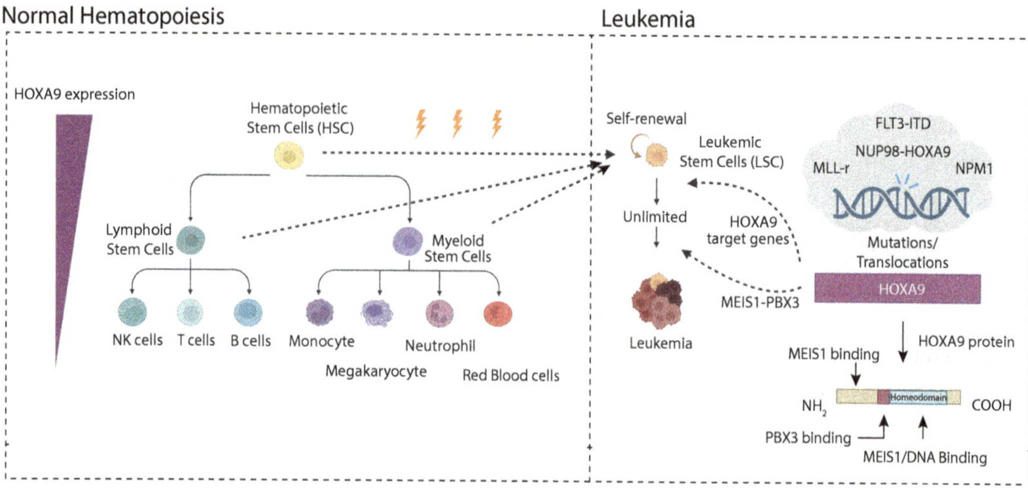

Fig. 18.1 An overview of the role of HOXA9 in normal and malignant hematopoiesis. The expression of HOXA is crucial in the creation and preservation of hematopoietic stem cells (HSCs). As HSCs undergo differentiation, there is a substantial reduction in HOXA expression. However, genetic alterations or mutations in upstream regulators, including rearrangements in the mixed-lineage leukaemia gene (MLLr), can cause an imbalance in HOXA regulation. This dysregulation is linked to a malignant phenotype, predominantly acute myeloid leukemia (AML) and acute lymphoid leukemia (ALL). Prior to the homeodomain in HOX proteins, there is a TALE (three-amino-acid-loop-extension) binding site. HOXA9 utilizes hexapeptide residues (HX) and homeodomain (HD) residues and interactions with cofactors such as MEIS and PBX. The figure has been designed using biorender.com

while HOXC and HOXD were not detected [45, 46]. HOXA gene expression levels could be strongly correlated with poor survival rates compared to patients with non-HOX or only HOXB expression levels [47]. Among HOX genes, HOXA9 is an essential TF that is expressed in HSCs and is pivotal for their self-renewal and differentiation [19]. Furthermore, HOXA9 has a defined role in the development of myeloid cells. Aberrant expression of HOXA9 has been documented in AML patients, and overexpression of this gene has been causally associated with leukemogenesis [14, 26, 48, 49]. It is overexpressed in about 70% of AML cases [26]. Overexpression of both HOXB4 and HOXA9, achieved through retroviral gene transfer, leads to a significant increase in hematopoietic stem cell proliferation but without noticeable abnormalities in peripheral blood or lymphoid development. Despite their similar effects on stem cell expansion, HOXA9 overexpression was linked to AML in mice, while HOXB4 did not indicate such an association [50, 51].

Functional research focused on deciphering the oncogenic role of HOXA9 demonstrated that the fusion of HOXA9 with nucleoporin 98 (NUP98), the overexpression of wild-type HOXA9, and its cooperative activity with MEIS1 could lead to the transformation of primary bone marrow cells and precipitate leukemia in murine models [14, 52–54]. The suppression of HOXA9, achieved through a histone methyltransferase that disrupts HOXA9 gene expression or via short hairpin RNA (shRNA), impeded the proliferation and self-renewal capacity of AML cells, resulting in a deceleration of disease progression [55, 56]. The pathogenesis of leukemia mediated by HOXA9 has been linked to its interaction with promoters—regions proximal to the 5′ untranslated region (UTR) of a gene—and enhancers, which are distal to the 5′-UTR. This interaction influences the transcription regulation of crucial oncogenes [57]. Additionally, HOXA9 can establish new enhancer regions implicated in the transcription of malignant genes by modifying histone methylation states [58]. However, an optimal therapeutic strategy targeting HOXA9 is currently absent, necessitating the development of novel approaches through the comprehensive understanding of HOXA9 expression mechanisms and HOXA9-dependent transcriptional processes [59, 60]. In mixed-lineage leukemia rearrangement (MLLr) leukemia, heightened levels of HOXA, inclusive of HOXA9, have been associated with the operation of epigenetic modulators such as DOT1L/KMT4, which are drawn to the fusion gene at the promoter site of HOXA genes [61]. In cytogenetically normal AML (CN-AML) patients harboring nucleophosmin (NPM1), one hypothetical model proposes that the mutation-induced relocation of NPM1 function from the nucleolus to the cytosol impacts MLL-driven HOXA gene expression [61]. Contrastingly, Claudal-type homeobox 2 (CDX2), a gene completely absent in healthy adult hematopoietic cells, exhibits high expression in CN-AML patients, acting as a direct upstream regulator of especially HOXA genes in AML [62, 63]. As an alternate mechanism, upregulation of HOX genes could be facilitated by long noncoding RNAs (lncRNAs) at the transcript level or via microRNAs (miRNAs), whose influence on posttranscriptional gene expression such as HOXA9 is diminished [64–66].

MLL fusion proteins play a crucial role in leukemia pathogenesis primarily via the direct upregulation of HOXA gene expression [26, 67]. Within the HOXA cluster, the most critical downstream targets of MLL fusion proteins are HOXA9 and its cofactor, MEIS1 [54, 68, 69]. In fact, among 6817 examined genes, HOXA9 has been identified as the single most significantly associated gene with unfavorable AML prognosis [70]. Moreover, it ranks as one of the most highly expressed genes in HSCs, indicative of its critical role in HSC function [29]. This was highlighted in an investigation where the overexpression of HOXA9 in *Hoxa9*-deficient murine models resulted in a partial rescue of hematopoietic stem/progenitor cell (HSPC) activity in vivo and reinstated the expression of multiple dysregulated genes [37]. A substantial overexpression of both HOXA9 and MEIS1 is required to induce and sustain MLL-rearranged leukemias [26, 68]. Consequently, a crucial reliance of HOXA9 to trigger and initiate AML in HSPCs has been documented [37].

18.3.1 HOXA9 and Its Association with Clinical Features

The expression level of HOXA9 and related mutations have been linked with several clinical features of leukemia, establishing it as a potential diagnostic and prognostic marker. Several studies have shown a correlation between high HOXA9 expression and poor prognosis in AML. High HOXA9 expression is also linked with specific AML subtypes like those with MLL rearrangements, which are known for their aggressive nature and poor response to therapy. In the context of age-specific manifestations, the dysregulation of HOXA9 carries unique implications. Notably, dysregulation and overexpression of HOXA9 are linked to distinct age demographics, particularly among pediatric AML patients who demonstrate elevated HOXA9 expression levels in comparison to their adult counterparts. This discrepancy in HOXA9 levels may play a critical part in the disparate disease behaviors observed across different age groups. Moreover, an elevated HOXA9 expression is connected to the development of acute leukemia (AL), with children diagnosed with AML exhibiting significantly higher expression levels than those with ALL. An observed positive correlation exists between the level of HOXA9 expression and the risk of childhood leukemia, with higher HOXA9 expression signifying a more unfavorable prognosis [71].

In AML, the bimodal expression of HOXA9 relates to clinical characteristics like age and white blood cell counts. Additionally, it aids in distinguishing patients into specific French American-British (FAB) or molecular subtypes [72]. In fact, the HOXA9 gene plays a complex role in various biological processes, and its expression levels can have diverse implications. For instance, a high HOXA9 expression is associated with an increased white blood cell count. It is also strongly linked to the presence of unfavorable karyotypes and mutations such as partial tandem duplication (PTD) of the KMT2A (MLL) gene and enhancer of zeste homolog 2 (EZH2). Conversely, the French American-British M3 subtype, a category of AML, exhibits a negative correlation with HOXA9 expression levels. In other words, higher HOXA9 expression is typically less prevalent in patients with this subtype of AML.

Interestingly, HOXA9 expression levels also influence patient outcomes. Patients with higher levels of HOXA9 expression tend to have lower rates of complete remission. Furthermore, they generally have poorer survival outcomes, both in AML overall and specifically in cytogenetically normal AML, compared to those with lower HOXA9 expression levels. This underscores the critical role of HOXA9 expression in disease progression and patient prognosis [73].

In conclusion, the HOXA9 gene holds substantial potential as a diagnostic and prognostic marker in leukemia due to its association with several clinical features. However, a comprehensive understanding of the diverse roles and implications of HOXA9 expression is crucial for its effective use as a biomarker. Further research is warranted to elucidate the exact mechanisms by which HOXA9 influences disease outcomes and to validate its utility in predicting treatment response. Such insights will pave the way for the development of more effective therapeutic strategies and personalized treatments for patients with leukemia.

18.3.2 HOXA9 as Biomarker in Leukemia

HOXA9 might also function as a predictive biomarker for treatment response. The expression levels of HOXA9 could potentially forecast responses to therapies that target its function or downstream effectors. For instance, a study on 91 Egyptian patients newly diagnosed with AML and 41 healthy controls discovered that HOXA9 and MEIS1 gene expression levels were significantly higher in AML patients [74]. Despite these findings, there was no association between these genes' expression levels and disease-free survival (DFS) or overall survival (OS) rates. Therefore, although HOXA9 and MEIS1 could be helpful as diagnostic markers, their role as prognostic markers remains unclear. Moreover, a case-control study involving 60 newly diagnosed AML patients and 40 healthy individuals revealed a significant rise in HOXA9 expression levels among the AML patients [75]. The most elevated levels of HOXA9 were associated with specific AML subtypes, older patients, those with hepatomegaly and splenomegaly, patients with unfavorable cytogenetic criteria, and those who did not achieve complete remission. The study concluded that high HOXA9 expression might be associated with a poorer prognosis and reduced response to chemotherapy. This suggests that HOXA9 could be a potential target for future therapeutic strategies.

18.3.3 HOX Proteins and TALE Cofactors in Leukemia

HOX proteins, including HOXA9, are widely acknowledged for their DNA-binding abilities and their role in gene expression regulation, working in tandem with a select group of cofactor proteins. Among these cofactors, the TALE family proteins, inclusive of *Pbx1–Pbx4*, *Meis1–Meis3*, and *Prep1–Prep2*, are particularly significant. The HOXA9 protein employs a mix of a HOX-specific hexapeptide (HX) motif and two specific residues from the HOXA9 HD to interact with its TALE cofactors PBX1 and MEIS1. These interactions, spanning from context-independent to diverse and context-dependent, exemplify the multifaceted ways HOX proteins can form associations with their TALE cofactors. These complex formations further enhance the binding diversity and specificity of HOX proteins [76]. Such formation of homo- and heterodimers by HOX proteins augments the diversity and specificity of their binding. Intriguingly, a sequential binding process is also observed, where TALE factors first link with regulatory elements, priming the stage with the appropriate chromatin marks. This initial interaction is succeeded by the recruitment of HOX proteins that instigate efficacious transcriptional activation [77].

Despite the known enhancement of DNA-binding and transcriptional activity of HOX proteins through direct binding with PBX proteins, the functional significance of this interaction in the context of leukemia remains enigmatic. Recent studies have shown that overexpression of a quartet of homeobox genes, specifically PBX3, HOXA7, HOXA9, and HOXA11, predicts poorer survival rates in patients with cytogenetically abnormal acute myeloid leukemia (CA-AML). Particularly noteworthy is the consistent coexpression of PBX3 and HOXA9 across different subtypes of CA-AML, suggesting that PBX3 could potentially act as a pathological cofactor of HOXA9 in this context. Further supporting this theory, study shows that reducing endogenous *Pbx3* expression significantly inhibits cell transformation mediated by MLL fusion, and coexpression of PBX3 and HOXA9 significantly promotes cell transformation both in vitro and in vivo. Interestingly, a small peptide, HXR9, specifically designed to disrupt interactions between HOX and PBX proteins, selectively eliminates leukemic cells overexpressing HOXA/PBX3 genes [78]. Collectively, this implies that PBX3 is a crucial cofactor for HOXA9 in the development of leukemia, suggesting that interrupting their interaction could be a viable therapeutic strategy for treating therapy-resistant CA-AML, such as MLL-rearranged leukemia, where HOXA/PBX3 genes are overexpressed.

18.4 Mechanisms of HOXA9 Dysregulation in Leukemia

HOXA9 dysregulation in leukemia can occur via various mechanisms, including chromosomal translocations, mutations, and epigenetic modifications [26].

18.4.1 Chromosomal Translocations

A common cause of HOXA9 dysregulation in leukemia is chromosomal translocations [79]. Translocations involve the rearrangement of parts between nonhomologous chromosomes, which can result in the fusion of two genes. In the context of HOXA9, notable examples include the NUP98-HOXA9 fusion due to t(7;11)(p15;p15) translocation and the MLL-HOXA9 fusion, both of which result in oncoproteins that promote leukemogenesis by deregulating HOX gene expression [80]. Such fusion proteins are potent oncoprotein that promotes leukemogenesis by deregulating HOX gene expression [81]. From a clinical perspective, the presence of the NUP98-HOXA9 fusion gene in AML is associated with a poor prognosis and ineffective outcomes using conventional induction regimens [82]. Two case studies by Fang et al. in 2022 revealed the challenging clinical management of AML patients harboring the NUP98-HOXA9 fusion gene. The gene fusion, a result of the t(7;11)(p15;p15) chromosomal translocation, is associated with poor prognosis. Both patients, one with AML-M5 and the other with AML-M4, failed to sustain remission after multiple induction chemotherapy regimens. The study also found that while the NUP98-HOXA9 fusion gene is typically found in younger Asian women with M2 AML, it can present diversely in older patients with different AML subtypes. The findings highlight the need for urgent hematopoietic stem cell transplantation in NUP98-HOXA9 bearing AML patients after achieving remission, if physically feasible. These cases underscore the urgency for more research into effective treatments for this particularly aggressive AML subtype [83].

18.4.2 Mutations

Another cause of HOXA9 dysregulation in leukemia involves mutations in other genes that influence the regulation of HOXA9. In this regard, a recent study demonstrated that DNA methyltransferase 3 alpha (DNMT3A), a member of DNA methyltransferases that is frequently mutated in AML, plays a significant role in HOXA9 regulation. It was shown that DNMT3A mutation interacts with PRC1, which in turn causes transcriptional silencing of *Hoxa9* in a DNA methylation-independent manner [84]. Moreover, this interaction between mutated DNMT3A and PRC1 leads to a block in the differentiation of HSCs and leukemic cells, thereby promoting monoblastic transformation. Notably, this aberrant interaction becomes particularly leukemic when combined with HOXA9, which further underscores the role of HOXA9 dysregulation in leukemogenesis. Also, DNMT3A mutations have been found to be associated with higher expression of HOX genes including HOXA9 [85]. These findings also shed light on the molecular mechanisms of leukemogenesis and indicate potential therapeutic targets in DNMT3A-mutated leukemias.

18.4.3 Epigenetic Regulation of HOXA9

Epigenetic mechanisms, including DNA methylation and histone modification, significantly influence the transcription of HOXA9 and other genes by altering chromatin structure and, in turn, the accessibility of the genetic material to the transcriptional machinery (Fig. 18.2). It has been shown that HOXA9 interacts with mixed lineage leukemia 3/lysine methyltransferase 2C (MLL3/KMT2C) and mixed lineage leukemia 4/lysine methyltransferase 2D (MLL4/KMT2D) histone methyltransferase complex and is required for HOXA9/MEIS1 leukemia both in vitro and in vivo [58]. Epigenetic modifications, such as methylation or acetylation of the DNA or histones, also contribute to HOXA9 dysregulation in leukemia. Once targeted to specific sites, HOX complexes are thought to regulate downstream gene expression via their interaction with histone modifying machinery. Both HOXA9 and MEIS1 have been demonstrated to recruit the histone acetyltransferase E1A binding protein p300/lysine acetyltransferase 3B (EP300/KAT3B) and CREB-binding protein/lysine acetyltransferase 3B (CBP/KAT3A) to instigate the activation of downstream targets. Recent work has also revealed a pivotal role of the leukemogenic transcription factor HOXA9 in AML, interacting significantly with the histone methyltransferase G9A/lysine methyltransferase 1C (KMT1C). It suggests that inhibiting G9A, and therefore attenuating HOXA9-dependent transcription, could serve as a promising therapeutic strategy to impede AML cell proliferation and self-renewal [55].

18.4.3.1 DNA Methylation

DNA methylation, especially in the promoter region of a gene, is a well-recognized mechanism of gene expression regulation. In general, hypermethylation of a gene's promoter region represses its transcription, while hypomethylation permits or enhances its transcription [86]. In the case of HOXA9, hypermethylation of its promoter region

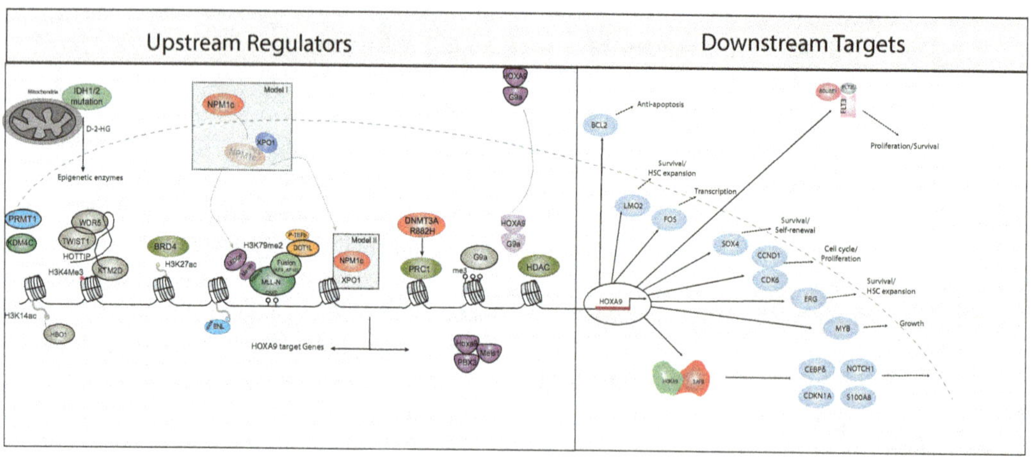

Fig. 18.2 An overview of the upstream regulators and downstream targets of HOXA9. A comprehensive depiction of the molecular interactions and pathways in which HOXA9 is involved. The right side showcases the significant upstream regulators, responsible for modulating the activity of HOXA9. The left side illustrates the downstream targets that are directly affected or regulated by HOXA9. Understanding this complex network is crucial for unravelling the functional roles of HOXA9 in various biological processes

has been linked to its overexpression, promoting cell proliferation and self-renewal [87]. Moreover, methylated HOXA9 is believed to be highly cancer specific and serves as a general diagnostic marker in certain cancers like ovarian cancer [88]. This pattern suggests that the methylation status of the HOXA9 promoter region can be a decisive factor in the determination of cell fate and the progression of disorders like AML.

18.4.3.2 MLL-MENIN-LEDGF Complex

MLL genes code for histone methyltransferase with a C-terminal SET domain responsible for methylation of H3K4 [89]. Leukemia-related KMT2A (formerly known as MLL) gene rearrangements, impacting a single allele of the inherent KMT2A gene, generate a fusion oncoprotein that persistently activates HOXA9 and the HOX cofactor MEIS1 [90]. While these gene fusions lead to the loss of the C-terminal SET methyltransferase domain, they retain the N-terminal domain crucial for interaction with chromatin cofactors like MENIN [91]. The wild-type MLL (KMT2A) is essential for normal hematopoiesis by maintaining HOX gene expression [92]. In MLLr AML (rearrangements involving the MLL gene, or KMT2A, leading to AML), the remaining wild-type MLL allele was considered necessary for HOX gene expression and leukemogenesis [93]. However, a newer study suggested that MLL2 (KMT2B), not MLL (KMT2A), is required for the cell growth of MLLr leukemia [94]. In other types of leukemia like NPM1c AML, the wild-type MLL continues to be important for leukemia development [95]. MENIN, a chromatin-linked nuclear protein, is crucial for the transcriptional regulation of MLL target genes and maintaining HOX gene expression by MLL fusion proteins [91]. A 5-amino acid sequence (RWRFP) near MLL's (KMT2A's) N terminus is critical for MENIN interaction [96]. This interaction plays a significant role in MLL's pathogenesis. MENIN might serve as an adapter protein, interacting with MLL (KMT2A) complex elements such as LEDGF [96], despite lacking known protein motifs. It has been found that MENIN, along with CDX4, coactivates the HOXA genes by binding to the HOXA9 locus' same regulatory region [97]. Additionally, MENIN inactivation leads to differentiation arrest and decreased oncogenic potential of MLL fusion [98]. MENIN can interact with both wild-type MLL and MLL fusion proteins to facilitate MLL-mediated gene expression regulation [99].

Lens epithelium-derived growth factor/PC4 and SRSF1-interacting protein 1 (LEDGF/PSIP1), another critical player in MLLr AML, is a transcriptional coactivator brought into the MLL (KMT2A) complex by MENIN [100]. Blocking the interaction between LEDGF and MLL (KMT2A)/MENIN can reduce MLL's target HOXA9 by impeding cell cycle progression and growth of MLL fusion-transformed human and mouse HSCs [100] (Fig. 18.3). HOXA9 regulation and pharmacological inhibition of HOXA9 in MLLr AML involves multiple epigenetic modulators such as the MLL-MENIN complex (involving KMT2A), DOT1L/KMT4 super elongation complex, H4-specific histone acetylase (HBO1) complex, and AF9 protein. Several small-molecule inhibitors have been created to target these epigenetic proteins, including MENIN inhibitors (MI-2-2, MI-463, MI-503, VTP50469) that block MENIN-MLL interaction; DOT1L/KMT4 inhibitors (SGC0946, EPZ-5676, EPZ004777) that inhibit DOT1L/KMT4 methyltransferase activity; HBO1 inhibitor (WM-3835), which directly binds to HBO1's acetyl-CoA binding site, inhibiting its acetyltransferase activity; and ENL inhibitors (SR-1114 and SR-0813), which disrupt ENL's interaction with histone acetylation. In NPM1 mutant AML, the HBO1, MLL- MENIN (KMT2A-MENIN), and DOT1L/KMT4 complexes also play a part in HOXA9 regulation. Additionally, several theories have been proposed regarding NPM1c-HOXA9 gene regulation. For instance, the AML-associated NPM1c mutant is transferred to the cytoplasm by the nuclear protein export receptor exportin 1 (XPO1). Pharmacological inhibition of XPO1 in NPM1c AML repositions NPM1c to the nucleus and suppresses HOXA9 gene expression (Model I). Recent findings also suggest that NPM1c binds to chromatin with XPO1 at HOXA genes, including HOXA9 (model II).

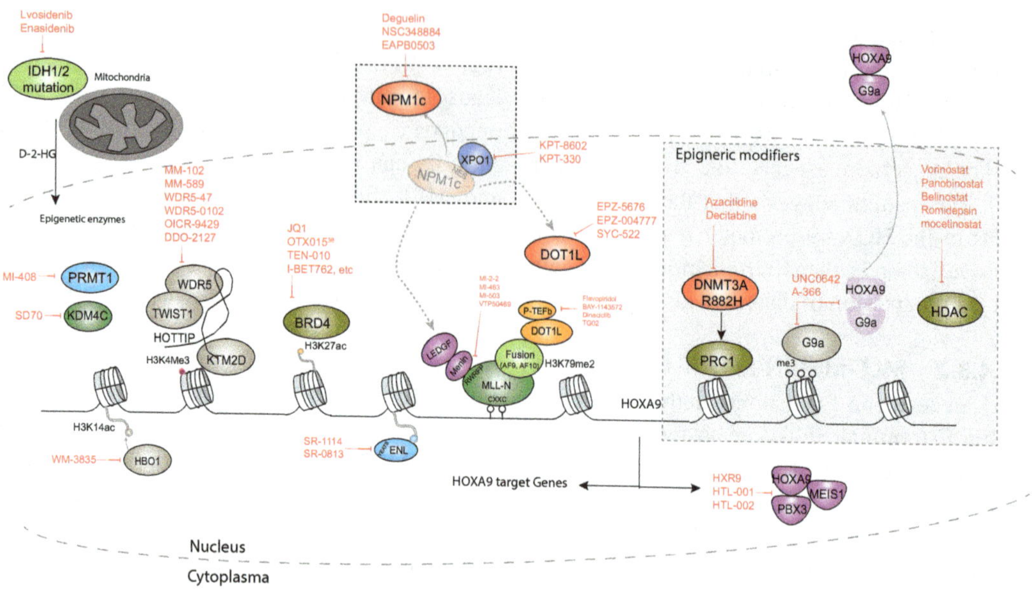

Fig. 18.3 An overview of approaches to target HOXA9. This systematic representation provides insight into the multifaceted ways to manipulate HOXA9 activity, offering potential therapeutic avenues for diseases where HOXA9 is dysregulated

18.4.3.3 DOT1L/KMT4 and Super Elongation Complex (SEC)

DOT1L/KMT4 is a histone lysine methyltransferase that methylates H3K79 [101]. DOT1L/KMT4 is associated with several members of the super elongation complex (SEC), which consists of RNA polymerase II elongation activators or coactivators, including positive transcription elongation factor b (P-TEFb), elongation factor for RNA polymerase (ELL) 1/2/3, ALF transcription elongation factor (AFF) 1/4, ENL, AF9, and ELL-associated factor (EAF) 1/2 [102]. Because the fusion partners of MLL (KMT2A) are often components of the SEC, the MLL fusion protein can recruit DOT1L/KMT4 to its target genes for aberrant H3K79 methylation modification [56]. This H3K79 methylation then facilitates constitutive activation of *HOX* genes and other oncogenes [19]. In addition, the aberrant methyltransferase activity of DOT1L/KMT4 is required for the leukemogenesis of several non-MLLr leukemias, including leukemias with *NPM1* mutation [103] and *DNMT3A* mutation [104, 105].

18.4.3.4 ENL

ENL belongs to the chromatin histone acetylation reader protein family with a distinct amino-terminal named YEATS domain-containing (YEATS) domain and a disordered carboxy-terminal protein-protein interaction interface [106]. ENL is known to bind to the acetylated histone 3 and colocalize with H3K27ac and H3K9ac on the promoters of genes essential for leukemia, hence promoting active transcription. Furthermore, ENL interacts with both MLL (KMT2A) fusion protein complexes (like MLL-ENL with SEC or DOT1L/KMT4 containing complex) and non-fused complexes to drive leukemia [107].

In the context of MLL (KMT2A)-rearranged leukemia, ENL has a vital role. MLL fusion proteins, like MLL (KMT2A)-ENL, interact with the EC and the DOT1L/KMT4-containing complex (DOTCom), which both have central roles in regulating transcription. The interactions between ENL and these complexes could provide valuable insights into the intricate network of transcriptional regulation in leukemia. ENL associates with these complexes and also interacts with both the fused (MLL (KMT2A) fusion/SEC/

DOTCom) and nonfused complexes [wild-type MLL (KMT2A)] to drive leukemia [108].

The regulation of HOXA9 is directly linked to ENL, a key component of MLL fusion proteins like MLL-ENL, which act as oncogenic transcription factors and are strictly dependent on the presence of 4-hydroxy-tamoxifen (4-OHT) for their activity. Microarray analysis has revealed that ENL, through the MLL-ENL fusion protein, upregulates HOXA9 among other targets, imposing a block on myeloid differentiation through aberrant activation of a limited set of genes, thus contributing to continuous cell proliferation and differentiation arrest in MLL-associated leukemias [69]. Importantly, ENL is not the sole actor in this scenario. Other proteins like TAT interacting protein 60 kDa/lysine acetyltransferase 5 (TIP60/KAT5), a MYST-type histone acetyltransferase, are essential for the expression of HOXA cluster and MEIS1 genes in MLL fusion leukemia. MLL-ENL fusion recruits TIP60 to the HOXA9 locus, where it acetylates the histone variant H2A.Z, thereby promoting HOXA9 gene expression [109].

18.4.3.5 NPM1c-XPO1

The *NPM1* gene, responsible for a multifunctional chaperone protein that cycles between the nucleus and cytoplasm, is subject to mutations in approximately one-third of AML cases. These mutations are AML-specific, usually impact exon 12, and are often associated with FLT3-ITD, DNMT3A, TET2, and isocitrate dehydrogenase 1/dehydrogenase 2 (IDH1/IDH2) mutations. Given its unique molecular and clinical attributes, NPM1-mutated AML is classified as a distinct leukemia entity. Notably, all NPM1 mutations lead to the creation of leukemic mutants that are misdirected into the cytoplasm of leukemic cells, significantly contributing to disease pathogenesis [110]. NPM1 is a ubiquitously abundant nucleolar protein that maintains genome integrity, DNA repair, and ribosome biogenesis [111]. In the context of normal physiology, NPM1 protein is localized in nucleoli, but AML-related mutations at its C-terminus result in abnormal cytoplasmic displacement [110]. This variant, referred to as NPM1c, is tied to aberrant activation of HOXA and HOXB cluster genes, including HOXA9 [45, 112]. Either the relocalization of NPM1c back to the nucleus or targeted degradation disrupts the oncogenic program via downregulation of HOX genes, promoting myeloid cell differentiation [41]. XPO1, a primary nuclear-cytoplasmic transport protein, collaborates with NPM1c to facilitate its transportation to the cytoplasm, and its inhibition can block this process, resulting in AML growth arrest and differentiation. Two potential mechanisms have been identified for the aberrant regulation of HOX gene expression by NPM1c. The first involves the nuclear relocalization of NPM1c in an XPO1-dependent manner, leading to direct binding to HOX cluster regions and thereby activating HOX genes. A second mechanism suggests that NPM1c AML also depends on the interaction between MLL and MENIN, a protein complex essential for maintaining normal HOX gene expression. Inhibiting this interaction reduces HOX gene expression and promotes leukemic cell differentiation.

18.4.4 Transcriptional Regulation of HOXA9

The transcriptional regulation of the HOXA9 gene involves a complex network of factors. Some transcription factors, such as CDX2 and CDX4, derived from caudal genes, have been shown to potentially increase HOXA9 gene expression. In some instances, CDX2 and CDX4 can form a fusion with the ETV6 gene locus, a process that has been observed in a subset of AML cases. Interestingly, ectopic CDX2 expression has been associated with accelerated leukemic development in mouse models, but this progression was observed without an associated upregulation of HOXA9 gene expression. Further, the N-terminal domain of the CDX2 gene has been found to play an essential role in maintaining HOXA gene expression. The deletion of this domain has been shown to result in decreased HOXA gene expression, indicating the critical regulatory function of the N-terminal domain. Likewise, the CDX4 gene has been implicated in maintaining HOXA9 expression during embryonic hematopoiesis, emphasizing the intricate regulatory network guiding HOXA9 expression. In addition to CDX2 and CDX4, the upstream stimulation factors 1 and 2 (USF1 and USF2) have been linked to HOXB gene expression during normal hematopoiesis. Of particular note, recent findings have identified USF2 as a novel positive regulator of HOXA9. This factor binds directly to a conserved motif at the HOXA9 promoter, and depletion of USF2 leads to downregulation of HOXA9 expression in MLLr leukemia cells.

18.4.5 Noncoding RNA Regulation of HOXA9

HOXA9 gene expression is controlled not just at the transcriptional level but also posttranscriptionally and epigenetically. MicroRNAs, notably miR-126, miR-145, and let-7s, are identified as regulators of HOXA9, as they target conserved sites within the *Hoxa9* homeobox. Forced expression of these microRNAs reduces HOXA9 protein levels and biological activity. Interestingly, miR-126 and HOXA9 mRNA are expressed in HSCs and downregulated during progenitor cell differentiation [113]. However, another miRNA, miR-708, when overexpressed, obstructs HOXA9-mediated transformation in AML cells, demonstrating the complex role of miRNAs in indirectly regulating HOXA9 during normal and aberrant hematopoiesis [114]. The expression of HOX genes, such as HOXA9, is regulated by lncRNAs. The regulatory mechanisms can be repressive or expressive. A crucial study by Rinn et al. established HOX transcript antisense RNA (HOTAIR), a transregulatory system where the noncoding RNA expressed from the HOXC locus regulates gene expression in the HOXD cluster. This regulation involves HOTAIR recruiting PRC2 member SUZ12 and increasing H3K27me3, leading to the repression of the HOXD gene clusters [115]. In addition to the possible regulatory action on HOXA9, HOXA transcript at distal tip (HOTTIP) has also been demonstrated to mediate HOXA9 to enhance the Wnt/β catenin pathway in pancreatic cancer stem cells, suggesting a significant role in regulating HOXA9. Furthermore, the lncRNA HOTAIR, which is frequently coexpressed with HOXA9 in high-grade gliomas, has been shown to have its promoter directly bound by HOXA9. This introduces a novel regulatory interaction, potentially allowing HOXA9 to regulate HOTAIR, thereby influencing HOTAIR's role in tumor progression [116].

On the contrary, lncRNA HOTTIP, as reported by Wang et al., functions via a cis-regulatory mechanism. Transcribed from the HOXA cluster, HOTTIP binds and activates HOXA genes. The activation of gene transcription is achieved through chromosomal looping that brings HOTTIP close to the target genes and promotes H3K4me3. Although a direct connection between HOTTIP and HOXA9 has not been established, the cis-regulatory action of HOTTIP implies that it may function similarly to regulate HOXA9 [117]. Moreover, HOTTIP has been shown to mediate HOXA9 to enhance the Wnt/β-catenin pathway in pancreatic cancer stem cells, reinforcing the concept that HOTTIP plays a significant role in regulating HOXA9 [116]. Together, these findings indicate that lncRNA HOTTIP is integral to the regulation of HOX genes, potentially including HOXA9.

18.4.6 Posttranslational Modifications (PTMs) of HOXA9

The multifaceted regulation of the HOXA9 protein through posttranslational modifications (PTMs), such as ubiquitination and arginine methylation, underpins its role in various cellular processes. The HOXA9 protein is regulated by PTMs, including ubiquitination by the Culin 4A (CUL-4A) machinery. This mechanism promotes the degradation of HOXA9, influencing the proper terminal differentiation of myeloid progenitor cells into granulocytes, thus revealing a novel regulatory mechanism of hematopoiesis through ubiquitin-dependent proteolysis [118]. On the other hand, protein arginine methyltransferase 5 (PRMT5)-mediated methylation of

HOXA9 plays a dual role in endothelial cell inflammation and cardiomyocyte hypertrophy. In the context of endothelial cell (EC) inflammation, PRMT5 is crucial for the tumor necrosis factor alpha (TNF-α)-dependent activation and methylation of HOXA9, which is required for the induction of endothelial leukocyte adhesion molecule (ELAM) expression, including E-selectin and vascular cell adhesion molecule 1 (VCAM-1) [119]. Conversely, in the regulation of cardiomyocyte hypertrophy, a decrease in PRMT5 expression and an upregulation of HOXA9 were observed. PRMT5, through its symmetric dimethylation of HOXA9, represses HOXA9 expression, ultimately impairing hypertrophy. These studies reveal the diverse roles PRMT5 plays in modulating HOXA9 activity in different cellular contexts, with potential therapeutic implications in controlling inflammation and cardiac hypertrophy [120].

In addition to ubiquitination and methylation, HOXA9 is also subject to regulation through phosphorylation by protein kinase C (PKC) and casein kinase II (CKII). This modification, particularly PKC-mediated phosphorylation of HOXA9 at S204 and T205 within a highly conserved sequence, can alter HOXA9's DNA-binding affinity and its cooperative DNA binding with PBX. Moreover, PKC phosphorylation is implicated in the process of myeloid differentiation, suggesting another layer of posttranslational modification contributing to the complex control of HOXA9 activity [121]. Thus, the regulation of HOXA9 through ubiquitination, methylation, and phosphorylation underpins its roles in hematopoiesis, inflammation, cardiac hypertrophy, and myeloid differentiation, emphasizing the potential therapeutic implications in these diverse biological processes.

18.5 Downstream Targets of HOXA9

Genomic and transcriptomic studies have discovered several transcriptional targets regulated by HOXA9 and MEIS1, including genes encoding transcription factor LMO2 [122], antiapoptotic factor BCL2 [123, 124], SRY (sex-determining region Y)-box 4 (SOX4) [124], and receptor tyrosine kinase FLT3 [125], STAT5 [48]. The HOXA9 protein functions as a transcriptional enhancer that regulates pathways controlling cell survival, often through activation of leukemia-associated proto-oncogenes, such as MYB and FLT3. Crucially, it has been found to control the expression of the BCL2 gene, which plays a vital role in suppressing apoptosis. Consequently, HOXA9-regulated *Bcl2* expression is significant for the survival of myeloid progenitors and contributes to the oncogenic effects of HOXA9 overexpression [123]. Moreover, HOXA9 functions as a transcription maintenance factor, playing a critical role in the regulation of genes like BCL2, SOX4, and IGF1, which have implications in leukemogenesis. Although HOXA9 is highly expressed in multiple cell lines, particularly those with MLL fusions, it cannot reactivate its target genes once silenced, indicating its role in preserving previously activated chromatin structures. Thus, high expression of HOXA9, BCL2, and SOX4 commonly appears together in leukemia patient samples, underscoring HOXA9's potential involvement in the maintenance of oncogenic chromatin states. The BCL2 and SOX4 genes, along with MYC, play a significant role in promoting leukemogenesis by reducing apoptosis or programmed cell death. When BCL2 or SOX4 were coexpressed with c-MYC, a considerable increase in the potential for leukemic transformation was observed. This effect was similar to the leukemogenic potential of c-MYC co-expressed with HOXA9. Furthermore, in addition to HOXA9, coexpression of BCL2 or SOX4 with c-MYC was found to alleviate apoptotic tendencies. Thus, BCL2 and SOX4, along with HOXA9, appear to be pivotal regulators in leukemogenesis, primarily through anti-apoptotic mechanisms [124]. HOXA9 gene plays a crucial role in regulating FLT3, a cytokine receptor essential for lymphoid priming and the generation of B-cell precursors. HOXA9 directly binds to the FLT3 promoter, with its presence leading to an increase in FLT3 transcription and expression. Conversely, the downregulation of *Hoxa9* leads to a significant reduction in *Flt3*, indicating that a failure to upregulate *Flt3* may underlie the lymphoid and early B-cell deficiencies observed in *Hoxa9*-deficient mice [125]. Recently, a study by Sahoo et al. has identified SCUBE1 as a novel target of HOXA9 and MEIS1, playing a critical role in the development of MLLr AML. The researchers found that signal peptide, CUB domain, EGF-like 1 (SCUBE1), which is highly expressed in MLLr AML cells but not in normal HSCs, promotes the activation of the FLT3-LYN proto-oncogene (LYN) signaling axis, essential for both the initiation and maintenance of MLLr AML. In addition, they have developed an antibody drug conjugation-based strategy targeting SCUBE1-expressing leukemic cells, effectively inhibiting leukemia growth. SCUBE1's potential as a therapeutic target is underscored by its role in activating the FLT3 signaling pathway, thereby reinforcing the oncogenic network downstream of HOXA9 and MEIS1 [126]. In conclusion, the importance of HOXA9 and its downstream targets, including BCL2, SOX4, FLT3, and the newly discovered SCUBE1, in the context of leukemogenesis, particularly AML, is underscored by our current understanding. The shared expression patterns between HOXA9 and these targets in AML patient samples suggest a potential coordinated function, which warrants further exploration. It is notable that SCUBE1, despite its absence in normal HSCs, exhibits high expression in MLLr AML cells, underpinning its potential role in the FLT3-LYN signaling axis. This discovery marks a significant step forward in our understanding of the molecular basis of AML and presents a promising new therapeutic target.

However, while these findings contribute substantially to our knowledge of the HOXA9 regulatory network, it is clear that more work needs to be done. For instance, the exact mechanisms by which these targets promote leukemogenesis and the potential crosstalk among them remain largely unknown. Unveiling these mechanisms might shed light on the complexity of HOXA9's role in leukemogenesis and the potential challenges that may be faced in developing targeted therapies. Future research should focus on further elucidating these mechanisms and exploring the therapeutic potential of these downstream targets, particularly SCUBE1. We are hopeful that these studies will lead to improved therapeutic strategies and ultimately better outcomes for patients with AML and other forms of leukemia.

18.6 Potential Therapeutic Strategies Targeting HOXA9 in Leukemia

HOXA9, due to its role in leukemogenesis, represents an appealing therapeutic target. As a TF, however, HOXA9 does not belong to receptor or steroid families that could be targeted by ligand mimetics through the binding to a binding pocket. Various strategies are being explored to disrupt its oncogenic activity in the context of leukemia (Fig. 18.3).

18.6.1 Targeting DOT1L/KMT4 Protein Interaction

DOT1L/KMT4 is aberrantly recruited at the MLL complex via interaction with MLL fusion partner moieties such as AF4, AF6, AF9, AF10, or ENL [127]. Several DOT1L/KMT4 inhibitors like EPZ-5676 (pinometostat), EPZ004777, and SGC0946 have been developed to compete for DOT1L/KMT4 binding to MLL fusion proteins, with pinometostat being the most advanced for therapeutic purposes [128]. Notably, DOT1L/KMT4 inhibitors have shown potential efficacy for MLL-rearranged leukemia and are currently under investigation in clinical trials [129]. In a homozygous deletion mouse model, it was observed that DOT1L/KMT4 deletion led to severe anemia, hypocellularity in the bone marrow, and depletion of hematopoietic stem cells. Further, using a hematopoietic cell KO model, it was confirmed that not all cells developed through hematopoiesis are dependent on DOT1L/KMT4 and that DOT1L/KMT4 contributes to MLL-AF9-mediated leukemogenesis through the upregulation of the oncogenes *Hoxa9* and *Meis1* [130].

These DOT1L/KMT4 inhibitors, including SYC-522, pyrimidyl-aminoquinoline derivative

9e; compounds 3 and 9; DC_L115; 2-chloro benzothiophene derivative 12; aza-benzimidazole derivative 13; phenoxyacetamide derivatives L01, L03, L04, and L05; and massonianoside B, are under development [59, 131]. They work by inhibiting DOT1L/KMT4 enzymatic activity by competing with the cofactor S-adenosyl methionine (SAM) [132]. Preclinical studies show that these compounds effectively reduce H3K79 methylation, causing a significant reduction in the expression of HOXA9 and MEIS1 and inducing apoptosis in MLL-leukemic cells [101].

Clinical trials with EPZ-5675, also called pinometostat, evidenced minor toxicities, making it safe to use at high doses. It was found that patients responded well to various therapeutic dosages. Only 2 of the 51 participants experienced complete remission at the end of the trial; however, one of those participants experienced an aggressive relapse following the study's conclusion. This suggests that continued DOT1L/KMT4 targeting treatment is required [133]. Since only minor therapeutic responses were observed when used as a monotherapy, it was therefore suggested that DOT1L/KMT4 inhibitors should be considered in combination with other therapies such as Ara-C or azacytidine in leukemia patients [134]. Currently, a phase Ib/phase II clinical investigation is ongoing to assess the safety and efficacy of pinometostat combined with chemotherapy in MLL leukemia patients. Despite the challenges, initial results of these trials are promising, indicating the potential of DOT1L/KMT4 inhibitors as part of future leukemia therapies [101, 127, 131, 133, 134].

The study observed a substantial correlation between pediatric AMLs with mutations in genes encoding for the subunits of the COHESIN complex and the genes downregulated following DOT1L/KMT4 inhibitor treatment. In contrast, gene expression changes in COHESIN wild-type AMLs were best correlated with genes upregulated following DOT1L/KMT4 inhibition. This suggests that both DOT1L/KMT4 inhibition and COHESIN haploinsufficiency impact a similar set of genes. Hence, using DOT1L/KMT4 inhibitors could be potentially advantageous in treating pediatric AML patients with HOXA9-upregulating mutations. The research also highlighted that cohesin-depleted cells exhibited reduced self-renewal when DOT1L/KMT4 was inhibited. Given the recent findings that AF10 acts as an epigenetic reader for unmethylated H3K27, researchers hypothesized that the global decrease in H3K27me3 associated with COHESIN loss could be accompanied by a global increase in the DOT1L/KMT4-dependent mark H3K79me2 levels [135].

18.6.2 Targeting MENIN-MLL (KMT2A) Interaction

The MENIN/MLL interaction interface is considered a promising therapeutic target for leukemias, particularly mixed phenotype acute leukemias (MPAL) [136]. MENIN/MLL inhibitors are in various stages of preclinical and clinical evaluation, with the goal to interfere with the interaction of MENIN and MLL fusion proteins, thereby reducing expression of oncogenes such as HOXA9 and MEIS1 [137–139]. Various compounds such as MCP-1, MI-2-2, MI-463, MI-503, MI-538, and MI-1481 are under preclinical evaluation [140]. MI-503 is especially promising due to its deeper interaction within the MENIN pocket and its stability. These compounds have shown to induce differentiation and display antileukemic activities in various AML cell models [141]. KO-539, developed by the Kura Oncology, is the most advanced MENIN inhibitor currently, with phase I clinical trials ongoing for relapsed or refractory AML. It was identified through high-throughput virtual screening and subsequent structure-based optimization. It has shown antileukemic effects in both the MV4–11 MPAL model and patient-derived xenografts from the NPM1c$^+$/FLT3-mut AML subtype. Other MENIN inhibitors, including ML227, MIV-6, and M-525, have been evaluated in combination with DOT1L/KMT4 inhibitors to restore differentiation in MPAL cell models. M-525 is particularly efficient in interacting with the MENIN protein and demonstrates higher specificity and efficacy in MPAL cell models [142].

Interestingly, DOT1L/KMT4L, a protein that interacts with MENIN, also collaborates with bromodomain-containing protein 4 (BRD4), another target for AML treatment. Several BRD4 inhibitors such as I-BET762, OTX015/MK8628, CPI-0610, FT-1101, BI-894999, BMS-986158, PLX51107, RO6870810, and GSK2820151 have entered clinical trials [59]. In patient models with AML subtypes having certain mutations, better responses have been observed with MENIN inhibitors. Compounds like VTP50469 and MI-3545 have shown promising results in preclinical studies by inhibiting proliferation, inducing differentiation, and promoting apoptosis [143]. Currently, KO-539 and SNDX-5613 are under clinical investigation for safety and efficacy in patients with relapsed or refractory AML and MLL-rearranged and NPM1-mutated leukemias, respectively. Preliminary reports indicate promising results [144].

18.6.3 Targeting NPM1c

NPM1 is a protein found in the nucleoli under normal conditions, where it plays an essential role in genome integrity, DNA repair, and ribosome biogenesis. In some cases of AML, NPM1 carries mutations at its C-terminus, termed NPM1c, causing the protein to aberrantly relocate to the cytoplasm. This relocation is a hallmark of NPM1c AML, a subtype of AML associated with the abnormal activation of genes in the HOXA and HOXB clusters, including HOXA9 [145, 146]. One way this abnormal gene activation is believed to occur involves the interaction between NPM1c and XPO1, a protein that transports various molecules across the nuclear envelope. In the context of NPM1c AML, XPO1 helps to maintain NPM1c in the cytoplasm, which has been identified as a crucial element in the development of this AML subtype. It is suggested that genetic or pharmaceutical inhibition of XPO1 can block NPM1c transport, leading to AML growth arrest and differentiation [41].

An alternative mechanism suggests that the abnormal activation of HOX genes might be a direct result of the NPM1c being relocalized to the nucleus in an XPO1-dependent manner. In this scenario, NPM1c binds directly to HOX cluster regions to activate HOX genes [145]. A relationship has also been observed between NPM1c and the TF PU.1, which may help keep NPM1 within the cytoplasmic compartment. In this case, overexpression of PU.1 is associated with decreased HOXA gene expression [147]. Given the essential role of the wild-type MLL protein in NPM1c AML, this subtype of leukemia is also influenced by the interaction between MLL and MENIN. Inhibiting this interaction with small molecules can reduce HOX gene expression and promote differentiation of leukemic cells. Thus, both these mechanisms provide potential targets for therapeutic interventions in NPM1c AML.

18.6.4 Targeting WDR5/MLL Interaction

Compounds such as MM-102, MM-401, MM-589, and others have been developed to target the interaction between WD repeat-containing protein 5 (WDR5) and MLL (KMT2A) proteins. They act by inhibiting this interaction, resulting in decreased HOXA9 and MEIS1 expression in specific cell models and inducing cell death in some cases [148].

18.6.5 Targeting HOXA9 Expression Through MLL-Independent Epigenetic Modifiers

A different approach focuses on preventing HOXA9 transcription in other AML subtypes by modulating the activity of epigenetic enzymes through metabolites produced by mitochondrial metabolism. This includes inhibiting enzymes like dihydroorotate dehydrogenase (DHODH), which is involved in nucleic acid synthesis, or IDH1/IDH2, which impact the normal catalytic activity and thus could affect the expression of HOXA9 [149].

18.7 Emerging Technologies Studying HOXA9 and Its Downstream Targets

Research into the HOXA9 gene and its downstream targets, particularly in normal and diseased cells, has been enhanced by recent technological advances. These emerging techniques provide greater depth and breadth of analysis, improving our understanding of HOXA9's regulatory network and its involvement in disease progression, particularly leukemia.

18.7.1 CRISPR-Cas9 Gene Editing

This revolutionary gene editing technology allows for precise modifications of specific genes in living cells. Researchers can use CRISPR/Cas9 to create specific mutations in the HOXA9 gene or its downstream targets to understand their role in cellular processes and disease progression. In addition, researchers can create models of disease in vitro and in vivo, enhancing our understanding of HOXA9's function in both normal and diseased states. Recently, CRISPR/Cas9 technology was utilized along with a HOXA9-mCherry knock-in reporter cell line to investigate the regulatory mechanisms of HOXA9. The use of CRISPR/Cas9 screens helped identify several TFs that control HOXA9 expression, notably revealing USF2 as a novel regulator. Depletion of USF2 resulted in a significant decrease in HOXA9 expression and subsequently impaired the proliferation of MLLr leukemia cells. Furthermore, when *Hoxa9* was reintroduced into cells lacking USF2, the leukemia cell proliferation previously impaired by the loss of USF2 was restored. This suggests that HOXA9 is an essential factor for MLLr leukemia cell proliferation and is directly regulated by USF2 [150].

18.7.2 Protein Degradation Tag Approach

This research involves the use of a unique degradation tag, specifically a mutated FKBP (FKBPF36V), which is used as a "degron" sequence. This sequence is attached to proteins of interest, in this case MEIS1, and allows for controlled protein degradation. This is achieved by the use of a proteolysis targeting chimera (PROTAC) named dTAG13. dTAG13 effectively bridges the FKBPF36V-tagged protein and the endogenous E3 ubiquitin ligase cereblon, prompting rapid proteasomal degradation of the targeted protein. By exploiting this system, researchers can exert control over the presence of specific proteins within the cell and observe the functional implications. In the context of the study, the researchers used this degron system to investigate the role of the transcription factors HOXA9 and MEIS1, both of which are involved in the pathogenesis of AML. While HOXA9 alone is known to immortalize cells, cooperation with MEIS1 is necessary for the full leukemia phenotype. Researchers found that MEIS1 primarily localizes to enhancers enriched for H3K27ac and H3K4me1 that have been preactivated by HOXA9. This association of MEIS1 with the chromatin requires the physical presence of HOXA9, and all MEIS1 DNA interactions were rapidly lost following HOXA9 degradation. Further, MEIS1 controls a gene expression pattern that is dominated by c-MYC, ribosome biogenesis, and ribosomal RNA synthesis genes. Even though c-MYC overexpression alone did not accelerate leukemogenesis, it contributes to the cell cycle-stimulating effect of MEIS1. Beyond c-MYC, MEIS1 also induces the transcription of ribosomal biogenesis genes, leading to elevated resistance against inhibition of ribosomal RNA synthesis and translation. Moreover, the researchers found that HOXA9 and MEIS1 proteins are stabilized by PTMs. Mutations of

HOXA9/MEIS1 phosphorylation sites or the inhibition of CKII led to rapid protein degradation, suggesting a potential pathway for pharmacological intervention in AML [49].

In conclusion, these emerging technologies have revolutionized the way we study genes and their functions. The application of these technologies to HOXA9 and its downstream targets will undoubtedly provide new insights into their roles in both normal cellular function and disease progression, thereby advancing our understanding of diseases like leukemia and opening up new avenues for potential therapeutic interventions.

18.8 Directions for Future Research in This Field

Given the complex nature of leukemia and the critical role that the *HOXA9* gene plays in its development and progression, there are several avenues for future research that could significantly enhance our understanding of this disease and subsequently improve patient outcomes. First, despite the comprehensive understanding of HOXA9's role in normal and leukemic hematopoiesis, the precise molecular mechanisms that cause HOXA9 overexpression and the subsequent development of leukemia remain somewhat elusive. Future research should strive to elucidate the genetic, epigenetic, and environmental factors that lead to aberrant HOXA9 expression, ideally paving the way for preventive strategies. Second, the intriguing synergistic effects observed between HOXA9 and other molecular players such as MEIS1 and PBX3 warrant further investigation. Detailed studies should aim to understand the structural and functional interplay between these molecules and how their combined aberrant activity leads to enhanced leukemic transformation. Elucidating the specifics of these relationships could uncover novel targets for therapeutic intervention, allowing for the development of more effective and personalized treatments. Third, while this chapter has highlighted the potential of HOXA9 as a biomarker for leukemia, further validation of this potential in larger and more diverse patient cohorts is needed. An improved understanding of the prognostic and predictive value of HOXA9 expression levels could enhance early detection efforts, risk stratification, and treatment planning. Moreover, the development of noninvasive methods for monitoring HOXA9 expression in patients over time could allow for more responsive treatment adjustments and better prognosis assessment. Lastly, the therapeutic strategies aimed at targeting HOXA9, while promising, present a number of challenges that must be overcome. Future research should focus on improving the efficacy and safety profile of these treatments. This may involve investigating novel drug delivery systems, refining existing drugs for improved selectivity and potency or developing combination therapies that can simultaneously target HOXA9 and its molecular partners. Furthermore, understanding the potential resistance mechanisms to HOXA9-targeted therapies could help in developing strategies to overcome such resistance. To summarize, while we have gained valuable insights into the role of HOXA9 in leukemogenesis, there remain considerable challenges and unanswered questions. Continued research in this field is not only crucial for increasing our understanding of leukemia biology but also for refining current therapeutic strategies and discovering new ones. The ultimate goal is to transform these scientific advances into clinical benefits, providing more effective and personalized treatments for patients battling this devastating disease.

Acknowledgments The authors thank the Lu Lab members Xinyue Zhou, Pengcheng Zhang, Lixia Zhang, Virginia Vasey, and Neha Srinivas at the University of Alabama at Birmingham (UAB) for helping with critical feedbacks. This work was supported in part by National Institutes of Health grants R01 CA259480. SA is supported by the American Society of Hematology's Graduate Hematology Award. RL is an American Society of Hematology scholar in basic science and a research scholar of the American Cancer Society (RSG-22-036-01-DMC). We apologize to researchers whose work could not be cited owing to space limitations.

References

1. de Bruijn MF, Speck NA, Peeters MC, Dzierzak E (2000) Definitive hematopoietic stem cells first develop within the major arterial regions of the mouse embryo. EMBO J 19(11):2465–2474
2. Canu G, Athanasiadis E, Grandy RA, Garcia-Bernardo J, Strzelecka PM, Vallier L et al (2020) Analysis of endothelial-to-haematopoietic transition at the single cell level identifies cell cycle regulation as a driver of differentiation. Genome Biol 21(1):157
3. Singh R, Soman-Faulkner K, Sugumar K (2023) Embryology, hematopoiesis. StatPearls, Treasure Island (FL). StatPearls Publishing Copyright © 2023, StatPearls Publishing LLC
4. Belyavsky A, Petinati N, Drize N (2021) Hematopoiesis during ontogenesis, adult life, and aging. Int J Mol Sci 22(17):9231
5. Zovein AC, Hofmann JJ, Lynch M, French WJ, Turlo KA, Yang Y et al (2008) Fate tracing reveals the endothelial origin of hematopoietic stem cells. Cell Stem Cell 3(6):625–636
6. Seita J, Weissman IL (2010) Hematopoietic stem cell: self-renewal versus differentiation. Wiley Interdiscip Rev Syst Biol Med 2(6):640–653
7. Cheng H, Zheng Z, Cheng T (2020) New paradigms on hematopoietic stem cell differentiation. Protein Cell 11(1):34–44
8. Kim CH (2010) Homeostatic and pathogenic extramedullary hematopoiesis. J Blood Med 1:13–19
9. Weiskopf K, Schnorr PJ, Pang WW, Chao MP, Chhabra A, Seita J et al (2016) Myeloid cell origins, differentiation, and clinical implications. Microbiol Spectr 4(5):1–18
10. Zhang CC, Lodish HF (2008) Cytokines regulating hematopoietic stem cell function. Curr Opin Hematol 15(4):307–311
11. Franco RS (2012) Measurement of red cell lifespan and aging. Transfus Med Hemother 39(5):302–307
12. Seifert A, Werheid DF, Knapp SM, Tobiasch E (2015) Role of Hox genes in stem cell differentiation. World J Stem Cells 7(3):583–595
13. Luo Z, Rhie SK, Farnham PJ (2019) The enigmatic HOX genes: can we crack their code? Cancers (Basel) 11(3):323
14. Bach C, Buhl S, Mueller D, García-Cuéllar M-P, Maethner E, Slany RK (2010) Leukemogenic transformation by HOXA cluster genes. Blood 115(14):2910–2918
15. Mark M, Rijli FM, Chambon P (1997) Homeobox genes in embryogenesis and pathogenesis. Pediatr Res 42(4):421–429
16. Grier DG, Thompson A, Kwasniewska A, McGonigle GJ, Halliday HL, Lappin TR (2005) The pathophysiology of HOX genes and their role in cancer. J Pathol 205(2):154–171
17. Slany RK (2009) The molecular biology of mixed lineage leukemia. Haematologica 94(7):984–993
18. Damante G, Pellizzari L, Esposito G, Fogolari F, Viglino P, Fabbro D et al (1996) A molecular code dictates sequence-specific DNA recognition by homeodomains. EMBO J 15(18):4992–5000
19. Collins CT, Hess JL (2016) Role of HOXA9 in leukemia: dysregulation, cofactors and essential targets. Oncogene 35(9):1090–1098
20. Banerjee-Basu S, Sink DW, Baxevanis AD (2001) The homeodomain resource: sequences, structures, DNA binding sites and genomic information. Nucleic Acids Res 29(1):291–293
21. Joshi R, Passner JM, Rohs R, Jain R, Sosinsky A, Crickmore MA et al (2007) Functional specificity of a Hox protein mediated by the recognition of minor groove structure. Cell 131(3):530–543
22. Breitinger C, Maethner E, Garcia-Cuellar MP, Slany RK (2012) The homeodomain region controls the phenotype of HOX-induced murine leukemia. Blood 120(19):4018–4027
23. Slattery M, Ma L, Négre N, White KP, Mann RS (2011) Genome-wide tissue-specific occupancy of the Hox protein Ultrabithorax and Hox cofactor Homothorax in Drosophila. PLoS One 6(4):e14686
24. Heinz S, Benner C, Spann N, Bertolino E, Lin YC, Laslo P et al (2010) Simple combinations of lineage-determining transcription factors prime cis-regulatory elements required for macrophage and B cell identities. Mol Cell 38(4):576–589
25. Salma N, Xiao H, Mueller E, Imbalzano AN (2004) Temporal recruitment of transcription factors and SWI/SNF chromatin-remodeling enzymes during adipogenic induction of the peroxisome proliferator-activated receptor gamma nuclear hormone receptor. Mol Cell Biol 24(11):4651–4663
26. Aryal S, Zhang Y, Wren S, Li C, Lu R (2023) Molecular regulators of HOXA9 in acute myeloid leukemia. FEBS J 290(2):321–339
27. Denans N, Iimura T, Pourquié O (2015) Hox genes control vertebrate body elongation by collinear Wnt repression. eLife 4:e04379
28. Hrycaj SM, Wellik DM (2016) Hox genes and evolution. F1000Res 5
29. Argiropoulos B, Humphries RK (2007) Hox genes in hematopoiesis and leukemogenesis. Oncogene 26(47):6766–6776
30. Steens J, Klein D (2022) HOX genes in stem cells: maintaining cellular identity and regulation of differentiation. Front Cell Dev Biol 10:1002909
31. Smith J, Zyoud A, Allegrucci C (2019) A case of identity: HOX genes in normal and cancer stem cells. Cancers (Basel) 11(4):512
32. Alharbi RA, Pettengell R, Pandha HS, Morgan R (2013) The role of HOX genes in normal hematopoiesis and acute leukemia. Leukemia 27(5):1000–1008
33. Sauvageau G, Lansdorp PM, Eaves CJ, Hogge DE, Dragowska WH, Reid DS et al (1994) Differential expression of homeobox genes in functionally distinct CD34+ subpopulations of human bone marrow cells. Proc Natl Acad Sci USA 91(25):12223–12227

34. Lawrence HJ, Christensen J, Fong S, Hu YL, Weissman I, Sauvageau G et al (2005) Loss of expression of the Hoxa-9 homeobox gene impairs the proliferation and repopulating ability of hematopoietic stem cells. Blood 106(12):3988–3994
35. Huang Y, Sitwala K, Bronstein J, Sanders D, Dandekar M, Collins C et al (2012) Identification and characterization of Hoxa9 binding sites in hematopoietic cells. Blood 119(2):388–398
36. Doulatov S, Notta F, Laurenti E, Dick JE (2012) Hematopoiesis: a human perspective. Cell Stem Cell 10(2):120–136
37. Lebert-Ghali C, Fournier M, Kettyle L, Thompson A, Sauvageau G, Bijl JJ (2016) Hoxa cluster genes determine the proliferative activity of adult mouse hematopoietic stem and progenitor cells. Blood 127(1):87–90
38. Alsayegh K, Cortés-Medina LV, Ramos-Mandujano G, Badraiq H, Li M (2019) Hematopoietic differentiation of human pluripotent stem cells: HOX and GATA transcription factors as master regulators. Curr Genomics 20(6):438–452
39. Ramos-Mejía V, Navarro-Montero O, Ayllón V, Bueno C, Romero T, Real PJ et al (2014) HOXA9 promotes hematopoietic commitment of human embryonic stem cells. Blood 124(20):3065–3075
40. Dou DR, Calvanese V, Sierra MI, Nguyen AT, Minasian A, Saarikoski P et al (2016) Medial HOXA genes demarcate haematopoietic stem cell fate during human development. Nat Cell Biol 18(6):595–606
41. Zarka J, Short NJ, Kanagal-Shamanna R, Issa GC (2020) Nucleophosmin 1 mutations in acute myeloid Leukemia. Genes (Basel) 11(6):649
42. Rugo HS, Jacobs I, Sharma S, Scappaticci F, Paul TA, Jensen-Pergakes K et al (2020) The promise for histone methyltransferase inhibitors for epigenetic therapy in clinical oncology: a narrative review. Adv Ther 37(7):3059–3082
43. Deshpande AJ, Deshpande A, Sinha AU, Chen L, Chang J, Cihan A et al (2014) AF10 regulates progressive H3K79 methylation and HOX gene expression in diverse AML subtypes. Cancer Cell 26(6):896–908
44. Rice KL, Licht JD (2007) HOX deregulation in acute myeloid leukemia. J Clin Invest 117(4):865–868
45. Spencer DH, Young MA, Lamprecht TL, Helton NM, Fulton R, O'Laughlin M et al (2015) Epigenomic analysis of the HOX gene loci reveals mechanisms that may control canonical expression patterns in AML and normal hematopoietic cells. Leukemia 29(6):1279–1289
46. Andreeff M, Ruvolo V, Gadgil S, Zeng C, Coombes K, Chen W et al (2008) HOX expression patterns identify a common signature for favorable AML. Leukemia 22(11):2041–2047
47. Chen SL, Qin ZY, Hu F, Wang Y, Dai YJ, Liang Y (2019) The role of the HOXA gene family in acute myeloid leukemia. Genes (Basel) 10(8):621
48. de Bock CE, Demeyer S, Degryse S, Verbeke D, Sweron B, Gielen O et al (2018) HOXA9 cooperates with activated JAK/STAT signaling to drive leukemia development. Cancer Discov 8(5):616–631
49. Garcia-Cuellar MP, Prinz A, Slany RK (2022) Meis1 supports leukemogenesis through stimulation of ribosomal biogenesis and Myc. Haematologica 107(11):2601–2616
50. Sauvageau G, Thorsteinsdottir U, Eaves CJ, Lawrence HJ, Largman C, Lansdorp PM et al (1995) Overexpression of HOXB4 in hematopoietic cells causes the selective expansion of more primitive populations in vitro and in vivo. Genes Dev 9(14):1753–1765
51. Thorsteinsdottir U, Mamo A, Kroon E, Jerome L, Bijl J, Lawrence HJ et al (2002) Overexpression of the myeloid leukemia-associated Hoxa9 gene in bone marrow cells induces stem cell expansion. Blood 99(1):121–129
52. Borrow J, Shearman AM, Stanton VP Jr, Becher R, Collins T, Williams AJ et al (1996) The t(7;11)(p15;p15) translocation in acute myeloid leukaemia fuses the genes for nucleoporin NUP98 and class I homeoprotein HOXA9. Nat Genet 12(2):159–167
53. Chung KY, Morrone G, Schuringa JJ, Plasilova M, Shieh JH, Zhang Y et al (2006) Enforced expression of NUP98-HOXA9 in human CD34(+) cells enhances stem cell proliferation. Cancer Res 66(24):11781–11791
54. Kroon E, Krosl J, Thorsteinsdottir U, Baban S, Buchberg AM, Sauvageau G (1998) Hoxa9 transforms primary bone marrow cells through specific collaboration with Meis1a but not Pbx1b. EMBO J 17(13):3714–3725
55. Lehnertz B, Pabst C, Su L, Miller M, Liu F, Yi L et al (2014) The methyltransferase G9a regulates HoxA9-dependent transcription in AML. Genes Dev 28(4):317–327
56. Bernt KM, Zhu N, Sinha AU, Vempati S, Faber J, Krivtsov AV et al (2011) MLL-rearranged leukemia is dependent on aberrant H3K79 methylation by DOT1L. Cancer Cell 20(1):66–78
57. Zhong X, Prinz A, Steger J, Garcia-Cuellar MP, Radsak M, Bentaher A et al (2018) HoxA9 transforms murine myeloid cells by a feedback loop driving expression of key oncogenes and cell cycle control genes. Blood Adv 2(22):3137–3148
58. Sun Y, Zhou B, Mao F, Xu J, Miao H, Zou Z et al (2018) HOXA9 reprograms the enhancer landscape to promote leukemogenesis. Cancer Cell 34(4):643–58.e5
59. Lambert M, Alioui M, Jambon S, Depauw S, Van Seuningen I, David-Cordonnier MH (2019) Direct and indirect targeting of HOXA9 transcription factor in acute myeloid leukemia. Cancers (Basel) 11(6):837
60. Lu Y, Yang Y, Zhu G, Zeng H, Fan Y, Guo F et al (2023) Emerging pharmacotherapeutic strategies to

overcome undruggable proteins in cancer. Int J Biol Sci 19(11):3360–3382
61. Yokoyama A (2015) Molecular mechanisms of MLL-associated leukemia. Int J Hematol 101(4):352–361
62. Rawat VP, Cusan M, Deshpande A, Hiddemann W, Quintanilla-Martinez L, Humphries RK et al (2004) Ectopic expression of the homeobox gene Cdx2 is the transforming event in a mouse model of t(12;13)(p13;q12) acute myeloid leukemia. Proc Natl Acad Sci USA 101(3):817–822
63. Rawat VPS, Götze M, Rasalkar A, Vegi NM, Ihme S, Thoene S et al (2020) The microRNA miR-196b acts as a tumor suppressor in Cdx2-driven acute myeloid leukemia. Haematologica 105(6):e285–e2e9
64. Wang LD, Wagers AJ (2011) Dynamic niches in the origination and differentiation of haematopoietic stem cells. Nat Rev Mol Cell Biol 12(10):643–655
65. Xu H, Wen Q (2018) Downregulation of miR-135a predicts poor prognosis in acute myeloid leukemia and regulates leukemia progression via modulating HOXA10 expression. Mol Med Rep 18(1):1134–1140
66. Huang F, Tang J, Zhuang X, Zhuang Y, Cheng W, Chen W et al (2014) MiR-196a promotes pancreatic cancer progression by targeting nuclear factor kappa-B-inhibitor alpha. PLoS One 9(2):e87897
67. Buske C, Feuring-Buske M, Antonchuk J, Rosten P, Hogge DE, Eaves CJ et al (2001) Overexpression of HOXA10 perturbs human lymphomyelopoiesis in vitro and in vivo. Blood 97(8):2286–2292
68. Roth JJ, Crist RC, Buchberg AM (2009) Might as well face it: MLL's addicted to HOX. Blood 113(11):2372–2373
69. Zeisig BB, Milne T, García-Cuéllar MP, Schreiner S, Martin ME, Fuchs U et al (2004) Hoxa9 and Meis1 are key targets for MLL-ENL mediated cellular immortalization. Mol Cell Biol 24(2):617–628
70. Golub TR, Slonim DK, Tamayo P, Huard C, Gaasenbeek M, Mesirov JP et al (1999) Molecular classification of cancer: class discovery and class prediction by gene expression monitoring. Science 286(5439):531–537
71. Jia XH, Zhu LP, Li JC, Wang CC (2013) Expression of homeobox gene HOXA9 in childhood acute leukemia, and its clinical significance. Zhongguo Dang Dai Er Ke Za Zhi 15(4):268–272
72. Talarmain L, Clarke MA, Shorthouse D, Cabrera-Cosme L, Kent DG, Fisher J et al (2022) HOXA9 has the hallmarks of a biological switch with implications in blood cancers. Nat Commun 13(1):5829
73. Gao L, Sun J, Liu F, Zhang H, Ma Y (2016) Higher expression levels of the HOXA9 gene, closely associated with MLL-PTD and EZH2 mutations, predict inferior outcome in acute myeloid leukemia. Onco Targets Ther 9:711–722
74. Abdelrahman AMN, Tolba FM, Kamal HM, Abdellateif MS, Ahmed HA, Hassan NM (2023) Evaluation of the HOXA9 and MEIS1 genes as a potential biomarker in adult acute myeloid leukemia. Egypt J Med Hum Genet 24(1):11
75. Kandel S, Ahmedy I, Tayel S, Mohamed A (2021) Role of homeobox-A9 gene expression in patients with de novo acute myeloid leukemia. Menoufia Med J 34(1):291–296
76. Dard A, Jia Y, Reboulet J, Bleicher F, Lavau C, Merabet S (2019) The human HOXA9 protein uses paralog-specific residues of the homeodomain to interact with TALE-class cofactors. Sci Rep 9(1):5664
77. Choe SK, Ladam F, Sagerström CG (2014) TALE factors poise promoters for activation by Hox proteins. Dev Cell 28(2):203–211
78. Li Z, Zhang Z, Li Y, Arnovitz S, Chen P, Huang H et al (2013) PBX3 is an important cofactor of HOXA9 in leukemogenesis. Blood 121(8):1422–1431
79. Kwan W, North TE (2017) Chapter Four—Netting novel regulators of hematopoiesis and hematologic malignancies in zebrafish. In: Sadler KC (ed) Current topics in developmental biology, vol 124. Academic, Cambridge, pp 125–160
80. Kroon E, Thorsteinsdottir U, Mayotte N, Nakamura T, Sauvageau G (2001) NUP98-HOXA9 expression in hemopoietic stem cells induces chronic and acute myeloid leukemias in mice. EMBO J 20(3):350–361
81. Oka M, Mura S, Yamada K, Sangel P, Hirata S, Maehara K et al (2016) Chromatin-prebound Crm1 recruits Nup98-HoxA9 fusion to induce aberrant expression of Hox cluster genes. eLife. 5:e09540
82. Hatano Y, Miura I, Nakamura T, Yamazaki Y, Takahashi N, Miura AB (1999) Molecular heterogeneity of the NUP98/HOXA9 fusion transcript in myelodysplastic syndromes associated with t(7;11)(p15;p15). Br J Haematol 107:600–604
83. Fang Y, Han X, Shen L, Hou J (2022) NUP98-HOXA9 bearing acute myeloid leukemia. Blood 140(Supplement 1):11614
84. Koya J, Kataoka K, Sato T, Bando M, Kato Y, Tsuruta-Kishino T et al (2016) DNMT3A R882 mutants interact with polycomb proteins to block haematopoietic stem and leukaemic cell differentiation. Nat Commun 7:10924
85. Jost E, Lin Q, Weidner CI, Wilop S, Hoffmann M, Walenda T et al (2014) Epimutations mimic genomic mutations of DNMT3A in acute myeloid leukemia. Leukemia 28(6):1227–1234
86. Dhar GA, Saha S, Mitra P, Nag CR (2021) DNA methylation and regulation of gene expression: Guardian of our health. Nucleus (Calcutta) 64(3):259–270
87. Zhou C, Li J, Li Q, Liu H, Ye D, Wu Z et al (2019) The clinical significance of HOXA9 promoter hypermethylation in head and neck squamous cell carcinoma. J Clin Lab Anal 33(5):e22873
88. Faaborg L, Jakobsen A, Waldstrøm M, Petersen CB, Andersen RF, Steffensen KD (2021) HOXA9-methylated DNA as a diagnostic biomarker of ovarian malignancy. Biomark Med 15(15):1309–1317
89. Malik S, Bhaumik SR (2010) Mixed lineage leukemia: histone H3 lysine 4 methyltransferases from yeast to human. FEBS J 277(8):1805–1821

90. Winters AC, Bernt KM (2017) MLL-rearranged leukemias-an update on science and clinical approaches. Front Pediatr 5:4
91. Yokoyama A, Cleary ML (2008) Menin critically links MLL proteins with LEDGF on cancer-associated target genes. Cancer Cell 14(1):36–46
92. Wang QF, Wu G, Mi S, He F, Wu J, Dong J et al (2011) MLL fusion proteins preferentially regulate a subset of wild-type MLL target genes in the leukemic genome. Blood 117(25):6895–6905
93. Ballabio E, Milne TA (2014) Epigenetic control of gene expression in leukemogenesis: cooperation between wild type MLL and MLL fusion proteins. Mol Cell Oncol 1(2):e955330
94. Chen Y, Anastassiadis K, Kranz A, Stewart AF, Arndt K, Waskow C et al (2017) MLL2, not MLL1, plays a major role in sustaining MLL-rearranged acute myeloid leukemia. Cancer Cell 31(6):755–70.e6
95. Kühn MW, Song E, Feng Z, Sinha A, Chen CW, Deshpande AJ et al (2016) Targeting chromatin regulators inhibits leukemogenic gene expression in NPM1 mutant leukemia. Cancer Discov 6(10):1166–1181
96. Caslini C, Yang Z, El-Osta M, Milne TA, Slany RK, Hess JL (2007) Interaction of MLL amino terminal sequences with menin is required for transformation. Cancer Res 67(15):7275–7283
97. Yan J, Chen YX, Desmond A, Silva A, Yang Y, Wang H et al (2006) Cdx4 and menin co-regulate Hoxa9 expression in hematopoietic cells. PLoS One 1(1):e47
98. Yokoyama A, Somervaille TC, Smith KS, Rozenblatt-Rosen O, Meyerson M, Cleary ML (2005) The menin tumor suppressor protein is an essential oncogenic cofactor for MLL-associated leukemogenesis. Cell 123(2):207–218
99. Thiel AT, Huang J, Lei M, Hua X (2012) Menin as a hub controlling mixed lineage leukemia. Bioessays 34(9):771–780
100. Méreau H, De Rijck J, Cermáková K, Kutz A, Juge S, Demeulemeester J et al (2013) Impairing MLL-fusion gene-mediated transformation by dissecting critical interactions with the lens epithelium-derived growth factor (LEDGF/p75). Leukemia 27(6):1245–1253
101. Sarno F, Nebbioso A, Altucci L (2020) DOT1L: a key target in normal chromatin remodelling and in mixed-lineage leukaemia treatment. Epigenetics 15(5):439–453
102. Ballabio E, Milne TA (2012) Molecular and epigenetic mechanisms of MLL in human leukemogenesis. Cancers (Basel) 4(3):904–944
103. Uckelmann HJ, Armstrong SA (2020) Chromatin complexes maintain self-renewal of myeloid progenitors in AML: opportunities for therapeutic intervention. Stem Cell Rep 15(1):6–12
104. Lu R, Wang P, Parton T, Zhou Y, Chrysovergis K, Rockowitz S et al (2016) Epigenetic perturbations by Arg882-mutated DNMT3A potentiate aberrant stem cell gene-expression program and acute leukemia development. Cancer Cell 30(1):92–107
105. Rau RE, Rodriguez BA, Luo M, Jeong M, Rosen A, Rogers JH et al (2016) DOT1L as a therapeutic target for the treatment of DNMT3A-mutant acute myeloid leukemia. Blood. 128(7):971–981
106. Hu H, Muntean AG (2023) The YEATS domain epigenetic reader proteins ENL and AF9 and their therapeutic value in leukemia. Exp Hematol 124:15–21
107. Krivtsov AV, Hoshii T, Armstrong SA (2017) Mixed-lineage leukemia fusions and chromatin in leukemia. Cold Spring Harb Perspect Med 7(11):a026658
108. Wilkinson AW, Gozani O (2017) Reading the future of leukaemia. Nature 543(7644):186–187
109. Yamagata K, Shino M, Aikawa Y, Fujita S, Kitabayashi I (2021) Tip60 activates Hoxa9 and Meis1 expression through acetylation of H2A.Z, promoting MLL-AF10 and MLL-ENL acute myeloid leukemia. Leukemia 35(10):2840–2853
110. Falini B (2023) NPM1-mutated acute myeloid leukemia: new pathogenetic and therapeutic insights and open questions. Am J Hematol 98(9):1452–1464
111. Rau R, Brown P (2009) Nucleophosmin (NPM1) mutations in adult and childhood acute myeloid leukaemia: towards definition of a new leukaemia entity. Hematol Oncol 27(4):171–181
112. Alcalay M, Tiacci E, Bergomas R, Bigerna B, Venturini E, Minardi SP et al (2005) Acute myeloid leukemia bearing cytoplasmic nucleophosmin (NPMc+ AML) shows a distinct gene expression profile characterized by up-regulation of genes involved in stem-cell maintenance. Blood 106(3):899–902
113. Shen WF, Hu YL, Uttarwar L, Passegue E, Largman C (2008) MicroRNA-126 regulates HOXA9 by binding to the homeobox. Mol Cell Biol 28(14):4609–4619
114. Schneider E, Pochert N, Ruess C, MacPhee L, Escano L, Miller C et al (2020) MicroRNA-708 is a novel regulator of the Hoxa9 program in myeloid cells. Leukemia 34(5):1253–1265
115. Rinn JL, Kertesz M, Wang JK, Squazzo SL, Xu X, Brugmann SA et al (2007) Functional demarcation of active and silent chromatin domains in human HOX loci by noncoding RNAs. Cell 129(7):1311–1323
116. Fu Z, Chen C, Zhou Q, Wang Y, Zhao Y, Zhao X et al (2017) LncRNA HOTTIP modulates cancer stem cell properties in human pancreatic cancer by regulating HOXA9. Cancer Lett 410:68–81
117. Wang KC, Yang YW, Liu B, Sanyal A, Corces-Zimmerman R, Chen Y et al (2011) A long noncoding RNA maintains active chromatin to coordinate homeotic gene expression. Nature 472(7341):120–124
118. Zhang Y, Morrone G, Zhang J, Chen X, Lu X, Ma L et al (2003) CUL-4A stimulates ubiquitylation and degradation of the HOXA9 homeodomain protein. EMBO J 22(22):6057–6067
119. Bandyopadhyay S, Harris DP, Adams GN, Lause GE, McHugh A, Tillmaand EG et al (2012) HOXA9

methylation by PRMT5 is essential for endothelial cell expression of leukocyte adhesion molecules. Mol Cell Biol 32(7):1202–1213
120. Cai S, Liu R, Wang P, Li J, Xie T, Wang M et al (2020) PRMT5 prevents cardiomyocyte hypertrophy via symmetric dimethylating HoxA9 and repressing HoxA9 expression. Front Pharmacol 11:600627
121. Vijapurkar U, Fischbach N, Shen W, Brandts C, Stokoe D, Lawrence HJ et al (2004) Protein kinase C-mediated phosphorylation of the leukemia-associated HOXA9 protein impairs its DNA binding ability and induces myeloid differentiation. Mol Cell Biol 24(9):3827–3837
122. Calero-Nieto FJ, Joshi A, Bonadies N, Kinston S, Chan WI, Gudgin E et al (2013) HOX-mediated LMO2 expression in embryonic mesoderm is recapitulated in acute leukaemias. Oncogene 32(48):5471–5480
123. Brumatti G, Salmanidis M, Kok CH, Bilardi RA, Sandow JJ, Silke N et al (2013) HoxA9 regulated Bcl-2 expression mediates survival of myeloid progenitors and the severity of HoxA9-dependent leukemia. Oncotarget 4(11):1933–1947
124. Miyamoto R, Kanai A, Okuda H, Komata Y, Takahashi S, Matsui H et al (2021) HOXA9 promotes MYC-mediated leukemogenesis by maintaining gene expression for multiple anti-apoptotic pathways. eLife 10:e64148
125. Gwin K, Frank E, Bossou A, Medina KL (2010) Hoxa9 regulates Flt3 in lymphohematopoietic progenitors. J Immunol 185(11):6572–6583
126. Zhou X, Lu R (2023) HOXA9/MEIS1 targets in leukemia: reinforced signaling networks and therapeutic opportunities. Haematologica 108(5):1205–1207
127. Lillico R, Lawrence CK, Lakowski TM (2018) Selective DOT1L, LSD1, and HDAC class I inhibitors reduce HOXA9 expression in MLL-AF9 rearranged leukemia cells, but dysregulate the expression of many histone-modifying enzymes. J Proteome Res 17(8):2657–2667
128. Cao M, Li T, Chen Y, Zhai X (2021) Nucleoside and non-nucleoside DOT1L inhibitors: dawn of MLLrearranged leukemia. Mini Rev Med Chem 21(11):1337–1350
129. El Chaer F, Keng M, Ballen KK (2020) MLL-rearranged acute lymphoblastic leukemia. Curr Hematol Malig Rep 15(2):83–89
130. Jo SY, Granowicz EM, Maillard I, Thomas D, Hess JL (2011) Requirement for Dot1l in murine postnatal hematopoiesis and leukemogenesis by MLL translocation. Blood 117(18):4759–4768
131. Zhang L, Chen Y, Liu N, Li L, Xiao S, Li X et al (2018) Design, synthesis and anti leukemia cells proliferation activities of pyrimidylaminoquinoline derivatives as DOT1L inhibitors. Bioorg Chem 80:649–654
132. Rudenko AY, Mariasina SS, Sergiev PV, Polshakov VI (2022) Analogs of S-adenosyl-L-methionine in studies of methyltransferases. Mol Biol 56(2):229–250
133. Stein EM, Garcia-Manero G, Rizzieri DA, Tibes R, Berdeja JG, Savona MR et al (2018) The DOT1L inhibitor pinometostat reduces H3K79 methylation and has modest clinical activity in adult acute leukemia. Blood 131(24):2661–2669
134. Yi Y, Ge S (2022) Targeting the histone H3 lysine 79 methyltransferase DOT1L in MLL-rearranged leukemias. J Hematol Oncol 15(1):35
135. Heimbruch KE, Fisher JB, Stelloh CT, Phillips E, Reimer MH Jr, Wargolet AJ et al (2021) DOT1L inhibitors block abnormal self-renewal induced by cohesin loss. Sci Rep 11(1):7288
136. Cierpicki T, Grembecka J (2014) Challenges and opportunities in targeting the menin-MLL interaction. Future Med Chem 6(4):447–462
137. Krivtsov AV, Evans K, Gadrey JY, Eschle BK, Hatton C, Uckelmann HJ et al (2019) A menin-MLL inhibitor induces specific chromatin changes and eradicates disease in models of MLL-rearranged leukemia. Cancer Cell 36(6):660–673.e11
138. Milne TA, Kim J, Wang GG, Stadler SC, Basrur V, Whitcomb SJ et al (2010) Multiple interactions recruit MLL1 and MLL1 fusion proteins to the HOXA9 locus in leukemogenesis. Mol Cell 38(6):853–863
139. Zhu N, Chen M, Eng R, DeJong J, Sinha AU, Rahnamay NF et al (2016) MLL-AF9- and HOXA9-mediated acute myeloid leukemia stem cell self-renewal requires JMJD1C. J Clin Invest 126(3):997–1011
140. Shi Q, Xu M, Kang Z, Zhang M, Luo Y (2023) Menin-MLL1 interaction small molecule inhibitors: a potential therapeutic strategy for leukemia and cancers. Molecules 28(7):3026
141. Borkin D, He S, Miao H, Kempinska K, Pollock J, Chase J et al (2015) Pharmacologic inhibition of the menin-MLL interaction blocks progression of MLL leukemia in vivo. Cancer Cell 27(4):589–602
142. Dafflon C, Craig VJ, Méreau H, Gräsel J, Schacher Engstler B, Hoffman G et al (2017) Complementary activities of DOT1L and menin inhibitors in MLL-rearranged leukemia. Leukemia 31(6):1269–1277
143. Swaminathan M, Bourgeois W, Armstrong SA, Wang ES (2022) Menin Inhibitors in acute myeloid leukemia-what does the future hold? Cancer J 28(1):62–66
144. Wang R, Xu P, Chang LL, Zhang SZ, Zhu HH (2022) Targeted therapy in NPM1-mutated AML: knowns and unknowns. Front Oncol 12:972606
145. Brunetti L, Gundry MC, Sorcini D, Guzman AG, Huang YH, Ramabadran R et al (2018) Mutant NPM1 maintains the leukemic state through HOX expression. Cancer Cell 34(3):499–512.e9
146. Federici L, Falini B (2013) Nucleophosmin mutations in acute myeloid leukemia: a tale of protein unfolding and mislocalization. Protein Sci 22(5):545–556
147. Gu X, Ebrahem Q, Mahfouz RZ, Hasipek M, Enane F, Radivoyevitch T et al (2018) Leukemogenic

nucleophosmin mutation disrupts the transcription factor hub that regulates granulomonocytic fates. J Clin Invest 128(10):4260–4279
148. Karatas H, Li Y, Liu L, Ji J, Lee S, Chen Y et al (2017) Discovery of a highly potent, cell-permeable macrocyclic peptidomimetic (MM-589) targeting the WD repeat domain 5 protein (WDR5)-mixed lineage leukemia (MLL) protein-protein interaction. J Med Chem 60(12):4818–4839
149. Sykes DB, Kfoury YS, Mercier FE, Wawer MJ, Law JM, Haynes MK et al (2016) Inhibition of dihydroorotate dehydrogenase overcomes differentiation blockade in acute myeloid leukemia. Cell 167(1):171–186.e15
150. Zhang H, Zhang Y, Zhou X, Wright S, Hyle J, Zhao L et al (2020) Functional interrogation of HOXA9 regulome in MLLr leukemia via reporter-based CRISPR/Cas9 screen. eLife 9:e57858